Recording Analysis

Recording Analysis: How the Record Shapes the Song identifies and explains how the sounds imparted by recording processes enhance the artistry and expression of recorded songs.

Moylan investigates how the process of recording a song transforms it into a richer experience and articulates how the unique elements of recorded sound provide essential substance and expression to recorded music. This book explores a broad array of records, evaluating the music, lyrics, social context, literary content and meaning, and offers detailed analyses of recording elements as they appear in a wide variety of tracks.

Accompanied by a range of online resources, *Recording Analysis* is an essential read for students and academics, as well as practitioners, in the fields of record production, song-writing and popular music.

Dr. William Moylan is Distinguished University Professor at the University of Massachusetts Lowell, where he is Professor of Music and Sound Recording Technology. He has taught recording, music theory and musicology, and critical listening for over 35 years. An active recording engineer/producer, William Moylan has worked with emerging and leading artists across the genres of classical and popular music. Dr. Moylan is the author of *Understanding and Crafting the Mix 3e.*

"William Moylan's authoritative work on the craft of record production has become a foundational resource for music scholars who seek to understand the myriad processes that shape records. This new contribution, *Recording Analysis*, offers a comprehensive interpretive framework for the serious listener, illustrating clearly how to account for the sonic materials and aesthetic values of recorded popular song. His analytic toolkit and practical approach will engage readers—interested in songwriting, musical performance, and recording technologies—who wish to explore their own affective connections to this powerful cultural medium."

—**Lori Burns**, Professor of Music, University of Ottawa

"In his most recent book, *Recording Analysis*, William Moylan's central concern is 'how the record shapes the song.' He provides a comprehensive, insightful, and refreshing exploration into the many dimensions that shape our experience of recordings. The book masterfully synthesizes work on popular music from across a number of disciplines, providing a useful overview that is richly supported with detail throughout. It is a must-read for any popular music scholar."

—**John Covach**, Director of the Institute for Popular Music, University of Rochester, and Professor of Theory, Eastman School of Music

"In *Recording Analysis*, William Moylan offers a substantial study of the qualities of recording which enable us to hear the music we hear. It is theoretically sophisticated but highly accessible throughout and, although of principal benefit to the student of recording, there is so much here that will increase the understanding, and pleasure, of anyone who enjoys the listening experience. And don't be put off by the term 'analysis' - Moylan's approach to the topic by means of everyday questions to which we can all respond greatly demystifies the process."

—**Allan Moore**, Professor Emeritus in the Department of Music and Media, University of Surrey, and author of *Rock: the Primary Text* 3e and *Song Means*

"Moylan's new book marks another large and important step in the literature of the evolving musicology of record production. This systematic approach to analysis creates a whole new toolbox for those practitioners and educators who are passionate about understanding how and why recordings work."

—**Simon Zagorski-Thomas**, Professor at the London College of Music, University of West London, and Founder of the Art of Record Production conference, journal, and association

Recording Analysis

How the Record Shapes the Song

William Moylan

Routledge
Taylor & Francis Group

NEW YORK AND LONDON

First published 2020
by Routledge
52 Vanderbilt Avenue, New York, NY 10017

and by Routledge
2 Park Square, Milton Park, Abingdon, Oxon, OX14 4RN

Routledge is an imprint of the Taylor & Francis Group, an informa business

Library of Congress Cataloging-in-Publication Data
Names: Moylan, William, author.
Title: Recording analysis : how the record shapes the song / William Moylan.
Description: New York : Routledge, 2020. | Includes bibliographical
    references and index.
Identifiers: LCCN 2019037213 (print) | LCCN 2019037214 (ebook) |
    ISBN 9781138667075 (hardback) | ISBN 9781138667068 (paperback) |
    ISBN 9781315617176 (ebook) | ISBN 9781317207160 (adobe pdf) |
    ISBN 9781317207146 (mobi) | ISBN 9781317207153 (epub)
Subjects: LCSH: Popular music—Analysis, appreciation. | Sound recordings—
    Production and direction. | Sound recordings in musicology.
Classification: LCC MT146 .M69 2020 (print) | LCC MT146 (ebook) |
    DDC 781.49—dc23
LC record available at https://lccn.loc.gov/2019037213
LC ebook record available at https://lccn.loc.gov/2019037214

ISBN: 978-1-138-66707-5 (hbk)
ISBN: 978-1-138-66706-8 (pbk)
ISBN: 978-1-315-61717-6 (ebk)

Typeset in Optima
by Apex CoVantage, LLC

Visit the companion website: www.routledge.com/cw/moylan

In loving memory of my mother,
Betty Lou (Peterson) Moylan

Contents

List of Figures

List of Tables

List of Praxis Studies

Preface

I really didn't want to write this book. I had a strong sense of what a large undertaking it would be, and didn't want to dedicate years to it. I was writing other interesting books and papers, making wonderful records, relishing my teaching and otherwise having a good life without being consumed by another big project—so I resisted. Still, this book was in the back of my mind for decades; numerous times I turned away from it, but always with a sense that this book would be important for the understanding of records. In April of 2015 I finally told my then current editor about this idea, and I offered to help find an author; Megan Ball's response was "You want to write that book." She persisted and soon we were negotiating a contract. I entered into the contract with trepidation, not really knowing where the boundaries of this writing would end, or the depths required to illuminate all facets of recording analysis. The length of this book is now more than twice what was proposed, and it has taken much longer to write than I imagined.

There are two reasons I decided to undertake this project now (actually four years ago, considering the time spanning proposal to completed manuscript).

First, studies of the art of record production and popular music analysis have steadily increased since the late 1980s. I am compelled to contribute to this momentum by compiling an in-depth examination of recording elements and of how they shape the artistry of tracks, and place them alongside the other central topics of popular music and song lyrics. My goal is to assemble all of these matters into a cohesive, bigger picture, one which emphasizes interrelations and that provides a framework for accessing and analyzing what is within records. The second reason is more personal; at this stage of my career, I am compelled to draw together many streams of my professional life into a more unified statement. This book spans my experience in the music and recording fields since the mid-1970s: my research (beginning in the early 1980s) exploring the dimensions of recorded sound and recorded music, and examining our myriad perceptions of the recorded sound qualities that augment the creative voice of records; and more than 35 years of teaching listening skills for recorded music, based on materials and techniques I developed and refined while carefully guiding, observing and teaching several thousand motivated individuals. Throughout my career I have sought to understand 'how the record shapes the song' by utilizing deep listening and open awareness. With this book, I am privileged to now share some of what I have uncovered.

I have examined and included a wide variety of perspectives and analyses in here, and have maintained a neutral tone throughout most discussions; there will be no critical reviews, only impartial presentations that open space for the readers' ideas (and growth). I also offer some new ground and my own ideas; I identify and analyze several important dimensions of records and recording elements that have not previously come into focus, and share what I hear within many tracks. This book is not about technologies or production techniques; it is about how the sounds of the recording process add dimensionality and richness to the musical and socio-psychological experiences of records. To this end, I have examined a large number of tracks, from a wide variety of genres and time periods—with records (and lyrics) by the Beatles and lyrics (and records) of Bob Dylan providing a thread of common context as many diverse topics unfold.

This book was written for those that study records, and for those that make recordings and records—and particularly for those who want to acquire these skills. It is also for those who are curious about what is within the recorded songs that speak to them. This book is a resource for these practitioners, academics and learners; it may be used for teaching at a variety of levels of study and for curricula in recording production, recording analysis, song writing, critical listening, and popular music studies; it has been written to promote informal, self-directed learning.

In recent years the pace of scholarship in the analysis of recordings has quickened, and it offers streams of evolving perspectives and ideas that ensure the future vitality of recording analysis. There is much more to discover. It is my desire that this book provide support for this broadening scholarship—those in research, those teaching, and those learning. Most importantly, my wish is that this book benefit all those who seek to understand, or to discover more about the recorded songs they hold dear.

EXPRESSION OF APPRECIATION

I have received the support of many people during the course of writing this book. The kind staff at Routledge, Taylor & Francis and Focal Press have assisted me greatly, and made this long process as easy as it could be. I want to especially thank Megan Ball, Shannon Neill, Hannah Rowe, Lauren Ellis, Jacqueline Dias, Claire Margerison, Lara Zoble, Zoe Meyer, Kristina Ryan, Peter Linsley and the many good people at Routledge that provided excellent support, but who were invisible to me.

My deep gratitude to M. Nyssim Lefford and to Christopher Lee, who read draft versions of each chapter and offered valuable suggestions and pertinent questions; their input brought greater clarity, focus and breadth to this writing. Thank you for joining me on this journey.

My sincere thanks to valued colleagues who have read one or more chapters and provided valuable feedback: Michael Millner, Anne Danielsen, Serge Lacasse, and Alan Williams. My gratitude to others who have been involved in the process in various ways: Adam Fergusson, Serge Lacasse and Sophie Stévance—and most especially, to Penelope Gooch, who closely observed the birthing of this book (and often heard of its challenges): your support, your sensitive listening and your loving presence and companionship are deeply valued.

William Moylan
June 21, 2019

Introduction

This book investigates how the process of recording a song transforms it. The process of recording constructs a rich experience that combines the music and lyrics of the song with captured performances; in addition, it imparts its own sonic signatures to deliver carefully crafted records. The recording is as much a part of the record's artistic voice as the music or the lyrics, and it is common for recording elements to add significant dimensionality to recorded songs. This dimensionality is at the core of this writing.

The goal of *Recording Analysis: How the Record Shapes the Song* is to articulate how the unique dimensions (elements) of recorded sound contribute to recorded music—how they add essential substance and expression to recorded songs.

In examining the central ways elements of recorded sound contribute to records, this book will analyze numerous tracks to examine specific traits of specific elements, and also the interaction of elements. It will also investigate how the record shapes the overall qualities of tracks down to its smallest details. *Recording Analysis: How the Record Shapes the Song* will explore other qualities that are contained within or elicited by recorded popular song to illustrate its larger context of ways all its parts might co-mingle.

'Recording analysis' is the study of the content of records; it can include (among other topics) the music, lyrics, social context, literary content and meaning, and/or the recording process. Studying records may take many forms, may be positioned from numerous vantages. For our purposes, recording analysis is the study of how the recording contributes to recorded songs, and to the content and character of the sounds of and in records. In *Recording Analysis: How the Record Shapes the Song* we will examine the other essential topics and content areas of records, though they are not its central focus; this broadening (to include music, etc. listed above) is necessary to accurately situate the recording elements in records and to recognize their contributions to the whole. Doing so will allow the reader to recognize how the recording interrelates with music and lyrics, affects and meaning, performance and expression, and other qualities that shape every track into its unique form.

Central topics and features contained in *Recording Analysis: How the Record Shapes the Song* include:

- Definitions and detailed examinations of each recording element
- Analyses of tracks examining how they are shaped by specific recording elements, functioning at various levels of perspective
- Developing listening skills to accurately hear and observe each recording element
- Developing listening skills to productively listen with intention and attention for a wide variety of purposes and types of perspective

- An analytical framework for approaching tracks as being unique from all others, utilizing listening with attention and intention to all elements as holding the potential of equal bearing on the track, and any level of perspective (detail)
- A process for analyzing tracks, emphasizing observation and evaluation of recording elements, that lead to discovery and conclusions
- Examination of music and lyrics, and of expression and affects within records

SOME PRELIMINARY DEFINITIONS: SONG, RECORD, RECORDING, TRACKS AND DOMAINS

Before moving further, some preliminary definitions of terms will clarify this discussion. These terms (most of which you will have previously encountered) will have specific meanings when used in discussions throughout the text.

Song is the core identity of the record; it includes the words, melody, chord changes, arrangement and structural design. It is the musical work that does not include the performance. In performance, all the items it includes can be modified without changing its identity. All that is critical to its identity could be contained in a lead sheet.

Recording represents the sound qualities (elements) and interrelations of sound qualities that are brought about by the recording process. 'Recording' is one of the three domains of the record.

Track is the finished work, it is the recording itself, the recorded song. For recorded popular songs, the track is comprised of the song (music and lyrics), the musical arrangement and its performance, and the sounds of the recording process—these are the domains of music, lyrics and recording with the performance of the parts intersecting all three domains.

'*Track*,' '*record*' and '*recorded song*' are used synonymously throughout this book. These are all meant to represent the finished musical work.

'*Track*' may also be a verb, that is used to mean to record a specific voice/instrument and to direct that recorded performance to a storage area that is isolated from others—a channel on a multitrack device or software. Occasionally this will be the intended definition.

The terms '*popular song*' and '*rock song*' are used synonymously, and also to speak generically. These terms refer to collections of musical styles and genres that are a product of oral and aural traditions, are created through recording, conceived in performance, and (most importantly here) that only exist in their complete forms as records (tracks). Springing from these generic terms are a myriad of styles and genres—a few examples are contemporary, rock, progressive rock, blues, country, western, dance, electronic, industrial, alternative, hip-hop, metal, grunge, soul, R&B, reggae, and so many more. The reader will recognize this list as grossly incomplete, and it is likely omitting some music that speaks deeply to you.

Lyrics is one of the domains of tracks. '*Text*' and '*song lyrics*' are used synonymously with lyrics in this writing.

Domain identifies the three distinct streams of information and expression that are within records. The three domains are music, lyrics and recording. Each domain is comprised of numerous, identifiable elements (for example, 'melody' is an element of music).

BOOK ORGANIZATION

An overall flow takes the reader from background information through the various parts of the recording analysis framework, to in depth analyses of recording elements as they appear in a wide variety of tracks. This takes place over ten chapters, that are divided into two parts (each containing five chapters).

Part One establishes the context and the requisite background for recording analysis and for the study of the recording elements.

Chapter 1 defines what a record or track is—its domains, expression and affects, and performances—and how outside disciplines (such as sociology or poetry studies) may provide insight into tracks; a cursory introduction to the book's analysis framework and process concludes the chapter. Chapter 2 delivers a thorough coverage of the framework and its principles and concepts; syntax, materials, perspective and form/structure are explored. Chapter 2 concludes with the first detailed discussion of listening with intention and attention, and how listening is personal.

Chapter 3 articulates the domain of music, and explores the appearances of music elements in popular music. The role of performance in tracks is encountered for the first time here, and bring focus to interplay of the lead vocal and accompaniment. Chapter 4 presents the domain of lyrics, as the voice of the song (and the track); structure of lyrics, message and meaning, dimensions of stories lead to a recognition of the central figures of recorded song: the lead vocal. The persona of the singer and the sound qualities of their performance are the focus of the second half of the chapter.

Part One concludes with Chapter 5, presenting approaches and challenges for observing the elements within music and lyrics domains; observation is the first step of the analysis process. This chapter includes transcribing and describing music elements, lyrics and vocal lines; timelines at various levels of perspective are first encountered here.

Part Two is dedicated to the recording elements: rhythm/time, timbre, frequency/pitch, spatial properties (lateral position and size, distance position, and environments), and loudness.

Chapter 6 provides an overview of the elements; it includes rhythm and time as recording elements, collecting information and transcribing recording elements, typology and syntax, and talking about sound. The chapter concludes with detailed coverage on track playback and listening systems, and ways they can impact what is heard and impact one's analyses.

Chapter 7 is dedicated to timbre and its link to pitch/frequency; the importance of timbre to the track (its embodiment of sound sources, its presence at all dimensions, and its influence on other elements) is emphasized. The content of pitch density and importance of timbral balance are introduced, along with acousmatic listening. Chapter 8 provides in depth coverage of the dimensions of space in records, and collecting observations of each spatial element. Its coverage extends through listener perspective, angular direction and image width, and distance positioning of sources to define the sound stage; the sounds of places—environments of tracks, of the sound stage, and of individual sound sources—are observed for and defined by their timbre and time characteristics.

Chapter 9 concludes the coverage of individual recording elements with its exploration of loudness levels, contours and relationships as recording elements; it then explores the confluence of domains in tracks, and how complex interrelations establish and co-function within tracks. Timbre as confluence, timbre of the track, crystallized form and deep listening comprise the second half of the chapter, and prepare for Chapter 10. Chapter 10 presents the final two stages of the analysis process (evaluations and conclusions) and expands on analysis goals discussed earlier. The majority of Chapter 10 contains detailed analyses of each recording element at several levels of perspective; numerous and varied tracks are examined in these analyses. Interrelations of elements and of domains are also examined to bring a sense of engaging the great confluence within records.

SUPPORTING MATERIALS

A set of individual praxis studies have been devised and are included in Appendix A. These studies are designed for the reader to undertake self-learning of the elements and the processes covered in the text. Individual studies can be followed in isolation or as part of a sequence; the reader can reorder the studies as they wish, though topics are offered in what has proven to be the most effective sequence.

Some eResources have been assembled to support this text; they are available at:

https://www.routledge.com/9781138667068

The eResources include:

- PDF files of select tables, graph templates, and various diagrams to aid the reader in analyzing recording elements and in developing listening skills
- Spectrographs of selected tracks
- Audio files to support praxis studies
- Guidance and audio files to assist sound system design, set-up and calibration

ESTABLISHING AN ACCURATE PLAYBACK OF TRACKS

In order to accurately analyze a record, one needs to hear the record accurately. This book can lead you to acquire the knowledge, develop the listening skills and engage pertinent listening experiences to hear a record's content with a higher degree of accuracy. A disconnect may occur here, though. Your playback system has the potential to change the sound of the tracks it reproduces; this change can be substantial without you knowing it, and can bring your analysis to be inaccurate because you are not listening to the track with the traits that the artists and other creators intended, and differently from other listeners.

Chapter 6 will provide much detail on playback systems, and the webpage will provide instructions and audio files to assist you in assessing your system and in making adjustments. Some of the topics that will be explored are:

- Components and specifications for a high-quality playback system to provide minimally altered, detailed sound
- Loudspeaker placement and loudspeaker/listening-room interaction
- Level-matching and spectral balance calibration of the system
- Listener distance and orientation to loudspeakers for accurate spatial reproduction
- Listening at appropriate monitoring levels to hear full-bandwidth frequency response

A professional quality system is not needed to establish an acceptably accurate listening condition, producing minimal changes in the sounds of the track. It is possible to assemble a listening system and physical arrangement within modest financial means and typical household logistical constraints. The goal is to reproduce the track with as few alterations as is practical for one's situation—and, importantly, to learn the sonic consequences of those limitations. It should be understood up front, headphones (and especially earbuds) are not a suitable substitute for listening over loudspeakers. Headphones and earbuds will distort most recording elements in numerous ways—they change the very objects of our study.

The playback system is the reader's access to the sounds of the track. Its characteristics will be inherent in all you hear.

As you navigate the topics of this writing, may you experience new dimensions and qualities within the tracks you know, and discover dimensions of sound you have yet to experience. The purpose of this book is to bring you to appreciate all that there is in records, emphasizing the recording elements. The journey of this book will take many readers into topics and concepts not previously experienced; the journey has been thoughtfully charted for you, though. It has proved rewarding for others in the past; I am hopeful you, too, will find it enriching.

PART ONE

CHAPTER 1

Recording Analysis: Domains, Disciplines, Approaches

This chapter sets the context for the study of recording analysis, and introduces the process. A sizeable portion of this context has been explored in the field of popular music studies. Those writings have provided valuable information and models, especially pertaining to disciplines outside the domains of the record. Popular music studies and popular music (and rock music) analysis topics will appear often in this chapter, and interspersed throughout this writing. Recording analysis emerges from these fields as a broader and more inclusive examination of the recorded song; recording analysis owes much to these origins of the study of popular music.

The record reshaped the listening experience and influenced the music itself—profoundly. The record is a finely crafted performance; it reframes notions of performance, composition and arranging around a production process of added sonic dimensions and qualities. It shapes spans of time and captures spontaneous moments. The record's influences are broad in scope, and worldwide in their impacts. A general accounting of them here helps to establish a meaningful point of reference.

Further establishing this context is an examination of the central issues surrounding analyzing each of the three domains of the record: music, lyrics and recording. The domains are examined separately to identify central concerns and issues to be engaged in their analysis. This provides background information and an overview to inform the upcoming chapters on each of these domains, as well as chapters pertaining to the analysis process and framework.

Disciplines outside the domains of the record have contributed significantly to popular music studies and analysis, some since before the field formally began. Others have recently been applied to the analysis and study of popular music. The most pertinent disciplines are identified and introduced here, with some indication of their scope, focus and potential application. These topics might serve to establish a context for recording analysis, depending on its goals.

Recording analysis and related concepts are introduced. A framework of four principles and several concepts, and a flexible, functional process, forms the approach to recording analysis that is presented throughout this writing. Important to the approach is its flexibility and transparency. The framework establishes a general scheme of inquiry from which the process may forge or explore many paths— paths that are unique and appropriate for the individual recorded song. From these paths, pertinent conclusions may emerge to illuminate the track.

RECORDS AND RECORDED MUSIC

Since its development over 100 years ago, the record has changed music ("music" being used in the broadest sense) in a great many ways. The record as a commodity emerged and has had significant, worldwide commercial influence and varied and broad social impacts. The record is a musical voice of the people, of all people—any person from the common and the elite, the sacred and the profane, and all between or beyond. It has changed how we relate to music and how we engage music, and it has changed the content of music. Much writing from many directions and disciplines has covered aspects of these topics and much more.[1] Our focus here is on the ways the record has shaped music and the experience of listening to the recorded song, of which the following section is but a glimpse, serving to set some context for our study of recording analysis.

The record shifted the ways we engage music. Before the record, music listening was a communal, group activity; it took place in concert halls, churches, pubs and other public places. Music was performed live and in front of the audience, it took place in real time and was created spontaneously, and once the performance was over all that remained of it were the memories of those in attendance. In the early days of recorded music—when listening to records (or perhaps cylinders) was largely a parlour or living room experience—the record (plus phonograph) had become the chamber music of the twentieth century; this soon evolved into something else. With the record one can have any music, any time, any place. Music has become personal (we own the records we choose from millions of options, and we can identify deeply with them) and portable (we take music with us wherever we wish and listen whenever we wish). Music listening has become largely solitary (Eisenberg 1987) as listening is often an activity individuals do when alone; increasingly, music listening is isolating, as we close ourselves off from the world with our earbuds.

Along with changing these processes of engaging music, the record changed the substance of music. It spawned types of music that could not have been possible without it—and continues to do so. Rock music, and much popular music, is created and disseminated as records; the medium and the music are fundamentally joined. This is music that is both created and heard, with musical qualities that are the result of the recording process, of the media that stores and delivers it, and of the reproduction methods that presents it to the listener.

Composition, performance and recording production are linked and entwined. Performers and performances generate musical ideas in "the same sort of considered deliberation and decision making as musical composition" (Zak 2001, xii). Recorded performances are intentionally crafted with great attention to details, and performer interpretations are filled with nuance and precision—all intentionally captured, then carefully combined into the sound of the record. There is a team of skilled individuals that contribute to the record—in addition to performers there are songwriters, arrangers, lyricists, engineers, producers, all of who might have overlapping roles. Those that shape the record are called "recordists" here, despite their varied roles of producer, engineer, mix engineer, mastering engineer, tracking engineer, and the host of other formal and informal positions and titles.

The composition of the music might continue or might even occur during the tracking of various performances, where the sound qualities, performance techniques and expression of the performers become wed to their performances and interpretations of the musical ideas to make something larger, more substantial. The traditional approach to the musical setting or arrangement shifts to the more immediate and subtle to incorporate the qualities of the performance, and also to embrace the techniques and sounds of recording. These performances and the arrangement next receive their final shaping, are combined and mixed into the final form of the track itself. In this process all stages influence one another, and they function inseparably. The composition is the record and the performance. The record is the composition that emerges from the captured, created and crafted performance that is guided and realized through recording technologies.

This brings us to recognize that the record is not only a performance and a composition, but something else as well. It is also a set of sound qualities and sound relationships that to a great extent do not exist in nature and that are the result of the recording process. These qualities and relationships create a platform for the recorded song, and contribute to its artistry and voice.

The sound qualities of the recording, then, contribute fundamentally to the artistry and sonic content of the record. The recording's sound adds qualities not found in nature, and reconfigures those qualities the listener already knows. The result is a reshaped sonic landscape, where the relationships of instruments defy physics, where unnatural qualities, proportions, locations and expression become accepted as part of the recorded song, of its performance, of its invented space. Sonic reality is redefined within the impossible worlds of records—and accepted without hesitation by listeners. Records can establish a "reality of illusion" (Moorefield 2005, xiii); a crafted world for the record that is uninhibited by the reality of natural sound relationships, where illusion is reality, where everything that can be created technologically is accepted as reality for the individual recorded song.

Given the high degree of control and scrutiny available in the production process, we should believe that what is sonically present on a record is what was intended, even if seemingly arbitrary or flawed. What is present is part of the record's artistic voice, and it is integral to the primary text of the track, "constituted by the sounds themselves" (Moore & Martin 2019, 1). Should seemingly unintended sounds be present, those 'flaws' were either not heard during the making of the record, or a choice was made that allowed them to remain. What is on the record are those sounds and performances that were selected and combined from all that were recorded or that were generated by, created within or captured through the recording process. The process of making a record is exacting, and the sounds of the final record are carefully crafted; what is there was put there, or was allowed to remain. The record can be crafted such that a flawless performance results—a performance perhaps perfect in its flaws, as 'flawless' is in relation to conventions and artist intentions, and therefor open to the subjectivity of context. Related, the record may be perceived as flawless, though by some accepted norms it might be considered technically flawed or the performance may have qualities that might be perceived by some as imprecise (an example might be the often criticized vocals of Bob Dylan).

The record then comes to represent the definitive version of the song and the performance of the song (in as much as the two can be separated). The record represents the correct version of the song, and any covers of the song by others or future performances of the artist are gauged in reference to the original. Of course exceptions to this occur, but when the original is well known, the record becomes widely accepted and recognizable in this singular form. Here the record becomes a permanent performance, a performance frozen in time along with all its nuance of expression in dynamics, timbre and time captured—and along with everything else the individual listeners or society associates with it.

Differing from live performance, this recorded performance (that is permanent) allows repeated listenings and reflection, deeper examination and more personal interpretations by the listener, and discoveries of the subtleties of the music, the lyrics and the recording. Just as the listener finds something new in music with repeated hearings, and lyrics can reveal something new to the listener upon reflection or new listening, well-crafted recordings can reveal "new facets and nuances on playing after playing" (Gracyk 1996, viii). Albin Zak (2001, 156) shares: "one of the great delights of listening to a well-mixed record lies in exploring aurally its textural recesses, its middle ground and background, and discovering that it offers levels of sonic experience far beyond its obvious surface moves." The sonic dimensions of the recording will reward the listener's attention with further enrichment of the recorded song; the experience of the record is enhanced and supplemented by the recording, and the listener's interpretation is further shaped by it.

ANALYZING THE RECORD'S DOMAINS

Analysis can demystify the record. It can provide clarity to its materials and their relationships, illustrate its structure and shape, and make its movement and energy more coherent to the listener/analyst; it may even allow one to peer into the content of its meaning. Analysis can provide some objectivity and depth of substance to listener interpretation, or to cultural analysis. In studying popular music, writings from numerous fields have examined various factors within and external to the song's music and lyrics, and a few approaches also incorporate aspects of the record (see below). Analysis has the potential to illuminate how the recorded song works, and what it communicates. This study can allow the analyst to develop an intimate knowledge of the song and the track, the artist and the genre; it can make the record intelligible.

Music analysis may bring benefits to understanding music, and thus to understanding the recorded song. So can an evaluation of the lyrics. Music and lyrics might be most easily analyzed individually, isolated from one another's influence. Within the song's artistic statement, however, they are not so readily divided, with their interactions and interrelationships playing key roles in the experience of the song. The interactions and interrelationships include the sounds of the recording; these greatly shape the experience of the record—potentially as much as the music and the lyrics. Thus, an analysis of the recording adds another layer of materials, contributing fundamentally to the multidimensionality of the record. Recording analysis embraces these three domains equally.

Popular Music Analysis

Analyzing music has a centuries long history, with established traditions, conventions, theories and methods. Some of this will be in play here, but much traditional music analysis seeks information and relationships that simply are not relevant in popular song, and especially in the popular song since recording became prominent. Traditional music analysis has long looked to explain music through study of its pitch and rhythmic materials, and their generation of melodic, harmonic and formal relationships. Traditional analysts sought (most still seek) the sources of unity and coherence in musical works through examination of harmonic and formal structures, (Cook 1987, 4) perhaps including thematic materials and their melodic and rhythmic elements. Less interested in the other dimensions of music, the traditional analysis process centers on score study. "The tools for analyzing 'serious' music assume a definitive written score which is regarded as the record of the 'composer's intentions.' Those distracting elements that 'creep in' during performance can be ignored as irrelevant: the music is structure, pitches and rhythms" (McClary and Walser 1990, 282). Traditional music analysis is bound to the score and its content. With no small irony, such score-based analysis may be entirely generated through sight, without accessing or taking into account the actual *sound* of the music.

Music analysis for popular music has many differences, and many challenges. Traditional approaches to music analysis are mostly inadequate and some are irrelevant; this makes sense, as the music and factors around production of popular music are significantly different. To start, three central matters are:

- The music is the recorded performance, bringing performance aspects to be central to musical content.
- As this recorded performance is not notated, the analyst engages the sound of the music itself, and is brought to decide on transcription matters.
- Musical styles are markedly different from Western art music, and there is an absence of tools and techniques for the analysis of popular song.

It is obvious but significant: musically, popular music's styles are considerably different from art music, and also from one style of popular music to another (though perhaps less of a difference than to 'serious'music). Any analytical method suitable for popular music needs to be flexible enough to meaningfully examine those elements central to the individual song and its particular musical style. The principles of traditional analysis are often of limited use here. The emphasis on harmony in traditional analysis will yield little information in many popular music genres, as the harmonic language of much popular music is limited to a few chords in recurring patterns—though certainly some styles and songs contain more involved and even quite complex harmonic relationships and structural designs.[2] Instead, it is typically other elements that create interest and carry expression—not the least is what Albin Zak (2001, 22) has simply called the record's "sound."[3] The skills learned for traditional music analysis are to be redirected and augmented to be relevant and useful for examining popular music—choices of what is worth studying will shift to that which shapes the identity of the individual work, and away from universal theory. Indeed, a theory for popular music seems well beyond possibility, not only because of the breadth of styles, but also because so few songs within any style have undergone meaningful, rigorous examination (Moore 2012a, 2). With this shift of approach comes the need to develop appropriate analytic tools and methods for these overlooked elements, and to construct a vocabulary and theoretical models with which to adequately address them (McClary and Walser 1990, 282).

Among these 'overlooked elements' are the qualities the performance brings. The matters of performance that are not reflected in the pitches and rhythms of the written score are integral to popular songs. For example, expressive shaping of dynamics and intensity shifts of timbre of individual lines and composite textures can take on prominent roles in a track's movement and expression. When performances don't comply with the rigid Western metric grid and pitches don't conform to diatonic tunings, for instance, this is part of the essence of the performance; it is part of the essence of the music. This is a matter of substance and style, and also an element of expression; it is not a reflection of inability or poor training. All this is especially prominent and integral to vocal lines.

A significant issue arises here. How can one study these musical materials, relationships, and performance qualities when they are not notated, when they are represented only within the listener's fallible memory, experienced from one fleeting moment to the next, and perceived differently, subjectively, by every listener? The music exists only as captured and crafted performances, on record. Should the analyst now try to transcribe the performances—and seek to do so without distorting their content? Can the analyst attend to evaluating all of the nuance and detail of the music from memory, or perhaps from copious (or sketchy) notes, but not notation? Some sort of transcription is often helpful to examine materials deeply or to illustrate points of an analysis, and as Peter Winkler (2000, 39) notes, "a transcription can make a performance 'hold still' so that we can observe it—or some traces of it—in detail." Still, one significant quality of popular music is that it resists conformity to Western notation, in fact it can defy it—and this is one of its characteristic qualities. This leads me to be a reluctant advocate of the strategic use of transcription, because it is especially useful to bring those materials and passages that present or generate the identity of the track into focus. Transcription as an exercise pulls the analyst into the music, and introduces them to detail and nuance they might not otherwise engage. Transcription brings the analyst to become intimately aware of what is truly meaningful within the sound itself—whatever the element. Every sound of the music does not need to be transcribed for an analysis to be relevant and successful. The challenge is to identify what deserves the focus and detailed attention of transcription. Which parts does one transcribe, and what level of detail must one capture to adequately inspect essential qualities? And, at what point does transcription into notation distort the qualities of the part, and thereby the track?

An absence of analytical tools to evaluate popular song's materials, an absence of a musical score to make materials readily accessible for critical examination, an overwhelming amount of sonic detail that is vital to the music, and no universally applicable systems of organization for musical syntax or structure, are some of the music analysis challenges of popular songs. These are challenges most analysts with traditional preparation in music theory and musicology may be unprepared to engage—and this only begins the list of challenges in analyzing records.

Lyrics in Popular Music Analysis

Popular music is overwhelmingly 'popular song.' Song lyrics bring added qualities of realism, everyday experiences, tangible concepts and communication to popular music; the songs communicate directly and in readily understood ways, using the language and imagery of everyday conversation, presenting the topics and events that mean the most to common people. Historically, this dimension is not prominent in 'serious' music—especially in the largely dominant instrumental forms, where music is abstract, absolute, isolated within its own patterns of organization, not connected with worldly matters, and often intended to speak to an elite, educated audience.

The language of popular song is colloquial—everyday words and expressions, often grammatically incorrect, and often slang. The language supports this sense of intimacy with the speaker/singer, and also a sense of casual and informal, direct and undisguised communication. The language may reflect a sense of social standing, a sense of culture and more; this reinforces the interconnectedness of the listener, language and the song itself.

While art songs, opera, liturgy and choral works offer certain text-setting conventions and bring some concepts to bear in analysis, popular song is starkly different. Dai Griffiths (2004, 24) has noted:

> [A] distinction needs to be made between 'song writing' as we've come to understand it, and 'text setting', a very different practice and tradition. In the German *Lied*, the French *mélodie*, or English [art] songs . . . the words are given as poetry, and in a sense the whole point of the 'setting' is to make the words into music, so that the emphasis is clearly still on the musical manifestation of words already given.

In contrast, lyrics are written to be sung, and communicate differently than poetry. Popular song speaks directly to the listener, at times perceived as to them alone. Its voice can be up close, at an interpersonal distance the listener allows few to enter in real life, and it can seem personal and private—this is in stark contrast to being detached from the voices projected from the stage or emanating from the choir loft. Popular music's singer is the intimate stranger.

Lyrics provide sound qualities and rhythmic elements that enliven the meaning of words, and may also function as musical elements. Indeed, in some genres and tracks, the sound of the text—the tones of the voices themselves—are themselves more meaningful and more significant to the listener than the words that are sung.[4] It is not uncommon for sung lyrics to be only partially understood, and perhaps not understood at all. No matter if lyrics are clearly understood or not, the role of the vocal line is central—for its sounds and rhythms, its dynamics and expression. These are contained within the lead singer's performance—the message it presents and for the character it projects, as well as the delivery of the song—that is the dramatic core of the song. The persona of the singer carries drama and enhances the performance and the song's content. A duality of the singer and the accompaniment— the individual above all those who are in the background—is established. The singer of popular music exhibits great subtlety of sound qualities and expression that are fundamental aspects of its message and musical content, as well as power of expression—with nuance only audible in recordings. The vocal line begs for transcription, as it contains much that is important—with its subtleties of expression in musical elements, language sounds and nonverbal sounds it is intricate and filled with significance.

In addition to the sonic qualities of the singer's voice, lyrics' sounds originate from their text. Conventions from poetry are often reflected in lyrics—in the structures of their stanzas and lines, and use of poetic devices. These may provide guidance for analysis of sound and rhythm, in addition to poetic/ literary devices, structure and content.

The literary content of lyrics will typically support the song's structure (integrating the structures of the music and the lyrics). In the interdisciplinary study of popular music, literary analysis of lyrics has been around since the beginning—in fact it preceded meaningful music analysis of popular music. Journalists and academics evaluated lyrics in many ways, such as according to literary principles, as poetry, related to meanings, related to current social influences, and related to how the song might reflect on the personal life of the artist, to identify a few. Interpretations of lyrics take a great many forms, as well as degrees of focusing on the text itself. The literary content of the lyrics certainly has a place in analysis. A danger is to place too much weight on the text, at the expense of the music—or the reverse. It can be a challenge to determine an appropriate balance of music and lyrics for the individual song—a balance that illuminates the song's message and its music.

Lyrics communicate the message of songs; they deliver narratives, stories, and more. The text can elicit meaning on many levels. Some songs plead for literary analysis of richly intricate content, and others scream their social positions. Audience members are eager to conceive their personal interpretation of lyrics. The analyst, too, might engage these realms—as will be explored below.

Analyzing the Recording

The recording is a third layer of the track (a third stream of data), and its qualities are examined equally with music and lyrics. This is the central topic of this book. The sound qualities of the recording add dimensionality, substance, and character to the record—they transform the song into the track. They are integral, just as any musical element and qualities of the lyrics. They shape listener response and the expressive qualities of the record; they contribute fundamentally to the experience that is the recorded song.

That recorded song is sonically different from a live, unamplified performance might be recognizable to the listener—or the experienced music analyst—without the differences being known, or identifiable. The factors that shape the recording are largely outside normal listening experiences—in fact, perceiving some of the elements of recordings actively contradicts listening processes that have been learned and utilized since birth (some perhaps before). Many elements of recordings have likely never been experienced fully by listeners, and the reader likely is unaware of some of the elements. That the elements are previously unknown, have never been consciously experienced and are often subtle, presents the necessity of learning—learning what might be present, learning how to listen for the elements, learning how to hear the elements' qualities and activities. The reader will be introduced to these qualities and guided in learning listening skills to address these challenges.

The recording is examined in ways mostly different from music. Though the concepts of structure, patterning, typology, movement, shape and others may be consistent with music analysis, the content of the recording's elements is obviously different. These concepts are adapted for insight into recording's elements. The elements of recordings are varied, and each requires a different approach. In general, elements exist in two basic states—elements that are active (that exhibit change over time) and elements that are stable. Stable elements are defined by their qualities and characteristics, as they exist throughout a song or within certain substantial portions. All elements of the recording have the potential to exhibit changes of state or value, and create events of shape (often in patterns) and speed (sometimes rhythm)—how much change, in what direction, at what rate, etc. This brings significant potential for widely varied qualities in the recording, and significant changes and characteristics in the recording's elements—a state that allows the recording to have an equal potential to be as significant as the other domains.

Listeners/analysts listen for conventions; we recognize conventions, or relate what is heard to conventions we know. Some conventions have evolved for certain elements, and some characteristic qualities have emerged in other elements. These help to define the 'sound' of various types of music, and their production styles. A systematic, orderly arrangement for some of the recording's elements has loosely coalesced in some styles of music—though constantly evolving, and at times suddenly shifting. Syntax is apparent within most records; it appears in the ways characteristic sounds and relationships of sound qualities (elements) are defined, and in the progressions of sound relationships that are present in various styles of music. These bring a sense of organization, syntax and style to the elements, and to the record. Evaluating the recording's elements requires devising suitable tools, approaches and techniques.

It is important to remember here, the sound qualities of the recording do not exist in notation. A collection of quasi-notational diagrams and X-Y graphs has been devised to transcribe the recording's elements into a written form. This collection serves as a tool to observe, collect, and bring to visual form elements' activities and characteristics, and to 'hold them still' for study. The graphs and diagrams are created in a transcription process—very similar to transcribing music. Emerging from this process will be some vocabulary for describing sound and the sound qualities of recordings. Through this vocabulary the elements can be described and discussed without relying on the typical personal, subjective and inexact ways we are inclined to use to talk about sound (Moylan 1992 and 2015). The analyst will be led to engage the subtleties of the recording, and to compile information on the elements that can be studied and compared to the other domains. Before this can happen, the analyst needs to accurately hear the elements. Engaging the listening process, and learning to hear the qualities of records, will bring the reader/listener to accurately compile information on the recording—a necessary first step toward evaluating them (Moylan 2017). Praxis studies engaging these skills are identified when topics arise, and appear in Appendix A.

OUTSIDE DISCIPLINES IN POPULAR MUSIC ANALYSIS

Outside the domains of music, lyrics and the recording there is much else to consider. While traditional approaches to music analysis might typically not extend far from the composition itself, there are disciplines and subject areas outside music and lyrics that are integral to the study of popular music. Popular music is not autonomous, self-contained from external influences, or transcendent of social interests as 'serious' music is inclined to be perceived (by artists, musicologists, audiences, etc.). These other disciplines that follow can contribute to or provide rich understanding of a different sort. The perspectives of these fields can add a great deal to an analysis of the record. Rock and popular music studies originated with scholars in disciplines outside music.[5] Adam Krims (2007, 185) explains:

> [T]he historic failure of music theorists and historians to engage seriously with "unserious" music has left a vacuum happily filled by scholars from such disciplines as communications, sociology, media studies, area studies, gender studies and geography, to name just a few.

Scholars from numerous fields have examined popular music, exploring various factors both within and external to the song's music and lyrics. Many of these studies emphasize either contextual or meaning matters, and some address both. Disciplines examine the record from their unique perspectives, often not engaging the actual substance within the song; perhaps understandably, such studies often focus on the issues within their discipline—disciplines such as sociology, social psychology, philosophy, literature or others. Other disciplines adapt existing theories or modify analytical methods that were originally designed to study different fields—such as semiotics, with its origins in linguistics. Together, this is a disparate group representing widely varied branches of knowledge. Table 1.1 provides a summary listing of these fields.

Table 1.1 Subject areas of popular music analysis outside the three domains of the record.

Disciplines	Subject Area(s)
Sociology	Social impacts, social behavior, cultural contexts and structures; crime, religion, family, racial divisions, social class; human action and consciousness related to social structures
Cultural Studies	Cultural analysis; cultural practices related to wider systems of power; ideology, class structures, ethnicity, gender, etc.; view of culture as constantly changing; semiotics
Philosophy	Aesthetics Phenomenology
Psychology	Cognition Perceptual psychology Psychology of music Phenomenology
English Studies, Poetry and English Language and Literature	Literary criticism; informed by sociology and cultural studies; journalism
Communications and Media Studies	Communication theory; mass media; journalism
Semiotics	Making meanings; application to music, musical sounds and issues of musical discourse
Hermeneutics	Theories of interpretation; conceptualization of interpretations by analyst, by listener, by a more universal audience.
Subjects studied in some way by most all of the above disciplines	Emotion in music, affects of feelings and moods, emotional response to music; Energy of music's motion, expression and emotion; Expression in music

What follows is a brief coverage of the significant subject areas that might be incorporated into a recording analysis by appropriating their methodologies, should their perspectives and knowledge base align with the goals of the recording analysis. Within the following chapters several of these areas will be engaged in greater depth. An adequate presentation and examination of these subjects is beyond the scope of what can be included here—despite the value and richness offered by some of these fields. The intention here is to identify other, potential areas the reader might seek out for more detailed study, initiated by following source references. With this broadened study, the reader will determine which, if any, external disciplines might enrich their individual analyses.

Popular music makes deep connections with listeners, and stirs listeners in individual and collective ways. It reaches the visceral and the physical, speaks to the cultural and the social, the personal and the collective, it engages reason and mood, touches heart, mind and soul. It does all this to a wide and varied audience; a single recorded song might resonate deeply within millions of people. How, then, might analysis illuminate the ways this music both influences society and reflects it, and the means by which those associations and connections are made? How might an analysis examine the record as interpreted by the individual, by the analyst, or by a more universal audience? How might an analysis reveal the meaning of the lyrics, or recognize how the track's sounds stir and move the listener?

This brings us to what might be of greatest interest to general readers (and music lovers of all types), and also to those of us fascinated with studying and creating recorded songs: an analysis can provide acknowledgement that the record reaches inside the listener. Our analysis might identify the source of what infuses the track with energy and brings it to penetrate into the bodies and psyches of listeners. We can explore how *this* track elicits the sensual, the disturbing, or the ecstatic—whatever the recorded song's core character might be.

Sociology and Cultural Studies

Sociology was among the first disciplines to begin examining popular music below its surface. It is the study of human behaviour and society, how human action and consciousness are shaped by and shape social and cultural structures—including social class, race, gender, religion, crime and economics. The interrelationships between music and society (creating some social context for the analysis) are central to popular music. This is articulated by Ian MacDonald, who frames "the main feature" of his *The People's Music* (2003, vii–viii):

> . . . lies in its espousal of this temporal aspect [that individual cult figures are products of their time] and the proposition that popular music is a product of society first, and rebel festivity second (and always in passing). . . . it is to stress the contextual factors that contribute to the rock revolt— everything from political backgrounds and technological innovations to reactions against incumbent styles.

Sociology's methods can bring insight into the social impacts of records, their social context(s), and the social influences on records—to a genre or artist in general, or in relation to specific tracks. By incorporating this discipline into one's approach, social meanings, influences and significance might formulate and emerge. Sociology is linked closely, perhaps symbiotically, with cultural studies. These two intersected fields largely informed music critics and music journalism, especially in the early years of popular music studies.

Cultural studies emerged from a confluence of various disciplines including (notably for our purposes) sociology, politics, media and communication studies, literary studies, geography and philosophy. It is a field engaged in critical analysis of cultural practices of cultural phenomena—particularly those related to power and control (especially political dynamics) often directed toward studies of ethnicity, gender, generation, nation issues, sexual orientation, ideology and class structures. Seeking meaning generated from analysis of social, political, economic realms, its critical approaches include diverse fields, including those pertinent to popular music analysis: semiotics, media theory, film/video studies, communication studies, literary theory. Keith Negus (1996, 4) defines the "general point" of his *Popular Music in Theory* "that music is created. Circulated, recognized and responded to according to a range of conceptual assumptions and analytical activities that are grounded in quite particular social relationships, political processes and cultural activities."

Philosophy and Psychology

Philosophy has two branches of particular significance to records: aesthetics and phenomenology. Aesthetics is traditionally defined as the theory of beauty, perhaps together with the philosophy of art—or the study of beauty and preference. It is a large field at its fullest. Aesthetic concepts, aesthetic value, aesthetic judgements, blend with other topics that might be brought to task in interpreting and evaluating tracks. The aesthetics of music concentrated on art music until recently—this position articulated its placement at the top of the higher forms of music and its advanced modes of response. The aesthetics of popular music can be studied in great detail in its own right (Gracyk 1996). While some studies of music aesthetics are closer to sociologies of music, aesthetic concepts can provide insight into the record, and also appreciation for the record's qualities. In a manner of blending these subjects, Simon Frith offers:

> [M]usic gives us a way of being in the world, a way of making sense of it: musical response is, by its nature, a process of musical identification; aesthetic response is, by its nature, an ethical agreement. The critical issue, in other words, is not meaning and its interpretation—musical appreciation as a kind of decoding—but experience and collusion: the "aesthetic" describes a kind of self-consciousness, a coming together of the sensual, the emotional, and the social as performance. In short, [popular] music doesn't represent values but lives them.
>
> (Frith 1996, 272)

Phenomenology studies the subjective experience, subjective phenomena and structures of consciousness, especially the first-person perspective (a common perspective in song lyrics, but most importantly: the listener's perspective). It studies how it is that humans are able to empathize with one another's experiences, and find ways to communicate about them, by engaging the mechanism of 'intersubjectivity.'[6] Phenomenology asks one of the most vexing questions in analysis (especially important with an analysis based on sound alone): 'Is my experience the same as yours?' In doing so it seeks to recognize what is universal or generalizable across humans engaged in the same experience; these are the objective aspects of an experience despite the differences across human subjects. Thomas Clifton (1983, 38) explains phenomenology is "more interested in describing rather than proving or predicting." Before extensive detailing of the manner of describing, he notes: "[D]escription of phenomena is something other than a blow-by-blow account of technical trivia . . . and also other than dreamy-eyed embellishments around a musical object pressed into the uncomfortable role of a narcotic." For the interested analyst, applications of phenomenology might hold strong potential to define shared musical experiences, communication and meaning between humans and groups. Phenomenology straddles philosophy and psychology—some refer to this intersection as 'phenomenological psychology.'

Psychology is a complex field, and is sub-divided into a natural science and a behavioural, human science. As a natural science it studies psychological phenomena as natural phenomena (tending toward biology and chemistry), and as a human science (behavioural) it accounts for experiential, social and cultural phenomena. For much of our use is in psychology's study of aspects of the conscious and unconscious mind and thought, including the study of behaviour. Psychologists explore many concepts central to our work, including: perception, cognition, attention and emotion. Perception and cognition of sound, of language and of music (and of recording) are central to music listening and analysis processes. They are woven into the framework for recording analysis presented here, and are strongly represented within other recent approaches.[7] Lawrence Zbikowski provides some insight into music cognition within his examination of a discussion of Robert Johnson's 1936 "Cross Roads Blues":

> Cultural, social and historical context cannot, by themselves, explain the origin of our affective response to the song, for our broad agreement on the effect of Johnson's music transcends these implements of high theory (even if they have a profound influence on what we do with "Cross Road Blues" once we have heard it). I propose that exploring the way cognitive structure informs our understanding of music gives us a way to account for the source of our broad agreement on the affect that pervades Johnson's blues and can help us understand better the way culture, society and history reshape musical practice.
>
> (Zbikowski 2002, xi)

The psychology of music is concerned with the psychological processes involved in music listening, composing and performing music.[8] Music psychology seeks to understand cognitive processes related to music and music behaviours of memory and attention (two central aspects of analysing music from records), in addition to perception and cognition. It can encompass emotion and meaning in music, as well as provide psychological analyses of perceptual, affective and social responses to music. We know music consistently elicits emotional responses in listeners, although it is a complex process and the nature of the affect is not consistent. Music psychology studies are helpful in examining this relationship between human affect and cognition (Sloboda 2005). This brings about interpretation as related to emotion and expression, as will be examined below. First we will look at interpretation related to meaning.

English Literature/Poetry Studies and Communications

Poetry and literature analysis and interpretation can be engaged to explore message, meanings and significance of and within song lyrics. While song lyrics vary considerably from 'high' poetry and literature in form and content, in their embrace of ordinary language (and defiance of correct grammar), and most importantly that they are written to be sung, they have received much attention from scholars and academic programs. Their analysis tools and methods can be usefully applied to the literary content of lyrics and the meanings of the lyrics, on various levels. Simon Frith demonstrates the potential of a sophisticated approach to lyrics as poetry as he explains Aidan Day's description of Bob Dylan's lyrics in *Jokerman*:

> Day starts from the assumption that although Dylan's lyrics are written to be sung, they nonetheless contain the "poetic richness of signification" and "density of verbal meaning" that characterize modernism; Day is therefore concerned primarily with "the semantic properties of the words of the lyrics," which "excludes consideration of the expanded expressive range belonging to performances of those lyrics." He concentrates on "obscurities in structure and verbal texture," on the "indirectness of language in the stanzas," on the "density of allusion and pun," on "lyrics whose framing of the multiform and bizarre energies of identity constitutes Dylan's most distinctive achievement."
>
> (Frith 1996, 177)

Of course, the lyrics of Nobel laureate Bob Dylan are not typical, and Day's observations are particular to Dylan. The point here is this type of examination can be pertinent to the study of the track, and can shed light on the content, meaning, and artistry of the lyrics.

Communication studies is concerned with the processes of human communication—written, verbal and nonverbal. It is concerned with how messages are interpreted within the contexts of their political, cultural, economic, semiotic, hermeneutic and social dimensions, and as they appear within the full spectrum of platforms from face-to-face conversation to mass media outlets (i.e. radio, television), internet and social media. Media studies is often incorporated into the study of communications, with and without broadcast and technology-related dimensions (media editing and designing, on-air talent and acting, production roles and content authors, web design, etc.). Journalism is a central component of the mass communications discipline. Traditional journalism is based on objectivity, ethics and standards that give high importance to eliminating bias. Advocacy journalism has become increasingly prominent (facilitated by blogs and social media), and can project extreme bias without documenting sources; there is also tabloid journalism intent on unearthing the sensational, or writing fictitious accounts; and a more socially-oriented form of journalism characterized by less depth of substance, that is intended to peer into the lives and homes of the famous and the fortunate, and thus entertaining a curious public sector. Knowledge-based, researched journalism generating factual reporting of content and context can prove highly relevant to popular music studies—and the study of the record. This is in contrast to entertainment journalism's infatuation with artists' lives and intimate secrets, that at times includes personal matters of context relevant to the track (but often not); this accounting has been called "a fake idea of truth" (Griffiths 2004, ix) and is of little use in to understanding the record. Emerging from all these approaches is the 'music journalist'; journalists who report on music and artists, review records, write analyses. At its best music journalism engages social commentary, the lyrics, or other fields, though even at its best it only rarely addresses the *music* itself;[9] at its worst it represents superficial and sensationalized writings of little value to understanding the record.

Semiotics and Hermeneutics

As a discipline, semiotics is the study of meaning-making and meaningful communication; including the study of communication through symbols and sign processes, and the study of non-linguistic sign systems. This points to the state of having meaning outside or beyond itself—a situation that 'absolute' music avoids through its 'abstract' concepts and relationships. Semiotics as applied to music looks deeply into meaning-related aspects of music and musical sounds, a process particularly relevant for popular music. Philip Tagg (2013, 145) provides this explanation: "The semiotics of music, in the broadest sense of the term, deals with relations between the sounds we call musical and what those sounds signify to those producing and hearing the sounds in specific sociocultural contexts." This begins his exhaustive coverage of the semiotics of popular music in *Music's Meanings*; therein he formulates theories, typologies and analytic approaches for popular music through semiotics that may prove useful to some analysts and analyses. In describing the complexity of music and the challenges of writing about it, McClary and Walser (1990, 278) observe how sounds carry meaning and signification to create

> a wide open semiotic dimension that can make us think we hear sincere remorse or bad-ass sassiness, that can produce the image of 'the authentic working-class hero' or 'the virginal slut,' that can engage associatively with anything from Louis XIV's Versailles to street gangs in the South Bronx.

Semiotics holds promise for the analyst to engage meaning in popular music, though the challenge is significant; the meanings of sounds are specific to social groups and contexts, therefore what is understood in one group may be understood as something entirely (or slightly) different in another. Cross-cultural or universally understood musical sounds and 'codes' are rare, and are mostly related to biological response to sound (Higgins 2012); aspects of music structure, musical functions and tension/repose also "share far fewer common connections to extramusical phenomena from one culture to another" (Tagg 1987, 286). Further, the large number of highly varied genres of popular music and widely varied production tools and techniques expands the typology of sounds and sound qualities, and their associated meanings; defining a set of characteristics and their associative meanings is a task that will likely continue to be ever more complex as music evolves. Incorporating methodologies from cultural studies and sociology may lead to defining this issue of sociocultural connectivity to musical meanings solely for an analysis of an individual record.

Hermeneutics, in the classic sense, is the theory of interpretation; originally for text interpretation (mostly to sacred texts), it evolved to include methodology of general interpretation. Modern use includes both verbal and nonverbal communication, in many diverse fields. Hermeneutics is sometimes framed as 'the art of interpretation.' It has come to be applied to music, and is more of a central matter to popular music than to art music—although it does have some relevance to traditional music analysis.[10] This makes sense when one remembers 'serious' music sought to be self contained, rational and objective, where it is not a central concern to question how music's expression of moods and feelings affects the listener and shapes their relation to the world.

Interpretation exists at a number of critical points in the experience of the record (or a musical performance), and it is ultimately linked to meaning. Hermeneutics recognizes that what exists as the track (the record), is a limited expression of the artist's view. Further, it recognizes the listener's view also affects their understanding of the record. The artist's view is more substantive than what is in the record; in other words, there is meaning in or behind the artistic statement of the record that has not been fully articulated. The listener's cultural context of time and place, and their experience, education, expectations and state of mind contribute to their receptiveness to the artist's view and influence their understanding. Here 'artist' is artists: lyricist, composer, arranger, performers, recordist, or producer; 'listener' is listeners: analyst, audience, or the individual listener. An important concept is the hermeneutic circle, which adopts the notion that as the whole is understood in reference to the individual, the individual can only be understood in reference to the whole; a circle between subject and object/event. Transferring the principle, the listener is constantly reassessing their conception of the record (the whole musical event), and with each reference (each new listening, or individual assessment of what has been heard) the interpretation is constantly being informed (and transformed, becoming more substantive or enriched) by the record itself.

Interpretation is inherently subjective, in all areas it is encountered or enacted. Recording analysis is itself an act of interpretation. Julian Horton (2001, 357) articulates this point: "[W]e never write the immediate, but an analytical strategy masquerading as immediacy," a strategy (analytical process, approach, tool, or methodology) that is already influenced by our subjective position; in this way, an analysis is already influenced by the analyst's predispositions and biases, as well as experiences and assumptions. From the others listed above, the interpretations of performers and others involved in the creative processes of creating and crafting the record are separate from our analysis concerns. In the application of hermeneutics to recording analysis we are concerned primarily with the receiver not the generators of the message—in other words, we are interested in the interpretations of the track itself, by the listener, or a more universal audience, or perhaps in the analyst. We might find ourselves in any of these three roles. In *Song Means*, Allan Moore articulates a methodology for *Analysing and Interpreting Recorded Popular Song* based on a "hermeneutic superstructure." Moore (2012a, 1–2) explains the book is "about *how* they [recorded popular songs] mean, and *the means* by which they mean. . . . The reason I focus on the 'how' is that I believe that, as a listener, you participate fundamentally in the meanings that songs have." Thus, he identifies meaning as being fundamentally linked to the listener.

The listener makes the meaning, not the speaker of the message (or the recorded song).[11] It is the act of listening that transforms a stream of sounds into meaning—the listener interprets the sounds to arrive at meaning. As we have noticed, it is not isolated sounds that formulate meaning, but how they relate and what they represent, communicate, elicit, and much more.

Affects: Emotion, Energy, Expression

The understanding of how a track's different elements come together and produce certain effects on the listener is distinct from those of interpretation or meaning. This sense of the expression, the energy and/or the emotive content of the record elicits response, reaction and sensation at the intersection of body and emotions. This might be conceptualized as the record's poetics—following the notion of suggesting "poetic feeling, intuition and imagination."[12]

Exploring, identifying and discussing what and how music emotes and what it communicates on kin-aesthetic and visceral levels can quickly become highly personal and subjective—yet these are clearly part of the musical experience, a part clearly attractive to listeners. The highly personalized experiences and subjective interpretations might be the most important consideration of the lay listener, and may be perceived as highly significant by the beginning analyst—after all, we consider ourselves at least partially defined by the music we use, listen to, and the meanings we find in it. Allan Moore (2012a, 1) articulated this: "The chances are that who you believe yourself to be is partly founded on the music you use, what you listen to, what values it has for you, what meanings you find in it." If our goal is a meaningful analysis that explains the listening experience (the track) beyond oneself—though, this is slippery ground. The personal can distort a productive collection and evaluation of information on the record, in contrast to a more detached process; the listener can become prone to seek information to validate or support subjective experience, rather than exploring deeply to discover more significant ideas. Passive and "pharmaceutical"[13] (or mood modulating) listening instead of active and engaged listening may play a role in this type of perception.

To explain what makes the record work as a coherent communication (or how the recording shapes the record) requires some intentional steering away from the personal, away from the affective and emotional experiences of the individual and interpretations that are largely subjective—and toward the universal, to the objective aspects of an experience as perceived by one's collective group of similar listeners who share some commonality of background, and to interpretations that are as objective as the human condition allows. This is not to imply listening experiences are less 'significant' during those times when listeners may (with or without intention) open themselves to the personal; this perspective is natural and common, and may well be deeply rewarding to the individual. This acknowledges such forms of understanding have strong potential to benefit only the individual, and have the potential to generate interpretations so highly personal they are shared by few others. To contrast, in seeking what may be the universal, one is more prepared to engage the unexpected or unknown and allow for the potential of any experience to be significant and to communicate uniquely (and in the realm required for analysis).

This is perilous territory to navigate. Most everyone talks around this, or addresses it in very vague terms. It is difficult (perhaps impossible) to generalize about personal experiences that hold deep meaning for listeners—individuals, groups, cultures, social clichés, etc. Emotion, energy, and expression will quickly tend to defy adequate description. The tools to engage these are few, and our vocabulary can quickly seem too much related to the senses than to analytical processes. Music has always defied verbalization to a considerable extent, and analysis can struggle for vocabulary and to bring concepts into language. Although music has a level of syntax, it lacks a consistent lexicon or semantics that can embrace or engage the many genres ('languages') across its breadth. Indeed, some believe it is simply not possible to adequately describe music, or the experience of music. Moore (2012b, 101) provides a response to this reluctance to engage what might be "always beyond precise verbalization," by recognizing "to avoid the attempt is to render our evaluations worthless."

Some of the above disciplines can provide access here. The 'intersubjectivity' and revealing meaning in semiotics (Tagg 2013, 195–228), hermeneutics' guidance in formulating and conceptualizing interpretation, seeking the objective qualities in shared experience through phenomenology, the lyrics' poetics and meaning engaged through literary methodologies, and exploring affect through the psychology (and sociology) of music all open different forms of access to this territory.

The questions to understand the record's emotion, energy, and expression are perhaps simple to frame, but are vexing to solve. A start might be:

- How might we seek largely universal descriptions of 'what provides energy' to the tracks, and 'what is the energy' of the record, 'what feelings, emotions, expressions are present,' *what* (by a quality, reference, an object, etc.) is it that elicits them,' and 'how (i.e. through an event or process) are they created?'

- What tools are of use to determine transfers of meaning and affect across groups (social, cultural, economic, etc.) of listeners?
- What is cultural, what is personal, what is biophysical within the qualities or meanings that are perceived?

Incorporating Other Disciplines Within an Analysis

These disciplines and subjects can be incorporated into the recording analysis in a myriad of ways. Many topics and disciplines might inform an analysis. In addition to those just described, the established research stream of sound studies might augment the context or meaning qualities of an analysis, and the recently emerged stream of ethnography can add dimension (related to context, interpretation or meaning) to the creative voice and process.

Ethnography has roots in ethnomusicology; it has been seen as an 'urban ethnomusicology' treating mediations and interactions as normal aspects of culture (Walser 2003, 18). Central to popular music studies is a stream of research examining studio-based ethnography.[14] This stream sits at an intersection of a number of communications and social disciplines, with a degree of focus on ethnomusicology research interests and methodologies. Roughly, this is the study of the cultures, interactions of people (often power-related), creativity, technology influences, and the creative processes within the recording studio environment—or of recording processes, as might seem more appropriate. Though related to ethnomusicology, research is rarely focused on the record itself (the artefact of the recording process), but rather on the social, cultural, functional, anthropological (and so forth) aspects of recording studio practice in creating the record. The process can shape and impact the record in profound ways, and can enrich recording analysis at times; this emerging field examines these, and more.

Sound studies also sits at the intersection of many disciplines, with a focus that can shift depending on the subject of the study. Human impact on the acoustics of the world is at the core of this discipline. The range of disciplines includes ethnomusicology, acoustic ecology, nature, acoustics, psychoacoustics, perception, anthropology, philosophy, sociology, media and cultural studies, film studies, urban studies, architecture, arts, musicology, music theory, and performance studies. Sound studies concentrates on the history of audio media, the nature of sound and listening, and the roles of sound and listening within the modern experience. Sound scholars look at perception and the effects of sound, the ways humans interact with sound and how sound interacts with the world around it; some study change and reactivity. Some streams hold significance to recording analysis as they inform understanding of recording's elements, their perception, and the human condition.[15]

As with incorporating other disciplines into an analysis, these interdisciplinary streams might be of interest to some analysts, and relevant to certain records, or enhance analytical methods; they can inform the analysis, but are not analytical tools. It bears reminding here: the record itself—the artifact that is the message, substance and content of the creative voice—is at the core of the analysis. Exploring and identifying the ways the recording shapes the track, and the track itself, are the central topics of recording analysis. Variables here are the goals of the analysis and the subjects the analyst wishes to illuminate, and the balance between the record itself and the outside subjects pertinent to understanding its content and message.[16]

Every record is unique and demands analytical techniques appropriate to it, and an analytic framework that is accepting, relevant and useful for the study of them all. The framework functions directly as a broad structure, inviting the analyst to apply their own questions and methods of inquiry within its process. Appropriate depth and breadth of coverage appropriate may be adjusted for the skill level of the analyst—from more general analyses to the meticulously detailed, graduate-level studies or a beginner's exercise, work by a novice or an established scholar will all approach recording analysis with a different sense of purpose, a different skillset and their own desires, experiences and preferences. When to incorporate disciplines outside the domains, which to incorporate, and how to assimilate all perspectives and the information those subjects offer depends on the analyst. No single approach is promoted here; rather, what is proposed is a sense of what is possible, so the analyst can seek to identify what is important for the individual track, and steer the analysis to directly engage it.

The lucid analysis will clearly define the degree to which it engages these subject areas—and which to evaluate. To be relevant, any conclusions generated from these areas will relate to the substance and sounds of the track. The content of the individual song will play a role here; for example, a song with expressed social views will be examined differently than one with materials and sounds that make connections with dance and romance. A number of these disciplines can generate studies in themselves—they can inform our analysis of the record, or they may stand alone.

Advocates for any one or particular combinations of these subjects will note their approaches are indispensable for the understanding of the track. They would be right, both from their vantage point and also for the completeness of an analysis (which might demonstrate what is relevant to the track, and what is not). This brings us to a need to recognize the content of the individual analysis is not only what it seeks to do, to discover, to evaluate and to understand, but also what it chooses to omit from detailed or direct consideration, or even from superficial acknowledgement. A lucid analysis will define its coverage within the track's domains, and those outside—and to what purpose—to best serve the individual record.

FRAMEWORK AND PROCESS

An approach to recording analysis is offered here, directed to the recordist as well as the analyst. This approach is intended as a framework and a flexible, functional process. It is intended to be practical and accessible, and intended to lead one to explore the contributions of the recording to the track. These contributions may include its materials, relationships, concepts, characteristics, expression, and more. The framework does not prescribe a methodology, a set of instructions, a philosophical or aesthetic basis, nor a theoretical premise. Rather, the framework establishes a point of reference and departure for a process of exploration—a process that is adaptable to the individual record, a process guided by inquiry into the record itself, beginning with the recorded song itself, a process based on a framework of a few simple and direct principles and concepts.

Table 1.2 Subjects that might be incorporated into a recording analysis.

Music
Lyrics
Recording
Performance
Affects: Emotion, Energy, Expression
Outside Disciplines: Context
Outside Disciplines: Interpretation
Outside Disciplines: Meaning

This approach is designed to guide readers to perceive and collect information with an intention of objectivity and accuracy. It offers direction for the evaluation of information to identify the defining qualities of the individual work; to arrive at conclusions of relevance and significance for the track studied. The framework establishes a general scheme of inquiry from which the process may forge or explore many paths—paths that are unique and appropriate for the individual recorded song—so pertinent conclusions may emerge.

This is a malleable, broad framework that does not define how it is to be used; one within which methods of examination appropriate to the individual record might be applied (or emerge) and function.[17] The process of the framework can incorporate or be adaptable to most theories or methodologies related to the analysis of popular music or the analysis of songs (and records). No single approach, theoretical construct or analytical methodology, can be the most effective portal for the study of all recorded songs. Nor might any meet the needs of all the goals one might have in analyzing a track—including embracing streams of knowledge outside the sonic content of the record. Therefore, the most appropriate approach and the most appropriate analytic tools are determined independently—and these can only be identified once within the process, once some pertinent information about the record has been uncovered. Within the process of recording analysis, the analyst will be required to determine the method(s) most appropriate. The recorded song itself will reveal the methods required to access its character and message.

With but a few important exceptions, existing methodologies do not examine the qualities of the recording, or the influences of the recording on the recorded song.[18] Rather, most approaches are focused elsewhere—be it aspects of the music, the message or meaning of the lyrics, circumstances around the track, social or personal interpretation, stories surrounding the artist(s), and other areas.

Herein may be the beginnings of a methodology for examining the qualities of the recording, which might supplement those analytical methods that cannot otherwise address the recording. Some unique approaches to the elements of recordings are offered, and are important aspects of this approach to recording analysis. The elements of recordings are defined and explained, and the reader will be guided in learning how to recognize the subtleties of their qualities. Tools for evaluating their characteristics and activities have been devised, including an approach to a vocabulary for describing their qualities. Their contributions to the record are explored, and will be central aspects of the framework.

Framework's Analysis Sequence

The framework's analysis sequence has two stages: a preliminary, preparatory stage and a stage encompassing the three-step analysis process.

The preliminary stage sets the goals of the analysis and establishes parameters for its content. The goal of the analysis identifies the purpose of the analysis, what it wishes to learn about the track.

Table 1.3   The framework's two-stage analysis sequence.

Preliminary Stage	3-Step Analysis Process Stage
Setting the Goals of the Analysis	*Observations* of the record's elements and materials, actively collecting information/data
Establishing the Content of the Analysis: 1. Subjects/Topics to be Included 2. Scope of Subject Coverage	*Evaluations* of the information that was obtained through observations
	Conclusions formulated by studying and analyzing evaluations

Establishing content articulates the subjects or areas that will be the focus (topics of study) of the analysis and the scope of coverage (its breadth, level of detail, etc.) each subject will receive. The topics to be explored, the object(s) of study, problems to be solved, tasks to perform, etc., are established. Domains to be examined or emphasized, outside disciplines to be included, specific traits to examine, and much more are embraced within the goal, focus and scope of the analysis. One can expect focus and scope to evolve during the course of studying the track—some subjects outside the record may be added, some subjects deemed uninteresting and discarded, some topics emphasized, some areas minimized, and so forth—though the goals will typically not change, but will likely be refined. As one learns more about the record, one recognizes what warrants further study or greater detail, how inquiry needs to expand or focus, and what one needs to uncover. This sense of scope and focus allows the process to unfold productively and with a sense of purpose that is defined and guided by the goals of the analysis.

At its simplest, the three-step analysis process follows the path:

- Observations of the record's elements and materials, and outside disciplines and sources;
- Evaluations of the information obtained through observations;
- Conclusions derived through study and analysis of evaluations, etc.

This general three-stage process has some similarity to traditional style analysis,[19] and some distinct differences. The process is designed to function fluidly for recording analysis; it can flex to embrace typical structure-oriented processes and style analysis approaches, and will open to other approaches as well—especially those that might be devised by an analyst to best serve the track being examined. The process is malleable and is general enough to remain consistent from one analysis to the next. One might pursue any level of detail at each stage—adapting to various skill levels, the purpose or goal(s) of the analysis, or the needs of the individual recorded song.

The observations stage is simple data collection of 'what is there' within the recording, within the lyrics and within the music. Any method that does not bring evaluation in to play might be utilized (at one's considered choice) in order to collect this information on elements and materials. The important matter is to collect information before interpreting or evaluating the information—limiting confirmation bias as much as might be possible.[20] This minimizes the influence that evaluating information might have on collecting information; when one begins to sense organization or content, one can involuntarily begin to perceive 'what is there' as 'what I have found' based on 'what I wish to find,'—or believe is there, or what I perceive as being present based on limited experience or on insufficient exposure to the materials. Some overlap of observation and evaluation is inevitable, as observations continue within the analysis process as one continues to discover each time one engages the record; the point is to delay evaluation and conclusions until sufficient information on the track has been collected so it can speak for itself. Thoughtfully sequencing this 'data collection' might help here; when one is inclined to make assessments of, say, the sonic characteristics of the text perhaps one might, instead, collect data on the melody that presents it, or the dynamic shaping of the line, or the interpersonal distance of the vocalist to the listener. This might seem a diversion, yet in practice is productive and informative, and will perhaps bring light toward identifying qualities that might otherwise go unnoticed. Observations originating in the materials and methods of other disciplines might be included here.

Within the evaluations stage various analytical theories and methodologies might be applied to examine the syntax of each element and the formulation of materials, to identify functions and to determine the relationships of materials, and much more. Formal theories and specific methods are not the concern here; rather, the focus is on revealing the content of the record.

At the 'conclusions' stage, the definitions of theories and methodologies might take more precedence, as per the position of the analysis. This stage is the broadest, most complex and most individualized by the analyst and/or conforming to the requirements of the track. The multidimensional interactions of all elements, and all materials of the three domains are examined, considered and insights formulated, tested and finalized. Here is where connections and discoveries are made, information synthesized, and where the analysis is collated and formulated. The goals and breadth of the individual analysis might include a careful examination—perhaps a central examination—of social context and impact, semiotics and interpretations, and expressive qualities and meanings, in addition to the three domains. The complex confluence that embodies the individual record is deciphered and revealed here.

Framework Principles and Concepts

As the framework seeks to be conceptually simple, it is guided by a few basic principles and concepts. These are principles that provide context for the framework, and concepts that provide points of reference for recording analysis. While the framework embraces these and allows them to provide a sense of guidance and coherence, this is not a methodology, but rather a simple approach that establishes a 'frame of mind' for freely and deeply exploring the record. It is deliberately nondescript in analytic method, and it intentionally avoids any one purpose or goal of analysis.

These principles guide the process, while seeking to be transparent in their influence on any outcomes. By these few principles the framework seeks to be open to information of all sorts, with a minimum of bias from past experience and knowledge. The framework seeks to impose limited influence on the analysis process and on what is revealed. The four principles are:

- Every record is unique
- Equivalence
- Perspective
- Listening with attention and intention

A primary position (principle) of this writing, and of its framework, is that every recorded song is unique. This might at once sound obvious, perhaps even trivial. The premise is quite radical, though, if considered in light of how often widely accepted analytical methods can bring vastly diverse pieces of music to look the same, or very similar, once they are analysed. Methods of analysis may easily function as "the equivalent of a sausage machine: whatever goes in comes out neatly packaged and looking just the same" (Cook 1987, 2). An analytical method might be applied with the intention of illuminating the work, but instead the work is forced into one of the few predetermined, model structures the method makes available. Analytic methodologies can thereby limit investigation of the individual piece; an examination with few options limits the opportunity for unique qualities to emerge, be recognized. The analysis might validate the theory (however loosely), but does not provide insight into the work. Fully embracing the notion that every recorded song is unique is indeed radical, then, as it throws conventions aside and opens one to the possibility of something new in each record, the possibility of something new that might embody each track. Each record is a world unto itself, and this position allows this world to be most fully explored. To recognize the uniqueness of the individual recorded song is one of the guiding principles here; it will be realized most accurately utilizing a process capable of revealing the record's unique materials and qualities.

The second guiding principle is that all elements are of equal value, and have an equal *potential* to be significant, to be used as a basic element in a musical idea (Tenney 1986, 8), and to shape the message and the artistry of the recorded song. This is the principle of *equivalence* that identifies any element of any of the three domains has an equal potential to contribute to the uniqueness of the recording at any moment, and at any individual level of the multidimensional texture of the recorded song—this includes the elements of the recording. This provides the opportunity whereby pitch, for example, is not necessarily more significant that any other quality or element; that the music or the text may not be more significant than the recording. 'Significant' here does not mean most important, primary or prominent; significant means that the element plays a crucial or substantive role. This principle is central to many stages of the analysis process, as well as to the listening process; it is integral to discovering the unique qualities of any record.

Perspective is the third guiding principle; it represents the act of observing and engaging specific level of detail. Perspective might be conceived as occupying a particular vantage point, or position of observation. Knowing one's perspective as a listener, allows clarity of what information is being observed and at what level of detail is the focus of one's attention. Perspective allows one to make marked shifts during observations and evaluations; for example, it might establish awareness to the activity on a specific structural level, or might focus attention within a sound's timbre. This principle embraces and acknowledges the multidimensionality of the record, and provides guidance in navigating its conceptual levels; it makes comparisons within the same level or across levels possible. Perspective assists in articulating intention and focus to the listening process, and it lends clarity to the analysis process of engaging specific structural strata, and their elements and materials—and much more.

Listening with attention and intention is the fourth guiding principle. Listening is the only way to access the record. The deep listening needed to access the workings, materials, messages and affects of the record requires an engaged and deliberate listening process. Directed attention is contrasted with open awareness; these are guided by intention in alternate listening experiences. Within the intention of listening a balance is sought alternating between seeking specific information and observing the sonic profiles of the record for what is significant and of interest. Attention holds the perspective intact, without intention perceptually salient features will grab attention and divert or distort the listening experience. Evaluating all dimensions of the record is facilitated through approaching listening with intention, and systematically holding focused attention throughout the experience. These concepts shape the process of collecting information from the listening experience and they shape the quality of the experience. This is not to say one cannot listen openly, seeking notable features that grab attention naturally; the analysis process requires hearings of this sort. One can establish a context of open listening, of being receptive to the unknown, of being prepared to hear the unexpected.[21] It is being open to the unknown and being receptive to the unexpected that allows discovery of the uniqueness of a record.

Interrelated to these principles, the framework divides form and structure into separate and distinct concepts. The two are distinguished by characteristics of time, dimension and other inherent qualities. Table 1.4 outlines the guiding principles and the general form and structure concepts of the framework. The concepts of form and structure will be explored in detail in the next chapter.

Table 1.4   Guiding principles and central concepts of framework.

Guiding Principles	Central Concepts
Every record is unique	Three states of form
Equivalence	Sound as object (stilled time)
Perspective	Shape, syncrisis, crystallized form
Listening with attention and intention	Hierarchies of structure
	Sound as event (unfolding over time)
	Movement

In conclusion, within this framework, any viable approach to analyzing the disciplines and domains of the recorded song might play out and reach its goals—if, of course, utilized with adequate skill. Methods of analysis and focused objectives many be used singly or in combination to best illuminate the unique record. Robert Walser (2003, 24) articulates this as he identifies that the success of an analysis "is relative to its goals, which analysts should feel obliged to make clear." Clarity within an analysis is formulated through clearly defined goals of subjects and disciplines to be examined, and success in illuminating the track is achieved through defining goals appropriate to the individual record.

CLOSING THOUGHTS ON ANALYSIS

This recording analysis framework and process, with its guiding principles and concepts, will facilitate controlled exploration and open discovery by not being tied to formal methodologies or theories. Controlled exploration and open discovery are essential for an analysis of a recording, and each present their own challenges. A suitable balance of the two is sought—a balance determined as much by the skill level of the analyst as it is by what is appropriate for the analysis. Later chapters will suggest ways of engaging and balancing these. When options are many, the processes to choose, to limit, to focus, to define gain significance—this becomes part of the process itself. Walser (*ibid.*) offers:

> Any analysis presupposes a host of choices that have been made by the analyst: what is worth studying? Which of its features are constitutive of its generic identity, uniqueness and efficacy? Why and how do the music and the analysis matter?

Some features of the record are common within the genre, some are unique, some features follow established known practices of language and syntax, and so forth. Deciphering between these conditions is part analysis, part learning the record, musical genre, context, etc.

When each record requires its own approach to analysis, the challenge arises to search it out, to uncover it, and to recognize it once it reveals itself. Knowledge, experience, skill and more support this process—and underlie at least some of the choices made by the analyst, and thus generate their evaluations and conclusions. In this way, the analysis is an interpretation. Allan Moore (2012a, 5) writes: "To analyse a popular song is, of its very nature, to offer an interpretation of it, to determine what range of meaning it has, to make sense of it." The analysis is the product of the analyst's complete background (education, experience, knowledge, ability, etc.), and their engagement with the record; the relationships and meanings they perceive are the ones life has equipped them to recognize.

In this is a tendency to emphasize what we can already detect and identify, to explore more deeply the subjects of our areas of specialty and our interests. Even in setting goals for the analysis there might be inclination to focus on a single discipline, and to ignore others—to focus on the music over the text, for instance. If done with intention, and the analysis does not present itself as more inclusive, perhaps the limited goals of engaging the record can be realized. Still, in the literature much has been written about rock and popular music that emphasizes disciplines outside the domains of the record, and that does not address the music itself—a situation that draws concern from many popular music musicologists. Among these are the observations of Robert Walser (2003, 21–22): "[A]ny cultural analysis of popular music that leaves out musical sound, that doesn't explain why people are drawn to certain sounds specifically and not others, is at least fundamentally incomplete." Allan Moore (2012a, 5–6) offers the perspective: "Too often in the literature, whether academic, journalistic, fan posting, or whatever, interpretations are made without adequate anchorage in the details of an actual aural experience of the song." The balance of information about the record and information from outside disciplines needs to be determined while setting the goals for the analysis, and sometimes adjusted during the analysis process; this is what will be emphasized in the analysis, and includes which outside disciplines might be used to explain, illustrate, support, supplement, find meaning in (and so forth) what the analyst has uncovered within the record. This balance frames one's interpretation of the record—the interpretation that here is called 'analysis.' This balance establishes the objective of the analysis, and the success of any analysis might be recognized against its stated goals.

It is difficult to discover materials and relationships, and to recognize qualities we have not previously experienced. We rely on what we know, and what we have repeatedly experienced, to make assessments. This is at least partially why so many analyses of popular music were once (a few still are) pitch-centric, reliant on functional harmony and structural, tonal pitch relationships when the music does not warrant it. Lori Burns (2008, 64) illustrates this point, as she explains the common problem of analyzing modal patterns with tonal expectations and analytic tools:

> Is such music truly nondirectional, or does our analytic system (that is, Roman numerals and linear analysis) simply fail us? After all, analysis is not just the act of hearing, but the act of applying a theoretical model. . . . I would argue that there is an implicit tonal bias when we analyze any harmonically conceived music. When we identify a chord as "I" or "tonic," this evokes a host of associations of tonic as the generator of harmonic activity, the goal of directed motion, and the source of unity. . . . If we cannot locate dominant-tonic patterns, we are likely to assert that the music is static, nondirectional, or unconventional.

This quote also calls out some of the impacts of the bias of our musical educations being based almost exclusively on 'serious' ('art,' 'classical') tonal music; our perception of musical experiences can be distorted, and our point of reference (knowledge) will lead us to seek to pull that information from whatever we encounter, whether to do so is appropriate or not. We perceive what we most readily recognize, what we confidently know, what we expect, and are prepared to experience. But all popular music and rock music do not follow such traditions; some are modal in their harmonic language, others approach harmony as sound or timbre, and more. What gives a song motion (a sense of departure, speed of direction, a point of arrival), unity, coherence, balance, etc., is not universal in the records we encounter daily. To understand these records, our approach must conform to the materials being studied, and not rely on inappropriate theoretical models to seek to discover the qualities of each record. Growing knowledge and experiences deepens our capacity to recognize new musical languages, materials and relationships, unique elements of the lyrics and to more fully experience and recognize the qualities of the recording. It is in the elements of the recording that many readers will encounter challenges—as many of their qualities may not have been experienced previously, or experienced enough to acquire requisite skills (and adequate knowledge) to hear nuance and to make accurate observations.

Ultimately recording analysis leads to talking about the record—its substance and perhaps its broader context. We present our conclusions and illustrate them with examples from our evaluations, and support it all verbally. Talking about music effectively can be difficult, and talking about the recording can be truly vexing. Music study is filled with technical terms that add a level of abstraction, making music less immediate and meaningful—and that is when the terms actually apply to the music. There are common alternatives to describing the affective experience, the journey of the unfolding song, the meaning of the text, personal interpretations or to describe association elicited by the recording. These all can carry one away from the substance of the record, its sounds, message, motions, meaning, and resultant affects. There is need for vocabulary and analytic methods to make a meaningful examination of popular music clear to the reader (and the writer), a means to explain precisely what is going on in the music and how it communicates to the listener. An alternate tendency is to "write in an impressionist manner that may capture something of the way the music feels and is received but that does not address the details that contributed to making the effect" (McClary and Walser 1990, 280).

Some small advance might be made within these pages toward establishing ways to talk about sound, and especially to talk about recorded sound. It is intended that the framework and process will lead toward clarity on the substance of the recording, with a goal to do so without overcomplicating the material, or its analysis presentation and description—though the result is ultimately dependent upon the reader, not this writer.

Remembering an analysis is an interpretation, the process and framework that leads to the analysis must also be an interpretation that is inherently reflective of the originator; thus in the spirit of disclosure, I admit to being predisposed to downplay the personal and to emphasize what I perceive to be more universal—though within the framework I have attempted to allow a breadth of space for all facets to be explored as equally as one might wish. The framework responds to the questions the researcher/analyst/listener is trying to answer, and opens one to explore and discover their own interpretation.

Finally, I remind the reader of the book's subtitle: "How the Record Shapes the Song." While significant coverage of music and lyrics appear, they are present so we can learn what they are so that we might recognize how the recording impacts them, interacts with them, and combines with them to form something else. The focus of this book is on the recording and its shaping of the record. Production techniques and artistic creativity continually bring new 'sounds' to music genres that are largely reflected in the sound qualities of their recordings. The broad concepts and intentions of the framework should allow it to remain relevant as records, genres and styles change, and as the listener/analyst engages unknown, unimaginable sounds and relationships in each emerging generation of records. This book will open the reader to discovering their own pathways through recording analysis—and through any individual track—to reveal their own engaged interpretation.

IDENTIFYING SAMPLE QUESTIONS FOR RECORDING ANALYSIS

The intent of the analyst might be quickly defined by asking, as clearly as possible: 'what is the information I will now seek?' Questioning 'how' to proceed follows. The act of proceeding is more deliberate and focused by asking 'where' to look, 'how' to find, and 'what' qualities might be contained in the object of inquiry. All the while, one uses prior experience of what *might* be present coupled with the knowledge and open expectation that the unknown awaits.

This is intentionally over-simplified. From this simple construct rigorous examination will emerge with ever-greater clarity. As one cycles queries between simplifying and seeking ever-greater levels of detail, an appropriate degree of reduction and amount of nuance and breadth appropriate to the track will be revealed.

Each chapter will end with a set of "Some Questions for Recording Analysis." The questions are not intended to be exhaustive of the chapter's materials. They are a point of departure to guide the reader to navigate those materials within analysis processes, but are not all-inclusive. At the end are some pertinent questions that have emerged from discussions in this chapter. Several are starting points, several frame the intent of the analysis as a whole, and others initiate streams of inquiry that can only be initiated after some groundwork has been laid.

Analysis is guided by inquiry. Asking the appropriate question, or framing a question appropriately, will lead to meaningful answers and uncover pertinent information. Learning to ask the right questions takes, well, learning—learning from failures more than successes. Individual songs may require uniquely tailored questions. As questions are pursued and perhaps do not yield results—or yield unexpected results—new and more insightful questions will be formulated, leading the reader through the analysis in greater detail and in increasingly individually crafted ways. In this way, the analysis process conforms to the qualities of the song—not the reverse.

Some Questions for Recording Analysis

How does the music work? *What makes* the music work?

What in the song is worth studying?

What elements shape the track's materials and their characteristics?

How does the music shape the recording?

How do the sounds of the recording shape and enhance the music?

What is actually going on in this recorded song?

What does this record communicate—on its surface, under the surface, within its core?

How do the music, the recording and the lyrics coalesce to support, project, or communicate the song's message?

How are the powerful moments, and the events that move the listener accomplished?

What makes this song so meaningful, to so many?

Why does it speak to me, reach me?

NOTES

1 See Michael Chanan's *Repeated Takes: A Short History of Recording and its Effects on Music,* especially the "Record Culture" chapter, for an initial entry into these topics.

2 A number of popular music analysts have demonstrated a significant depth in the harmonic language and harmonic structures and relationships within some styles of rock. Notable (but not exclusively so) among the (quickly growing number of) scholars are Lori Burns, John Covach, Walter Everett and Stan Hawkins. Some of their important writings are referenced throughout.

3 Albin Zak (2001, 22): "The record's most essential character, its sound, remains, even at this late date, its least-talked-about aspect."

4 In this reference Frith (1996, 159) is identifying: "we actually hear three things at once: *words*, which appear to give songs an independent source of semantic meaning; rhetoric, words being used in a special, musical way, a way which draws attention to features and problems of speech; and *voices*, words being spoken or sung in human tones which are themselves 'meaningful,' signs of persons and personality."

5 The culture of college/university music schools, until recently, held that popular music studies were deemed inappropriate for music scholars to pursue, as that music had little sophistication, and therefore limited value for study. One who wished to advance in the profession needed to study and write about more sophisticated, significant and serious topics.

6 Clifton (1983, 41–42) explains this: "The knowledge to be gained from description is neither objective nor subjective, but intersubjective. Phenomenological description tries to avoid description based on personal attitudes which vary with empirical circumstances. . . . Phenomenological description is subjective in the sense that its terms are subject-dependent and subject-related."

7 Especially noteworthy is Allan Moore's *Song Means* (2012a).

8 Deutsch (2012a) *The Psychology of Music* provides entries to this subject's many areas.

9 This prompted Susan McClary (1994, 38) to write: "the study of popular music should also include the study of popular music."

10 See Savage (2015) *Hermeneutics and Music Criticism* for hermeneutics within the context of art music.

11 This is a reference to the three stages of communication: emitter, channel and receiver. Here the emitter is the performers, producer, lyricist, composer, recordist; the listener is the receiver, and the channel (vehicle) is the record, called "the sounding object" by Tagg (1987, 290).

12 This is one of several definitions of 'poetics' offered by Albin Zak in explaining the title of his *The Poetics of Rock* (2001, xv).

13 Sloboda (2005, 319–320) uses this term to discuss a form of passive listening experienced by subjects within laboratory listening tests that are designed to study listener response to music stimuli. This term is used here to describe a type of escapism, in using music listening to modulate mood or transform the effects of life experience, inferring a replacement for drugs that might serve these purposes.

14 A growing number of scholars are delivering important work in this area; among them are Alan Williams, Paul Thompson, M. Nyssim Lefford, Phillip McIntyre, Louise Meintjes and Eliot Bates.

15 R. Murray Schafer's pioneering writings in this area explore the relationship between humans and the sounds of the environment, especially as reflected in spaces and soundscapes: "The soundscape of the world is changing. Modern man is beginning to inhabit a world with an acoustic environment radically different from any he has hitherto known. These new sounds, which differ in quality and intensity from those of the past, have alerted many researchers to the dangers of an indiscriminate and imperialistic spread of more and larger sounds . . . Noise pollution is now a world problem" (Schafer 1994, 3). Important resources are *Sound Studies* (M. Bull, ed.), a significant collection of important historical writings as well as current research, and *The Sound Studies Reader* (J. Sterne, ed.), especially relevant for studies on noise and silence, listening, and form and meaning of sound across cultures, contexts and centuries.

16 An example of an analysis that substantively incorporates disciplines and engages relevant issues from *outside* the record while also providing keen insight *into* the track is "'Joanie' Get Angry: k.d. lang's Feminist Revision" by Lori Burns (1997).

17 While I've stated this subtly before, I want to be clear: I do not intend or wish to offer a competing theory or method on the analysis of rock or popular music. Rather, this framework is an approach and a process within which more formalized theories and methods might be applied and function—as one might wish, or deem appropriate. Just as importantly, one might choose to incorporate portions of various methods, to modify existing methods, or to devise their own, as might seem appropriate for the individual track or some larger context.

18 Allan Moore's *Song Means* (2012a) and "Beyond a Musicology of Production" (2012b) are two important methodologies that embrace aspects of the recording as being integral to the track; Philip Tagg (2013) provides a method that incorporates semiotics, and offers a detailed examination of content and meaning in popular music that includes analysis of all aspects of the track, including the recording; in *The Musicology of Record Production* (2014, 1) Simon Zagorski-Thomas pursues "the ontological question of how recording changed music and how that change needs to be incorporated into its study" which is of great relevance here.

19 This process owes some of its origin to Jan LaRue's *Guidelines for Style Analysis* (2011). The process I propose has been substantially reshaped and redirected to be pertinent to the record. Especially important is its realignment to popular music, lyrics and the recording, and thus the nature of the data being collected and its typologies, as well as the methods and principles for evaluations and conclusions.

20 Confirmation bias is the common act of seeking data to support what we already believe, or believe to be present. (Lewis, 2016)

21 Still relevant, in *The Principles of Psychology* (1890, 402) William James wrote of attention: "My experience is what I agree to attend to. Only those items I *notice* shape my mind—without selective interest, experience is an utter chaos. Interest alone gives accent and emphasis, light and shade, background and foreground—intelligible perspective, in a word."

CHAPTER 2

Overview of Framework: Principles and Concepts, Materials and Organization

This chapter adds detail to the three domains (music, lyrics and recording), and will engage several of the framework's central principles and concepts in greater depth. The three domains are each examined for their unique syntax and language, and the materials they may generate. The various functions which elements and materials might assume will follow. Principles and concepts of the framework are clarified by examining:

- Perspective
- The delineation of the various concepts of form
- The relationships of form and perspective to structural hierarchy
- Listening with intention and attention

The framework's guiding principles of each record's uniqueness and of equivalence are imbedded in the discussion of domains and syntax. The principle of perspective is investigated as the concepts of form and structure are distinguished. Here the passage of sound in time as events and the stilling of sound and time formulating sound as object are contrasted; this distinction is clarified by the concepts of shape and movement.

Listening with intention and attention is the fourth principle of the framework. Accurate listening is central to successfully perceiving the qualities of the record; the confluence within records easily obscures its individual elements as well as their subtle features. The highly personal nature of listening brings challenges to identifying universal information and understanding personal bias and our unique perceptions and understandings. Listening with intention (supported by knowledge and sense of awareness), cultivating attention, and listening without expectations can bring the listener/reader to hear and recognize the record's qualities—those qualities that are actually present—in ever-greater precision, clarity and detail.

THE THREE DOMAINS: MUSIC, LYRICS AND RECORDING

As briefly discussed in Chapter 1, the recorded song is comprised of three streams of distinctly different information: the domains of the music (and its performance), of the text (lyrics), and of the sonic qualities of the recording. These three realms form the materials of the record—and the record's content, its message and meanings, and its expression. They merge in confluence, blending synergistically into a single multidimensional sonic tapestry.

Despite this intricate confluence, the three domains exist individually with their own unique elements and characteristics, and may function independently as well as in tandem with other domains. Each of the three domains contributes its individual materials and imparts a characteristic imprint on the track; each may be created independently, though they may be crafted to complement another or both others. On some level, they are each crafted and shaped individually, in their own distinct voice.

We engage all three domains within the recording analysis process, seeking to understand their materials and organization, and their contributions to the recorded song. We examine how they interact to form intricate relationships—interactions that add significantly to the overall depth and complexity, richness and subtle detail of the song. An emphasis on the elements of the recording lights an examination of how the 'record shapes the song.'

The content of each domain is shaped by its elements. Their elements function with purpose, and carry the potential to contribute fundamentally to the character and content of the song. Next is a brief outline of the elements of each domain, to set the context for the discussions that follow. These elements are explored more deeply later, in chapters dedicated to each domain.

Elements of Music

Elements of music in the record are more than the traditional score, or even the live performance and the score; the elements represent a specific performance, captured in a specific time and place. The performance is intrinsic to recorded song, and embodies the musical elements. Here the music and the performance have potential to contribute equally to the musical idea. Most significantly, they combine collaboratively into a single sonic and musical statement.

The traditional elements of music in the 'music' domain are shown in Table 2.1. In much of recorded song, it has been practice to engage music with a focus on pitch relationships—with successions of single pitches forming melody and groupings of simultaneous pitches creating chords, thereby enacting harmony and tonality. Rhythm manifests the temporal to become integral to and to propel melody and harmony through organized time. Traditionally pitch-based elements and rhythm have been enhanced by dynamics and timbral constructs of instrumentation and arranging, and tessitura and range. Ultimately all coalesce into states of texture and fabric.

Table 2.1   Elements of music.

Rhythm	Pulse, meter, tempo, durations, rhythmic patterns
Melody	Succession of individual pitches, grouped into a line
Harmony	Simultaneous pitches grouped into chords, progressions, key, tonality
Dynamics	Loudness levels, loudness relationships, gradual changes of levels
Tone Color, Timbre	Selection and scoring of instruments and voices, orchestration and arranging
Range and Tessitura	Span of pitch levels of a line, an instrument, or of any pitched aspect of the musical materials or structure
Texture	Number of individual parts and the type of materials being presented, such as monophonic (a single melody)
Fabric	Composite sound of the music, includes all elements

Melody and harmony are commonly examined in traditional music analysis. The other elements found here are typically less often evaluated, and appear less explicitly (if at all) in the musical score. These will receive detailed attention alongside melody, harmony and rhythm as we examine them within recorded song.

In the approach presented herein, all elements are examined for their contributions to the music's character, flow and substance. Each element is examined with equal attention, and all elements are examined for their place in the interrelationships that are formed in many ways and on many levels, as a complex web of the interaction of elements is integral to even the simplest music.

In contrast to traditional, score-based analysis processes, analysis here will emerge from the sounds of the music itself; evaluated directly from the record, where the song is complete.

Subtleties of music performance are integral to the musical materials, musical expression and other musical qualities of the record. In the record, sonic aspects of the performance are elements as well. The energies and sounds of performance meld with the ideas of composition to create a larger experience—especially once transformed by recording processes. Albin Zak (2001, 51) recognizes this as he articulates:

> The performances that we hear on records present a complex collection of elements. Musical syntactic elements such as pitches and rhythms are augmented by specific inflections and articulations, which include particularities of timbre, phrasing, intonation, and so forth. Furthermore, the inscription process captures the traces of emotion, psyche, and life experience expressed by performers. That is, the *passion* of the musical utterance is yet another element of a record's identity.

The performance is about intensity and expression, as much as any other element; these are present in the nuance of timbre or sound quality most strongly, but are also reflected in other elements, such as dynamics or subtle flexing of time, silence and pitch.

All of the sound qualities of musical performance blend into the substance of the music. They are woven into the complex tapestry of the record, just as fundamentally as the traditional elements of music. These performance characteristics include a "flexible molding of pitch, rhythm and timbre" (Winkler 1997, 186) that establish fluidity of the line, and considerable added substance. This substance becomes inherent to musical material; it brings the performance to be inseparable from the composition.

To not study the performance, then, is to omit significant substance as well as the music's living experience, diminishing full understanding of the content of the record. Our approach to understanding the music domain must extend to encompass both musical materials and their shaping by music performance, and acknowledge their inherent contributions to the record. Though not a simple undertaking, the dimensions and qualities of the performance are part of the analysis of the record's 'music.' Every detail is part of the whole; every subtlety of the captured performance shapes the musical idea and the musical experience, though not momentarily as in a live performance but in perpetuity as a recording. Following and examining this detail, so important to the music, is rife with challenges; our music notation system is not designed to effectively depict this information, and our ear has a tendency to distort, simplify or ignore details of what we are hearing in order to allow a passage to conform to our limited notation system. We will work to address this matter; as "those elements which listeners tend to find most interesting in popular music [records] and which most nearly capture the music's particular strengths (rhythmic and pitch nuance, texture and timbre) are impossible to notate accurately" (Moore 1997, x), a window into understanding these materials is needed.

Table 2.2 outlines some of the elements generated by a musical performance.

Table 2.2   Elements generated by music performance technique and expression.

Timbre-Related	Sound quality (tone color) of instrument/voice, changes in sound quality, rhythmic movement of sound quality changes, performance intensity cues
Rhythm-Related	Rhythmic placement of notes, accuracy of rhythms and rhythmic placements; Use of time for shaping musical idea vs. surface expression: stretching and contracting time
Dynamics-Related	Shaping of line; accents; tremolo; Generation of performance intensity
Pitch-Related	Intonation; pitch inflections, bends, glissando; rhythmic placement of pitch inflections; vibrato
All Elements	Changes and characteristics generated by expression, performance style and performance techniques

Table 2.3   General dimensions of lyrics.

Message and Story	Theme and subject; narrator and protagonist; Voice and persona; story and narrative; Figurative language devices; References to external sources
Structure	Stanzas; lines; patterns and relationships; structural devices.
Sound: Timbre, Rhythm, Pitch and Dynamic Inflection	Word sounds and patterns of sounds (such as rhyme); Rhythms (word, line, patterns between lines and stanzas); Diction; Performance techniques and qualities

Lyrics and Text

Even the 'unrecorded' song contains much more than music. Lyrics provide a core dimension of the song, and are readily recognized as equal in significance to the music (and at times more so). From earliest known songs, the lyrics and the music interact synergistically and establish a larger whole; generating great richness of interrelated and complementary materials, communication and expression for the song.

Typically the song's music provides a platform upon which the story, theme, narrative or drama of the lyrics unfold. The text generates the song's message and communicates its meaning. As the vocalist delivers the lyrics, the persona of the storyteller is revealed; the performer and the lyrics become enmeshed; a myriad of performance aspects enhance and enrich each sound of each word to further the presentation, drama and expression of the lyrics. The text communicates the theme, and defines the song's character; it creates motion in time and a sense of intrigue, and it provides the narrative of the song. Lyrics may contribute strongly to the song's movement in time, through time, or as shifting time, and may present a message of direct simplicity or interwoven complexity.

Lyrics also generate sonic qualities when they are sung. Timbral sound qualities and rhythmic patterns, tonal inflection and stress emphasis points are integral elements to the lyrics. Structural organization of sounds, theme and story, concepts of delivery and persona provide additional elements. Great richness of ideas and sounds, word play and meanings, expression and contemplation, and much more may be instilled in the text.

Table 2.4 Elements of the recording.

Spatial Properties	Sound stage dimensions, stereo or surround phantom image location and size, distance location; environments of individual sources, of the sound stage, and space within space; ambience, echo, reverb
Timbre	Crafted sound qualities and timbres of sound sources, component parts of timbre/sound quality; altered performance intensity and performance techniques; timbral balance, pitch density
Dynamic Levels and Relationships	Loudness balance of voices and instruments; dynamic contour/shape of individual parts, dynamic contour of individual sounds; dynamic contours of reverberant energy; reference dynamic level; dynamic contour of overall program
Rhythm- and Time-Related	Tempo, durations, patterns of durations (rhythms) in all artistic elements; time functions; timbre of time units; rhythms of reflections
Pitch-Related	Pitch areas; pitch registers; pitch/frequency placement, range of sounds/sound sources; pitch density, timbral balance

In short, lyrics provide the message, sonic and the structural elements in the text domain. These will be examined in Chapter 4.

Elements of Recording

The recording process and reproduction medium of the record impart sound qualities and sonic dimensions to the record (Moylan 1992). This creates the elements of the recording—sonic qualities that are unique and distinct dimensions of the recorded song.

These elements are condensed in Table 2.4. The elements' qualities and relationships shape each track uniquely, just as each song's music and lyrics establish it as unique. The elements of recordings transform, enrich and enhance musical materials and the lyrics; they may also provide core substance to the song (Moylan 2015).

Some elements—such as dynamic contours, pitch areas, timbral balance, and others—are conceptually linked to music and music performance elements; while associated with music-related elements, these elements function differently and appear in different dimensionalities in recordings. Other elements such as spatial properties of lateral location, distance and environments are elements that provide qualities that clearly delineate the uniqueness of records, and the unique experiences of recorded music.

These interactions establish a rich set of relationships where elements may reinforce one another, may contrast with or complement others, may balance or may delineate any of the other elements. In much the same way as traditional musical elements interact to create larger concepts like implied harmonies in melody, the elements of recordings interact to establish larger concepts such as the sound stage. Chapters 6 through 9 will detail these elements.

THE LANGUAGES OF DOMAINS

A language is a system of communication. Most often we conceive 'language' to be human communication related to linguistics—perhaps existing in both spoken and written forms. As such, language is a set of words and a system of rules for combining them (grammar and syntax), through which a community can communicate. The concept of the syntax of language might be extended to materials and organizational systems in music, and perhaps to communication and to meaning in music. To a lesser extent there is a language created by usage of the recording's elements; recording's language and its syntax of defining materials and building structure is in its beginning stages (Moore 2012a; Tagg 2013).

Syntax in Each Domain

The use of the term 'syntax' and its concepts in linguistics are adopted here to conceptually represent relationships and materials within each domain, especially those in music and in the recording. With this, 'syntax' creates a means to link all three domains in terms of basic-levels, structures, materials, and more.[1] This allows insight into how the domains are contrasted and how they interact.

In the field of linguistics, the syntax of a language is generally agreed to be a set of rules, principles and processes governing the structure of sentences. Syntax dictates how words, clauses and phrases are combined into well-formed sentences, into sentences that conform to governing rules of structure in a language. In different terms, syntax provides the basis for coherent ordering of words (of various types) and phrases (with their different characteristics and functions) into properly constructed sentences, and how smaller units might combine into larger to result in fully formed sentences. From a finite set of rules for connecting words, an infinite number of sentences can be generated; each meaningful to one knowing the vocabulary and syntax of the language. Grammar, in contrast, dictates agreement between words and other aspects of the sentence, in how clauses and phrases communicate; syntax is a subset of grammar. It is common for people to be able to speak a language fluently with no formal knowledge of the language's grammar.

These concepts of language and syntax are extended to organizational systems in music (Cogan and Escot, 1984), as well as related to meaning (Sloboda 2005; Tagg 2013). Syntax in music is generally agreed to refer to rules and conventions that shape musical materials and that guide its structure. Music has a syntax that is conceptually and (in some ways) functionally comparable to linguistic syntax in its hierarchical organization and complexity of variables—though, of course, containing its own substance.

Syntax of Music

Dominant theories of musical syntax are grounded in the principles and conventions of tonality; most particularly, the tonality reflected in the procedures of eighteenth-century European art music (reflected in its scale degrees, chord structures, key structure). Thus, traditional theories and related methods of evaluating musical syntax will have varying degrees of relevance (or irrelevance) to the analysis of the music of an individual popular song. The harmonic language of a song may well have little in common to the art music conventions.[2] What might be considered a 'universal law' in the syntax of a formal language (or perhaps in any established musical tradition such as any European art music of an era predating the twentieth century) (Lerdahl and Jackendoff 1983) is in popular music a convention or a matter of musical practice slowly resulting in loose guiding principles. Conventions are established as musical practices are repeated and refined, though conventions are inherently a matter of artistic choice to conform to a common set of sonic qualities and principles of organization; conventions are invented constructions, not an inherent and universal 'law' of organization. A tremendous variety of music and production qualities is found in the tracks of today, coupled with all those stretching back to the beginning of recorded songs. There are few (if *any*) governing rules or principles outside individual genres (although the 'words' of music's language may be borrowed or adopted across genres), and I venture to offer there is very little *universal* across them all. In reality, while (as with language) some syntax must be intact for music to be coherent, it is not rare for popular music artists to seek to ignore or wilfully counter anything that could be recognized as a 'rule.'[3] Think of how quickly one might recognize that a new track is by a known artist simply by recognizing telling qualities of the 'sound' of the track, or by language of the opening bars of the music, and when the text enters, if the syntax of the lyrics is not telling, the vocal timbre removes all doubt.

The ethereal language of musical style in popular music, the evolving language of the artist, and the language of the individual song are all in play in the syntax of a popular song. Examining syntax, it should be possible for every song to be recognized as unique in its materials, syntax and language of musical elements. Similarly, artists develop unique inclinations leading to a personal syntax within the language of the style. In some styles more than others, the individual piece of music may be based on a language or syntax that is uniquely its own (Zbikowski 2002, 138).

Just as with verbal language, we do not need to know explicit rules of grammar to understand music—or to perform music, or compose music. Music makes sense to us as we recognize its construction and materials—even if that recognition is not conscious (or non-discursive or nonverbal) and calculating of the matter and materials of construction (syntax and language) (White 1994, 55). At some inherent level, music communicates to a listener who is at least somewhat acquainted with its syntax. What it communicates to the listener, though, can vary—as *how* it makes sense can be individual, and *what* sense it makes might be immensely personal. As a musical style's principles and conventions are known across a social group, the music communicates more broadly, and is understood similarly throughout a community and between its individuals. John Sloboda (2005, 179) offers: "It seems the mere exposure to the standard musical culture is enough for children to build grammatical structures."

Syntax of Recording

There is an emerging language represented in the structure, content and usage of the recording's elements; not surprisingly its depth and sophistication is much looser and simpler than what appears in lyrics and music. The typology and organization of recording's elements establish its relationships of structure and its character. These can be conceptualized as syntax, and what is perceived as the recording's sound qualities and their evolving appearances is generated by production style. Because recording is a young language (or languages if separately considering stereo, surround, VR, mono, etc.), its syntax is not as thoroughly developed, and its activities are largely static for extended time periods and are rarely temporal at the level of surface rhythms.

Listeners have few production-related expectations of sound qualities of recordings, although some generalized practices and conventions have become established that create an expected sense of relationships (such as the lead vocal located in stereo's center). Recording's syntax provides little sense of directed motion and of what should logically follow next, however—the grammar and syntax of succession that creates the tensions, expectations and resolutions of directed motion are latent, and remain in a process of being defined and established. One such example, though is the shifting presence of the lead vocal that often occurs in an established pattern of qualities between verses and choruses. The elements of recordings are still understood by the listener on a nonverbal level, however; the elements 'make sense' and function in ways that are recognized and processed by the listener, even in the absence of grammar in the language of recordings.

Still there is the basis of language and syntax. The attributes of elements shift through various values, and create patterns over periods time in their own ways (from microrhythms through macrorhythms); they form simultaneous groupings of attributes with and between other sound sources, establish relationships within elements. Further, recording's elements substantially determine the character of the track's overall sound. These all are leading to conventions of usage for consistently producing sonic qualities and relationships resulting in shape and motion within an element, but in ways vastly different from the elements of music or of lyrics. These have brought stylized production principles within genres of music—principles that might be explored through examining typologies.

Approaching elements by typology can provide some access to their content and activities. The approach to typology adopted here blends aspects of type 1 categorization and prototype effects explained by Lawrence Zbikowski (2002, 36–49), with principles of various computer programming techniques for data collection and analysis, and the concepts of a spectra of characteristics and contrasts of musical elements and materials employed by LaRue (2011).

Typology of recording's elements can follow some of the identifying principles as music's elements—though they are all unique in content.

Typology (as used here) is a flexible process of categorization, a way to organize data and make available pertinent examination of its content. The unique qualities of recording's elements can be reflected in a typology that is set up to categorize its unique attributes and values. A typology might collect the following information (these are examples, not an exhaustive listing):

- Collection of identifiable attributes (or variables) and their values (levels, etc.), activities, states, or qualities within each element
- Spectrum or inventory of all values (levels, characteristics, etc.) within an attribute
- Range of values, levels or qualities within each attribute (such as the highest and lowest level)
- Increments between the values assembled in the 'spectrum of values' to identify the smallest and largest increments in use, and, if present, the lowest common value (this might be conceived as synonymous with the half-step in pitch as its smallest increment)
- Preferred or predominant values of each attribute or element
- Relationships of values between successive sounds, and/or between other sound sources that are simultaneously present
- Speed (tempo) of change of values (levels or qualities)
- Amount of change of values
- Static attributes (unchanging characteristics)
- Patterns of activity in and/or patterns of characteristics throughout events
- Any number of other attributes might most appropriately serve the goals of the analysis

Typology establishes the lexicon of the individual qualities within elements, and perhaps their organization. This set of data is (to extend the association with syntax) the alphabet and a budding vocabulary of the elements of recording. This will be of great use as we later evaluate the content and characteristics of the track.

Any element might generate materials that can create larger ideas by combining smaller ideas, resulting in more complete artistic statements, or in sound objects; and materials might likewise be divided into similar divisions. In this way the structure of linguistics' syntax tree can be conceived as branches of structural relationships, and is context dependent—such an example would be space within space (see Chapter 8). With this approach it can be evident that the syntax of recording is both functionally and conceptually present, and will be approached as such in our explorations of the recording's materials and organization.

To summarize this section, the concept of 'syntax' has been adopted in order to observe how 'the language' of each domain functions uniquely and to identify any similarity. It serves as a tool through which each element of each domain might be observed and evaluated, compared and synthesized with others. Thus far discussion has focused on the individual domains, and on the unique language and content of each.

Ultimately we engage the confluence of the three domains as the single overall sound—a multidimensional texture. Continuing the association with language and syntax, with this overall sound a 'dialect' of the record is formed. The dialect results from the three domains of the record communicating with a single voice. The unique qualities of individual tracks are reflected in this syntax of confluence, as well as in the ways its materials are constructed (typology) and the ways they are organized (structure) within the individual streams of each domain. Through examining syntax, the unique qualities of the individual record will begin to emerge.

FUNCTIONS AND MATERIALS OF DOMAINS AND THEIR ELEMENTS

The multidimensional texture created by the confluence of the three domains can seem a kaleidoscopic array of activity. Elements within all three domains contribute to the song in any number of unique ways, at every moment in time. In doing so, the domains and their elements acquire function within the materials they shape; they come together in various ways to create vocal lines, instrumental parts, accompaniments, and much more.

Elements contribute to and present the musical, lyric and recording materials and ideas that are infused with the expression of the song. Each element is in a defined role; each at a certain level of significance. Domains, their elements and the materials of the song they establish all function in various ways within the song, and often change roles as the track unfolds.

Elements can play a primary, supportive, ornamental or contextual role (function) in shaping materials. For instance, traditional approaches to analysis will identify the melody of a lead vocal as being comprised of pitch and rhythm, and perhaps harmonic implications of the line. There are many other elements that appear as well; each is of concern to us. A few might appear as a slowly changing vocal timbre shaping performance intensity of the line, dynamic shaping of individual notes and/or of the entire line, unique alignments of pitches and vibrato against the metric grid, subtle expansion of the phantom image, distance placement, and more. These elements all have a role in shaping the line, some primary, others in another role (but all sharing an equal potential to function at any other level of significance), as they coalesce into a single idea or material.

The materials of the recorded song and the sources that perform them also assume one of these same four functions, though in their own way. Table 2.5 outlines a potential hierarchy of performers/ instrumentation and their materials based on function in a hypothetical song. The table identifies the significance of each part based on function. The dominant element(s) that shape those parts follow. The table represents typical parts and instrumentation of a track, but does not represent a model; an

Table 2.5   A single track's hypothetical hierarchy of the functions of musical materials and performers' parts, including select elements of lyrics and recording.

Primary Materials & Elements	Supportive Materials & Elements	Ornamental Materials & Elements	Contextual Materials & Elements
Lead vocal – song's persona; delivery of text; melody/pitch, lyrics/ sound quality, rhythm	Rhythm guitar – harmonic accompaniment; harmony, rhythm, sound quality	Backing vocals – verse ornamentation; sustained harmonies, sound quality, register	Drum set – groove, pulse; rhythm, pitch registers, sound quality
Lead guitar – melody/ pitch counter melody, solo section; chord progression theme	Piano – harmonic accompaniment; harmony, melody, sound quality	Ancillary percussion strikes – rhythmic reinforcement; rhythm, sound quality	Bass – groove, pulse; melody/pitch, register, rhythm, implied harmony

(Continued)

Table 2.5   (Continued)

Primary Materials & Elements	Supportive Materials & Elements	Ornamental Materials & Elements	Contextual Materials & Elements
	Strings – melodic accompaniment; pitch, rhythm, sound quality	Brass section – rhythm, register, sound quality	Synth pad – harmonic pedals; harmony, sound quality, register
Lyrics – images, story, subject, message; word sounds, expression; paralanguage	Backing vocals – chorus response; melody, harmony, lyrics/sound quality	*Lyrics* – literary devices, word-play	*Recording* – sound stage dimensions and its environment
Recording – timbres of sources and performances	*Recording* – positional location and size of sources	*Recording* – sound source environments	

appropriate reduction of possible collections and relationships of parts and materials will emerge from the track itself to illuminate an analysis.

Functions

The four fundamental functions are primary, supportive, ornamental and contextual functions. These functions are roles, elements and materials (musical, text or recording materials) that, respectively, serve to deliver primary ideas, support or reinforce the primary material, embellish material, or create a sonic context or backdrop.

Elements, domains or materials that carry the weight of content and expression are substantive; they present the central ideas of the song represent the primary function. Examples of these in the music domain might be the 'material' of main melodies of the song—typically the lead vocal line is shaped primarily by the musical element of 'pitch,' of course. The primary materials might also include other musical elements: a dominating beat, groove or a recurring rhythmic pattern occurring within the element of 'rhythm,' or a 'striking chord progression';[4] elements of the text can typically be primary elements as well, or elements of the record might serve a primary function. Parts with primary functions are essential to the song, and deliver its core themes and ideas, character and characteristics.

Materials, elements and domains in a supportive role provide reinforcement to those in a primary role. In providing reinforcement, they add critical dimension(s) to the core character and are also essential to the fundamental qualities and nature of the record. They add character to the primary materials and support that material in contributing to fundamental shape and motion.

Elements (or groups of elements) that dominate primary materials of a track (or a specific section or structural level) are controlling elements. Controlling elements are those elements that contribute most significantly to structural organization and/or to the over-riding motion of the track. Controlling elements drive movement and/or are the most substantive contributors to shaping the primary materials and concepts of the track.

Those elements with an ornamental function provide enhancement to the essential—whether supportive or primary. An ornamental element adds character by providing decoration to the essential qualities of materials. Materials/elements in an ornamental role embellish the substance found in others; they are not essential to the track's core, but provide important nuance and adornment. Ornamental treatments are often important characteristics of musical style and performance technique.

Elements or materials with contextual functions establish a sonic context or backdrop that is more or less consistent throughout the entire track or major sections. They might present an ambience or some specific dimension to an overall sound, against which all other materials and all other functions play out. Contextual functions are inherently stable, and create a unifying aspect to the track.

In typical recorded songs, primary materials are dominated by the elements: melody and pitch relationships. The supportive roles might be comprised of rhythmic activity and pitch relationships providing harmony, and some parts emphasizing sound quality; ornamental roles might be assumed by dynamics and the timbre changes of interpretation; contextual roles might be reflected in the sound stage dimensions or perceived performance environment, and others. Obviously, these examples are highly simplified, incomplete and are not universal. The track is much more complex.

Recognizing this concept of interrelationships of elements and functions applies to each domain and between domains, a myriad of possible combinations results. This vast number of potential interactions point to the complexities within even the simplest of records—and the number of potential ways a song might be shaped. These many interrelationships shape and create its many dimensions and structural levels—this stratification is explored below.

Domains also assume these roles of function—at a higher dimension. Any domain can serve the primary function for the entire song, or might switch to a supporting or an ornamental role. Using text as an example, that domain might provide the controlling elements and primary materials throughout much of the song, and assume a secondary role when only backing vocals delivering only vowel sounds are present during a certain section. Switching of roles for domains is typically apparent between sections, but might occur at any moment in the record.

Still, in recorded song, the lead vocal nearly always plays the central role, and assumes a primary function. This is perhaps the only universal rule or principle of recorded song.

The conventions[5] that establish any musical style bring certain relationships of elements (creating stylized musical materials) and their functions to dominate, or be most common. Conventions may also be reflected in typologies and characteristics of the materials. Individual songs in any musical style will deviate from convention at least somewhat in order to provide novelty, interest or variety, though certain intrinsic relationships of materials and elements will remain evident.

In all types of song, along with the lead vocal (with its musical elements and its text) any of the elements of music might be ornamental, supportive or primary at any moment, or throughout a major section of the song; any element or group of elements may be its controlling element. Similarly, any of the elements of the domain of the record hold equal potential to assume any function in shaping the materials of the track, and in relation to the musical elements, or to the text. All elements of all three domains share the potential to be equally significant (or provide context, support or ornamentation) in shaping and providing content for the song.

This concept of equivalence is the recognition that each element has an equal potential to assume any function at any moment in time—no matter its domain. Any element has the potential to be the central carrier of the artistic statement at any moment in time; any element may also be in any other role, and that function might shift at any moment (Tenney 1986). This is a guiding principle in listening as well as in analysis.

This discussion of functions of elements and of formulations of conventions in the record brings a way to access and recognize the contributions and qualities of the recording's elements, typologies and concepts.

Inside the Elements and the Materials of the Song

Reviewing Table 2.5, common allocations of materials are apparent. Some individual parts/lines are performed by a group of instruments, and others are performed by a group of voices; many materials are presented by a single instrument or voice, while some instruments may present several musical ideas simultaneously (common in piano parts). Each of the musical materials may be defined by a number of elements that are actively shaping or otherwise significant to those materials. Identifying the activities and functions of the elements that shape and characterize the musical materials is a central concern in recording analysis.

By considering George Harrison's lead vocal in the first verse of "While My Guitar Gently Weeps" (*LOVE* stereo version, 2006), we might recognize some of the contributions of elements to the track's materials.

From the entry of the vocal we readily identify a single line of pitches, organized by rhythmic patterns and patterns of intervals, delivering a text exhibiting a variety of sound qualities and rhythmic elements, and performed with dynamic shaping of the melodic line and precisely placed vocal timbral qualities. This statement is purposefully vague to illustrate it could apply to any vocal line, and that it is without distinguishing detail—it does not allow us to learn about the line. To bring clarity to the content of the vocal part we might examine (as appropriate): (1) the melodic structure and pitch content of the vocal line, and its phrasing, (2) implied harmony and other patterns of intervals, and repeated phrases or repetitions of partial phrases, (3) the rhythmic and dynamic elements that further shape the phrases, (4) the text sounds and rhyme scheme, text rhythm and the repeated lines and words. This is only a start.

Upon further examination of these text and musical elements in Harrison's vocal, we identify the significance of various pitch relationships within the context of the song; the placements and stresses of the rhythms related to the meter, phrasing, text structure; the timbral modifications of the text to bring expression and emphasis, and clarify meaning; the tessitura of the line, placement within the singers range; and more. There are intentional qualities to the recording, too; the proximity of the vocal, distance of the guitar, and the open areas of lateral space between Harrison's vocals and the strings. The exposed nature of the vocal allows ready access to these observations, including intentional recording gestures.

With further listening, one becomes increasingly aware that the lead vocal is more than a melody of pitches, rhythm and words. It is the result of the performance, where pitches do not always conform to scales, and rhythms flex for expression or to more closely reflect speech inflection. Word timbres morph in quality without changing vowels, all the while communicating intensity, a certain level of tension and an expression of melancholy—and more.

Harrison's first verse vocal is also shaped by the elements of the record. The vocal has a width of approximately 8-degrees and is centered in the stereo field; it is located within close proximity to the listener, and while there is just a short length of distance between Harrison's voice and the perceived location of the listener, there is also some instability to distance cues generating some ambiguity and motion. The vocal's dynamic level in the mix is markedly louder than the moderately soft intensity level at the performance's beginning. Finally, there is a negligible influence of environment cues with only the environment's barely noticeable accentuation of timbral characteristics within the vocal's frequency range.

As these observations demonstrate, methods used to evaluate the lead vocal's melody, harmony and rhythm are of no use to evaluating the record's elements. Nor are they applicable directly to evaluating the lyrics. Elements from each domain are unique, and can be vastly different from one another. As we have just observed, the record's elements are equally important to hear and understand as the music elements, or the content of the text. This gives rise to questions of concern on how the elements, materials and domains interact. A beginning might be:

- How does the music present the lyrics, enhance them, support them?
- How are the lyrics and the singer's voice quality utilized to present the musical ideas and materials?
- What does the recording contribute to the character of the music, of the lyrics, of the track?

FORM, STRUCTURE AND PERSPECTIVE

Within our human experiences of time and space, internal or external, we grasp to make sense of our perceptions—of all kinds. The record is no exception. As we experience materials and their component elements, they coalesce into shapes and events, into sequences and patterns, into hierarchies and structure, into concepts and form.

Recorded songs have form, in three types. First, form is an overall shape comprised of recurring large sections, defined by their materials (Cook 1987). Second, form might be framed as a global quality, whereby the record crystallizes into a singular core essence (Moylan 2015). Third, form can be conceived to reformulate the recorded song from a temporal experience to a realization, or conceptualization of out of time (Tagg 2013).

Herein 'form' and 'structure' are defined and used as separate entities and concepts. While many sources and common convention might use them synonymously, a distinction is made here so these terms might bring some clarity to important opposing, but complementary ideas.

In this approach, form is the record's shape, frozen in time. It is the recorded song crystallized into a single sound object, one unified and multidimensional entity or conception. Form does not evolve as an event; it is a memory of the song fully formed, as an object of many dimensions. Crystallized form, like sound object, embraces the notion of music as a memory, crystallized into a single representation of overall qualities comprised of rich detail and existing out of time. Crystallized form is distinct from the concept of sound object (Schaeffer 1966), however; a sound object is isolated from the context of the track, whereas crystallized form *is* the context of the track.

This contrasts with structure. Structure provides hierarchies and the relationships that bring form to its overall shape. The recorded song's structure is the architecture of its materials. As such, structure is comprised of many component parts that represent the materials of the song, at various levels of scale and significance. These materials establish interrelationships as the track progresses throughout its duration; its many section, phrase, sub-phrase and combination of section time-spans establish a complex hierarchy and stratification of structure—and the materials of the recording are also present within the structure, as are those of the lyrics.

Structure is dynamic and continually unfolding. Its materials are sound events. Events begin, progress and conclude; they appear experientially while unfolding over time. Structure recognizes motion and movement, departure and arrival, tension and resolution. The sound event draws attention to the experience of the existence of sound and music over time, as unfolding structure.

Structure encompasses the characteristics of all materials (and all elements that shape them) coupled with a hierarchy of their interrelationships, as they function and provide motion. The structure of

Table 2.6   Defining and contrasting concepts of form versus structure.

Form	Structure
Shape	Movement
Large Dimension Formal Divisions	Song Structure
Syncrisis, 'Now Sound'	Linear, Temporal Direction
Sound Objects (stilled time)	Sound Events (unfolding over time)
Large Dimension Qualities of Elements	Structural Strata and Architectural Framework
Crystallized Form	
Abstract Concept and Global Qualities	Functions of Materials and Elements

the music will be the basis for examining the structure of the track; in other words, the song's structure will be used as the reference against which the structures of the three domains are related to one another. The elements of all domains function to provide the musical materials with their unique character. All musical materials, their functions, content, and time-spans can be related within the hierarchies of structure.[6]

Time Perception and Pattern Perception

Two areas of perception are central to experiencing the track and its structure and form: time perception and pattern perception. These also impact the perception of all temporal aspects of the record, its domains and its elements.

We might understand the track more deeply by examining time perception and how our perception of the passage of time differs markedly with and without engaging the metric grid. Recognizing our innate and our learned organizational abilities for pattern perception, will provide some clarity when engaging groupings of materials of all elements and dimensions. The motion and dramatic effects of rhythm and its organizational force will become clearer as examination of each element unfolds.

Time perception not only factors into the perception of rhythm, it also factors into the perception of time units for evaluating subtle qualities of environmental characteristics and timbre. It is also central to the listening process and in recognizing the global qualities of form. The following discussion is applied in observing all elements, refining listening skills and in conceptualizing the track.

Time perception is distinctly different from duration perception. We make judgments of elapsed time based on the perceived length of the present. The length of time humans perceive to be "the present" is normally two to three seconds, but might be extended to as much as five seconds and beyond. We do not perceive the passage of time as such, but rather what we do perceive, we perceive as the window of time of our existence, the present time—as what is going on right now. This 'right now' is a window of time within which occurrences are perceived as occurring during the moments of active, conscious awareness; the tipping point between the past and the future within which our lives play out. The present has "fuzzy edges" (Snyder 2000, 51) that are limited by the type, speed and amount of information that occur within short-term memory (STM)—averaging 3–5 seconds, and may reach as long as 10–12 seconds.

> The focus of conscious awareness could be thought of as the 'front edge' of STM. The focus of conscious awareness has an even smaller capacity than STM: three items at most. This is the cutting edge experience as it is happening; it is what is in immediate awareness.
>
> (*ibid.*, 50)

From this we recognize, while we are hearing in the present, we are hearing and understanding as a function of memory. This front edge also contains the briefest of the three memory types, echoic memory; its duration is maybe 2 to 3 seconds and it functions for perceptual processing of various sensory memory systems (Zbikowski 2011, 185). The maximum length of the experience of hypermeter seems to be influenced by the amount of information (lengths of measures, number of measures, amount of stimuli within measures) that can be held in STM (Kramer 1988, 372). Working memory is a buffer of around 10–15 seconds duration "within which information provided by perceptual processing can be evaluated" (Zbikowski 2011, 185); it is important for many processes, one ready example being the processing of language, which requires holding an amount of information, evaluating it and then determining its content and what to do with it.

This is the psychological present; it is our window of consciousness. It allows conscious awareness with which we perceive the world and listen to sound. We are at once experiencing the moment of our existence, evaluating the immediate past of what has just happened and anticipating the future (projecting what will follow the present moment, given our experiences of the recently passed moments, and our knowledge of previous, similar events).

Humans do not perceive the passage of clock time accurately; judgments of clock time are inherently imprecise. The time length of a piece of music (or any temporal art form, such as a motion picture) and all temporal aspects within the metric contexts of the record are separate and distinct from clock time. A lifetime can pass in a moment, through the experience of a work of art, and its aesthetic and dramatic statements. A brief moment of sound might elevate the listener to extend the experience to an infinite span of existence. Just how the experience shapes time perception is complex, and is not consistent between listening sessions to the same work, let alone between individuals. Time judgments are strongly influenced by the listener's attentiveness and interest. If the material is stimulating, the event might appear shorter in duration; desirable experiences typically seem to occupy less time than would an undesirable experience of the same (or even shorter) length. Expectations, interest, contemplation, mood, energy and even pleasure caused by music can alter the listener's sense of elapsed time.

Within musical contexts, humans have the potential to accurately gauge clock time. The metric grid can be used to calculate clock time. The process is to recognize a tempo meaningful to the listener, and to transform that tempo into a metric pulse; the pulse may then be transposed into a tempo of a pulse per second or half second. This will obviously work more effectively for some tempos than others. With this sense of recognizing clock time, time units can be recognized, allowing observations of timbre and spatial dimensions to be practical. Time units as short as a few milliseconds will become recognized as timbres of time, containing unique sound qualities; this allows access into identifying and understanding several important dimensions of recordings. This skill will also be useful in hearing the nuance of subtle rhythmic fluctuations of performances, and will assist in defining the characteristics of materials; praxis studies for developing an awareness of small increments of duration as timbre-related percepts and as related to clock-time are contained in Appendix A.

Listening to records is facilitated with an awareness of the window of the present. This is an understanding that what is being heard 'now' will soon be past, and will be replaced. It sets intention, heightens and cultivates attention and focus.

An event occurring over time might be perceived as an aural image, its qualities simultaneous instead of sequential. The temporal aspects of perception are suspended and awareness shifts to the global qualities of a piece of a recorded song, and toward appreciating the broad and textured shape of form as 'stilled time.' The entirety of the record is thus represented as auditory image (not unlike a visual image), where information on the work has coalesced within long-term memory into a single object; such a perceptual representation is not time-dependent and is pre-verbal.[7] Our creating auditory images of music might explain why many aspects of music's meaning and expression defy verbal explanation (Snyder 2000, 23). The qualities of crystallized form appear largely dependent on suspending the temporal experience of music—and broadening the perception of the passage of musical materials across the listener's focus of conscious awareness. In effect, this allows the listener access to the entirety of the piece of music within the elongated window of the present.

Pattern perception is a deeply ingrained cognitive process; it is central to how we perceive our complex environment. We use pattern perception to process the complex information that arrives through sight as well as sound. At its most basic, structure (and the activities of all elements, especially as related to rhythm) is comprised of patterns of activities and their durations—durations that are organized into patterns within perception (Sloboda 2005). Pattern perception carries to pitch, where melody and harmony are understood by their patterns of pitches; patterns can be established in all other elements, as well.

These patterns are deciphered from the mass of sound (the acoustic wave) arriving at each ear through groupings of acoustic events "formed by their basic similarity and proximity to each other" (Snyder 2000, 264), and "the tendency for individual elements in perception to seem related and to bond into units" (*ibid.*, 259).[8] Here we interpret 'elements' to refer to activities occurring within each element of the track, bonding into an event (such as a musical idea or a language statement). Grouping may be innate (happening automatically) or learned (requiring the engagement of long-term memory), and thus can conform to changing contexts—individual records (Sloboda 2005, 135–146). Grouping can occur at multiple levels, and results in the breaking of the acoustic wave into separate events (typically any number of simultaneous events) (Handel 1993, 185). This is stream segregation that brings the perception of individual temporal ideas (including language and music), where each event becomes a separate stream—even if overlaying or overlapping. Streams may be sound sources, musical lines and materials, lyrics, performances, recording's gestures—anything that can be identified as a separate event. This is relevant for us because each event stream unfolds in time, and has its own iterations, its own rhythmic patterns, its own structural relationship with other streams and to the whole.

These rhythmic patterns are created differently within each individual element in each domain, and (as we have already encountered) the elements in play all fuse to formulate the track's materials and also multi-dimensions in elements of music such as melody and harmony. The ways patterns are established within individual elements will be examined as each is presented.

Hierarchies and Structure

Within the hierarchy of structure, the song branches out into structural layers and sections of time units. These time divisions are reflected in the sections of the song, and used as a reference herein; larger sections may be verses or choruses, groups of verse-chorus combinations; smaller sections are compound phrases within song-sections (such as verse or chorus), phrases within song-sections, sub-phrases, beat patterns, and so forth. In this way, every musical material is a subpart of other, higher-order musical materials; all time spans of beats, measures, phrases, etc., are imbedded within other higher-order sections.

Further, this hierarchy of the musical structure organizes musical materials into patterns and patterns of patterns; contrasting and complementary materials; repetitions and variations; etc. In this way, relationships are established between the subparts of a work and the work as a whole. The hierarchy is such that any time span may contain any number of strata of smaller time-spans, and may be contained within any number of strata of larger time spans; musical material at any level may be related to material at any other structural level. This musical structure can serve as the basis for organizing the materials and elements of the recording, and a template upon which the structure of the lyrics can be displayed.

Figure 2.1 presents the structural hierarchy of the Beatles' "Every Little Thing" (1964). This diagram allows us to recognize the common phrase length between all sections is 2 measures in duration, and the patterns created by these phrase-level materials. In this example, phrases are labelled using lower case letters; numbers following the letter denote changing lyrics, tick marks denote changes in musical materials. Compound phrases within song sections are upper case letters, and song sections are named. Groupings of song sections are bracketed, allowing quick recognition of the verse-plus-chorus groupings that reside at a higher structural level than the sections.[9]

These temporal divisions represent passage of time. The materials within the divisions chronicle the flow of movement to the song—gauged against the underlying pulse of the metric grid. This allows observation of processes that draw music through time. Movement is created by directed motion of tension pushing or leaning toward resolution, temporary relaxation of arrival points, cycling oscillations of activity that provide minimal forward momentum, and more. Created by musical materials, their

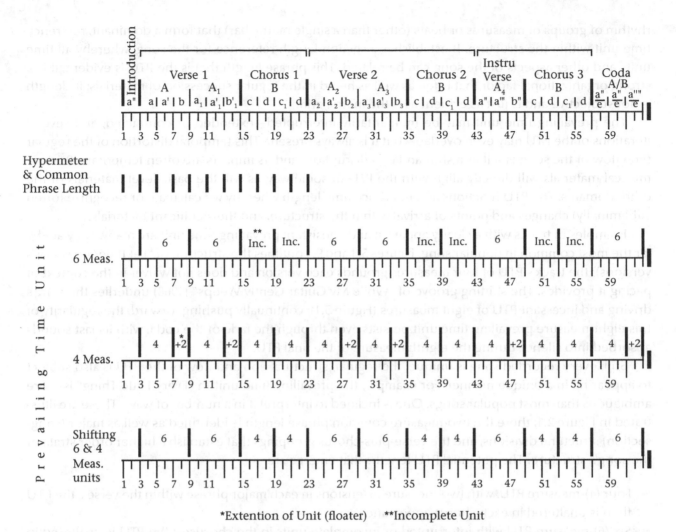

*Extention of Unit (floater)   **Incomplete Unit

Figure 2.1 Structural hierarchy and overall shape (form) of "Every Little Thing" by the Beatles.

interactions, and stylistic tendencies, movement delivers the moment-to-moment experience of structural materials and relationships unfolding. Movement's journey through time—embodied in a myriad of speeds, amounts and intensities of motion from the slightest, most restrained crawl of complacency to the immense, most extreme burst of urgency—is propelled by any of the elements of the record, the intensities of the performance, the drama and unfolding story of the lyrics, scene changes between invented worlds in the recording, and all that is the record.

Within this sense of motion and rhythm of time passage, there exists an underlying dominant section length. Each song will have this time module of activity—a phrase-length or section length that establishes a fundamental characteristic of the song. This phrasing drives or propels the song along with a regularity of pacing; it forms a steady, cycling, pulsating wave of forward motion. This is the 'prevailing time unit' (PTU). It resides in the middle dimension of the song—its length is typically aligned with the 'basic-level material' (explained below) or a multiple of that length.[10] It is the hypermetric grouping that organizes the passage of materials into a fundamental recurring structural division. This is a syncrisis unit—manifesting here as a perception of a coherent temporal grouping with a duration that repeats incessantly—as will be discussed below.

Thus, the prevailing time unit is a length of metric time that repeats throughout the song. It establishes an underlying, regular wave of time—a consistent pulsation—within the structure. It is a regular rhythm of groups of measures or beats (other than a single metric bar) that form a dominant, reference time unit within the structure. It establishes a section-length reference for the song, whereby all time units and other aspects of the song can be related. This phrase length that is the PTU is evident; it is a strong organizational factor that is felt as well as heard in the regular succession of materials; its length is not subjective and is just as evident as is the metric grid.[11]

The prevailing time unit may be momentarily interrupted—extended or contracted, and several iterations of the PTU may even overlap—but it is always present. This temporal distortion of the regular time flow of the song is felt as a shift and/or a disruption, and its impacts are often temporary. Primary musical materials will directly align with the PTU in some way, as will the basic-level materials of the other domains. The PTU functions as a reference time length whereby we calculate or recognize (often subliminally) changes and points of arrival within the structure, and the significant materials.

Examples of tracks with a clearly apparent and consistent prevailing time unit are many. They are by far the most common in popular song. Figures 5.3 and 9.4 outlines the structure of "Let It Be" (in various versions); the track's PTU of four measures grounds each version and does not waiver in the context of pacing it provides. The shifting groove of "While My Guitar Gently Weeps" (1968) underlies the song's driving and incessant PTU of eight measures (Figure 5.1); continually pushing forward, the regularity of this eight-measure prevailing time unit persists even through the fade of the coda, with its last sounds absorbed into silence during the eighth measure of the final PTU.

Still, every recorded song is unique; the regular, structural ebb and flow of the PTU is also subject to appearing in a unique manner. For example, the prevailing time unit in "Every Little Thing" is more ambiguous than most popular songs. One is inclined to interpret it in a number of ways. These are illustrated in Figure 2.1; there the two-measure common phrase length is identified as well as major (song-section) structural divisions, and the verse-plus-chorus groupings that establish a higher order strata of the song's structure. The potential PTU lengths that emerge are:

- Four (4)-measure PTU, with two-measure extensions in each major phrase within the verses; the PTU then is unaltered in each chorus appearance
- Six (6)-measure PTU with interrupted or incomplete units in the choruses; the PTU is unaltered in each verse
- Shifting of the PTU between verses and chorus appearances

A four-measure PTU is clear within each chorus appearance. Applying the four-measure PTU throughout the song requires a division of each six-measure phrase of the verses, bringing the PTU to be heard as extended by a two-measure phrase that completes the 'aab' sequence; the verse divisions are structurally organized and are propelled by movement of three two-measure phrases that form a 4+2 grouping. Walter Everett (2009a, 187) identifies this structural concept as an "expanded prototype" and calls this added two-measure phrase a "floater"; a floater is usually two measures in length and is attached to the front or back end of a four-measure phrase. The "floater" extends the four-measure phrase to six-measures; this would temporarily disrupt or shift the PTU.

A six-measure PTU might also be considered, as that is the length of the defining grouping of each half of each verse (each 'aab' sequence). This requires each half of choruses to be heard as being an incomplete presentation of the PTU, or heard as one that has been interrupted; neither of these fit the material, as the phrases are complete, coherent materials with their own structure and movement. While a six-measure PTU tends to be reinforced by the solo and the coda, the chorus appearances do not have structural or movement characteristics of being contractions or interruptions of six measures. The chorus is stable as it appears, and so is the verse. While the two measure phrase length is common between both sections, it does not reflect the musical movement of either the verse or the chorus—where materials clearly articulate closure at the end of either four- or six-measure phrases.

While a shifting prevailing time unit counters the definition of an underlying, consistent wave of time throughout the work, some works exploit ways to interrupt, shift, expand, contract, or bring ambiguity to it. Songs like "Every Little Thing" are the exceptions that confirm convention by offering contrast. The verses are clearly 6+6, and the choruses 4+4. The PTU might be interpreted as shifting between verses and chorus; a changing span of syncrisis. This contrast is not typical, but not uncommon. The Beatles used wide ranging surface rhythms, asymmetrical meters in songs such as "Within You Without You" (1967), and many examples of freely mixed meter as in the repeated 4/4 and 3/4 alternation in "All You Need is Love" (1967). Their creative use of phrase rhythm brings larger-scale patterning generating hypermeter and PTU that can exhibit instability, but also movement (notice the PTU shift in the middle section of "Here Comes the Sun" in Figure 9.2). Walter Everett (2009a) explores the Beatles' free phrase rhythms in considerable depth, identifying six characteristics of these irregularities that impact PTU. The Beatles used what might be heard as a shifting prevailing time unit numerous times, and perhaps more than other songwriters. A shifting PTU of contrasting lengths is clearly present in "You Never Give Me Your Money" (1969) and "We Can Work It Out" (1965). The shifting meters between and within sections of "She Said She Said" (1966) and "Here Comes the Sun" (1969) bring PTU and hypermeter adjustments as well. Metric modulations between sections in "Lucy in the Sky with Diamonds" (1967) (illustrated in Figure 9.3) and transitioning into the final phrase of "The End" (1969) provide coherence to changing tempo, meter and PTU; the result is a shifting of the prevailing length (both metric and clock time) of the unit that depicts the time span of movement as well as surface rhythm.

Dimensions of Structure

The hierarchy of structural divisions can be grouped into three basic dimensions: large, middle, and small (Moylan 2015). These allow the examination of structure and musical materials at clearly defined levels of detail, and allow levels to be compared in many ways. All strata (levels) of structural hierarchy are contained in one of these three dimensions—dimensions that are sometimes conceived as micro-, middle- and macro-levels in music analysis (White 1994, 24–27).[12] Note, these three dimensions have no relationship to the Schenkerian analysis concepts of foreground, middleground and background.[13]

Large dimension concepts concern the entire song. Should one be examining an entire album, this large dimension might be a group of songs or the entire album. Large dimension may also be considered as the uppermost level and highest, simplest grouping(s) of major sections/time-spans of the song. Large dimension relates to language at the level of the entire work (poem, story, book), or major divisions such as the parts of a book that group chapters into a few key subjects (roughly the equivalent to the movements of a symphony).

Small dimension recognizes the smallest ideas and relationships. Small dimension materials are related to sub-phrases, melodic motives, rhythmic cells, riffs, beats, timbres of individual sounds, and so forth; in some songs shorter phrase lengths may be small dimension. The smallest, subtlest details of musical materials and performance, of music elements, the elements of the record, or the text's elements are represented in the small dimension. Much detail will be discovered here. The small dimension detail functions to embellish or combine its groupings to create middle dimension materials.

The small dimension relates to language at the level of words and syllables, where complete ideas are not formed. The middle dimension is related to language—structural levels of clauses, phrases, sentences, paragraphs, sections, and chapters. The middle dimension is the most varied and complex of the three dimensions.

The multi-layered middle dimension holds the majority of the information and activity of the song. It is comprised of strata of mid-level activity, and is where the primary ideas and concepts of the track are present and evolve. The majority of individualizing characteristics of a song and its characteristic materials and internal workings are resident in this dimension. Songs differ in complexity, and many individual levels can exist between the small and large dimensions, or there may be just a few, depending on the individual song.

In this stratified middle dimension, we find some references by which we calculate workings of the song, as materials and elements carve out and establish their places. The prevailing time unit will exist somewhere within the middle dimension; this precise level of where the PTU is present is the 'characteristic dimension' of the recorded song. This is the level of dimension that inherently reflects the motion of the song, and which serves as a structural reference against which other levels of the structural hierarchy are gauged.

As mentioned above, the primary musical materials of the song directly align with the PTU in some way, as will the materials of the other domains. These all are resident in the middle dimension at the level related to the idea of the 'basic-level' of categorizing materials. 'Basic-level' is a mid-level perception/cognition that is central to categorization and to our structural grouping. It also pertains to delineation of sound sources and their materials; in future discussions we will observe how the basic-level is the perspective at which humans interact, and sound sources (performers) exist as the categorization of reference. Discussing the categorization of objects, Lawrence Zbikowski (2002, 33) explains:

> The basic level is the highest level whose members have similar and recognizable shapes; it is also the most abstract level for which a single mental image can be formed for the category. The basic level is also the highest level at which a person uses similar motor actions for interacting with category members. The basic level is *psychologically* basic: it is the level at which subjects are fastest at identifying category members, the level with the most commonly used labels for category members.

These basic-level shapes reside in the middle dimension, and represent individual concepts within a larger context (middle dimension within the large). It is in the middle of the dimension, the most useful for categorization and also clearly presents the central, primary materials and concepts of the song—these materials are complete (rather than parts of whole thoughts, statements, ideas), and are perceived by shape as much as content. Therefore, in this one specific level of the middle dimension we find the primary musical ideas, the primary statements and concepts of the lyrics, and the central characteristics and relationships of the recording—a level that provides a critical reference to the song.[14]

Perspective

This division of structure into three dimensions is related here to perspective. Perspective is one of the guiding principles of the analysis framework.

Perspective is the act of perceiving at a specific level of detail, as if from a defined vantage point of observation. Eric Clarke (2005, 188) has called this the "scale of focus" of attention, "one of the remarkable characteristics of our perceptual systems, and the adaptability of human consciousness, is to change the focus . . . of attention."[15] This may be applied to listening at any level of the structural hierarchy—attention focused, observing materials/elements functioning or appearing at the same level of structural activity. Perspective allows the recorded song to be observed from various but specific levels of detail, and with attention directed to musical materials, specific elements, any groupings of activities within that structural level, etc. Clear use of perspective relies on the listener's ability to bring focus of attention to specific materials at specific levels of detail.

Each level of detail represents a unique perspective from which the material can be heard, and the song examined. Perspective allows the listener to observe different qualities, content and activities of the sound material and its elements. At times it is helpful for perspective to be considered as a conceptual distance between the listener and the sound material; the nearer the listener to the material, the more detail the listener is able to perceive. Simply, perspective is the level of detail at which one is listening; focus brings the listener's attention to a specific level of perspective. (Moylan 2015, 92–95, 112)

Critical to utilizing perspective is the ability to consider elements/materials in relation to one another clearly and accurately. Elements/materials are compared at the same level of detail in the structural hierarchy; there each has equal opportunity to assume any level of significance; further, any element might have a different function in any dimension, or level of perspective, simultaneously.

Levels of perspective and related levels of detail for the recording are outlined in Table 2.7. Observing this table, we might identify the overall qualities and shapes created by the recording's elements and materials. On the other extreme will be the subtle details of individual sounds, materials, elements, etc., that take place below the surface; such information demands heightened attention to detail within sounds. As in discussion of middle-dimension structure, a significant amount of information and activity takes place in the middle dimension perspective. Two primary perspectives are central to understanding the recording, they are (1) the texture established through the interactions of all sound sources within an upper level of the middle dimension (this is the 'composite texture' where all sources appear as an equal presence), and (2) the level of the characteristics and activities of the individual sources. These similar but distinctly different perspectives allow the sound sources (instruments and voices) of the recording to be accurately perceived and accurately compared (Moylan 2017).

The structural hierarchies of the three domains might now be recognized as functioning in parallel. The strata of structural hierarchies function similarly within each domain; each domain functions according to its unique syntax, materials, elements and modes of expression. Structural levels can be compared across domains by observations at the correct perspective, from the lowest through the highest levels.

Table 2.7 Levels of perspective and related levels of detail.

Level of Perspective	Level of Detail
Highest Level, Large Dimension Overall Texture	Overall, primary shape Overall character and characteristics Big picture, overview Experience of all individual sounds coalescing into one blended sound
Upper-Level Middle Dimension Composite Texture of the presence and interactions of all individual sound sources	All sound sources recognized as being of equal importance Detachment from individual sound sources Level where balance relationships can be accurately observed
Middle-Level Middle Dimension Performances, materials, and sound qualities of individual sound sources	Overall characteristics of individual instruments or voices and their musical materials Typical perception, surface-level detail of life experiences Relationship of self to the world, and to others Perceived relationship of sound sources to the rest of the mix, which inaccurately places prominence on the source
Lowest Levels, Small Dimension Characteristics of and within individual sounds Small-scale activities of elements	Detail below surface level of sound sources and musical materials Detail is heightened, exaggerated Focus on the characteristics of sounds Microscopic details

While at the highest level of perspective each domain may be heard individually as an overall quality (as a fabric of its unique elements, character and characteristics), within the experience of the record the domains fuse. All blends into a single complex, multidimensional quality that is more than the sum of all of its parts—as elements complement, support, enrich and add meaning to other parts, across domains, across levels of perspective and structure. In an all-inclusive "weave of fabrics" (Erickson 1975, 139) the three domains create a multidimensional texture.

Arriving at this highest level of perspective and the upper stratum of structure, we encounter the traditional concept of form as shape.

Form: Shape and Stilled Time

Form is a term used here for three separate concepts:

- Traditional shape. Comprised of the succession of major sections.
- 'Now sound' of syncrisis. The intensional, synchronic sonority within the window of the extended present.
- Crystallized form. The mental representation of the work's essence, conceptualized as a single manifestation, heard simultaneously, crystallized out of time.

In its traditional usage, form is the shape created by the progression of the song's major sections—the sections at the highest level of structure. Form is the memory of movement. The sections are labelled and arranged as they move in succession, in passing time, until the whole is represented. This approach to 'form' or 'formal structure' (structure at the highest, 'form' level) is still useful to illustrate the shape of the song as produced by structure. These highest order sections allow the 'shape' of the work (song) by the relationships of major sections to be recognized.

After acknowledging this traditional notion of musical form, Philip Tagg (2013, 385) has identified "aspects of form and signification bearing on the synchronic, intensional, arrangement of structural elements inside the extended present." In this concept of form, which he calls syncrisis and 'now sound,' form is "created through the arrangement of simultaneously sounding strands of music into a synchronic whole inside the extended present."[16] The vertical aspect of musical expression (intensional) within the limits of the extended present (*ibid.*, 591) is often called 'texture,' but there is further richness here. "Syncrisis involves the *simultaneous* combination of elements and can also be considered in terms of shape, form, size, and texture of a scene or situation" (*ibid.*, 417) within the time perceived as 'now' extended to encompass the experience (incorporating aspects of short-term memory functions).

This provides a window for the immense complexity of the track to be observed with greater clarity. Further, the notion of the extended present related to musical materials allows the individual sound to be a 'sound object,' and is also important for recognizing musical structures (such as the prevailing time unit) and materials (such as 'pitch density') where sound activity over a period of time is linked, or for chunking (memory process identified in psychology) which binds the pieces into a meaningful whole, into a unit that can be recognized and remembered.

Syncrisis also acknowledges the 'scene' as a span of time (a window of the present) in which materials are grouped and shaped, ideas and drama play out; "a pop song . . . is more likely to [derive interest] in batches of 'now sound' in the extended present" of which the "3.6 seconds of guitar riff accompanied by bass and drumkit in *Satisfaction* (Rolling Stones, 1965) is a textbook example" (*ibid.*, 272). Metaphorically the composite 'now sound' can be considered as a scene; as part of the whole song, with its own contained sound and message. It can also "connect with patterns of social interaction specific to the culture in which they are produced"; in this way

> distinctly social sense 'scenes' represented in music consist of simultaneous strands of sound . . . that are organized in different ways for different purposes to produce different effects in patterns that make sense to members of society in and for which the music is produced.
>
> (*ibid.*, 417)[17]

Within the track, syncrisis can manifest the prevailing time unit as a scene (or rather a consistent pulsation of scenes), can bring musical gestures to be coherent sound events or sound objects. Chunking helps moments to be linked into scenes of greater breadth, but with a singular impression; it brings us to experience syncrisis. For the analyst, syncrisis opens considerations of characteristics and qualities of different sorts—ones at different structural levels, different conceptualizations, and more.

The third concept of form—crystallized form—is central to the analysis framework. It is integral with shape, reconceived to acknowledge the piece of music can be perceived synchronously, all at once, as a single large-scale percept or conception. This is somewhat akin to how we are able to engage a work of visual art (a painting, photograph or sculpture) in its entirety, as a single impression of many dimensions, all at once. In observing a sculpture, for instance, its elements and form (composition in the visual arts) of their presentations do not change while one is examining the work—though one's gaze may linger at any point, view the object from any angle, examine it up close or from afar, or even see the work in a different setting (context), with different lighting, or under different circumstances. The many aspects of composition related to the sculpture are as rich in detail and variety as the elements of the track—except they do not unfold over time.

This brings us to define this sense of 'form' as a conceptualization of the song, as if perceived in total, all at once.[18] The record's form is conceptualized in an open consciousness, perceived as an object like a sculpture—but now a sound object, existing out of time. This object is multidimensional, faceted, textured; the record is crystallized into a fundamental shape that represents all that makes it unique, along with its essence, its core impression; its time stilled, time passage simply becomes but one of its qualities. It is a conceptual or mental representation of the whole that contains what unfolded temporally but is not chronological, or episodic memory, and that contains all different types of patterns, at all structural levels, in all three domains, but of no assigned significance. The significance of materials and message, energies and expressions that establish its core impression are unique to the individual track.

The record might now be conceptualized, observed and understood as if by processing all of its variables all at once, into a single expressive, aesthetic object. The record becomes a single, global, multidimensional object or manifestation. The recorded song becomes one, as if all is sounding simultaneously, and in an instant; the shape of the song and the essence of its character and content are recognized within a single, broad, rich understanding; a manifestation of what the song 'is.'

The idea of the sound object helps us understand music out of time, crystallized into a singular singular whole, into one complex aural image. All of the song's multidimensional characteristics can be examined while considering it as a whole, all at once as if it were a physical object. The complex composite sound itself is observed as if suspended from time, crystallized. In this concept of 'form,' the sound object is the perception of the whole recorded song in an instant, with time stilled.

This manifests in a held consciousness of the song, reflecting on its purest core embodiment, or essence; reflection not of working memory during listening, not of episodic memory recalling events, but reflecting on the whole, as if all occurs at once. This state is observed with intention to be unencumbered (as much as this is possible) by personal inputs of interpretation, preferences, cultural influences, past experiences of the song, personalized outside associations with the song, and more. Further, the resulting mental representation of the record appears in nonverbal form, but it is fixed and definite. It is nonverbal because one may only describe/verbalize the whole by considering the characteristics of its parts—and the parts are not the objects of attention here, it is the singular coherent whole. The core essence of the record is held nonverbally, but still accessed and appreciated for all its great richness of detail and dimension comprising the whole. One comes to a place of 'knowing' the record for its core substance and for its broad-reaching concepts and affects that define its context; this 'knowing' recognizes its dimensions of expression, energy, core concept(s), as well as its materials and structures.

This 'knowing' is important functionally as well. It allows recognition of large-dimension characteristics of the recording that are essential aspects, though they may be difficult to recognize and understand. These aspects can supply important sonic references to establish context for the record, such as the holistic environment of the track and its reference dynamic level. This way of knowing can likewise impact understanding of the music, the track's performances, and perhaps of the lyrics. This knowing and suspending time allows ready consideration of the record's many dimensions simultaneously or instantaneously, and at any moment.

Joachim-Ernst Berendt describes this conception of form as the total being perceived and understood in an instant of 'inner hearing.'

> Outer hearing is bound to the unfolding of a piece of music in time, but inner hearing which . . . hears everything simultaneously is independent of perception through our sense organs. . . . Inner hearing perceive[s] everything simultaneously . . . independent of the dimensions of space and time.
>
> (1992, 55)

In "Burnt Norton" from *Four Quartets* (1943, 19), T.S. Eliot expresses this experience in the passage:

> *Words move, music moves*
> *Only in time; but that which is only living*
> *Can only die. Words, after speech, reach*
> *Into the silence. Only by the form, the pattern,*
> *Can words or music reach*
> *The stillness, as a Chinese jar still*
> *Moves perpetually in its stillness.*
> *Not the stillness of the violin, while the note lasts,*
> *Not that only, but the co-existence,*
> *Or say that the end precedes the beginning,*
> *And the end and the beginning were always there*
> *Before the beginning and after the end.*
> *And all is always now. . . .*
> (T.S. Eliot, "Burnt Norton," *Four Quartets*)

Song Structure

Conventions of structure have evolved in all genres of music, in the music of all cultures. Some conventions of structure are complex in the number, sequencing and nature of materials, some are simpler in the amount of contrasting materials and the nature of the contrast. Recorded song has evolved into certain conventions as well—though a great diversity of genres and practices exist.

Songs typically have any number of contrasting sections, with sections having associated functions and containing materials that contrast and/or connect with others. The sections within song structures establish patterns that create variety and unity, and establish some measure of balance; the musical materials themselves, and the elements that comprise them also contribute to this sense of unity and coherence, variety and balance. Structures can provide familiar content and shapes to songs—shapes that can establish a sense of comfort in being able to correctly anticipate (mostly unconsciously) the next material, or that may generate novelty and listener interest by modifying content or introducing unexpected materials.[19]

Within the upper tiers of middle dimension, song structures can be viewed as a sequence of complete sections; these are major divisions in the structure that are typically similar in length. Many of these sections have recognizable names: introduction, verse, chorus, refrain, bridge, middle eight, pre-chorus, solo or break, coda.[20]

Sections of songs have musical materials and a character that are distinct from others. Some sections will repeat with materials slightly altered (such as verses), others nearly exactly as before (such as a chorus or a refrain). Sections are typically clearly delineated by lyrics, and by music as reflected by primary and supporting (accompanying) musical materials (i.e. melodies and harmonic progressions, etc.), by their instrumentation (instrumentation may change with different sections), and all other music elements. Sections are also delineated by recording's elements; sections typically have different mixes, with a different arrangement of the record's elements contributing to delineating sections. Each section functions within the song's structure, with each type of section serving its unique role; this function is defined by the text as much as the musical materials (Moylan 2015, 405–414).

Through learned norms, standard couplings of chorus plus verse sections are recognized, and deviations from these couplings and the characters of other sections are noted. This patterning of sections brings the recorded song its shape and its structure. Most importantly it helps us to recognize what makes a track unique, and how it differs from convention and from all others.

LISTENING WITH INTENTION AND ATTENTION

Let us recall here: the record can only be experienced, engaged, or studied through listening. Indeed, it only exists in its complete form as an aural experience—as *our* personal listening experience. Any understanding of the record we can ultimately obtain will be an experiential understanding; an understanding based on our experience(s) of listening to the record. From this, it becomes apparent our perception of the record is shaped by our listening skills and our knowledge of the dimensions of recorded songs, and by our physique and personal experiences; further, our predispositions of all types factor into our perception of the record. These can all play significant roles in the success of recording analysis.

'Listening with intention and attention' is one of the guiding principles. Listening with intention requires knowledge of what to listen for (what might be present) and the skill to identify and observe the characteristics and activities of what is present. Listening with attention requires the discipline to remain on task and the ability to focus one's attention as desired. Intentional attention is listening with focused attention to specific aspects of the record—such as a specific musical idea at a specific level of perspective. Understanding how to listen and acquiring knowledge of what might be present and of how our bodies and experiences are unique and shape our perception will help guide the reader through our process.[21]

Listening to the unique qualities of recordings and of the record requires listening in unique ways, and hearing qualities not found in nature and therefore not previously encountered with regularity, if at all. Core listening skills unique to recognizing the qualities of recordings are woven throughout Chapters 6 through 10. Listening concepts for hearing the qualities of recorded performances (of music and lyrics) and for engaging various facets of the framework are introduced throughout this book. Appendix A contains numerous praxis studies to engage these listening and analysis tasks, and guidance for refining listening skills. All these approaches are successfully brought into practice by listening with intention and attention.

Listening is Personal

Listening is a highly personalized activity, as many aspects of the listening process are unique to each individual. As we cast a cursory glance at one another, we recognize each individual's unique physique, unique experiences and knowledge, unique formations of cognition and perception, and more. We all have our own way in which we perceive the world, our own perceptual framework (Clarke 2005). Our perception filters our experience and shapes the way(s) in which we listen, the way(s) we hear, the way(s) we approach and engage listening, the way(s) through which we learn what is within a track, the way(s) we process the world of sound around us, the way(s) we make sense of sound, music and language—and more. Our individualized filters are the product of many things— our cultural experiences, our assumptions and attention, our preferences and inclinations, our skills and more; we are also shaped by our physiology and by biology, and by the way our individual mind directs attention toward the world. Our unique physical qualities shape sound itself, and our perception of it.

Beginning with the dimensions and shape of the head, of the shoulders and pinnae, and continuing through the hearing mechanism, sound is modified by the body of the listener in ways that are subtly different from all other people. The sound qualities reaching the hearing mechanism in one individual are not precisely the same as those reaching others; nor are the workings of the hearing mechanism and the transduction of sound from mechanical energy of the hearing mechanism into neural impulses the same in us all. These differences increase as we factor in physical and functional differences between the two ears and other physical attributes of our hearing mechanisms.

Further, we all have our personal set of experiences. These prior listening experiences uniquely shape our capacity to recognize and engage language, sound and musical materials—and to make sense of what we hear. Our listening skill levels continuously improve with more experience and increasing exposure, especially once the process has been informed, guided and directed with the knowledge of what might be present to be 'heard.'

These experiences also shape how we process sound and perceive sound. We each have our own relationships with sound, our own ways of conceptualizing sound (beyond language), and of understanding or of identifying the qualities of sound (Sacks 2008). We all gauge pitch in our own way, we all tend to recognize some pitch levels more deeply than others whether or not we have pitch recognition (Sloboda 2005). We each are able to calculate time relationships accurately by using our individual memories of the tempo of songs we have found meaningful. We all can recognize thousands of timbres, many unique to our personal lives. Further, we react physiologically to sound in ways common to all, and in ways that are subtly unique to each of us (Roholt 2014).

We can learn to access our unique ways of perceiving and understanding sound. Through this, we can establish a personal listening practice that is both effective and relevant.[22]

Modes of Listening

Listening processes can be divided into two general modes: active and passive listening. Active listening is engaged, with the listener focused and absorbed in the listening process. One's attention is fully concentrated on listening. To extract the significant detail and large amount of information present in the record requires active listening (and a willingness to undertake repeated listenings). Active listening is central to the analysis framework. Within active listening we can engage the modes of critical listening and analytical listening, and listening with intention and with open awareness.

Passive listening occurs when the listener's attention is elsewhere. In music listening for much of the lay audience or in most everyone's daily situations, passive listening is common—and perhaps necessary to avoid sensory overload. Music is being played back and sounds are being produced in many segments of our daily lives; often we do not actively bring the center of our attention to many of these. Setting intention in listening can allow us to choose to be a casual listener; we can intentionally allow sound/music to drift by and not occupy our focused attention. Music, language and sound within passive perception is not engaged consciously, is not fully within one's attention, is in the background and ambient in one way or another.

Analytical listening and critical listening allow us to approach active listening in two different ways. Much of analysis is accomplished using analytical listening, supplemented by critical listening to extract certain observations and to process certain evaluations. Analytical listening might be considered functionally aligned with structure and structural relationships, and critical listening more conceptually aligned with form and with sound objects at any structural level. (Moylan 2015, 90–94)

Analytical listening engages the content of the record as it unfolds in time. It looks at the track's materials and how they are formed by various elements, at the sounds and presentations of the lyrics, the structural relationships of materials, at the functions of materials that produce motion—all within the context of its artistic voice. All observations and evaluations drawn from analytical listening seek to understand the materials (music, lyrics and recording) within the context of the recorded song. This includes how the ideas of the record relate to other ideas, get divided, form larger ideas; how they move over time, are shaped, and how they are combined with other simultaneous sounds. Analytical listening looks at all the characteristics and the relationships of the recorded song's materials as they unfold over time. This unfolding over time is an event. Analytical listening observes and evaluates events; it engages the record as containing sound events of many types.

Critical listening pulls sounds (individual or groups of sounds, etc.) out of the context of the record. Such evaluations examine materials or sounds for their specific content, without relating them to other materials or sounds within the record—without relating the material(s) or sound(s) to the musical context of the song, or the message of the lyrics, or the qualities of the recording. In this way ideas can be isolated and studied for their own unique characteristics, without concern for how their characteristics relate to other materials or how they contribute to the record. After study, that information may be applied back within context; this creates a loop of observing materials out of time and out of context to identify pertinent information, then applying that information within the track's context. In this way critical listening can enhance analytical listening. The important distinction is pulling sound or material from context for isolated evaluation or recognizing context while evaluating the sound or material as it relates to all other sounds or materials as well as its qualities over time. A sound object is created by pulling the sound or material out of context and observing it without its unfolding in time—observing the sound as an object, with all parts available simultaneously.

This is similar to the concept of crystallized form (described above), but at a different hierarchical level of structure or level of perspective. The sound object "is the perception of the whole musical idea (or abstract sound) in an instant, out of time; it is understood as the qualities of the sound itself in its many variables and as it exists as a global quality [with rich dimensional detail] or 'object'" (*ibid.*, 454). In this way the sound object is without relation to other sounds, or to musical context; it is "independent of its cause and of its meaning" (Chion 2012, 50). Their temporal aspects can be observed as a dimension of the objects without distorting the idea that they are pulled out of context and observed as objects;

this is most effective when the object occupies a duration within the window of conscious awareness and the perception of the 'present' (Snyder 2000, 8–11). Individual sound objects can be conceptualized for many aspects of the record. Of particular value is the opportunity to explore sound quality and timbre of individual sounds or groups of sounds; this concept may be extended to concepts of sonority, pitch density, timbral balance and texture. This approach is aligned with Pierre Schaeffer's original concept of *l'objet sonore* (Schaeffer 1966). Conceptualizing sounds as objects can prove helpful in defining aspects of the recording; for example, environmental characteristics, phantom image size and placements, and distance location placements of sound sources.

Sound objects allow for comparison with other sound objects for abstract qualities, outside the context of a track or tracks. For example, the Gibson J-200 acoustic guitar George Harrison performed on "Here Comes the Sun" (1969) was also used on "While My Guitar Gently Weeps" (*The Beatles* "White Album" version, 1968); the two sound qualities of that instrument may be examined out of context as sound objects, evaluating their timbral qualities and recording elements. The two objects (different guitar sounds) might then be compared, bringing to light their differences, and thereby extending the sound object concept. As a next step, each instrument might be re-examined within its original context for its contributions to the track, in an analytical listening process.

Listening Deeply

In order to compare sounds, one must recognize their unique qualities to identify their differences. These qualities and differences exist *within* sounds, and in the subtle details of the overall qualities of sounds. Listening within sound is an activity that is rarely, if ever, engaged in everyday listening. In fact, since our earliest listening experiences we are taught to listen to overall qualities of individual sounds, and that the subtle information within sounds is important in as much as it shapes the overall sound. Listening for subtle qualities is also quite unnatural, except for a few experiences—as examples, the tuning of an instrument, or attention to vocal intonation can pull one into the simplest experiences of deep listening.

Deep listening is a focus of attention to extreme detail—detail at any level of perspective. It may be hearing extreme detail at upper levels of perspective, such as the overall sound, composite texture of individual sources, or even the basic-level sound; here, deep listening brings attention to minute changes in level, rhythmic placement or other activity. Deep listening brings attention to extreme detail at all dimensions. Deep listening is also hearing within sounds, at the smallest dimensions; hearing what is happening under the basic-level surface (where much human experience happens) is not a common or natural activity, and must be cultivated for the skill to be acquired.

Deep listening *to* sounds and *within* sound is central to understanding records, though. Listening at extreme levels of detail is common for those involved in making records. It is also common—in one way or another—for instrumental performers and singers to be aware of the inner workings of the sound qualities of their instruments, and of their performances. Listening within sounds allows access to hearing the subtle qualities of performances (music and lyrics) and the subtle qualities of the recording—and thereby the subtle qualities of records.

The term deep listening has connotations from the compositions and the musical aesthetics and theory writings of composers Pierre Schaeffer, John Cage, Pauline Oliveros, R. Murray Schafer and others of that era.[23] While the use of the term here embraces some of their philosophies, it is also used to provide intention for the reader/listener to access extreme depth of listening for musical events (analytical listening) and sounds that are pulled from context and their causal origins (critical listening). In this extreme depth and detail are the subtleties of performance and of the recording that substantively shape the record. Recordists develop skill in deep listening at these levels of perspective and detail to shape the record in important ways. The reader, analyst, listener also will engage these important qualities, and these atypical ways of listening, in order to fully understand the substance of the track.

Listening with Intention

The process of listening begins with intention. When intention is brought to listening, a sense of purpose and a direction for attention is established, directed by forethought of what one is seeking to find. Awareness is brought to specific elements, materials and all other aspects of the record—at specific hierarchical levels or levels of perspective. Listening with intention functions in two ways: (1) it can direct a sense of a specific focus of attention, or (2) it can establish a position of neutrality and an approach of searching for information and holding an open awareness in scanning the record's entire texture.

Listening with intention allows planning of what information to seek, and provides opportunity to dedicate one's attention to that task solely. A great many types of information can be extracted from what is heard at any moment; intention brings one to recognize all of that information cannot be extracted all at once. When each listening has a specific purpose, attention can be directed toward that specific aspect of recorded song and information extracted with greater fluidity and accuracy. Listening for one set of information provides the luxury of limiting attention to only one aspect of the track, ignoring all others; intention allows directing mental resources to a specific task. Focusing on a single topic will more quickly and directly allow recognition of that element and its activities, and it will also develop the disciplines of focus and attention. This act also embraces the reality that it is not possible to hear several things simultaneously; intention acknowledges there will be repeated hearings of the material, and opportunities to bring focus on other materials. Listening repeatedly to the same material reveals new results, as observations are checked and refined, or as the listener's intention shifts to another aspect of the track.

When each listening experience begins with establishing an agenda of intention, the experience can be guided towards more complex situations than a focus on one aspect of the record. Attention can be broadened to engage two or more aspects of sound simultaneously, intentionally alternating focus. This process does not seek to hear the qualities of several elements simultaneously. Instead, a deliberate but rapid alternation between elements brings control to the process and greater awareness of the qualities and the interactions of materials.

Intention brings conscious choice to the listening process. One chooses to listen in a manner of either a controlled exploration with open awareness, or one chooses to listen with intentional focus on a single aspect of the track. These two processes may alternate depending on what is needed, but are largely exclusive activities.

In listening with open awareness, one is exploring the materials and the experience of the track. This is an open listening field, engaged in actively scanning and listening to discover aspects of the record. Open awareness holds a broad net open within which the activity of the track can be observed. This is an intention to recognize and follow certain traits of the record, or to be open to discovering the unexpected, or unique or important sound qualities or sound relationships. Listening with open awareness puts into practice the principle of equivalence, and opens to the concept of listening without the bias of expectations. It allows us to assume the most objective position possible to recognize the unique qualities of the track—with a minimum of personal interpretation and bias from our personal predispositions. Listening in this way is hard to do.

When we open to an understanding that all elements of each of the three domains have an equal potential to serve any function in the recorded song, our approach to hearing, learning and analysis are fundamentally shifted; the depth of our understanding potentially deepens and broadens. We turn away from a narrow focus on pitch's melody and harmony, the text and some rhythmic materials; we adopt an awareness of the breadth of activities within the song and the potentials of the medium. We open to the possibility of recognizing the unknown, to hearing what has not previously been experienced, and to discovery and realization. Equivalence brings us to embrace the possibility that the unknown and the unimaginable may happen in any way and at any time. It brings us to understand that any aspect of any element in any dimension may be significant—not necessarily most important, but significant—and thus worthy of our attention and evaluation.

Listening Without Expectations

Successful listening requires a willingness to accept and to engage as relevant all that comes along—without expectations or personal desires. This listening from a neutral vantage point includes not having preconceived ideas of what one will hear in the lyrics, music, or recording. It is atypical to listen in this way—our tendency is to seek what we know and what we expect to find. When listening for what we believe may be present, we can begin to imagine it even when it is not there; worse, this state will usually bring us to miss what actually *is* present.

It is common to listen *for* something, and to miss what actually occurs. We can do this in conversation, when we so intensely hope to hear someone say what we want to hear, that we hear it although it is not stated; or do not hear it when it is not stated precisely as we had hoped, or expected. In music this can happen similarly, and the recording can be misheard from misdirected expectations as well.

Listening with an open awareness of intention works in concert with listening without expectations. The two approaches are mutually supportive.

Listening without expectation allows one to be receptive to the unexpected, and the unknown; ready to engage things never before experienced just the same as those that are known, and those expected. It allows for accurately hearing and understanding something new, something unexpected—as well as recognizing and observing what is familiar. This being open to all aspects of sound will allow things to be detected and embraced that otherwise might have gone unheard.

Cultivating Attention

The deep listening required of recording analysis takes concerted energy, undivided attention, and focused concentration. Remaining focused on listening requires vigilance in noticing when our attention begins to wander. It is part of our human nature and conditioning to constantly shift our attention; we do this in a great many ways, and this often serves us well—as we suddenly remember to do something important, just in time. It also challenges listening processes.

Here, let us remember that sound is a memory, music is a memory, communication involves memory. We observe sound through the window of the present, and process it and its content and messages within short-term and long-term memory.

While we *listen* during the passage of time (aware of activities within our window of consciousness), we *hear* backwards in time (remembering and considering what has happened). Riding our window of consciousness, we do not know the duration of a sound until it stops, we do not know the shape of a melodic line until it has cadenced, we do not know the message of a song until reflection following its conclusion. Accurate listening requires not making judgments or reaching conclusions until having listened completely, and considered fully what has happened. Accurate listening occurs from a position of neutrality, engaging equivalence. This happens as a by-product of engaging an awareness of what is happening *now*, and retaining (making a memory of) that experience to form larger experiences and to compare one experience to others.

Attention can be improved with practice; it can be focused and disciplined by intention. The conscious mind can become unmoving, and solely aware of observing the record as it passes—in this way attention can be improved and held for ever-longer periods, with attention given solely to the record.

'Listening with intention and attention' brings guidance to the listening process. It improves endurance and promotes the discipline of staying on task and seeking specific information, or in exploring the texture to engage whatever comes along from a neutral, unbiased position and to discover unknown qualities. This simple principle provides a vehicle for improvement of listening skills to all elements and materials of the record. 'Listening with intention and attention' opens the listener to encounter the detailed, impossible worlds and the unique artistic expression within a record.

Some Questions for Recording Analysis

What is the typology of the sound stage in the verses and choruses? How does this evolve and support the function of the primary materials?

What is the prevailing time unit that drives the motion of the song?

At which level of structure do the complete primary musical ideas occur?

What is the controlling element for the chorus of the song?

How do the smallest details relate to and contribute to the large dimension structures?

How do materials in a lower structural level contribute to the functions of the materials on the next highest structural level?

How does equivalence shape the significance of elements throughout the record?

Have I experienced the crystallized form of this record? Am I able to perceive its singular presence? Its dimensional qualities?

Am I able to retain a clear focus on the lead vocal throughout the record—holding it in the center of my attention without distraction? Am I able to accomplish this with less significant instruments? Less prominent instruments?

NOTES

1 In *Music's Meanings* (2013, 603) Philip Tagg provides a summary of his three definitions for syntax related to texts, semiotics and form: "SYNTAX = order, array [1] (general) the study of principles and rules governing 'texts', including written and spoken language, musical works, recordings, etc.; [2] branch of semiotics focusing on the formal relationship of signs to each other without necessarily considering their meaning; [3] *mus.* ordering of events in sequence rather than simultaneously, particularly inside a phrase but also inside an episode (motifs, phrases, harmonic progressions etc.). The ordering of events in sequence ('long-term syntax') is referred to as DIATAXIS."

2 Walter Everett's "Making Sense of Rock's Tonal Systems" (2007a) provides a significant resource that defines many (if not all) of the central ways tonality has been approached in popular music. There is a great breadth of approaches to tonality in popular music, a good number of which are fundamentally counter to art music's tonal organization practices.

3 This frequently used quote by Claude Debussy is relevant here: "Works of art make rules, but rules do not make works of art." Stretching boundaries is part of the creative process, and as boundaries are stretched rules get redefined, or perhaps become irrelevant.

4   Here the 'primary' idea or material is used more broadly than the traditional view of 'theme' used in the study of musical form. Acknowledging the limitations of traditional formal analysis' focus on melodic themes ('tunes') and shortcomings of its approach to sectional functions that do not contain thematic material, Cook (1987, 9) explains: "[Theme] refers to some readily recognizable musical element which serves a certain formal function by virtue of occurring at structural points. A tune can be a 'theme' in this sense; but so can a striking chord progression, a rhythm, or indeed any kind of sonority."

5   The term 'convention' can be rife with contrasting applications and interpretations. In *Conventional Wisdom* (2000, 2–6) Susan McClary provides a significant exploration of the concept of musical convention: "By 'convention' we usually mean a procedure that has ossified into a formula that needs no further explanation. . . . Why do pop ballads end with fade-outs? Convention." Therein she articulates its contrast with "purely musical" procedures, the aversion to convention by some as: ". . . interpret reliance on convention as betraying a lack of imagination or a blind acceptance of social formula." Here I use 'convention' in ways closely aligned with her statements: "genuine social knowledge is articulated and transmitted by means of shared procedures and assumptions concerning music." ". . . the procedures we regard at different moments as 'purely musical' count rather as the most crucial set of conventional practices." ". . . conventions always operate as part of the signifying apparatus, . . . it is not the deviations alone that signify but the norms as well." And "the apparently universal laws of syntax . . . were 'merely' conventions."

6   Lerdahl and Jackendoff (1983) identified a number of hierarchical structures in various aspects of music, all directed toward listener understanding of music. Most relevant to recording analysis appear to be 'grouping structure' hierarchy of non-overlapping groupings of motives, phrases, sections; 'metrical structure' the hierarchy of strong-weak pulse patterns; 'time-span reduction' hierarchy relating the time spans of the most predominant element on each level as the head element to be represented at the next higher level imbedded within that level's head element, creating a tree-like hierarchy; 'prolongational reduction' hierarchy corresponding to time-span reduction it attempts to reflect the flux of harmonic tension and resolution of the piece of music; 'reductional hierarchy' is similar to a Schenkerian approach wherein the hierarchy moves through higher and lower levels that progress from the most detailed lowest level from which only the most important elements (materials, etc.) survive in the highest levels. It is significant that the authors identified the rules guiding the formation of these hierarchies were *preferences* of the listeners, and were not 'universal rules.'

7   Long-term memory occurs in parallel rather than serial fashion; this parallel process allows us to access and select from many different memories simultaneously, rather than in succession (Snyder 2000, 216). In this way, the record can be perceived as an object out of time.

8   In addition to similarity and proximity, Diana Deutsch (2012a) notes grouping may also follow the principles of 'good continuation' where elements that follow each other in a given direction are perceived together, and 'common fate' where elements which move in the same direction are perceived together.

9   John Covach (2019, 285) offers a more detailed examination of the structure and materials of "Every Little Thing." His analysis defines verses as being six measures in length, contrasting with the 12-measure verses (divided into two six-measure phrases) offered here.

10 The prevailing time unit (PTU) is often aligned with the "basic level categorization" relationships explored by Zbikowski (2002, 31–36)

11 The PTU is a phrase length and time unit. It is in contrast to the typical rhythmic grouping as described by Cooper and Meyer (1960, 9): "There are not hard and fast rules for calculating what any particular instance the [rhythmic] grouping is. Sensitive, well-trained musicians may differ." Here, the phrase length is evident in the structural organization, not in the spinning out of rhythms of materials.

12 I have chosen the concepts of three dimensions instead of the concepts micro, middle and macro (that are widely adopted in music theory) because the concepts of dimensions are more readily applied to the domains of lyrics and recordings, and more adaptable to the concept of crystallized form.

13 At its most basic level, the reductive linear analysis of Heinrich Schenker (1979) is a formalization of the intuition that music should be understood in terms of larger-scale shapes and patterns rather than a series of notes. It proposes music is comprised of various layers, the surface being the foreground and the deepest layer the background, and the elaboration layers of the middleground in the details between foreground and background. The approach is based on reductive hierarchies of layers of detail, focused on common-practice harmonic

motion and relationships—voice leading, melody and harmony. This almost singular focus on common-practice harmony and melody is irrelevant for some popular music, but can give insight into pitch/harmony matters when applied to others. For example, Walter Everett has effectively used techniques from Schenkerian analysis to illustrate pitch-related qualities in many popular songs, including deep studies of Beatles songs in numerous writings (see for example Everett 1986, 1999, 2001, 2004) and Lori Burns has adapted it effectively in her writings—see especially "Analytic Methodologies for Rock Music" (2008)—that also embrace modal languages in popular music.

14 It seems important to acknowledge that Lawrence Zbikowski recognizes "that music comprehension begins at the motivic level, for such a level maximizes both efficiency and informativeness" (2002, 34). The context of this study is art music, however, and the example he chooses is the famous four-note motive from the first movement of Ludwig van Beethoven's Symphony No. 5; in this context the 'motive' as the 'basic-level unit' makes sense. Shifting this concept of basic-level categorization to popular music analysis, it is clear the 'fastest identified subjects' of primary substance are rarely so short; rather four- or eight- or two-measure units typically contain the fastest identified subject in popular music.

15 Eric Clarke (2005, 188) continues: "what might be called the 'scale of focus,' of attention—from the great breadth of diversity of awareness to the sense of being absorbed in a singularity. At one moment I can be aware of the people, clothing, furniture, coughing, shuffling, air conditioning and lighting of the performance venue, among which are the sounds and sights of a performance of Beethoven's string quartet Op. 132 and all that those sounds specify; and at another moment I am aware of nothing at all beyond the visceral engagement with musical events of absorbing immediacy and compulsion."

16 The extended present "is a duration experienced as a single unit (*Gestalt*) in present time, as 'now' rather than as an extended sequence of ideas" (Tagg 2013, 588). It can stretch from one to eight seconds of clock time. Events separated by more than an average of 3–5 seconds are not automatically part of the conscious present and not perceived immediately, but must be recollected, and are perceived in retrospect (Snyder 2000, 69).

17 Philip Tagg has offered significant approaches to popular music analysis and music semiotics. His chapter on syncrisis (2013, 417–484) incorporates 'social anaphones' and other observations outside the three domains of the record that can hold much relevance to some analyses.

18 This is not significantly different from how we recall complex events, manifestations resulting from and potentially containing rich detail and emotions, thoughts and contributions of others in long-term memory (Snyder 2000, 15).

19 Covach and Flory (2015, 10–17) offer a concise and informative overview of typical formal types in American popular music that includes 12-bar blues, verse and verse-chorus forms, and AABA form.

20 Walter Everett's *The Foundations of Rock* (2009b, 134–156) examines song sections in depth and with many examples to demonstrate the vast richness of these sections, and ambiguity that can arise when trying to define them.

21 Portions of the sections 'Listening is Personal,' 'Listening with Intention,' 'Listening without Expectations,' and 'Cultivating Attention' have been adapted from the author's "How to Listen, What to Hear" (Moylan 2017). They have been reframed here in this broader context.

22 See *Understanding and Crafting the Mix* (Moylan 2015, 85–125) for more detailed explanations of these concepts and practical exercises for building foundational skills.

23 A source for foundational writings by these composer/author/philosophers and others is the collection *Audio Culture: Readings in Modern Music* (2004), edited by Cox and Warner.

CHAPTER 3

Domain and Elements of Music

This chapter presents an overview of the elements of music, as commonly found within the context of recorded popular song, and some analysis methods or techniques that might be particularly relevant to recording analysis. To more deeply understand how the record shapes the song, an understanding of what comprises the song (which is in large part 'music') is central. Within any recording analysis, an examination of music elements (in any level of detail deemed appropriate) may be within the goals of an analysis. Traditional analytical techniques, theory and common practice may or may not apply or be relevant, depending on the individual song. Popular music has many styles; its materials have many places of origin and roots of style; those who write such music come from innumerable backgrounds and levels of training. Further, intuition can drive creativity much more deeply than traditions, or an understanding of conventions and theoretical principles. For the analysis of popular music—and analyses incorporated into recording analysis—analytic methods may need to be devised, or established methods may need to flex in order to address the music in meaningful ways.

Contained here is an overview of the variables that are present in popular music, of basic concepts of their content and organization, and of how they might contribute to or impact the track and how it is perceived. Some commonly found practices and a few underlying traits will be identified and explored. The elements of the music domain will be presented as:

- Rhythm
- Melody
- Harmony
- Dynamics
- Timbre
- Arrangement
- Texture

This is not a book *about* the analysis of popular music—about how to analyze or to make sense of the music of an individual track. I do not seek to cover the analysis of popular music substantively, but rather to open the door *toward* an understanding of 'how the song shapes the record,' which is the mirror of this book's central focus. The subject of the musical materials and practices of recorded song is very broad, and eludes generalization necessary for broadly applicable analytical methods. Some notable experts from music theory and musicology fields have dedicated years of study to covering popular music analysis extensively; the breadth of their work cannot be adequately condensed into this book, let alone into one chapter. Many of these sources (that do not always speak in the same voice) are referenced throughout; they should be consulted to provide guidance in understanding the vast potential materials and practices of popular music, and to provide additional vocabulary and methodology to facilitate analysis.[1]

While existing approaches and methods of music analysis may produce some guidance for entry into understanding a specific piece of music, it is important to

> avoid casting popular music as if it were 'just like' art-music. The challenge then becomes the investigation of popular music along traditional musicological lines while maintaining a careful sensitivity to how popular music may differ from art-music in its specifically musical dimensions.
>
> (Covach 2001, 466)

Few popular songs are substantively aligned with the structures and syntax of art-music, for which existing analysis techniques were devised to address. The qualities that characterize individual popular songs are traits that often push the envelope of existing popular practice—let alone art-music constructs of centuries past. Creative voices seek unique ways of expressing themselves, as well as creating the new substantive ideas within the song—this encompasses all participants in the creative process. These 'unique ways' are often reflected in the syntax of the elements, and in the new combinations of syntax content when elements are combined.

The goal of this chapter, then, is to identify and discuss the elements of popular music and its potential characteristics and practices. Through this, additional framework concepts and points of reference will be introduced here (and more will appear in later chapters). Using the concepts that appear here—pursuing an approach of inquiry based on established methods (as appropriate) and based on conforming methods/approaches to the materials, language and musical practices of a particular style or song—an appropriate approach to analysis might coalesce to allow discovery of the unique qualities of an individual song. In this process of examining specific elements, it is helpful to remember they do not operate in isolation. Allan Moore (2001, 33) has framed this:

> One of the major difficulties in discussing music lies in its multi-dimensionality. The stream of sounds a listener hears is composed of rhythm *and* harmony *and* melody *and* instrumental timbre *and* lyrics and, quite possibly, other elements as well. These basic elements are distinguishable one from another in the abstract, and on reflection . . . but they conspire together to produce the music we hear.

This chapter begins with popular music's most visceral element: rhythm. Rhythm is also the element that functions with or exerts influence on all others. Further, rhythm is also a factor in the elements and characteristics of lyrics and the elements of recording.

RHYTHM AND TIME

Rhythm is a driving force in much popular music, and records. Popular music's strong, regular beat is a characteristic trait, and hard to ignore. Rhythm is immediately evident sonically, and felt in the body, or perhaps "through the body" (Roholt 2014, 2). Rhythm is temporal; its substance is time. Simple, prominent and pervasive, rhythm provides the point of reference for the experience of the record.

Discussion of all individual elements begins with rhythm because it is the organization of and the perception of time; rhythm's sequential durations appear within the many sound qualities of the record, and provides the platform for their activities. While rhythm is presented here as a musical element, all of the elements in music, lyrics and recording take place in time; their materials and changing values have the potential to establish rhythmic patterns. Rhythm is generated by iterations of sounds, bringing a repetition or change of an element's characteristics—of any characteristic of any element, such as (as example) a changing pitch, or a repetition of a drum sound. What is discussed here under rhythm applies to all elements, and is adapted to all domains and at all dimensions. Concepts of rhythm will reappear often within discussions of all elements (over the next few chapters). Its many applications and subtler characteristics will be revealed throughout.

Pulse, Rhythmic Layers and Metric Grid

Rhythm is the pattern of durations, groupings of discrete impulses that underlay most forms of communication. Durations are unified by a common point of reference: an underlying recurring pulse. This underlying pulse is an ever-present time reference; a time reference reflected in an established speed of its repetitions (tempo). Meter organizes this incessant pulse as a recurring pattern of strong and weaker beats (pulses), and thus establishes the reference pulsation of the metric grid. It is against this metric grid that the durations of rhythm can be accurately calculated and perceived.

Rhythm can be conceived in three layers. Each level of the multidimensional structure carries its own rhythmic expectations—whether at the pulse level, measure, phrase level, hypermeter, or larger levels. This is reflected in all different structural levels; it can appear as the 'rhythm' of major sections, as the waves of the prevailing time unit and of the hypermeter, within the recurrent patterns of phrasing (such as 4+4+4 measures), as consistent divisions of the beat into two or three equal parts, and so forth.

Hypermeter plays a central role in popular music. Hypermeter is the song's meter of measures. It is a large-scale meter where measures act as beats, where several measures are combined into one unit, a hypermeasure. In popular music groups of bars are normally four measures, and often the measures have different levels of stress similarly to stronger and weaker beats in a measure (Moore & Martin 2019, 42). The pattern of hypermeter establishes the prevailing time unit—the two may be one in the same at this structure level—and combinations of hypermeasure groupings can functions on higher hierarchic levels (Kramer 1988).

The first layer is below the surface; this layer contains the underlying pulse and its divisions and the larger units it forms. The point of reference is the metric grid and the "hierarchy of expectations and implication" (LaRue 2011, 90) of motion and rhythmic patterning it generates. The incessant pulsation of some tracks—especially evident in dance musics—although blatantly present, resides below the surface in this layer.

The second layer is surface rhythm. This includes all relationships of durations in the record's elements and materials at the level of perspective of the individual sound source, or musical idea. This is the level at which sounds are most visibly represented in the flow of traditional music notation. This is the rhythm of melody and accompaniment patterns, and of changes of harmony and drum parts, as examples.

The third layer is comprised of the rhythms created as elements (and their materials) interact. Rhythms may be generated by cross-referencing the activities of elements as they relate to one another (such a relationship is sometimes found between a lead vocal and backing vocals); in this instance they fuse into a single expression. A grouping of the elements must be established for these interactive rhythms to be perceived; some similarity or connection must be present to create cohesion, otherwise the elements (materials) will tend to separate into independent streams (Cooper and Meyer 1960, 9). The instruments of a groove will often exhibit this type of interactivity, and establish rhythmic gestures between sound sources.

The traditional idea of the metric grid defines the pulse as "one of a series of regularly occurring, precisely equivalent stimuli. Like the ticks of a metronome or a watch, pulses mark off equal units in

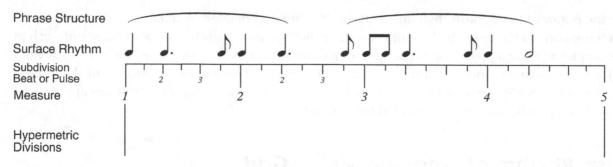

Figure 3.1 Underlying metric grid, surface rhythmic pattern, hypermeter and phrase structure.

the temporal continuum" (*ibid.*, 3). In popular music, the pulse is often not so rigid; it is alive and it can breathe. In contrast, some styles have a mechanized, electronically produced and exacting beat; "some artists . . . are attracted to generating grooves of the square feel of quantized evenness and an exaggerated tempo" (Danielsen 2010a 2). The exacting precision of digital and electronically generated pulses will reflect this traditional definition, though many other styles are fluid. Most exhibit a slightly relaxed pulse, and some reflect great elasticity in the pulse and microrhythmic deviations from the metric grid;[2] this can be especially noticeable in lead vocals.

Tempo (the speed of the recurring pulse) is as pervasive as the beat. Although it can often flex in some styles, in others it is mechanically constant. Tempo provides a consistency and point of reference in contributing to the metric grid. Gradually increasing speed (accelerando) or slowing (ritardando) tempo—the entire ensemble moving from one tempo to another, appearing quite regularly in art music—are not common in popular music "outside a more sensitive rock music, such as in retransitions to verses in Paul Simon's 'Something So Right' (1973)" (Everett 2008, 131). Tempo and meter can 'modulate' into another tempo and/or meter, usually relying on some common beat division to serve as a pivot relationship; though not typical, a metric modulation of this sort is clear in the Beatles' "Lucy in the Sky With Diamonds" (1967).

The stolen time of rubato provides a flexible pulse that can be subtle or pronounced. Rubato's influence often appears as a rhythmically flexible solo melodic line accompanied by an ensemble holding the steady pulse of the metric grid (though perhaps not rigidly), or with accompanying instruments (especially common for drum parts and related reflected in the 'groove'). These are common in records, where there are much deeper flexibilities of pulse and meter found in much popular music that in art music. This might go so far as to establish layers of different rubato streams between lead vocal, groove instruments and other independent vocal or instrumental parts.

The human qualities of performing rhythms can use duration relationships to bring expression to the line. Performing styles, techniques and expression influence rhythmic patterns and the beat. The pulse might be dragged at one moment and pushed the next, or consistently pulled by one part and pushed by another to establish equilibrium; rhythms of vocals flex to interpret the sounds and message of sentences, words and syllables. Perhaps no other element is shaped by performance as much and as subtly as rhythm, and in so doing rendering the rigidity of traditional notation increasingly inadequate in its reflection of rhythm's subtle motions.

These subtle undulations characterize many records, and are often loosely addressed as 'the groove' of the song:

> In a musical groove, a musician, dancer or an engaged listener has a similar feeling of being pulled-into a musical "notch," guided-onto a musical "track," buoyed by a rhythm, being lifted up and

carried along. . . . musicians . . . go to great lengths not only to accurately perform one rhythmic pattern or another but to perform rhythms in such a way that they acquire various qualities of groove, specific qualities of "pushing," "pulling," "leaning forward," being "laid-back," being "in the pocket," and so on. Musicians achieve this by playing certain notes ever-so-slightly early or ever-so-slightly late (in addition to the subtleties of dynamics, timbre, etc.). Loosely speaking, a groove is the *feel* of the rhythm.

(Roholt 2014, 1; emphasis in original)

The groove establishes a rhythmic context for the record. It may function as a continuous point of reference, or as a subtle backdrop; in some songs the groove is a dominant driving force and a primary material. It is the regularity, subtlety and the nuance of the slight deviations from the metric grid that characterize the individual groove, and that contribute to the unique character of the track. Consider the rhythms of the piano, acoustic guitar, bass drum and hi-hat within the introduction of "While My Guitar Gently Weeps" (1968). Careful listening will recognize rhythmic notation would be pressed to capture the subtle pushing and pulling of the pulse within the piano and the rhythmically moving timbre of the hi-hat; while the pulse is grounded by the acoustic guitar and bass drum, there is elasticity of pulse within each as well. Notice, too, the rhythms established by interactions of these sources.

Similar to a groove, rhythmic ostinatos and beat patterns pervade recorded songs. These are incessantly recurring patterns that may be part of a groove, or separate material; ostinatos can be short riff-like patterns, a single measure drum groove or drum kit pattern, or longer phrase-length patterns. The 'beats' of hip-hop are typically a combination of groove and ostinato patterns, often emphasizing backbeats.

Remembering that elements fuse with other elements to create more complex materials, we can understand no element functions in isolation. For example, 'melody' and 'melodic lines and materials' result from patterns of pitch in conjunction with patterns of rhythm; these patterns that may not remain

Table 3.1 Variables and characteristics of rhythm.

Tempo	Speed(s) (beats per minute)
	Number of tempos
Pulse/Beat	Underlying recurring pulse
	Surface rate (notes per minute)
Metric Grid	Organization of pulses (strong and weak beats)
	Sub-divisions of pulses (i.e. two or three divisions of the beat)
	Reference for duration patterning
Rhythmic Patterns	Duration patterns in all elements
	Between sources, materials or elements
	Exist at all structural levels
	Pattern groupings: patterns of patterns, sub-patterns
Performance Alterations	Stretching and compressing pulse
	Rubato
	Groove
Hypermeter and Structure	Prevailing time unit and characteristic phrase lengths
	Meter of measures
	Hypermetric groups at all structural strata
	Regularity of hypermetric unit lengths
	Surface rhythmic patterns of elements
	Rhythmic interaction of elements and materials

aligned between the elements, and that in their union will create another set of characteristics (a typology) that combines the activities and states of pitch and rhythm of that particular melody. The melody will involve other elements as well, as its emphasized pitches imply harmony, and as it is delivered by an instrument or voice, it is shaped by its dynamics and timbre. The web of relationships is intricate, and elements fuse in perception—though distinguishable upon study and reflection, and as abstract activities. All of the record's materials are comprised of more than one element, whereby a change in one element brings a change to the whole.

The musical elements of harmony and melody are organized by the conventions and principles of tonality. More precisely, melody and harmony adhere to (or perhaps define) the varying principles of function within one of the various tonalities found in popular music.

TONALITY, HARMONY AND MELODY

In its simplest definition, melody is the linear succession of pitches in rhythm. When experiencing music, we hear melody as a single percept dominated by pitch and temporal relationships in some type of successive motion, but also with other dimensions. We engage melody as intervals of pitch and time between the notes, as organized into patterns that are perceived as shapes, and as an overall contour and impression that also contains qualities of performance (dynamics, timbre and other sound qualities).

Relationships of pitch receive the most attention in many forms of analysis. Walter Everett (2008, 111) offers the position "that pitch relationships are of central importance, forming the core of the structure, the identity, and even many of the expressive qualities of pop-rock music." While equivalence opens our perception to potential of all elements, we also recognize which elements present materials of central importance—and pitch is often, though certainly not always, central. Pitch relationships are based on the assembly of pitches in use, organized into scales or modes; the functions of the pitches are reflected in the song's tonal system. Such functions are established by convention, as relative levels of stress or significance, and more.

Harmony plays a crucial role in the song's tonal system. Harmony is often framed as the vertical aspect of pitch (with melody the horizontal); with its chords of simultaneously sounding pitches, built on pitches of the scale degrees of the scale or mode in use, and containing various intervallic content (though based on diatonic thirds). Harmony is also linear. In ways somewhat similar to melody, its chords carry functions and levels of tensions largely in parallel to the scale degrees of melody. Chords form patterns that are the sequences of harmonic progressions. Progressions are stylized within various types of tonalities, and carry conventions of use within those tonal systems; they also represent an important characteristic of the song.

Thus, harmony is comprised not only of vertical sonorities, but also of their horizontal movement from one chord to another. The influence of harmony is felt even in the absence of clearly articulated vertical sonorities.

Counterpoint appears at the intersection of melody and harmony. In popular music it typically manifests as two or more melodic lines with clearly independent identities—such as a vocal and a bass line. The lines are interdependent harmonically, but are separate ideas in rhythm, shape and function. This harmonic interdependence is the underlying chord progression—which can be veiled at times and clear at others. These streams of independent melodic lines produce harmony (chords and progressions) by their harmonic suggestions and melodic motion. Voice leading links the notes of one chord to the next with melodic motion and melodic lines' implied harmony; conventions of tonality carry expectations of how some pitches are to resolve into the chord that follows.

Tonality

Engaging tonal systems gets complicated. The tonal system plays a defining role in the song's pitch language—language that communicates through the song's scale material(s) and chord choices and successions. This language generates the syntax for motion and relationships of tones, and establishes expectations within the listener—expectations generated from learned convention as well as from materials of the song itself. Before progressing forward, we need to define some matters of tonality.

At a base level, a tonal system (or tonality) is a hierarchical system for organizing pitch. In a tonality one individual pitch has the greatest stability; the remaining tones are related to that tonic pitch. Those remaining tones are the aggregate of the scale in use, and they are defined by scale degree and their relationships to the tonic. With relationships come functions (such as the common-practice functions of tonic, dominant and subdominant) inherent within the system; this establishes a hierarchy of perceived relationships and functions between the pitches of the tonality and between the chords built on the pitches of the tonality. These relationships of pitches and chords carry expectations inherent in the tonal system, and induce stability and tension, directed motion and points of arrival, and so forth. The cadence is a sequence of harmonies that bring musical motion and tension to a point of rest or resolution; this point of arrival can have varying degrees of strength and stability depending on the chord functions in the tonality, and other factors of the musical texture. This generic definition of tonality opens to include the many tonal systems found in recorded song.

Tonal systems take many forms. Popular music tonalities incorporate a wide variety of modes or scales and related chords and progressions. Further, with the inventiveness of popular music composers (and perhaps the creative impulse to stretch or even to defy conventions), it is not unusual for an individual song to have a tonal system that combines modes, or that is otherwise unique in some subtle but nonetheless significant way. A very large percentage of popular music is tonal, likely more than 98% (Everett 2007b, 303). With so many songs and such a variety of tonal systems, the details of tonality in rock and popular music have received numerous interpretations. There are any number of informed and somewhat diverse concepts and approaches in current and recent studies; this plays out most noticeably in identifying which existing analytical tools might appropriately be used, and in how relevant tools might be devised. The many tonal systems within popular music nearly all differ from the common-practice major and minor tonal principles of art music in some way—some slightly and some profoundly. All recorded songs will benefit from an examination that conforms to their unique system, materials and qualities; an examination in which the analyst ignores biases of all types and origins as much as is possible, and one that allows the song to reveal how it can most easily be understood (Cone 1989, 54).[3]

Modes and Scales

While major and the forms of minor scales (modes) are reflected in the majority of popular recorded song, a significant number of other modes are commonly found. Diatonic or church modes, pentatonic major and pentatonic minor modes, and blues scales are all typical; each generates unique organizational features and inherent qualities and characteristics.

Of these modes, Dorian, Mixolydian and Aeolian are quite common in popular music; see Figure 3.3. Mixolydian is very common, and is used in the Beatles' "Tomorrow Never Knows" (1966). Also from *Revolver*, "Eleanor Rigby" provides a clear example of combining modes within a song, as it alternates Dorian verses in contrast with an Aeolian chorus. The influences of the remaining modes are certainly found in popular music,[4] though pure, unaltered Phrygian and Locrian are rare; "Wooden Ships" (1969) by Crosby, Stills & Nash contains a clearly Phrygian verse, and the lowered 5th of Locrian is strongly evident in "Army of Me" (1995) by Björk. The Beatles' "Blue Jay Way" (1967) represents a clear example of a song in the uncommon Lydian mode.

Peter Winkler (2000, 30) notes:

> [P]entatonic melodies are extremely common in popular music, for a number of reasons: they contain no half-steps or dissonant intervals, they are clearly centered around a triad, and they easily accommodate variable intonation—portamentos, blue notes, speech-like inflections, or just plain bad singing.

Pentatonic scales, with minor and major third scale degrees, contain five pitches; notice these scales contain only the intervals of major second (M2) and minor third (m3). The pentatonic minor scale may be the most common pop-rock scale after major, and it also often occurs through substantive melodic inflections within major mode (Everett 2008, 158). The most common forms of pentatonic and pentatonic minor scales appear in Figure 3.2.

The blues has had a significant impact on rock and popular music genres. Scales are often altered to incorporate 'blue notes,' which usually appear as the lowered third, lowered fifth and/or lowered seventh scale degrees. 'Blue notes' can be incorporated into scales, or superimposed on major modes creating "the possibility of blue alterations of the third scale degree in pieces with major tonic triads: in many rock styles, it is common in melodies to tune the third scale degree so low that it forms a m3 [minor third] with the tonic" (Stephenson 2002, 37). A blues scale will typically appear in one of three forms: (1) a hexatonic, six-note scale that consists of a minor pentatonic scale plus the ♭5th scale degree, (2) a seven-note conception is the major scale with lowered third, fifth, and seventh degrees, and (3) a chromatic variation of the major scale incorporating a flat third and seventh degrees which alternate normal third and seventh scale degrees (in context this can manifest as the Dorian mode occurring simultaneously with major mode, its minor quality superimposed over the major-key chord changes).

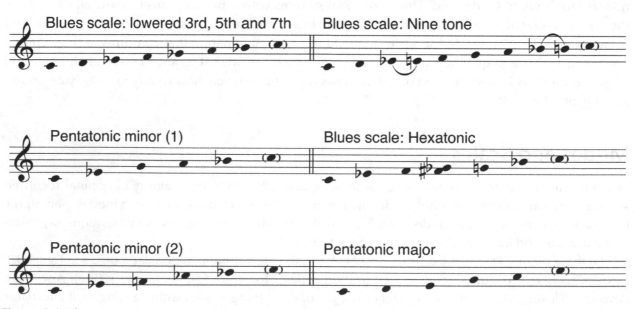

Figure 3.2 Blues, pentatonic and pentatonic minor scales commonly found in rock and popular music.

These blue notes appear to conform to equal temperament when considered as incorporated into these scales. Musical practice can change this substantially for any instrument capable of bending pitch (such as stringed and wind instruments) and for singers. Blue notes often appear as microtones, flattened by a variable amount smaller than a minor second (half-step). This 'worried note' is sung or played slightly differently from conventional tuning, both for expressive purposes and for tonal effect. This alteration differs among performers and styles (jazz, blues, etc.); though it is often a quartertone or somewhat less, it is noticeably and distinctly flat. These microtonal blue notes are typically integral to the substance of melodies, though they may instead provide ornamental qualities.

Blue notes are just one example of microtones in melody. Microtones are often incorporated into vocal performance styles and melodic lines. There they can provide or support speech-like inflection of lyrics, embellish pitches or add tension and expression. As flat pitches create tension of anticipation that the performer will reach the target pitch, or a sharp pitch can represent the stress of perceived over exertion, embellishing the line by intonation and microtones can add direction and motion to melody. Thus, being 'out-of-tune' is not necessarily poor performance technique, but rather an integral part of performance technique, interpretation of the line, and its musical materials. The current widespread use of auto-tune erases microtonal inflections; the resulting inhuman pitch accuracy (described by many as mechanical) is a core characteristic of other popular styles.

Recorded song can take great liberties with tonalities and scales. The unique qualities of tracks can reside in modes that morph between forms, bend pitches ambiguously, as well as countless other activities. Blues or folk music origins can influence mode usage in certain genres significantly. The recording medium often accentuates the presence of shifting modes' subtle changes.

Chords and Progressions

Chords and harmonic progressions are formed from the pitch resources of modes. The conventions of tonalities are reflected in the qualities of chords and the sequences of progressions. Chords built on the scale degrees of modes reflect the unique qualities of the mode's sequence of whole- and half-steps. These unique qualities are imbedded into the chord sequences (progressions) that characterize the music written within or utilizing the mode. Figure 3.3 presents the triads built on the successive scale degrees of Dorian, Aeolian (natural minor), Mixolydian, and major (Ionian) modes. Chord letter names and the nomenclature often found in popular music are listed above the staff; chord quality is indicated by uppercase alone for major 'C,' for minor 'Dm,' and for diminished B°. Below is the system (with roman numerals) that relates numbered roots of chords to illustrate their relationship to the tonic; this system also provides indication of chord quality.

Both forms are provided here to draw attention to identifying the names (based on the pitch of the root) and qualities of chords as a separate step from identifying the functions of chords— functions that can only be identified after the tonality has been determined. Further, Roman numeral analysis places the chord within the tonality and within the context of the song; simple chord names identify the chords themselves and not their functions or relationship to tonic, allowing ambiguity to be embraced and explored. Both approaches have value at different times. Tones added to these basic chords are reflected in chord symbols. Chord symbol notation and nomenclature has many variables. It can become very involved quickly; all nomenclature used herein will follow one of these accepted systems. A source that lists and explains systems of nomenclature will prove valuable.[5]

Notice there is a second line of roman numerals under Mixolydian; these denote how the chords on those scale degrees relate to their counterparts in the major mode. Some quotations below will use this approach. Notice all roman numerals are in uppercase and a ♭VII chord is present; this signifies

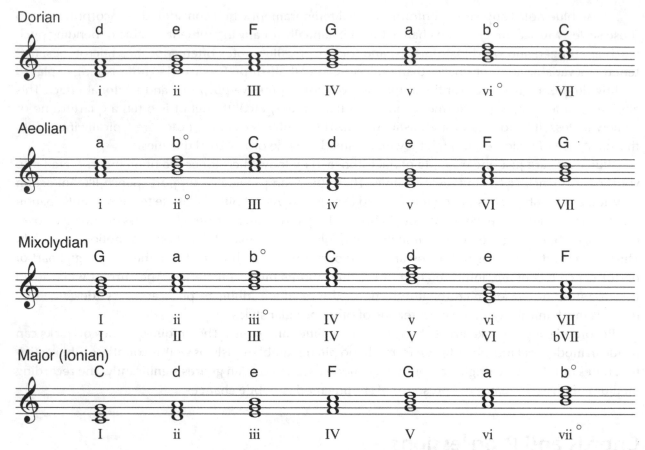

Figure 3.3 Chords built on the scale degrees of Dorian, Mixolydian, Aeolian and major modes, with designations of chord qualities and scale degrees.

the chord is one half-step flat from the seventh scale degree of major. This system is a clear example of how a well-meaning approach to use traditional major/minor scales as a reference to characterize modes might (depending on usage) undermine the mode's perceived status as a norm in the context of rock and popular music.

Harmony is the element of music that has been most thoroughly systematized. Analyses of music often emphasize harmony, and dedicate the most space to its materials. Early opinions of most academics identified pop-rock's chord types and functions, and the content of harmonic vocabulary, as simple and limited; even those most accepting of popular music framed their observations with a clear implication that art music should be the point of reference, as it is somehow superior, more complex and more sophisticated (Mellers 1973). We now know this is far from accurate, though aspects of this position still persist.[6] An abundance of general tonal systems are used, as we have seen from the diverse scales, above. Each of these general tonal systems can generate a unique harmonic syntax for the individual song—even the well-used 'major' tonality can spawn an array of harmonies and harmonic relationships that may create a unique vocabulary for a track—especially with the addition of chromatic harmonies. Further, the ways a tonal system is used within a critical mass of songs can generate subsets of styles and establish conventions that are widely shared, and that establish specialized tonal systems with certain characteristics.

Walter Everett (2007b, 304–320) has organized the richness of rock's harmonic vocabulary into nine 'classifications of rock's preeminent tonal systems' that "progress from traditional major-minor and modal systems through pentatonic patterns, ultimately ending in chromatic relationships that may bear little resemblance to any normal established tonal centricity." Of these, four classifications are major-mode systems, one classification each of minor-mode systems and diatonic modal systems, and three classifications related to pentatonic systems. A review of the classifications confirms the striking differences of harmonic languages and other characteristics, and their levels of complexity. Tonal systems are not only diverse, explaining them can become very involved. Identifying chords and the actual function and classification of those harmonies can become complicated or ambiguous (indeed, sometimes this mystery is part of the allure of music that can be 'heard' in different ways upon repeated hearings, though it can frustrate analysis). This can become further complicated when songs move between tonal systems—more specifically, between major or minor tonal systems and modal or pentatonic systems—or are resident in a modal tonality. Fixation on how a song fits into one or another system, though, can miss the point: analysis is to identify what is going on in the song, not to identify to what tonal family the song belongs. Often a song can seem to defy clear classification, not because it is somehow flawed, but because its unique language is not clearly reflected within existing classifications—and this may simply be part of the process of the artist pushing boundaries, though perhaps the methodology is inadequate or has been misapplied, and other possibilities certainly exist. Every song sets its own context, defines its own rules—even if some of those rules are borrowed (or adapted) from conventions.

Classifying and describing works within major or minor tonalities are the focus of traditional music analysis (steeped in common-practice conventions that originated in a different century). This approach to analysis is not always useful, or at all pertinent, for some tracks—or for recording analysis. In some songs, harmonies difficult to reconcile within major or minor tonalities might be explained in the context of modal usage. Moore (2001, 55) identified, "this obviates the consideration of some chords (such as the Mixolydian VII) as aberrant. Mixolydian VII is far more common in rock than Ionian VII." Similarly, a major IV chord within a minor tonality evokes the impressions of Dorian influence. Still, our established theoretical/analytical methodologies are typically misleading for the songs that are clearly modal—that contain no tangible presence or influence from major or minor tonalities. Modal scales and progressions have unique qualities—often reflected in chord root movements by seconds, and sometimes thirds; their characteristics cannot be fully recognized or appreciated within the context of tonal analysis (that emphasizes concepts of harmonic relationships based on fifths). These unique qualities are fully formed within the contexts of individual songs, as well as within modal systems; that modal songs do not have the tonalities' directed motion of fifth relationships (which may well be more of a learned association than an inherent tendency) does not make them without tension or direction, these are simply achieved by other means. It is clear that modal progressions are unique and stand on their own—with their characteristics and intrinsic tendencies. Given this, it seems most relevant for valid understanding of the song that they not be considered as substitutes for major/minor, or even aberrant in their nonconformity. It follows that "modal progressions might themselves be elevated to the level of paradigm or norm"; that "a knowledge of modal idioms comparable to our knowledge of tonal idioms is valuable as one attempts to analyze popular music" (Burns 2008, 67–68).

Clearly, accessing the tonal and harmonic language of the individual song may not be a simple process.

Characteristics of Harmony

Tonality is a system of organization at the large dimension. Tonality provides the relationships of pitches and chords that generate harmony's characteristics in nearly all popular songs. A governing tonality is the defining tonality of the song and its key; the tonal center is the point of reference for all of the pitch and chord relationships, and any new keys that might appear. The characteristics of the governing tonality—which, to be clear, includes modalities—include its system of pitches, their organization and relationships. It also includes the aggregate of influences from any additional tonal centers and their modes. In a major or minor song, the degree to which it conforms to common practice principles could be a factor of significance. Modes may be superimposed or appear successively; the resulting tonality is comprised of the qualities of both when recognized at the large dimension level. There is vast tonal/modal richness available to popular music.

The relationships of some tonalities (especially common-practice major and minor, and some blues traditions) are fraught with a "particular cluster of conventions" (McClary 2000, 63) all their own. This extends to include how tones and chords are to function in harmonic progressions and cadences, how tonal centers are to relate, and how harmonic tension can directly create motion (departure, travel, arrival). This motion generates and directs tensions (of varying degrees of strength) toward full and partial resolution and stability. The tonalities of rock and popular music may carry some of these common-practice expectations, though its listeners are easily accepting of their absence—and willingly embrace popular music's wide palette of replacements. Further, the conventions of tonality carry a deep ideological stigma due to their history within European classical music; its principles and terminology carry loaded definitions, relationships and expectations into popular song that often imply (or overtly promote) it is the accepted way against which all other systems must be compared.[7] References to tonal systems in popular music need not carry this weight, though it is easy for an analyst to unconsciously find oneself in this position.

Modulation is the movement between tonal centers (such as from the tonic key to another). New tonal centers are identified by their modality/tonality, tonal center, and their relationships to the tonic; traditionally these might be subdominant, dominant, or parallel major/minor relationships, though popular song may modulate as it wishes, with mediant movement common and second relationships not unusual. Structural use of tonal areas is common in popular song, with it being common for a new key to be introduced in a bridge or a middle eight, a modulation to occur between verse and chorus, and so forth. Lori Burns (2008, 86–89) explains the shifting tonal focus in the four main sections of Tori Amos' "Crucify" (1992) as verse in G♯ Dorian, pre-chorus in B major, chorus in G♯ Aeolian, and bridge in G♯ major/minor; this demonstrates modulation as taking place between modes as well as between tonal centers, with tonic shifts to relative major and parallel major relationships to G♯ Dorian. The large dimension governing tonality of "Crucify" reflects these modes, and the upper strata of the middle dimension organizing the sections by tonal centers as well as acknowledging the individual modal scales.

Modulations occur as harmonic pathways between the keys. A path of modulation might be a common chord between the two keys, a slightly altered chord in one key that pulls into the new, movement between relative keys common to the same scale, paths that are ambiguous in common-practice but 'work' within the language (or modality) or unique context of the song, or any other means. Songs may even oscillate freely between keys with no hesitation or effort to reconcile the two by common tread; such a 'double tonic' relationship with keys a tone apart can allow the flat seventh scale degree to be prevalent (Mellers 1985, 39–40). Brief excursions outside the key are ornamental in function—especially those with duration of less than the prevailing time unit. These are transitory embellishments, providing a 'flavor' of another tonality without its established presence. Modulations that bring a new tonal area to an entire song section are structural; a new pitch level is tonicized, and perhaps a new mode established. Number of tonal centers, duration of each center, modulations path(s), relationships of the keys(s) to the governing tonality, structural placement and frequency of returns of keys, are some of the characterizing factors of tonal areas and modulation.

Harmonic progressions are, simply, sequences of chords. They are central to establishing and reinforcing tonality, reflect or produce the governing characteristics of the song's mode, and the chord patterns they establish and reiterate are typically defining characteristics of songs. Harmonic progressions are stylized within the specific song, and might reflect a certain style of rock or popular music—such as chord motions by descending perfect fourth (that reverse the usual functional progression of common-practice harmony) and parallel shifts by step that frequently occur in rock songs (Koozin 2008, 267). As we have encountered, many languages of harmonic progressions (based on modes and tonalities) exist, and their content and contexts can be very complex.

Here I wish to make clear, popular music is extremely forgiving about harmonies (chords) that do not conform to any 'language' of harmony that might be in play. When intuition is the guide for creativity—as is often the case with recorded song—any 'rules' of harmony (or conventions of style) are irrelevant. Indeed, even the vertical sonorities of chords can morph to resemble something closer to timbre than superimposed discrete pitches. The guitar line within the introduction of U2's "Bullet the Blue Sky" (1987) is one such example; this line has tension, directed motion, a sense of departure and of arrival, movement between established points—all qualities of harmonic progressions.

Progressions play a significant role in musical movement within the majority of musical styles. Their qualities can be characterized in this way. Progressions contribute significantly to the track through the types of motion they provide—speed of motion, strength of motion (resulting from tension), amount of motion (separation from the tonic note), among other attributes.

The characteristics of harmonic progressions that establish motion and tension are typically directed toward some type of resolution or stability. Tension does not need to resolve, however (Everett 2008, 145–149); some chords, contexts and songs exploit this, hanging tension without the release of resolution. The tension of harmonic relationships that creates movement is intrinsic to language, some modes providing less obvious movement than others. This may be a result of root movement between chords, as the fifth movement within major/minor creates (for some listeners, in certain contexts) more of a sense of direction than the modal conventions of movement by a second or third. Musical languages with their built in tensions are cultural and learned (Sloboda 2005, Handel 1993). Tensions that may be clearly heard by a skilled listener well versed in common-practice principles, might well be heard differently by a skilled listener that conversely is experienced in the modalities of folk music; one might recognize little directed motion with a divergence from common-practice, where the other might notice the tensions within a root movement by thirds with only one common tone. This is central to calculating and understanding the continuum of maximum tension to unwavering stability that characterizes progressions and the chords they contain.

Thus, progressions are characterized by type and strength of motion. The unique syntax of an individual tonal/modal language provides the organizational cohesion and coherence for progressions; the same progression can function differently—mean something different—in different songs given a different context of syntax. The listener brings experience, listening skill, knowledge, attentiveness and more to the listening experience; thus a degree of uncertainty and subjectivity is present in the *functioning* of tonal systems. The interpretation of the analyst can often reflect this, intentionally or not—as the analysis itself is an interpretation.[8] The possibilities for chord progressions in popular music, then, might be seen as broad, with numerous ways to make sense of them—some ways are specific to the song, some are based on the conventions of the past (perhaps preceding common-practice), some borrowed between conventions (such as major borrowing from minor pentatonic), some particular to a style within popular music, and new ways are likely currently emerging.

Progressions are also characterized by chord content—chord types in use, which chord types and roots dominate, and how they are sequenced. Triads and added note chords, seventh chords and ninth chords, chords built on fourths or fifths, and chords without thirds all appear in progressions—and this list is sorely incomplete. These chords all function within the movement of the progressions—or not, depending solely on usage and context. The dominant seventh chord is perhaps the chord that demands the most attention and has the most stylized function within much of our culture. "Function" has been determined largely by common-practice principles, and a chord like the dominant seventh has a deeply engrained function—its sound quality demands resolution, movement in a preordained voice-leading path to a singular goal.

This motion is typically directed to a cadence of some varying strength—cadences are points of arrival and of closure, of resolution and of stability—usually near the end of a hypermetric unit. These are often marked by slowing rhythmic drive, a pause, and descending melodic motion, and are strongly reflected within harmonic progressions. Cadential progressions immediately precede such points of arrival. Certain cadential progressions have emerged in common practice that are deeply engrained and stylized—these are progression patterns ending with V, V–I(i), vii°–I(i), or V–vi. When these formulae appear they may be adapted to popular music by voice leading or rhythmic emphasis. Modal progressions can also generate cadence types; one common cadence is the ♭VII to I (Moore, 2007a). In widespread use in the late 1960s and continuing onward, the stylized double-plagal cadence (♭VII–IV–I) is unique in the many forms this pattern has generated in popular music (Everett 2008, 154–155). Allan Moore (2007a, 193) offers: "Clearly, the VII–I cadence does not have the finality of the traditional V–I, although it is articulated as a full close . . . In terms of poetics, it seems to me to qualify the certainty of V–I with 'nevertheless'."

Principles of common-practice function simply do not apply to some (perhaps many) popular songs;[9] chords and progressions in these instances are characterized in other ways. Chords will not always have a function, or predisposition of how they must move. Some chords will be found that do not fit into constructs of triad-based sonorities, or within the construct of having a role within a progression. Power chords and color chords are examples.

Power chords do not contain a third, and are comprised of solely a perfect fifth or fourth (typically sounded by two adjacent guitar strings, often open); this staple of metal, new wave, grunge and punk can generate progressions of parallel-motion fifths, often moving by seconds. As Robert Walser (1993, 43) describes, "Power chords are manifestly more than these two notes, however, because they produce resultant tones. An effect of both distortion and volume, resultant tones are created by the acoustic combination of two notes." The two-note chord generates an additional tone at the frequency that is the difference of the two tones; for example, two pitches, one at 147 Hz (D_3) and one at 110 Hz (A_2) would generate the pitch 37 Hz (D_1) which is below the range of the guitar but will be audible. Further, the typically distorted guitar amplifier that plays the chord generates other frequencies, many unrelated to chord tones; this distortion adds considerable harmonic complexity to the sound.

Color chords typically function for their abstract sound quality (they can be a type of sound object inserted into the flow of an event); their sonority can vary widely, and may not be comprised of stacked thirds; the chord will be unique in some way. Color chords might make tension, but not resolve fully or expectedly; they can bring either instability or stability, to varying degrees. An example is the opening chord to the Beatles' "A Hard Day's Night" (1964) with its D in bass and piano, and F-A-C-G on electric 12-string guitar. Examples of non-functional chords abound. Non-functional progressions can be found as well.

As noted, modal progressions exploit the unique qualities of the mode. They are functional in their own unique ways, and their characteristics place emphasis on those unique relationships to create tension and motion—in ways distinctly different from major/minor. Modal progressions and modal systems, of course, have a rich history dating back to the Middle Ages and the Renaissance. Modally derived harmonies and progressions are the primary tonal language of the sixteenth century and earlier— and have survived and thrived in the folk musics of Europe and the Americas, reappeared in concert music from the late nineteenth century onward, and exist in a great many forms throughout the world, including the musics of Africa whose influences are deeply felt in popular music.[10] Modal

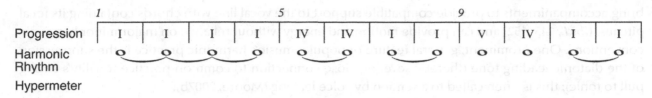

Figure 3.4 A typical twelve-bar blues pattern indicating harmonies, harmonic rhythm and four-bar hypermeter.

conventions have new roots in today's rock and popular musics—whether intentionally imported and adapted or intuitively uncovered. The characteristics of all modes have been explored in a wide variety of contexts; three examinations of Aeolian mode will serve as examples here. Alf Björnberg (2007) examines numerous works with Aeolian harmonic progressions; he notes the harmonic progressions the mode can generate, including the cycling i– ♭ VII – ♭ VI– ♭ VII Aeolian progression of Bob Dylan's "All Along the Watchtower" (1967). Lori Burns (2005) demonstrates how the verses in Sarah McLachlan's "Ice" (1993) use the F Aeolian progression "i–VII–VI–V";[11] further, voice leading within the Aeolian mode plays a central role in the analysis as a whole. Finally, Stan Hawkins (2000, 58–65) examines how the Aeolian voice leading, chords, and derived chords appear in Prince's "Anna Stesia" (1988); he presents a 'macrocosmic cellular sequence' of how the various scale degrees of C Aeolian establish a sequence of tonal emphases that function structurally throughout the song—providing a set of relationships that clearly maps out how these tonicized areas within the governing Aeolian modality are akin to a 'key rhythm' (see below) for the song.

A progression does not always elicit directed motion—simple chord patterns might simply repeat, instead of progress (Mellers 1973, 36). Cycling chord patterns is a characteristic of harmony that also brings structural and duration (hypermeter-related) aspects into play. The incessantly repeating twelve-bar blues harmonic pattern is a common example, and one that is commonly modified in pop and rock. Note the pattern's three-chord vocabulary; a vocabulary of three or four chords (or fewer) has appeared in a wide variety of songs. This principle of circular motion with repeated harmonic progressions has been explored as various types of 'chord loops' by Philip Tagg (2016, 401–414), and identified as 'repeated circles' of chords by Richard Middleton (1990, 113). These patterns may create and resolve tension in waves by their repetition rather than (or perhaps in addition to) any functional characteristics the chords might carry.

Voice leading's melodic movement from one chord to the next reflects the harmonic relationship between simultaneous melodic lines; it is the synergy of horizontal melody and vertical harmony. This motion articulates or implies harmonic progression and its chord functions, and may bring directed motion of increasing or wavering tension, but is resident within and responsive to the character of the melodic line. These and more can shape and be reflected within voice leading, and are responsive to the governing tonality and its conventions. The voice leading and contrapuntal relationships between the vocal line and bass is especially significant, as noted by Lori Burns (2008, 67): "[T]he specific way in which a vocal line is staggered contrapuntally with the bass is a very important feature of popular music performance and one that contributes in large part to the distinctive sound of a given song or artist."[12] These lines and their relationships typically appear in the recording in ways that provide them with space and clarity, allowing them to be apparent and readily appreciated. Voice leading and counterpoint can serve other important functions. All harmonic activity in single chord songs—or most activity in songs with two chords as in Bruce Springsteen's "Born in the U.S.A." (1984)—are generated by counterpoint, such as "rap vocals . . . laid over tracks involving contrapuntal ornamentation of a single harmony" (Everett 2008, 145). Voice leading can establish non-triadic added tones to chords, in major/minor tonalities it can bring successions of chords that are outside common-practice relationships, can bring accompaniments to provide compatible support to the vocal line with chords containing its focal pitches, (*ibid.*, 145–152) and can provide motion and energy without reliance on major/minor harmonic conventions. One prominent, general feature of popular music's harmonic practice is the suppression of the diatonic leading tone (thereby severing close connection to common-practice tonality's strong pull to tonic); this is often called to attention by voice leading (Moore, 2007b).

Like cycles of chord patterns, other harmonic characteristics contain temporal elements, falling into rhythmic patterns. Harmonic rhythm is the rhythm of chord changes. As seen in twelve-bar blues Figure 3.4, harmonic rhythm can establish a rhythmic pattern of chord changes that can recur; patterns

Table 3.2 Variables and characteristics of harmony.

Tonality	Governing tonality: tonal center/key
	Governing mode
	Aggregate of modes in play: number, how related
Tonal Center(s)	Governing tonal center
	Number of tonal areas
	Relationship of tonal areas to governing tonal center
	Structural locations(s) of tonal centers
	Key rhythm
Modulations	Structural modulations: durations and key
	Ornamental modulations: durations and key
	Relationships of modulation to governing tonal center;
	Path of modulation: harmonic progression, transitional process
	Repetitions of modulation paths
Chord Types	Chords types in use, which dominate
	Recurrence of specific chords (characteristic of song)
	Unique chords, color chords, power chords
Progressions	Chord content of progressions
	Recurring characteristic chord pattern(s) throughout the song
	Placement of unusual or characteristic chords within progression(s)
	Chord types in use: appearance of unusual chords
	Characteristic chord root movement
	Root movement and common tones between chords that dominate
	Relationship of progressions to modal and tonal conventions
	Tonal progressions: degree of divergence from common-practice
	Number of different sequences
	Variations within recurring progressions
	A specific characteristic chord within a recurring progression
	Patterns of relative intensities (stability to tension)
	Progression as theme
	Harmonic rhythm
Cadences	Location of cadences related to hypermeter, phrases, and sections
	Stylized cadential progressions: used repeatedly,
	Cadence progressions used at important structural points
Voice Leading	Melodic resolutions at cadences
	Counterpoint of voice and bass
	Enhancing or replacing harmonic motion
Harmony in Rhythm	Harmonic rhythm
	Key rhythm
	Chord rhythm

might also shift between song sections, or may vary (develop) in recurrences. Chord rhythm is a pattern of iterations of the same chord, common in many rock styles, and is a strong stylized trait in others. In songs with modulations, especially songs that modulate frequently between sections (such as when verse, chorus and bridge are all in different keys), 'key rhythm' can be significant; key rhythm will be at a structural level in the upper middle dimension or large dimension, and will manifest as a rhythmic pattern established by the durations of tonal areas.

It may now be evident, for some songs, harmonic language might be a minor, insignificant or even irrelevant factor. Pop/rock benefits from tonal ambiguities, chord progressions unrelated to traditional harmonic structures or progressions. Further, it is not at all obliged to progress from a propulsion of harmonic tension and to resolve harmonically like art music. Motion in the music can be a product of any musical element—and any element of the recording or the lyrics.

Characteristics of Melody

Melodies carry many characteristics, utilizing the grammar of the scale and resulting tonal influences. Melody can be characterized by its contour, melodic framework, structure and phrasing, and (for vocal lines) how the melody presents the lyrics. The characteristics of melody are summarized in Table 3.3.

Melodic contour characterizes the shape of the line that is created by pitch levels over time. Shape is the overall pattern of the contour, and any division of that pattern into shorter patterns—whether

Table 3.3 Variables and characteristics of melody.

Contour	Shape of the line
	Direction and motion
	Pattern of peaks and lows
	Range, register,
	Stepwise or leaps
Melodic Framework	Tonality
	Mode(s): governing, others
	Intervals, activity, articulation, duration
	Density of pitch changes and rhythmic durations
	Pitch and interval content: patterns, focal pitches
	Articulated and implied harmonies
	Rhythmic flow and patterning
	Compound melodies
	Sequences
	Short-duration pitch/rhythm patterns: riffs and melodic ostinatos
Linking Melody and Lyrics	Number of pitches per syllable, or syllables per pitch
	Syllabic and melismatic
	Speech-like, linking of melody with speech
Structure and Phrasing	Rhythmic effects of phrase lengths
	Regular phrase lengths and sequences
	Irregular phrase lengths
	Relationships of phrases
	Repeating, variation and completion phrases
	Contrasting phrases

interrelated or contrasting. Contour variables include direction, interval change (changes in ascending or descending movement), frequency of direction change and the timing of direction changes. Contours may be classified as ascending, descending, pendulous (in moving in one direction then the other), oscillating (melodies that alternate between two structural notes) terraced (melodies in which entire phrases move between structural pitch levels), axial (melodies that circle around a structural pitch), static or chant (melodies that remain fixed on a repeated note); these are general classifications that might often be elaborated upon.[13] Range is the interval (or distance) between the lowest and highest pitches of the melody; an important consideration is the distribution of pitch materials within this range, whereby the melody concentrates its activity within an area—or register. In this way the pitch range emphasized by the melody can be identified. Related is the rate of pitch change, and any variation in the speed (or rate) of pitch change.

Melodic framework is established by the interval content of the melody (the predominance of steps, skips, especially leaps of specific intervals), the speed and amount of rhythmic activity, patterns, the use of silence to articulate pitch activity, and the relative length of the melodic ideas and how they are sequenced. It is also defined by focal pitches within the melody—pitches that are emphasized in some way, whether by metric accent, relative length, or some other means) and their tonal functions. This gestural outline might provide directed motion to the melody, establish pitch terraces (registers) for the phrases of compound melodies, or be reflected in sequenced melodic lines (repeated melodic patterns transposed to other pitch levels) (Everett 2009b). A short melodic or rhythmic pattern that is repeated over and over is often called a riff; a riff can function as a melodic ostinato. Changes in other parts will take place at the same time as the riff repeats unchanged; a multi-level process is established where "sections of riff-based circularity are *set against* sections of cadential closure" (Middleton 1990, 195).

Related to this melodic framework is the relationship of lyrics and music. The vocal melody dominates songs. This is a by-product of the singer's persona, language grabbing the listener's attention, and of the many other dimensions of the lyrics. While these are covered in detail in the chapter on lyrics, the connection of melody and lyrics is important to note here.

Lyrics influence vocal melodies. The rhythms of speech can influence the rhythm of melody, and the tonal inflections of speech might often be reflected in pitch. These can be structurally incorporated into the melody, appear within the pitch, dynamics and rhythmic materials of melodic shape. These might also appear in performance as nuance of pitch and rhythm. Range and register also assume particular importance within the characteristics of a vocal melody, as the singer's voice sounds and communicates differently in different areas. It is common for singers to shape lines subtly, to breathe the definitions of words and phrases into melody; shaping the sounds of words and in so doing shaping (interpreting and communicating) the meaning of the lyrics. Singers also add body sounds, vocal gestures and other nonverbal sounds around the lyrics, within their performance and the lyrics.[14]

The complexity of the vocal melody tends to conform to the nature of the lyrics, often reflected in the number of pitches per syllable, or syllables per pitch. When content of the lyrics is important (telling the song's story), syllabic rhythms dominate, and often on or focused around a single pitch; this may be taken to the point of the vocal line appearing nearly speech-like, creating a connection between speech and song. More ornamental and even melismatic settings might present recurring ideas in the lyrics, as in the chorus sections. The manners in which syllabic and more ornate melodies appear and are shaped relative to speech is a defining characteristic; speech inflections and intonations provide important insights into this link between the lyrics and the vocal melody. In performance, inflections of pitch, dynamics and durations provide substantive characteristics of the line. Hip-hop and other genres take this in a different direction, as lyrics are rapped with varying degrees of indefinite pitch in the lead vocal; pitches are suggested, and they become more apparent in their repetition.[15] There will be more on all this later.

THE REMAINING ELEMENTS: DYNAMICS, TIMBRE AND ARRANGEMENT

Dynamics, timbre and performance-related alterations of timbre, and arrangement and texture are the remaining elements of music. All aspects of music other than rhythm, melody and harmony have been grouped, here, into one of these three elements. These elements are elevated in significance in recorded popular and rock music, especially pronounced in comparison to live classical music. This is largely due to two primary matters related to the record being a permanent performance.

First, these characteristics are fixed and unchanging upon completion of the recording. Therefore, they are integral to the performance, and fused with its qualities and expression. The interpretation of the song is fixed, and how these elements contribute to that interpretation has been carefully evaluated and crafted to best suit the recorded song. Remembering the record not the score becomes the canonical version, the recorded performance emphasizes relationships and qualities beyond the capability of notation.

Second, these elements can contribute to the musical materials of the track in ways not possible in live unamplified performance. Records provide the opportunity to reveal these elements more clearly, and to draw the listener into them. These elements can assume any of the four functions carefully crafted to contribute substantively to the track.

These other elements are also most closely linked to the elements of the record. There will be much interplay between these elements of music and their complements in the elements of recording. This is reflected in the ways the recording transforms dynamics, sound qualities of sources and groups of sources, and the track's arrangement; the elements of the recording transform these traditional concepts with added characteristics—from subtly to profound.

The many styles of popular music make any generalized statement suspect, and anyone that has listened carefully for any length of time will be bursting with exceptions. The statements that follow will be riddled with examples that do not conform.

Dynamics

Loudness and dynamics can be a powerful element in some rock and popular music. Dance music, rap, heavy metal, some rock music, whatever music is current for adolescents, and many other popular music genres will find substantial loudness as a primary element within the listener's overall experience. Dynamics are loudness in musical context. Dynamics can be a dominant element at the highest dimension, strongly shaping the listening experience—it can make music scream. Dynamics may also bring some sounds to be barely present within a sparse texture, other sounds vibrant with subtle complexities of dynamics within their timbre, and rich expression to musical materials.

Dynamics appear as loudness levels and as relationships between levels—just as pitch has levels and the ear processes the relationship (interval) between pitch levels. Dynamic relationships are typically comparisons of one dynamic level to another, allowing one to recognize one musical part as being louder than another. Dynamics in music can be significant, and will function independently in all dimensions and structural levels.

Traditionally it has been used in supportive or ornamental roles for expression and subtle shaping of a line, supportive for the process of balancing lines, parts or instruments with other, and in a more central role for dramatic purposes. Dynamics can serve an important syntactic function to articulate and shape patterns, and to structure at all strata (Meyer 1973, 35); by itself it can create a cadence for a single line or for all.

Recorded song retains these functions; in addition, the dynamic subtleties of performance take on more relevance with the performance meticulously shaped and its expression is permanent. Dynamics in recorded song have many qualities that are the result of the recording. Those qualities will be considered within the elements of recording and discussed later, and many of the traits described below will carry over in dynamics in recording.

Traditional analysis techniques rarely approach the impacts of dynamics in understanding the work; if discussion appears it is likely related to drama. Dynamics may add direction to the line, create tension, and more—though rarely is such activity apparent in the musical score. Dynamics are largely implied in scores—by the number and ranges of instruments for example—or identified in vague and incomplete terms. Musical dynamics are manifest in great subtlety in performance; general instructions and relative levels are resident in the score, even in the detailed scores of modern concert music, give little direct indication of the *experience* of loudness. The score represents the experience of pitch and rhythm in some detail, everything else, not so much.

Dynamic accents regularly occur in meter. These subtle changes in loudness establish a pattern of stressed and un-stressed beats. In performance, dynamic accents take many forms within musical lines—syncopation, backbeats, marking a moment's significance, etc. The strength and placement of accents may help characterize a record or a line.

Changes in dynamic levels may be immediate or gradual. Crescendo and diminuendo are obvious examples within middle dimensions—whether the entire group changing together, the rhythm section as a unit, or the vocal line. Dynamics function at all structural levels, and these gradual and immediate changes can be traced throughout all dimensions. The overall dynamic shape of the song may exhibit sudden changes in level or gradual increases and decreases of loudness; these exist at the small dimension as well.

The subtleties of the groove are reflected in dynamic changes as much as rhythm changes, as meter is the result of varying stresses of beats. This is a clear example of subtle changes of loudness. Performances in records are carefully crafted for their musical materials and for the expression; dynamics plays central roles in shaping and in communicating both. Much dynamic shaping of musical lines happens in performance, and that shaping can be studied as important to the content and character of the musical material. This is especially important to vocal lines. Subtle, sudden and gradual shifts in loudness are difficult to describe—and difficult to notate. Our vocabulary to describe dynamics, and our notational nomenclatures to write them, are inherently insufficient; both struggle to reflect what happens in the music, within the *sound* of the music.

Dynamics is the musical application of the perception of loudness. Dynamics and loudness are difficult to engage conceptually and practically, for good reason. The way the ear processes loudness does not readily allow loudness levels to be compared accurately; we do not recognize specific loudness levels, and our sensation of loudness varies with the sound's frequency level. Loudness perception is complicated and nonlinear. Further, loudness is readily confused with prominence—we are prone to assume that if something grabs our attention it is loudest.

Our weak vocabulary for all things related to sound is especially pronounced with dynamics. The indeterminate language we use with dynamics makes analysis difficult. Dynamic levels are conceived and discussed in very imprecise terms: loud, soft, moderately soft, or moderately loud; very soft or very, very loud, and very, very, very soft. Discussing the relationships of loudness levels is not better: we are apt to say 'louder than' or 'softer than,' but are at a loss as to 'how much.' 'Twice as loud' is meaningless, as we do not know the starting level of 'loud' in any meaningful way. This is largely the way we talk about dynamics and evaluate dynamics, though a deeper examination can be possible.

This imprecision leads to more general observations, grouping many perceptions of loudness into a few categories of piano, mezzo-piano, mezzo-forte, forte, and so forth. These loudness 'levels' have become loudness 'areas' in use. We have many graduations of 'moderately-loud' (for example); while all can be categorized as 'moderately-loud', each graduation is readily recognized at a different loudness level—a difference that may be pronounced as well as subtle. This establishes a continuum of innumerable 'loudness-es' within these 'dynamic areas' that provides great extremes of loudness to exist within each. Silence is the lowest dynamic level; it is one that can stop motion, suspend it with anticipation, be a recurring musical presence, and much more.

Dynamics are not only about loudness. Dynamics also carry a sense of energy, intensity and expression—traits that are reflected in timbre, not in loudness. Loud is significant energy being expended, and soft significant energy being withheld. Robin Maconie (2007, 71) framed this as: "the sound of a musical instrument combines force and resistance." 'Moderately' then is a matter of degree of loudness and a degree of intensity. 'Moderately loud' asserted with some force and 'moderately soft' restrained with some resistance.

Loudness and performance intensity (the energy of the performance) do not follow in parallel. It is within the experience of many performers that a 'high mezzo-piano' may have a higher sound-pressure-level loudness than a 'low mezzo-forte.' The difference is in the intensity of the sound's timbre; its sense of urgency expending forward or sense of restraint and holding back. The threshold or tipping point that separates mezzo-forte and mezzo-piano might be conceived as a level of energy, intensity, loudness that is neither pushing forward nor holding back; a level of energy that has no effort (as withholding energy requires effort) and that might be able to be engaged indefinitely. This can provide some tangible way to process dynamics and intensity, and bring some clarity to subtle changes within and differences between dynamic areas. The question becomes a matter of degree, moving from 'moderately soft' to a more substantial restraint within 'soft;' just in to the area of 'soft' or more towards the threshold where 'soft' transitions to the more extreme restraint of 'very soft.' Loudness and intensity (the degree to which energy is exerted or restrained) fuse to become our impression of dynamics—an impression carries complex information not only of loudness, but also of the expression and energy provided by performance intensity (and manifest in timbre, see Chapters 7 and 9). We find engaging our perception and understanding of dynamics into any meaningful discussion is thwarted, because we have no point of reference on which to base discussion, by which we might gauge levels with any precision, or begin to calculate changes in level.

The reference dynamic level provides such a reference. It is one of the dimensions of crystallized form that embodies the entirety of the track; within this perception and recognition is a specific level of intensity, energy, and expression. The track's form and essence has a dimension that *is* the energy, drama, level of urgency and quality of expression of its singular level of intensity; this is a significant quality of the global, sound object form—crystallized form. Every work can be conceived as having a single, overall reference dynamic level (RDL); this is a specific dynamic level reflected by a track when its global form is envisioned. The RDL is a single, specific dynamic level that represents record's intensity and expression—the record as a complete entirety, as realized in an instant of conceptualization. RDL is the inherent spirit of the recorded song as a level of exertion, expression, mood and sometimes message combined into a single concept.

The RDL does not change within a piece of music. The reference dynamic level is an unwavering reference. It remains constant throughout the track, and serves to unify all of the dynamic levels and relationships of the record. Conceptually this is similar to the decibel. The decibel is a ratio; its level is meaningless until it is identified in relationship to an established reference level. In dynamics, levels can change at any moment and by any amount, and just like the decibel, this changing level does not change the reference level. The reference dynamic level serves to unify all of the dynamic levels and relationships of a track because it is unchanging and constant. The RDL does not vary even when the nature of the music changes; it allows very contrasting ideas (and even large sections) to relate to the singular concept of the track.

In the most fundamental of ways, everyone who has performed a work they knew deeply and were able to shape with artistry, incorporated into their sense of 'knowing' the work was the sense of the expression, energy, intensity of the piece they were seeking to reflect in their performance—the level that held together their performance into a coherent and cohesive statement and expression. Their first gestures in performance that generated a dynamic level, performance intensity, and expression, are calculated against the RDL they conceive. In all likelihood, such an impression was intuitively held. Performers (and producers, arrangers, recordists, and conductors) can and do intuitively tap into the RDL; in fact, within the processes of composing, producing and recording a project the RDL is held as a reference. At stages in projects where more than one person is shaping the artistry of the track, sharing the same impression of the RDL can make the difference between working seamlessly together and significant disconnect.

Relying on intuition does not guide analysis, though. Arriving at this understanding of RDL requires becoming well acquainted with the track; many aspects impact RDL, and all aspects impact every song differently. RDL might first appear to be an ethereal dimension of a recorded song; still, you know it once you have discovered or experienced it, though it can be illusive at all times prior. Awareness of RDL can be cultivated through remaining focused on the large dimension. Determining how all the qualities of the track come together into a single impression can be illusive. The tendency to focus on what grabs attention (such as the lead vocal), the most prominent parts, what is perceived as significant and many other factors all serve to confuse. The RDL is a large dimension quality that is *influenced* by lower dimension activities. The listener is drawn to the middle dimension's basic-level, and followings the experience of unfolding sound events. To calculate or understand RDL requires a different way of listening, perceiving; attention must settle on the experience and of the mental representation of the track as a whole. This representation is established by listener experiences of all the sounds of materials and expression of structure at all levels (and in all domains) as they coalesce into a single impression, as one aspect or dimension of crystallized form. Just how this happens, and in what proportions these elements come together, is one of the factors that make every song unique.

Perhaps a bit more detail will clarify. The reference dynamic level *may* be influenced by timbre, tempo, meter, the energies of instruments or dominant instruments, the expressive qualities of the lead vocal and its dominant characteristics, key dramatic moments of the song or its general context, affects and all other aspects of the musical fabric—in relation to their contribution to the overall character of the track. The arrangement plays a key role in shaping the RDL, as most of music elements are molded and balanced there. The other two domains also shape the RDL of the track. Lyrics can play a central role in shaping the track, and thus the RDL; tone, message, story, drama, pacing, sound qualities (and much more) all contribute to intensity and expression. The recording has the potential to equally contribute. Revealing the reference dynamic level will be covered more deeply in Chapter 9. The RDL, once identified, functions in the domains of music and recording at all dimensions except the individual sound (explained in Chapter 7).

Every song will have a unique balance of these factors, and will undoubtedly contain others, some perhaps unique to that record. It is important to remember these are middle dimension materials, and the RDL is a characteristic of the largest dimension: the singular crystallized form of the track.

The RDL is subjective and relative to the individual performer/listener/analyst only in as much as the interpretation of the song is personalized and defined by the quality and nature of the analyst's unique experiences. Within the set of similar experiences of a cultural group (or of skilled analysts), the RDL will largely be a common experience—listeners will arrive at the same point given enough exposure to learn the track deeply, adequate attention, acquired listening skills and requisite knowledge.

RDL will prove helpful in the evaluation of dynamics in all dimensions, within the domains of both music and recording. Dynamics is also a significant element of recording. In this way, the domains of recording and music can fuse into one percept, a single gesture. Recording carefully shapes loudness balance, the shapes of individual lines, the dynamic shapes of individual sounds, and more—at a point between the performer and the listener. In live performance these are entirely generated by performers. From subtle aspects of internal dynamics of sounds to the overall dynamic contour of the track, recording and music function jointly. These will be explored within the elements of recording.

Timbre and Performance

The musical element of timbre is most apparent in the sound sources (instruments and voices) that present materials. The timbres of instruments and voices (also called tone color as well as sound quality in some sources) are very significant in records. The timbre of the instrument, the musical materials, and the energy and expression of the performance all fuse into a composite sound. This fused sound

Table 3.4 Variables and characteristics of dynamics.

Roles and Functions	Functions in all dimensions
	Potential to assume primary, supportive, ornamental or contextual roles
	Articulates structure at all dimensions
	Shapes patterns and materials
	Can generate motion and tension
Variables and Qualities	Levels and areas: general, indeterminate language
	Range of dynamics of the song
	Dynamic ranges of sections
	Changes of level between sections
	Relationships between simultaneous or successive sounds
	Reference dynamic level
	Silence
Accents	Metric
	Stress accents and exaggerated metric accents
	Stylized: syncopation, backbeat, groove, other
Activities of Dynamics	Dynamic contour of the song
	Shapes and inflections of lines
	Gradual shifts: crescendo, diminuendo
	Immediate shifts, terraced levels
	Implied levels and implied changes of level
	Dynamic shapes of individual sounds
	Dynamic shapes within sounds
Interactions with Elements Across Domains	Allows all other elements to be audible
	Integral to performance: dynamic nuance and shaping
	Timbral modifications from loudness
	Distortions of loudness and prominence
	Activities of recording elements easily confused with dynamics

is at the core of timbre in recorded song. Timbres of instruments and voices are also altered by performance techniques, and each register (portion of the range) of an instrument or voice has a characteristic sound. These all combine to create a rich palette of timbral variables for nearly every instrument and voice. Of significance is the level of energy (performance intensity) placed into the performance; this can take a voice from the qualities of a whisper to a scream without changing the lyrics phonetically, yet transform the timbre immensely.

Timbres in records are selected, created or shaped to most appropriately present the musical ideas and character of the line. Their characteristics are deliberately chosen, as they will forever exist in this way. Changes of timbre in performance, especially vocal performances, can bring a significant display of nuance in timbre and timbral changes (timbral changes are how phonetic sounds turn into words, into phrases). This selection is integral to the character of the song, and the sound of the record, and the qualities of the timbres of the sources are thus worthy of examination.

Individual performers bring unique sound qualities to their parts. This is especially noticeable in lead vocal and instrumental solo lines, though it exists at all levels. Performers bring their unique sound quality, performance technique or style, and creative expression to the song. They contribute their own creative ideas (improvising, offering embellishments or alternative materials) and performance talents; admittedly more in some tracks than others, their own musical sensibilities will infuse their performance, its intensity and its timbre substantially, and in uniquely personal ways. (Moylan 2015) The ways in which the individual performer's timbre and expression contribute can be significant to understanding the song.

Specific instruments are often chosen for specific materials, as well; their unique sound qualities selected to reflect the desired character of the song. The coupling of the timbre presenting the musical ideas becomes inseparable, and becomes increasingly vital with timbral changes brought by performance intensity. Those parts essential to the song might receive acute shaping in this way, but decisions have been made for all sounds in the texture. The instrumentation for the musical lines is inseparable from the sound of the performance.

Instrument selection might also carry electronic modifications of its timbre. Amplification, distortion, or signal processing might be used to establish enhancements that are integral to the sound of the instrument (or voice). How an alteration to the original source timbre is accomplished is typically not important for recording analysis; the resultant sound qualities are—as it is the sound quality that presents and shapes the musical idea and its expression.

As a parameter of musical expression, timbre can work in four general ways. First (1), we have been examining the qualities of the performance. Of these, performance intensity is the impact of the energy and expression of the performance on the sound source's timbre. This can be understood as a direct correlation between the timbre of the instrument and an appropriate dynamic marking to represent that timbre. It brings a connection of timbre and dynamics through the energy of performance as reflected in the timbral modifications of the instrument or voice. Many other performance-induced timbral qualities exist. These are a primary concern in recording analysis; understanding them can be approached through examining the physical properties of timbre, or through describing their characters. The impact and qualities of performance can be understood through examining modifications of the source timbre itself.

The other three ways timbre contributes to musical expression manifest in musical affects, in eliciting connections of music and characteristics from outside the musical context, and associations of timbres with particular musical styles or other cultures. Second (2), timbre can establish associations with or representations of musical affects (which can include intensity and urgency); this can lead to a host of subjective terms that are perhaps as close as we can get to defining affects. Third (3), timbres can elicit moods and general character traits "that resemble sound, touch or movement that exist outside musical discourse" (Tagg 2013, 308); as such, listeners are drawn to connections between the sound's timbre and experiences outside the piece of music, or outside music entirely. Fourth (4) brings connection of the timbre to a particular musical style or a musical genre, such as the piccolo trumpet sound in the Beatles' "Penny Lane" (1967) connecting with the music of a bygone time; this may also take the form of producing connotations of some other culture or environment, such as the swarmandal (on the flip side of the 45 rpm single) in "Strawberry Fields Forever" (1967).[16] While there are places in analysis for these observations and the conclusions they bring, the timbre of the sound source itself (including those nuances provided by performance intensity and expression) is typically the primary object of study, here, in recording analysis.

When we discuss sound, we are prone to use language from our other senses, from associations with other sounds, or from our subjective impressions of the sound and/or its context. None of these communicate information that is common with the perceptions of others; nor do they provide adequate detail about the sound to assist our understanding of how the timbres of sources contribute to the musical idea. We do not have a process or the language to describe sound and its specific content; this simply has not been part of the way we approach discussing sound. Instead, our custom is to describe sound by analogy and by using terminology from our other senses. We grasp words such as 'warm,' 'dark,' 'crisp,' 'bright,' and a great many others. It is nearly impossible to accurately imagine a timbre's substance and character (its *sound*) described with such a vocabulary (Moylan 2017, 28). Yet this is the state of common communications about sound. We can observe, describe and characterize timbres with intention to hold observations to shared cultural norms, or conventions; using a vocabulary that is as objective (or as neutral related to personal interpretation) as one can craft, and that stresses the shared experience, the character of timbre may perhaps be more effectively described—effective, in that it communicates the same information to many people within a culture.

We will learn later that by listening at lowest levels of perspective we can access the subtle details of timbre. It is possible to talk about the sound itself; to recognize the characteristics of timbres in and of themselves, as opposed to relating them to something else. We can evaluate an instrument's timbre through making observations of its dynamic envelope, spectral content and spectral envelope. We can also observe, describe and characterize timbres with intention to hold observations to shared cultural norms or conventions; using a vocabulary that is objective in that it stresses the shared experience rather than personal interpretation. These approaches can lead to a rich understanding of timbre, and how it contributes to the recorded song.

Thus far we have been considering timbres of voices and pitched instruments. Certain instruments do not have a strong sense of pitch, or a strong fundamental frequency. Rather, these timbres contain a pitch-like quality; their spectrum is divided into frequency bands instead of the (more or less) discrete frequencies of pitched sounds. (Moylan 2015, 153–164) This sensation results from a strong interaction between pitch and timbre. These timbres contain bands of frequencies, often in harmonic relationships; the most prominent bands provide a pitch-like quality. Drums and cymbals are examples. These instruments have aperiodic, noise-like waveforms but with resonant areas of periodic wave activity. Thus we have drums that are higher and lower than others, some are broader in the frequency band they occupy than other drum sounds and so forth. The range of some cymbals might span several octaves, others a narrower range of perhaps a fifth; certain registers within their ranges will have more prominent frequency activity than others.[17] This sense of the width of the pitch/frequency content of non-pitch sounds allows us to understand how these sounds 'fit' into an arrangement, although the pitch-related sonic content of cymbals and drums is never notated into the musical score.

Timbre on a higher structural level than the individual sound or the individual sound source (instrument or voice) is the result of combinations of instruments and voices. This is central to the track's

Table 3.5 Variables and characteristics of timbre and performance intensity.

Selection of Voice or Instrument Source	Inherent sound qualities of the individual voice or instrument
	Qualities relative to character of song
	Sound source selection for lines, parts
	Specific performer for timbre and musical contributions
	Specific instrument (make and model) for musical part
	Specific of performance techniques, style
	Integral modifications to instrument timbre: amplification, distortion, signal processing, etc.
Performance Intensity	Relationships of performed dynamics and resultant timbre
	Shifts and gradations of timbral qualities
	Perceived energy and expression
	Timbres of performance techniques and style
	Sound qualities of instruments' range and register
	Tessitura of voices
	Nuance of timbre for expression, communication
Expression	Performance intensity
	Musical affects
	Inter-sensory traits or mood connotations from outside musical context
	Connection to other musical styles, genres, cultures
Evaluation	Timbre's *Gestalt*
	Listening within sounds
	Physical dimensions of timbre
	Evaluation of physical dimensions
	Objective vocabulary
	Identifying sources by order of appearance
	Identifying sources by defining characteristics

arrangement, and is typically largely determined by the producer and executed by the mix engineer—a process built upon a host of collaborators, resulting in an overall texture.

Arrangement and Texture

The elements covered thus far have been related to the song—its melody, harmony, rhythm and dynamics. We've examined certain qualities of the performance—dynamics, rhythmic nuance, timbral shifts—which will become essential to the track. Between the song itself and the qualities of the final track sits the arrangement. The arrangement is a process in creating the record, and it is also an element of the relationships of sounds and parts, containing variables that get shaped and that contribute to the track. Antoine Hennion (1990, 187–188) has proposed "The song is nothing before the 'arrangement,' and its creation occurs not really at the moment of its composition but far more at the moment of orchestration, recording and sound mixing." This notion is central to this writing, and also connects orchestration with the recording (and the processes that generate the recording).

The process of arranging typically involves shaping a song, or a reconceptualization of a previous work. In making a record, arranging bridges songwriting and recording the track. It shapes the song and can be part of the composition process; it establishes sound qualities that are realized within or created by recording. Arranging is a significant part of the entire creative process. "As long as its basic features remain intact, a song can be reconfigured in various ways and still be recognizable" (Zak 2001, 25). This discussion is not concerned with the process of arranging, but rather in the arrangement as an element of music—an element with characteristics and variables. In recording analysis, we examine the finished artifact, what it *is*, not how it got there. "The arrangement is a particular musical setting of the song" (*ibid.*, 24), and as such has characteristics to explore in analysis. An arrangement itself can be strikingly different from the original song—consider the difference between Joe Cocker's cover version of "With a Little Help from My Friends" (1968) and the Beatles' original (1967)—and can encompass a reworking of all musical elements. Here we will introduce arrangement's variables not previously covered: texture and orchestration.

The characteristics of arrangements include instrumentation (selecting timbres, as just covered), musical parts, rhythmic groove and so forth—that all come together in combining the instruments and materials. The results of combining bring blending of voices, instruments and parts, while setting some apart; clarifying structure and establishing structural relationships of parts; combining materials into texture; and the timbre of the track (its pitch density and timbral balance).

The arrangement centers around assigning musical parts to various sound sources best suited to present them. Here timbres are observed in their presentation of musical ideas (as approached above), at the low-middle dimension. Combinations of sound sources add complexity to the overall texture, but also to the individual lines—or groups of lines that work in tandem, such as might be found between the bass guitar and bass drum. The number of instruments, number of parts, various combinations of sounds and their voicings, parts shared and traded, interactions of groups, the sonic

Table 3.6 Variables and characteristics of arrangement and texture.

Assigning Timbres to Parts	Instrumentation in use (timbres)
	Instrumentation sound qualities
	Delineation of musical parts/materials
	Number and types of parts, instruments & voices
Combining Timbres	Combinations of instruments within musical parts
	Formations of groupings of instruments
	Sonic layering of materials and lines
	Interactions and connections of layers
	Dimensional strata of parts and combined instruments
Pitch/Frequency Range/Spectrum;	Range of sonic spectrum,
'Pitch Space'	Registers of most and least activity
	Pitch density: musical lines plus timbre
	Vertical density of chords and timbre
	Linear density of rhythmic iterations
	Timbral balance: layers of pitch densities
Structural Usage	Delineates materials
	Delineates structural strata
	Articulates cadences at all structural levels
	Delineates structural divisions by changes of instrumentation
	Characteristic combinations of song sections
	Characterizes appearances of important materials
Texture	Consistent throughout song
	Varied, especially between sections
	Number of textures, structural locations
	Melody and accompaniment
	Monophonic
	Homophonic, chordal
	Contrapuntal, polyphonic
	Contextual layer: drone, groove, rhythm section

layers established by combinations and interactions all create a complex web within the sonic spectrum of the song's overall range. The width of this range, the registers of heightened and lesser activity, the closest voiced chords and lines, the sparsest, and so forth all coalesce to bring the timbres and materials of instruments and voices into a characteristic set of relationships and sounds that make the song unique.

An important aspect of the arrangement is how sources exist in frequency bands, or pitch ranges. Anne Danielsen (2006, 51–52) and Allan Moore (2001, 121–126) both engage this as a vertical pitch-space within their similar conceptions of a soundbox.[18] These relationships are often vital parts of the essential character of a track. Musical materials appear as occupying a bandwidth of pitches, that may or may not change over time or in range, and within a relevant time period; they form gestures that have vertical and horizontal dimensions in pitch/frequency over time, as well as relationships with the pitch-spaces occupied by other sources.

The musical materials the sound source presents are fused into a single idea, combining the substance of the musical idea with the dominant timbral characteristics of the source; this brings a sense of the source occupying a pitch space or bandwidth for a certain segment of time. This is pitch density.

These gestures of pitch density establish a layering, sometimes with separation of sources and sometimes overlapping; the full collection of these gestures by all sources present may be conceived as a spectrum, where the pitch densities of individual sound sources/materials that can be conceived as contributing to the timbre of the track. The individual sounds of the track and the sources that present them contribute as partials to this large dimension spectrum we will later examine as timbral balance. This concept of 'partials within the large dimension spectrum of the track' can be framed as pitch density (Moylan 2015, 271–274). Alternatively, each source might be conceived as a chord tone within the 'harmony' that is the overall sound (timbral balance) of the track. Pitch density appears on the levels of perspective of the individual sound source or of the groupings of sound sources where they can be perceived as equal in significant. This situates it in the middle dimensions, where the pitch ranges of materials and the sound sources that deliver them can be identified.

Pitch density allows some access into this complexity in the arrangement and orchestration. Pitch density conceptualizes the area of the hearing range occupied by musical materials and their timbre (the frequency content above its fundamental frequency, and also any sub-tones and sub-harmonics it might contain) as a 'pitch space.' Sound sources (represented with timbres) fuse with their materials to occupy an area in this space (Moylan 2015, 42–43). The length of the musical material may be identified as the phrase length, hypermeter or prevailing time unit—as might be appropriate for the content. This time is related to syncrisis, or the "extended present" (Tagg 2013, 272–273); this duration has a relationship to short-term memory (or working memory), and it is between one and eight seconds depending on the complexity of the materials; it produces a chunk of 'now sound.' Pitch density combines the linearity of musical parts (melodic, harmonic, rhythmic, etc.) with the vertical dimension of timbre. Here it might be helpful to note that timbres are 'chords of harmonics' that exist on the lowest levels of the small dimension, within the individual sound. The vertical dimension has density between the spectral components and the simultaneously sounding pitches of the part, the linear dimension has density comprised of rhythmic iterations, or spacing of within the musical materials. With pitch density, the pitch relationships of non-pitched sounds (such as drums or cymbals) can be placed and evaluated against those that are pitched.

The layering of all the pitch densities of all the various materials and parts establishes the complete pitch range for the track. This is timbral balance; the balance of the vertical character of all sounds and their musical materials. Balance here means distribution of frequency information throughout the hearing range; places of sparse or no activity, places of considerable activity, intense activity, and so forth.[19] If observed as the song progresses, one can evaluate how pitch densities interact between each other throughout the song; the 'shape' of the changing vertical dimension can become apparent at the highest dimension. This is important, as it can be a characteristic feature of the track—a characteristic evident in the Beatles' "A Day in the Life" (1967). Timbral balance and pitch density work in complement, at different levels of dimension.

It is common for timbral balance and arrangement to have a relationship to structure. Changes of instrumentation often coincide with changes of section; sections typically reflect their characteristic sound qualities created by groupings of sounds, and the voicings and relationships of those sounds. Much variation emerges from the timbres of sound sources and various groups of sources, and their relationships to musical materials—variation that can serve as a source for musical movement and shape, and can characterize materials, sections or entire songs.

Texture is the number of musical parts present within a song, or an identifiable section of the song. Types of textures are defined by the characteristics of the parts and their relationships. Textures can be thin, with only a few (perhaps two or three) parts present, or thick, containing many parts of individual instruments and voices. Textures are traditionally identified as monophonic, homophonic, polyphonic or melody and accompaniment. Related to popular music, a drone layer or basis for texture is important.

Monophonic texture is simply a single, unaccompanied melodic line. Homophonic texture sees different musical lines moving simultaneously and in the same rhythm; a chordal texture of one chord moving to another, unadorned is a more bare homophonic texture. Its opposite is counterpoint, polyphony.[20] Polyphony is the texture of two or more simultaneous melodic lines or parts that clearly differ in pitch and rhythmic content; these lines are perceived as equally important.

Contextual layers establish a point of reference against which other parts of the texture work polyphonically. Drone layers contain a continuously sounding pitch, timbre or chord (such as a synth pad) that sustains or is repeated throughout a song or a significant portion thereof. They will create a context—a reference point or a backdrop—against which the other parts change. A drone is usually (not always) the lowest sound; it is often sustained, though it may have a rhythmic character. The rhythm section and the groove (often generated by the rhythm section) can also function as a drone—a quite active and dynamic one that establishes the rhythmic, dynamic and energetic context that might contain harmonic and melodic functions as well.

Melody and accompaniment is not only by far the most dominant texture found in song, it is the inherent texture of song. Vocal melody against all else (all other instruments functioning as accompaniment) is one of the fundamental characteristics of song. The accompaniment can be complex or utter simplicity; it often contains a contextual layer of some type. The next section bridges the elements of music and some of the content of lyrics—the subject of the next chapter.

INTERPLAY OF LEAD VOCAL AND ACCOMPANIMENT

A duality between the lead singer and all the remainder of the musical parts plays out in many ways. This interplay emerges from the arrangement and contributes greatly to texture. It can also generate central characteristics of song structure.

In a pop record musical parts happen around the lead singer, and that happening may not seem correlated, related or supportive. It can often appear as a misalignment of phrasing of the vocal melody and the song's hypermeter, as lyric structures and music structures do not necessarily correlate fully. Lyrics can play a major role in this. The streaming lines of lyrics fused with the lead vocal melody might be conceived as running simultaneously with the other, accompanying musical streams; the streams of the band supporting, countering, providing groove and context, and making space for the vocal and lyrics. There is much room around the vocal for all this.

"Tonal music's phrasing creates spaces which the words in performance occupy: we can visualize the combination of consistent phrasing and words producing *lines*, the line being a feature of pop songs to an extent share with poems." With this Dai Griffiths (2003, 43) identified 'verbal space' where "the words agree to work within the spaces of tonal music's phrases, and the potential expressive intensity of music's melody is held back for the sake of clarity of verbal communication." Verbal space will be explained in greater detail in Chapters 4 and 5, as it is related to vocal phrasing and analysis of lyrics.

The vocal/lyric conforms to available musical space, with melody often subservient to the lyrics' sounds and messages; stratification of materials, domains, momentum and expression results. These parallel streams flex, and the alignment of vocal phrases with the regular measure groupings of hypermeter can shift.

The phrase structure of the vocal line relates to the lyrics, with phrases typically articulated to a line of the lyrics, a grouping of lines, or a portion of a line. The lengths of the phrases are often consistent, but need not be; even slight changes in lyrics can result in notable changes. Further, line lengths and lengths of phrases need not be consistent between song sections (as between verse and chorus). This contrasts with the incessant momentum of hypermetric groupings. Vocal phrases often begin just before a grouping, and can sustain to stretch the phrase beyond the end of the hypermeter and into the next; vocal phrases may delay their start until the hypermeter is underway, or conclude before the hypermeter has run its cycle. This can bring phrase groups of irregular lengths, irregular phrasing in instrumental parts as well as the vocal phrase, and perhaps irregular rhythms; all of which might interrupt or slow forward momentum or merely impart character.

While hypermeter's rhythm of measures provides an underlying structural reference and momentum, it too can be interrupted and extended. Hypermeter can be temporarily disrupted by contraction or extension, and this disruption is very noticeable when it occurs; its underlying pulsation of rhythm of measures remains felt, if not in the listener's consciousness, and the sense of interruption can be a means of validation of the length and strength of hypermeter.[21] Allan Moore (2012a, 60) explains, "at [the] hypermetric level we frequently find cuts, elisions, and extensions. . . . hypermetric groups at the end of sections may have 3, 3 1/2, 4 1/2 or even 5 bars." An example of vocal phrase and hypermetric groups being extended and contracted is illustrated in the Yes cover of "Every Little Thing" (1969); irregular hypermetric unit lengths are evident in the nine-measure verse and seven-measure chorus sections and their contrasting successions. The irregular phrase rhythm of David Bowie's "Changes" (1972) is explained by Everett (2008, 130–1), where he notes how "metric irregularity at the hypermetric level . . . is often quite expressive."

FUNCTIONS AND RELATIONSHIPS OF LEAD VOCAL AND ACCOMPANIMENT

The very nature of song elevates the individual singer to a role of storyteller, protagonist, narrator or observer, with music functioning in relationship to the singer's presentation. It often assigns all other aspects of the song to providing a vehicle to aid the communication of the song's lyrics and message. This topic is complex; its details are woven throughout Chapter 4.

This primary function of a lead vocal exists in nearly all songs. Still, a possibility exists for the lead vocal to assume different functions. It might serve a supportive function during certain passages or sections, when other musical parts assume primary roles—as appears in the coda of Yes' cover of "Every

Little Thing" (1969). In some songs a lead vocal can take on an ornamental function, though this is not common. When the lyrics are presenting nonsense syllables it is far more likely that the vocal line can assume an ornamental function, than when it is communicating a message.

All this might bring an image of the singer in the spotlight with everyone else (and their musical parts) in the shadows. While this relationship might occur, the relationship of the lead vocal to aspects of the accompaniment is complex, and is shaped by stylistic conventions between popular music genres and by cultural influences. In practice, it is not typical for the singer's message, musical materials and persona to be so starkly isolated in character and content from all others—though this is a central stylistic trait of country music, that emphasizes the storyteller (and persona) and the story (lyrics) above all else, to the point the vocal may be segregated from the accompaniment's support and context. Highly varied degrees of separation between singer and accompaniment exist between styles and sometimes between songs within styles; this can play out in conflict, competition, interconnectedness, and innumerable other relationships. Related, Simon Frith (1996, 182) has offered: "The best pop songs, in short, are those that can be heard as a *struggle* between the verbal and musical rhetoric, between the singer and the song." All songs carry this dichotomy between the lyrics and the music, and the tensions they create are central to the song—the types of interactions that result in the 'best' songs is not the issue here. An examination of this relationship can benefit our understanding of the song in numerous areas.

The lead vocal nearly always delivers primary material and any message of the song; lyrics establish the singer as a presence in the center of the song. A typical lead vocal delivers the lyrics using melody as the primary element for delivery, with rhythm, dynamics, and timbre as supportive or ornamental elements. Some lead vocals utilize the elements of music differently; common are vocals emphasizing rhythm and sound quality of the lyrics over melody of the song (perhaps a sung recitation), in sections or in total. The lead vocal (with its performed materials and the image projected by the performer) is typically set in a duality relationship with all other sound sources, and their perceived performers. The lead vocal is the focal point of the song, presenting its text and all that entails; the other instruments/ voices providing other materials that can have a variety of functions.

This melody and accompaniment relationship is a defining texture of songs. Other textures are possible, but rarely appear. While chordal, contrapuntal, monophonic and other unique textures are commonly found in liturgical settings, choral works and in instrumental, art music genres they rarely appear similarly in songs—whether popular song or art song. The traditional concepts of texture often do play out within the accompaniment or between the voice and accompaniment. The contrasting duality of primary and supportive roles and materials is clearest in song, and is reflected in "melody and accompaniment" and the individual voice of the singer/narrator set against all others—a duality that regularly adds drama to the story.

The basic roles of the song's accompaniment are (1) to provide a stable pitch and metric reference, (2) to establish tonality and deliver a harmonic vocabulary (no matter the harmonic language), (3) to contribute to or to articulate structure, and (4) to produce the energy and demeanor of the song through its expression, instrumentation (sound qualities with range and register), harmonic content, and rhythmic materials. These will have differing functions and qualities from song to song; while they are typically consistent within songs, they might change with sections (such as between verse and chorus), or for brief passages.

There are five potential functional states for the accompaniment:[22]

1. Contextual function with supportive traits. Supportive only in that it provides essential basics of pitch and meter reference stability and the harmonic/modal language (this is the central role of all accompaniments). It establishes the context of base sound qualities within the musical style; this

accompaniment role is also ornamental in that it does not reflect the song's emotive character or contribute to its message.

2. Mostly supportive function with some contextual and ornamental traits. Supportive in that it establishes the context of tone or demeanor for the song, and the singer confirms and conforms to this overall emotive character; ornamental in that some materials are decorative and nonessential, and the accompaniment does not participate in communicating the meaning or message of the song. This is the most common state of accompaniments.

3. Function clearly supports the vocal narrative. This accompaniment actively supplies qualities that reinforce or perhaps illustrate the song's meaning; carries and may enhance the song's character and contributes to the narrative by using tools such as text painting.

4. Primary function with support traits. Enacting the drama or amplifying the meaning content of the lyrics, it provides information beyond the text in order to augment the song's subject or theme during (sometimes large) sections or (at times brief) moments of significance; some interaction is common, at other times the accompaniment remains supportive of the narrative similar to number 3.

5. Primary function. Accompaniment is independent in some traits; can counter the text's tone and/or meaning with contrasting, nonverbal commentary; accompaniment presents an alternative point of view and perhaps a contrasting character, perhaps delivering opposing musical ideas and expression; clear interaction and perhaps alternation can be characteristic.

This relationship between the accompaniment and the lead vocal brings many variations of and between these potential states. It should now be clear, a potential for interaction, support, ornamentation or dominance of either are present. These common relationships play out within the above hierarchy of musical materials, and the functions of the elements of each domain. The typical dominance of the voice in this relationship does not at all diminish equivalence, though. Within the song, no matter the relationship between vocal and accompaniment, any element has the potential to be significant.

CONCLUSION

Many readers might have music analysis as their primary interest in engaging this book.

This chapter might be used as a starting point to analyze a popular song without engaging the recording. Such an analysis might include lyrics, as discussed in the following chapter, or focus solely on the song's music. Analysis might also include other disciplines, as introduced in Chapter 1, and explore rich social, cultural and perceptual dimensions of the song outside of, represented in, or elicited by the music.

The analysis of popular music requires unraveling a few issues of musical language and syntax, of how much to rely on traditional approaches to analysis and to common-practice principles and concepts, of how to engage the recording's sounds and music aurally without distorting them by personal bias, of how to treat all elements with equal attention in order to fully understand the individual song. These are not small matters individually, let alone collectively.

Pitch materials and relationships immediately jump to the spotlight in discussions about music. Melodies immediately attract the listener, and harmonies contribute to motion and fabric in mysterious ways largely unnoticed (at least consciously) by the listener; the intricacies of tonal/modal systems (and their harmonic language) engage the fascination of analysts. While this is overly generalized, my point is pitch is central to our understanding of music—no matter how strong the beat, how loud or distorted the guitars—pitch receives some significant attention by all because its contributions to the song are significant. This relevance of the listener's experience as well as the analyst's unraveling is important—and in many ways reaches beyond the sounds of the music.

In "Confessions from Blueberry Hell, or, Pitch Can Be a Sticky Substance" Walter Everett offers this perspective:

> I believe that purely musical effects—nearly always connected in some way to matters of pitch relationships—contribute to any composer's or listener's appreciation, regardless of training or superficial awareness. If the masses believe they are attracted only to rhythm or loud volume and "can't hear" the pitch or have no conscious understanding of functional tonal relations, I say they are merely unaware of why, for instance, they become more excited by expanded dominant-seventh retransitions enhanced by added uncontrolled dissonance than they do in the face of less tonally valent alternations of weak III and VI chords. (Where, musically, did the Beatles and their 1964 listeners shake their mop-tops and shriek most fervently? Follow the retransitional dominants!)
>
> For me, pitch is a sticky substance. It is hard to avoid discussing, no matter what the central musical topic might be, because pitch relations are the matter that is colorized by timbre, that is shaped by formal design, and that is measured by rhythm. . . . Even those who agree on that basic point might disagree about just how pitches relate to each other—in the myriad tonal systems combined in the multitude of styles of rock music, the situation can be, well, sticky.
>
> (Everett 2000, 269–270, 336)

While this perspective might sit very differently with different theorists and musicologists in the details of perception and harmonic languages—as well as with lay listeners and budding analysts alike—it articulates several key issues very clearly. Its central point of the connection of pitch to music and to musical effects must be acknowledged as relevant and significant. Pitch relationships clearly matter; it is in the 'how so' that it all gets sticky. Also important, is 'when' musically and structurally does shrieking occur, as this might lead the investigation as to 'why.' Yes, certainly the learned tensions of harmonies are strongly in play here—though some could take the position other tonal systems might well embrace III and VI chords differently, where they may not be so weak within another context, and others could focus on the ways other elements are contributing to 'uncontrolled dissonance' within this build and bursting release of tension. But what else is in there that caused such a reaction of many thousands of people? Those reactions were—still are—quite profound and almost surreal. There are certainly social/cultural factors in play. Restricting our comments to matters within the song, what other musical factors contributed—even if in a less significant or evident way?

Albin Zak (2001, 25) observed, "Words, pitches, rhythms, chords, arrangements, can be modified without changing the song's essential identity. As such, the song is easily separated from any particular recorded rendering." As any record under study will be such a rendering, just what is it in the 'song' that is so malleable and yet so distinctive?

While this book is about 'how the record shapes the song,' it is apparent and must be acknowledged that 'the song shapes the record.' 'What is in the music that shapes the song' has been the underlying current of this chapter—a current that is much richer than the content here.

Under it all, the song is an 'essential identity' of the record. The song resides within the track's core experience. The recording can add substance and dimension, elaborate its ideas and qualities, or provide a platform for its drama—but the song is what makes the track, that forms its core experience, that is its essential identity.

The song is much more than music, though. The song communicates something; it is often 'about' the story. Music is the way the story is told. Lyrics provide the message and meaning, the drama and the story told within the song. This story and message of the song—with its language and other elements—is the subject of the next chapter.

Some Questions for Recording Analysis

What is the length (in number of measures) and pattern of hypermeter? How does the vocal phrasing align with hypermeter?

Do changes in hypermeter occur? How do they impact the track's momentum?

Is the groove of "Tomorrow Never Knows" contextual or primary in its function? Does its function shift or remain stable?

What is the governing tonality and mode? How are these developed between song sections? Does a key rhythm emerge?

What is the phrase structure of the vocal line, and how does that phase structure align with hypermetric divisions?

What are the harmonic progressions that characterize various sections of the song? Are characteristic chords, progressions, or color chords prominent within the song?

What are the melodic contour, intervallic content and motion, and the phrase relationships of the verse melody?

What are the dynamic relationships of the lead vocal, and the supportive and ornamental parts in the chorus? How do these differ from other sections, especially from verses?

How does dynamic contour shape the unfolding, large dimension of the song?

What are the inherent timbral qualities of the lead vocalist, and the timbral qualities of the performance of the vocal lines? How does the timbre of the voice and vocal performance complement or contrast with the character of the melody and the text?

How do rhythm, pitch and dynamics conspire in the lead vocal's shaping of lines?

What are the sound sources (voices and instruments) present in each successive song section? How does this impact texture, timbral balance and pitch density throughout the song?

How is the lead vocal situated within the musical fabric?

What is the relationship of the accompaniment to the vocal?

NOTES

1 Acknowledging the omission of any number of significant scholars and writings upfront, I refer readers to review the following authors and writings (full titles and citations in this chapter's references, and other writings appear in the full Bibliography): Richard Middleton's *Studying Popular Music* (1990), *Music's Meanings* (2013) by Philip Tagg, Alan Moore's *Rock: The Primary Text* (2001) and *Song Means* (2012), Walter Everett's "Pitch Down the Middle" (2008) and *The Foundations of Rock* (2009), "Analytic Methodologies for Rock Music" by Lori Burns (2008), and Simon Zagorski-Thomas' *The Musicology of Record Production* (2014).

2 In "Feel the Beat Come Down: House Music as Rhetoric," Stan Hawkins (2003) presents a deep examination of beat organization and metric phrasing, mechanizing the beat, listener dance response, DJ roles, and much more, using the track "French Kiss" by Lil' Louis to establish focus in the study. Note his charting of cellular groove patterns, hypermetric units and formal structure, and the use of sonogram to capture spectral content and to relate it to rhythm and texture. This writing can contribute much to our understanding the content and effects of rhythm under consideration here.

3 I attempt to take no 'sides' related to this diversity of thought, except to underscore the framework principle that every song is unique.

My own opinions will likely emerge to the reader, though I cannot and will not position myself for or against any of these respected scholars. I sense a fully considered solution is still out of reach, for reasons of culture and detachment of time. I am willing to concede I am too close culturally and chronologically to be unbiased; I am inadequately aware of the literature of popular styles that are culturally foreign to me, to be able to accurately include them in an inclusive assessment of principles and practices. I doubt I am alone in this regard.

What is essential here is that the song be understood. The song is the object of attention not the method of analysis or a definition of the tonal system to which the song may belong, though both may prove necessary at some point. The goal is to illuminate the song, not to "recompose" it to uphold the theory (Burns 2008, 67)—any theory or method.

4 This is demonstrated by Dominic Pedler (2003, 239–291), and throughout Everett's two *The Beatles as Musicians* volumes (1999, 2001) and especially in *The Foundations of Rock* (2009).

5 A source that presents and explains chord types, chord spellings, chord symbols and nomenclature can be a useful reference, especially for those with limited experience or when encountering distant harmonic relationships. Walter Everett's *The Foundations of Rock* (2009, 401–405) provides such a resource, though there are many other sources.

6 See Covach (1997 and 2001), McClary and Walser (1990), Middleton (2000a), and Tagg (2000) for several diverse yet reinforcing perspectives and discussions on the dimensions, depths and implications of these issues. While several decades may have passed, certain perceptions and opinions persist—change can be slow.

7 Middleton (1990, 104) points out that many of musicology's "terms are commonly ideologically loaded. 'Dissonance' and resolution immediately suggest certain *harmonic* procedures, and a string of technical and emotional associations. . . . These connotations are ideological because they always involve selective, and unconsciously formulated, conceptions of what music *is*. If this terminology is applied to *other* kinds of music, clearly the results will be problematical."

8 In *Song Means*, Allan Moore (2012a, 5) offers: "To analyse a popular song is, of its very nature, to offer an interpretation of it, to determine what range of meaning it has, to make sense of it."

9 I recognize some theorists may contest this statement. I clearly hear the workings of any number of modal songs to be free of reference to tonal relationships.

10 An interesting angle on the latter is *Sound of Africa! Making Music Zulu in a South African Studio* by Louise Meintjes, where music, culture, history, politics, studio dynamics and much more intersect.

11 This nomenclature is as quoted from Lori Burns' writing. An alternative nomenclature for this progression is "i–♭VII–VI–V".

12 Richard Middleton (1990, 118–120) also brings attention to the significance of the bass as an organizing force in popular music, and recognizes its connections to music past: "Similarly, the descending scalar bass-lines so popular as 'grounds' for passacaglias and sets of variations in the sixteenth and seventeen centuries survive in twentieth-century popular music."

13 See Burns (2008, 69), Middleton (1990, 201) and Moore (2001, 49–51).

14 Serge Lacasse (2010) has explored paralanguage of vocal timbres, effects of the body's vocal production, vocal noises and nonverbal vocal differentiators. His study will be summarized in Chapter 4.

15 This is much aligned with the concept of *Sprechstimme*, somewhere between speaking and singing, that is associated with the Second Viennese School, notably Arnold Schoenberg and Alban Berg. Of course, the actual origins of Hip-Hop are undoubtedly found within very different cultural and musical contexts.

16 Philip Tagg (2013, 305–308) discusses how timbre may be used "anaphonically" and "synecdochally" as forms of musical expression. Here he also offers a broad list of "synaesthetic-aesthetic descriptors" of the type I caution against within the practice of analysis—or whenever one wishes to communicate meaningful detail about timbre. Such terms are inherently subjective and hold limited universal meaning that is specific enough to add substance to analysis or discourse; still, they may privately prove useful to the analyst within limited processes that are clearly defined as one's own impressions.

17 Range is the interval between the lowest and highest pitches an instrument can produce; register is a specific segment, or interval of pitches within the range.

18 One will encounter 'sound-box' and 'sound box' variations of this term. I am using 'soundbox' as it is the form used by Allan Moore in *Song Means*—his most recent extensive writing containing this topic.

19 This timbral balance is largely controlled by the engineer, and in this way the engineer functions as arranger.

20 Here a distinction is not made between counterpoint and polyphony, as rarely is this distinction relevant to popular music analysis. Some may find it useful or relevant to define differences, though; if so, see Philip Tagg (2013, 338).

21 See Stephenson (2002, 7–27) for examples of hypermeter lengths and ways it can be interrupted.

22 The concepts of this outline are explored deeply by Allan Moore (2012a, 188–204) in his discussion on 'Persona/Personic Environment.' Therein he observes a re-conception of 'environment' as occupying five 'positions' defined by the singer's persona, text setting, textural matters of accompaniment, harmonic and modal/tonal vocabulary, and "formal setting or narrative structure." These positions have been slightly reframed here to also reflect function, and are understood and presented here as 'accompaniment.' I will be addressing 'persona' in Chapter 4.

CHAPTER 4

Domain of Lyrics: The Voice of the Song

This chapter examines the domain of lyrics.[1] An understanding of the content of lyrics (its structure, story and sounds), and how lyrics are transformed in performance (connections with melody, vocal qualities, intelligibility, style, etc.) will open one to a deeper recognition of the ways the record shapes the song. The recording can significantly shape the sounds, meanings and impressions of the performed lyrics; these will be introduced here, as an introduction to what will be presented in considerable detail later. This chapter seeks to set a context for engaging lyrics, and to define some fundamental concepts; it is far from an exhaustive study of song lyrics.

This chapter might seem incomplete to some readers. Established methodologies and approaches to examining the content of the lyrics are noticeably absent in this chapter. Further, there has been no coverage of how the lyrics are situated in culture, and the host of ancillary concerns. Many of the numerous disciplines introduced in Chapter 1 might be included in the examination of song lyrics. In fact, many of these disciplines have research methodologies that have been used in examining popular music, and some resulting important literature in this field. Lyrics may be studied from the vantage point of poetry and verse, literature and literary criticism, sociology and cultural studies, psychology and philosophy, and innumerable other disciplines and sub-disciplines. The strengths of each have been used to examine the lyrics, to attempt to study the social forces that produced them, and for a diversity of purposes. The reader is encouraged to engage those disciplines directly, from within their specializations.

The goals of an individual recording analysis may bring significantly different types of emphasis to the lyrics. The flexibility of the framework can be applied for different analytic purposes with lyrics, just as it can for other areas. Deconstructing the materials and interrelationships of music and lyrics may be central to some analyses, and the recording and lyrics central to others; in some analyses these may be more of a peripheral consideration, and some analyses might steer clear of lyrics entirely. The performance of the lyrics is shaped by the singer's sound and style, and by the content of the lyrics; the recording captures and can enhance the performance. Social, cultural, philosophical and many other connections can emerge from lyrics, and be rightly incorporated into an analysis—as can the lyrics' affects. The unique qualities of the individual track reflect these. Again, the recorded song itself will determine how it might be most effectively examined—lyrics and vocal performance included.

The role of the song's lyrics immediately seems clear, and mostly obvious. Lyrics speak to the listener; communicate the message or the story of the song. Yet there is more to lyrics; song lyrics are performed, and the performance is part of the communication. The singer not only communicates the lyrics, the singer also communicates *through* the lyrics. Lyrics reveal the persona and narrative of the singer's dramatic role, and more. These become apparent in the contrast between performances of "All Along the Watchtower," the original by Bob Dylan (1967) and the cover by Jimi Hendrix (1968). Meaning and story, ideas and emotions, word sounds and rhythms, listener memories and associations, listener interpretation and countless more, all blend into the expression and message that is heard.

This message can be direct and simple, with all that is intended clear to the listener—evident in many early Beatles songs such as "I Want to Hold Your Hand" (1963). As found in many songs, messages can be complex in content, containing layers of meaning encrypted in a maze of inter-textual references, symbolism and metaphors; all to some extent subjective, perhaps highly inter-subjective. Bruce Springsteen's "Born in the U.S.A." (1984) communicates directly through some of these devices and in layers of meanings. In each of these, the singer's voice might express a singular thought and emotion, then move quickly between various states of feeling, all the while expressing different views of the subject, and providing the listener with "not only comprehension, but comprehension accompanied by felt experience" (Oliver 1998, ix).

The message of the song is imbedded in the text. Message and its meanings are revealed as the structure's organization unfolds; the lyrics' language and poetic devices shape its content, word meanings, pacing, rhythms, rhymes and sound qualities. The text is further enriched by its delivery. The singer provides voice to the song, through their persona and their performance of the lyrics; their "recorded form embodies both singer and persona" (Lefford 2014, 56). Added expression and sound qualities enhance meaning, shape the message, and enrich the felt experience.

The voice of the song is manifest through the lyrics and in the singer's performance. This holds whether or not the lyrics are accurately heard, understood, or the focus of listener attention. The message may be delivered regardless of whether the words are intelligible in performance or in the recording. The text does not need to be understood for the persona's attitude and character to be recognized; the voice projects intensity and sound qualities as no other instrument can, and also exhibits the image and dramatic presentation of the main character. The human voice attracts attention above all other sounds, and in unique ways. "Pop songs celebrate not the articulate, but the inarticulate, and the evaluation of pop singers depends not on words but on sounds – on the noises around the words. In daily life, the most directly intense statements of feeling involve just such noises: people gasp, moan, laugh, cry, . . ." (Frith 1983, 35).

Acknowledging this, we will start discussion from the position of seeking to understand the literary and poetic content of lyrics. Following, we will examine how the lyrics are presented and are often enhanced in performance. The last section offers a cursory view of ways the recording brings drama and dimension to performed lyrics; this material will be explored again in Chapter 10. All this will be undertaken knowing at times the lyrics may not be understood by a listener, and may be misunderstood, or may contain few intelligible words and communicate little actual verbal meaning. The information on lyrics is related to:

- Message and meaning in the song
- Themes and stories, subjects and ideas
- Structure of the lyrics
- Contributions of poetic devices and references to outside sources
- Sound elements of the text (timbre, rhythm, pitch and dynamic inflections)
- Delivery of the lyrics
- Paralanguage and nonverbal vocal sounds

Since there has been song, music and lyrics have complemented one another. In interacting, bending to reinforce the other, contributing their individual unique qualities, and fusing into one, they create something much different and richer than what each represent separately. In the song, many expressive and communicative qualities emerge through the innumerable materials and interactions of lyrics and music—and from their performance.

The interaction of the song's sung melodic materials with lyrics is crucial to the song's overall character and message. A multitude of other characteristics are generated from the song's sung text (in all its dimensions) within its musical setting (and its many elements). The song is a message or a story on a musical journey—its lyrics elevated, enriched or expanded by its musical context.

It is deeply apparent: the song contains much beyond its music. The recorded song—the record—is richer still. A glimpse of the interactions between the record and the lyrics will follow.

In engaging lyrics and the voice of the song, the analyst is drawn to question the contributions of the lyrics to the recorded song. Contributions that are sonic (rhythmic, dynamic, timbral), structurally related to sound and to ideas, concepts, drama, and more are potentially relevant. In this examination, an appropriate guiding question might be: How do the lyrics communicate by their substance and expression, and perhaps what they communicate—though 'what' lyrics communicate is (as we shall learn), in the end, our own interpretation.

LYRICS: MESSAGE, STORY AND STRUCTURE

Song lyrics often have some similarity to poetry. Both utilize a robust sense of language, and seek to engage the reader/listener on both emotional and intellectual levels. Many song lyrics carry strong traits of lyric or metered poetry, some only hints; other lyrics are more like prose. Dai Griffiths (2003, 42) recommends: "first and crucially, that we stop thinking that the words of pop songs *are* poems, and begin to say that they are *like* poetry, in some ways, and that by extension if they are not like poetry then they tend towards being *like* prose." Perhaps some greater clarity of the qualities of song lyrics might emerge with a comparison to poetry, whereby we might seek the "borderlines which can and do become blurred" (*ibid.*) between the two.

Allan Moore (2012a, 113) offers: "Although lyrics are not poetry, and the two categories of expression should not be confused, some technical poetic devices can be found in lyrics, and can add a certain expressive quality." The guiding concepts and principles of poetry, and poetic devices of all sorts, might provide a 'point of reference' whereby one might identify how a song's lyrics 'do and do not' conform, to better observe and identify their qualities. Some of the terms and concepts used here, as we explore this connection, will reflect those of literature, and some will be unique to song lyrics.

The theme and subjects of poetry are often more abstract and complex than song lyrics, though this is certainly not always true. Song lyrics may have a strong message, tone or theme, but this is not necessary; they may have a story to tell, or song lyrics might convey little substance. Lyrics have the support of the musical setting to help communicate its theme and subject matter; some song lyrics do not communicate as well without music. Poetry speaks by itself; it stands on its own content to deliver its message and ideas. Poems are written knowing the reader may stop and reread sections as desired. Readers can study poems; they can look up terms and explore imagery and ideas, and more. Poems can be examined deeply outside real time.

Conversely, song lyrics must speak to the listener immediately. Some lyrics will yield a depth and complexity to reward additional listening, study or thought, though this is often not the case. Song lyrics must also quickly capture the listener's attention in order to inspire their further engagement. Song lyrics pass in real time, incorporated into the performance of the track. Listener engagement and communication of message are often the function of the lyrics' chorus or refrain; there the song's idea coalesces, exemplified as a line from the Rolling Stones' song that is also its title: "(I Can't Get No) Satisfaction" (1965).

Both lyrics and poetry are structured, incorporating word sounds and rhyming; poetics are often found in song lyrics, as well as point-of-view perspectives and tone. In contrast, the structure of song lyrics typically establishes a recognizable departure from literary poetry. Poetry can be of any length and structure, while song lyrics tend to be concise, efficient in their use of language. The prominence of recurring phrases, topics, words, sounds and rhythms of many song lyrics are simply out of place in the context of many forms of poetry. Further, song lyrics integrate into and/or complement the rhythm and structure of the music. Though some poems benefit from being recited, most can be read silently and be successful; lyrics are intended to be sung—they are conceived as sound events.

Structure of Lyrics

Many lyrics may share certain traits with metered poetry. Similarities can be visually apparent in looking at the layout of structure, as song lyrics are often clearly divided into stanzas, just as poetry. The first stanza in metrical poetry establishes an initial design of lines; the design is comprised of a recurring metrical pattern, a line length or lengths, and a rhyme scheme, or system of rhyme. This likewise is common within song sections.

Lines within the stanza have rhythm. Within the lines there exist metrical patterns, called feet. A foot is either two or three syllables in length, organized in patterns of two or three light and strong stresses. Scansions notate the metrical pattern of a poem, and can be applied to lyrics. The notation accounts for each syllable of every word. Syllables receive heavy or light stresses within meters; these are marked by strokes on heavy stress syllables and by curves on those that receive light stress (see Figure 4.1). A recurring metrical pattern is established as the feet are repeated within each line. The metric patterns of the feet generate a recurring rhythm that is similar to music's metric organization based on the measure—a metric grid for the poem. (Oliver 1998, 7–28)

Scansions reveal prevailing patterns and their variations of meter within the poems by reading the line as naturally as possible. The meter of spoken text does not necessarily need to align with the meter of the music, though. The stresses of spoken or recited lyrics may appear differently within the performances of song lyrics—the stresses of the melodic line, musical expression, meter of the music, dialect, and so forth, may bring heavy or light stresses to syllables that differ from what would be heard in a reading of the line. This brings us to recognize, there is a subjective component of interpretation involved in determining these patterns; this is especially prominent when no strict metrical pattern has been established—such as in "Strawberry Fields Forever" (1967).

Figure 4.1 presents the scansion from the chorus and first verse of "Strawberry Fields Forever."[2] Each of these stanzas has five lines. The opening chorus presents the two two-syllable feet (light+heavy and heavy+light) throughout. Exceptions to the two-syllable feet are the lines containing "Strawberry Fields" and "nothing is real." Thus, a metric connection is made between them, which also links their content. The number of heavy stresses per line varies: lines one, two and five contain three heavy stresses, line three disrupts this pattern with two heavy stresses (omitting one heavy stress similarly to the fourth line of verse one), and the fourth line has a rather unusual heavy stress on the final "a-bout" which brings the line to contain four heavy stresses. Also within that fourth line, the words "to get" are an example of stresses that differ between the spoken word and the sung; speech would emphasize "get," but Lennon sings "to" with the heavy stress.

All lines of the chorus end on a heavy stress, except the last; the word "forever" ends on a light stress, and the final line remains without closure (both metrically and philosophically)—forever. This is significant because it leaves the line and the chorus open-ended, aligning with the meaning of the line—structure and function supporting each other. There being no true rhymes in this stanza (the song's chorus) also resists closure and emphasizes openness.

The verse (second stanza) presents a fairly regular four-beat line (one that is close to iambic pentameter) with the exception of the fourth line, though its stresses and syllables are somewhat elaborate.

The many heavy stresses and syllables of the first three lines give them a meditative quality. In general, in poetry the more stresses there are, the more thoughtfully complex and reflective the line seems to become. The word "someone" at the end of the third line is rather difficult to interpret; it may be heard by some as a heavy stress on each syllable (giving the line five heavy stresses) or a light stress followed by a heavy stress (as notated in Figure 4.1), with the last syllable slightly more substantial than the first. The line "It all works out" contains three-beats of heavy stress amid the four-beat lines; this disrupts the meter of the verse, and gives a sense that something is missing; it also is less thoughtful, and though rather thrown out it makes an impact among the other lines that are more complex in meter and content. A pattern of light+heavy dominates all but the first line, which contains a rather floating ("living is easy") impression from its heavy+light+light pattern that repeats until the line's final word ("closed") stops this patterning. The final line creates a sense of closure; it does this from its very regular four-beat light+heavy stress pattern, and from its rhyme—the only rhyme in these two stanzas, and the first rhyme of the song. These combine to make a strong sense of closure to both the line and the stanza.

Recurring metrical patterns typically dominate flow throughout a poem, as they often do in song lyrics. Though flow might be interrupted from time to time, and a different dominant meter may exist in different sections, an underlying flow and patterning serves as a reference. A prevailing metric pattern is commonly established that is similar conceptually to the prevailing time unit in music; though much more prone to disruption, it is a reference against which we are inherently aware of variation. In "Strawberry Fields Forever" we see patterns shift regularly, though an underlying pulse is discernible. The patterning of "Here Comes the Sun" (1969), once established, shifts much less. The number of feet per line is part of the structural patterning of the poem. Line length functions to organize the syllables of the line into a meter for the poem. Patterns of line lengths establish another layering of structure within the stanza. Notice the patterns of line lengths of "Here Comes the Sun" as shown in Table 4.5.

The four-line stanza quatrain is a common stanza length and line structure. Quatrain stanzas allow a variety of end-rhyme patterns. This stanza may have one pair of rhyming lines and one non-rhyming pair (abcb), or two rhyming pairs of lines (abab). Any number of other rhyme schemes combining the four lines are possible: (abba), (abac), (aabb), (abcc), and so on. Rhyme systems need not be this simple. Internal rhymes allow words within lines to connect with line endings, thus connecting words at different points in line and stanza rhythms. Off rhymes (or slant rhymes) are words that do not rhyme exactly; these bring a different sense of connection to words, sounds, and ideas. Examples of these devices in song lyrics are very common. While all these rhyme schemes may be formalized in metric poetry, they often appear with considerable flexibility in song lyrics.

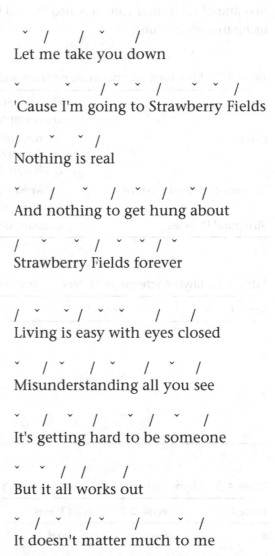

Figure 4.1 Scansion notating the chorus and first verse of the Beatles' "Strawberry Fields Forever" (1967). © 1967 Sony/ATV Music Publishing LLC.

Turning to the structure of typical song lyrics, the similarities with metric poetry may be readily apparent. In most songs, text stanzas are comprised of a fixed number of lines in each repetition of the same section; the lines will similarly have fixed lengths, an established meter will be present, and a recurring rhyme scheme is common. Stanzas establish regular divisions of the lyrics. Stanza lengths and their line patterns (number of lines and number of syllables within lines) tend to repeat. Deviations from these line lengths and line sequences are notable departures that tend to be noticeable in their disruption of patterning. Contrasting stanzas are common in song lyrics—as verses and choruses are regularly at least slightly different in structure. Structural devices are commonly employed and alter metric, rhythmic or line length patterns in some way.[3]

Woven into these stanzas may be regularly repeated sections. These may work on a number of structural levels—a phrase, a group of lines, or a complete stanza. Some will represent refrains. Other repeated lines or sections might also be the song's hook and/or repetitions of the line containing the song title; these anchor together the lyrics and music, and contribute significantly to song structure. The line and song title "If not for you" provides such an anchor in Bob Dylan's song. The song title is the first line of each stanza and it is also the last line in each stanza—the first chorus is the single exception, using the word "true."

Table 4.1   Structural components, patterns and relationships within lyrics.

Stanzas	Number of stanzas, lengths (number of lines), topics, repetitions, structural areas emphasized
Lines	Number of syllables and words, line lengths; feet, metrical patterns and rhythms; rhyme schemes: rhyme, near or half-rhyme, internal rhymes, deliberate non-rhythm in a rhymed setting
Patterns and Relationships	Between stanzas, within stanzas; between lines, within lines; recurring topics or words; recurring placements of rhymes or poetic devices; feet
Structural Devices	Caesura, elision, enjambment, free verse

Table 4.2   Rhyme scheme of "If Not for You" by Bob Dylan from *New Morning* (1970).

Verse 1	Verse 2	Chorus 1	Chorus 2	Verse 3
a	a	a	a	a
b	c	d	d	e
b	c	a	a	e
a	a	d	d	a
a	a	a	a	ea
	a	a	a	a

Table 4.3   Rhyme scheme of George Harrison's cover of "If Not for You" from *All Things Must Pass* (1970).

Verse 1	Verse 2	Chorus	Verse 3	Chorus repeated	Verse 3 repeated
a	a	(a) d	a	(a) d	a
b	c	a	e	a	e
b	c	d	e	d	e
a	a	a	a	a	a
a	a		a		a

The rhyme scheme of "If Not for You" (1970)[4] by Bob Dylan (2016, 257) illustrates connectivity between the song's verses and chorus. The sections all begin and end with the same rhyme; verses all end with the song's title—a recurring line throughout the song—except section. The fifth line of the third verse ends with "rings true"; the line rhymes with the first line of the stanza, while the word "rings" rhymes with the second and third lines.

George Harrison covered "If Not for You" on *All Things Must Pass* (1970). He adapted the verses to follow Dylan's first verse phrasing and rhyme scheme, modifying the lyrics of verses two and three slightly to create the pattern of the first verse. Through his performance, Harrison runs the first two lines of Dylan's chorus into a single line, and thus shifts the rhyme scheme; as he also modifies the last line of the chorus, the phrase structure is now a more traditional four-line stanza, that contrasts with the five-line verse. Harrison's structure is a more typical rhyme scheme and phrasing; further, he separates the presentations of the two choruses by the third verse, and repeats the third verse at the end. Dylan's version disrupts or suspends typical phrase rhythms and rhyme patterns, especially in the fourth and fifth lines of the verses and the chorus. Here we can clearly observe Dylan's ability to shape the structure of lyrics in unusual ways, bringing unusual qualities and variety to both lyrics and phrasings; this is in striking contrast with Harrison's version that distils the song into a more common structure, and with lyric pacing that is also more readily anticipated by the listener.

Sting's "Fields of Gold" (1993) represents a somewhat typical quatrain patterning of stanza lengths and line patterns; each stanza is four lines, deviations are very slight and are the result of repeating the last line (the middle eight is five lines with its last line is repeated, the final stanza is six lines with its last line is repeated twice). In terms of structure and content, though, the lyric is uncommon; its refrains are embedded within the verses. No rhyme scheme is evident within the stanzas, though a distinct patterning of line lengths is present. The recurring images of barley and of fields of gold anchor structure; these occupy the second and fourth lines of each stanza (except for the middle eight) and provide the character of a refrain. The first and third lines resemble an unfolding of story; through them time shifts and romance evolves, children run and seasons change; though these lines comprise only half of every verse these present the bulk of the song's action.

It is not unusual for repeated phrases or words, syllables or word sounds to contribute to structural coherence. These might establish end or internal rhyme schemes. Though repetitions of words or sounds may function independently from a stanza's established structure, they most often bring coherence to lyrics. Even lines, phrases or sections comprised of nonsense syllables work in this way—as in the coda of "Hey Jude" (1968).

Timbre, rhythm and tonal inflection of lines, words and syllables all contribute to the character and the sound quality of the lyrics. This may reach beyond message, often touching into the realms of the visceral or the aesthetic; still, a connection to the human voice will remain. Word play between characteristics of the words and other sonic elements commonly appears, and expression can be enriched or transformed; the sounds themselves are often manipulated in performance, potentially establishing patterning of word sounds. "The Boy in the Bubble" (1986) by Paul Simon exemplifies word play, shifting word meaning (meaning play), repetitive word sounds and more; much is packed within this short

Table 4.4 Sound elements of lyrics, and word sound devices.

Word Sounds	Alliteration, assonance, consonance, rhymes and near rhymes, nonsense syllable phrases
Tempo and Rhythms	Line meters, phrase rhythms, word rhythms, rhythms of images, etc.
Pitch and Dynamics	Diction and inflection
All Factors Combined	Performance techniques and sound qualities from performance

excerpt from one stanza: "It's a turn-around jump shot/It's everybody jump start . . . Medicine is magical and magical is art/The boy in the bubble/And the baby with the baboon heart."[5] Contrasting with patterning of conceptual ideas into "rhythms of verbal images" (Bradby and Torode 2000, 223), or a type of 'image play,' discussed below.

It is common practice for sections of lyrics and sections of the music to coincide, though "lyric structures do not necessarily correlate with musical structures" (Middleton 1990, 238). Griffiths (2000, 197) provides a clear example of rhyme scheme in opposition with melodic structure found in "The River" (1980) by Bruce Springsteen: "The rhyme scheme is aabb: across it the melodic structure cuts chiastically, ABBA."

Song structures typically move through sequences of verse/chorus combinations of various sorts—perhaps broken by repetitions of either section, or by inserting a middle eight or bridge. These stanzas might form patterns establishing units larger than the sequence of individual stanzas of verse and chorus materials—an example grouping verses and choruses appears in "Here Comes the Sun," Table 4.5. Its verse+chorus patterning emerges when one recognizes the first chorus is linked to the introduction and the final chorus is linked with the coda; between these are three verse+chorus pairings, with a 'middle section' interjected between the second and third verse+chorus pairs (between Verse2+Chorus3 and Verse3+Chorus4). Between all three verses only the second line is different; as verses accumulate the rhyme scheme of verses is abac, adac, aeac, if labelled cumulatively. In the table the labelling of materials begins anew in each stanza section; melodic lines are present for comparison. If melodic material was labelled cumulatively from beginning to end and across sections, we would be able to recognize the second phrase of the chorus ("do, do, do, do") and the first phrase of the verse is the same melodic motive.

The overall shape of the song lyrics emerges in recognizing the relationships and combining patterning of stanzas. Patterns of stanzas combine in groupings, establishing ever-larger sections until arriving at the least number of combinations. An overall structure is thereby established at the highest hierarchical level, just as in musical structure. This structure of stanzas is articulated within line lengths, meters and rhymes—in combinations and in divisions.

Free verse texts, and free sections within the lyrics, are certainly found, though these open structures are not as common in songs—Bob Dylan's "Brownsville Girl" (1986) provides an example of stanzas of varying length (four, five and six lines). Less common still are sections or stanzas of text that overlap different musical settings, though this holds the potential for pre-chorus/chorus coupling.

Table 4.5   Outline of "Here Comes the Sun" stanza sections and sequence of lyrics, with syllables per line, rhyme scheme and relationships of melodic materials.

Syllables Per Line	Stanza Section	Rhyme Scheme	Melodic Structure
	Introduction		
4 + 4 + 4 + 3 + 3	**Chorus 1**	abacd	ABACD
4 + 9 + 4 + 8	**Verse 1**	abac	ABAC
4 + 4 + 4 + 3 + 3	**Chorus 2**	abacd	ABACD
4 + 9 + 4 + 8	**Verse 2**	abac	ABAC
4 + 4 + 4 + 3 + 3	**Chorus 3**	abacd	ABACD
3+3, 3+3, 3+3, 3+3, 3+3	**Middle Section**	ab, ab, ab, ab, ab	AB, AB, AB, AB, AB
4 + 9 + 4 + 8	**Verse 3**	abac	ABAC
4 + 4 + 4 + 3 + 3	**Chorus 4**	abacd	ABACD
4 + 4 + 4 + 3 + 3	**Chorus 5**	abacd	ABACD
	Coda		

In some songs, if one is able to recognize the text of a pre-chorus and chorus as being a single stanza, perhaps an extended stanza section, their movement through different musical materials could serve as an example of this type of activity. An example of this occurs in the pre-chorus (of eight measures, beginning with "You just gotta ignite the light . . .") and chorus extended stanza of "Firework" (2010) by Katy Perry.

Poets and lyricists (songwriters, playwrights, film writers, etc.) place concepts, ideas, activities, etc. at specific structural places to shape and control the unfolding narrative. When lyrics are considered for this literary shape and content, an overall structure of the lyrics' concepts might emerge. The structure of the lyrics might be reshaped when factoring both the sonic and conceptual. "Rhythms of verbal images"—that is rhythms of ideas, concepts, and any literary aspect of the song—might be formed in the strata of the middle dimension, as well as the large dimension. The song structure might remain sectional (verses, choruses, etc.), though articulated within its subject matter or some other aspect of the text (perhaps different characters present in various stanzas, a different topic appearing at differing points in the text, perhaps defined by recurrent and differing imagery or symbolism) will be a structure of the unfolding story in all its dimensions. There is a potential for the structural shape of the lyrics to differ from the sequencing of literary content within that structure. Such a divergence often appears as a progression of some sort, typically driven by the narrative.

Bob Dylan's "A Hard Rain's A-Gonna Fall" (1963) provides a clear example of rhythms of imagery; each line presents a different image rarely connected with those around it. Each stanza is framed by a two-line refrain at the end and with a two-line inquiry from the protagonist's mother at the beginning; a question-answer antiphony between mother and son is established (Dylan 2016, 59). The internal lines answer the inquiries, and vary in number between stanzas. Most lines are complex and contain several images, and either add surreal detail to its image, present images of characters or of actions within an image of a scene, or multiple images. The number of images per line varies; though most lines contain two connected images, some contain a single image and others three or four. A pattern of speed of changing images is established between the beginning and end of each line, between lines and between stanzas. A number of images of the first stanza reappear in the last (mountains, forests, oceans); most images contrast with others, but pertain to the mother's queries of "where have you been," "what did you see," "what did you hear," "who did you meet," and (finally looking to the future) "what'll you do now."

Subject matter within stanzas may be another organizing factor. Patterns of topics, of scenes of action, or of repeating refrains between stanzas reinforce the verse and chorus identities. The repetition of text for refrains and for choruses, and the common unfolding storylines of verse narrative solidify the linkage of the musical materials and the poetic topics. The unfolding musings of "Strawberry Fields Forever" verses are grounded by its chorus refrains; though it contains some rhyme, its poetic voice dominates and organizes its montage of imagery and ideas.

Message and Meaning

A difference between "meanings intended by the writer and those inferred by the analyst . . ." (Everett 2009b, 364) can be expected, though there are other points of view as well. Three vantage points of interpretation can seek the meaning of lyrics; each is capable of generating a different meaning. These interpretations are that of the author (lyricist, songwriter), of the analyst and of the listener (the intended audience of the song). Lyrics have surface meanings, and at times deeper meanings (both intended or by chance); searching for deeper meaning can be appropriate at times (or within certain tracks), and not at other times—a significant portion of the countless writings seeking to 'identify the meanings' within John Lennon's or of Bob Dylan's lyrics attest to how convoluted this can get.[6]

Unless the author is forthcoming with the 'meanings' that were meant to be present, any discussion of author intent is speculative. Songwriters are rarely so revealing. Analysts or lay listeners might be aware of the artist's history and life situations, and thus inclined to read them into the lyrics; this knowledge does not, however, mean those events are present within the text. Such action seems akin to projecting one's own interpretation onto the lyrics, into what the artist 'was feeling or thinking' at the time the lyrics were written. Even if a lived experience is present in the lyrics, the songwriter may willingly reshape reality. Further, a songwriter might understand an experience only after writing about it, or may not put the full experience into a verbal or sonic form; expression transforms the experience, and also the author; "songwriters . . . are often surprised at what they create and often only retrospectively comprehend what they were attempting to communicate" (Negus and Pickering 2002, 184). Even when the songwriter overtly shares, it can be through implication, with a more abstract poetic voice, or in any other way avoiding explicit communication; the lyrics wilfully open to listener interpretation, to make the lyrics somehow their own story. Of course lyrics can be fictional, and not at all autobiographical; a song might simply be a story.

The interpretation of the audience is highly individualistic; listeners seek meaning in lyrics with great flexibility, in a great many ways. It is important to recognize this is an intricate matter, and exploring this thoroughly is well beyond our scope. To identify a few central issues though, first we must recognize the listener can identify deeply with a song; Allan Moore (2004, ix) has noted "an unintended word can have life-long consequences . . . all these details can become part of listeners' lives and identity." Lyrics can invite listener interpretations to complete their narrative; Richard Middleton (1990, 173) proposes that in some songs it is necessary for listeners' experiences to be used to complete the song's meaning. Further, lyrics need not be taken at surface meaning, at face value, and such flexibility invites personal interpretation of phrases, scenes, story and moods by individual listeners; what the song means to one is likely not fully consistent to another. Lyrics (in tandem with their performance) speak to them personally, and what it says to them is based on their own experiences and perceptions. The recognition of what is personal, and what is cultural and shared with others is important to identifying an appropriate interpretation.

The vantage point of the analyst will be cultivated as we continue to explore the domain of lyrics. Here there may be some cultural universality, though even between analysts there will be justifiable differences; as we each can only interpret meaning from our own vantage point, "a perspectiveless perspective is impossible" (Moore 2012a, 330). In addressing the general audience listener (although he may as well have been addressing a fellow scholar), Allan Moore (2004, ix) has offered: "And there is no reason why mine [his sense of the meaning of certain lyrics] should be more plausible than anyone else's."

As is now evident, for the analyst, the process of determining the message and/or meaning of a text is fraught with potential subjectivity. Navigating this boundary between what is present in the text and what is personal in our hearing/reading of it can be challenging. Some subjectivity is inherent in the analysis process, as we bring our own life experiences, personal perspectives and sense of the world—and even misperceptions and misunderstandings—to the analysis. Conversely, some universality can be sought and a certain amount expected, as topics, ideas, concepts will generate common messages and meanings amongst listeners of the same cultures. Within the goals of the analysis is setting the balance of the universal and the personal in interpreting meaning.

Sub-topics, concepts, ideas, references, images contribute to shaping the lyrics and its overall concepts and meanings. Paul Simon's *Graceland*—both the song and the album—are filled with examples. Figurative language devices pull the listener into more intricate relationships with the text, and connections with external sources (texts, images or ideas) push the listener outside the confines of the lyrics and the music to bring the work greater breadth and perhaps more substantive meaning.[7]

These can all work together, pulling the lyrics to establish meaningful connections, while generating imagery, feelings, and messages well beyond their few words. Lyrics may have "symbol-rich meanings that cannot be expressed in conventional words and thoughts" (Everett 2009b, 370). Indeed, lyrics can be crafted that bring the listener to engage or recognize abstract concepts, aesthetic dimensions and philosophical ideas—all generated from the sparse wording of the song. In some instances, meaning may appear concealed and unclear. Dylan's "All Along the Watchtower" is one of many of his lyrics that can leave one wondering which of a myriad of possibilities to accept as its intended meaning; intention cannot be known, though (unless the artist shares it). Some lyrics open themselves to, or even invite, a wider interpretation.

The subject of these concepts, sub-topics and topics, ideas, images (etc.) may reflect or be placed within a specific time period (present, past or future). Popular song is often rooted in the time of its creation, reflecting the then current culture with references, language or topics of the day. The song "San Francisco (Be Sure to Wear Flowers in Your Hair)" (1967) is unmistakeably rooted in place and time; Joni Mitchell's "Woodstock" (1970), while also rooted in place and time, speaks to a broader context and with a more universal voice about an event she did not attend (Whitesell 2008, 33–38). A lyric might also present a message and ideas that are more timeless—a more universal context void of topical or cultural references, establishing theme and subject outside the constraints of time. Songs such as Bob Dylan's "Blowing in the Wind" (1963) and John Lennon's "Imagine" (1971) "while advocating political causes, employ nonspecific, even mythicized images and a rhetoric of metaphysical questioning rather than activism" (ibid., 47).

Dimensions of the Story

A common approach to theme in song lyrics is the narrative. The following are intended to assist interpreting the narrative song's story, content, subject and perhaps message, in as much as might be present within an individual text.

Some dimensions of the narrative story (setting, situation and plot) are summarized here:

The plot is the activities, action and events in the story. The plot is often organized toward a particular end; it typically incorporates unfolding drama and time lapse, placing the narrator or other characters in situations. Situation refers to the state of affairs present at a given moment in the story, and can refer to a character's circumstances at any given moment. The setting frames the context for the song; it is the scene or the backdrop, in front of which the story unfolds—a memorable barroom setting opens Bob Dylan's "Hurricane" (1976). Setting can establish a point of reference, and include a specific time

Table 4.6   Components of message and story.

Theme and Subject	Topics
Narrator and Protagonist	Person(s), animated creature(s), object(s)
Voice and Persona	Singer(s)
Story	Plot, characters, situation, setting
Narrative	Point of view, narrative time, language style, tone
Cognitive Devices and Figurative Language	Connotation, denotation, simile, metaphor, personification, hyperbole, onomatopoeia, pun, riddle, irony, oxymoron, paradox, imagery
External to the Lyrics	Literary, historical, cultural references; symbolism, intertextual references, intertextuality, hypertextuality

and place in which the story takes place; it is also the physical environment in which a story or event takes place. Setting may also establish an atmosphere within which the story exists.

Stories typically include characters or imaginary persons (or other personified figures), including the narrator and the protagonist. These are the participants or actors of the drama and story; their presence typically provides central interest and meaning to the text.

The story's narrator or speaker is integral to all aspects of the lyrics. If appearing as a character within the story, the narrator is a participant. A non-participant narrator is an implied character, or perhaps an omniscient or semi-omniscient voice or presence that conveys the story to the audience, though such a voice is not involved in the activities or story. The narrator gives voice to the lyrics. Through singing, rapping, speech, non-language sounds and more, the voice of the narrator may take many forms.

A narrator might project personality characteristics and traits of an individual through the singer's performance; this narrator might be provided a history or some other backstory. In other lyrics the narrator may be a non-personal voice—an anonymous, detached, or stand-alone entity. The narrative of many lyrics is a voice within which a listener might project themselves (Durant 1984, 203). The significant majority of song lyrics are presented from the point of view of a narrator. This person's presence is also known as persona—a concept explored in detail below.

A protagonist is typically (though not necessarily) the main character of story, and often the storyteller. The protagonist is at the center of the story; this character is making the difficult choices and key decisions, experiencing the consequences of those decisions, relaying thoughts and emotions of those situations and decisions. The narrator of a story can be the protagonist only when the story is told from the first person point of view.

The point of view of the lyrics establishes a vantage point from which the story is told. The narrative might be told from a first person, second person, or third person point of view. Each have unique vantage points that shape the listener's experience of the text/lyrics, and may have a profound effect on the message of the lyrics, and the many dimensions of the recorded song. In this way, one step of the analysis might be to consider the pronouns one encounters. Who are the characters and how are they related? There is a level of distance or intimacy indicated, as an audience or an individual is the intended recipient of the discourse.

A first-person narrative is always a character within the story—as in Joni Mitchell's "The Last Time I Saw Richard" (1971) or Bob Dylan's "I Want You" (1966). Narrators refer to themselves using variations of "I" (the first-person singular pronoun) and/or "we" (the first-person plural pronoun). This brings the listener to experience the point of view of the narrator, such as their opinions, thoughts and feelings, etc., but not those of other characters.

In second-person, the narrator addresses "you," "your," and "yours"; the narrator assumes the position of speaking to the listener directly (though exceptions do exist)—as in Bob Dylan's "Like a Rolling Stone" (1965). It can communicate an alienation or distance (this may also take the form of a detached formality) from the events being described, or from the listener/reader. Second-person perspective may be used to guide the reader/listener,[8] or to address the audience directly. The second-person form is found with great regularity in song lyrics.

Alan Durant (1984, 203–204) identified four individual distinctions for "you" in rock music. Suggestions through second-person pronouns with unspecified or general direct addressee provide immediacy, and allow it to transcend limitations of formalism:

1. Addressed to one specified individual (clear in the context of the song),
2. Directly addressing *any* singular individual,
3. An address extending this sense to a general or universal listener,
4. Addressed to an addressee that is determined by the listener herself or himself, upon occupying the imaginary position of the singer.

The first- and second-person pronouns in rock songs bring forth a unique possibility of identification. Listeners may superimpose their person on the "I" of the singer; effectively the rock singer is singing out on the audience's behalf. Alternatively, the listener might occupy the position of second-person addressee, to be the one addressed by the "I" of the singer (*ibid.*).

In third-person narrative, the narrator refers to all characters as "she," "he," "it," or "they." The narrator is an uninvolved person or an unspecified entity that conveys the story but is not a character within the story. In an omniscient third person point of view, the narrator knows the feelings and thoughts of all characters, and full knowledge of the situation and setting—as in Bob Dylan's "Hurricane." When the narrator only knows the thoughts and feelings of one of the characters—or in some way has limited insight into the situation or thoughts and feelings of some characters—the text is in the semi-omniscient (or limited) voice.

Songs often contain an alternating person point of view—Bob Dylan's "Tangled Up in Blue" (1975) continually shifts between the first person and third person narration. This shifting is often reflected in a third-person commentary of chorus or bridge sections, contrasting with the unfolding of the story in verses that are written in a first-person voice. In this way, alternating person narratives will shift from one point of view to another; this is often present between the direct communication between narrator and listener that is quite common of verses, and the more detached and universal position of a chorus' commentary.

More unusual in songs, some lyrics have multiple narrator points of view. Such songs illustrate storylines containing several narratives, and bring a more complex and non-singular point of view of the subject. The mother and son antiphony previously noted in "A Hard Rain's A-Gonna Fall" is one example. "All Along the Watchtower" is another example; a different speaker presents each of its three stanzas: a joker, a thief and an omniscient narrator. In Bob Dylan's "Boots of Spanish Leather" (1964), a man and a woman alternately exchange points of view of one leaving to travel and one staying behind—though it is ambiguous whether the man or woman speaks first, as the one who is leaving (Ricks 2003, 404).

Shifting can also happen in other dimensions of the text. Similar to alternating person, narratives might instead shift narrative time, language style and tone—as in "All Along the Watchtower." Shifting from one form to another may occur in several of these aspects simultaneously or each might change separately. These changes are most often found between sections—one stanza to another—but shifts might also appear between lines. This provides different sections with the potential of different point of view, narrative time, tone, or language—as is often found between verse and chorus, a middle eight, or a bridge.

Narrative time is the temporal setting of the narrative. This may be simply the point in time of the narrative, or refer to chronological, historical or cultural aspects of the text, or surrounding its events. The time of the narrative fixes the plot in either the past (occurring sometime before the time at which the narrative is expressed to an audience), in the present (events occurring 'now,' spanning an arc of real time explained as it seemingly takes place), or in future (events of the plot are set to occur at a later time). Time may progress in real time, speed up (action summarized), be stretched (action slowed down), be paused (story comes to a stand-still while a commentary is presented), or time shift (i.e., the storyline suddenly leaps ahead 20 years, or flashes back a century). The time shifting of Sting's "Fields of Gold" provides several temporal settings for narrator and protagonist.

Language style of the lyrics communicates information about the narrator and possibly the singer; it also identifies the time period and cultural context of the subject or story. Popular song often represents a conversation and is mostly written using informal language. In using slang, and regional/cultural dialects (often grammatically 'incorrect'), the text is highly reflective of the social origins of the narrator/performer or of the subject matter. Language—especially informal language—can bring people and groups of people to unite and connect. It may also divide or distinguish groups, cultures and nations from others. Everyday language is brought into the poetic, and toward the potentially profound, within song lyrics.

This is in contrast to proper language of literature and formal interpersonal situations, which might project greater literacy and social correctness. Formal language minimizes group identity within cultures, and can establish a sense of detachment, as in a conversation between people who have just met. Language shifts can provide stark contrasts within lyrics, instantly transporting the narrative in setting, place and culture, and perhaps through time. Shifting language is evident in the lyrics of "All Along the Watchtower"; Bob Dylan juxtaposes conversational characters and an observational narrator, by using grammatically incorrect, colloquial speech and slang, contrasting with archaic word usage and precise formal language.

Tone presents an overall mood of the lyrics, enmeshed with an overall sound quality; it is an intricate web of feelings that stretch throughout the text. Tone is the lyric's feeling or attitude as an over-arching quality. The narrator's demeanour, or perhaps the narrator's attitude toward the subject or the audience will also contribute to the lyrics' tone or mood—the Rolling Stones' "Sympathy for the Devil" (1968) has a (strangely) unique tone and narrator. In this way, tone also exists as an overall quality of the song, present throughout, and it is a dimension of the track's form.

Tone is also present in more subtle levels, where it can embody the lyrics' feelings from one moment to the next. This mood is rarely static. Relating it to structure, a tone might be fleeting and last only momentarily in the words and phrases of the small structural dimension, or it may be reflected throughout middle dimension materials and sections, or shape the character of major sections. Following the unfolding narrative, tone might change as the song progresses—perhaps gradually, often suddenly. David Bowie's "Changes" (1971) displays these shifts ('changes') of tone clearly within the lyrics, the music, and within the singer's persona.

Persona: The Messenger, Story Teller, Narrator

The individual singer and their personal identity are entwined with the role they assume in delivering the song lyrics.[9] Within this role, the individual that is the singer delivers the song by stepping inside a persona, perhaps assuming the lead character inside the song.[10] This is clearly exemplified by David Bowie and his Ziggy Stardust persona.[11]

'Persona' originated in early Latin as the term for masks that were used in Greek drama. It referred to the situation of an actor being heard through a mask, which represented a character within a play. Central to this concept, the actor's personal identity could be recognized by the audience, the public; with their voice projected through the mask's open mouth they became another—the character within the play (Perlman 1986). The mask provided persona. Here in the song, the singer assumes the role of a character in the song; the mask of Greek drama is transformed into a "vocal costume" (Tagg 2013, 360) that is the persona. "Song characters, then, live through the singer's voices . . ." (Lacasse 2005, 12). While the performer is an individual outside the text, they establish a persona that exists within the narrative and gives it voice; the persona is the identity of the singing voice (Moore 2012a, 180–181).

The persona might speak from two basic positions in the song; these are vantage points related to the narrative points of view. A persona may either participate in the narrative, or be an observer of the narrative and perhaps the messenger of its subject. In order to better define the persona's position within the lyrics, Allan Moore (ibid., 181–183) identifies its range of options around three questions:

1. Does the persona appear realistic, or overtly fictional? A realistic persona is perceived as emanating directly from the singer; a direct address from the singer. A fictional persona is much the same as an actor assuming a role; the singer unambiguously becomes a fictional character.
2. Is the situation described in narrative itself realistic or fictional? The realistic situation is one that might reasonably occur any day, either to the listener or to the larger community of which both the

listener and the singer are a part. The fictional situation is beyond the experience of the listener and the singer; perhaps it is an imaginary, mythological world, or an historical context that provides the situation for the narrative.

3. Is the singer personally involved in the situation being described, or acting as an observer, external to the situation? The persona is inside and participating in the activities of the narrative, or is outside and observing the situation and plot.

These three positions are not all-inclusive; records are complicated and can defy clear categorization. These categories do provide a valuable reference for situating the persona within the greater narrative, though. They allow a meaningful examination of how the persona is placed and interacts with the narrative to emerge, and begins a definition of the image projected by the persona. Moore (*ibid.*, 183) goes on to identify a 'bedrock' normative position (framework) that is very common for songs—a typical set of characteristics that are most common. The 'bedrock' position contains the attributes: a realistic persona; persona is in an involved stance; a realistic, everyday situation; the plot takes place at the present time; and explores feelings, thoughts, events of the moment, or a momentary situation. This normative position might establish a point of reference for identifying 'who is talking' and from 'what position,' as well as their relationship to time, setting, situation and plot. This connects what the persona contains and portrays to basic details of plot and situation.

Stepping outside the context of the individual recorded song for a moment, our sense of persona can be generated from our perception and knowledge of a wider perspective than the individual song. Entire albums (such as *Sgt. Pepper's Lonely Hearts Club Band*) may exhibit qualities of a guiding persona that provides a consistent voice throughout. Further, a sense of persona of an artist or group might emerge from an historical reflection on their catalog—perhaps a number of consecutive releases, perhaps their entire catalog (such as Alice Cooper or Kiss). The sonic signatures of certain producers have generated personas that extend over any number of projects, over any number of artists. These greater contexts may be found useful to scholars, especially in examining stylistic decisions used by some artists/producers (such as Phil Spector or Brian Eno). This has much to do with a sense of tone—tone as an overall sound quality to which the recording contributes significantly.

The persona utilizes tone to project attitude; tone shades the text into an attitudinal position. The singer's tone of voice communicates an attitude toward what is being said, and toward the subject of the lyrics. This attitude projects into all structural levels, from the topic of the moment through the overall concept of the text. One can ask, is the persona serious, reverent, glib, mad, pained, satiric, humorous, hostile, detached, intimate, loving, playful, ambivalent—or other? Is the tone in this moment, exhibited throughout this section, or an overall mood? The tone of the persona is an interpretation of the tone of the lyrics; these are two different opportunities for establishing mood. The tones within the lyrics and around the performance may be aligned or not; may be complementary or oppositional. Tone is manifest in subtle or pronounced timbre modifications, dynamic and pitch inflections related to linguistics and articulation of melody's pitch and rhythm, and also in extra-musical vocal sounds. Tone is a product of the delivery, or performance of the lyrics—it is timbre and attitude, sound qualities and demeanor.

Sounds of Language

The International Phonetic Alphabet (IPA) might be used as a tool for recording analysis; it can engage an array of language sounds and dialects, and much more. IPA can represent the sounds of language(s) and other utterances, and allow them to be deconstructed and opened for more thorough examination. It can also represent the nuances of the voice in a performance—nuances of language, abstract sounds or paralanguage and non-linguistic sounds.

Word sounds may function as abstract sound, and be separated from meaning. This can occur whether by nonsense syllables or by the sounds within words, and within phrases. The sounds of language, and other non-language vocal sounds, add substance to the record, if not also the lyrics. The sounds of language are quite complex, and the sound qualities of spoken and sung languages are richly varied.

It is a challenge to delineate the subtle sounds of language, given their great variation. Phonetics can provide some guidance to examine language sounds directly. Phonetics engages the sound qualities of the text itself. This represents the sounds and sonic structure of the lyrics, before it is transformed by the song's setting of the text and the singer's performance.

Since this book is written in the English language, within the conventions of English as used in the United States (and acknowledging a great many dialects exist), this is the point of reference for discussion.[12] Though other languages have different sounds, this section might serve the lyrics of other languages as well.[13] Several phonetic alphabets and systems of symbols exist. The International Phonetic Alphabet was devised by the International Phonetic Association to bring the sounds of oral language into some standardized form. It has its origins in the late nineteenth century, and is meant to be adaptable for studying any language; this universal application is an advantage. The North American Phonetic Alphabet (or Americanist phonetic notation) is another system of phonetic notation that is in common use; it was originally developed for Native American and European languages, making the name 'Americanist' misleading. The IPA symbols will be used herein because they may reach a wider audience; either system is valuable for studying the sounds of language, as accurately as that might be possible.

The human voice is capable of making many distinct sounds, though only a select few contribute to constructing words in any language—and some different vocalizations will appear in individual languages. Utterances and stringing of select vocal sounds combine to form unique languages worldwide.

Words are the smallest meaningful element of language. They are used in defined grammatical structures, with rules of convention to formulate and communicate their meaning. In this section we are concerned with sound alone (the sound quality of words and language), not with meaning—though meaning does define the separation of words.

There is little universality here, even within the same language. Language sounds are inexorably linked with dialect; language sounds can differ widely while retaining meaning. The same words may have the same meanings but widely varied sound qualities between geographical areas, between different social groups, or ethnic groups—they may even be profoundly different between adjacent neighborhoods. The qualities that bring these distinctions are often clearly present in popular song; they may even be exaggerated. The context and meaning of the lyrics might bring one to extract meaning from dialect as well, though this is not our concern here. Here we are focused on the vast palette of sound qualities within and between the words themselves and how they are shaped by the fuller sound sequence of the sentence.

Words are built in combinations of phonemes, or segments of syllables. Phonemes are abstract units of sound, and the smallest part of a language that can serve to distinguish between the meanings of a pair of minimally different words, a so-called minimal pair. Notice how the words *bat* [bæt] and *pat* [pʰæt] form a minimal pair; the "b" and "p" sounds differentiate the two words (Chomsky and Halle 1968). The sound of the text using the International Phonetic Alphabet appears within the brackets. This represents a standardized representation of the sounds—a type of notation—of oral language. IPA symbols are composed of one or more elements of two basic types: letters and diacritics, which provide description of the letters. Depending on the level of detail one wishes, IPA symbols for the sound of the English letter "p" may be transcribed with a single letter [p] or with a letter plus diacritics [pʰ] that signifies the "p" is aspirated.

Phonetic Symbols

		Consonants	
1	/p/	as in	**pen** /pen/
2	/b/	as in	**big** /bɪg/
3	/t/	as in	**tea** /ti:/
4	/d/	as in	**do** /du:/
5	/k/	as in	**cat** /kæt/
6	/g/	as in	**go** /gəʊ/
7	/f/	as in	**four** /fɔ:/
8	/v/	as in	**very** /'veri/
9	/s/	as in	**son** /sʌn/
10	/z/	as in	**zoo** /zu:/
11	/l/	as in	**live** /lɪv/
12	/m/	as in	**my** /maɪ/
13	/n/	as in	**near** /nɪə/
14	/h/	as in	**happy** /'hæpi/
15	/r/	as in	**red** /red/
16	/j/	as in	**yes** /jes/
17	/w/	as in	**want** /wɒnt/
18	/θ/	as in	**tanks** /θæŋks/
19	/ð/	as in	**the** /ðə/
20	/ʃ/	as in	**she** /ʃi:/
21	/ʒ/	as in	**television** /'telɪvɪʒn/
22	/tʃ/	as in	**child** /tʃaɪld/
23	/dʒ/	as in	**German** /'dʒɜ:mən/
24	/ŋ/	as in	**English** /'ɪŋglɪʃ/

		Vowels	
25	/i:/	as in	**see** /si:/
26	/ɪ/	as in	**his** /hɪz/
27	/i/	as in	**twenty** /'twenti/
28	/e/	as in	**ten** /ten/
29	/æ/	as in	**stamp** /stæmp/
30	/ɑ:/	as in	**father** /'fɑ:ðə/
31	/ɒ/	as in	**hot** /hɒt/
32	/ɔ:/	as in	**morning** /'mɔ:nɪŋ/
33	/ʊ/	as in	**football** /'fʊtbɔ:l/
34	/u:/	as in	**you** /ju:/
35	/ʌ/	as in	**sun** /sʌn/
36	/ɜ:/	as in	**learn** /lɜ:n/
37	/ə/	as in	**letter** /'letə/

		Dipthongs (two vowels together)	
38	/eɪ/	as in	**name** /neɪm/
39	/əʊ/	as in	**no** /nəʊ/
40	/aɪ/	as in	**my** /maɪ/
41	/aʊ/	as in	**how** /haʊ/
42	/ɔɪ/	as in	**boy** /bɔɪ/
43	/ɪə/	as in	**hear** /hɪə/
44	/eə/	as in	**where** /weə/
45	/ʊə/	as in	**tour** /tʊə/

Figure 4.2 Phonetic symbols for English vowel and consonant sounds. Adapted from the IPA Chart, http://www.internationalphoneticassociation.org/content/ipa-chart, available under a Creative Commons Attribution-Sharealike 3.0 Unported License. Copyright © 2015 International Phonetic Association.

The symbols of the International Phonetic Alphabet include 107 letters that represent consonant and vowel sounds, the 31 diacritics that are used to modify them, and 19 additional signs that indicate the suprasegmental qualities of length, tone, stress and intonation.[14]

There are approximately 44 phoneme sounds in the English language. The 26 letters of the alphabet represent these 44 phoneme sounds, individually and in combination; their sound qualities will undergo some variation with accent and articulation. The phoneme sounds are divided into two major categories: consonants and vowels.

Vowel sounds are produced with an open, unrestricted vocal tract; there is no build-up of air pressure above the glottis, and the tongue does not touch the roof of the mouth, teeth or lips. Five or six letters ('A,' 'E,' 'I,' 'O,' 'U,' and sometimes 'Y') are used to represent 20 vowel sounds—'y' can be either a vowel (as in 'sky' or 'fly') or a consonant sound (as found in 'yellow' or 'yesterday'). Consonant sounds (the remaining 20 or 21 letters) are produced by a partial constriction or a complete closure at some point in the vocal tract; these sounds are produced with the lips [p], with the front of the tongue [t], with the back of the tongue [k], in the throat [h], by forcing air through a narrow channel [f, s], or by air flowing through the nose [m, n].

TONES AND WORD ACCENTS

LEVEL			CONTOUR		
e̋ or ˥		Extra high	ě or ˄		Rising
é	˦	high	ê	˅	Falling
ē	˧	Mid	´e	˧˥	High rising
è	˨	Low	`e	˩˧	Low rising
ȅ	˩	Extra low	˜e	˩˧˩	Rising-falling
↓		Downstep	↗		Global rise
↑		Upstep	↘		Global fall

SUPRASEGMENTALS

ˈ	Primary stress	foʊnəˈtɪʃən
ˌ	Secondary stress	
ː	Long	eː
ˑ	Half-long	eˑ
˘	Extra-short	ĕ
|	Minor (foot) group	
‖	Major (intonation) group	
.	Syllable break	ɹi.ækt
‿	Linking (absence of a break)	

DIACRITICS   Diacritics may be placed above a symbol with a descender, e.g. ŋ̊

̥	Voiceless	n̥ d̥	̈	Breathy voiced	b̤ a̤	̪	Dental	t̪ d̪
̬	Voiced	s̬ t̬	̰	Creaky voiced	b̰ a̰	̺	Apical	t̺ d̺
ʰ	Aspirated	tʰ dʰ	̼	Linguolabial	t̼ d̼	̻	Laminal	t̻ d̻
̹	More rounded	ɔ̹	ʷ	Labialized	tʷ dʷ	̃	Nasalized	ẽ
̜	Less rounded	ɔ̜	ʲ	Palatalized	tʲ dʲ	ⁿ	Nasal release	dⁿ
̟	Advanced	u̟	ˠ	Velarized	tˠ dˠ	ˡ	Lateral release	dˡ
̠	Retracted	e̠	ˤ	Pharyngealized	tˤ dˤ	̚	No audible release	d̚
̈	Centralized	ë	~	Velarized or pharyngealized ɫ				
̽	Mid-centralized	e̽	̝	Raised	e̝ (ɹ̝ = voiced alveolar fricative)			
̩	Syllabic	n̩	̞	Lowered	e̞ (β̞ = voiced bilabial approximant)			
̯	Non-syllabic	e̯	̘	Advanced Tongue Root	e̘			
˞	Rhoticity	ɚ a˞	̙	Retracted Tongue Root	e̙			

Figure 4.3 Listing of diacritical and suprasegmental IPA symbols. Adapted from the IPA Chart, http://www.internationalphoneticassociation.org/content/ipa-chart, available under a Creative Commons Attribution-Sharealike 3.0 Unported License. Copyright © 2015 International Phonetic Association.

Vowel sounds normally form the nucleus (the center core) of syllables; consonants typically form the onset (beginning) of syllables and will form the coda, or end of the syllable, when one is present. This concept of segments of syllables is especially important when considering sung lyrics. Vowel sounds are the sounds most often sustained by the singing voice; the open vocal tract used to advantage for extending the vowel with minimal stress and effort to the singer's body. These open vowel sounds are rich with potential for morphing between various forms and tones of the same vowels, and to alter pitch, dynamics, timbres and intensities of the sounds.

Consonant sounds typically articulate the first and last sounds of syllables and words, and punctuate word rhythms. Some consonant sounds can be sustained when singing (such as the "l" sound in "table"), and this aspect of vocal technique is common in records. In song, use of these consonants might imitate vowels, creating new syllables such as when John Lennon sings "fields" as "fi-ldz" in the song title appearing in the chorus' last line: "Strawberry Fields forever." In such cases, the sound of the recorded singing voice differs from normal speech and real-life experiences. Constrained consonant sounds do not project well in live (unamplified) settings, but are readily captured by a microphone; Frank Sinatra's expressive shadings of 'm' and 'n' sounds are clear examples. While sustained consonant sounds typically cannot be altered as substantially as vowel sounds, they have the potential to change with unique characteristics; consonant usage is significant and worthy of attention in evaluating the sounds of lyrics.

The phonetic symbols for English using IPA symbols for vowels and consonants appear in Figure 4.2. To use the International Phonetic Alphabet, word syllables are divided into phonemes and then transcribed into appropriate IPA symbols. IPA symbols also exist for suprasegmentals, tones and word accents, and diacritics; these are added to the vowel and consonant symbols to further define word sounds. Especially relevant for lyrics analysis are vowel sounds as modified by suprasegmentals and diacritics, as appropriate. Diacritical marks are symbols added to letters; some diacriticals indicate a different pronunciation of a letter is in effect. There are instances when suprasegmental symbols might

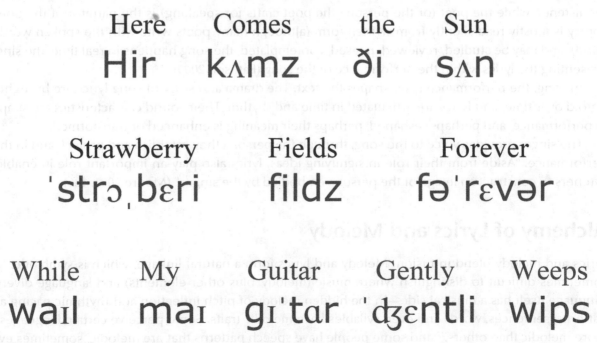

Figure 4.4 Song titles transferred into the International Phonetic Alphabet.

be incorporated to provide some useful information on the lyrics' sound qualities. Symbols for tones and word accents are also incorporated in use of the IPA in language analysis. A listing of diacritical and suprasegmental symbols appears in Figure 4.3. A further defining of these symbols and their functions is beyond the scope of this writing but could be valuable to the reader; these are thoroughly covered in sources listed here.[15]

In using a phonetic alphabet, a significant amount of sonic detail can be extracted and notated. Detailed phonetic analysis can be a valuable supplement, allowing one to recognize sonic traits of the text.[16] This may prove especially valuable in identifying or transcribing the sounds of sung lyrics and of paralanguage sounds.

Examining the sound of the lyrics can clarify how sound of the language has been utilized, and how the lyrics are structured related to sound. The sound qualities of the lyrics can be examined without the added layer of meaning. This examination can assist in identifying sonically connected words; word sounds that appear and reappear at structurally significant locations; word sounds with connections imbedded in lines and phrases; and syllables in repeating patterns of many types. Other sonic aspects of the text may likewise emerge, where detailed attention to the lyrics' timbral qualities is given.

With this section, we have drawn a distinction between the sound of lyrics as written (read silently, originating internally), and the sound of the lyrics as sung (performed, originating externally). This distinction becomes clearer as lyrics and melody fuse into a single expression.

LYRICS IN PERFORMANCE: GIVING VOICE TO THE SONG

Obvious but significant: song lyrics differ from poetry in that they are written to be sung. This difference takes many forms. The singing voice is substantially different from the narrator of verse; the singer is external to the listener, the poem's narrator speaks from within the reader. The singing voice represents an individual (and their personal interpretation) interjected between the author of the lyrics and the listener, while the poet (or the persona the poet crafts for speaking) is the narrator of the poem. Poetry is mostly read silently from written form (although many poets write with the spoken word in mind), and may be studied, reviewed, paused, contemplated; the song happens in real time, the singer presenting the lyrics within the performance of the song (Pattison 2009).

In song, the performance itself shapes the text. The drama and story of song lyrics are brought to unfold over time, and lyrics are articulated in time and rhythm. Their sound characteristics are shaped in performance, and perhaps reshaped; perhaps their meaning is enhanced or transformed.

The singer provides voice to the song, through the persona they establish and project, and in their performance. "Aside from their role in signifying ideas, lyrics also play an important role in enabling listeners to construct an image of the persona embodied by the singer" (Moore 2001, 186).

Alchemy of Lyrics and Melody

Lyrics and melody blend in fusion. Melody and lyrics form a natural linkage, which is symbiotic. It is sometimes difficult to distinguish where music (melody, plus other elements) and language diverge. Language itself has a musical side—in the hidden melody of pitch inflection and rhythmic pacing and patterns, sentences, words and even syllables have melodic traits. Some perceive certain languages as more melodic than others,[17] and some people have speech patterns that are melodic, sometimes even 'singsong-y.' All these traits might be amplified within a vocal line, as melodic contours can seem to mirror the intonations of verbal phrases.

Especially common in song verses, melody can have linguistic pacing and inflection that reflects a narrative character; recitative lines are common, allowing the narrative or story to be emphasized over the music; in such cases little might separate speech and the sung lyrics. Here, the text can largely minimize pitch changes in melody, and bring melody's content to be largely rhythm with dynamic and pitch inflections. A context of "speech rhythm" results, where the sung phrase appears much as it would if it were spoken (Bradby and Torode 2000, 214). At the other extreme, melody itself may take the music-traits of language to great sophistication; instead of mirroring speech, melody can magnify it. Melody can exaggerate the sounds, rhythms and phrasing of text; it may take language traits and rework them into musical effects, or into substantive materials.

While melody and lyrics inherently have great connectivity, in many contexts melody and lyrics will establish traits independent of the other. The two are just as likely to appear with different content and characters—from slightly varied to distinctly different to contrasting. Melody and lyrics may be contrasting or even oppositional in their content and characters, though they are most often complementary in some way, and to some degree. An effective song is likely to have a melodic setting of lyrics that together produce something wholly different than their independent contributions; melody and text working off one another, sharing and enhancing characteristics, bringing the concepts of the text into the activity and character of the melody. Lyrics are elevated by music's content and context; music is elevated by the sounds and the meaning of the lyrics; together they fashion a more meaningful and more sophisticated voice.

Middleton (1990, 231–232) outlines a three-pole model to this melody and lyrics interaction that identifies how "the hybrid practices of actual popular songs come into being." The model distinguishes between story, affect and gesture.

In the 'story,' words are in narrative form; the voice tends toward speech and can be prose-like; the lyrics direct rhythmic flow and harmonic movement. The text merges with melody in 'affect,' where words contain expression and the voice tends toward song, intoned feeling. Words functioning as sound qualities define 'gesture'; as gestures, words are prone to being absorbed into the music, and the voice may resemble an instrument. While individual songs might not clearly fall into any one of these three practices, one trait is likely to dominate within any section or song. These concepts might help guide an understanding of the relationship(s) of lyrics and melody.

Dai Griffiths has identified the concept of 'verbal space' as an alternative approach to evaluating phrase structure of lyrics within the track. Verbal space is the result of "the pop song's basic compromise: the words agree to work within the spaces of tonal music's phrases, and the potential expressive intensity of music's melody is held back for the sake of the clarity of verbal communication" (2003, 43). The idea is the music's phrasing creates spaces, and the performed words occupy a portion of that phrase. The approach seeks to establish a 'word consciousness' to systematically explore how the words and lines of the lyrics work. Verbal space is explored more deeply in Chapter 5's observations of lyrics.

Phrasing and the words produce lines, just as with poems. The prevailing time unit and hypermeter establish the clear dominant phrase length within which lines of text are contained. This phrase length is the basis for this examination. Griffiths conceptualizes pillars at each end of the phrase, with the lines of words unfolding in time, from left to right. Between these pillars there is a point where the words begin and where they end; this establishes the boundaries of the line of text.

Within these lines, phrases can be evaluated for their word content. Lines can be full or empty, and words can be positioned at various points on the line. How the words occupy the space is the central issue; compiling information on syllable count and placement of words allow the relative density within and between each line to be revealed. This 'syllabic density' (or syllable count) reveals important information of the word-presence in the line. This density is impacted by word rhythms and tempo, by the speed of a song and the way it is phrased. The approach offers a way to understand the relationship between lyrics and the relative speed of their delivery.

Changes in density as the song progresses can be telling; Griffiths (*ibid.*, 47) notes a doubling syllable count in a middle-eight is common in some pop song styles. Patterns of density might emerge, along with other aspects of syllable density and rhythms. Verbal phrasing and musical phrasing will come together in many aspects of rhythm, as music and words trade off each other's rhythm. Other characteristics will be readily apparent in the data collected here, such as a change in syllable count shaping a stanza, or propelling motion within a line.

This concept of verbal space has considerable potential for the collection and evaluation of other lyrics-related relationships, as well. Repeating words or word sounds might be identified in lines; literary information of story, subject matter. Drama and pacing of action can be collected and made evident as well; Griffiths notes "the return of the pillar can be a moment of some drama" (*ibid.*, 43). This concept of verbal space might be broadened to determine what information is to be collected (beyond syllable count and placement); we can develop ways of evaluating and talking about the proportional relationships within the verbal space of any particular song that begin with syllable density and perhaps encompass word sounds and rhythms, meaning and story, and more.

Just as a prevailing time unit might flex, verbal space can be malleable. Verbal space can change position, both within the time unit and shifting to overlapping musical phrasing. It can extend or contract in time as the line is shaped. This can be a valuable concept, as one can readily observe how activities shift with and without the presence of the vocal, as well as observe central characteristics of the performance of the lyrics—and 'anti-lyrics.'

Within the anti-lyric, the "emphasis shifts away from sonorous rhythm towards the detail of its statement, away from rectitude of rhyme and rhythm towards the novelty or interest of words and ideas" (*ibid.*, 55). More prose-like lyrics loosen language and sound; here Griffiths describes Patti Smith in writing: "[T]he words are free to lose the markers of lyric, rhyme and syllabic consistency, in favor of a looser relation more akin to prose forms: less poem as analogy, then, and more short story, novel, letter, confession, manifesto" (*ibid.*, 54).

Sound Qualities of Sung Lyrics

The singing voice and its musical line introduce characteristics and dimensions, and add many sound qualities to the lyrics.

To start, all human voices are unique; each contains a vocal timbre and speaks (sings) with a vocal style like no others. The unique sound of each voice is a one-of-a-kind musical instrument. A singer's unique sound adds qualities to the song lyrics that are unmatched by any others. Each voice contributes in its own way to support the presentation of the text, and add to the character of the persona. In production, this brings great significance to matching the song (with its lyrics, music, persona) with the sound and character of the individual singer's voice—and to the singer's performance style and persona.

In the recorded performance the persona becomes more fully developed. It manifests as a synergy of the content and expression of musical material, the literary and dramatic content of song lyrics, and the expressive qualities of paralanguage, non-linguistic vocalizations and the real-life qualities generated by physical sounds from the singer (such as breath and mouth sounds).

We recognize the music domain of the sung line is most closely linked with melody, while other musical qualities (rhythm, dynamics, register, timbre) can be significant. The lyrics domain continues to carry the sonic substance of linguistics—the rhythms, timbres, inflections, accentuations and diction of speech—and layers of communication and literary meaning. This traditional connection plays out in the shaping of individual words and word sounds, the individual lines of text that are fused with lines of music, and the lyrics and the vocal line as a whole. The persona is deeply enmeshed with its delivery of the lyrics and its performance of the melody.

Returning to the singer, we recognize the sound qualities of sung lyrics are a blend of the timbre of the individual's singing voice and the qualities of the language. This blend establishes the singer's voice as a unique musical instrument, and the singer as a unique individual, capable of projecting a unique persona. Additionally, the individual singer has an inherent vocal technique; technique is often influenced by speech dialect, and will exhibit personalized characteristics of performance style and sound quality.

Everyone has a unique voice, owing much to physiological differences. Greatly simplified: vocal sounds are produced from a sequenced mechanism starting at the lungs and progressing through the larynx, resonated by the chest and head cavities, and articulated by the tongue, palate, teeth and lips; each individual has unique vocal folds (commonly called cords) and vocal tract. As small differences in human physique are reflected in different sound produced, it becomes clear how each individual has vocal qualities like no other. These physical differences create formant regions of accentuated frequency patterns; this acoustic resonance greatly distinguishes individuals by the tonal quality of their voice. Closer examination allows recognition of spectral envelope characteristics of the vocal formants.

Differences between singers' voices are often clearly evident within particular word sounds. Distinctive timbral characteristics will appear most prominently in different phonetic sounds, and will vary between individuals. As vowel sounds are the phonemes most often sustained, the differences will either be quickly distinguishable or be rather blended. Often the most unique voices will have prominent formants in sustained vowels. Some singers have cultivated the technique of activating a formant region that is not naturally present in their voice. Further, use of falsetto and other vocal techniques extend the voice's timbral characteristics and range of expression, while the persona remains identifiably the same.

Vocal timbre can impact the qualities of persona. For example, as timbre communicates a sense of intensity and energy when sound is produced, this can cast an impression of degree of urgency upon the persona, as presenter of the narrative. The vocal qualities of individuals may inherently project a sense of calm, grace, tension, aggressiveness, panic and so forth. These not only shape the persona, they impact the message or meaning of the song.

Lyrics as Mosaics of Moments and Tone, and Unintelligible Lyrics

Even after this lengthy coverage of lyrics, we must acknowledge lyrics are not always the first thing written in the song. Further, the content of lyrics is not always the primary consideration of songwriters, performers or producers. Ian MacDonald (2005, ix–x) noted:

> [T]he lyrics of The Beatles were wrought in a collage spirit, line by line. The group rarely thought of them as overall structures, still less as of any real emotional consequence. . . . only bothering about overall concepts in formula terms (e.g., using the second or third person, aiming towards a punchline, and so on). . . . Lennon and McCartney wrote their lyrics to create a mood or tone, so as not to get in the way of the effect created by the music and the sound.

The sound and tone of the lyrics, blended with vocal qualities and melody took precedence over story, for the Beatles; "Eleanor Rigby" (1966) is rare among their songs in its "sustained line of thought and expression amounting to a poem" (*ibid.*). Perhaps such lyrics might be conceived as mosaics comprised of disparate moments, connected by mood, sound, ideas. Whether or not lyrics are finely wrought hardly seems the point to popular song; just a few words can speak volumes when sung. Clearly, the Beatles' lyrics (and those of many others) have communicated something special to many, through the moments, images and moods within the lyrics' collage—aided by much else.

The Beatles' lyrics are not unusual for having their "word-technique driven by sound" (Griffiths 2003, 53); such an approach to lyrics is common in popular song. Emphasis on content of lyrics certainly varies considerably from style to style—pop, country and rap might place greater emphasis on lyrics and what they say than perhaps rock and metal—and from artist to artist, and song to song, within genres. Even Bob Dylan has down played the importance of lyrics; in a 1984 radio interview he stated: "When I do whatever it is I'm doing, . . . It's not in the lyrics."[18] In all this, song lyrics are clearly not poetry. While perhaps overstated, there is likely some bit of relevance in the view Theodore Gracyk (1996, 65) offers:

> To be blunt, in rock music most lyrics don't matter very much. Or, to be more precise, they are of limited interest on the printed page, divorced from the music. We emphasize the wrong thing if we think that profound lyrics are in any way superior to "Wop Bop A Loo Bop"' as something to sing.

Some observers have extended this position to mainstream rock/pop lyrics as well, relegating them to banality and worthlessness (Shuker 2002, 181).

Richard Middleton (2000b, 163) identified "the axis can be seen as running from the idea—so popular in the sixties—of lyrics as 'poetry' to the argument that actually listeners pay no particular attention to words at all." On the extreme opposite from Gracyk, of course, are opinions and writings based on the exact position that songwriters *are* poets (Goldstein 1969); still nearly all songs fall within the continuum between the two positions, as lyrics communicate within and are enhanced by the song. In reality, song lyrics are not intended to be poetry, and fundamental differences exist between the two. Simon Frith (1996, 182) notes:

> Good lyrics by definition . . . lack the elements that make for good lyric poetry. Take them out of their performed context, and they either seem to have no musical qualities at all, or else to have such obvious ones to be silly.

The richness of popular song holds no single position related to the meaning, communication or significance of lyrics. Frith (1988, 121) suggests the critical question is "how do words and voices work differently for different types of pop and audiences?" As with all other aspects of the record, we can reach beyond the context of the song's genre and culture to recognize each song is unique and carries its own relationship with words; the task becomes to identify the 'poetic status' of the lyrics, in order to approach the record appropriately.

The importance of words and lyrics also plays out in the intelligibility of lyrics, and the listener's ability to understand the message (or language). A great many of us have had meaningful listening experiences without understanding the sung text. Unintelligible lyrics can result from a song with lyrics in an unknown language or dialect, whether a foreign culture or not; such lost words might also be the result of musical texture, of the mix, of the singer's diction, or our own inabilities to decipher the lyrics (each might be caused by any one of a multitude of reasons).

At the end of the first verse of "Purple Haze" (1967) many listeners believe they hear Jimi Hendrix sing "'Scuse me, while I kiss this guy"; his articulation of "the sky" is often misheard as "this guy". Understanding lyrics depends on the listener; all interpretations are influenced by the listener's experiences, attitudes, knowledge, skills, culture, etc.—and their perceptions. With unclear words, listeners are prone to use their frame of reference and experiences to quickly process meaning; with words and their meanings moving by quickly, and with the added information of a competing fabric of musical setting, little opportunity (mental capacity) is available to question and explore the uncertain. In these instances, communication relies heavily on the listener, perhaps more than on the singer. This difficulty in understanding song lyrics leads to misinterpretations by listeners, and sometimes to rather personal interpretations as listeners apply their own words to indistinguishable sounds using the tone of the voice and context of the music as guides. As just illustrated, misunderstood lyrics—even a single word—can radically shift meaning.

Diction and enunciation can play a key role in intelligibility of lyrics, and also vary between types of popular music. Gracyk (1996, 104) observes: "rock singers are notorious for injecting degrees of incoherence into their vocals . . . there may be no attempt at clear articulation." We have all experienced singers who articulate clearly and those that do not. Cultural factors can enter here as well, as some people struggle to make sense of some genres of music that are clearly perceived by others.[19] While singers play a role in intelligibility, the clarity of the vocal may result from the arrangement or the treatments of the vocal within the recording's mix. The vocal (and its lyrics) might compete for its place in the track, and the listener's perception—and not emerge clearly enough to be intelligible.

Unintelligible lyrics function as sounds in rhythm, perhaps with the sound of the voice functioning as an instrument without words—whether or not unintelligibility is intentional is irrelevant here. The lyrics of "Louie, Louie" (1963), as performed by the Kingsmen, have eluded generations of listeners, who were otherwise engaged with the song; this is a clear example of words not as important as rhythm and sound, of the voice communicating by its gesture and expression. Even when lyrics are unintelligible, or when lyrics are not correctly understood, songs are able to communicate directly. A "musicalization of the words" (Middleton 1990, 228) takes place that provides the vocal line with significant depth of meaning well beyond all other instruments. The persona of the singer remains, and the vocal's sound qualities communicate a great deal beyond all instruments. Though language content of lyrics may be missing, masked or very largely unheard by listener, the lyrics' rhythms, expression, energy, performance intensity, sound quality/timbre are perceived and communicate in place of words. The human voice travels deeply, even when void of language.

Paralanguage and Nonverbal Vocal Sounds

Certain physical experiences, particularly extreme feelings, are given vocal sounds beyond our conscious control—the sounds of pain, lust, ecstasy, fear, what one might call inarticulate articulacy: the sounds, for example, of tears and laughter; the sounds made by soul singers around and between the notes, vocal noises that seem expressive of their deepest feelings because we hear them as if they've escaped from a body that the mind—language—can no longer control.

(Lacasse 2010a, 225)

The recorded song embraces and elevates nonverbal vocal sounds. Daily-life vocal sounds are commonly incorporated into the performance; they become integral to the recorded vocal line. Other 'noises around the words'—sounds generated by the physical act of singing and breathing—are also incorporated into the persona, musicality and drama of the performance, and thereby, the substance of the track. This is significant, as "non-linguistic indicators . . . are as efficient, if not more efficient, than language for expressing meaning" (Lefford 2014, 46).

Paralanguage—vocal components of speech that are non-phonemic and assist communication—is an important dimension of the vocal performance. With popular music anchored in recorded performance, its manner of delivery is central to its communication. Non-linguistic vocal sounds, are widely varied—some will be interpreted relative to cultural convention and some more universal—and are surrounding elements that "participate in the conveying of emotions in everyday communication activities" (Lacasse 2010a, 226). Popular song seeks ways of enriching language; at the same time it is deeply rooted in everyday language and human interaction and "popular singing acquires most of its inspiration from everyday speech" (*ibid.*). This is one of popular music's great strengths for expression and communication, and connections with the listener; oohs and ahhs, exclamations and much, much more, all become integral, assimilated within the communication and language, the performance, and the track. Including these non-linguistic sounds within a particular analysis of lyrics or vocal performance might reveal unique qualities of the track.

Serge Lacasse (2010a)[20] has examined the very humanizing, everyday sounds of paralanguage. To offer a starting point for developing a theoretical framework for popular singing, he has adapted the work of linguist Fernando Poyatos (*Paralanguage: A Linguistic and Interdisciplinary Approach to Interactive Speech and Sound*). Paralinguistic features are classified in four categories: primary qualities, qualifiers, differentiators, and alternants. Any of these might be used within the delivery of the vocal line.

Primary qualities are predetermined by the physical, biological features that individuate voices and how those voices are modified. These are the timbral qualities that allow us to recognize an individual solely by the 'sound' of their voice. Other primary qualities include the musical elements of resonance, loudness, tempo, pitch and intonation, rhythm and duration, etc., but are "approached from a slightly different angle, according to whether they are considered to be predetermined (either for biological or compositional reasons) or altered by the singer" (*ibid.*, 229). The distinction here is between what is within the music and what is inherent to the voice itself. Primary qualities establish a point of reference whereby we might witness: "the most interesting aspect of performance resides precisely in how a singer will alter these predetermined parameters" (*ibid.*).

Qualifiers modify the voice itself. The modifications are noticeable and may be controllable or uncontrollable, and are heard related to established social values, though some (such as the intimacy of whispering) are more universal. Many qualifiers exist:

- Breathing sounds (such as breathy voice, whispers)
- Laryngeal effects (such as falsetto)
- Pharyngeal control (such as opening the voice),
- Nasal sounds
- Sounds produced by the tongue, lips or jaw

Differentiators appear in conjunction with verbal language (such as yawning while speaking) or can appear as independent vocal sounds. These include laughing, crying, shouting, panting, gasping, coughing, throat clearing, and sighing (among many others)—all of which can be produced while speaking or on their own. Alternants is a broad category of vocal sounds that can only occur in isolation; they are set apart from speech or singing. Though alternants might appear within a vocal line, a demarcation will exist between these sounds and the music/lyric gesture; a few examples are hisses, moans, slurps, sniffs and snorts.

Table 4.7 lists some of these sounds around words. All of these sounds contribute to the singer's image and persona. Critically, they also contribute substance and expression to the vocal performance and the vocal line. They can supply a deep connection to the performer and to their communication. When 'inarticulate' sounds are incorporated into the vocal they can provide a personal statement to or an intimate connection with the listener. They punctuate rhythms, add dimension to persona, clarify meaning and enliven language. Further, nonverbal vocal sounds provide a tangible representation of act of performance—lifelike qualities can thus be instilled within the performance, within the record.

Table 4.7 Select non-linguistic vocalizations, body-generated sounds and sample paralanguage.

Non-Linguistic and Emotional Vocalizations	Body-Generated Sounds
Gasps, moans, groans, growls, giggles, laugh, coo, cry, cough, sniffle, sobs, etc.	Breath sounds: Inhale, exhale, wheeze, blowing, etc.
Wordless expressions from feelings: surprise, disgust, happiness, anger, sadness, fear, pleasure, pain	Mouth sounds: pops, tongue clicks, lip smacks, gurgles, swallows, etc.
Sounds from feelings: ughh, mmmm, ow, aahaa, hmmmm, oooo, yeoow, etc.	Language-initiated: implosive consonants, sibilant sounds, etc.
Alternants: sighs, hisses, moans, groans, grunts, gasps	Alternants: voluntary throat-clearing, clicks, inhalations, exhalations, throat and nasal frictions, sniffs, snorts, slurps, pants

The 'sounds around the words' are significant when they appear. Their existence is nearly always calculated and intentional, not accidental. The majority of these sounds can only be part of any vocal performance through recording; these qualities are typically exactingly crafted to contribute to the record.

Vocal Style

Vocal style shapes the fusion of lyrics and melody. How the line is sung is potentially as significant as its content. Serge Lacasse (2010a, 226) observes from watching his young daughter singing along to recordings: "[O]ne might be under the impression that the true content of the music she likes . . . lies not so much in the lyrics or in the melody, but rather in the way they are sung." Material and its delivery will find a balance within the individual track; these too will become entwined as a single expression. As "in popular music, questions about melodic structure cannot be separated from questions about vocal style" (Winkler 2000, 39).

In performance, subtleties of sound qualities of both melody and language emerge (and merge) in vocal style. Those elements of music performance technique and expression that were identified in Table 2.2 are augmented here. The sung vocal exhibits additional activities and elements idiomatic of the unique qualities of the voice and those generated by the lyrics' qualities.

Table 4.8 contains numerous qualities and techniques, many identified by Walter Everett in "Pitch Down the Middle" (2008, 118–123); there he provides brief, clear examples (encompassing many timbre-related qualities) and acknowledges his listing is necessarily incomplete. Everett notes at the end of his listing: "all sorts of approaches to articulation, phrasing, portamento, note bending and glissandi, dynamic shadings and matters of stylistic ornamentation (with Janis Joplin unmatched for her variety of high-voltage techniques applied to multitudinous contextual relationships, exhibiting both talent and control . . .)." Of further interest to style is his observation that "A singer's level of ease or tension is palpable and an important conveyor of expression" (ibid., 119).

Allan Moore (2001, 45) has noted a void in how a multitude of factors are not embraced in characterizing vocal style and the voice's presence. He has proposed four factors that should be taken into account in any analysis: (1) range and register, (2) degree of resonance, (3) the singer's heard attitude to pitch and (4) the singer's heard attitude to rhythm. In Song Means (2012a, 102–108) he frames these as "four positional aspects" of the singer's voice. These are intended to allow the analyst to "determine whether the singer is conforming to the apparent meaning of the lyrics in the way they are delivered, or is perhaps clarifying them, whether the singer is equivocal about the lyrics . . ., or is even subverting them (ibid., 103)."

Table 4.8 Variables of vocal qualities and performance technique.

Timbre-Related, Sound Quality-Related	<u>Vocal timbre</u>
	<u>Resonance:</u> body cavity being used (throat, diaphragm, etc.)
	<u>Sound quality of the text</u>
	<u>Performance intensity cues:</u> shifting vocal timbre from dynamic change; Shifting vocal timbre from emotive intensity change
	<u>Non-linguistic vocalizations, body-generated sounds</u>
	<u>Vocal articulation</u>
	<u>Rhythms of sound qualities:</u> rhythmic placement of phonemes within lines, syllables and words.
Rhythm-Related, Time-Related	<u>Rhythmic placement of notes:</u> shaping 'within' the beat, anticipating the beat, or slightly behind; compressions and expansions of the beat; disguising starts of notes: attacks can be smooth and gradual
	<u>Vocal articulation</u>
	<u>Meter:</u> syncopation; polyrhythmic interplay between the pulse and divisions, stresses and phrasings; rubato
	<u>Emphasis by using time:</u> elongating notes; delaying attacks (back of beat); anticipating the beat (ahead of beat)
	<u>Manipulating time:</u> shaping musical idea, vocal space, shaping surface expression, stretching and contracting pulses, phrases
Dynamics-Related	<u>Shaping:</u> contours of lines and phrases, individual notes
	<u>Accents</u>
	<u>Performance intensity:</u> energy, effort, and expression of performance reflected in loudness
Pitch-Related	<u>Vocal intonation:</u> precisely in tune to pitch highly erratic; tension in unresolved, non-conforming intonation; blue notes
	<u>Pitch inflections:</u> microtonal inflections; speech-like inflections; bends, glides, or portamento; pitch-glide beginning as note is attacked; other forms of pitch-glides and bends; blue notes
	<u>Rhythmic placement of pitch inflections:</u> in front, during or at end of note
	<u>Vibrato:</u> continuums from narrow to wide, slight to pronounced, controlled to highly variable, and fast to slow
	<u>Range and registers of voice</u>

Peter Winkler performs a deep examination of Aretha Franklin's pitch inflections and intonation, and intricate beat subdivisions and inherent rhythmic flexibility (1997, 186–191). Moore views attitude to rhythm in relation to singing exactly on the beat or ahead or behind it; attitude to rhythm might also include singing in a rhythm that approximates everyday conversation, or that might conflict with linguistic syntax. The singer's heard attitude to pitch relates to singing 'in tune.' Singing perfectly in tune, flat or sharp by a slight or greater amount project attitude and vocal style. These performed pitch characteristics might be inherent to the singer; dramatically placed, the "approach may vary not only from phrase to phrase, but sometimes from moment to moment" (Moore 2012a, 103).

Currently, a prominent convention in mainstream popular music is to autocorrect vocal tuning; the natural singing voice is almost avoided in some styles, with only the beginnings and endings of pitches left unaltered and the sustain auto-tuned (pitch corrected). Pitch intonation is left unaltered from the performance in other styles (and before auto-tune became available and fashionable); it can appear in many forms from carefully in tune to rough approximations of pitch. Intonation might even be (somewhat) disguised by a vibrato's wavering pitch. For many instruments pitch conforms unambiguously to scale degrees and to tuning. Pitch in vocal performance may appear precisely in tune from its first moment and throughout its duration; this creates an impression of exactness and formality. Exacting pitch intonation is not typical in popular song, though; in keeping with the casualness of popular music, precision of tuning is out of character, and often undesirable. Pitch gliding is a common part of vocal style—sometimes subtle, often somewhat pronounced; this reinforces the linkage with speech inflection. A few common shapes and uses of pitch glides[21] (or bends) are:

- Pitch-glide beginning as note is attacked to resolve; glide is typically upward.
- Pitch is sustained from beginning of note, then gliding upward to resolve
- Glide between two pitches within the line, with the second pitch sustained
- Glide at the end of phrase, with the final note heard but not sustained
- Glide at the end of phrase, with the final note not clearly articulated

Alterations of pitch as much as a quartertone above or below the target pitch are common. These glides bring many qualities and functions to the music, and heighten the interplay of being 'in tune' with the affects (and the effects) of 'out of tune-ness.' For example, a sustained pitch that is slightly flat from the perceived, intended pitch creates tension that continues to increase the longer it remains. This un-resolved pitch is a suspension of sorts; considerable tension can be generated with this technique, as the pitch may never arrive at its destination, creating a sense of incompleteness, or may arrive after much time and establish a sense of finally having arrived. Other instruments, of course, can control pitch as exactly, but no others incorporate intonation quite so intrinsically into technique and performance style.

There are many nuanced aspects of vocalized lyrics. These subtleties appear in timbral qualities of phonemes, rhythmic placement and transitions between text sounds and nonverbal sounds, accents and dynamic shaping of individual syllables or words, and pitch inflection—to name but a few. As we seek to examine the vocal line and text fully, we observe and recognize great subtlety in the performance. Examined at a close perspective, we hear many dimensions, of the domains and elements in action; a multidimensional experience is presented by the vocal performance. The performance itself adds sounds and enhances the sound qualities of the lyrics; it adds layers of emotive and attitudinal interpretation that can transform its message; it adds dimension and drama with an unfolding of the text in time; and the performed vocal line becomes a musical idea, with all the elements of music present.

To illustrate some of these concepts, Allan Moore describes Joe Cocker's performance of "With a Little Help from My Friends" (1968) as

> a fine study of wringing meaning out of a song. He sings with obvious effort, with rather way-ward tuning, in an uncomfortable high register, and with a huskiness that suggests a voice raw with shouting or crying. All of these connote a personal authenticity or integrity—they signal that we should, or at least can, trust the singer's displayed emotions. . . . Cocker takes an immense amount of time within the beat. It is almost as if he is unaware of where the beat is.
>
> (Moore 2012a, 107)

So far discussion has emphasized individual elements and traits of the vocal line. Shifting to the perspective of the vocal's overall quality, there is a 'unified impulse' of music and lyrics domains—alchemy of voice and melody at the core of the song. In describing Aretha Franklin's voice in her seminal "I Never Loved a Man (The Way I Love You)" (1967) Peter Winkler observed:

> [I]n separating out these discrete 'dimensions' or 'aspects' of Aretha's singing—pitch, rhythm, dynamics, timbre, text setting . . . As she sings, I hear, not an assemblage of discrete ingredients, but a single entity, the result of a unified impulse. When I listen to that singing, I can feel that impulse, but it is not anything I can describe with any precision.
>
> (1997, 192)

'A unified impulse' is a striking description of the sung vocal line. Its many dimensions blending into an overall quality, the vocal is where the individual contributions of pitch and rhythm, dynamics and timbre, word sounds and expression dissolve; no longer discrete elements that built the melody-plus-lyrics vocal line, but rather a whole so well integrated one aspect does not stand out. While the masterful performance of Aretha Franklin is central to the richness of this integration described by Winkler, all vocal lines present some level of integration and a clear sense of an overall quality. All vocal lines present an integration of interpretation, pitch, intonation, rhythm, time, text, diction, drama, timbre, dynamics and intensity into a single coherent whole, into one 'unified impulse.'

Examining Lyrics Within the Track

Interpreting, or making sense of lyrics is uniquely challenging. Performed song lyrics have many levels and dimensions, beyond message and meaning. Lori Burns (2010) has developed a detailed approach to examining aspects of the lyrics as they appear *in* song—that is, as they appear in the *experience* of the song. This approach may prove valuable to many analysts engaged in the complexities of lyrics.

In "Vocal Authority and Listener Engagement" Lori Burns proposes an "interpretive framework for the integration of musical strategies with lyrical narrative perspectives" (2010, 155) to study female pop-rock expression. It serves to bring considerable depth and clarity to her analyses. Her framework might be successfully applied to a wider array of pop songs to bring a functional approach to—or at minimum a meaningful guide towards—exploring lyric context (including meaning and message) and its relationship to the record.

Burns' 'narrative-theoretical framework for the interpretation of the voice' incorporates theories of narrative authority, musical persona and agency, "the subjective layers of protagonist, character, and artist, defining those roles within narrative theory, . . . [and] expressive strategies of the female singer-songwriter" (*ibid.*, 158). Recognizing the study's "critical objective is to explore female vocal authority" is well achieved and significant; its more universal value is in what follows: "by examining strategies of narrative voice and vocal expression in four selected songs. The analyses will demonstrate how the strategies contribute to the sociomusical communication from artist to listener" (*ibid.*, 156). The framework unfolds in four stages, from artist to listener, by way of implied author, narrator, narratee (the person "inside" the text to whom the narrator is speaking), implied listener:

1. Narrative agency: author-implied, author-narrator. In this first stage the 'implied author' establishes the norms of the narrative, and the distance (such as emotional, moral, intellectual) between the narrator and the implied author; the implied author is the sensibility (opinion, knowledge, intelligence, feeling) that accounts for the narrative. Through the way the author crafts the narrative voice the listener/reader forms an interpretation of the ideologies, values and authority of the persona that guides the narrator, which is the implied author. This is especially important in popular songs in which the artist uses personal voice

2. Narrative voice includes the narrator, character, and persona. Of central concern is 'narrator status,' which is informed by 'narrator identity' and communicated through 'voice': the identity of whose story is being told and by whom; contained in the identity is the reliability of the narrator, their honesty and sincerity. Narrator status might be deduced through questions such as: Is the narrator an observer or a character, involved in events? How does the narrator come to know the story? Narrative stance, or point of view, is the relationship of narrator's personality and values (social norms) against the story. Voice is projected from this position. Forms of voice that are common are:

*   Authorial—third person and public narratives, omniscient narrator (reading to an audience), narrator and narratee both exist outside the story and are not humanized by the events
*   Personal—first-person, narrators telling their own story
*   Communal voice—narrative is a collective voice, or a collection of voices

3. Modes of contact occur between the narrator and the narratee. The critical questions about the narrator's communication to the narratee are recognized with the assistance of narrative theory. Framed as opposites, these include:

*   Address: public (aimed at a broader readership) versus private (aimed at a specific character inside the narrative)
*   Communication from the narrator to the narratee: direct and active (with "I/you" communication) versus indirect and passive (through an accounting of actions, events and feelings within the narrative)
*   Expression: sincere versus oppositional (such as an ironic expression toward the narratee)

4. Listener engagement links artist to reader/listener. Using Andrea Schwenke Wyile's (2003) concept of engaging narration, central "elements of narrative that develop the relationships between and among characters (including narrator and narratee)" (*ibid.*, 165) emerge:

*   Proximity—a narrator can seem so close to the character that the listener might conflate narrator and protagonist, blending into one in many listener minds, though the 'I' who narrates and the 'I' who experienced the events are not one and the same
*   Sincerity—narrator relaying events and character experiences earnestly; a shift in engagement or perspective represents oppositional viewpoints, and might present two or more competing voices
*   Temporality—consonant (having no temporal interruption between events and their telling) versus dissonant (resulting from time-shifting techniques such as retrospection)

Table 4.9 presents a narrative-theoretical framework for "Strawberry Fields Forever." Narrative agency is clearly first-person, addressing a second-person narratee, with a sense of detachment (perhaps emotional) separating them. Lennon (as narrator) is telling *his* story, and it is personal; he is a character in the story and his personality is enmeshed with what is shared; there is a sincere desire to share the experience and place of Strawberry Fields with the listener, narratee. The modes of contact see the narrator and listener in private interaction, directly communicating (though Lennon's persona is hesitant, uncertain at times), and the narrator appears sincere, with some avoidance and oppositional positioning. Some distance (perhaps emotional) appears to separate the narrator and listener; the narrator is not engaging reality earnestly, so relaying of events may be distorted; activity takes place out of time, with no temporal sequencing.

To these examinations of narrative, this approach explores musical expression and interpretation to reveal how narrative elements are developed in the musical realm. Interpretation focuses on the details

Table 4.9 Narrative-theoretical framework for "Strawberry Fields Forever" by the Beatles.

Stages	Description
Narrative Agency	First person narrator, second person narratee. The narrator wants to share their feelings, mood, experience, and perhaps a metaphysical place or idea with the narratee, although there is a sense of detachment between them.
Narrative Voice	Narrator is telling their own story, and it is personal. The narrator is a character, the central character.
	Narrator implies a wish to connect with the listener (a sincere desire to take the listener down a place with nothing to get hung about).
	The narrator appears wistful, weary and self-doubting; living with anxiety and loneliness of being misunderstood; reality and fantasy may be blurred.
Modes of Contact Between Narrator and Narratee	Address: private, narrator and narratee are together alone
	Communication: direct, with some hesitation and uncertainty
	Expression: mostly sincere, with some avoidance and oppositional positioning
Listener Engagement Links Artist to Reader/Listener	Proximity: while the narrator and narratee are alone they are separated by some distance (of connection if not space); listener is invited to 'follow' narrator in each chorus
	Sincerity: narrator is not engaging reality earnestly; seems preoccupied, conflicted, practicing avoidance
	Temporality: non-temporal, happening outside of time

of vocal expression bound to the four narrative elements, and is supported by music analysis of relevant materials. The musical elements in her interpretations are: vocal quality, vocal space, vocal articulation, texture and recording techniques. Burns notes: "For each of these musical layers, the listener interprets the extramusical connotations of musical gesture within the context of style and genre" (*ibid.*, 166). This flexible yet focused process represents a tangible framework for interpreting the lyrics' narrative, and its unfolding story, perhaps revealing stated and implied concepts within the lyrics—one that will fit well with our recording analysis framework.

LYRICS IN RECORDED SONG

The recording represents an additional layer of sonic dimensionality and interpretation to the song's lyrics and vocal performance. Recording elements can provide lyrics with drama and motion, they can support meaning and expression, and they can add color to and create a context for the lyrics. This section is intended to provide some grounding of the lyrics in the track by introducing some examples of how the recording might enhance lyrics and enriches its story and communication; notice that these discussions largely emphasize timbre and spatial qualities. The recording, lyrics and the performance are in confluence within track; they fuse into a single statement, into one complex voice.

This confluence is apparent in Bruce Springsteen's narrator in "Born in the U.S.A." (1984); a title he brings "to sound like both a sentence of doom and a hopeful declaration of optimism" (Himes 2005, 19). The protagonist speaks his story directly to the audience, in a voice that often is as much a yell as singing. The story is not of overt patriotism; it is rich in ambivalence. Woven against the backdrop of working-class struggles and limited opportunities, are the bleak social challenges that greet the narrator upon returning from the Vietnam War—no jobs, a Veterans Administration unable to provide support; his story arrives at "nowhere to run" and "nowhere to go"; with more than a little sense of defiance he stands his ground, still seeking the life he deserves (*ibid.*, 9–35). The recording supports this, as Springsteen's character struggles to be heard above the noise, struggles to establish his own place in the track (to carve out his own place in life and country). The mix and arrangement place the lead vocal at a distance from the listener and at a lower loudness level than others. With a snare drum processed to create explosions of dense sound, a piano that is distinct in its clarity, and synthesizer sounds creating timbres in detail that assert a dominating repeating pattern; even within the sparse texture of the first verse+ chorus sequence the vocal must exert itself to compete with the others, and in the mix. The extreme performance intensity of the lead vocal is not the loudest in the mix, other sounds playing at lower levels of physical exertion (though they may be high, as the drums, they are not higher than Springsteen's vocal) are louder, some only at times. Further, the lead vocal is the sound furthest from the listener; the piano and synthesizer sounds contain more timbre detail and are more immediate to the listener, the snare drum is somewhat ambiguous, though its density of sound information brings it a sense of being nearer than the protagonist. Once the second verse starts, the vocal must compete harder, and with the third verse harder still, as throughout the track the instrumentation and mix become increasingly dense and complicated, and the sound stage gradually becomes more complex until the track obtains its full texture. This relationship of the lead vocal to the ensemble is the result of crafting loudness levels of sources in the mix, and also of modifying timbres to create a sense of depth of sources; here, recording shapes and enhances vocal qualities and its relationships to other sounds. Recording provides a context against which the vocal asserts itself against many forces and challenges—the narrator never backs down, nor does he triumph.

The stereo image of David Bowie's character Major Tom fittingly travels through space in "Space Oddity" (1969). The lyrics' first line is presented in the center of the stereo array and with considerable width. The second and third lines are sung by two Bowies, one hard right and a higher doubling on the left. Each image is narrower (occupies less space) than the vocal image of the first line; the effect is that these two voices are not more substantial than the single voice that preceded. Line four returns to a single voice centered, with characteristics like the first line, as the spoken countdown begins in the left channel (with an image similar to the second and third lines). The lead vocal is located in the right channel for lines five and six, with the same qualities as lines two and three; the left channel continues the dispassionate countdown (image unchanged) until "lift off" during the sixth line. The chorus follows with left and right channels performing the same line roughly in unison, with slightly different interpretations. Verse two opens with a single centered lead vocal, which then resumes two right and left channel vocals—this time with the lower line in the left channel. The result is not having a fixed point of reference for the lead vocal; the vocal is moving in space, it is surreal. Major Tom is moving, his presence is changing sizes and locations; at times he seems duplicated and split, though in mismatched proportions. These changing physical locations and sizes of image all support the story and the drama of the lyrics.

In contrast, the listener is transported between two distinctly different places within the first line of "Hello" (2015) by Adele. The first word, "Hello" contains contradictory spatial information—a reverb tail of a large space and an early time field of near walls that provide a sense that the voice is coming from a different place, perhaps a different time; the spatial character of the word provides a sense it is directed to the listener from a rather detached (but still close) position within the listener's personal distance. Immediately we hear "it's me" arriving from another place—it is a smaller space, noticeably different, less reverberant (though clearly reverberation remains) and more naturally proportioned; one may imagine this voice in the same room as the listener, with the protagonist leaning in and entering the listener's intimate space—breaths and mouth noises are audible. While the first word "Hello" has larger substance, the following words "it's me" are clearer and piercing. The remainder of the verse continues the character of these two words, in a voice that seems to penetrate deeper than the first; it is more immediate, and it cannot be ignored. After the second verse, a pre-chorus brings a shift to the voice and a more urgent intensity builds from a slightly more detached distance; the chorus erupts "from the other side" with a high intensity vocal, from a place that is more distant than the verses but not the place of the first "Hello." In the chorus all words are within a consistent distance location, and clearly at a more polite social distance that is well outside conversational speaking range.

This shaping of distance and environments is not at all unusual; "Hello" is filled with the drama, motion, direction and expression provided by the recording that has carried many tracks, though it (of course) has unique qualities. The verses have a different spatial, loudness and timbral profile from the chorus (plus, in this case the pre-chorus)—verses more intimate, chorus noticeably more detached, pre-chorus builds energy and provides a different and intermediate set of spatial relations, the return to the verse after the chorus is like turning the page at the end of a chapter, and back to a familiar scene. Qualities of the performed lyrics very often shift between verses and chorus, and support the content and expression of the lyrics; recording elements reinforce structure and fuse with the performed lyrics (voice) as it shifts from one set of relationships to the listener to another—here these are a sense of intimate verses, a more detached chorus and a sense of movement from one to the other.

"You Oughta Know" (1995) by Alanis Morissette follows a very similar shape of intimate verse, more detached and more animated pre-chorus, and an emotively intense chorus with increased separation between the listener and the singer's persona. The qualities of the voice and how the intimacy and distance translates into the communication of the lyrics are different between these tracks. The deeper breath sounds of the first verse carry an anger and energized restraint, where Adele had vulnerability and perhaps regret. Morissette's vocals utilize image width changes and voice in two channels that (similar to "Space Oddity") provide an unsettled place from which the lyrics emanate—exacerbated by distance locations that also shift unpredictably. Image size and proximity to the listener are used to emphasize certain words and feelings. At times the text can appear menacing, in stark contrast to the emotional detachment of Bowie's track. A rich set of connections are present in the way the recording enhances the performed lyrics and also non-language vocalizations; the vocals in the 16-measure bridge provide shifting distances from very close breath sounds and some from greater distances, and vocalized sounds that exceed the distances of human interaction.[22]

Vocal performances do not need pronounced recording-element qualities for their lyrics and expressions to be transformed. In Joni Mitchell's "The Last Time I Saw Richard" we hear her delivery coming from a realistic place, and the narrator does not move in distance or position. We do hear changing energy in her performance and the shifting ranges of her voice that result in loudness changes, though the changes are not proportional to the performance—in other words, the recording process shapes the loudest and softest parts of the performance into a smaller dynamic range, reducing the loudest sounds and increasing the softest ones. Further, we hear Mitchell's voice up close, with tongue clicks, saliva sounds, cracks in her voice, unpitched sounds in the back of her throat—all sounds that shape the timbres and expression of the lyrics, and that would be utterly inaudible in live, unamplified performance.

Subtle qualities of recording elements are also found in Bob Dylan's "Simple Twist of Fate" (1975). Like Mitchell's track, Dylan's vocal is closer than a live performance could be and all instruments carry similar space information as the vocal. Loudness levels vary with performance intensity, like Mitchell's, though the performed dynamic range is narrower—that is, until the peak point of each stanza. At those dramatic moments the voice suddenly increases in loudness and in performance intensity and the room's reverberation is excited; a new sense of much larger space is introduced. Out of an unforced, fluid presentation the end of the last phrase within the fourth line of each stanza quickly grows in loudness to provide strong emphasis of the words "straight, freight, gate, relate, wait, late" (Dylan 2016, 334). Those six words generate a significant shift in reverberation as well as loudness and timbre; they are the highest pitch and are the words that set up the rhyme that connects the verse lines to the song's refrain that closes each stanza. These words set apart by reverb, loudness and performance intensity close the story, action and concepts of each verse. The significant amount of room sound and the size of that space add unmistakable emphasis and substance to those words; their reverberant sound contributes drama to these lines. The reverb tail brings the words' impact to linger after their sounds cease; as the words dissipate and return into the original performance quality—into the final line of the stanza—each of these words introduces a different take on the song's core idea and its song title.

This final section has introduced select qualities of recording elements to help the reader connect this chapter on lyrics to the core topic of this book—the recording elements. The recording elements and related concepts introduced in these few paragraphs will be more fully examined in Chapters 6 through 9—along with the remainder of recording elements. Chapter 10 will present detailed analyses of recording elements and will explore ways these elements shape lyrics, and the song as a whole.

CONCLUSION

This chapter has explored how lyrics exist within the song, as well as how they communicate. It has sought to articulate how lyrics are fundamentally linked to performance; that musical materials and vocal performance styles shape the content and character of lyrics—and that the recording participates.

The framework for analysis offered throughout this book can embrace methods such as Lori Burns' approach to the interpretation of the voice and all it generates (2010), as well as other approaches looking at the analysis of lyrics as they are situated within the track. It can also embrace the sonic qualities of the lyrics and the performance, the energies and expression of vocal lines and their impacts. Existing methodologies from disciplines that specialize in the examination of lyrics and poetry, of how lyrics are situated in culture, and the host of ancillary disciplines that can look deeply into the language and communication of lyrics, and their social and psychological dimensions may also find space in the framework that is offered. Any of these might be used to supplement, enhance or add considerable substance to the analysis of lyrics within a track, and of the track itself.

In addition to all this, we remember: song lyrics are not poems. Lyrics are meant to be sung. They are written to communicate with their sounds and through their performance as much as their language—sometimes more so. What lyrics communicate and how lyrics communicate differ from poetry.

> Lyrics . . . let us into songs as stories. All songs are implied narratives. They have a central character with an attitude, in a situation, talking to someone (if only to herself). This . . . is one reason why songs aren't poems.
>
> (Frith 1996, 169)

Some Questions for Recording Analysis

Who is the persona singing the text? To whom are they singing?

What is the subject of the song? What is the story of the song?

What is the theme of the song?

What is the narrative point of view?

Is something being told or described that is below the surface of the lyrics?

Is there a metrical pattern in the lyrics? How is it created? Is it interrupted?

What patterns of numbers of syllables between phrases and sections are present?

What rhyme patterns are present? How do they contribute to structure? How do they recur?

What words or topics are emphasized by structure, by the meter, or by rhymes?

Are there connections of word sounds other than rhyme? (Alliteration? Assonance? Onomatopoeia?)

What is the ordering of ideas within the lyrics? Is there a progression of time?

How are important sections of the text emphasized in the setting?

Do specific words or word sounds occur and recur at structurally important locations?

What line rhythms are contained in the lyrics? Do they remain consistent?

What subjects or topics recur in the text, or are grouped together in different sections?

Does each stanza cover a separate topic, or is there some on-going activity or element throughout?

NOTES

1   I gratefully acknowledge the input provided by my colleague Michael Millner—with whom I share an affinity for Bob Dylan's lyrics and music. He generously gave of his time to review several iterations of this chapter, and to suggest a good number of the examples used. Given his expertise as a professor of English, I very much appreciated his being selective and gentle in his comments about my writing.

2   Numerous observations in this section, including those pertaining to these first stanzas of "Strawberry Fields Forever," were offered by my colleague Michael Millner.

3   Definitions of the common structural devices listed are: caesura is a pause within a line of the lyrics usually formed by the rhythm of speech (typically in the middle of the line, it may also occur near the beginning; elision is the omission of a syllable or other word sound in speaking (used in lyrics to mirror speech); enjambment is the running-over of a sentence or phrase from one line of the lyrics to the next; free verse poems do not contain a regular meter or rhythm, rhyme scheme, or patterning of line lengths (they can seem more prose-like than poetic).

4   All Bob Dylan lyrics discussed in this book have been examined in their official forms, contained in the collection: *Bob Dylan: The Lyrics 1961–2012* (NY: Simon & Schuster, 2016).

5   Accessed on 22 September 2019 from the Paul Simon official website: https://www.paulsimon.com/track/the-boy-in-the-bubble-6/

6   No, I will not provide examples on this. The reader will not need to search far to find representative examples.

7   Serge Lacasse (2007) has articulated a distinction between two concepts that are useful here: intertextuality and hypertextuality. The theory of hypertextuality has its origins in literature, studying and characterizing relationships between different works. This distinction can assist in defining how quoted, borrowed and modified sources appear in and add dimension to lyrics. While presented here in lyrics, Lacasse's study explored how these concepts are equally pertinent to popular music, and its musical elements and the recording—where an

easily recognized example exists in sampling. Intertextuality is the actual presence of a text within another; it is used to define various types of quotation and allusion, where new works are produced that re-use material from another source (this might take the form of sampling). In hypertextuality a new work is created from the content of another. The 'borrowed' material (music, sound, or lyric) is transformed, renewed or modified; this generates parody, remixes, cover, copy and other useful categories such as translation, pastiche and travesty, as well as others less central.

8 Some sections in this book provide instructions and guidance for performing certain tasks. I will at times use the second-person form to address the reader as 'you,' to encourage your actions and activities during those sections.

9 Exceptions to this might be (1) the star performer, where it is "through the voice that star personalities are constructed . . . The tone of the voice is more important in this context than the actual articulation of particular lyrics" (Frith 1987, 143) as quoted in Moore (2001, 185); and (2) a singer who is not adequately skilled.

10 Moore (2012a, 180–181) clearly articulates a distinction between the performer, persona and protagonist, and establishes the significance of each.

11 See *The Rise and Fall of Ziggy Stardust and the Spiders from Mars* (1972).

12 The 'English' language obviously originated in the British Isles of 'England,' and certainly not in the United States. Further, it is the de facto national language or the official language of a good many countries well beyond the current United Kingdom, including Australia, Canada, New Zealand, South Africa, and others.

13 It is hoped what is offered here might be in some way adaptable to other languages and their unique sound qualities; this will be a more reasonable hope for some languages than for others.

14 Complete symbols of the International Phonetic Association's alphabet may be obtained from their website at: https://www.internationalphoneticassociation.org/content/ipa-chart

15 Sources for explanations of IPA diacritical and suprasegmentals symbols are: Patricia Ashby (2011) *Understanding Phonetics*, New York: Routledge and Mehmet Yavas (2016) *Applied English Phonology*, 3rd edition, West Sussex: Wiley-Blackwell.

16 Several websites contain tools to convert English into IPA symbols. Two are: http://lingorado.com/ipa/ and http://upodn.com both accessed July 21, 2016.

17 Exactly which are 'most melodic' is a cultural assessment and subjective, and will not be approached here.

18 Quoted in Paul Williams (1990), *Performing Artist: The Music of Bob Dylan, Vol. 1*, 261.

19 You may notice I am not providing examples here. I believe further clarification of these points is likely counter-productive, as each reader will have experienced each state, though from their own set of experiences. I wish to avoid inserting my own personal and cultural biases between the reader and their own observations.

20 Musicologists such as Rob Bowman (2003), David Brackett (2000), Walter Everett (2008), Serge Lacasse (2010a), Richard Middleton (1990), Allan Moore (2012a), and others have discussed aspects of vocal performance, vocal production and vocal dramatic qualities in popular music. Lacasse (2010a, 227) states: "most of these discussions are dispersed throughout literature and leave out certain important vocal features."

21 Types of pitch glides adapted from Peter Winkler's "Writing Ghost Notes" (1997, 190).

22 Serge Lacasse (2000a, 193) observed Morissette's "You Oughta Know" opens with the singer/protagonist in the same room as the listener, stating that: "the most likely interpretation of an acousmatic voice [a voice detached from its origin] displaying no reverberation would be to hear that voice as sharing, or as sounding inside, the listener's own environment."

CHAPTER 5

Observing Elements of Music and Lyrics in Records

With this chapter we begin to engage the actions of putting the framework for recording analysis into practice. The observations process is defined, and then put into practice by examining this initial stage of analysis for the domains of music and lyrics in popular, recorded song. Practical matters will be found alongside more conceptual issues. The analyst's playback system can impact observations of music and lyrics; while a detailed discussion of playback is postponed to Chapter 6, its presence is felt throughout this chapter.

The focus of this chapter is to bring about the start of the analysis process from a position of intention in defining the goals of the analysis and a position of awareness and attention in collecting observations.

The three stages of the 'observations' process are presented in detail. Guidance is provided for engaging listening for the purpose of performing aural analysis of all domains within records, and for enhancing requisite listening skills. For those who wish closer guidance, relevant Praxis Studies are identified in side bars throughout the remaining chapters, directing the reader to Appendix A for specific studies and exercises.

The many issues concerning transcription are approached, including when transcription is appropriate, when it might not be necessary, and when transcription might adversely influence data collection, materials and an analysis. Approaches to transcription, notation conventions, assisting tools and devices, devising workable notation nomenclatures and replacing notation with relevant descriptions provide alternatives to traditional notation and a recognition of the simple values of accepted systems—no matter their limitations.

Some analyses will benefit from transcriptions; others will require some materials of the track be transcribed. Engaging and organizing observations comprises the second half of the chapter. Timelines of various sorts are constructed with goals of making structural divisions and hierarchies, and temporal divisions visible. Select approaches to collecting information on the elements of music and on the performance of lyrics are examined; these will provide a glimpse of what information from observations was deemed pertinent within these analyses. Adding information collected from observations to timelines is explored in various formats.

THE PROCESSES OF RECORDING ANALYSIS

An analysis of a record is directed toward a goal—a goal of what *level* of understanding is sought (and will be communicated), a goal of what *type* of understanding is sought (what one seeks to learn or discover).

This goal may be directed towards understanding the song, and its lyrics, music, and expression. The objective may be directed towards understanding how the track is situated culturally or socially, what and how it communicates, or any number of other goals outside the record. While a worthy goal is to illuminate the track, to bring clarity to its essential qualities and how the other qualities contribute, this is only one approach, one objective.

For instance, the underlying objective and goal of this writing is to bring (provide a path toward) understanding of how the recording impacts the song—the affects of the recording process. The recording is a central concern, and the object of detailed examination. The track's musical materials, lyrics' sounds and content, message, structure, etc. are important to the analysis in that they need to be understood at some level in order to recognize how the recording impacts them, delivers them, enhances them, etc. How the qualities of the track's recording establish relevant relationships to areas outside its sounds might also be pertinent or of interest. Still, the focus is on the elements of the recording; other domains and other disciplines are, well, 'other'—they are examined to understand the qualities and influences of the elements of recording.

One can define the objective of an analysis of a recording in many other ways. One can focus on the entire track, or just a portion; one can focus on the lyrics, or on their interactions with accompaniments; one can explore the intricacies of the music, or its affects or expression. The choices are without number.

Even when an objective is well defined, it might shift due to what the analysis reveals. Such a shift occurs as one learns the track better, makes discoveries and connections, and recognizes what needs to be explored more deeply and what other topics need to be included to understand and present what makes that track unique. Still, a clear objective creates a clear path for collecting information—a path which can be broadened and deepened as the process unfolds.

Just as important as the subject matter is the level of detail sought—breadth and depth of coverage. At some point information collection must stop and evaluation begin; at some point the subtle qualities of the record are either central to understanding the track or superfluous, and so forth. The breadth and depth of the analysis is part of defining its goal. This might be the level of detail that is needed to adequately reveal the essential characteristics of the record, or the level adequate to support the needs of the individual analysis—to identify just two of many possibilities. For instance, the analysis might be directed toward a specific audience and need to conform to what they might reasonably be expected to understand.

An interpretation of the track results from the analysis. The interpretation is generated by and reflects the analyst—their skill, interests, knowledge and more. It also represents what they wish to learn about the record, and how they choose to communicate about what they learned. Defining goals that recognize the interests, strengths and skills of the analyst can be beneficial.

Overview of the Framework

To begin this overview of the recording analysis process, it can be helpful to recall the guiding framework's four principles:

1. *Every recorded song is unique*, with unique essential traits. Those essential traits are supported, ornamented and provided context in unique ways, and are organized in a manner that will also have unique qualities, even if subtle, and speak in a unique syntax and language. An appropriate analytical approach is needed to reveal the track's essential traits.
2. These essential traits are reflected in all dimensions of structure. *Perspective* allows the analyst to navigate these levels of detail and of hierarchy to collect, evaluate and synthesize information within all levels of detail, and to compare those diverse sets of data.

3. The elements that form the essential traits—within and between its lyrics, its music and its recording—contribute in unique ways to every recorded song. While conventions might bring certain elements to dominate a style, every element is in play in shaping the record, every element impacts the recorded song, and every element holds the *potential* to contribute significant and essential information. *Equivalence* reminds the analyst to examine all elements, as each has a role and each may provide any function, at any time, at any (every) level of perspective.

4. The concept of equivalence also helps guide the listening process that is at the core of analyzing the record. *Listening with intention and attention* establishes an approach to access all the record's traits and materials, structural levels and dimensions of form, and so forth. Further, listening with intention and attention brings coherence to the process of making observations of music's elements, the sound qualities of lyrics, and of the elements of recordings. It also facilitates navigating the challenges of transcription in each domain.

The differences between analyzing music from a score and from a record are many, though differences all tend to lead to one being perceived through sight, and the other through sound. The fundamental challenge of accurately analyzing a record is to hear what is there, and to hear it accurately. While not a simple task (especially at first), all paths of the analysis process rely heavily on deep, intensive listening—this is our only access to the materials and expression that draw us to it.

First Step of Process: Observations

The three stages of the analysis process might be reduced to collecting information, evaluating information, and bringing together conclusions. With a multiplicity of materials and elements, domains and dimensions, the track is complex. Dividing the process into three clear steps brings some clarity to engaging this complex texture. "Observations" (collecting data to analyze) is the first stage, and will be discussed here; the stages of "evaluations" and "conclusions" underlie Chapter 10.

There are three primary activities in the observations stage. These three activities do not need to be accomplished in a set sequence. This stage is dedicated to collecting raw data—information on what is present within the track. No explorations of functions of materials, patterns of organization, how the materials work or evolve and such will appear or be determined within this observations stage.

First, the track's sound sources are identified, along with the basic-level materials they present. These are the surface materials in the music, along with the sound sources that present them (such as accompaniment and bass line, piano or perhaps groove and bass line, electric bass guitar).

Second, several basic characteristics of structure are defined in observations. This includes identifying the length (number of measures) of the recorded song, which requires recognizing the meter and tempo (and any changing meters or tempi). The large dimension timeline can now be established, as in Figure 5.1. It is important to note, the details of this figure contain information that is revealed from evaluations of materials—the song sections and the eight-measure prevailing time unit. This figure is configured to illustrate the overlying, inexact palindromic arch of the song's structure; notice measure 57 is marked to identify the point where the structure reverses; before and after this point is a verse+bridge+verse sequence. The arch is inexact because the introduction and coda are not proportional; Clapton's solo substitutes for a verse, and Verse 3 slightly modifies Verse 1's text.

Third, information or data on each individual element in each domain are collected. The depth and detail of this information is intrinsic to the record (including recording elements). Data collection embraces all that is of concern to the analyst, and can be extended to pertinent stylistic conventions (once evaluation is engaged). Here the musical materials become better defined. Their content is more

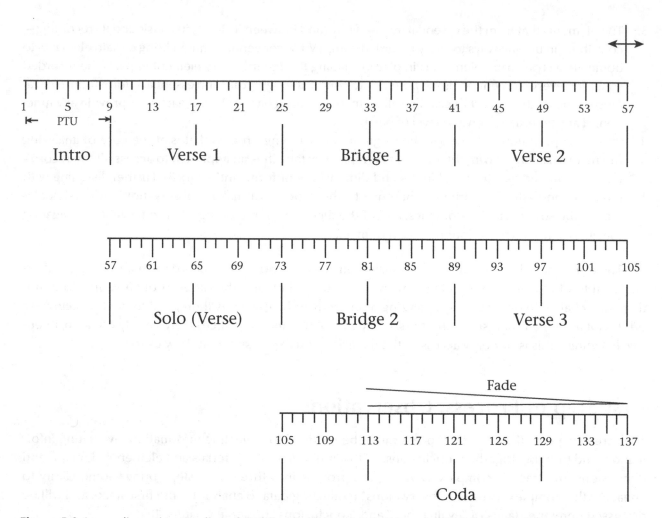

Figure 5.1 Large dimension timeline of "While My Guitar Gently Weeps" (*The Beatles*, 1968).

clearly delineated, but not evaluated. Transcription, descriptions and notation in some form will take place in this stage—if it is to be undertaken. We will learn transcription and description are interpretations, and interpretations are often based on evaluations of sorts.

Observation, Interpretation and Bias

Interpretation of information begins at the 'evaluations' stage; with all (or much) data available, patterns will emerge. Once in the evaluations stage, it is common to return to 'observing' to confirm information or to collect additional information. Being aware of the separation of collecting and evaluating brings clarity to the process, and can keep the analysis from being distorted.

It is very easy for data collection (assembling basic information) to be distorted by interpreting that data (even certain portions of it) before collection is complete (or sufficiently complete for accurate evaluation to emerge); for example, one cannot identify the tonality of a song from the materials within its first moments, and one cannot determine the functions of chords without learning how a progression concludes, or the nature of the tonality in which it is situated. Once one begins to interpret information, a natural tendency is to seek information to support that interpretation. This tendency can be so strong that determining the information that is present is transformed into causing the new information uncovered to conform to what one suspects might be present or wishes to be present to confirm one's theory.

Scientific inquiry has taught us to separate data collection from data analysis, for good reason. For observations to be unbiased (as much as this is possible) the principles and theories explored within evaluation should not appear here, as they will cloud and distort this collection. The notion of neutral or purely unbiased observations is utopian; we approach any material within any track from our unique position of pre-understanding. Our different experiences with music, our listening skills, our personalities, ethnicity, and so on, all unconsciously influence what we perceive. While we seek a neutral observation—seek to discover what is present not what we wish to find—this is not ever completely possible. Significantly, this connects with setting the goals for an analysis, where deciding what data to collect already implies some kind of interpretation. While the ideal is to be unbiased, we need to be self-reflective in order to avoid being biased—it is easy to be biased without being aware of it.

Similarly, there can be a curiosity to begin to evaluate observations early in the process. Some have proposed transcription can be aided by some early evaluation of materials (Winkler 1997, 174–181), and the experienced analyst might instinctively engage evaluation early in the collection process. While this may work (to some extent) for some experienced individuals, it is not an unbiased position; it is clear that in order for the song to be engaged from a neutral position—the position that allows one to most readily and accurately recognize its unique qualities—evaluation is most effective later (and as a separate stage) in the analysis process.

Other forms of notating the track's sound qualities can take place in observations. Functioning like music notation X-Y graphs and other diagrams for the recording, and phonetic sounds of lyrics can be transcribed. This stage of the process can become quite involved and detailed; it is common for this to happen. A middle dimension timeline, or perhaps a separate timeline for each section of the track, can assist in establishing some order to observing elements and materials. Through timelines, elements may be observed more systematically, thorough data collection can be ensured, and observations organized to facilitate the evaluations to follow.

These observation activities are accomplished through listening to the record. This is an activity that will be repeated a great many times to study a single track. In each successive listening, new information will be identified or extracted; noticing different features and attributes each time, successive hearings access deeper levels of structure, greater detail and/or shift to focusing on different instruments or other aspects of the track. This process will become more efficient as auditory memory improves. To avoid holding all of that information in active memory—and to allow the pool of information to grow—it is helpful to write it down. Transcribing the music and its performance, the performance of the lyrics, as well as the qualities of the recording will greatly assist the analysis, but are rife with issues that will be explored later.

Outside Sources

Reaching beyond the track itself, should the analysis seek to include other disciplines, information in those areas might be collected at this stage. The most pertinent and productive way to engage those areas and concepts will vary between disciplines. Forethought when defining the goal of the analysis will play a central role in when and how to collect that information, and when to begin evaluating it.

A literature review, collecting relevant background information (etc.) will influence, and potentially frame, an analysis. These may bring new awareness to the analyst. They may also skew their direction of inquiry and their perceptions. Engaging outside information should be approached with an awareness of purpose and significance of what is sought, and how that information is used in the analysis going forward.

LISTENING, TRANSCRIPTION AND DESCRIPTION

Listening for the purpose of analysis is a distinctly different type of listening activity than one encounters in music listening, performance or analysis following a score—aural analysis is atypical in a great many ways. Many readers will carry a significant depth of listening experience in some areas to apply to recording analysis. Many of us have learned a considerable amount of music from listening to records; some have even learned to play instruments by listening. This is an extension of the aural/oral tradition that once passed musical traditions (especially popular music traditions) between individuals and generations; the record can be a teacher by virtue of the easy access and repetition of its content.

> **Praxis Study 5.1.** Developing auditory memory (pg. 490).

Within the process of observations, listening with attention and intention play central roles. It can take two basic forms in observations. One approach is intentionally focusing the listening experience (attention) on a certain element or idea—at a specific level of perspective, and within a certain time unit. The second approach brings the listening experience to intentionally and systematically scanning the track for information. This is a contrast of controlled exploration searching for salient features, and of bringing focus to certain features at the exclusion of all else.

Learning to listen to tracks is multifaceted—it is just as multifaceted as the texture of the record. Each domain requires a unique set of skills—the chords of a harmonic progression and the timbral modulations of a vocal line require different skills and abilities to recognize. While there are some aspects in common between domains, each has unique qualities; those qualities require listening skills that are unique in some way to access their information fully.

We learn from examining the elements of recordings that the record holds sonic experiences not represented in nature. Therefore, it is expected that a good many listeners (readers) will have little prior experience with some of these elements and materials—especially without prior knowledge of the elements that might exist in the record. While equivalence can assist the listener in being open to possibilities, the listener needs to be aware of what those possibilities 'sound like' in order to engage the act of searching the sonic landscape to experience the sounds of the record. We rarely see something for the first time, but when we do it is obvious—such as a friend's new hairstyle—and is nearly always something within a known category. Within our senses of touch, smell and taste we rarely encounter something completely foreign or new. In sound, encountering something new is rare in most contexts, but not in some others. The sounds of unknown languages can be readily encountered, and while perhaps not understood the sounds are still recognized as being within the category of language, and connected to the voice and to communication. The nature of recorded sound, however, contains qualities that defy adequate verbal description and therefore listeners cannot be adequately prepared to perceive the qualities; access to perceiving such qualities is thwarted without a clear indication of what one is trying to hear, how one needs to listen to hear it, and where such a quality might be situated within the track.

It can be a challenge to discover sound qualities that have never before been experienced. Awareness of the possibilities, trusting that what has never been experienced is actually present, and the perseverance to keep trying is needed to begin to hear unknown dimensions of sound. Some qualities are more apparent than others; some qualities require a more developed awareness or a more concerted effort before they are perceived.

Learning to listen to tracks in multiple ways extends to perspective and dimension. Observing elements in records requires one to focus attention on many different levels of detail—individually, one at a time. As each level of perspective and dimension contributes to the track, each requires separate attention to its unique level of detail and activity. It can also be a challenge to listen at various levels of detail. Table 2.7 presented the levels of perspective and related detail:

- Highest Structural Level or Large Dimension; the Overall Texture
- Upper-Level Middle Dimension; the Composite Texture

- Middle-Level Middle Dimension; Basic-level, individual sound sources
- Lowest Levels, Small Dimension; Individual sounds, and activities within individual sounds

Individuals are inherently pulled to the perspective of the individual sound source and musical materials, the basic-level; as organisms we have adapted to instinctively identify sources in our environment. This also represents the surface-level detail of life experiences, and the relationship of self to the world that is reflected in the lead vocal of the song. Listening successfully at other specific levels, though, can be elusive—while listening to the overall sound one can be pulled into details of the middle dimension's lyrics, listening to the blend of the middle dimension one can be pulled toward an individual sound or instrument, and listening for the subtle workings of a timbre can seem impossible. The mind gets distracted and pulled away. The four primary levels of perspective listed are central to understanding a track; information about the track will be collected at each of these.

Identifying the activities of various levels of perspective requires holding ones attention there. Intention of listening at a specific level, and holding one's attention there requires continual monitoring of information, to ensure one has not shifted focus to another level of dimension. An intentional shift of focus and perspective is part of this process. The intention of where to focus, and the act of shift from one dimension to another is a valuable listening-related skill that can be developed; it will aid recording analysis in numerous ways. With competence in intentionally shifting attention from one specific dimension to another, other dimension shifts can be engaged; this can lead to learning to shift attention between elements as well as between dimensions. Ultimately listening deeply into sounds (as in the smallest dimensions) and listening to the overall quality of the texture (in the large dimension) might be engaged as comfortably as bringing attention to the basic-level of the vocal line (within the middle dimension).

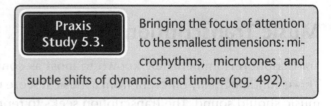

Praxis Study 5.2. Developing focus of attention and intentional shifts of focus and attention (pg. 491).

Praxis Study 5.3. Bringing the focus of attention to the smallest dimensions: microrhythms, microtones and subtle shifts of dynamics and timbre (pg. 492).

The lowest levels of dimension contain much nuance and subtle detail; these are the small dimension and the lowest levels of middle dimension. Few people come to recording analysis with much relevant listening experience listening at the lowest levels of perspective. This type of listening may be periodically experienced in practicing performance techniques, within recording production processes, during careful attention when listening to sounds, and more. Still, holding one's attention at the lowest levels of the small dimension is an atypical activity, although it is one to be cultivated.

The large dimension, or overall texture, is a strikingly different listening experience that is also not common for most listeners. Similar to the lowest dimensions—though conversely—the overall qualities of the track may also present challenges when one seeks to hold that experience within attention. Perception of this structural level brings awareness to qualities like the dynamic shape of the track or its evolving timbral balance.

We are most experienced engaging sounds within the middle dimension levels. Here sounds can be framed as either (1) a focus on an individual performer and/or the materials they present, or as (2) the interactivity between individual ideas and performers; these are directly related to common life experiences. Distinguishing between these two middle dimension levels allows for the examination (1) of the individual (middle-level middle dimension, basic-level) and an examination (2) of how the individuals relate to one another without placing more significance on any one (the upper-level middle dimension, composite texture). Conceptually, the composite texture is a way to perceive all of the sounds of the basic-level in a relationship of equal prominence. If any material, element or instrument is held in the focus of attention, that object of focus will be inherently prominent. To recognize how all of the sounds relate without this distortion, one listens at a level of perspective just above the basic-level of individual sound sources and materials; at this level all can be held in attention in equal significance. Only then can their relationships be accurately heard in the existing balance. This distinction is central to many analysis observations.

Transcribing Lyrics

Lyrics have typically been included with records, within its accompanying documentation, since *Sgt. Pepper's Lonely Hearts Club Band* (1967). Recently, if not included in documentation, lyrics are often found on an official website of the artist, their label or their publisher. As more artists are self-releasing material, official lyrics are less consistently included. Relying on third parties, such as websites authored by fans, may not prove reliable.[1]

Transcribing the words, lines and stanzas (etc.) of the lyrics as they appear on the record is the goal here; this is separate and distinct from the lyrics as performed, which is part of the sung vocal explored later.

Compared to transcribing music, transcribing lyrics appears straightforward; often it is, at times it isn't. The performance and arrangement can make words difficult to confidently identify, unknown words (or words with several meanings or spellings) might appear, line and stanza structures are commonly disguised, punctuation is assumed or goes unrecognized, some words may be unintelligible or easily misunderstood, and many more challenges will arise. Some interpretation of the sounds and structure of lyrics is to be expected.

Still, transcribing lyrics oneself is the most direct and active way into the words of a song. The process is simple listening and writing, re-listening and correcting. With the aid of digital editing and playback this process can be direct and productive—to the extent words are audible, articulation and punctuation clear, and structure detected. Some techniques of music transcriptions may be pertinent or valuable to unraveling the track's lyrics.

Music Transcription

Transcription is immediately known to most as converting the sounds of music into music notation. This is fundamentally different from a musical score meant for performance, which represents how the music should sound. The transcription seeks to represent visually what is present in the record (in as much as that is possible), how it sounds.[2] Any need we might have for transcription here is to serve the analysis—to make the information available for more careful study. Transcribing to assist analysis and transcribing with a goal to capture all of the aspects of sound are different objectives. Our objective is not to create a document that reflects the complete content of the record, and certainly not to create a musical score that can be used for performance.[3]

Transcription can aid in collecting information to bring it to hold still for study. One can hold the idea in one's hand, look at it carefully and study it; one can consider it without relying solely on memory, and process it more deeply. Materials can be compared to others, out of sequence, out of time, at different levels of perspective. Much more detail can be collected, as listening repeatedly refines observations, adds richness and accuracy.

In recording analysis each of the three domains may be transcribed—each in their own way—to still them for study. Transcription here can include the sound qualities of the recording and the lyrics as well as the music—as per the needs of the individual song, and the objectives of the analysis. Though the discussion that follows will emphasize music notation, these factors also apply to the other domains. Specific approaches and concerns of transcribing the recording, music, lyrics, and the vocal line will each be covered separately later.

Transcription within the three domains can allow one to find out what is going on within the track. The very act of focused listening, discovering and transcribing brings a level of familiarity with and a connection to the track's substance that is difficult to duplicate from listening alone. The looping sequence of listening, writing, listening, correcting, listening to verify can draw one into the analysis process deeply. This exploration and deciphering can be enlightening, and lead to an intimate connection that could not be experienced without exploring its subtleties, and notating them in some manner.

Allan Moore (2001, 35) offers: "[N]otation can act as a memory aid, enabling the aural experience to be (re)constructed." When something is written down it does not need to remain in active memory; the material can be edited, corrected, added to and referenced at any time in the future. Notation is also an efficient way to isolate instruments and parts, structural units and individual elements in the musical stream, to draw them close for study and to commit them to memory.

Peter Winkler (1997, 172) observes:

> Notation enables us to transcend the evanescence of music. Its effect is to neutralize time—to *kill* time. When we write music down . . . we apprehend it independently of the temporal stream. Thus we can focus on a particular musical event for as long as we want, scan instantaneously back and forth in time, make side-by-side comparisons of temporally distant moments. . . . viewing music from outside time, free of the constraints of the past, present and future.

The sound of the track is ever unfolding. To freeze its ideas and elements for more careful examination is in some ways liberating. When listening there can be a sense of urgency to absorb and remember as much as possible before the fleeting moment has passed; notation can free one to ignore all but specific materials, knowing all others can be included later. In this way, the listener can seek to find what is going on deep in the track, as details emerge that a casual listening could not reveal—details that are integral to the track, but that might otherwise evade discovery.

This all, of course, assumes the transcription actually reflects the sounds within the track; a state that is not entirely possible.

Perils

Transcription creates challenges and brings difficulties as well. "Transcription is not an innocent activity" (McClary and Walser 1990, 282). Its results can bring severe consequences that may not only limit the effectiveness of the transcription; it may actually distort the material, and thereby diminish the analysis. Indeed, sometimes it can be best not to transcribe materials, as "fixing a piece in notation may be not only unnecessary, it may be an actual impoverishment" (Winkler 1997, 173). Transcribing is a reductionist activity; it summarizes an experience of the specific performance captured on the recording in order to have it conform to a notational system (whatever the system). Nuance and detail is lost, and perhaps its essence distorted or its substance transformed.

Notation systems themselves are inherently flawed and incomplete. Western music notation has great limitations in presenting the sounds of music, but it was not designed for this purpose. Wilfrid Mellers (1973, 15) observes:

> [I]n reference to Beatle music (and to most pop, jazz, folk and non-Western music) it [music notation] may be not only inadequate but also misleading: for written notation can represent neither the

improvised elements nor the immediate distortions of pitch and flexibilities of rhythm which are the essence (not a decoration) of music orally and aurally conceived.

Western music notation at its core is an incomplete set of performance instructions; it most clearly represents pitch and rhythm, and which conform neatly to precise grids within written form. Tempo, dynamics, arrangement, expression, instrumentation, variations in timbre from performance and all others are represented with arbitrarily assigned symbols and verbal indications, and are less precise representations. Thus, the notation does not reflect the sound of the performance. The act of bringing a performance into staff notation

> shoves its pitches into Western diatonic patterns—with important inflections showing up as irrational deviations; and to try to "capture" its rhythms in the square metric system of Western classical music usually either winnows out whatever was interesting about the piece in the first place or reveals the piece as a hopeless tangle of ties and switching meters.
>
> (McClary and Walser 1990, 282–283)

Western music notation cannot easily flex to match the nuance of pitch and rhythm represented within much popular music performance. Capturing some aspects like the flexing of the beat and interactions of parts in the all-important rhythmic feel or groove may well be impossible at times. The aspects of the record Western notation does not directly address are some of the most important aspects of popular music; "those elements which listeners tend to find most interesting in popular music and which most nearly capture the music's particular strengths (rhythmic and pitch nuance, texture and timbre) are impossible to accurately notate. . ." (Moore 1997, x).

There is "an absence of a standard, easy visual representation" for vocal quality and timbral qualities of any instrument; "conventional music notation is helpless here" (Moore 2001, 35). The matter of vocal and instrumental timbres is a significant one. While an approach to a visual representation of sound quality and an approach to discussing it is offered within the domain of the recording (one which could be applied to the other domains as well), it is not easy to accomplish and by no means standard; still 'sound' is at the core of popular song, and begs to be engaged—visually represented or described. Describing the materials might seem a better choice to notation, though this, too, has its issues.

It must be obvious by now that transcription can take considerable time and effort to accomplish—even if only making incomplete sketches. One's skill levels in hearing within the various elements and domains will impact the accuracy of the transcription, as well as the time involved. Knowledge of and experiences with music, the elements of recording and the sound qualities of lyrics also impact accuracy and time/effort factors.

It is often difficult and sometimes nearly impossible to hear inside the fabric of the record. Real perception issues exist, such as trying to determine the notes present in a piano chord within a complex musical texture. Psychoacoustics can inform some of these difficulties, but does not provide easier access to the materials. There can be uncertainty about which notes are being played, and where they are placed against the metric grid; pitch and rhythm cannot be determined with absolute certainty. Instruments playing in unison with others have timbres fused and disguised; this can happen when harmonically related as well. These are but a few examples of many, and similarly impossible perceptual situations abound in the recording as well.

Then the question becomes what is possible to hear that is being missed by the analyst, and what must be accepted as ambiguity; whether brought about by blending, combination tones, masking, or something else may be helpful to recognize, but might not be worth the effort. Ambiguity within the sound, the skill-level and experience of the listener all conspire against detail in the transcription. Peter Winkler (1997, 193) has observed these challenges:

[E]ven the most scrupulously detailed transcription is full of guesses, suppositions, and arbitrary decisions. . . . 'What notes are being played and sung?' can never be answered definitively. It is a mistake to think of a transcriber as a scientist, objectively recording aural phenomena.

Lastly, music notation can be unwieldy to use, even for those with much experience and skill in music dictation. One should expect to find other approaches to music notation, and the ways to notate the recording and the lyrics at least equally involved and uniquely complicated. Just as an excellent control of music notation can be valuable, so too will be a sense of timbre, as noted by Allan Moore, above; timbre analysis and descriptions have many applications in the recording, in vocal phonetics and sound qualities, and in music.

Tools and Devices, and Inventing Notations

The most valuable tool for transcription is a well-trained and informed ear. Knowing how to listen effectively, knowing what one is seeking to hear, and knowing the options and mechanisms of how to accurately write down what was heard are part of this trained and informed tool. Experience taking music dictation is valuable; it is a skill that can be applied toward engaging the elements of the recording as well. Some practical techniques have been devised to assist dictation and transcribing music. Pitch matching to decipher melodies, searching out chords and progressions on a piano or guitar, tapping or clapping out rhythms and singing to replicate vocal sounds can all supplement or carry the process; other performance and deductive techniques can be discovered by individuals (what works for them individually) to identify what is going on in the sounds—and to simultaneously improve their listening skills.

A digital audio workstation (DAW) or sound editing software can likewise be used as a tool to improve listening skills as well as to aid transcription. It can 'stretch time' (*ibid.*, 174) to allow materials to be heard more clearly, especially helpful when slowing the playback speed without altering pitch or timbre; details will go by at a slower pace, altering tempo without substantially altering pitch or timbre. Portions of the audio range may be filtered (diminished or removed) to reveal a particular element or material by reducing interference and/or masking by other sounds. It can isolate sounds, sections, phrases, etc. for study; it can loop them for ease of examination; even reorder materials from throughout the song to place them side by side. Numerous other uses will arise. As a tool for educating the ear and improving listening skills, it can be used to aid in recognizing and transcribing many recording elements and their activities as well as music and lyrics. DAWs can highlight the spatial elements of the recording: allowing one to become aware of timbral detail and definition, provide a sense of lateral spaciousness and discrete boundaries, and might provide ways to examine environments. A DAW can provide access to the inner workings of timbres—a skill central to the analysis of many elements of the record, across its domains.

A potential concern with using the DAW for learning and transcribing is a diversion from a focus on sound; it is very easy for one to become overly engaged and reliant on sight and what is on the screen, and to lose focus on hearing the track. Further, one might see items that one cannot hear—and that have no place in the sound of the track. This concern also applies to spectrograms, waveforms, and other visual representations that are directly generated from audio through sound analysis devices and software, though they, too, can be of assistance.

Spectrograms can be "effective in refining and focusing the listening experience" (Cook 2009, 225). They represent sound in three dimensions: time (left to right), frequency (bottom to top), and amplitude (with color or degrees of shading in black and white). Spectrograms appear in many different forms, depending on formatting and sensitivity settings of the device or software; though these three dimensions are unchanged, different qualities can be generated.

David Brackett (2000, 27) used a spectrum analyzer to "freeze" the musical surface into photos of the spectrum of the sound over time as an aid in transcribing "the most prominent aspects of the musical surface and to comment on the melodic process, rather than to search for hidden relationships between different components of the musical texture." "The spectrum photos record all the sounding physical vibrations present in the recording" (*ibid.*, 65), and in effect display the timbral balance of track. These photos were able to provide some visual clues as to the content of vocal sounds, that might not be possible with other musical textures. The photos were used effectively to examine and to illustrate the performance styles and vocal qualities of Bing Crosby's and Billie Holiday's recordings of "I'll Be Seeing You." Among the pertinent information discovered,

> The spectrum photo also reveals many differences not captured in the transcription to staff notation; this includes the great variety of vibrato and pitch-bending employed by Holiday in contrast to the almost constant wide vibrato of Crosby. Although Crosby does occasionally "scoop" up to pitches, he usually holds the pitches after the initial attack; and he employs a far smaller number of vocal inflections than Holiday.
>
> (*ibid.*, 67–68)

Serge Lacasse (2010a, 2010b) has incorporated spectrograms into several detailed studies of language and vocal sounds within recordings. He offers: "In the absence of any comprehensible system of notation for paralinguistic features, an aural interpretation will be relied upon primarily, though with recourse to spectrograms which offer a useful visual representation of some of the paralinguistic features encountered" (2010a, 231). A spectrogram can illustrate timbral qualities of vocal sounds, as combinations of this frequency and amplitude information over time. Lacasse uses this to illustrate and analyze paralinguistic sounds, serving to engage the emotions and expression they produce. Spectrograms serve to replace notation; they are able to represent visually exactly what is happening acoustically and allow access to the sound itself (though, again, it is important to remember what is represented visually often does not translate accurately into what is heard). Nicholas Cook observes

> [T]heir attraction is that in principle all aspects of sound are present in them; the downside is that in practice it may be hard to extract the information you want. They are most useful for homing in on the details of performance—the unnotated nuances that are responsible for so much of music's meaning.
>
> (2009, 226)

In some instances, this approach may not be effective in bringing clarity to the content of some vocal lines, or of various sounds or sections; spectrum analysis of records is not a universal source of information (and Brackett clearly did not present it as such).

Unless one has access to the source files/tapes that generated the record, the visual representation will contain all the track's sounds; this is the track's mix. Seeking information about an individual instrument (even a prominent lead vocal) may yield limited results—these results may provide some important insight or information, as we have just learned, but may also be misleading or difficult to interpret. This is entirely dependent upon the context of the track. The process of trying to interpret sound through a visual image rekindles thoughts of the musical score, and the difficulties in imagining sound from the visual—and how one's tendency turns to evaluating what is seen instead of what is heard; this holds for the X-Y graphs that will be offered for recording elements as well. Approached with caution and remembering our goal is to understand the experience of sound, spectrograms can be useful. "When they are integrated into the working environment of studying recordings . . . they help to transform listening into analytical interpretation" (*ibid.*).

Waveforms are generated by software programs and within DAWs, and illustrate dynamic alternations of frequency over time. These have been used successfully to assist in observing sounds. Anne Danielsen (2010b, 2012, 2015) has made extensive use of waveforms to transcribe microrhythms, as they can be very helpful in parsing out dense percussive textures. In Stanyek (2014) she notes:

> [W]aveform representations . . . are very useful for identifying things that you're not able to grasp because they pass so quickly . . . new software tools are *very* useful in order to kind of freeze time. You can map sound to the visual representation and figure out what kind of pattern it is, or where it is actually placed in time, or what kind of structure it has.

Waveforms may be used alone or in combination (aligned in time) with spectrograms images as a relevant option; the timbral information of spectrograms contrasts with that of the waveform representation to reveal sound multi-dimensionally.

Automatic transcription appears within the discussion below, in making nuanced observations in rhythm. It is not a replacement for listening, but can add a different perspective of pitch and rhythm. It does not add substance to observations for other elements (dynamics, timbre, texture, etc.). Some approaches of automatic transcription for pitch and rhythm will have predetermined interpretations of these elements; the authors of those transcription algorithms have made choices that add another layer of interpretation. This predetermines the set of choices available when specific characteristics of the sound are translated and transferred into notation. Choices are made by the process that define the raw data against a few sets of preconceived possibilities—not unlike the limited options of some traditional music analysis methods.

Some analysts have devised their own approaches to notation. Among these are the graphic renditions of gestures offered by Richard Middleton (2000a, 111) that are reflected in Figure 5.4, and also the arrows showing alterations in rhythm and pitch incorporated by David Brackett (2000, 63) that are incorporated into Figure 5.9. New or invented notations can be helpful, effective and relevant, when carefully conceived, and the analyst might choose to explore this matter themselves.

Descriptions

Description can often replace notation, or function independent of notation. The reasons for this are many. Some aspects of sound defy notation, and an analysis may well include dimensions other than sound. At times description alone can reveal the essential character of the materials of music and lyrics; at others notating material is unwieldy or distorts the materials. Perhaps it is a choice related to the skill level of the analyst, or that the texture makes transcription difficult or impossible. Incorporating disciplines outside the track into an analysis will assure descriptions will need to be used, in order to embrace their substance and benefits to the analysis. Notations of music, lyrics or recording all have limitations, and simply might not provide a suitable platform to represent certain materials—especially when domains and disciplines intersect and overlap.

Bringing language to the track is an interpretation of its sounds and materials, from the listener/speaker's unique perspective—their own 'subject-position.' Eric Clarke introduces this concept from cultural studies and the context of film. Every viewer has a unique interpretation resulting from the "individual's particular circumstances, experience, background, and aesthetic attitudes, as well as the specific . . . occasion . . . But . . . there is a limit that can be attributed to properties of the film itself—understood within a certain shared cultural context" (Clarke 2005, 92–93). The personal and the cultural impact our interpretation, as well as skill levels at assessing sound also influencing the success of talking about sound. With our limited vocabulary specific to sound, our attempts to adapt or adopt terms from other modalities or experiences are fraught with difficulties and distortions.

Though often desirable, describing lyrics, music and sound can be difficult. Many words are needed to replace what might be represented in a visual; writings can quickly become repetitive and tedious; descriptions themselves can be misleading, filled with imprecise language. Descriptions can color the materials with thoughts and language better suited to creative literature or a personal blog than to an academic writing that seeks some level of depth, with some level of commonality and objectivity. It can be difficult to explain the substance of sound, of musical materials or of the lyrics without including one's impressions *about* them; yet those impressions are exactly what one wishes to avoid while collecting details (making observations) of what *is* present. This is the duality of content (what is present) and of character (interpretation of what is present). The techniques of observing timbre and pitch areas (in Chapter 7) might open one to this substance of 'what is present,' and enhance understanding to enrich descriptions. Describing without distorting the sound or without relying on personal impressions is a challenge that will be engaged often throughout the remainder of the book.

'Sound' is central to the recording, and to the recorded popular song. Talking about sound and describing sound will be covered in greater depth in Chapter 6, and timbre analysis in Chapter 7. Those concepts can be applied to the 'sounds' of lyrics and of musical instruments. Descriptions of musical materials and their activities and of the lyrics are separate matters from timbre itself. Descriptions are based on 'what is happening' and 'how' they are pertinent—and relate to evaluations and conclusions explored in Chapter 10.

No matter how skilled the writer or speaker, to put a perception into words is to filter its richness. Description is reductive, and description is an interpretation. Language summarizes an experience, eliminates detail and dimensionality. The resulting description loses something, perhaps much, of the original experience. Depth, detail and character are likely to all be diminished, if not erased. Experiences of sound defy language—by their very nature, and from our limited vocabulary. The experience is lost within the description—and, yes, also within notation. To describe an experience is to define it; we define it by what we perceive, what we understand and what we are capable of putting into words.

Linked to Analysis Goals

To be valuable, a transcription should provide information that aids the execution of the analysis, and that facilitates and illustrates its discussions. The ideas, conclusions and discoveries one wishes to offer might benefit from illustration in notation. These can include information central to the goals of the analysis that are suitably explored with notation. There should be a purpose to transcribing materials: from exploring elements to determine their relevance and materials to examining in detail materials clearly identified as the substance of the track.

For example, an analysis that seeks to define the content and sound qualities of lyrics may pay little attention to some musical qualities, and an analysis that focuses on pitch and harmonic relationships may pay little attention to lyrics.

An analysis that brings its attention to how the recording contributes to the record will seek a significant breadth and depth of understanding of the elements of recording. This can bring one to create detailed graphs or charts of the elements; to explore their characteristics, shapes and motions, etc.

The analysis could also seek to identify how those elements deliver the lyrics, and their impacts on the character and context of its performance, potentially bringing a need to transcribe areas of lyrics. The impact of the recording on the musical materials might also be objects for examination; perhaps this will require some transcription of musical materials, perhaps not. A clear objective of the type of understanding and level of detail of the analysis will guide decisions related to transcription—how much detail to uncover, and which areas to examine.

All sounds need not be transcribed for an analysis. An analysis can be thorough, meaningful and effective without converting all sounds into some type of notation.

Certain features are pulled from the sound of the record that are of particular interest for careful examination. This occurs once evaluation has begun, and transcription and analysis are no longer separated. Prior to seeking this level of detail and understanding, transcription is used to gather basic information: what chords are present (not their functional relationships to a tonic), what are the basic rhythmic patterns (not the subtle aspects that will be sought later), what pitch do I believe I am hearing in this performance (ambiguity acknowledged, for follow up once evaluation begins) and so forth. Once some evaluation has established such things, a transcription can include "looking for specific things . . . and I found them. . . . As the transcription process went on and I focused on more specific questions . . ." (Winkler 1997, 194). It is important to recognize that this establishes a circular function of sorts, where more transcription may take place after evaluation has started.

COLLECTING OBSERVATIONS

To collect observations is to establish the pool of information that will be evaluated in the analysis. These observations are inherently the analyst's preliminary or first interpretation of what is heard. Mindful of one's biases and skill level, one might be able to identify what is present without distorting it; one might look beyond what they expected to find, be open to recognizing unknown qualities, and so forth—though this is a challenge that cannot be entirely achieved. Collecting information on the track must reveal the unexpected and the unknown, if its essential qualities—what makes the track unique—are to be found. Holding a sense of one's inclinations and biases might minimize distorting the data, and may keep data collection as neutral as can be possible.

Before collecting observations begins, some basic questions will be engaged to provide some guidance. These can establish a meaningful way to determine the elements and level of detail to be engaged. One should expect the path to collecting observations will take unexpected twists, and lead to exploring some elements/materials more than expected, and some less. The questions might be:

- What is the goal(s) of the analysis, and what must be examined to meet them?
- What is pertinent to this stream of inquiry, and what is going to be of limited significance?
- What materials/elements need to be explored to understand the essential qualities of the song, as they pertain to the goal(s) of the analysis?

What speaks most prominently from within the track? Those features that draw one's attention are very likely to be significant. The obvious is typically important, and might be explored first as a gateway into the more concealed or less known.

Expect this observations process to turn in to a circular process after a critical mass of observations has been established. As one observation leads to another, one will be drawn to other areas and collecting information on unexpected elements/materials; these might lead to some adjustment of the analysis. As initial evaluations gain depth, it is common to determine other materials need observations, more detailed observations are required of some elements/materials and some might have already been examined adequately, perhaps excessively.

Just as every track is unique, every analysis process will be unique (in some way) if it is responsive to the track. Observations within each domain will also be unique in some way. The next sections will identify the steps to collecting and organizing observations that, if approached without bias, will reveal the essential characteristics of the track:

- Establishing appropriate large dimension and middle/small dimension timelines
- Identifying sound sources, with a general identification of materials and lyrics
- Processes and concerns for collecting observations within each domain, with music and lyrics explored separately, and recording explored in the succeeding chapters

Constructing Timelines

Structural timelines can be important organizational and analysis tools. These are distinct from the timelines that will appear in recording element X-Y graphs. Here, timelines are their own graphs or figures; they display information about the track against its structure (as time). Data recognized in observations is added to a timeline, organized without evaluation simply by placing the data against time; the information has no meaning or function, it awaits later evaluation. Once evaluation starts, patterns might emerge visually or from examining data; functions can become evident as large spans of time can be condensed, large amounts of data can be observed rapidly, if not simultaneously.

The 'X' axis of the timeline is always time, unfolding left-to-right. The axis may be divided into measures, groups of measures, or beats within measures. Once established, the resolution of this time axis remains constant, so as not to distort the consistent speed of unfolding time.

Structural timelines may take many forms; they can be applied to all perspectives, and can incorporate both observations and evaluations of elements in all domains. The vertical, 'Y' axis will dedicate space (a level of the graph above or below the line) for each aspect of the track included; this will not move, so the graph has consistency and one knows where to find information. The 'Y' axis will be labeled at the start of the graph as to what it includes at each level of height.

Timelines may be dedicated to large dimension structure alone, as seen in Figure 5.1. They can display structure in general terms, or present great detail. Structural divisions at or below the measure level and extending up through the highest strata might be visible simultaneously. Figure 2.1 illustrates several structural levels of imbedded strata within "Every Little Thing" (1964).

Timelines have great potential as analysis tools. As observations are collected, the nature of that information may be noted on the timeline; alternatively, when information might be more complex (such as details of a melodic line), the timeline may reference notes or notations on the material in an outside listing or outline. The large amounts of diverse information available (and visible) in timelines aids evaluation processes directly. Information of all sorts can be related, as they are located vertically against structure. Materials from one section or level of perspective can be compared to others in any structural level or location. This ready access to large amounts of information will be useful in a great many ways.

Adapting Timelines for Each Dimension

Timelines can be dedicated to specific perspectives, or structural levels, such as large dimension, middle dimension or combining middle and small dimensions. Within these dimensions, a timeline might be dedicated to a single domain and its elements, or may combine domains.

A large dimension timeline will, by definition, reflect the track as a whole. The time units displayed will be measures or groups of measures, as appropriate to clearly show the major structural divisions and any large dimension materials it contains. Elements at this large dimension will be listed in the 'Y' axis; they will reflect a single domain being observed, or may mix elements between domains (such as including lyrics within a music domain timeline). Elements may have fixed levels for the entire track; for example, a single tempo of the track would be a fixed element, as would the track's perceived performance environment. Some elements will exhibit variable activities. In some instances, this means a few changes over the course of the track, such as key changes between major sections; these tonal areas can be readily shown within the major structural divisions. Other variable elements might require a higher degree of detail in the timeline even at the large dimension; for example, the track's pitch density may continually change and require a resolution that makes changes clear within each measure.

The divisions of lyrics into stanzas might be noted on the large dimension timeline, identifying the first line of text on major structural divisions. Defining the functions of sections—such as verse or chorus—is best left until the evaluations stage; assessments of any type are typically most effective when delayed until all information has been collected. The large dimension structural divisions, those that separate major sections (such as at the verse/chorus level, or higher-level division) should be noted when they are audible (and they typically are) by a vertical division in the timeline.

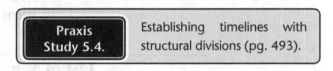

Praxis Study 5.4. Establishing timelines with structural divisions (pg. 493).

A middle dimension timeline can contain activities at one or more levels of perspective; it might be appropriate to locate composite texture information above the timeline and sound-source level materials and activities below. It can also be valuable for a timeline to display middle and small dimension materials in a similar way. Middle dimension timelines will be divided into measures or half-measure, as at least some of the material they graph or represent will have the potential to change often, or to start/stop at any point within a measure. The timeline attempts to depict where activities take place; the exact nature of those activities is most effectively articulated on accompanying pages—especially within the working timeline that compiles observations.

Middle dimension timelines might represent a single section or several adjacent sections (such as an introduction and the first verse). It might encompass the entire record. The individual sections of a track-length timeline might still be compared with little effort. The timeline in the middle/small dimension is most useful for data collection. Figure 5.2 illustrates how melodic phrases, chords, sound sources (timbres and arrangement), verbal space, lyrics and structural divisions might be placed on a single middle dimension timeline; the elements contained in the graph can be changed to best suit the track.

Timelines are most useful when time divisions remain consistent once they are established; any inconsistencies distort the rate of time passage and confuse data. This is disorienting to the reader of the analysis (perhaps confusing the analyst, too). Moving between timelines is preferable to changing a timeline's 'X' or 'Y' axis. Additional timelines can be incorporated into data collection or analysis as the work unfolds. The information they contain, their perspectives and lengths can vary markedly one to another, but should remain fixed once established.

Timelines can also be used to illustrate evaluations and conclusions (this will be explored later). In these instances, timelines may have a more unique format in order to most clearly present materials and relationships under discussion.

Identifying Sound Sources and Materials

It makes good sense to begin our analysis by using the perceptual process that allows our auditory system to sort out and differentiate sounds from the complex mixture of acoustic energy in our natural

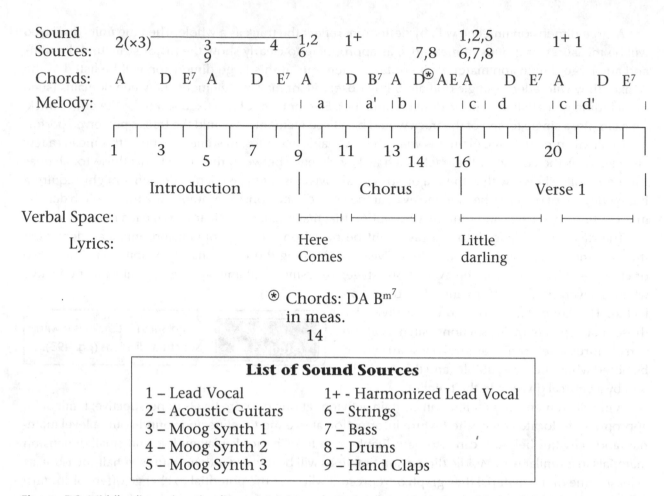

Figure 5.2 Middle dimension timeline containing select elements of music and lyrics; beginning sections of "Here Comes the Sun" (The Beatles, 1969).

(or created) listening environment. Auditory stream analysis studies how our brain can separate sounds from the mixing of all sounds at once; which sounds, from which source, identified as connected in a unique stream allow the listener to track single lines within the texture. Otherwise, if, for example, we heard several people talking at once we would combine the separate streams into new words and sentences (Bregman 1990). Identifying the track's sound sources and the basic-level materials they present not only pulls the listener deeply into the track, this process is also our natural tendency, and one of the basic ways we understand what we hear.

These processes provide an opening into the essential qualities of the track, without encumbrance of seeking to identify detail, function, meaning and so forth. The sound sources in records typically have unique characteristics and warrant individual attention; these include all individual instruments and voices present, as well as extra musical sounds and effects. The result will be a listing of the sound sources of the track, and some sense of the materials they present.

Depending on analysis goals, this listing of sources may need to be complete and detailed, or selective in identifying only prominent sources, significant instruments/voices, or some other limited number that serves the particular analysis. Identifying all individual instruments in groupings (such as all cymbals and drums in a drum kit) might be needed for certain observations, and not for others. Listening for and identifying sound sources can often be readily accomplished in textures with a few instruments or mixes that leave sources exposed. Depending on how sources are combined as well as the

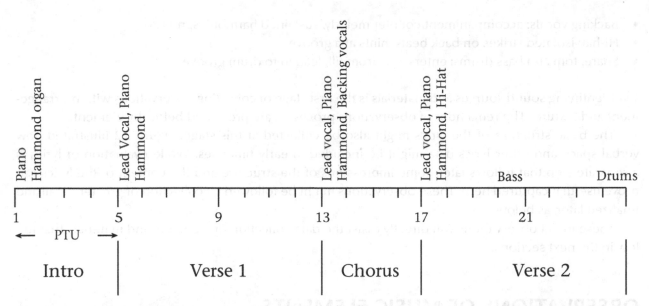

Figure 5.3 Timeline of the opening sections of "Let It Be" (The Beatles, *1*, 2000 version) with sound sources identified in each section.

number of sources, instruments and voices can fuse, blend, mask, bury or otherwise disguise certain sources. It is not unusual for a source to be 'discovered' only after many hearings of the track.

Naming sources is sometimes direct and simple—such as identifying an instrument that appears in only a single form, or recognizing the lead vocalist by name; this is not always so. It can become confusing to identify sources as more instruments are added and groupings of sources are formed; then it can be challenging to name them. Many instruments can be identified by their unique sounds (such as 'Hammond organ'), by the type of instrument (electric bass) or by the instrument plus the type of material presented ('acoustic rhythm guitar'). With more than one independent instrument of the same type, sources can be numbered by order of appearance ('synthesizer 1' being the earliest), by an effect on a source ('flanged electric guitar') or by range with the highest instrument most commonly the lowest number ('tom 1' being the highest). A group of sources functioning together as a single idea might be labeled similar to 'background vocals' or 'string section.' If a source cannot be recognized, 'unknown sound #1' would be appropriate (noting timing of the sound's appearance) until it has been successfully identified.

The general content of materials presented by sound sources may be noted now. This should not be detailed here, but rather provide a bit of definition between sources and delineate their interaction. This will give some guidance as to which materials are primarily melodic, rhythmic, harmonic, rhythmic/groove or sound related—while recognizing these observations might later be refined or revised.

Figure 5.3 is a timeline of the opening sections of the *1* (2000) version of "Let It Be" divided into song sections, and showing the main structural divisions and the four-measure prevailing time unit. Sound sources are placed against the timeline; sources present throughout a section are listed at the beginning of each section, and sources that enter well into a section appear near the point of their entry. A general idea of each source's materials might appear either on the timeline or in an accompanying listing, where descriptions such the following (pertaining to the Figure 5.3) might be clearer:

- Piano: accompaniment, rhythmic chords and bass line
- Hammond organ: contextual accompaniment, sustained chords
- McCartney, lead vocal: primary melodic materials and text

- Backing vocals: accompaniment/counter melody, sustained harmonies, no text
- Hi-hat: isolated strikes on back beats; hints at a groove
- Snare, tom and bass drums: enters as a drum fill, lead in to drum groove

Identifying sound sources and materials is the first stage of collecting observations within 'arrangement and texture.' The remainder of observations processes are presented below by element.

The basic structure of the lyrics might also be collected at this stage. Figure 5.2 illustrated how verbal space and other lyrics data might be included in early timelines. While evaluation of lyrics is a separate step that follows later, some impressions of the structure and the content of the lyrics can prove useful if captured here. These observations might be followed by collection of greater detail and finalized later, as below.

These initial observations will directly assist the data collection of elements and materials that follow in the next sections.

OBSERVATIONS OF MUSIC ELEMENTS

Following will be six descriptions of analyses of popular songs, or of writings that contain song analyses. These examples are offered to guide one in determining what to evaluate in the music to meet the goals of an analysis. Further, these might provide some direction to the acts of identifying what must be transcribed and of choosing what will be transcribed in order to support the observations process or to articulate an analysis.

David Brackett (2000) analyzes many dimensions of Elvis Costello's "Pills and Soap" (1983). In his analysis he includes a transcription into staff notation of the song; in it the synthesizer, voice, piano and percussion parts are clearly identified. The analysis covers underlying harmony, harmonic motion (chord progressions and implied tonalities), materials and developments in the vocal melody, its octatonic pitch collection, ostinato lines, percussion parts, vocal qualities (aided by spectrum photos), range and tessitura, grooves, and melodic gestural shapes. Observations needed for these evaluations identified musical instruments and parts, structure, chords, rhythms, instrumental melody lines (bass, synthesizer and piano), vocal part and vocal timbres, percussion rhythms and cross rhythms of grooves, pitch areas and ranges of parts, and likely more. The transcription's detail does not include rhythmic and pitch nuance of the performance, though discussion addresses timbral qualities of the vocal performance in some depth.

In "Analytic Methodologies for Rock Music" Lori Burns (2008) focuses her evaluation on the governing harmonic progression and on the bass and vocal lines (to examine voice leading). These evaluations are central to her analytic method, which is the topic of the writing; the analytic method is demonstrated through her analysis of Tori Amos' song "Crucify" (1992). Observations that would have been performed to lead to these evaluations would have included identifying chords, notating the bass line and the vocal line—though likely much more actually occurred. The level of detail of these transcriptions did not need to include the nuance of the performance to meet the defined goals of the analysis; the transcriptions gave rise to presenting the normative progression, voice-leading graph and reduction staff notations that are central to the method (ibid., 74).

Stan Hawkins (2000) examines harmonic materials and organization of Prince's "Anna Stesia" (1988). Observations required for this primary goal include chords and chord sequences, structural divisions and melodic lines. The full breadth of the analysis, however, extends well beyond 'harmonic analysis' and presents evaluations of phrase structure, texture, bass lines, vocal timbre and expression, dynamic contour, and more. Included in his study are sonic qualities of the production: vertical density, recording's elements related to signal processing and mix elements, and other subtleties within the production. Few transcribed materials exist in the presented analysis; it is obvious a considerable amount of detail was extracted from the track that was not presented in his writing.

Observations of pitch materials' melody and especially harmony form the backdrop of Walter Everett's (1986) analysis of "Strawberry Fields Forever" (1967). Though "Fantastic Remembrance in John Lennon's 'Strawberry Fields Forever' and 'Julia'"[4] examines other aspects of the song in some detail, considerable attention to the harmonic vocabulary provides strong cohesion to the analysis. Other areas covered include the 'fantastic remembrance' basis of the article, the sound qualities of the recording and production techniques, orchestration, voice leading, connections of text and music, and meaning. Some sizeable sections of the song are transcribed in considerable detail; included as outgrowth of the transcription process are Schenkerian graphs of middleground and foreground structures. Performance nuance is not included in the score; though rhythms are detailed to provide clarity of materials, performance nuance is not central to the goals of the analysis.

Contrasting "Fantastic Remembrance" with Everett's discussion and analysis of "Strawberry Fields Forever" in his *The Beatles as Musicians: Revolver through the Anthology* (1999, 75–84), we witness a shift of breadth and focus. The discussion broadens to include the compositional process and the development of the lyrics and some musical elements. A more detailed examination of making the recording follows, that discusses vocal and instrumental sound qualities as well some qualities of the recording. The musical elements are not explored in great detail, but the coverage is all pertinent and insightful; also acknowledged are the sound qualities of instruments and vocals, melodic ideas and lyrics (at times acting as a unit), harmonic progressions and ambiguities of surface harmonies, tonal areas and arrival points, and bass lines. Observations undoubtedly included pitch collection for melody, bass lines and chords, rhythms, instrumentation, percussion rhythms, and structure. Other observations certainly informed what is present, though they did not become topics of focus in the writing. Many transcriptions were included, all of the compositional drafts that provided support for his discussions of the compositional process and sequence; these do not contain performance nuance, as that is not relevant to his discussion.

In "Determining the Role of Performance in the Articulation of Meaning: The Case of 'Try a Little Tenderness,'" Rob Bowman (2003) looks deeply into four records of this Tin Pan Alley song.[5] These tracks are performed by Bing Crosby (1933), Aretha Franklin (1962), Sam Cooke (1964), and Otis Redding (1966), and contain many performance technique and stylistic differences. The analysis carefully studies the vocal lines of the four performances for melodic interpretation (relationship to the original sheet music), pitch, rhythm, timbre, "timbre, dynamics and playful voicedness" (*ibid.*, 115–118) and speech singing. Other aspects examined include structure, harmonic setting and instrumental accompaniment, and, of course, meaning. A significant number of vocal transcriptions are present, some showing nuance in performance related to rhythm and pitch. Observations would have included transcriptions of some vocal lines, and collecting information on chords, rhythms, vocal timbres, dynamics, accompanying instrumentation and musical parts, structural divisions of timeline, and likely more. The evaluation process also utilized the original sheet music—a step not common in popular music analysis, but of relevance here as it is the original medium for this song.

Melody

At the observations stage we attempt to access the melodic materials without becoming overly detailed. Without engaging the function of pitches, information will be examined and collected on the vocal melody and bass line. The crucial vocal melody and the bass line jointly establish the voice leading that is central to many songs (Burns 2008, 67). Riffs and ostinatos, and other melodic materials such as solos and background vocals, may also be engaged, as they are often significant.

A detailed transcription of pitch and rhythm of the lead vocal and bass line is often useful, but is not always needed for the analysis. Instead the melody might be conceived as a gesture, with shape and contour over time. This simplified approach of capturing some of the essential characteristics of melody might be useful here. Melodic contour might be sketched; using beginning and end pitches, highest and lowest pitches and sustained pitches in the phrase, a general shape can be established that can then be filled with further detail—for example any emphasized pitches, implied or outlined chords, and repeating shapes. The shape of melodic contour unfolds, and can increase in detail or transform into transcription, as desired.

Melodic phrasing of the lead vocal is typically important to collect. It can be marked on a middle dimension timeline to establish phrase structure, against the structural divisions. Phrasing of the lead vocal will coincide with verbal space (as in Figure 5.2), as it is a different way to observe the data percept.

By collecting information on melodic contour, melody might be engaged without excessive detail. This can be helpful, especially as a beginning point. Gestures of melodic contour might be mapped phrase by phrase. A reference of the highest and lowest pitch levels of the middle dimension section (defined above) might be identified. Central or significant pitch level(s) might be identified to serve as references for mapping the contour. The shape (from very general outline to a more detailed contour) of melodic movement can then add detail between these reference pitches without concern for specific rhythms, intervallic patterns, or pitches.[6] The gestural shapes of contours can illustrate tessitura, shapes and ranges of melodic gestures, indication of melodic complexity (patterns, speed, density), and so forth. Perhaps a tonal center will be perceived by experienced analysts, but tonality will be examined later in evaluations and the goal here is to collect information without concern for function. Figure 5.4 illustrates general melodic contours of Lennon's vocal in the first chorus and the first verse of "Strawberry Fields Forever"; each contour is a phrase, separated from others by silence. This figure could be more precise (showing specific pitch levels) if warranted.

Repeated pitches can be identified and incorporated into the contour, and can illustrate qualities that might later be found related to recitation or lyricism. Completing this process, melodic materials that might be recognized as significant—a prominent bass line or the main ideas of the vocal are typically readily identified—may be notated. These core melodic materials can be transcribed in a general way, to capture their basic interval sequencing and fundamental rhythms, without engaging evaluation. The nuances of performance will not be incorporated at this time; more detail can be added later

Chorus

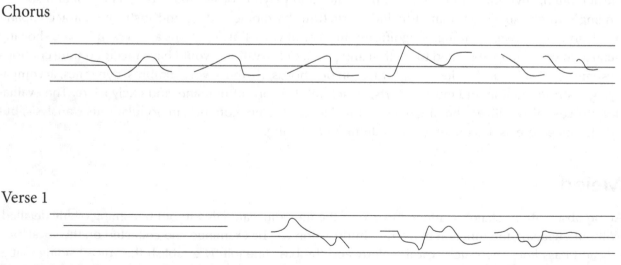

Verse 1

Figure 5.4 Melodic gestures illustrated as general melodic shapes and contours. John Lennon's lead vocal from the first chorus and first verse of "Strawberry Fields Forever" (The Beatles, 1967).

as needed. The purpose of this final step is to begin to accumulate more detail on melodies, without subtle characteristics or functions. All this activity occurs at the middle dimension, and some of the observations are towards the small dimension.

Melody may have implications related to tessitura, shape and gesture over larger spans. 'Observation' collects the unbiased information of shape, range, rhythm, intervals or pitches, etc. These and other qualities may be worthy of attention as well, perhaps listing variables and qualities in a typology table, or with melodic transcription or gestural mapping. Melody can have a shape that unfolds throughout a section of a song, or throughout the entire song. Attention might be brought to recurring ideas that seem to evolve to get a sense of information that may require attention. For example, in "All Along the Watchtower" (1968), Jimi Hendrix makes use of the introduction and four interludes to shape an over-arching motion of his guitar solos. Albin Zak (2005, 634–636) explains how Hendrix builds tension in his solos by climbing from C_5 in the introduction, through E_5, G_5 and an ornamented G_5 in the first through third interludes, and finally reaching C_6 at the end of the song (the guitar's highest pitch), and establishing the song's dramatic and triumphant peak of arrival of Hendrix's cover version. We learn in Zak's 'evaluation' that a dissonant 'D' pitch continually appears and is resolved throughout these sections; while this evaluation is clearly ahead of ourselves, here within observations we clearly sense the tension of this pitch; our collection of melody data will reflect its presence, though at this stage would not necessarily identify its dissonance function and instability. This building arc is only evident when one brings attention to evaluating at the large dimension, though this evaluation cannot happen without some collecting of information on pitch content, peaks and shapes of the guitar solos.

Harmony

Much information about harmony is wrapped in function—and function can only be determined later, by evaluating harmonic content and its inherent directed motion (and the significant contributions of melody). In making observations, we are concerned with chords: chords by letter names and chord types/qualities, perhaps also including inversions. Chord information often prominently comprises guitar parts and keyboard parts. These sonorities might be determined by pitch/chord matching, alternating listening to the record with searching for the right chord, perhaps including the right voicing. Noting chord voicing in the correct octaves will help acquire the information needed for pitch (vertical) density observations, and later allow timbral balance evaluations. Guitar tablature[7] might provide helpful shorthand for some to collect guitar chord data, though transferring it into staff notation may at some point be needed for other analysis steps. Depending on the texture, this task has various levels of difficulty.

There are advantages to using letter names only as one does not need to engage the evaluation process to identify a key within discussions. Chord type and letter name removes all doubt about which chord is being observed; progressions can still be observed and noted without engaging chord functions or identifying mode.

Lloyd Whitesell (2008, 117–147) uses only chord name/qualities within his detailed discussion of Joni Mitchell's 'harmonic palette.' The absence of chord functions and of typical emphasis on tonal centers brings a sense of flexibility to the discussion: specifics on chord names and harmonic rhythms and progressions, without concern over tonal centers and chord functions that are not intrinsic to his explanation of Mitchell's harmonic language. This seems to especially work well in his discussion contrasting Joni Mitchell's original version of "Woodstock" (1970) with that of Crosby, Stills, Nash & Young (1970) (*ibid.*, 33–39).

There are practical concerns here as well. Using chord letter names allows more information to be collected in order to determine what key is in play. The roman numerals of harmonic analysis indicate a scale degree of a key and imply (or define) tonal function; calculating the key (tonal center) is not possible until study of the song has advanced and evaluation is in process. Separating observations from evaluations is simplified by merely naming chords; this can help minimize skewed evaluations drawn from incomplete or hastily interpreted observations.

Information is also collected on the rhythms of chord changes (harmonic rhythm). This is the simple pacing of chord changes; specific rhythms or rhythmic patterning will not be sought here. By placing chords on the timeline where they begin (or roughly so), this pacing is noted; patterns, speed and so forth will emerge later.

Cadence locations, or points of arrival or release of harmonic tension might be noticed during this process as well—although there should be no effort to identify function (which requires modal/tonal evaluation). These are typically apparent, and can be located without analyzing specifics. Finding these articulation points allows one to recognize harmonic movement and the phrasing they establish. Chord phrasing and the movement of departure and arrival should be treated loosely, as this will be refined significantly in evaluations, once tonality/modality has been established.

Arrangement and Texture

Collecting observations on the arrangement and texture begins by delineating the sound sources and the sounds of the recording, and recognizing areas of harmonic fusion or a composite rhythm creating a fusion of sources. Included in observations is relationships of source timbres of musical materials; this includes the timbral qualities of the performance and the pitch spaces occupied by the sound sources joined with materials. A dataset is generated that will later allow evaluations of how sources interact in these dimensions, and how they shape the track. These are among the fundamental observations of the arrangement. There is much alignment between arrangement and texture in music, and timbral balance in recording.

With identifying sound sources (as outlined above), observations of the arrangement began. Here, information on the presence of sources is verified, and the list of instruments/voices throughout the track is refined. Sound sources that are present within the structural sections are noted on timelines.

Connections of instruments/voices to general materials they present are identified next. Some sources were engaged during observations of melody and harmony. Remaining sources might be explored here; information on these other sound sources may be revealed by tracing how sources form groups and interact, and by the materials they present. All previous observations are refined to bring more focus to sound sources and their materials, and other pertinent sources identified. The sources are organized by general functions, among which are likely:

- Primary melodic parts—lead vocal and instrumental lines
- Groove elements—dominant rhythmic parts and often bass line
- Secondary melodic parts (such as backing vocals)
- Accompanying harmonic parts (often keyboards, guitars, string sections, etc.)
- Secondary rhythmic and timbral parts
- Thematic, riff or other defining gestures of sounds—whether melodic, harmonic, rhythmic or timbral

The final determination of a source's place in the musical fabric can be delayed until the evaluations stage, as will finalizing the number and types of musical parts in the fabric. Some general observations are possible to make here, though, and they will prove helpful. An observation of the type of texture (melody and accompaniment, chordal, contextual groove, etc.) within each section (or sub-section) can be added to the timeline; this provides a sense of the number and nature of materials, and a general sense of how the sources are used within the fabric.

The combinations of sources that result in an arrangement will be apparent by the listing of sources in each section. Groupings of sources play a significant role in the timbral qualities of the ensemble; it can be a valuable observation to identify those instruments/voices performing the same or similar parts, sources that are synchronized in some way, or those that are interacting in some way. Often a set of dynamic relationships between the different sources contributes to this ensemble timbre; these general levels and relationships could be noted here, and lead to further observation within dynamics. An example of combining a variety of different sound sources to sync or interact would be the instruments that comprise a groove (see below).

The concept of sources occupying an area of pitch that extends over a defined period of time is central to the arrangement and the texture it produces. Pitch density allows access to understanding many qualities of an arrangement in the music domain, as it will later for the recording domain in Chapter 7. Pitch density exists at the perspective level of the individual sound source and the musical materials a sound source presents. These are fused into a single percept, combining the substance of the musical idea with the dominant timbral characteristics of the source. The result is a sense that the source (its material and timbre) is occupying a pitch space or bandwidth (within the full pitch-range of the record) for a certain, defined segment of time.

Pitch density unfolds as 'scenes' of gestures created by the source's materials. It is mapped out as a series of snapshots of time, passing successively from one to the next. These time units are a syncrisis unit. They typically relate to the prevailing time unit or the phrase structure of materials.

Collecting observations on pitch density in the music domain uses a similar process to the pitch density as a recording element:

- Materials are observed as occupying a pitch space (a pitch area comprised of its material and timbre)
- Pitch space is identified by its boundaries and the interval between: (1) the lowest pitch present, or melodic contour, and (2) the highest significant partial of its timbre
- The interval between these two boundaries establishes its range.
- Changing levels of lower boundary over time (melodic motion)
- Changing upper boundary caused by movement of the lower boundary or by a change of timbral characteristics (such as those caused by performance expression, technique, intensity, etc.)
- Time span of the 'scene'

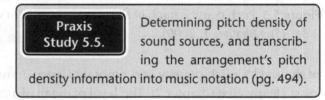

Praxis Study 5.5. Determining pitch density of sound sources, and transcribing the arrangement's pitch density information into music notation (pg. 494).

Collecting the basic information of pitch density begins with identifying pitch material; this may have been observed when examining melody and/or harmony, depending on the sound source. Attention is next drawn to the timbre of the source to identify the interval spanning from this pitch material to that of the last prominent partial within the timbre; this defines the lower and upper boundaries of the gesture. Pitch area and timbral quality observations will assist this process; these will appear in Chapter 7. Other variables might include shifts of the pitch density area brought about by changing pitch level of the materials, or shifting upper boundary brought about by changes in the timbre of the source (such as those caused by increased or decreased performance intensity).

As a recording element, pitch density is plotted against pitch registers on an X-Y graph. This practice can be applied here within the music domain as well. In practice, observations of pitch density as a recording element can replace observations of the arrangement in the music domain.

Alternatively, one could attempt to notate these pitch areas in staff notation; a quasi-score of more than one system of staves will often be needed to clearly show most tracks. This will generate a score

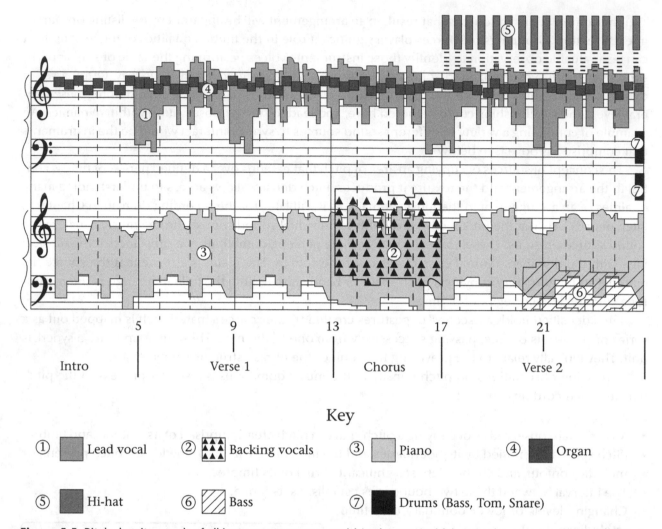

Key

① ⬛ Lead vocal   ② ▲▲▲ Backing vocals   ③ ⬛ Piano   ④ ⬛ Organ

⑤ ⬛ Hi-hat   ⑥ ▨ Bass   ⑦ ⬛ Drums (Bass, Tom, Snare)

Figure 5.5 Pitch density graph of all instruments present within the musical fabric in the opening sections of "Let It Be" (The Beatles, *1*, 2000).

of the track's pitch density and timbral balance. In many instances an X-Y graph presents the materials most clearly (see Chapter 7). The level of detail of a pitch area's boundaries, densities and changes over time can vary with the goals of the analysis. Notations of pitch density content might be coordinated to appear as a tier on a data collection timeline, should one wish.

Within the evaluations process, pitch density observations will generate large dimension considerations of the arrangement's use of range and registers. It will also facilitate upper middle dimension comparisons and interactions of pitch and timbre between sound sources.

This pitch density information allows the frequency and pitch range of the overall fabric to be identified, and provides the data for timbral balance evaluation. This can take varying forms including recognizing the highest and the lowest pitches of each section (any section) or of the entire track. It might also be a variable that can be traced from the beginning of the record to the end, representing the changing shape of the range of highest and lowest pitches, and the amount and location of activity within the range. Bringing this information into evaluations, ranges of most and least levels of activity can be identified, as well as the distribution of pitch materials throughout the pitch space of the track. The observations of pitch density can reveal much about the arrangement. There is no indication of dynamics within these pitch area relationships, though; these are revealed in the loudness levels and dynamic relationships of musical balance.

All this is explored more thoroughly in Chapter 7. There pitch density is explored as a recording element. As a recording element, each source is conceived as a level of pitch density, a level of strata. These strata of all sources establish timbral balance.

Dynamics

Dynamics can take place in any dimension. Information on dynamics might be collected in any dimension for individual sources or groups of sources that are in some way linked. Dynamics also functions at the large dimension, shaping a dynamic contour of the track, as well as at the smallest dimension within sounds and sound qualities.

Dynamics of musical materials and of recording are also closely linked in significant ways. The various graphs of the recording can examine the overall loudness levels and contour, and the dynamic contours of individual sound sources; these materials need not be collected here—unless there is a specific purpose to do so. It might be desirable to observe dynamic contour of certain lines, of musical balance of certain instruments or certain groups of instruments, to trace the dynamics of the entire musical fabric for a section or throughout the track, or to examine individual notes within a line—all instances at different perspectives, and many more exist.

Observations of central concern to dynamics will be loudness levels of sources, loudness levels within the sounds and the materials of sources, and loudness levels representing a balance of loudness between sources, or between groups of materials. These middle and small dimension activities might be related to metric and syncopation accents, crescendos and decrescendos, dynamic shaping and inflections in lines, and similar qualities. Musical materials interact with dynamics in a great many ways; observations of dynamics will typically be tied to elements and/or the sound sources that present them. Groups of instruments—such as those that might form the groove—will have their own overall dynamic shape, as well as the dynamic contours within the group. Often great dynamic complexity exists within grooves—and contributes to their character and drive.

Making written notes about dynamics might make use of traditional dynamic markings, though these are general at best. Individual lines and instrument dynamics can be concerned with:

- Performed dynamic level at any moment
- Shape of changing loudness level (contour)
- Dynamic relationship of a source to others (louder or softer than, and by how much)

The dynamic areas adopted for recording elements in graphs such as the loudness balance graph covered in Chapter 9 as a recording element might be of greater assistance. At this stage of collection, general observations may be adequate, and will lead to deeper examination of some materials once evaluations begin. More detailed examination requires the reference dynamic level (RDL), which can only be determined in the later stages of evaluation or within conclusion; with that knowledge deeper observation and evaluation of dynamics are possible. Typology tables examining the types of dynamics and their values and characteristics, or their changes, may be used to collect unbiased observations on dynamics. This is covered in Chapters 6 and 9.

Character, Timbre and Performance Qualities

The loudness levels of sound sources (instruments and voices) carry an inherent and characteristic timbre. Changes of loudness (dynamic level) cause shifts of timbral qualities. These shifts of timbral qualities themselves are dynamics-related, as portions of the sound's frequency spectrum become emphasized or de-emphasized as the energy and intensity of the performance shifts. This connection between how a sound is performed and the resultant timbre plays out in analysis through observations of the character of the sound (or sound source) and the timbre of the sound source. This duality of character balancing color might also be gauged as the interpretation of 'affect, expression, energy, intensity (and more),' that is contrasted against an observation of the substance of the sound (what is physically present).

Describing the Character of Sound Qualities

Descriptions of the character of materials, and even the musical character of instruments or voices can have value within an analysis. Descriptions that directly address the qualities of expression of lines or materials—the levels of intensity or, conversely, the degrees of passivity of a performance—can have relevance. Wrapped in with intensity, is the energy of the performance and the resultant sense of character, tension, motion, drama, stability, and much more.

The 'character' portrayed by a sound (timbre) is often linked to semiotics, and in variable ways to meaning attached to the character of a timbre. This is fundamentally entwined with 'interpretation' by the individual listener, but also by 'meaning' within the listener's broader collective culture. Thus, identifying the character of a timbre is subject to some interpretations (or impressions) being personal, while others speak widely within a common culture—and are open to variation between listeners and cultural groups. "The semiotics of music . . . deals with relations between sounds we call musical and what those sounds signify to those producing and hearing the sounds in specific sociocultural contexts" (Tagg 2013, 145).[8] Here this definition is extended to the timbre itself. It can be related to what Philip Tagg (*ibid.*, 525) has called "genre synecdoche" where sound sources bring associations outside the track; sounds and musical structures that "connote paramusical semantic fields—another place, another time in history, another culture, other sorts of people." Such outside elements might be sounds of other cultures (highland pipes, tin whistle, didgeridoo), or from the same culture, but a different context (such as a gospel choir within a rock song).

The expression and affective qualities that bring character to a timbre have some inherent qualities (qualities that are culturally learned), some qualities related to context within the track, some qualities personal to the listener, and qualities that might originate elsewhere or from any combination of factors.

Here is perhaps where subjective connotations might be used and found useful, when approached with clear communication of a common experience in mind.[9] This subjective position is not overtly personal; rather, it is a softening of restricting observations to the universal that can be objectively quantified and qualified, to a position that might also embrace the expressive qualities of timbre, the affective qualities of the track and its impacts on the meanings of timbre, and on the energies and intensities sounds carry that sit squarely within the timbral qualities of sound sources. Embracing the character of timbres will be especially useful using the broader, more 'universal' or 'cultural' perception as a starting point for description, then weaving more personal interpretation as might be appropriate. How a timbre might bring the listener to feel on a personal level, or the energy, intensity or meaning that the timbre might elicit is difficult to navigate, and can often generate descriptions that are meaningful only to the individual. Using language to describe the character of a timbre is often tenuous ground; problems of precision of matching term and timbre are to be expected, as definitions of terms between individuals and cultures (and geography) can shift. These communications are never fully adequate or accurate, though they are often as good as it gets.

We have few words in our vocabulary to describe 'how sounds sound,' and do not have an objective, unbiased approach to describing the content and character of sounds. Still, talking about the sound of the record is a vital part of analyzing the record, the music, lyrics and the recording. Describing sound is also central to observing the characteristics of timbre, expression, performance qualities, and more. In our attempts to describe sounds we rely on terms from other senses, analogy, associations, feelings, affects, and a host of other well-intentioned but ineffective (or counterproductive) terms. Without careful attention, it can be difficult to avoid terms that do not present quantifiable information, and do not represent the experience shared between others. In the absence of terminology and methods, we easily fail to make meaningful assessments (Moylan 2017), and our analyses fail to communicate clearly.

The literature contains no small number of descriptions of tracks, sounds and musical materials that are filled with words that communicate little more than an author's feelings or personal impressions, and do not address either the content or the character of the material within the context of the track. I leave it to the reader to recognize examples of such writings; you may have already encountered some. Descriptions effective in communicating tangible and relevant observations (and subsequent evaluations) about the qualities of materials are certainly possible, and a process and several resulting descriptions will be offered in Chapter 10.

With the timbre imbedded within the context of the sound stream, it is heard and presented to the listener within context of the track. The timbre is situated within the track; its character and the meaning of its character are formed (at least partially) by this larger context. The timbre is not isolated from context (as a sound object), but rather holds a character that has meaning *within* that context. As we are starting to recognize, the track's context is rich and complex. The timbre of the sound is not only sound quality of the source and its musical material, it is also language and performed lyrics, and the persona of the performer(s). As we will encounter later, it is also the timbral definition that establishes distance and the sound qualities of environments, among other qualities of the recording.

The impressions of character are thus the result of many factors interacting into an overall quality—a quality that might be of the individual sound or of the sound source (and the materials it presents and the qualities it holds), a group of instruments or perhaps the entire work. Character is thus identified within the 'conclusions' process, where evaluations of elements are considered and the richness of their confluence might be recognized. This will be explored in greater detail in Chapter 10 under conclusions.

Recognizing the Content of Timbre

Defining the timbre of a sound is different from describing its character. Talking about the content of a sound requires engaging the sound itself—noticing physical properties as well as their affect. Describing timbre is defining the substance of the sound, not the characteristics that make a psychological impression on the listener—or that bring an interpretation of the sound or material within the context of the track.

Listening for timbral qualities requires listening inside the sound. Our natural approach to listening to sounds is to recognize overall quality; for example, we recognize a person by the sound of their voice, but are not aware of what makes that voice different from all others. Listening inside the sound brings awareness of its subtleties, and to the variables within timbres. We clearly detect minute differences, but are not practiced at identifying the nature of those qualities; instead we identify what those qualities elicit. We hear very subtle changes in the timbre of voices, and we recognize a change in meaning, a change in expression, perhaps a change in mood, but we are not aware of the content of the change. Observations of the content of timbre are most accurate and relevant when recognizing and describing the activities of the sound's inner workings.

Timbre is one of the two dominant elements of recording. Chapters 7, 9 and 10 will cover timbre analysis in detail; there this contrast of physical content with character will be explored in depth. Those concepts can also be applied here to music and lyrics, when suitable. The following table presents a more general and simplified set of variables for examining the content of timbre that may prove more workable within the contexts of music and lyrics; it will often be adequate for addressing the concerns of timbre in these domains.

Table 5.1   Variables for observing the general, physical characteristics and content of timbre.

Variable	Values or Characteristics
Pitch or Pitch-Range	Defined by the pitch-range of the change (for example, C5 to F5), or by identifying a pitch or pitches within the range and the approximate interval of the range
– Type of Change	Increase or decrease of loudness level of the pitch-range
– Degree of Change	Description in general but objective terms such as: minute, slight, noticeable, moderate, substantial, pronounced, extreme.
Dynamic Shape or Envelope	Defined by speed (related to pulse or clock time) and attack level (related to sustain level or previous sound) and decay shape (contour and duration of levels)
– Type of Change	Increase or decrease of speed, increase or decrease in attack level, increase or decrease of the level of the sustain and of the initial decay
– Degree of Change	Description in general but objective terms such as: minute, slight, noticeable, moderate, substantial, pronounced, extreme.

Rhythm, Gesture and Groove

Rhythms and rhythmic patterning exist in all elements and in many structural levels. Microrhythms on the small dimension, middle dimension surface rhythms, and the macrorhythms of large dimension traits exist in all styles and genre. Most importantly, these rhythms at different structural levels function to shape those styles and genres as a (more or less) unique attribute. Discussions presented here should be taken as one example from the diverse popular music catalog. Electronic music contexts often treat rhythm, gesture and groove differently than more performance-based popular music. It is toward performance-based genres that many comments are directed—and for each comment, there will be exceptions, and perhaps more exceptions than tracks that conform. These following comments are not rules, but rather general observations, as rhythm—at all strata—might be the element that most deeply characterizes genres and styles of popular music.

Collecting information on rhythms might occur within observations of another element (such as harmonic rhythm, for example) or within musical materials (such as a vocal line); these are examples of surface rhythms (the rhythms of materials and sources). Within any musical style, rhythm data might uniquely emerge from any material or sources, and at any dimensional level.

Propelling and organizing surface rhythm is tempo and meter; tempo's elasticity in some tracks and styles of popular music will warrant observation. Tempo and meter are the references for calculations of rhythm in most contexts of tracks. Structural rhythms, such as the rhythms of verse/chorus exchanges, or of hypermetric phrases are macrorhythmic and proportional extensions of meter and/or pulse; these tend to be evaluations that are recognized after several levels of structure have been identified.

Figure 5.1 reveals macrorhythmic groupings and the prevailing time unit's regular pulse. The prevailing time unit and various structural levels of hypermeter can be recognized within the stages of establishing a middle dimension timeline. These structural aspects of rhythm may be important additions to the middle dimension timeline, as they may serve as reference time units. Further, they lend a sensation of musical movement and thereby significantly assist observations of other elements (as just seen with pitch density), and later evaluations in all elements and domains.

The rhythmic qualities of popular song are perhaps foremost (along with timbre) among the qualities that establish its essential character and sound. Rhythmic qualities often manifest within a groove's tight patterning and/or the rhythm section's more flexible layers. These slightly different concepts each establish a textural accompaniment containing percussion's rhythmic and timbral pulsation, harmonic/rhythmic materials from guitar or keyboards, melodic/rhythmic patterns by a bass, and perhaps other contributors such as winds. In grooves of tight patterning, even the voice might participate, as Anne Danielsen (2006, 108) identified in James Brown's productions from the late 1970s: "[T]he lyrics acted first of all as part of the groove and/or as comments on the qualities of the groove." Both the contrapuntal layering and the tight patterning may take many forms, and each form contributes to delineating one genre and style from others. Electronic music contexts generate grooves that differ from those that are performed, though the sources themselves need not be the factor that determines microtiming and microrhythmic deviations and patterning, and interplay of parts; rather than a dichotomy between two types, though, there is a continuum of groove types with more than two poles of variables. Indeed, not all grooves emphasize microtemporal deviations from the pulse, some grooves (especially electronica-related styles) are generated with all rhythmic events aligned with the metric grid, with the square feel of quantized evenness and an exaggerated tempo (Danielsen 2010a, 2). What is important here is the sense of approaching grooves as rhythm+pitch+timbre+dynamics textures (perhaps performed by many individuals, though perhaps not) in a way that might be applied to popular music styles that have yet to emerge.

The groove is situated in the middle dimension, at a level higher than individual sound sources where sources interact. While multilayered with timbres of different instruments—typically drum kit, bass, keyboard or guitar, though this can expand or contract—it functions as a single musical statement, and is appropriately observed as such. While each instrument contributes something substantive, the groove is a composite of rhythmic activity, timbral qualities, texture, dynamics and energy. The groove engages an interaction of musical elements as well as an interaction of instruments (and as we will discover, it can also engage recording's elements). Its complexity can vary widely between types of music, and can even shift within the track; while we immediately grasp the groove as rhythm-related, the interactions of dynamics, timbre and pitch-register also contribute and make the cross-relationships (and cross-rhythms) of elements significant to its overall quality (Danielsen 2006, 50–52; 2010a, 10). This section will seek to recognize the rhythms of the groove instruments, along with some approaches to getting it on paper as appropriate. Observing the rhythm section's groove can bring one to engage great nuance in the performances and materials, as well as broader gestures of rhythm, motion and energy. Collecting information can take a variety of forms.

Richard Middleton (2000a, 105–112) underscores the musical 'gesture' in relation to the 'performance' and the experience of somatic movement. The starting point for his 'gestural modeling' of songs and song-types is the rhythmic 'groove'; interactions of drum kit, bass line and other instruments (perhaps guitars, keyboard and horns) produce a gestural center. The rhythmic groove represents a 'given' around which many popular songs are oriented, and which vary between types of music. He offers a graphic of the gestures within the groove in Madonna's "Where's the Party" (1986). This concept of gestural modeling and graphic notation illustrates the various layers of texture with shapes depicting motion of elements or materials. Middleton defines the groove by identifying the rhythms and characters of its parts (instrumental lines) and their interactions; some materials are transcribed, but much is left undefined, save for general graphic shapes extending over time. Analysts may find value in this general data collection, especially for initial observations of some materials—in as much as this might align with the track and the goals of the analysis.

The groove, however, often calls for careful examination to understand its energy and rhythms. Anne Danielsen (2006, 100–103) has examined funk grooves in substantial depth, and demonstrated her findings with notated examples capturing much nuance. Her analysis of the two grooves in James Brown's "Funky President" (1974) provides clear examples of its rhythmic relationships and patterns, cross-rhythmic tendencies and pick-up gestures; as an example, the song's underlying feel of four against the first three beats of the bar within its first groove is clearly evident in notation that illuminates the rhythmic complexities upon listening. This approach is in stark contrast to the graphics of Middleton's gestures; it will be appropriate to some tracks, and others perhaps not as much.[10] Danielsen (2006, chapters 3–7) substantively embraces the concept of rhythmic gesture, but differently. Danielsen (2010a, 6):

> used the terms 'actual sounding events' and 'virtual reference structures' (nonsounding schemes that structure sounding events) to describe this interaction, exploring the microrhythmic features . . . In addition to metre, pulse and patterns of subdivision, I introduce a reference structure at the level of the figure, and a related notion of the gesture as the figure's sounding counterpart. My distinction between figure and gesture . . . the figure is the virtual reference structure behind the gesture, while the gesture is a sounding actualization.

An important observation of 'gesture' is offered by Danielsen (2006, 47):

Gesture names a demarcated musical utterance within the fabric of rhythm. It might be a riff or a vocal phrase, or part of either, or a group of beats, or just one beat, as long as it is perceived as forming an entity, a sounding gestalt. A musical gesture includes in principle every aspect of this entity—that is, the actual as well as the virtual. Even though one parameter often tends to be the primary characteristic aspect of the gesture—it may be shaped by, for example, timbre, rhythm, or melody—the gesture transcends any traditional division into analytical parameters. In the case of funk, the primary shaping aspect is most often rhythm, but timbral shaping is also common, as with James Brown's remarkable shouts.

Danielsen (2010a, 10) further defines gesture: "Whereas a stylistic figure is no more than a preliminary condition for musical performance, the gesture is the music as performed for someone."

The notion of gesture can be explored and represented in great detail. Gesture is inherently rhythmic; though this rhythm may manifest in any element—as with the timbre of James Brown's shouts. Gestures are also shapes—as with Madonna's harmonic gestures creating shapes of repeating chord-sequences in (Middleton 2000a, 112). Gesture may manifest in a more general temporal, dynamic, timbral, registral shape; it may also have great precision in its material, or subtlety of expression.

Peter Winkler (1997, 180–186) brings attention to the groove's subtle shadings of the subdivisions of the beat and the shadings of different beats within the measure, along with the often-found backbeat—the accentuation of metrically unstressed beats. He identifies the rhythmic pushing and pulling of the beat by the rhythm section resulting from the bass riff patterns, and the microrhythmic deviations from a 'metronomically exact center.' With the use of an automatic transcription device[11] he is able to calculate deviations with great precision, revealing unexpected differences in the durations of beats and points of attack—though Western notation cannot capture these essential traits. Upon studying the lead vocal, Winkler makes some notational suggestions that will be presented below.

Making observations of the groove pulls one into recognizing activities of the drums and the bass, perhaps with guitar(s), keyboard(s), horn section, or more. Activities of rhythms, placements of the beats against the metric grid, subdivisions of beats, dynamic accents and timbral changes, are all part of the groove. Further, not all instruments may be involved in approaching microrhythms in the same way (or at all). Minute rhythmic tensions may emerge *among* instruments and parts, and *between* them;

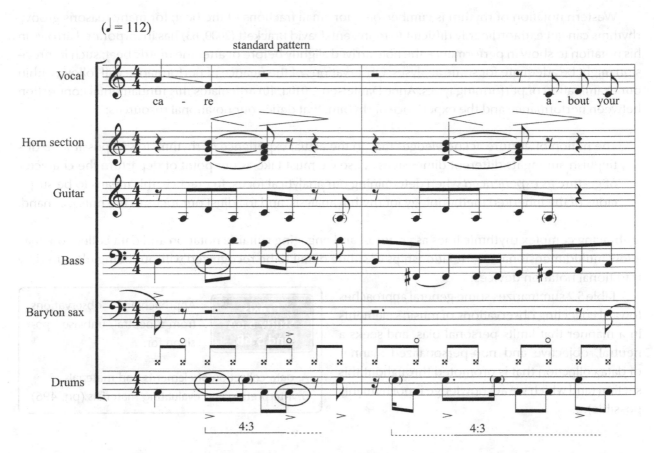

Figure 5.6 Basic groove in "Cold Sweat" (1967) by James Brown (standard pattern indicated in circles). Figure from *Presence and Pleasure: The Funk Grooves of James Brown and Parliament* © 2006 by Anne Danielsen. Published by Wesleyan University Press. Used by permission.

the subtle complexities of grooves can be quite intricate. Transcribing the groove nearly always brings percussion into notation and some significant rhythmic complexities are often imbedded.[12]

Figure 5.6 is from Anne Danielsen's *Presence and Pleasure* (2006, 74–76), and shows the groove from James Brown's "Cold Sweat" (1967). In the transcription the standard pattern of the groove is circled, though its elaborating elements are intricately complex. The drumming is woven together with the rest of the band (does not appear as an independent layer), and presents accents, syncopations and performance techniques to bring the sixteenth note as the smallest division present. Horns play to a different density referent (the pulse of melodic and rhythmic shaping as the shortest possible duration) of the quarter note, guitar to eighth notes, bass mostly to eighth notes, and saxophone and drums to sixteenth notes. The track's phrasing and subdivisions imply different durations that do not fit, and provide its funky style.

The goal of a groove transcription might be to recognize and capture the groove pattern (or some semblance of it), which typically repeats for significant portions of the song. Alterations of the groove might occur, and those can be noted by how the groove has been modified, and where on the time-line modifications occurred. In some tracks a second groove can be present, and its traits might be observed likewise; "While My Guitar Gently Weeps" (1968) is an example of a track with grooves alternating between song sections. To hear inside the groove, some transcription can be helpful; how much is needed, or if it is needed at all, cannot be anticipated until one is engaging the record itself. Much subtlety of microrhythm is common, and is integral to the groove. While these microrhythms define the groove's character and energy, they often elude clear and accurate notation.

Western notation of rhythm is cumbersome for small fractions of the beat; for many reasons groove rhythms can get extraordinarily difficult to represent. David Brackett (2000, 63) has incorporated arrows in his notation to show in performance the note arrived slightly before or after the metric beat; such imprecision might be adequate for some analyses. Observations of the nuance of performance will follow, within our examination of performing lyrics. Anne Danielsen (2010a, 10) articulates this fundamental connection between performance and the experience of rhythm, that defies our notational resources:

> The notion of gesture acknowledges that, in the actual experience of rhythm, it is impossible to distinguish among its different dimensions . . . so we must take as our point of departure the character of gesture *as experienced* when determining our analytical focus. In practice, this means to be suspicious of the inherited methodology for rhythm analysis, and to adjust our focus to the music at hand.

Embracing complex rhythmic lines as gestures, and engaging suitable notation and data collection, may allow multidimensional interrelationships of subtle detail to be recognized without being obscured by traditional notation devices.

Table 5.2 summarizes some general approaches toward collecting observations of music's elements in a manner that limits personal bias, and seeks a neutral, objective and non-personalized manner of data collection that is promoted throughout this section (and which we acknowledge is not humanly possible).

Praxis Study 5.6A and 5.6B. Approaching observations from a neutral, unbiased position for:

A. data collection of elements and materials
B. interpreting and evaluating materials (pg. 496)

Table 5.2   Summary table of general approaches to approximate neutral, non-personalized, objective, unbiased observations for music elements.

Sound Sources (Listing of Timbres)	Identify and name prominent instruments and voices; groupings of instruments; individual sources within groups; define musical parts by instruments/voices
Structure	Tempo and meter; identify primary song sections; PTU and section groupings, and section subdivisions might emerge
Melody	Gestures: contours and shapes against meter; transcribed melodic lines, riffs, motives, bass lines, and vocal lines
Harmony	Chord names, rhythmic placement and durations of chords; harmonic gestures (sequences of chords); harmonic themes and characteristic progressions; chord voicings
Arrangement and Texture	Identification of surface-level materials, and sources presenting them; pitch density of sources and their materials, placement of sources in registers, denote overlapping ranges; gestures of materials plus timbres; identification of texture type
Dynamics	Gestures: dynamic shapes, contours; terraced dynamic changes; dynamic relationships of sources
Characteristics of Timbre and Performance Intensity	Non-personalized observations or impressions of expression, affect, intensity and energy for, and any cultural meanings of individual sounds, sources, gestures, parts, lines, etc.
Rhythm	Gestures: speed and contours of any element against time; transcribed rhythmic patterns, cells, riffs; groove: gestures and/or transcription of individual parts/sources, and cross rhythms generated by interactions

OBSERVATIONS OF LYRICS AND VOCAL LINES

Lyrics observations fall into two broad categories. First is the literary text itself, its characteristics of structure, use of language, and ideas. The second is lyrics as they are performed in the record; the vocal line's presentation of the lyrics is a complex fusion of speech and singing, melody and sound qualities, musical expression and dramatic representation, voiced utterance and paralanguage, and more. The types and qualities of relationships between the vocal and the music are significant in the record, and are seen in the alignments of their phrasings and in the level of interplay of their materials.

Literary Analysis of Lyrics

Data collected on the lyrics is used to understand its content, ultimately as it relates to the vocal performance and material. Examining the lyrics in printed form, their most apparent structural dimensions are laid out before the reader. Structurally, stanzas are in the upper middle dimension, and individual lines of the lyrics are at or around the basic-level in the middle dimension.

The observation process identifies the number and ordering of stanzas, and some fundamental characteristics of each. Stanza structures contain a hierarchy of number of lines, line lengths (number of syllables within each lines), rhyme schemes (of various sorts) between lines within stanzas, and any other devices. Patterning of stanzas (as in verse – verse – chorus successions) and stanza groupings (as in verse+chorus pairings) will emerge. Reading the lyrics internally and aloud (as one would poetry), rhyme types, devices such as alliteration will be revealed, as well as rhythms of line lengths and internal feet patterns (inherent rhythms and meters within lines), and more. These aspects will become objects of attention and of careful examination at the 'evaluations' stage, where their interrelationships with the structural strata of musical setting become of interest.

Structural qualities of both the poetic and prose contexts may emerge from this process. Observation can then shift to the content of lyrics. The content of lyrics and their structural design, their meter and prosody, all contribute jointly and individually to the flow of motion and the generation of shape — these observations might ultimately lead evaluations to identify 'how' these are accomplished.

Lyrics often have "formed the basis for the interpretation of popular songs" (Brackett 2000, 192). Observations on the content of the lyrics might seek to identify the topics in Table 5.3. This table might establish a rudimentary start to exploring lyrics' content; it is by no means exhaustive.

The individual traits of the lyrics will bring unique twists in exploring these matters. It seems too apparent to state (and too important to ignore): the content of song lyrics is deeply original to each track, fundamental to its communication and message, and immensely varied. Following are brief summaries to provide a bit of illustration. Collecting information for observation and considering it in evaluation blend into a single activity most distinctly in lyrics; bearing in mind when one is holding objectivity, and when one is forming ideas, may allow the process to evolve most efficiently. Certainly in interpreting lyrics, one may be quick to assume a position of personal experience, rather than to formulate an objective reading

Table 5.3 Basic observations of the story.

Theme and subject	Story
Narrator	Protagonist
Narrative vantage point (first-person, etc.)	Audience and/or narratee
Temporal setting	Language style, dialect
Tone	Voice and persona
Location, place	

of the intent of the artist or of a more universal meaning—and some might propose the interpretation of the listener (analyst) might be more relevant, more pertinent, or otherwise more valuable.[13] In this writing we will continue to seek more universal (or culturally shared) interpretations with direction towards some commonality within recording analysis; the following provide a few examples.

Lloyd Whitesell (2008, 78–116) dedicates an entire chapter to thematic ideas in the songs of Joni Mitchell. His observations are refined into thematic threads running throughout her career's work. The threads include the themes of 'traps,' 'quests,' 'talent,' 'flight,' and 'bohemia'—quite removed from the most common themes of popular song. Whitesell focuses "on 'voice' in its literary-technical sense, indicating the vivid fictional characters and implied speaking presence in Mitchell's poetry" (ibid., 42) to more deeply explore the illusions of fictional voices and the persona types characteristic of her work (ibid., 41–77).[14]

Returning to Elvis Costello's "Pills and Soap," David Brackett (2000, 192–194) explores its lyrics and extracts Biblical allusions and clichés from nursery rhymes, recurring figures in various guises (accumulating meaning) and the effects of an intricate use of pronouns. He notes the lyrics "do not convey a straightforward narrative, . . . They express neither loss nor desire in any overt way. They do fall loosely into what might be termed the 'social criticism' genre" (ibid., 192). His complete analysis presents an interpretation of the lyrics, with supporting evidence from the musical setting.

Brackett also examines the lyrics of Hank Williams' "Hey, Good Lookin'" (1951); it addresses some different concerns that will transition us to the next section. This analysis includes aspects of the performance of the lyrics; it notes the sustained syllables in each line and their relevance, and the differences and interactions of durational and sonic stress on words within the verses. The stresses change meaning and mood. He offers: "[T]he performance of the lyrics . . . affects their meaning. We should first note how the performance stresses certain words through duration and sonic (rhyme) so that they reinforce the rough formal outline . . ." (ibid., 82). Brackett's detailed analysis extends through analyzing the lyrics' content, to situating them culturally, and more (ibid., 75–107).

Qualities of Performed Lyrics and the Vocal Performance

Performance situates lyrics within the time passage of the track. The lyrics' lines occupy sung phrases. These phrases represent 'verbal spaces' (Griffiths 2003)—the spaces of time occupied by lines of lyrics—and also by the word spacing (rhythmic density and distribution) of lyrics within lines and (thereby) within stanzas (Griffiths 2013). These are convenient concepts, allowing the words to be mapped out against the song's timeline, as in Figure 5.7. The phrases formed by each line are articulated on the timeline, showing beginning and ending points. Lines of text can be displayed, as well as the general rhythm of the line; this allows the syllabic density of the text to be visible. Proportional relationships between lines, words and stanzas can be made apparent, allowing one to evaluate their significance and contributions. Should the track warrant, other qualities of the text might be incorporated into the graph as well; some of these qualities might be rhymes, repeated words, emphasized (stressed) words, rhyme scheme, number of syllables per line, or significant word sounds, etc. The graph might readily accommodate expression characteristics and non-language sounds as well as other characteristics of vocal style.

Vocal style shapes the sound qualities of the lyrics in fundamental ways. Style might be considered in the following ways:

- Inherent qualities of the singer's unique voice
- Performance style idiosyncrasies of the artist
- Use and sounds of language
- Expression characteristics
- Use of paralanguage and non-linguistic sounds

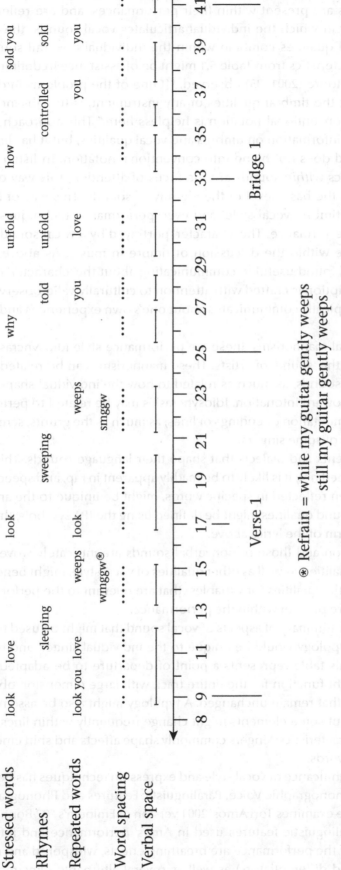

Stressed words		look		look		why		unfold		how		
Rhymes	love		sleeping		look		told			controlled		
Repeated words	look you love	weeps	sweeping	you look	weeps	you	unfold you	love	you	you	sold you	sold
Word spacing	⁝ wmggw⊛	⁝	⁝ smggw	⁝	⁝	⁝	⁝	⁝	⁝	⁝	⁝	⁝
Verbal space												

8 9   11   13   15   17   19   21   23   25   27   29   31   33   35   37   39   41

Verse 1   Bridge 1

⊛ Refrain = while my guitar gently weeps
               still my guitar gently weeps

⊛ = while my guitar gently weeps, within Verse 1 and Bridge 1 of "While My Guitar Gently Weeps" (The Beatles, 1968).

Figure 5.7 Select observations of lyrics incorporated into timeline, within Verse 1 and Bridge 1 of "While My Guitar Gently Weeps" (The Beatles, 1968).

All singers have voices that are unique, just as all speaking voices have unique characteristics. These inherent qualities are present within their performances, and are reflected in the timbre of their voice. The manner in which the individual articulates vocal sounds, the singer's natural resonance, and other added qualities combine within the individual's overall sound. The variables for observing timbral characteristics from Table 5.1 might be of assistance in defining some observations on vocal quality. Allan Moore (2001, 35) observed, "[O]ne of the problems in discussing vocal quality, indeed in discussing the timbral qualities of any instrument, is the absence of a standard, easy visual representation. Conventional notation is helpless here." This approach to describing timbre can provide meaningful information on timbres and vocal qualities, but it has limitations; it is not fast and simple to create and does not blend into conventional notation. In listening to vocal style and timbre, the characteristics *within* sounds are the focus of attention; this way of listening is seeking information underneath the basic-level of the identity of sounds. This way of listening holds for all observations related to timbre, vocal style, and even performance techniques and transcription of the subtleties of vocal performance. The 'character' portrayed by a vocal sound or a vocal line might be approached as above within the discussion of timbre in music. As above, subjective connotations might be used and found useful in communicating about the 'character' of a sound to a broad readership—if the description is crafted with attention to culturally valid observations, and an awareness that the personal typically communicates about one's own experiences and not often the experiences of others.

Artists also have vocal mannerisms; these are performance style idiosyncrasies that contribute to and often largely define the 'sound' of artists. These mannerisms can be related to expression, enunciation or non-linguistic sounds, as much as related to how the individual shapes vowel sounds, uses vibrato or their attitude toward intonation. Idiosyncrasies may be related to performance style such as characteristic pitch ornamentation or endings of lines, as much as the grunts, screams or breathing that define the performances of some singers.

Individuals have accents and dialects that shape their language sounds. This may influence their singing style, though it need not; it is likely to be readily apparent in rap, and speech-singing of any type. Sounds of language, often reflected in specific words, might be unique to the artist and therefore the track. In observations, sound qualities might be defined using the IPA symbols; they may then be noted and related within the form of the lyrics above.

Qualities of expression and those of nonverbal sounds are intricately woven into performances. Observations of these qualities, as well as other qualities of vocal style, might benefit from an organized approach to assembling the qualities (or variables) that are relevant to the performance and the values (or characteristics) that are present within the performance.

Table 5.4 represents a summary of aspects of vocal sound that might be used to assemble a typology of the vocal style. This typology would be unique to the individual singer, and to that singer's performance on the record; this table represents a point of departure to be adapted, not a definitive listing. Such a typology might function for the entire track with large dimension observations, or middle dimension observations that remain unchanged. A typology might also be assembled for sections that vary. It should be apparent some elements might change frequently within lines, and demand a more detailed assembly of characteristics; singers commonly shape affects and shift emotions within phrases, sometimes even within words.

An example of the significance of vocal style and expression techniques has been provided by Serge Lacasse (2010a) in "The Phonographic Voice: Paralinguistic Features and Phonographic Staging in Popular Music Singing." In it he examines Tori Amos' 2001 version of Eminem's "97 Bonnie and Clyde." Lacasse examines the many paralinguistic features used in Amos' performance and interpretation. Incorporated into the essence of the performance are breathing effects, whispered and murmured voice (acting as both qualifiers and differentiators) as well as several alternants—lips moistening, swallowing,

Table 5.4   Typology table for defining vocal style.

Variable	Values or Characteristics
Range of voice	
Register(s) of voice	
Degree of resonance and placement	
Degree of diaphragm support	
Attitude toward pitch	
– Intonation	
– Types of inflections	
– Rhythmic placement of inflections	
– Vibrato	
– Glides and bends	
Attitude toward rhythm	
– Note placements relative to beat	
– Precision of subdivisions	
– Relationship to spoken word	
– Rubato or free vs. polymeter	
– Microrhythms compressing or stretching the beat	
– Manipulation of time, pulse	
Inherent timbral qualities	
Inherent performance style	
Tremolo	
Language sounds	
Expression	
Affects	
– Frequency shifting states	
– Degree of shift	
Paralanguage	
– Non-linguistic vocalizations	
– Wordless expressions of feelings	
– Body-generated sounds	
Level of ease of the singer	

sighing—all with much subtlety. The paralinguistic features' sonic qualities and contributions to the performance and to the track are made apparent; illustrations of the paralinguistic features allow the reader to clearly recognize their significance and roles in the track's artistry and communication. Lacasse incorporates spectrograms into his observations in order to identify the frequency content of these sounds, and to supplement aural interpretation—noting the absence of any system of notation for paralinguistic features. Observations of paralinguistic features might be noted within the vocal line, incorporated within the timeline as alternants (stand-alone sounds), or collected as part of the qualities of lyrics. As evidenced here, exploring the effects and the timbral qualities of paralanguage features may provide significant insight into the sounds and expression of some tracks.

Transcribing Vocal Lines and Performed Lyrics

Observing the qualities of vocal lines is filled with challenges, owing much to the importance of nuance in the lines. A number of approaches might be appropriate to the analysis, at one time or another. Observations to collect data might engage transcription to hold those subtle qualities still. Such a transcription might generalize pitch and rhythm (perhaps include expression and dynamics (if addressed at all). Others may base observations on the content and sound qualities of lyrics and presented by the vocal melody, with only cursory transcription of portions of melody. Incorporating a typology of vocal style might help define expression and performance techniques enough to characterize a vocal performance to adequately subsidize a transcription of strategic passages in generalized (pitch and rhythm simplified and adjusted to conform to notational conventions) melodic notation. Other transcriptions may be filled with pitch, rhythmic and dynamic details, with rhythmic placements of precise phonetic qualities of the lyrics, and with speech-singing, expression affects and paralinguistic sounds.

The matters surrounding notation covered earlier illustrate how problematic vocal transcription can be; even the most accurate and detailed transcriptions will only present a close correspondence between notation and what is actually heard on a recording, and only within a limited number of elements. This is the best one can hope for when using a system of performance instructions to represent sound. Still, transcription can be useful for many purposes, and might be flexibly applied; few other options are available to make the record hold still for evaluation. The level of detail of the transcription can be part of the decision-making process of the goals of the analysis, and what information needs to be thoroughly examined to achieve those goals.

The microtonal inflections in pitch are difficult to capture in traditional music notation. Quartertones can be noticed in the techniques of many singers, as well as smaller deviations of pitch. Figure 5.8 presents a notation for quartertones, a precise pitch that resides between the half-step intervals of the chromatic scale. Figure 5.9 contains arrows above note heads representing a slight raising (upward arrow) or lowering (downward arrow) of pitch; these intervals are not as large as a quartertone, and are

Figure 5.8 Quartertone notation demonstrating enharmonic spellings using one- and three-quartertone symbols.

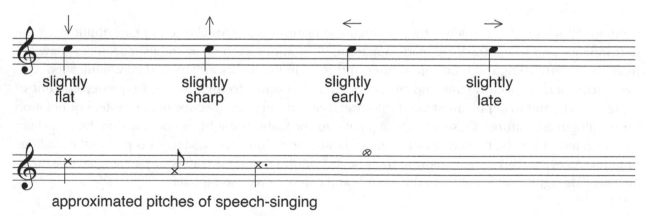

Figure 5.9 Notation for an indeterminate notation for microtone and microrhythm deviations from conventional notation values, and for speech-singing.

a slight but undetermined degree of sharp or flat. These arrows can be defined in several ways; they can represent a slight raising (upward arrow) or lowering (downward arrow) of pitch, or they might represent an undetermined degree of sharp or flat. These options allow the transcription to address intonation matters with varying degrees of detail or accuracy.

Microrhythmic inflections are equally difficult to capture in traditional music notation. Notation of microrhythms can quickly be an unreadable mess of ties, with short duration notes and rests tied to others. This results in a complex notation, that is difficult to identify, a challenge to notate and cumbersome to read. Rhythm is extraordinarily complex in the vocal line. The rhythm might flex with speech inflection, it might assume a relaxed relationship to the beat similar to rubato, or it might compress or expand the beat in such a way to establish a polyrhythmic relationship to other musical parts. These subtle shapings are just a few that are possible. In listening closely to singers, one becomes increasingly aware that portions of a word's sound fall in specific places rhythmically. As examples, some words will start with a consonant preceding the beat, others are slightly late; some internal consonants might be sustained, others punctuate the rhythm by precise percussive placements; ends of words are often as important as beginnings in terms of rhythmic placement and audible articulation. The arrows above the pitches in Figure 5.9 designate that the note began slightly earlier than written (left pointing arrow) or the note was slightly delayed (with the right point arrow designating the beginning of the sound pushed after the note head); the amount of this displacement is, however, indeterminate.

Rhythms of and within words and lyrics are part of the vocal line. Syllables and words may be contained within the melody as clearly articulated syllables on each note. Words are often much more nuanced in performance, and vocal style brings much attention to the timbres of syllables, and in the ways syllables morph into others within words. Drama can play out within words, as lyrics unfold and are interpreted by the singer. The rhythm of the vocal might reflect any of these, in addition to any rhythmic placements generated by paralanguage sounds, nonsense syllables and other singer-produced sounds. All of these might be of concern in collecting observations.

Writing observations of the text can include unusual divisions of words, denote prolonged vowel sounds, or the placement or sustaining of consonants; unusual word usage may bring important qualities to the track. Recognizing extremes, some vocal styles emphasize or enunciate articulation and endings of words, while some blur language by slurring words; these same effects might appear as temporary vocal techniques to support the context of the track. The sound qualities of word usage within the text therefore might warrant observation. IPA symbols can aid in identifying, notating and evaluating some language and non-linguistic sounds; others might be best recorded as letters or in some letter combinations denoting a sound ("zzzzzzk"). Some of these observations might be clearly added to a timeline (as Figure 5.7), others might be more appropriately added within a transcription of the vocal line (as Figure 5.10).

Timbral qualities are central to vocal quality, to language and to performance's shaping of the vocal line; the timbral qualities of the voice and the timbral qualities of words and syllables may blend into unique sounds. Any of these may generate an observation central to the track, and thus be worthy of close examination.

Here, again, it is important to acknowledge and separate 'character' of timbre and its 'content.' Describing character, as above, may be pertinent and appropriate to the track or sound source, though the content of the sound is often the actual object of discussion. Timbre's content is difficult to observe—in order to collect data, describe and explain—and requires deep listening. Further, we have few tools to collect this information, and they all are quite complex, which is needed to address the complexities of timbre itself. Listening to the inner workings of timbre and engaging approaches to observe and evaluate timbre are outside the prior experience of most analysts. Depending on the goals of the analysis and the skill set of the analyst, any one or combination of these following options might be pursued:

- The process of observing the characteristics of timbre from Chapter 7 and/or Table 5.1
- Using one or more of the fields in the typology listing of Table 5.4
- Incorporating IPA symbols into a description (see Chapter 4)

Finally, performances of popular songs can shift between singing and speaking; sometimes the singer is neither singing nor speaking. Speech-singing allows the spoken voice to have pitch qualities of singing, while maintaining some quality of spoken word. Figure 5.9 also includes an accepted notation for speech-singing that is similar to standard pitch/rhythm notation but substitutes an 'x' for standard note heads.

Figure 5.10 illustrates the four performances of the line "Nothing is real" from "Strawberry Fields Forever." Within this figure many of the topics from previous pages are applied. Notice the complexities of the rhythms, the use of arrows to designate indeterminate raising of pitch, the quatertone intonation (providing expression, interpretation and motion), the use of IPA and other approaches to word

Figure 5.10 The line "Nothing is real" from "Strawberry Fields Forever" (The Beatles, 1967) as it appears in each chorus.

sounds, and speech-singing. Above all, note how each is unique; for instance, the word 'real' appears in front of the beat once, and a sixteenth behind the downbeat the three other times, and its intonation in each instance is slightly different. John Lennon's performance of the word 'nothing' varies very little over the four presentations of the line.

Observing the Relationship of Vocal and Lyrics to the Musical Fabric

The opposition and interrelationships between the vocalist and the other performers (and the musical accompaniments they generate) can represent a central concept in the track. Observations pull out information on how the vocal line—including the persona of the singer, the content of the lyrics, and the performance of the line—is situated in the song relative to the musical accompaniment (and all that it contains).

Data on this interrelationship has been collected through explorations of the musical materials and their sound sources, and the arrangement. That initial data collection will be revisited at the evaluation stage to explore this relationship, seeking to reveal more nuance of some materials or adding other sources to those explored. These evaluations are only possible with more directed observations that seek to understand the relationship of the accompaniment and the lead vocal.

The five states of accompaniment discussed in Chapter 3 will provide a point of reference to these observations, and warrant review. With the vocalist in the primary role of the song, the accompaniment can take on the following functions:

1. Contextual function with supportive traits
2. Mostly supportive function with some contextual and ornamental traits
3. Function clearly supports the vocal narrative
4. Primary function with support traits
5. Primary function

These functions can change at any time, though changes typically happen at section changes. The function of the accompaniment might differ from introduction to verse to chorus to middle-eight to coda; it may also establish a grounding and reliable context for the vocal that stretches throughout the track. Alternately, any backing material may emerge as significant, and assume the primary role for moments or for sections of the track; they may enter into dialog with the track or contrast. Options for interaction are innumerable. David Brackett (2000, 92; emphasis in original) engages this relationship as well the quality of the vocal performance when he observes an "important element in Williams' singing is the *rhythm* of his vocal line—the way it either emphasizes the underlying pulse of the band or strains against it."

This interchange is often evident within or supported by relationships of the phrasing of the vocal line against the accompaniment. The alignment of the vocal's verbal space with the accompaniment's phrasing can be significant, as can the prominence of the accompaniment's phrases in drawing attention away from the voice. The strength of the prevailing time unit and hypermetric units can add to or limit the strength of phrases within the accompaniment and its layers or parts. Observing these interrelationships can contribute to collecting pertinent information of the functions of accompaniments, and of the characteristics of how the vocal is situated within the musical fabric. There is no one approach to observations that can establish the groundwork for these evaluations. Each track's accompaniment establishes a more-or-less unique relationship to the lyrics and the expression of the vocal; the five functions of accompaniment merely serve as guides for the data collection of observations that might ultimately reveal this interrelationship.

CLOSING REMARKS

Observations for lyrics and music can rely heavily on traditional approaches to music analysis, in the steps that precede determining function and organization. The processes of evaluating observations toward discovering the conclusions of how the song works will appear in Chapter 10. Our focus will now shift to the recording, and away from coverage of music and lyrics—though they will not be forgotten.

The chapters that immediately follow cover the elements of recording, and the unique challenges found in engaging them fully. Collecting observations for the elements of recordings will be woven throughout those chapters. The information on recording's elements is collected through creating X-Y graphs, sound location diagrams and/or typology tables—observations take place in unique ways for each of the elements.

Some Questions for Recording Analysis

What is the purpose of the analysis, and what are the goals of the analysis?

How many measures are present in the track?

What is the smallest time increment of the middle dimension timeline? Of the large dimension timeline?

Where on the timeline are the divisions between major sections?

What instruments and voices are used in the track? Where do they appear?

Are you able to maintain attention and concentration while listening to specific materials? Listening throughout the track?

What transcriptions of parts are needed to achieve the goals of the analysis? How much detail is required within those transcriptions? Does this vary between parts?

What tools are available to assist transcription? Do they enhance hearing the material, or replace sound with visuals?

How can one approach each element under consideration to collect observation data from an unbiased and objective position?

How can one collect information on the track's melodic and harmonic materials without knowing tonality?

What chords are present throughout the song, and where are they placed in time?

What are the pitch densities of accompaniment instruments, as they shift from phrase to phrase? How can these areas be placed accurately against the timeline and pitch registers?

How can the variables for observing timbral characteristics be applied to understand the guitar sound?

What elements are in play in the groove (timbres, pitch registers, dynamics, rhythms)?

What are the layers of the groove? Do the layers interact? How?

How do rhythmic gestures shape and propel the performed groove?

"How do timbre, dynamics, and texture shape our experience of the temporal distance between events, or the friction that can arise between asynchronous rhythmic layers?" (Danielsen 2010a, 10)

How does the verbal space of the vocal align with the phrasing of the groove?

What characteristics are present in the vocal style?

How are the performed lyrics enhanced by non-linguistic and expression vocalisms?

What level of detail is needed to capture the nature of the performed vocal?

What is the relationship of the musical fabric to the vocal?

Who is speaking? Who is being addressed? What is the topic?

Peter Winkler (1997, 170) offers these four questions:

What happens when I represent recorded sound in graphic form?
What is the relation of my transcription to the actual recorded sound?
What does transcription help me learn and discover?
Are there things that the act of transcription obscures or minimizes?

NOTES

1 Looking for sources of transcriptions by third parties brings one to rely on the perceptions (and biases) of others who have no first-hand knowledge of the lyrics; the web is filled with lyrics collected (and sometimes deciphered based on personal interpretations) by well-meaning people with no connections to the song; people representing widely varied skill levels for transcribing lyrics.

2 See Brackett (2000, 27–29) for a more detailed distinction between "prescriptive" and "descriptive" notation.

3 Jason Stanyek convened a "Forum on Transcription" (2014) that examines many dimensions of transcription in more depth than can be covered here; its bibliography will lead the reader further.

4 This writing also appears in *Critical Essays in Popular Musicology*, edited by Allan Moore (2007a, 391–416).

5 These tracks are cover versions of the Tin Pan Alley standard "Try a Little Tenderness" written by Reg Connelly, Harry Woods and James Campbell. The song was first recorded in 1932.

6 Richard Middleton (2000a, 114) provides a perspective on mapping melodic contour. His approach to melodic contour is demonstrated by depicting several important gestures in "(Everything I Do) I Do It for You" (1991) by Bryan Adams.

7 Robert Walser provides a detailed explanation of guitar tablature in *Running with the Devil* (1993, 91).

8 Incorporating semiotics into the substance of a recording analysis is a different matter than I am suggesting here, in recognizing the connection of semiotics to articulating the character of sounds/timbres. Allan Moore (2012a, 9) views "the principles of semiotics are opposed to the principles I shall develop here"; thereby recognizing his approach to interpretation (cultivating the reader's interpretation) is distinct from the semiotic approach to musical meaning.

9 For those who have read any of my previous work, or who have heard me speak over the past decades, this statement could seem contradictory to matters of sound evaluation I have previously proposed. This position I am taking here is, rather, a softening of those positions to acknowledge the expression, affect, intensity, energy, and more that form a significant dimension of timbral quality.

10 Anne Danielsen 2006 and 2010b are important resources for examining performed and digital grooves, the harmonic, timbral and melodic aspects of grooves, microrhythm and microtiming, and other related topics.

11 "An automatic transcription device should not be thought of as a replacement for aural transcription. They perform different but equally justifiable functions. The primary value of automatic transcriptions would be to throw light on what we do *not* 'hear,' what we change in the process of 'hearing,' or what we take for granted. They can also provide an insight into some of the extremely subtle elements of music which we cannot readily distinguish aurally, but which might nevertheless influence our perception of the music on a subconscious plane" (Jairazbhoy 1977, 270).

12 Percussion notation conventions might arise during transcription and can be found in most jazz arranging or orchestration texts.

13 See Moore (2012a, 1–7 and Chapters 7–10) for discussion on interpretation by both listener and analyst, and "finding out what we make of the song" (*ibid.*, 286).

14 Whitesell's (2008) examination of lyrics as poetry is particularly relevant in this context, given the significance of Joni Mitchell as a poet, deeply versed in the craft of writing poetry.

PART TWO

CHAPTER 6

The Recording Domain: Elements, Listening, Notation, Rhythm

With this chapter we begin detailed coverage of the recording domain. This chapter is divided into six major sections. In a broad sense, this chapter progresses from reintroducing recording elements and rhythm/time, issues around transcribing and describing recording elements, and the challenges of hearing and reproducing recording elements.

The first section explores "Engaging the elements of recording." The qualities and interrelationships of recording elements are discussed. The brief introductions of elements found in earlier mentions are expanded upon to set a context for the chapters that follow and also for a short introduction to contributions of recording elements within the multi-domain texture of the track. The individual recording elements are distributed over the following three chapters. Chapter 7 covers timbre (one of the recording's primary elements) in most of its many of its guises, as well as its related element of pitch/frequency. Chapter 8 is dedicated to the other primary element: spatial properties and relationships. Chapter 9 concludes coverage of the elements with dynamics and loudness, before moving into other topics.

The second major section presents rhythm and time as a recording element. Both rhythm and time appear with different attributes and syntax when functioning with the other recording elements. Rhythm patterning and durations occur within the materials of each recording element, though conventions differ from music and lyrics. Macrorhythmic relationships and microrhythms largely replace the surface rhythms that dominate music and lyrics as the temporal units that characterize their activities. Time is important for several central qualities of recording elements—where time durations and relationships (functioning outside the metric grid of rhythm) establish important qualities. The timbre of time will be an important concept as our studies progress.

The next two sections are interrelated. First, data collection and transcription of recording elements is introduced, along with a 'notation' for elements of recording. Specific X-Y graphs and several diagrams are created for each element, some are at specific levels of perspective, others illustrate several levels; these will be explored deeply within discussions of individual elements throughout the following chapters. This notation constitutes data collection of the attributes and activities of elements. In the next section we explore how these same observations may be directed toward organizational typologies for recording elements; typologies can categorize and organize elements at any level of perspective or dimension. This information provides some raw data from which we might recognize syntax of recording elements. Data collection, organization and display may make use of typology tables alone, be represented by graphs or diagrams, or these might be combined.

The fifth section explores hearing the elements of recording. The X-Y graphs and diagrams of the notational system, and also typology tables, provide some basis for discussing the elements of recording, and recognizing their content. Still, like music notation, they will not be easily accessed and useable for immediate observations, especially initially when learning the processes and notations. While these representational systems may not be immediately intuitive, they are very flexible once learned. One of the potential uses is for describing or talking about sound.

Effectively 'talking about sound' is important for discussing recording elements (as the record exists only as sound)—though little vocabulary unique to sound-in-general exists, let alone the sounds of recording elements. Common ground between listeners and a shared vocabulary are challenge descriptions of sound. These are explored in a section dedicated to talking about sound; the concepts introduced here are further explored in Chapters 7, 9 and 10 as they pertain to various specific elements.

Many readers may not be aware of most qualities and subtleties of recorded sound, let alone practiced in the types of attention and awareness needed to recognize their presence and dimensions. Even seasoned listeners are encouraged to review the praxis studies woven throughout the coming chapters; these studies are designed to pull the listener into understanding and perceiving recorded sounds for their unique individual characteristics, relationships, and the sound qualities they form when combined. The studies allow each element to be thoroughly explored in a singular focus of attention; bringing a deeper understanding and higher level of awareness than is otherwise possible.

Listening to the track and playback variables comprise the final section of this chapter. It can be no surprise by now that listening to records is necessary for recording analysis. This 'listening to records' is the basis of the analysis; it establishes the sound qualities from which an analysis is developed. This listening requires a playback system, and the system itself is part of the methodology of an analysis. The playback of the track contains numerous choices for the analyst—from selection of storage media, playback format, loudspeakers, the listening room, and much more. Each variable of the playback system will shape what is heard within the track (the qualities of sounds and their balances), and how significantly the qualities of the track are transformed compared to the original sounds. The very substance of a recording analysis is established with what one hears—these are the sound(s) of the track, its primary text. The record is "itself a text" (Zak 2001, 41); our active listening to the record is the primary analytical text, and any model for functional analysis must look at one's personal text, or one's listening situations and intentions (Kennett 2003, 208–209); the playback of the track is its reading.

An analysis might have greater relevance to others if what is heard is most reflective of the original recording; the track itself, and its sound qualities, are the common ground between the analyst and those they address. Selecting playback variables is part of the interpretation and analysis process; it can have a profound impact on what the analyst hears and what information is experienced, observed, evaluated and so forth.

These quite diverse topics have been placed in a single chapter because they interconnect and mingle together in various ways. All carry a degree of interpretation and potential subjectivity, and all function within the sequence of experiencing the track, and making sense of it. We will explore this (inter)subjectivity in experiencing and interpreting the track next, before beginning this chapter's major sections.

THE INTERCONNECTEDNESS OF CHAPTER TOPICS, AND INTERSUBJECTIVITY

Playback systems, notation of elements (and transcription), syntax and communication about sound are all fraught with subjectivity. At any moment their level of subjectivity can be anywhere on a continuum from slight (highly culturally shared) to profound (immensely personalized)—and may shift in the next moment. To these topics, we can add variables and the potentially subjective impressions we have already covered, such as socio-cultural context, personal experience and the skills of the listener/ analyst. These are all intermingled, interconnected. This chapter will seek to parse them out a bit, and also allow them to blend, as they do within our daily practices of understanding the world about us.

Our task in analysis is to balance the objective against the subjective—the shared experience contrasted with the personal, what can be measured opposing what cannot, the experiential balancing the conceptualized, and so forth. The science in our approach to analysis seeks out "things that are intersubjective, that pertain to the music [or recording or lyrics] itself and not to anyone in particular. . . . something [attributes of elements] with an independent and objective existence . . ." (Ford 2019, 19). Attributes of elements (and the values and activities of those attributes) may be discovered within the track and identified in a manner whereby others might recognize this content as reflecting their own perception; the notational graphs and diagrams and the typology tables that are offered will provide assistance in collecting and displaying this information. This all, however, occurs within the experience of listening. An aural experience of a track

> is always the experience of the individual. It is always a physical vibration happening at some specific place in time in some specific ear canal. If I wish to write an account of such concrete experiences, I cannot avoid putting myself in the picture.
>
> (*ibid.*)

This accounting of the experience requires technical data (observations of the track focused on content that is collected in a manner that is as unbiased as possible) in order to establish a balance between the personal and the shared (within one's culture). This accounting is also needed to establish some access to what is artistic expression within the experience, and its accompanying affects and energies. That which can be measured is divided from those that defy measurement—the human, artistic element that is inherently subjective, but that even within the subjectivity has some social norms of intersubjectivity, "a shared subjectivity at the end of a communication process . . . that arises when at least two individuals experience the same thing in a similar way" (Tagg 2013, 195–196). Some level of intersubjectivity is necessary if we are to compare analyses with one another—or if the comments another makes about their experience reflect our own.

The intermingling of listening observation and intersubjectivity whirls and spins. This cycling of objective observations of elements (that are interpreted when diagramed or graphed) within the listening experience (that is uniquely personal in many ways) with its subjective tendencies and influences, and with the syntax and sonic conventions that establish some semblance of intersubjectivity, just begins to reveal how complex (and fragile) accurate listening to recording elements can be.

We seek a balance of recognizing what is culturally shared affects and expressions with observing the elements' attributes and their values; wary that one may distort the other, wary that the human may bias the empirical, wary that objective data cannot reveal the motion and kinetics (and more!) inherent in popular music. We experience the record—the primary text of the track—through our hearing mechanism; we perceive physically its acoustic energy (in the form that was determined by one's unique playback system, that includes one's related conditions and equipment choices); as neural impulses are generated at the end of the sequential hearing apparatus (with many opportunities for it to become unique to each of us—and also to each ear) we begin the process of observation, of engaging our knowledge and skills, and of building and shaping our interpretation(s). This conceptualization of tracks might be divided into areas that contribute to objective content and those that contribute to intersubjective character (seeking to minimize the personal); these two concepts will return continually throughout the remainder of this book.

This all might appear overwhelmingly untamable. Yet, it is at the core of the magic that allows tracks to communicate similarly (though not precisely the same) to millions of people.

In practice, we might find a balance between the subjective and the objective by using objective observation of the physical content of elements as the basis for subjective characterization and description (that inherently carry some level of bias). As determining what data will be collected inherently restricts those areas that will be examined, and determining the parameters of observation (such as level of detail, breadth of examination, qualities to be examined, etc.), it is necessary for one to regularly assess one's choices as the observation process unfolds—this act that might pull evaluation into the observation process if one is not cautious. The act of deciding what to observe (and how) is based on personal bias (the magnitude of this influence is a personal variable as well); to minimize these influences, one may seek to exhaust options for data collection (what data to seek and how to observe it) by intentionally observing data that is outside one's biases—for instance, one might intentionally observe timbral balance changes in density and its overall contours throughout a track, although one's natural tendency would be to not consider that element. In the end, it is one's responsibility to know how their personal biases lead their analysis process, and how their biases become incorporated into an analysis—and to balance those with what is (mostly) objective within their observations.

ENGAGING THE ELEMENTS OF RECORDING

The elements of recording are outlined in Table 6.1. In broadest terms, these elements are timbre, space, pitch/frequency related percepts, loudness/dynamic levels and relationships, and rhythm and time. Each element appears in a different form on each level of perspective. The recording elements are complex in how they shift form and function at each level of perspective. They each contribute something markedly separate and unique to each level, yet are intricately and inextricably interconnected.

The elements of recording do not generate sound, as do the instruments that create music or the voices that form lyrics. The elements of recording are qualities that are applied onto instruments, voices, the overall texture (and so forth) in the processes of recording them, or are qualities that result from crafting the relationships of what has been recorded. The elements of recording transform and reshape the sounds and materials of the other domains, and they all work synergistically. In doing so they can add, enhance or create new qualities of the music and lyrics, and establish other dimensions and relationships of those sounds and materials.

The Elements of Recording

While all elements hold equal potential to be significant at any moment, the inherent nature of recorded sound itself and its applications in production convention[1] manifests most profoundly in the elements of timbre and space; they are "organizational in their own right" (Camilleri 2010, 200). They can present significant variety that can fundamentally shape a track's unique character and content. Timbre and space are situated as primary qualities and play decisive roles in shaping the 'sound' of the track. Both also play fundamental roles in how humans understand and interface with the world around them. In this way, they can pull outside connotations into the track, as well as add a sense of realism, an altered reality, or an invented world.

Common Traits of Timbre and Space

Timbre and space are each multidimensional, appearing in various forms and having various distinct dimensions. They have different identities at different levels of perspective. Timbral and spatial qualities are complex elements resulting from innumerable combinations of frequency, amplitude and time relationships. Neither element is entirely independent, though. Some qualities of space are determined by timbral characteristics, others by dynamic shape and relationships, other qualities are manifest in time characteristics, and more. Timbre is closely linked to pitch and frequency, to dynamic shapes and

Table 6.1 The elements of recording, with summary of their states and functions at all hierarchical levels.

Element	States, Manifestations, Qualities
Timbre	Timbral balance
	Pitch density
	Crafted sound qualities and timbres of sound sources
	Performance intensity, expression and performance techniques
	Component parts of timbre/sound quality
	Timbral qualities of distance cues
	Timbral qualities of environments
Space	Sound stage dimensions (width and depth of overall stage)
	Sound source positions and relationships: stereo or surround phantom image location and size, distance location
	Sound source positions and relationships: listener to sound source distances, and distance relationships of sources
	Environments of individual sources (qualities of spaces, including depth)
	Environment of the sound stage (holistic environment)
	Space within space relationships of holistic environment and individual source environments
	Ambience, echo, reverberation
Pitch and Frequency Related	Pitch registers and pitch/frequency placement
	Pitch areas and range of sounds/sound sources
	Pitch density
	Timbral balance
Loudness Levels and Relationships	Dynamic contour of overall program
	Reference dynamic level
	Loudness balance of voices and instruments (musical balance)
	Loudness/dynamic contours of individual parts
	Dynamic contours of individual sounds
	Contour/shape of dynamic envelopes
	Contour/shape of spectral envelopes
	Dynamic contours within reverberant energy
Rhythm and Time	Tempo, meter, and metric durations
	Patterns of durations (surface rhythms) in all recording elements
	Time units, durations
	Microrhythms: timelines for activities within timbres and environments of individual sounds
	Timbre of time units
	Rhythms of reflections
	Reverberation time duration and density characteristics
	Macrorhythms: changes that occur between sections that are greater than hypermeter or prevailing time unit
	Patterns of changing characteristics between major sections

levels, and dynamic and time relationships within its internal components. The multidimensionality of these two elements cuts across all levels of perspective in the track. As Table 6.1 illustrates, timbre and space function at all structural levels.

Timbre and space are each perceived as unified wholes, comprised of internal dimensions that are subject to change. Timbre is comprised of three basic component parts (dynamic envelope, spectrum, spectral envelope), each of which contain a significant number of variable attributes (see Chapter 7). Space similarly has a number of distinct dimensions (providing virtual representations of physical spatial relationships), one of which—environmental characteristics—is also comprised of three basic component parts (covered in Chapter 8). This multidimensionality of both timbre and space is often generated by their relationships to the elements of pitch, dynamics and rhythm—although often most directly understood by the frequency, amplitude and time characteristics that form their substance.

Pitch, Dynamics and Rhythm Reframed as Recording Elements

Some recording elements contain qualities that are conceptually linked to music and music performance—such as dynamic contours, pitch areas and pitch density (among other elements). While associated with music-related elements, these elements function differently and appear in different dimensionalities in recordings. As elements of recording, rhythm, pitch and dynamics are reframed from their states and functions in music; they not only provide unique qualities to recording, they function differently and appear in different forms.

Pitch as a recording element is closely associated with pitch density (in registral placement) and timbral balance. In practice, pitch in recording is not as discrete as the pitch levels and relationships of music. The registral placement of pitch provides crucial information for characterizing a track, often more so than its specific frequency or pitch level. Pitches in recording function more independently, more as isolated objects. They are individuated, as are frequencies, and each level is unique; indeed, pitch as a recording element can often be most clearly defined as frequency and not as a pitch within equal temperament. This, of course, is in contrast to pitch in music, which functions in context of a system of relationships and is understood by how it relates to other pitches and the organizational system. Pitch as a recording element does not relate to the music's tonal system; rather, the harmonic series provides a unifying presence in some instances. Should a source's recording-element pitch be in relationship to other pitch levels, the harmonic series establishes any relationship between them (relationships that need not conform to equal temperament). They are heard by how they relate to an implied harmonic series. Pitch perception/recognition is transformed to a conception of a bandwidth or area within the texture; there it functions more like a component of the timbre of the texture, than as an independent element. Indeed, it is hard to know where timbre and pitch diverge—as timbre morphs into pitch space, and pitch space becomes timbre.[2] And this can manifest at any specified level of perspective. Timbre and pitch are closely linked in recording, and are in some ways inseparable.

Dynamics (loudness) can play a substantive role in shaping the recording—not only the sounds of the track, but also within the recording process itself. The term 'loudness' has been chosen to allow the element of recording to be more readily identified from the element of music; within musical contexts are relative 'dynamic' relationships of parts; within the context of the recording loudness can be measureable as amplitude and it often functions on perspectives other-than the basic-level (of sound source dynamics). This activity of adjusting loudness and dynamics in the recording process creates sonic qualities and relationships—relationships that might manifest in other elements, such as timbre. Loudness levels and contours have the potential to appear in a vast array of states and activities, and at all structural levels—such as contours of overall sound for the track, for individual layers or lines, and individual sounds. Loudness manifests as relationships of sound source layers and relationships within and between layers of materials; this balance delineates sounds and blends others. While dynamics is often assumed to be the most significant element in shaping the mix, in practice (and sonic result), dynamics is typically supportive in function—just as its mirror musical element. Dynamics (and loudness) are often confused with prominence; this causes changes in other elements to be misperceived as changes in loudness. As a recording element, dynamics is closely aligned and interactive with timbre (again) and to the spatial elements of environments, sound location, reverb, echo and more; changes in these and other elements often bring an *impression* of loudness shift, where none exists.

Rhythm, too, appears differently as a recording element. Surface rhythms of recording elements are atypical; they rarely exhibit change at the basic-level of individual sources. Though some elements are mapped against the metric grid in certain X-Y graphs, surface rhythms do not often drive the elements of recording. Instead, rhythmic activity functions most strongly at other structural levels; for recording elements, temporal activity is common at macrorhythmic (at or longer than hypermeter or the prevailing time unit) and microrhythmic (within individual sounds) levels. Rhythms created by changes in element relationships and characteristics between alternating verses and choruses establish macrorhythms, and recording element rhythms of structural units larger than the prevailing time unit are common. This is explored more thoroughly below.

Interdependence of Elements

There are many levels of interrelationships between the elements of recording; elements rarely function independently. This has some similarity to the elements of music, the multi-dimensionality of which Simon Zagorski-Thomas (2014, 11) describes: "We do not . . . interpret pitch or harmony in isolation from rhythm or timbre." The multi-dimensional features described here in music function in the recording just the same. Zagorski-Thomas explains how conceptual blending, cross-modal and cross-domain mapping have strong application to recording's elements (*ibid.*, 7–13). Intermodal associations (here between motion, vision and audition) are especially prevalent related to spatial cues and physical gestures of performance intensity reflected in instrument timbres and in vocalization.

The multi-dimensionality of recording's elements is clear as pitch, dynamics and rhythm play substantive roles in creating timbral qualities; timbre, loudness/dynamics, pitch/frequency, rhythm/time all play roles in shaping various spatial dimensions. This can establish intricate associations between elements, and cause perceptions to be easily misunderstood—especially prior to some study of the sounds of the recording.

Recording's Elements and the Multi-Domain Texture

To briefly revisit an important concept: a multidimensional, multi-domain texture exists at the highest dimension of the recorded song; this is the overall sound quality of the track. All three domains contribute substantively to this texture that represents a confluence of all their elements and materials.

We have just explored the contributions of music and lyric domains to this multidimensional texture. To recall, music contributes layers of melody (lead vocal, bass line, other melodic parts), of harmony (keyboard, guitars, etc.), of the groove (drum kit and other rhythmic parts), and of various parts that straddle these functions or otherwise contribute to accompaniment. Lyrics add other layers; these interact significantly with the layers of melody and also provide the richness of language and its sonic qualities, as well as drama and expression. All these together weave an interconnected hierarchy of structure, that builds the overall, multidimensional texture.

This concept of 'layers' can be helpful in working toward observing and evaluating the stratification of domains and elements, the strata of materials and their functions, and of all their transformations—all, ultimately, observed for their contributions to the overall texture. In broadest terms, the multidimensional texture's parts might be delineated as:

- Music: layers of materials and their functions (pitch centric, with metric pulse and surface rhythms)
- Lyrics: text and its performance (convergence of melody, language and expression)
- Recording domain: Timbre (timbral balance and matters of sound source timbres; integrated with pitch/frequency and percepts of dynamics)

- Recording domain: Space (sound stage and positional relationships of sources; environments and related spatial/timbral cues; integrated with percepts of dynamics)

While we have conceived of the record being divided into the three domains and their elements, the overall texture of the popular music record is fundamentally connected to the recording's timbral and spatial qualities and relationships, which have the potential to contribute significantly to the materials of the music and lyrics domains. Allan Moore (2012a, 19–49) has arrived at a similar set of qualities to the overall sound. He introduces his approach: "Observations focus on what the instruments sound like, how they work together, and how they appear to be situated in the recording. . . . three broad categories. . . *functional layers*, the *soundbox* and *timbre*" (*ibid.*, 19). The terms he has chosen and his approach have some important differences to what is offered here.[3] The functional layers he proposed contain the layers of the song's beat layer, bass layer, melodic layer and harmonic filler layer; the primary melodic materials of the vocal melody are included in the melody layer, though the lyrics themselves and their performance are not addressed within this context. The 'soundbox' contains many (though not all) of the space attributes outlined above, and his use of 'timbre' is also somewhat different as it is directed primarily to the qualities of the individual sound source; though source timbre is essential here, I seek to acknowledge timbre's contribution to the overall texture as a significant dimension of the track.

We will engage the dimensions of timbre and space as discrete concepts, and turn to observe how they are interrelated and even fuse in some ways. Together with the layers of music and lyrics (their materials and in performance), the layers of timbral qualities and spatial properties of the record comprise the four 'layers'; or parts of the track's texture. This does not imply the recording domain's two layers contribute more to the record than music and lyrics; they are identified in this layering to better illustrate the recording's role in shaping the track.

The density and registral placement of pitch/frequency information contributes directly to the overall texture as timbral balance. Contributing to this impression are all matters related to the timbres of individual sources: their inherent qualities, presentation of materials, expressive qualities and intensity levels of performance, and more—these originate from within the materials of music and lyrics, and receive final shaping as a recording element. Timbre is not only the sounds of instruments and voices, timbre is linked to the sounds of performance. Timbre is shaped in records with precision and in exacting detail. Timbre is linked closely with pitch/frequency, and also with loudness/dynamics; space and timbre can also be linked closely, as in the timbre of reverberant sound.

The space layer also plays a prominent role in defining the track, and contributes to its fundamental 'sound.' The multiple attributes of space breathe real world-related features into the track—even if that world is virtual, and impossibly unreal. Space is a largely independent element, though through distance location and the spectral characteristics of environments the elements of space and timbre are fused. The spatial dimensions of the sound stage, and the positional relationships of individual sound sources, and the environments from which individual sound sources emanate, as well as that of the entire sound stage, contribute substantively to the multidimensional texture.

RHYTHM AND TIME WITHIN RECORDING ELEMENTS

With rhythm and time applied within recording elements, the context for the study of the elements is established. As we examine the ways rhythm and time are unique in the substance and functionality of recording elements, their contrast with music and lyrics elements will become evident. Rhythm appears differently in recordings; this is especially noticeable as the elements do not exhibit surface level changes and rhythm relationships that are common in the other two domains. The elements of recording change over time at different relationships to the pulse and meter (and to the prevailing time unit and hypermeter) than do music and lyrics. Only a few of the characteristics and variable values of recording's elements are temporal at the surface level, and those changes are typically not rhythmic. Rarely do they contribute to the rhythmic drive and patterning of surface rhythms, which are so central to the materials of music and lyric domains—though with a notable exception, below.

Recording's elements exhibit pronounced activities and defining characteristics at macrorhythmic and microrhythmic levels, however. At these structural levels—levels where little competing activity is typically present within the music and lyric domains—recording's elements shape the sound of the track significantly. Microrhythms appear in recording elements in a number of guises; of particular note, some elements function in clock time, rather than in relation to the metric grid of the song's pulse.

In the recording domain, rhythm is most relevantly approached in three layers as: microrhythm, surface rhythm, and macrorhythm. This is in stark contrast with the three layers of rhythm presented for the music domain in Chapter 3. Rhythm as an element of recording will hold some similarities to these rhythmic layers of the music domain, though many of its most significant characteristics will be present at different levels of perspective.

Surface Rhythms

The surface rhythms that are integral to melody or to poetry and lyrics are typically given room to flourish, with the activities of recording functioning more strongly on other structural levels. Recording elements simply do not often generate rhythmic patterns at the basic-level.

When rhythm does function on the surface level as a recording element, it is most often functioning interactively with the timbres of individual sound sources, and often with their lateral (stereo or surround) locations. The drum solo that appears in "The End" (*Abbey Road*, 1969) is a clear example of a sequence of surface rhythms of lateral locations and of timbres; these function interactively and in parallel to the surface rhythm of the drum kit that, as a whole, presents the musical statement. These types of rhythmic patterns (phrases) established by repeating sequences of locations and/or timbres are common in the drum kit parts found in grooves, as well. This is "timbre melody"[4] coupled with sound location positions. When timbre sequences are repeated, they become established and anticipated; timbre sequences can create an ostinato, perhaps contribute to (or comprise) a groove—and each timbre is located in its own position, thus adding space to the gestural material. This is perhaps the most common way recording elements participate in surface rhythms, but not the only way.

Musical balance may also have subtly changing levels on the surface rhythm level, though such changes are rarely organized temporally into patterns, especially recurring patterns. Pitch density and musical balance may also couple to have a presence in surface rhythm in ways similar to those just discussed for timbre/lateral location. Patterning of changing loudness and of pitch density's bandwidth (representing music's melody plus the source's spectral content) are difficult (sometimes impossible) to perceive in discrete increments, and therefore are not typical at the level of the individual source or the composite texture. Musical balance and pitch density materials are typically gestural, and although present at the structural level of surface rhythms these materials typically do not contain rhythmic patterning. Their activities are related to the metric grid, however; musical balance will often exhibit change within the beat, and pitch density often exhibits durational values of phrase-lengths (or hypermeasure). An atypical example of sound location changing at the phrase length is the Moog sound at the end of the Introduction to "Here Comes the Sun" (1969).

With these examples we open to an awareness of how elements of recording might function similarly to music and lyrics on this basic-level of surface rhythm—though these are not typical. Rather, the substance of recording elements appears largely at higher and lower structural levels—some of these levels are within the beat and others perhaps at the phrase length or at the hypermeasure. Rhythmic activities at even higher and lower structural levels—straying even further from surface rhythms—are common in recording elements.

Microrhythms

Microrhythms can exist as subtle or microscale variations in surface rhythm. The same instance might also be observed at a lower level of structure as a sub-rhythm or as a modification of a simpler beat. Microrhythms of certain recording elements might also appear within sounds, as the temporal activity of one of its attributes; this level of perspective is within the sound itself.

Very small subdivisions of the pulse that appear between iterations of sounds, or that appear within sounds create microrhythms. These very short durations can appear as qualities iterated within recording elements, appearing within a single sound or appearing between successive sounds. These may also appear from stretching the beat—or a portion of the beat—or from contracting the beat. Microrhythms can take place at the surface level, or can function at (or even establish) the lowest levels of structure; in this context they will be referenced to the metric grid and its pulse. Microrhythms are common in the sound qualities of lyrics, and often within the timbral qualities of the voice; this can establish a cross-domain connection blending timbre (as a recording element), lyrics and musical rhythm.

Anne Danielsen (2006 and 2010b) has observed microrhythmic relationships exist throughout the range of musical elements—here I expand this notion to include the domains of lyrics and recording. Microrhythmic relationships not only occur within each element, at least as significantly (and at times perhaps more significantly) are the rhythmic gestures that emerge from the *interaction* of elements. While these microrhythmic durations and relationships between elements allow us to understand the nuance of performance and the intricate rhythmic interrelationships that take place in the groove, in the lead vocal, between layers of the musical fabric, microrhythmic durations also exist within the elements of recording; they appear within the sounds of environments and timbres, and they appear as the elements of recording interact between other domains and with other recording elements. With this we are brought to remember the three domains are in confluence. While we separate them to understand their subtleties, they cannot be entirely separated as they appear within the track in some degree of symbiosis and synthesis.

Another type of microrhythm exists within sounds, within their recording elements. These short duration iterations within the sound might repeat a value or present a different value; these iterations often form patterns of time, in time. These microrhythms within sounds are typically heard and calculated in clock-time increments (tenths of seconds, milliseconds, etc.); these temporal iterations are perhaps more accurately termed as 'microtiming'. Microrhythms are both perceived and their values calculated within the context of the song and the metric grid. Sound elements perceived for the microtimings of their attributes are perceived out of the context of the track, and their values are measured in relation to clock time. Rhythms of reflections—as the iterations of a sound that is an environment attribute—are such an example.

Echoes are at the intersection of metric relationships and clock time. They appear in various forms, though always as an audible re-generation of the original sound delayed by some time increment longer than 50 milliseconds. Echoes might relate to the musical texture as rhythmic reiterations of sounds, using the metric grid as a reference. They may also be perceived in relation to an environment, as a series of distinguishable reflections or as a single repetition of the direct sound; in these contexts, the iterations of echo are measured in units as clock time, as they are within of a self-contained sonic impression. The object of analysis here is the echo and the sound in which it is contained, not how the sound fits into and relates to the track—though it is common for the delay time to be of a duration related to the pulse of the metric grid (as in Les Paul and Mary Ford "Falling in Love with Love" (1955) using a semiquaver echo),[5] it need not be. This allows us to recognize microrhythm functions in analytical listening contexts (related to the sound's relationship to the track) and in certain critical listening contexts (where the sound is detached from its context within the track). In critical listening, 'microtiming' is more accurate, as the timeline reflects clock time.

Some common attributes within the elements of recording that make use of microrhythm or microtiming are:

- Delay echoes of sounds
- Early time field (of environments)
- Dynamic contour of reverb (of environments)
- Time- and amplitude-based rhythms of reflections (of environments)
- Dynamic envelope (within timbres)
- Dynamic contours of spectral envelope (within timbres)

As the time delays of echoes shorten and reach a speed of approximately 20 times per second, the iterations fuse into pitch perception—given a continued regular repetition lasting a sufficient duration. When sounds are followed by only one or just a few repetitions of similar amplitude, the iterations will fuse into a single sound object with a timbral quality; the result is a quite subtle, though distinctly unique timbre. A timbral quality unique to the time increment is established; in this way we are able to perceive a 3 millisecond (ms) time delay (reiteration of the sound) as having a unique and different timbral quality than (as examples) 8 ms or 5 ms, which will also have unique timbres. Recognizing these 'timbres of time' facilitates identifying the content of many microtiming attributes—for example certain qualities of environments. Snyder (2000, 125) describes this: "[T]one color . . . is the quality of the microrhythm of events that are too fast for us to perceive individually."

The short durations of microrhythms and microtimings often flirt with the thresholds of perception. It can prove valuable to reference microrhythm's metric grid against clock time. Table 6.2 provides a summary of related perception thresholds. It identifies the smallest microrhythmic unit we can reliably perceive is (at its best) 62.5 ms (or a 16th of a second) between the soundings of two clicks, or of two sounds that contain an abrupt attack—this is the duration of a sixteenth-note at 240 beats per minute (bpm), or a thirty-second-note at 120 bpm. With shorter durations between iterations, the sounds begin to fuse from individual events to a sensation of pitch (Roederer 2008). This rhythm to pitch transition zone is interesting as we move between ways of hearing and processing:

> [A]lthough the *only* thing that changes as we pass from below 16 cps to above it is our own neural processing, the nature of our experience changes dramatically . . . the ear and auditory cortex switch over from functioning as an event pattern processor to functioning as a more holistic waveform processor.
>
> (Snyder 2000, 27)

2 ms (0.002 seconds) is the threshold for hearing clicks as individual events (a threshold that may extend up to 5 ms for some individuals); with shorter durations between clicks we no longer perceive separate iterations—the two are experienced as a single tone, as if occurring simultaneously. This is important for our understanding of microtimings within sounds, especially those of environments.

Table 6.2 Time values and brain processes related to rhythm and pitch perception.

Time Value	Brain Process	Description
2 ms (2 milliseconds = 0.002 seconds)	Window of simultaneity	Threshold interval below which two tones or clicks are experienced as a single tone
3–25 ms $^3/_{1000}$ – $^{25}/_{1000}$ 0.003–0.025 seconds	Threshold of order	Time window within which two different events will be perceived as separate, though perception of event ordering is unreliable; above 25 ms ordering becomes reliable.
50 ms $^1/_{20}$ second, 20 Hz	Pitch fusion threshold	Threshold time interval below which chains (as opposed to pairs) of very similar events fuse together into a sensation of pitch
62.5–50 ms $^1/_{16}$ – $^1/_{20}$ sec 16 Hz–20 Hz	Rhythm-to-pitch transition zone	Time window within which the 'speed' of reiterating clicks transforms into a perception of the 'height' of pitch (rhythm transforms into pitch)
62.5 ms	Rhythm and durations	16th note duration at 240 beats per minute (bpm) 32nd note duration at 120 bpm 64th note duration at 60 bpm

While we are sensitive to these arrivals, our perception is that of timbre, not of discrete sounds; our perception of timbre becomes linked to time (Handel 1993). This helps define the limits of our capacity to perceive and observe the subtle workings of recording elements.

Skill in recognizing the timbres of microtime increments is important to perceiving qualities of environments (Chapter 8) and timbral content (Chapter 7) in perception and will be explored there. The first step toward engaging this area is to learn how one personally, innately and uniquely remembers tempo. We each have a personal, spontaneous tempo (Handel 1993, 385)—a tempo one carries as a preferred rate of periodic movements, such as foot tapping—and we each carry the potential for an absolute memory of tempo related to familiar songs (Levitin & Cook 1996). Our spontaneous tempo can be recalled viscerally with significant accuracy. Our ability to recall the specific tempo of familiar songs is equally capable of significant accuracy; for some of us experienced performers and listeners it may be more accurate to rely on the tempo of well-known records. Once there is facility at accurately recalling tempo consciously from either or both origins, specific tempos can be used as references to perceive and calculate clock-time increments—moving between the rhythmic context of tracks and the content of timbres of time increments that need not be related to rhythmic context. This important skill can be developed by working through Praxis Study 6.1.

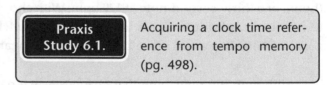

Praxis Study 6.1. Acquiring a clock time reference from tempo memory (pg. 498).

Macrorhythms

Rhythmic sequences and relationships using durations longer than phrase lengths are common for recording elements. These macrorhythms are perceived as changing scenes within the track, a rhythm of sections; sometimes patterns form, sometimes not. In this way, one or more recording element can exhibit a stable set of characteristics or qualities within a time length, and change to another set of qualities in another (or others). Changes at the structural level of sections and higher levels are common for certain recording elements.

An example is a change of distance relationships and qualities of sound sources that commonly occurs between a verse and a chorus. A set of distance relationships and qualities between the various instruments and vocals is established in the verse; this set of qualities and relationships is often changed to a new set in the chorus. Alternating verses and chorus establishes a shifting pattern. This type of change between verse and chorus can be reflected in distinctive differences in one or several other recording elements as well, bringing changes in audible characteristics that can alternate in a rhythm of sections. This may readily establish expectations within the listener for the pattern to continue.

More complex relationships are possible, especially as recording elements interact and interrelate. Rhythms of higher structural level sections might appear as recurrences of groupings of major sections (such as groupings of verse+chorus, verse+bridge, or prechorus+chorus) or perhaps broader patterns. Some of this activity might carry a sense of rhythm and timing, and some might appear as changing sonic qualities without a temporal dimension.

Typology of Rhythm and Rhythmic Patterns

Rhythm may appear within any characteristic of all recording elements. A potential exists for rhythm to be woven into the typology of most recording elements (see below). As rhythm generates patterns of duration that facilitate grouping of activities within elements, these patterns, too, are incorporated into typologies.[6]

Rhythm and rhythmic characteristics pertinent to recording elements might pertain to:

- Tempo and meter
- Divisions of the beat and number of types of durations
- Shortest and longest note durations (boundaries)
- Surface rhythms within phrases
- Recurring rhythmic patterns
- Lengths of recurring patterns
- Primary or significant rhythmic patterns
- Rhythmic patterns at the length of the prevailing time unit
- Division of recurring rhythmic patterns into sub-patterns, motives, gestures
- Rhythms of groups of phrases, sections and groups of sections
- Bending of pulse with microrhythms
- Common microrhythmic unit
- Usage of microrhythms and microtimings
- Use of rubato (and the ways it is used)

This is a partial list of attributes and is in no way complete. Any individual track may necessitate a unique approach toward the typology of rhythm as it pertains to one or more of its recording elements.

DATA COLLECTION AND TRANSCRIPTION OF RECORDING ELEMENTS

The elements of recording often exhibit rich detail. By their influence, they provide what may be the most fundamental qualities of the record.

Rock is a tradition of popular music whose creation and dissemination centers on recording technology.... In rock the musical work is less typically a song than an arrangement of recorded sounds. Rock music is both composed and received in light of musical qualities that are subject to mechanical reproduction but not notational specification.

(Gracyk 1996, 1)

The record itself is the art object, the expression. The elements of recording contain just as much substance and subtle information as music and lyrics. Describing these elements is just as perilous. Again we will turn toward examining what is present within these elements to observe the content of sound, and to collect data on its attributes and their values. As with the other domains, transcribing elements into a notation might help collect this information; we will also make use of typology to define and organize their variables and values.

Notating Recording's Elements

Notating some aspects of the elements of recording can at least partially circumvent the need to describe the unfolding activities of recording's elements and the substance of its sounds—just as notating music circumvents describing every pitch and duration in a melodic line. The challenge turns to devising a suitable system of notation to depict their sonic gestures and qualities. Such a system needs to capture qualities that exist on many levels of detail, and qualities that defy being blended into traditional notation systems. The system offered here does not pretend to be perfect, but it is accurate and detailed in displaying elements and functional in its use. Other approaches are certainly possible, and others (such as the soundbox) have been devised.

The notational system offered here is based on X-Y graphs. It can be used for writing the dimensions of sound, capturing the details of recording's elements, and for documenting elements of recordings; it assembles the data of observations and makes it possible to study the elements deeply. It is a notational system to represent recording elements, making their qualities visible and (at least relatively) clear for observation and analysis. Graphs may take many forms; forms are most often based on the formats:

- Plotting a single element within a single sound source
- Plotting multiple elements (in separate tiers), within a single sound source
- Plotting the activity of multiple sound sources within a single element (within one tier, or multiple tiers)
- Plotting the activity of multiple sound sources within multiple elements (with a tier of the graph dedicated to each separate element)

From these combinations of number of elements and number of sources, great flexibility results—allowing the method of graphing elements to bend to be suitable for the individual track (and thereby conform to and reinforce the framework) and the data being collected. Aspects of the sound itself are visually displayed; this is in direct contrast to music notation, where the notation is a set of performance instructions. In this system, the notation is a visual representation of the actual sound. Like a transcription of music or the writing of language, the X-Y graph brings the elements to 'hold still' for study and reflection, and provides a visual representation that can support an analysis' description of the track.

The notational system presented here is the result of some considerable refinement over several decades of use by colleagues in various production positions and in a variety of academic levels and institutions, by generations of diverse student groups, and by me (in production work, teaching, and recording analysis). It has evolved significantly since the early 1980s, when I began devising it and using it as an analysis tool and in composition and production work. During the early research of this work, Pierre Schaeffer generously offered some guidance that influenced my first drafts, an influence that continues still.[7] He offered that notation must distinguish between the sonic, natural and universal, and the musical, cultural and specific within a musical system; the first is a 'necessary law,' the latter more or less arbitrary (author's correspondence with Pierre Schaeffer in 1983).

I found helpful guidance in our correspondence; it led toward the principles of the notations offered here for recording elements, timbre/sound quality, and for spatial traits as reproduced over loudspeakers. Schaeffer offered that a theory and the notation that illustrates it should address two concerns: what is physical (imbedded within sound itself in its natural form, brought into existence and perception, and fixed) and what reflects the arbitrary system that represents it. Thus, the arbitrary choices of notation must serve the natural, universal sounds from their creation and into perception.[8] The strong pull I have felt to identify what is largely universal (or more specifically, that which is largely in common within a culture) within a record—music, lyrics, recording, and all they communicate and represent, together and separately—and between listeners, had some of its origin in our correspondence.

In the context of this notation, the 'theory' is recording's elements themselves; more specifically, how the sound qualities of recording are defined and divided into elements. The resulting approach to notation and analysis of recording's elements has been directed to sound itself in its natural form as reproduced in the record, as fixed within the record, and as perceived/interpreted by the listener/

Table 6.3   Primary X-Y graphs for observation of recording elements at various levels of perspective.

Element	Perspective	X-Y Graphs
Pitch	Individual Sound	Pitch Area [Graph]
	Individual Source	Melodic Contour
	Composite Texture *	Pitch Density
	Overall Texture	Timbral Balance
Loudness (Dynamics)	Composite Texture	Loudness Balance
	Overall Texture	Program Dynamic Contour
Rhythm	*All levels*	*Implied in all graphs that contain a time-axis*
Sound Quality	Individual Sound	Sound Quality (Timbre)
	Composite Texture	Performance Intensity
	Overall Texture	Timbral Balance
Spatial Properties	Individual Source	Host Environment Characteristics
	Composite Texture	Stereo Image Position and Size
	Composite Texture	Surround Image Position and Size
	Composite Texture	Distance Position
	Composite Texture	Sound Stage Imaging [Graph or Diagram]
	Overall Texture	Sound Stage Dimensions [Diagram]
	Overall Texture	Holistic Environment
Example Combined Elements *with* Timeline	Composite Texture	Loudness Balance and Performance Intensity
	Composite Texture	Loudness Balance and Pitch Density
	Composite Texture	Distance Position, Stereo Imaging and Pitch Density
Example Combined Element in Diagrams	Composite Texture	Pitch Density and Stereo Imaging
	Composite Texture	Distance Position and Stereo Imaging
	Composite Texture	Soundbox: Distance Position, Stereo Imaging and Pitch Density

* Represents clear presence of individual sound sources; sources may be observed individually or in a relationship of equal prominence with all other sources.

analyst. This notational system presents the content of each element by placing their variable values against time; the timeline is most often the metric grid—allowing the recording elements to be directly compared to those of music and lyrics. The notational system is intentional in its content, directed toward clearly and appropriately presenting each element, carefully devised for ease of use (as much as possible), and containing enough commonality to culturally accepted practices of notating ideas and materials to be accessible and readily learnable (not an easy task)—and the notational system is inherently arbitrary in its choice of the X-Y paradigm and other details, though intent on *serving* the content of recording's elements and their place in the track without distorting their content.

Creating these graphs has value in itself. The process draws one into the listening process directly, and deeply. It brings one to engage each element individually and to observe its activities. This leads to detailed discoveries into what makes each element unique, how they are shaped and ultimately how they contribute to the track.

The X-Y graph is highly adaptable, and can be made suitable for any of the elements of recording. Further, it can be used at any level of perspective, from showing the smallest detail to depicting the overall shape of any element's material. Table 6.3 lists the collection of graphs that are central to this approach; most graphs are dedicated to a single element, and each graph is focused at a specific level of perspective (though the 'composite texture' can be 'viewed' and evaluated at various levels of perspective).

Features of this notation include:

- Graphs use the X-axis for time, and thus mirror music notation's use of time movement from left to right; axis can be divided in increments reflecting the song's meter and measures or can use clock time as appropriate for the passage and element
- The normative timeline resolution is for each increment (division) of the timeline to represent a single measure, though the divisions can flex to reflect changing meter and tempo
- Any element can be plotted on the Y-axis against the same timeline resolution
- The music's structure can be displayed under the timeline to add clarity to the material and its relationship to the song
- The Y-axis readily conforms to the needs of individual elements, to reflect their unique content, and to present their variables and values with sufficient detail and clarity
- Elements are graphed individually and at specific levels of perspective
- Elements may be graphed in a high degree of detail and with great precision
- Graphs can demonstrate as much detail as the analyst wishes by modifying the resolution of either or both axes
- Several (to many) sound sources can be plotted on the same graph provided they are clearly labeled or otherwise identified
- Several tiers can be stacked against the same timeline (allowing different aspects of an element, or different elements to be readily compared)
- Two different elements might be placed on opposing X-Y axes to visualize their interaction at a given moment in time, or within an identified section of the track

Graphs are created through a transcription process: observing the activities of elements, and transferring those activities to the graph. The levels and changes within the element are transferred to the graph as they are recognized, typically located in time against the track's meter and measure divisions. Each element is graphed in a unique way, using a Y-axis specifically devised to portray its continuum and activities.

Figure 6.1 presents a sample graph that contains the material of three elements: loudness balance, stereo location and distance location. Note the graph is at the perspective that displays the activities of individual sound sources; it can be viewed at the perspective of the activities of individual voices/instruments or viewed from the vantage point of perceiving all instruments simultaneously and with equal

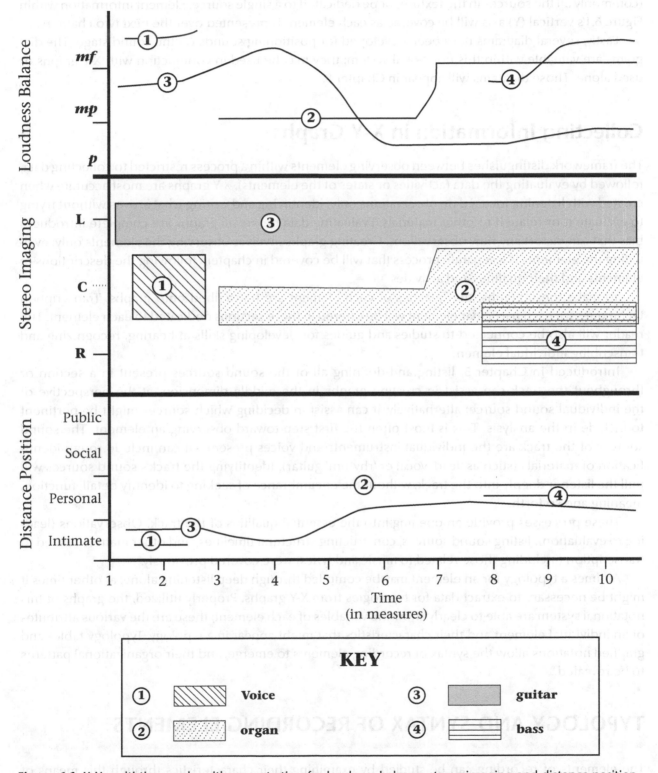

Figure 6.1 X-Y multi-tier graph, with separate tiers for loudness balance, stereo imaging and distance position.

prominence. The Y-axis is divided into three tiers, each incorporating a different set of divisions and boundaries unique to each element. Notice the four instruments are numbered in the key, and appear on each tier; their appearances on each tier are aligned against the timeline (X-axis) and their activities oriented against the Y-axis scales and increments for each element. This particular graph is somewhat unusual; most graphs will be dedicated to a single element, and they commonly contain more sources (commonly all the sources in the texture) or be dedicated to a single source. Element information within Figure 6.1's vertical (Y) axis will be covered as each element is presented over the next two chapters.

Lastly, several diagrams have been developed for positioning sounds on the sound stage. The diagrams are valuable within this notational system; they may be used in conjunction with X-Y graphs or used alone. Those diagrams will appear in Chapter 8.

Collecting Information in X-Y Graphs

The framework distinguishes between observing elements within a process restricted to collecting data, followed by evaluating the data (activities or states of the elements). X-Y graphs are most accurate when created with listening focused on observations only; identifying and writing what arrives without trying to evaluate it or relate it to other materials. Evaluating data before all graphs are complete introduces bias that can distort any new observations. Creating graphs involves observing the elements only; evaluating the elements is a separate process that will be covered in chapters following the descriptions of elements and their qualities and activities.

To make these graphs, the qualities of each element are transcribed onto graphs. Transcription techniques for each particular element will be offered in these sections dedicated to each element. The reader will also be connected to studies and guides for developing skills at hearing, recognizing and transcribing individual elements.

Introduced in Chapter 5, listing and defining all of the sound sources present in a section or throughout the track can assist in creating graphs in the middle dimension, at the perspective of the individual sound source; alternatively, it can assist in deciding which sources might be pertinent to include in the analysis. This is most often the first step toward observing an element. The sound sources of the track are the individual instruments and voices present—it can include rough identification of materials (such as 'lead' vocal or 'rhythm' guitar). Identifying the track's sound sources will pull the listener deeply into the track, without the encumbrance of seeking to identify detail, function, meaning and so forth.

These processes provide an opening into the essential qualities of the track. Observations (leading to evaluation), listing sound sources, constructing structural timelines, and other matters related to transcription (including those related to music and lyrics) were covered previously.

At times a typology for an element may be compiled through deep listening alone; at other times it might be necessary to extract data for typologies from X-Y graphs. Properly utilized, the graphs of this notational system are able to clearly show the variables of each element; these are the various attributes of an individual element and their characteristics that might appear in a typology. Typology tables and graphed notations allow the syntax of recording elements to emerge, and their organizational patterns to be revealed.

TYPOLOGY AND SYNTAX OF RECORDING ELEMENTS

The elements of recording can be studied by examining their characteristics through this means of categorizing and organizing observations.[9] Typology tables allow this type of categorization and organization—this is an important supporting concept of the framework. As we previously explored in Chapter 2, typology can be used to reveal the syntax of the elements of recording—their content and their activities, and ultimately how they function and their conventions of usage within a track. Along with the various graphs for recording elements, the data set of a typology can be utilized within the 'evaluations' and 'conclusions' processes that form a stream throughout much of Chapter 10. While typologies will be used here to observe the elements of recording, the concepts of typology might prove effective for examining elements of the other two domains (music and lyrics).

Typology is a means to classify and categorize recording element attributes, and to identify and observe their values.[10] The resulting collection of unique attributes and their values establish the unique typology of an element during a specific time period. Stated another way, a typology of an element identifies and organizes each of its attributes by the values and characteristics they contain. A typology might be detailed (with an element's attributes being defined to examine many variables, each of which generating values) or more general (with element attributes roughly defined) in the data it includes.

The specific attributes of recording elements will be covered throughout this and the following chapters. These attributes (or dimensions) within elements are variables, as they hold the potential to change their state or their value. Most elements have several attributes capable of change. Attributes can change in some more-or-less specific way, some reflecting certain levels of values particular to the element; levels are increments of a value (such as dynamic levels).

For example: if this were a discussion of the elements of music, and the element was melody, one attribute (variable) would be melodic contour, a value would be shape of the segment within the typology, levels would reflect the smallest change to within the shape (vertical change of pitch and horizontal change of time).

For recording elements, the terms and idea of 'levels' might be replaced with 'characteristics' or 'traits' or some other appropriate term for a particular element; values might be pitch-areas, locations, dynamic levels, durations, or any other specific 'data unit' (such as an area in space) pertinent to the elements of recordings. What is important to understand, is that 'levels' (or its substitutes) are the means necessary to define 'how' the attribute 'appears' out of the number of its possible variable states.

In discussing typology, "activity" refers to changing values within an identified dimension or variable (of an element), these can be gestural in nature. "State" identifies the attribute's values are fixed (not changing during the time under examination). This implies a typology might only apply to a specific period of time. This is important, as a typology may be compiled for any span of time. A typology

Table 6.4   Typology content and organization.

Attribute	Value (characteristic, trait)
Variable	Level, location, area, duration, or other 'data unit'
– Activity (active)	– Activity reflects changing value(s)
– State (static)	– State is fixed, with unchanging value(s)
Data collection is specific to:	1. Time period (phrase, song verse, chorus, track, etc.)
	2. Structural level or level of perspective
	3. Specific element
	a. An individual attribute, with values identified
	or
	b. As many attributes as best serves the analysis, with each attribute separately categorized

compiled from verses will typically yield contrasting information when compared to a typology of a track's chorus; typologies of certain attributes throughout entire tracks might be pertinent to one analysis, and other analysis (or study) might seek to define a typology of, say, a producer's final albums. Typology can classify an element's materials within a short passage or a complete section, be restricted to the lead vocal or collected throughout the entire track. This points to the final primary consideration: just as elements function at different levels of perspective and/or structure, typologies are also collected at a specific level of perspective, or a specific structural level.

Typology Tables

Typology tables can be especially useful for organizing observations or categorizing data of recording elements. Just as with X-Y graphs, typology tables present information at a specific level of perspective, and within a defined period of time. Typology tables also clearly define the element(s) being categorized, and the specific attributes of the element(s) that are being observed.

A typology of an element will classify the values of its attributes; it establishes a way to access the details and workings of an element and to collect information without interpreting it. Through the process of collecting information and transferring it into a typology, an element's unique dimensions are observed and values are identified—seeking to perform an objective process to define the sound without interpreting it. The unchanging states and gestural shapes (activities, events) of those values (or characteristics) are noted, and data is collected on the element; this ultimately will lead to evaluation of the data, to understand how elements function and how their materials are organized.

A total spectrum of the activities and states results; "within this spectrum are the preferred and predominant types" (LaRue 2011, 16) of these activities and states. A spectrum of characteristics reflects the actual content of the sound; this is the characteristics of each attribute, each element in themselves. This contrasts with a spectrum of contrasts which identifies relationships between values; this allows the identification of information such as the amount of diversity of values (number of different types), range of values (highest and lowest, for example), common increments between values (how values might be related), and more. Seeking to identify predominant values or attributes, those that are most common would be an activity related to 'observations' and data collection. Seeking information regarding the preferred or most significant values—values that perhaps appear at important structural moments, or that support the primary musical materials—identifies information related to 'evaluations' and 'conclusions.'

To step through an example, Figure 6.1 contains a stereo imaging tier with four sounds. The sounds have attributes of positon (placement in the stereo field) and of width (span of area occupied in the stereo field); these are calculated relative to the center location, located by degrees left (+1° to +45°) or degrees right of center (-1° to -45°), with loudspeaker locations at 30° either side of center. Sources may appear anywhere from 15° outside the left loudspeaker to 15° outside the right; this establishes the potential range of location (within which sources may be located) at 90°. The unit of value for stereo location attributes is the degree, which marks the amount of angular deviation from center; increments are expressed as degrees left of center (to +45°) or degrees right of center (to -45°).

A typology table of Figure 6.1 appears as Table 6.5. The perspective of this typology is at the individual sound source (although Figure 6.1 and the data of the typology are also viewable as composite texture, comparing the individual sources); the time period is 10 measures. The typology also contains the attributes of amount of change, speed of change, and spectrum of width values (identifying all widths that appear in the passage). In Table 6.5 the stereo imaging attributes of location are listed for each sound (items one through four); next the widths of four sounds are listed separately (one containing two numbers: its initial width and the width after it changed). Both of these variables are expressed

Table 6.5 Typology table generated from the stereo location attributes found in Figure 6.1.

Attribute (variable)	Value (by level, location or other unit)
Image locations	Sound 1: 11°⇔ -15° (11°L spanning 15°R of Center)
	Sound 2: -3°⇔ -18° and 15°⇔ -18°
	Sound 3: 35°⇔ 17°
	Sound 4: -10°⇔ -33°
Range of locations	35°⇔ -33° (total area of the stereo field occupied)
Size of image widths	Sound 1: 26°
	Sound 2: 15° and 33°
	Sound 3: 18°
	Sound 4: 23°
Range of widths	15° to 33°
Spectrum of widths	All sounds: 15°, 18°, 23°, 26°, 33° (smallest increment of size: 3°)
Amount of change	Sound 2: 18° (width changing from 15° to 33°)
Speed of change	Sound 2: 2 beats to travel 18° (with steady change of level and speed)

in degrees. In this example, in the attributes of the changeable position of the location, range is the area of the 90°stereo playback field utilized (the extremes of the material present in the passage, not what is possible); for the attribute of width, range is the span between the smallest and largest widths. Spectrum, as used here, is the collection of all individual values within an attribute—here within the attribute of image widths. Typologies may also include attributes of change; here one of the images changes size. Figure 6.1 incorporates a source that changes width; the typology table categorizes the sound by the amount of change to its width (the change between beginning and ending width), the speed of change (the time period of the movement from the first to the second width). Other sounds may have attributes that exhibit different types of motion; typologies can incorporate types of motion, regularity of speed, regularity of change, time at which the change occurred, even rhythms created by durations of changing values establishing patterns.[11]

The typology of each element has unique variables and values/characteristics. Using typology for recording elements allows us to engage the content and activities of each, in an appropriate manner. Observations to compile a typology begin with creating the graphs and extend through adding detail to the graph to reflect the activities and characteristics of relevant values. Features from graphs can be categorized into a typology—observing the variables in play, and their values and characteristics, and engaging them as appropriate. Typology begins organizing the information (data) from the graphs without interpreting it. Typology is pivotal between collecting information (observations) and examining it (the evaluation process)—this is explained further in Chapter 10.

As demonstrated, typology can be constructed to extract information from graphs in order to identify information about elements. Questions for analysis can be formulated to explore some of the more common variables found and more common types of values within each element; some generalized topics follow that can be adapted for collecting data appropriate for an individual track, and lead toward a suitable evaluation process.

The following is a listing of generic attributes that are commonly employed within elements of recording:

- Sounds with attributes that are fixed in value, that remain in a steady state
- Sounds that exhibit activity within or change of values

- Values can be represented as 'data units'; such as levels (dynamic level or pitch level), area (pitch area, image width), positional locations (lateral location, distance location), density (pitch density, time density), and others
- Spectrum of values or characteristics (inventory of specific levels of values establishing a collection of all used; allowing identification of smallest and largest increments)
- Amount of variation of levels (any common size of increment, smallest and largest increments, etc.; this information is available through examining the 'spectrum of values')
- Range of levels (for example furthest left and furthest right in use)
- Rates of activity
- Boundaries of rates of activity (fastest and slowest changes, longest and shortest durations, etc.)
- Identifying durations of specific levels (perhaps most prominent, or structurally significant)
- Tracking changes of values (speed of change, amount of change, regularity of change, etc.)
- Identifying patterns of levels/characteristics that may be created by changing increments of values
- Patterns created by changing characteristics or states of values
- Rhythmic patterns created by changing values occurring in patterns of durations

The following are but a few values and characteristics that might be found:

- Increments of levels (such as dynamic levels)
- Increments of other values (degrees in location)
- Basic pulse or speed of change
- Duration of activities or appearances of values
- Characteristics of the values of stable elements, singly and in aggregate

To conclude, typology may be applied to identify the collection of variables for each of the track's recording elements. The element's dimensions are observed by their values and their characteristics within the track. Later within the evaluations process, patterns of activity, types and degrees of contrast, and any characteristic shapes, states, forms or qualities might emerge. Variables might even be identified as clearly structured into patterns of activities varying over time—if this is present. Thus, each element can be observed, characterized, understood and classified to reflect its unique qualities, and in the way(s) in which it might function, interact with other materials, and contribute to the record. Information collected with these typologies might be examined within existing analytical methods, or perhaps studied through a unique self-generated approach based on inquiry; uses of this raw data are numerous, and examinations are (stated once again) based on the goals of the analysis and the preferences of the analyst.

Syntax, Structure and Style

Syntax results from identifiable patterns, and "the ability to identify a pattern as similar although occurring at different times" (Snyder 2000, 200). Identifiable patterns are needed for syntax to be established; patterns based on standardized scales or comprised of proportional values (ones that can be categorized and remembered) create syntax in music. Learned systems, comprised of relationships and content that is "repeatedly identifiable across different experiences" (*ibid.*, 196) is central. The syntax of language governs its sentence structure sequencing of subject, verb and object; it is the patterns of formation of sentences and phrases from words. Syntax can include word choices and phrase orderings, etc., that ultimately result in a learned (or implied) set of rules, principles and processes.

Recording elements have a capacity to contribute content to the record, though standardized systems of organization barely exist and conventions of use are few (compared to music and language). They do more than support or shape a primary domain (music or lyrics), though. Allan Moore (2012a, 29) has noted:

> [F]or recorded popular music, secondary domains [recording elements] are able to do much more than simply "shape" content: indeed, they frequently *constitute* content, even if they do not embody syntax. . . . there is not much academic literature dealing with it, [secondary domains providing content] and the notion of theorization in this area is fraught.

That recording elements do not embody syntax in the traditional sense of music and language is clear; it is also clear a void of study in how 'secondary domains'[12] can constitute content exists, and that it extends to the nature of the content present in recording elements.

Syntax works differently for recording elements, compared to language and to music's melody, harmony or rhythm. Recording elements cannot easily be divided up into many clearly recognizable categories. They cannot be organized into scales, and do not generate surface level materials and relationships (though rare examples of surface activity does emerge in certain elements, in various tracks). They will not generate materials that resemble melody within another element, though this does not mean the activities of these elements are somehow deficient—recording elements are simply different in their content, and in how they function in current conventions.

Recording elements do establish patterns, though, and they do have identifiable values and characteristics. Most elements have values that cannot be scaled; a common, scalable value is not present for elements such as timbre, distance or dynamics. Yet timbres can be remembered and categorized in other contexts—such as everyday communication, bird sounds, etc.—in great precision, and dynamics has little to establish discernment between the thousands of loudness gradients one might perceive. A few elements can be scaled; one such element is stereo location that can be placed precisely at a trajectory from the listener, though listeners are likely not to be inclined (or not open to being induced by the material) to hear these materials as identifiable and repeating. Clearly the scales that create syntax in music's pitch cannot be the model for perceiving recording elements. Many recording elements will "present *different* kinds of possibilities for organization: gradients and simple contrasts rather than more complex patterns" (Snyder 2000, 200; italics in original). These are the attributes and values that are observed and categorized in the typology of a recording element.

Identifiable patterns can be simultaneous sounds as well as successive—we experience this with harmony. For example, the positional relationships of sources establish a pattern of distribution; other similar examples within other elements at the overall texture and at the composite texture are common. Thus, while identifiable surface-level patterning is uncommon by variations of recording elements, identifiable relationships between sound sources are common.

Syntax in recording elements function at different levels of perspective than lyrics and music; it also functions at different structural levels, in different time scales (structural units). The language of recording elements plays out over longer stretches of time, in macrorhythmic relationships. Philip Tagg (2013, 603) refers to this as "long term syntax" or diataxis—"the ordering of episodes throughout a whole piece of music into an overall sequence." Recording elements, functioning at higher levels of perspective, often shape choruses differently from verses, differently from introductions, and so forth; basic characteristic qualities are often established for repeating sections, setting up a sense of syntax that is broader in relation to time.

This can establish a sense of expectation for sound to change, and anticipation for what is next to arrive, what might next appear. These senses of expectations and anticipation contribute to (or may even establish) movement and motion within the track. These qualities of sections can be shaped by any element, and at any level of perspective; the overall texture's timbral balance and sound stage might play important roles; so can relationships of individual sources, or a recording element's qualities within an individual sound source (such as the timbre and distance placement of the lead vocal).

Syntax of recording elements also contributes to shape and style. The elements' materials (changing values) and characteristics (unchanging attributes or values) are reflected in the shape of the track, and are important components of its form. Timbral balance, sound stage, holistic environment (also called perceived performance environment) and perhaps other timbral attributes are present in the track's crystallized form. Recording syntax can have some consistent qualities across many tracks, and be an important component in establishing 'musical style' or fundamental qualities for genres of popular music. Bob Snyder (2000, 201) proposes: "[T]he rules of syntax may be part of a tradition, or they may be established within the context of a single piece of music." For recording elements, the short history of their presence in music records and listening has brought few entrenched conventions, and as such syntax within its elements is highly malleable for an individual record, or for an entire genre of music. This does not limit its ability to shape a track; arguably, recording elements are more flexible in their potentials to supply the original sounds, sonic relationships and platforms for expression that are so often sought in popular music.

As we end this section on engaging the recording elements, and as we prepare to experience them, we return to the principle of 'listening with intention and attention.' Introduced in Chapter 2, the principle will be put into practice as we examine the elements of recording, and learn to recognize their qualities—and as we graph them. Throughout the remainder of this book, as we engage the recording analysis process, that principle of the framework will be applied to all elements and domains. Listening with intention and attention will guide our observations of recording elements, and all other aspects of tracks.

The acts of creating typology tables, graphing elements and otherwise collecting data require detailed listening at all levels of perspective. For the domains of music and lyrics, the analyst had the advantage of considerable exposure to those elements and a considerable ability to discern the activities and materials of each element. For recording elements, many analysts will be learning to direct attention to qualities never before experienced, directing attention to levels of perspective rarely consciously engaged previously. Hearing the elements of recording may require refining conceptualization and listening skills.

HEARING THE ELEMENTS OF RECORDING

Discovering the elements of recording within the fabric of the track can be elusive, especially at first. Hearing them accurately typically is the product of guided study.[13] Many important traits of recording's qualities are subtle, and most listeners may not have previously been aware of their presence. The sounds of recording are perceived (heard) non-consciously and unconsciously by many music listeners—they may not be consciously engaged in auditioning them, but listeners perceive and react to recording elements. To hear the sounds of the recording requires listening in ways that are different from how we typically engage music and speech, and most all of everyday experiences—it also requires knowing what to listen for.

It is difficult to recognize a quality not previously experienced. Experiencing a new quality requires taking one's attention to a different place in perception—this place for some aspects of the recording is beyond experience and perhaps imagination. It also requires a commitment to trust that what is not known is actually present—that this sound quality that has never been perceived, or never been experienced, actually exists. When one has never experienced a certain sonic dimension, it is difficult to know how to listen for it; how to discover the dimension of sound and determine how to hear it. It is a challenge to direct attention to recognize a sonic dimension that is beyond imagination and experience.

In order to hear certain qualities of recordings, one is required to listen in unnatural ways—or, rather, in ways unlike how we listen day-to-day within real-life listening situations, like interacting and gathering information. For example, when we recognize a friend's voice we perceive the overall quality and know it is our friend; we hear subtleties within the voice that let us know if they are excited to see us or are upset, but we only engage those qualities in as much as they modify the overall sound. In listening to the recording element of timbre, we listen within the sound to hear the components of timbre that all together make up the overall quality. Spatial qualities will likewise bring one to listen differently, to be open to unanticipated and unknown qualities.

Listening at levels of perspective we rarely attend to consciously takes deliberate attention, and is something many readers may not have previously experienced. Deep listening within sounds is the gateway through which much detail and richness of recording's elements is accessed. This is countered by listening to the overall qualities of the track—the entire record as a single waveform. Both of these extremes are unusual experiences, and represent listening challenges (at least initially) for many. Also unusual is listening at the level of the composite texture, just above the level of individual sound sources; while recordists, conductors, composer/arrangers often engage this level of perspective, many musicians do not, and lay listeners are rarely aware of this way of hearing.

Some of the sound qualities within recording's elements do not exist in nature; this makes believing that we hear them harder, and makes analyzing them tenuous. We never encounter these qualities outside the record. While we may have intuitively or subconsciously 'heard' them within the record's context as they shaped its expression, they were likely at least partially (if not entirely) out of our attention and awareness. For example, in nature we do not 'hear' the size (width) of a car horn, though image size (for example the width of a lead vocal) is a significant factor within the record. Some qualities are subtle, and will only reveal themselves when the listener 'learns' how to discover them, how to hear them. In real life situations vision is nearly always used to verify or determine the precise locations of sounds around the listener—vision plays a conclusive role, after sound provides an initial, general indication of direction. In records, of course, sounds are invisible, void of visual cues; a keener awareness of the sound qualities of spatial perception is cultivated, and some spatial skills outside the experiences of worldly physics are needed. Guidance for embracing these challenges will be offered throughout.

Learning to hear recording's elements is particularly difficult. We are challenged to discover aspects of sound that defy our real-world experiences, to cultivate unusual and specific types of attention, and seek dimensions only known to us by descriptions. How can we be certain our interpretation of the sound is complete and accurate? In terms of our perceptions, we are all alone in our own minds. How, then, do we gauge that our perceptions are the same as another's?

Our language on sound challenges all this even further. How might we describe the aspects of sound we are learning to hear, when few pertinent terms are available? How can we communicate our interpretations and our perceptions to others? How might we at least partially 'de-personalize' our perceptions, interpretations, and observations—and thereby allow our analyses to be more meaningful to others?

Describing Sounds and Sound Qualities

Building on our prior discussions of timbre in music and in the language, expression and performance of lyrics, describing sounds in the recording offers some additional challenges. It also reflects some similar concerns. The following discussions might add some clarity to music and lyrics timbres, and offer tools for examining and describing sound quality in all three domains.

It is difficult to discuss sound with another person—to exchange information about sound itself, let alone the sound of the recording. Two factors that seem to contribute to this are the difficulty to identify what constitutes a shared experience and a weak approach and insufficient vocabulary for describing sound and sound's qualities.

It is difficult to describe an experience (and sound, music, lyrics, the track are experiences) without imposing a personal memory, association, or interpretation. To accomplish this takes an awareness of remaining unbiased, and the ability to do so. It is a challenge worth the effort, as this is just what recording analysis is about: discussing a record's substance, materials and sound qualities in such a way as to be of value to others, so they might better understand *their* own experience of the same track (an experience of the same, shared sound characteristics), not learn about *our* own *personal* experience. We can seek to describe the track as an independent object, one that is outside of one's self—something objective on the outside, not personalized within (Ford 2019, 18–20).

Distinguishing our personal biases from a more objective position requires self-reflection. To what extent does our personal interpretation distort the objective qualities of sound and the track? To what extend does any discussion of expression and affects rely on our own personal experience, rather than a cultural sense of perspective? Stated differently: how personal are the analyst's interpretations? Describing sound, and especially the elements of recordings, should be clear in its approach to personal impressions. This holds no matter the goal of the analysis.[14]

Attention is drawn to determining to what extent personal impressions guide choice of language and guide observations of the content of sound. When all of analysis is, in the end, an interpretation—one based on our own experience, prior experiences, personal biases, etc.—to what extent does one seek clarity between what is personal and what is cultural, between the extremes of seeking to describe the track as a *shared* experience with the greater audience of common background, or providing a personal interpretation (no matter how learned)?

The concept of a shared experience points to the elusive question from phenomenology: 'Is my experience the same as yours?' Turning this question for our purposes in recording analysis, the only experience of mine that might be the same for others is the record itself—but only the artifact of the record *before* it is experienced. In other words, the (hypothetically unaltered) track sounding in air before it becomes a personal experience by my (or your) unique physiology and experience, perception and cognition, personal predispositions and biases, and more. And this assumes the possibility of establishing identical sound qualities of the track reaching the analyst and the greater audience—a state entirely unrealistic, and yet needing definition to establish common ground. Once the sound or track becomes experience, intersubjectivity becomes the common ground.

Interpretation of a listening experience leads to personalized impressions and language that might often focus on the listener's feelings or reactions to what was heard. These interpretations may be meaningful to ourselves—and perhaps have great value to the person in the mirror—though such affective responses are often of little real value to others wishing to learn from an analysis. These are often driven by reactions to the experience rather than descriptions brought towards understanding its content. Eric Clarke (2011, 198) explains this further: "[T]he manner in which people report their musical experiences is necessarily a product of graspability of the musical materials, of individuals' capacities as listeners, of the limitations of language, and of the inaccessibility of unconscious and non-conscious processes."

Any analysis process requires, by definition, that we step outside—to the extent possible—our biases; tools for this are offered for each element, as they are explored in turn, throughout Chapters 7, 8 and 9. Analysts seeking a broader voice, one that speaks to as many others as might be possible—one that embraces the substance of the materials with some objectivity—are faced with the task of communicating with clarity. Still, we are often prone to talking about our impressions of the experience and what the experience elicits (associations, feelings, etc.)—but not sound itself. This points to the two approaches to observing timbre: for internal content and for overall character (these are explored in Chapter 7).

It may be difficult to articulate an interpretation or analysis with any kind of specificity, given our limited vocabulary, but that is what we must do. While we have few non-personalized ways to describe sound in general, let alone the elements of recording, we seek to communicate about the content of the sound or material, or about the shared experience. In the end, verbal descriptions become necessary for the study of the record, and to talk about it.

Finding Common Ground: A Shared Vantage for Talking About Sound

Our approaches to discussing sound do not allow us to address its substance, or the specific dimensions of its content. Nor does it allow for us to capture or define its character without imposing personal bias. We have no language to describe timbre, no timbre-specific adjectives; timbre is often merely "described in metaphor or by analogy to other senses, and this is true in many, many languages of the world" (Fales 2002, 57). We have no satisfactory way of describing what we are hearing with timbre—as well as with other qualities of sound, elements of the recording and numerous traits within the track.

As touched on before: our custom is to describe sound by using analogy and by using cross-modal terminology from other senses. We resort to words such as 'warm,' 'round,' 'fat,' 'mellow,' 'percussive,' 'edgy,' and a *great* many others—imprecise at best, and typically grossly inadequate and ineffective; often misleading, and commonly merely meaningless jargon. When we resort to describing sounds by mood connotations (i.e. 'somber'), the imprecision of language may get more extreme. Table 6.6 lists a small sampling of words recently used by some to describe sound qualities. This is to demonstrate how creative language can be, and at the same time how meaningless or inexact; I do not advocate their use, unless they are clearly defined within the context that a term is used. An example of the intent to share important information about sound with this vocabulary, Dean Nelson (2017, 132) has written: "Words I use to describe tone are warm, woolly, fuzzy, dull, dusty, woody, vibrant, shrill, metallic, glassy and crystal clear, just to name a few." The terms offered by Nelson are obviously meaningful to him, they are of personal value to his work, and reflect his understanding (and unique interpretation) of those sound qualities. These terms are almost certain to be interpreted differently by others—perhaps a subtle difference, though potentially a substantial difference. If one has never heard a sound or sound quality, it is nearly impossible to accurately imagine it by having it described in this way, or by any of the terms in Table 6.6. Still, without alternatives we continue to try to communicate in these ways, and certain terms even become widely used, though they mean different things to different people—they might provide an illusion of transferring knowledge or communicating about the experience of sound, the material or a track, but they fall far short.

Taking this further with one more example, the term 'punchy' is currently appearing in many contexts—describing types of sounds within tracks, performances, overall sounds of records, and even applied to certain styles of popular music. The term *might* communicate a sense of energy and intensity, and perhaps something of the character and affect of the sound, but communicates nothing about the substance, or 'sound qualities' of the sound itself or its content. Still, many might be quick to 'know' what a 'punchy sound' is—though each would define it differently given the means and skills to describe their idea of the qualities of a 'punchy' sound. Perhaps they have experienced recordings or performances that others have described as 'punchy' so they believe they know what 'punchy' is—or perhaps they have developed a personal impression of 'punchy.' In whatever way the definition was developed, it will be at least somewhat vague and largely personal, and prone to describing affect over

Table 6.6   Listing of terms recently used to describe sounds; chosen at random from a variety of sources. The list reflects a cross-section of common, current word usage.

aggressive	creaky	glassy	mellifluous	punchy	spacious	veiled
airy	dark	grainy	mellow	raucous	steely	vibrant
bassy	deep	grating	metallic	rich	strident	violin-like
blanketed	delicate	grungy	muddy	round	sturdy	warm
bloated	detailed	harmonious	muffled	seismic	sweet	weighty
boomy	dull	harsh	nasal	shrill	thick	wet
brassy	dusty	heavy	opaque	sibilant	thin	woody
breathy	edgy	hollow	open	silvery	tight	woolly
bright	euphonious	honky	piercing	sizzly	tinny	
clear	fat	juicy	plaintive	smeared	transparent	
closed	full	liquid	present	smooth	true	
cool	fuzzy	lush	puffy	soft	tubby	

substance. The term does not reflect the content of the experience several listeners might have heard while together, or have in common from separate hearings. That the term has been used to describe entire genres of music, individual recorded songs, instrument qualities, the character of loudspeakers, and more, is testament to its vagueness and imprecision, and the flexibility of its meaning.

It is clear these terms are largely subjective and they are personalized—with terms simply stated, without being defined. Meanings will vary between individuals, and communication will be imprecise (at best). We need different approaches to describing sounds.

To talk about sound, it can be most useful to begin by examining the sound itself. If there is any semblance of joint experience—and a common sonic point of reference—between humans, it is the sound itself. In as much as it is possible, considering the sound before it is transformed by perception and interpretation (or attempting to minimize their impact) provides an opening to a more objective understanding of sound. Allan Moore (2003, 7–8) explains this: [When]

> we hear a sound when our eardrum vibrates at, say, 440 Hz, having set in motion by the sound-waves vibrating at the same speed. They, in turn, have been set in motion by the vibration of some material, again at that speed, . . . This is an objective description of what happens. In saying this I mean that any person with normal or near-normal hearing from whatever culture they come, will have their eardrums vibrate at the same speed. They will not identify the "sound" that the brain receives in terms of speed . . . there is a code fairly widely accepted . . . which would call the sound "A above middle C" and, in choosing whether to use that code or some other, we are entering into an act of interpretation, but it is an interpretation of the cultural context of the sound, of how to understand it, rather than the sound itself.

Objective description is available through recognizing (actually extrapolating the state of) the waveform prior to its arrival at the listener.

In describing the physical qualities, a more objective, more readily understood and more functional communication about the content of sound is possible. For character, one might bring focus to an interpretation supported by cultural codes, and that provide more substance and universal application than personal views. Both the person talking and the person receiving the communication may discuss something about the sound itself; this may be in addition to (or in place of) an analogy or metaphor of it. A more meaningful interpretation and exchange of knowledge might be possible in this way, including the experience of the sound and its place in the context. Terms such as those of Table 6.6 might become meaningful when they are selected based on an interpretation that includes observations of the sound's content, as well as character—the substance of the sound balanced with what the sound evokes in the listener. Descriptive terms will have more value and communicate more clearly when chosen carefully to reflect the broader culture's view of the sound characteristics and the word choice explained or defined; descriptive words require definition or explanation to communicate with any precision. This act blends the content of the sound with the interpretation of its character.[15]

This process requires one to listen inside the sound and to recognize the physical dimensions of the sound. This necessitates some knowledge of the physical dimensions of sound, and some focused practice and skill to access them. Using this approach, communicating information about the sound itself is possible, though the approach takes attention to develop and some on-going discipline not to resort to old patterns. While this approach may not be readily adaptable for spontaneous dialog (especially at first), it can be valuable for recording analysis (Moylan 1992).

The physical dimensions of sound can be described by observing the values of its component parts, and changes to those values throughout its duration. Word choice can indicate a level of perspective and focus a description. For example, a dynamic envelope can be described by its contour and loudness levels over time; spectral content can be addressed in specific terms of what partials are present and when they are present, and the contours and levels of their individual amplitudes. All other recording elements—as well as the elements of other domains—can be addressed with equal precision and detail by describing their unique dimensions and activities against time (Moylan 2015). This should make obvious for us to recognize: substantive communication about sound—with sufficient accuracy and depth to be of use in recording analysis—requires one to identify what is present *within* sounds, in order to inform the *global quality* that results from the *coalescence* of components within sounds. Talking about the substance of sound can be accomplished with some preparation and background. This will receive more detailed coverage as this chapter progresses.

In the end, it is important to recognize that to describe any experience or object is to distill it. Descriptions by their nature filter detail and richness; they generalize, minimize, provide an overview. The very act of putting a sound into words diminishes its experience. Notation might retain some of the richness and detail of sounds—if appropriately devised to capture specific details and dimensions—though much is still lost.

Chapters 7 and 9 will broaden discussion to describing the character of timbre. Despite the great difficulty we have in talking about sound, verbal description is a necessary part of analysis. Within an analysis sounds (timbres and other qualities) need to be described in a way that minimizes misinformation, misunderstanding and confusion between individuals, and clearly presents the substance of the discussion.

Listening inside sound (content) as well as listening to its overall attributes (character) can bring clarity and enhance language and descriptions. These are supported and guided by the data collection of observation afforded by compiling typology tables of, or transcribing recording elements. These processes open the content and character of elements to the analyst.

LISTENING TO RECORDS AND RECORDING ELEMENTS

Consider each of these rhetorical questions for a few moments. They are interrelated, and some reframe prior questions to put forward another point of view:

> In your best listening situation, is what you are hearing what is on the record? In other words: Are the qualities you are hearing the same qualities that were instilled within the record—the qualities the artists and recordists shaped as its final state?

Are you hearing an unaltered presentation of the track's artistic statement, or some type of transformation?

To the extent your listening situation and playback system is altering the sound of the record, do you know the approximate or precise nature of those alterations?

To what extent is it important to a recording analysis for the analyst to hear the track in its original, unaltered form? To be analyzing the record with its intended qualities? To what extent is this important for *your individual* analysis?

To what extent do your listening practices bring the qualities of the track to vary between repeated listening sessions? Are there potential consequences to not hearing the same qualities during each hearing of the same material?

Without hearing a track in a way that does not transform the sonic content of the record—and that has consistent qualities with each re-hearing—what might one expect from an analysis of a record?

Is it important for the sound of the track you hear when performing an analysis to be similar to—contain some common ground with—the qualities heard by others experiencing the track?

If your playback of the track transforms its intrinsic qualities, what are you hearing? Can an analysis of the track that is meaningful to and reflective of the experiences of others be performed in this way?

You are likely reading this because you have some interest in learning more about 'how the record shapes the song.' We now encounter a situation wherein *your* choices shape the record as well. Your choices of how you listen shape your perception of the record, your experience of the content of the track. The analyst's choice and use of playback variables fundamentally shapes an analysis—fundamentally shapes the analyst's interpretation of the track. This returns us to the goal of the individual analysis. By acknowledging the fundamental impact on the content of the listening experience as it is shaped by its delivery to the listener/analyst, one may make choices in the playback variables that support the goals of the analysis.

For many central concepts of this and the following chapters to be fully and accurately experienced, one needs to hear the record with some confidence in the content of what they hear—with some at least cursory knowledge of the performance of their playback system, in all its variables. No ideal listening situation exists. We can, however, have a sense of how our listening processes and playback variables are shaping the record—are transforming it from its original form.

As we begin to study the elements of recordings, it is appropriate to consider how we listen to a record can significantly alter its qualities, and thus our experience(s), and our interpretations.

We would not study a painting—a masterwork by, say, Michelangelo or Monet, Degas or Dali—by viewing it through a veil, or from position distinctly closer to one side than the other. We would not place it in a shadow-streaked location, or in poor lighting that does not reveal nuance in its broadest shapes. Nor would we wear tinted sunglasses when studying it, or rely on a reproduced image from a magazine as our primary source for observing, defining and examining its subtleties of its colors, textures, lines, etc. We readily recognize to do any of these would be to compromise the study from its start. As absurd and contrived as any of these conditions appear, similar sonic circumstances may well be (unknowingly) established by analysts/listeners—generated by their playback system and listening conditions.

It is possible for even an experienced listener to be fully unaware of the veil of lost detail, of acoustic shadows creating dead zones absent of spatial cues, of unbalanced colorations of spectrum tinting timbral balance (and percepts in other dimensions), of playback formats storing and delivering an incomplete reproduction of the original signal, of unbalanced or mismatched loudspeakers, or of a misaligned listening position—and this is just the beginning, as very much more might distort the record.

This situation is especially possible for those who have never fully experienced these playback percepts or the elements of recording.

Those who might otherwise be astute listeners of popular music might be unaware that important substance and subtle sonic qualities are missing from their experience, or transformed into something subtly or substantially different. This is especially relevant to the elements of the recording. This all may be irrelevant for recreational listening, but this is a concern of recording analysis. Accepting how one listens is an analysis decision; it readily follows, the choice of the variables that reproduce the track should be appropriate to the goals of the analysis.

Analysis Goals and Listening to the Track

Reconciling the characteristics of playback of the track with the goals of the analysis sets the context for the analysis. In setting the goals of the analysis (see Chapter 1), the materials and concepts that will be the focus of the study are identified. The sound qualities generated from the playback system need to support the observation of those materials. Consider, again: there are no 'ideal' listening conditions. The concept of listening through a system that is utterly transparent in presenting the track's full palette of qualities, and that does not imprint additional qualities onto the track (or distort and eliminate others) is utopian, and wholly impossible. Even the finest recording studios imprint the sounds reproduced through their monitoring system (though often only very subtly). In this situation, what level of distortion of those materials might be acceptable? To reframe, how much might the sound under observation be altered before the goals of the individual study or analysis are compromised?

In discussing music and lyrics elements of the track, the subtle qualities of the playback can tint their substance. For recording elements, the qualities of the playback system *are* their substance. This is a decisive difference, and points to the significance that reproducing and experiencing the record plays in our perception and interpretation of it.

When all stages of playback systems (room, components, media) alter sound, how do we know we are discussing the same 'sonic experience'? Simply, on some level we will not be discussing the same sound. We all will experience the track and everything within it differently—even those in the same room at the same time will be at different locations. The issue is 'at what level' are we discussing the same sounds, and are we identifying the same sounds, as what we perceive and how we perceive will differ from the experiences of all others. To some extent (subtle to extreme), our perception is shaped by the technology through which we hear the track. In this way, we are experiencing the track *through* the unique voice of the technology. This is related to what Ragnhild Brøvig-Hanssen (2010, 2019) has called "opaque mediation," where the medium through which we observe an object (here the track) imparts qualities onto the object; further, "the technology . . . forces the listener to reckon with it. . . . the technological mediation [here, the playback system] has a voice in its own right . . . and insists on its role in the experiential meaning of the music" (Brøvig-Hanssen & Danielsen 2016, 5).

Defining this level of mediation—this voice of the playback system—will provide some point of reference. Through this context, others might find, or arrive at, some state of commonality or some understanding of the content of the perception of the other; through this, we might meaningfully engage the observations of another. The variables of playback system are many, though, and can profoundly transform the sounds of the track; these will be covered below. Selecting these variables, as stated earlier, represents an interpretive choice of the listener, and determines the content available to a recording analysis. This selection determines how the recording elements will manifest in the listening experience—most profoundly in timbral detail and timbral balance at all levels of perspective and spatial attributes in all dimensions. The playback system is our entry into the track.

There are some parallels between subject-position related to interpretation (see Chapter 5) and the positions of playback quality that might assist us. "Part of the listener's experience of music is the manner in which he or she is invited to engage with the "subject matter" of the music. . ." (Clarke 2005, 10). We might recognize the playback system as one's invitation to the track; shifting the concept of the quality of playback, the "listening perspective that the listener adopts" (*ibid.*, 62) is established by the playback system. The outer extremes of a playback-quality polarity might be similarly divergent to what Clarke (*ibid.*, 92) observes as "infinite pluralism which posits as many readings as there are readers, and an essentialism which asserts a single 'true' meaning." This concept of extremes applied to sound reproduction might be considered as (1) seeking to reproduce the original track accurately, void of any transformation, with a high-definition monitoring system and (2) freely altering the qualities of the track to conform to personal listening taste and preferences with a low-end consumer system, without concern for engaging the track's original qualities. As with most matters, the space between extremes is rich in variation, and is a continuum within which nearly all activity occurs.

This space between is a middle course, wherein the experience might be examined for its variables. Somewhere between the author-centered experience (ideology) that cannot be experienced or recreated, and the overtly personalized interpretation of an individual that cannot be experienced or recreated by another (and is potentially only relevant to that individual), lies some semblance of the shared experience. Here we might find some guidance through the intersubjectivity that "arises when at least two individuals experience the same thing in a similar way" (Tagg 2013, 196). While no two people hear precisely the same thing, cultural circumstances bring patterns of experiences to have representations (and perhaps meanings) shared by members of the culture, establishing a shared subjectivity within cultural groups (that may or may not transfer well to other groups) (McGuiness & Overy 2011). While perceptions may be

> grounded in certain sonic signifiers that have . . . embodied and ecological connections, [they are] also partly personal and arbitrary. In the larger scheme of things, an individual's subject-position will give them a unique viewpoint and therefore a unique interpretation, albeit one that may lie in a predictable large sector of viewpoints determined by generic and universal factors.
>
> (Zagorski-Thomas 2014, 215)

Just as interpretations differ and have personalized segments, so are the qualities of playback; each system is unique, each system imparts a set of transformations that are unlike all others. These transformations will have some broader, higher-order qualities in common with the systems of others, though their details that establish the subtleties of the traits of recording elements will almost certainly vary.

This commonality between listeners is reflected on the broader qualities of the track, and also within those qualities that most significantly define a track. The sonic details and dimensions of the track will be transformed by the unique opaque qualities through which the track is experienced. The shared representations can provide a basis for communication, and the content of the communication is generated from the details contained within that experience—details that are defined by and placed within the context and qualities of the playback system. This brings cause to reflect on how the track is being transformed, and how those transformations might be observed.

The track is observed from a listener position in several ways: hearing the sounds as being reproduced and qualities of the mediating technology, and hearing the track in itself (though it is presented by the mediating technology). Considering the listener's point of audition,[16] we recognize the significance of the listener's perspective—of their physical and perceived relationship to the track. This acknowledges "the way in which a recorded voice is [and all other sound sources are] presented to someone hearing the recording" (Lacasse 2000b) both as the listener hears them within the track and as the listener hears the track within the listening room. This placement of the listener is crucial for spatial properties of the recording, and also for the listener's perceived aesthetic, socio-psychological and physical relationships to the track (see Chapter 8). For hearing the qualities of the technology, this placement brings some sense of detachment and objectivity and provides a vantage from which the technology can be segregated from the program; the qualities of the technology—the variables of the playback system—might then reveal themselves. From these virtual and physical positions, the listener is an outside observer hearing the track reproduced in the room and mediated by playback; also the listener is experiencing the track from within, adopting a position from which all sounds contribute to a singular experience.

To conclude, a methodology that articulates the playback variables in relation to the goals of the analysis can set the context for the study. This allows the playback variables to support the goals of the analysis. For readers of an analysis, knowing the methodology presents opportunity to identify common ground with the study, or to adjust their listening situation to create common ground with those used for the analysis. This common ground allows a basis for exchange of knowledge, comparison of experiences, and further scholarship. Rick Altman (1992, 41) observes we are able to discuss film sound as we file out of the theatre: "In spite of the fact that we have literally, really heard different sounds, we still manage to find a common ground on which to base our conversation."

A cursory examination of playback variables follows, with some exploration of how their selection might support the goals of an analysis, and be reflected within its methodology.

Playback Variables: Reproducing the Track

How the record is delivered by the playback process is a substantial topic. As Table 6.7 illustrates, the process contains many stages, each with its own variable qualities; further, how one stage progresses to the next is also a part of the playback system, and often a source of diminished signal integrity. Each stage of the process contains areas that are highly technical, and some that are highly subjective. Some steps in the process can be measured, others are perceptual. Just how the subjective impressions of listeners and the objective design parameters correlate is not always known (Everest & Pohlmann 2015, 65). This has brought a perception in many people that a mysterious layer of unknowns or unknowable factors surrounds designing and evaluating audio systems. The reality is, should one wish to invest the time, it *is* possible to identify what is measureable, to determine what is within one's perceptual capabilities, and to account for factors that remain ill-defined. A common concern is assembling a playback system suitable for listening purposes is an arduous undertaking and likely unaffordable. Admittedly, the expense can certainly add up quickly, and assembling a system can be a puzzle. Working through the details of this quagmire is out of the scope of this writing. Resources are available that can provide some realistic guidance within a modest budget.[17]

A number of discrete stages establish the qualities of playback—reproducing the track. Here, we will briefly approach a few of the most decisive matters:

- Storage media and formats
- Loudspeakers and power amplifiers
- Listening rooms and loudspeaker placements
- Loudness level of playback
- Headphones

These topics and the remainder of the playback process deserve further research than can be offered here. Some technical and practical guidance on system set up appears on this book's webpage. The reader is encouraged to use the calibration audio files provided to examine the performance on your system.

Table 6.7   General variables of playback systems.

Stage	Options	Variables
Storage Media	Digital discs	CD, DVD-A, SACD, Blu-ray
	Server file formats	ACC, AIFF, ALAC, FLAC, MP3, Ogg Vorbis, WAV; PCM and DSD Hi-Res, MQA
	Vinyl	LP record
Audio Source	Disc players	Internal digital-to-analog converter
	Music servers	External digital-to-analog converter
	Turntable	Phono cartridge, tonearm & preamplifier
	Analog tape	¼ inch tape, ½ inch tape, audio cassette
Power Amplifier	Multichannel, integrated	One amplifier powering two or more speakers
	Bi-amplifying	Two amplifiers powering one loudspeaker
	Digital amplifier	PCM digital source drives output transistors
Loudspeakers	Speaker cabinet	Design (size, shape, ports, grill, internal structure), materials, enclosures, stands
	Cabinet types	Full-range, floor standing, 'mini' monitors, bookshelf, subwoofer, etc.
	Individual drivers	Number, size & types of drivers
	Powered loudspeakers	Internal power amplifier(s)
Matching	Loudspeaker to amplifier	Power, impedance, sensitivity, crossover, etc.
Listening Room	Acoustics related	Size, ratios, materials, reflective objects, etc.
	Speaker placement	Listener-speaker equilateral triangle, height, distance and orientation to walls
Playback Level	SPL meter	Nominal listening level, weighting
Headphones	In ear	Earbuds, earphones, in-ear monitors
	Over and around ear	Open-back & closed-back
	Enhancement options	Noise cancelling, crossfeed circuits & SVS
	Headphone amplifier	Dedicated amplifier & built-in to component
Cables	Interconnections	Speaker cables, digital & analog interconnects
Power	Conditioning	AC line noise in system & external

Digital Source Files

Compact discs contain pulse code modulation (PCM) digital files, formatted in word lengths of 16 bits, at a sampling rate of 44.1 kHz. This represents the standard against which other digital audio files are compared; 16-bit/44.1 kHz is typically the file format of the track at the end of the production process. Two file formats of uncompressed audio are commonly found: WAV (waveform audio file format) and AIFF (Audio Interchange File Format); CDs copied into these formats contain identical PCM data that was recorded on the CD. WAV files have no provision for metadata (that contains artist and track information, among other); AIFF files contain metadata tagging. Codecs that compress files and return to original form when decompressed are lossless formats; these files are typically half the size of WAV and AIFF files, but have bit-by-bit accuracy to the original data when decompressed. Apple Lossless Audio Codec (ALAC) and Free Lossless Audio Codec (FLAC) formats are lossless; the process is conceptually similar to Zip files on a computer.

Lossy compression schemes reduce the number of bits in the data stream by discarding information— information that cannot be recovered. This discarded information represents portions of recording elements. The result is the content of the track is diminished, and recording elements are audibly compromised. These changes are noticeable—anywhere from subtly to profoundly. The data that is deleted often contains the subtle characteristics of the recording we seek to study, acknowledging its significance. Examples of lossy compression formats are MP3, Ogg Vorbis, ACC and the Dolby Digital format encoding film soundtracks. Once one learns the subtle qualities of recording elements, the consequences of compression become audible. Zagorski-Thomas (2014, 218) notes that "in extreme cases the effect is . . . like an audio form of pixilation." These consequences of compression are especially detectable in the timbres of sources and the track, in the spatial qualities of image locations and size, and the stability of the lateral and distance imaging, shapes of dynamic contours and loudness balance relationships, and in pitch space and timbral balance of the overall texture.

Digital files can be stored and played in a variety of ways, from smartphones to dedicated systems. A music server can be as simple as a personal computer playing digital files stored on internal or external hard drives, perhaps controlled by an iPhone or iPad. In contrast, stand-alone music servers are available that are specifically designed to contain and provide immediate access to one's entire music library, and to download files from commercial sites; they contain unique control systems, intended for ease of use. Servers can store and play any file format, including high-resolution files (Harley 2015, 221–234). Hi-res files are those with a sampling rate of 88.1 kHz or higher, with a word length of typically 24 bits; common sampling rates for 24-bit PCM files are 88.2 kHz, 96 kHz, 176.4 kHz and 192 kHz. Direct stream digital files (1-bit formatting) have sampling rates of 2.8 MHz and 5.6 MHz. File sizes increase quickly with higher resolution formats; a 1TB drive can hold approximately 2,000 CD-quality albums of music (16-bit/44.1 kHz), approximately 350 albums of 24-bit/176.4 kHz hi-res, or approximately 250 DSD formatted albums (at 5.6 MHz). SACD, DVD (DVD-A), and Blu-ray discs support high-resolution audio files; they require a suitable disc player, and support surround sound audio as well as stereo.

Integral to the system is digital-to-analog conversion. The process transforms the digital file's data stream into an analog audio signal that can be fed to the preamplifier/amplifier. The DAC may be a stand-alone component or one internal to a disc player or music server; they vary considerably in how well they decode the data stream and in the content of the analog signal they establish. Audible differences exist between file formats and between various DACs—these differences can be significant in how they impact the sound of the system. Differences are largely oriented towards revealing nuance and detail of sounds, and the level of accuracy of reproducing all frequency levels equally. This is especially noticeable in tracks with exposed sources (where detail can be observed more readily) and those with activity in the extremes of high and low frequency ranges.

Loudspeakers, Amplifiers and Listening Room Considerations

Loudspeakers coupled with power amplifier(s) anchor the system, and are the components that contribute most clearly to the character of the system. Within a loudspeaker cabinet are separate drivers of various sizes dedicated to reproducing the signal of a specific range of frequencies; the signal is divided by crossover networks to deliver the identified frequency band to each driver. Cabinets vary in size depending on the number and size of drivers; size varies from large floor standing speakers to small bookshelf speakers. Monitors designed for audio production must have as level a frequency response as possible to mix accurately, and they tend to clearly expose the details within the track. Conversely, hi-fidelity speakers often seek to establish a pleasing sound by smoothing over detail (Winer 2012, 434). Consumer sound systems are often designed to impart sound characteristics to enhance the timbral qualities that define certain styles of music, to attract those segments of the listening public to their product. Audiophile speakers tend more towards the detail and accuracy of professional studio monitors, though differences remain. The coupling of amplifier and loudspeaker is symbiotic; much nuance of matching specifications compatibly can be involved. A direct way to work around this matter is to purchase powered monitors, such as those that have become common in home recording studios; they come in an array of sizes and price points. Powered studio monitors are designed for the loudspeakers, cabinet and amplifier(s) to work as a unit; each speaker cabinet contains an amplifier, and higher quality studio monitors will have an amplifier circuit for each driver in the cabinet. The important goal is for this pairing to reproduce all frequencies throughout the hearing/audio range equally well—a lofty ideal impossible to achieve, though some systems have but subtle deviations.

The loudness level at which one listens impacts the overall timbre of the track, and all the timbral elements and relationships within the track. As we do not perceive all frequencies equally well at all loudness levels, it is important to listen to the track at a sound pressure level (SPL) at which humans most readily perceive frequencies in equal loudness. Equal-loudness contours demonstrate perceived loudness (as phons) in relation to frequency and sound pressure level; the least deviation of perceived loudness with SPL at the widest frequency range exists between approximately 80 and 100 SPL; even in these curves, lower frequencies (below 200 Hz) require greater SPL to be perceived as equal in loudness to 1000 Hz (B. Moore 2013, 134–137). This is significant in that the loudness level of the listening experience determines the frequency balance ('spectrum') of the track. Listening to the track at a consistent, nominal level between 82 and 85 dB SPL with peaks up to approximately 90 dB SPL is a desirable listening level; it is a fairly loud home-listening level (that one's family may not long

| Praxis Study 6.2. | Cultivating sensitivity to loudness levels (pg. 500). |

appreciate), but it is a level that can be sustained throughout a work day with minimal fatigue (Moylan 2015, 370–371). An SPL meter set to 'A-weighting' will be indispensable for monitoring levels.

The listening room is part of the playback system, as the sound from the loudspeakers reflect off room surfaces before arriving at the listener location. These reflections can be controlled, minimized, and many problems circumvented through loudspeaker placement and room treatments—though realities of living conditions can often dictate or limit options appropriate for a living space used for listening. Figure 6.2 illustrates placing speakers with a clearance of 4 feet to side walls and 3 feet to the wall behind can minimize reflections arriving at the listener in a direct-field monitoring relationship (Moylan 2015, 363–370).

This arrangement may well be impractical for one's daily use within a living space. Here the analyst may establish a listening environment that might be useable for empirical data collection—to a reasonable extent—by careful (though perhaps temporary, but necessarily reproducible) adjustments to loudspeaker locations in relationship to room boundaries (walls, corners, floor and ceiling). Distances from walls are important for managing reflections and frequency response, though, and should be observed as much as feasible; it may be workable to move speakers into a listening position when needed, and otherwise tucked away in a position suitable for the lifestyle of you and your housemates. Other alignments of loudspeakers, wall relationships, and listening positions can be adapted to the geometry and volume of listening rooms; setting up a stable and accurate listening position is often a process of compromise, calculation and experimentation. Loudspeakers placed on stands can be relocated easily; further, the stands can be decoupled from the floor and from the loudspeaker minimizing several energy transference and resonance issues. Stands also allow the tweeter of the loudspeaker cabinet to be at ear height, approximately 32–40 inches from the floor (Harley 2015, 359–373). Hard reflective surfaces and resonances between parallel surfaces are concerns, as are furniture and other objects that can reflect sound. As the sound of the room is controlled, and the stability of spatial relationships, system frequency response, and time alignments and amplitude contours between drivers and speakers will be improved (Everest & Pohlmann 2015, 397–410). The room can become a positive

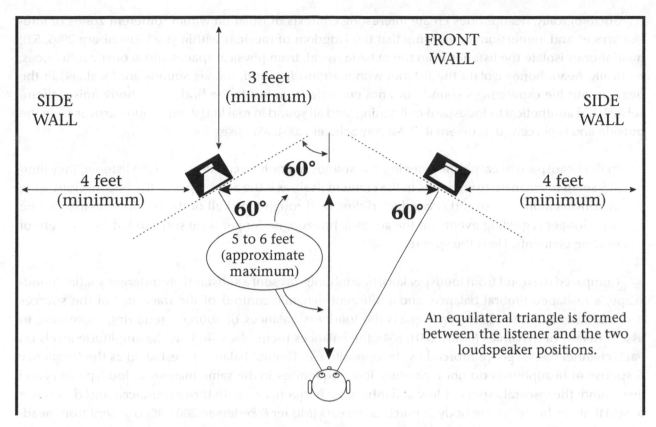

Figure 6.2 Direct-field loudspeaker relationships to the listening room and the listener location.

contributing factor to a sense of spaciousness and visceral experience of the track; with controlled reflections off ceiling and floor, reproduced sound becomes more of a real-life experience (Rumsey 2001, 120–127).

Headphones

It may be tempting to look to headphones as a means to simplify listening to the track; to do so may also save money, not disrupt the household, provide portability, among other attractive features. Headphone listening is not an equal substitute for the open-air playback system, though. Headphones and earphones (also earbuds and in-ear monitors) produce a different experience entirely; some qualities of the track are physically transformed, and others are perceived differently with headphone listening. This is a result of the closeness of the driver to the ear and the isolation of each ear—and also isolation of the listener.

The vast majority of tracks were created (tracked, mixed, assembled, mastered) while listening over loudspeakers. The track's final version is crafted into its distinctive and definitive state—its primary text—while listening over loudspeakers. The qualities of this final version are substantive and are elements of central concern to recording analyses—these are qualities that potentially shape the song in ways that characterize the track. To listen to tracks over headphones distorts these qualities and produces an entirely different sonic experience—one that does not reflect the primary text of the vast majority of tracks. Some might argue that with the proliferation of home studios more tracks are being produced over headphones; it is certainly true recordings over headphones are more common in some genres and markets, though this does not justify using headphones for tracks recorded over loudspeakers, which still represent a very significant majority.

Sociologically, headphones create interesting contexts of isolation within "different zones of interior space" and immersion "suggesting that the kingdom of music is within you" (Eisenberg 2005, 53); headphones isolate the listener from the outside world, from physical spaces and acoustic influences. Sonically, headphones isolate the listener within an *inside* world, too, as sounds are localized *in* the head. In our life experiences sound does not emanate from inside the body (save body noises themselves); it is antithetical to loudspeaker listening (and all sound in real life) where sound arrives from the outside and is perceived as external. R. Murray Schafer (2004, 35) observes:

> In the headspace of earphone listening, the sounds not only circulate around the listener, they literally seem to emanate from points in the cranium itself, as if the archetypes of the unconscious were in conversations. . . . sound is conducted directly through the skull of the headphone listener, he is no longer regarding events on the acoustic horizon; no longer is he surrounded by a sphere of moving elements. He is the sphere.

Compared to sound from loudspeakers, headphones present a substantially different spatial soundscape, a reshaped timbral balance, and a different dynamic contour of the track and of the sources it contains. Headphone listening impacts the loudness balances of sources, requiring more time to assess details of dynamics (Poldy 2001, 654); the balances themselves shift, as the amplitude levels on each channel do not get reinforced by the opposite ear. Timbral balance is reshaped as the frequency response of headphones do not reproduce low frequencies in the same manner as loudspeakers and live sound; the visceral aspect of low and infrasonic frequencies (both those produced and difference tones) that are 'heard' by the body as much as the ears (Møller & Pedersen 2004, 40) is absent from headphone listening.

Spatial information over headphones redefines the sound stage and environment cues. The elimination of interaural time, spectrum and amplitude cues "results in an unnatural stereo image which does not have the expected sense of space and appears inside the head" (Rumsey 2001, 59); the closeness of the drivers to the ears adds unintended and uneven emphasis to timbral details, which alters distance positions (Poldy 2001, 652); the absence of the individual's unique head-related transfer functions (HRTFs) and pinna cues impacts our perception of reflections and room cues (as well as source positioning). Beginning in early childhood, we each learn how the acoustic properties of our unique ear shapes, torso and pinna affect the frequency, amplitude and time information of sound arriving at the ear from different angles; the result is each person has a different filter of sonic properties, or different HRTFs (B. Moore 2013, 263–265). Listening to sound reproduced from loudspeakers within a room, we continue the opportunity to process room reflections and the sounds of spaces that are within the track to understand its spaces through these personalized cues; these cues are eliminated by headphones, except pinna cues which are altered by the close proximity of the driver and further reshape the percept. Earphones eliminate pinna cues completely, as the driver is within the ear canal.

The sound stage and positional localization are transformed in comparison to what is heard in space, emanating from loudspeakers. Without interaural cues, sounds are localized in the center and at each ear; headphone reproduction establishes voids or holes in the stereo field between the center position and far left and far right. Crossfeed networks (often found as a "Headphones" button on an integrated amplifier) are intended to compensate for the absence of these cues by adding interaural crosstalk; an electronic circuit mixes some left-channel information (attenuated and delayed) into the right, and likewise right into the left, to simulate interaural cues. Crossfeed cannot address missing HRTF cues generated by our unique physiology. Further, mixing channel information electronically often brings unintended sonic consequences. The networks may "reduce 'inside-the-head' imaging, but . . . at the expense of blurring imaging, softening the treble, reducing resolution, and muting dynamic contrasts"

(Harley 2015, 346). Crossfeed networks bring further reshaping of the elements of the recording; stereo image sizes and positioning, clarity and stability of source images, distance placement positions, and environmental cues are particularly influenced.

Headphone use is not without some benefit, though; the qualities they reveal might be used to advantage in a number of listening and analysis tasks. Keeping clear the content of those experiences are skewed from the original form of the track, observations over headphones can supplement those over loudspeakers by adding further detail to certain elements or tracks. Headphones' isolation can facilitate focused concentration by reducing perceptions of all sorts, and by bringing the sounds into the listener. Their enhanced detail allows one to hear into textures and into timbres that typically blend by design in the production process; this has many applications from identifying individual sound sources in a complex mix to experiencing the subtle content within sounds. Hearing this heightened detail also allows more ready perception of subtle changes of rhythm, pitch and timbre; here, headphones can assist in accessing these substantial dimensions of performances that are challenging to observe. Headphones bring clarity to finding specific points in the track, where sounds, sections, or other activities begin and end. These observations may be added to others acquired over headphones to bring another perspective, that must then be balanced into the analysis against those of the same percept experienced over loudspeakers.

Headphones Conclusion and Consistency Between Listening Sessions

Headphones and earbuds are certainly in widespread use for much of the public. In itself, this does not bring them to be a point of reference for auditioning the sound qualities of the track. To restate, a significant majority of tracks are created while listening over loudspeakers; those qualities represent the track's primary, original form. Alterations to the original form are unavoidable with any playback system; the analyst will need to determine to what extent it can be altered and still be its original form— unless, of course, one wishes to make the altered track the object of study. Widespread headphone and compressed file listening does not necessarily represent a cultural shift away from open air loudspeaker listening and more detailed high-resolution sound reproduction. Nor is this a shift in point of reference for discussion of tracks. What is present is that in the current state of consumer listening habits, many people within certain demographics find it more economical, convenient, or 'whatever'—given a viable choice, logistics of space and economic resources, they may well choose differently. Detail-rich, full bandwidth (or better) sound seems likely to remain available as download sales of hi-resolution and of 16/44.1 kHz albums are steadily increasing in parallel with internet speed.

In the end, headphone listening may be a preference of the individual. In making that choice it should be acknowledged: observations performed over headphones will have different qualities from those that will be present over loudspeakers—and spatial qualities are substantively different. This represents a decision that will bring their interpretation and analysis to be disconnected from the experience of those listening over loudspeakers. And of course, analyses performed over loudspeakers will be disconnected from the experience of those listening over headphones as well. This returns us to the notion that the goals of a study, an interpretation or an analysis need to guide the decisions of the most suitable playback experience—and that the variables of track playback be revealed within the methodology of a study.

The playback system determines and shapes the qualities one hears. When one studies aspects of sound that could be substantively altered by playback, one should choose carefully and learn the consequences of that choice. For example, one study of spatial qualities, in particular, was fundamentally flawed when headphones were used to evaluate spatial positioning; the study did not obtain the goals intended (and asserted), because inaccurate data was collected.[18]

One last significant matter will conclude this topic: consistency of listening. Small changes to the playback system or the listening situation can shift the traits of the reproduced sound. Settings or performance of all system components, arrangement of the listening room, listener location, loudness levels, and perhaps other factors all impact the heard experience of the track. If re-hearings are not consistent in reproducing sound qualities, an attentive analyst will recognize inconsistent observations; an inattentive analyst's observations and findings will be inherently skewed from one listening session to the next. A developing listener and analyst may be confused and frustrated—or simply unaware. Inconsistent listening practices will deliver inconsistent results and observations; this does not serve the goals of analyses of recording elements.

Playback Systems Used by the Author for Observations and Analyses

This section reveals the playback system (actually systems) used for listening to the examples and analyses in this book. It represents a portion of the methodology of observing and evaluating the tracks discussed throughout. Knowing information on the equipment, rooms and other pertinent matters may allow readers to evaluate the graphs, observations, evaluations and descriptions that will be (and have been) encountered. Readers can compare these findings to their own experiences on their own listening systems and playback conditions. Comparisons can reveal similarities and differences—the source of differences might be related to playback systems, or any number of factors (such as listening skill level).

The playback conditions for this writing are not typical, though, and they are well beyond what most readers can replicate. I have worked intimately with sound reproduction systems all my career, and have gradually acquired suitable and high-quality systems for my home listening room, my living spaces and my campus office. I also have ready access to a professionally designed and constructed critical listening room, as well as recording studio control rooms. These systems represent a great diversity—from very high quality professional systems to more affordable consumer systems. I use my home listening system (with high quality studio monitors) and campus critical listening room as primary references; the two are generally quite similar in broad characteristics, with the critical listening room being substantially more detailed and accurate at reproducing nuance. I listen on the other systems to compare results and to hear what they might reveal from their own inherent qualities of mediation; this allows me to verify the content of my perceptions, and also to hear what other, differently designed systems might reveal—allowing me to predict what others might hear on similar systems. My final observations balance all information collected (privileging my two primary systems) against my knowledge of performance characteristics of each system (rooms included); this is the result of some considerable experience in listening to records and the characteristics of sound systems, and an acquired knowledge of sound system variables and the specific components of each system.

First in my playback chain, and most easily replicated by readers, all tracks were digital files (CD-quality or better) heard directly from CD or DVD-A (for surround); some tracks were accessed through SACD and DSD file format (both stereo and surround). Discs were used instead of stored digital files for both convenience (in my situation) and to not confuse multiple versions of the same track on different versions of the same album. The DACs I use are within either a very high-quality Blu-ray disc player or a high-end commercial CD player, and vary between systems.

Second, an SPL meter is kept near. Loudness level is checked when changing discs, and when resuming an interrupted session. This ensures listening levels remain consistent at a nominal level at or slightly below 82 to 85 dB SPL A-weighted; the meter will retain maximum and minimum levels, allowing one to identify peaks levels. I will raise the nominal level slightly if the minimum SPL is below 78 dB for more than a few moments, and this will become my new loudness setting for listening to the track on that system. When changing systems, I seek to keep a track at the same SPL. Once set, the levels are not adjusted and are recalled (from my listening session notes) for the next listening session. Slight deviation of level is common when changing between tracks; differences can shift considerably between albums (discs).

The following are details on the playback systems I used to examine tracks. The first two systems (critical listening studio and home listening room) are the ones I use primarily; the systems incorporate ribbon speakers for high frequencies, and are both spaces dedicated to listening. The systems in my office and in my living room I use to verify information and to add a different perspective on tracks; significant differences are these systems have dome tweeters and these rooms also serve other purposes (living, working, hosting guests, etc.); the influence of these rooms on the track is minimized (as much as possible) by repositioning speakers for listening sessions; the impact of these rooms on the track cannot be neutralized. Again, by using four playback systems, observations were cross checked, and some systems revealed qualities that were obscured on others. I own a set of headphones; I very rarely use them, and then only for limited specific tasks (such as for locating sounds in time, and for listening within textures or sounds to reveal details). I use the audiophile-quality surround-sound audio system in my car as a source of curiosity and entertainment, never for data collection—and I am mindful not to be a traffic hazard.

The critical listening studio is an extraordinarily accurate listening and monitoring environment for both stereo and 5.1 surround playback. Each of the five monitors have a high-performance planar ribbon high-frequency transducer, two 6.5-inch midrange drivers and two 12-inch low frequency drivers that are designed for exacting accuracy and speed to keep transient performance synchronized between all drivers, resulting in great clarity; each speaker pairing is amplified separately with electronics designed to match the speakers and crossovers with mastering grade electronics; they have an operating range of 34 Hz to 30 kHz. A significant subwoofer extends response into infrasonic frequencies. Speakers are integral to the walls, and the room is designed to direct sound within a focused area of audition for accurate and detailed listening in both stereo and surround. The room is isolated from the building and built to exacting standards acoustically. This space has 16-foot ceilings with various acoustic treatments and the walls contain various surfaces to control reflections; the area of audition is located from 15 feet 5 inches from all five monitors to 17 feet 2 inches from the stereo pair. A professional grade Blu-ray/DVD/CD player with exacting external DAC plays discs. I use this system for all surround sound analyses.

Home listening room is in a 20-foot by 21-foot family room that has a vaulted ceiling that is 15 feet at the peak; an open balcony is part of the considerable space behind the listening position (60% of the room volume). Loudspeakers are separated by 6 feet and form an equilateral triangle with a sofa listening position 21-feet wide, and located 3 feet from the front wall and 7 feet from side walls to minimize the role of the room. Loudspeakers are full-range two-way (active crossover at 1.5 kHz) world-class powered studio monitors with discrete digital amplifiers (270 watts and 50 watts); each front-ported cabinet contains an 8-inch driver and a wide dispersion planar ribbon high frequency transducer (same manufacturer and similar design to the above system); the loudspeakers deliver accurate transcient response and clear timbral detail, and its cabinet design eliminates signal diffractions from the sides of the cabinet; operating range is 36 Hz to 40 kHz.

My work office is in a 14-foot by 15-foot room. Speakers are arranged near-field to utilize as much direct sound as possible, and to minimize the input of the room; the speakers cannot be placed symmetrically to room surfaces, as the room is a working office space where I host visitors daily. Loudspeakers are superior quality near-field powered monitors; each cabinet is a two-way system that is bi-amplified with its 1-inch soft-dome tweeter coupled with an 80-watt amplifier and a 200-watt amplifier powers its 8-inch driver (active crossover at 2.6 kHz); the design incorporates two front ports. Their operating range is 35 Hz to 20 kHz.

My home living room is arranged for reading and relaxation, for entertaining guests and for everyday living. When I want to use the system for active, attentive listening I must move a chair out of the room and move each loudspeaker from its next-to-the-wall out-of-the-way location to a specific location on the area rug; this takes little more than a minute, as I know how to precisely position each speaker stand on the rug pattern. The speakers are not symmetrical in the room, one is 2.5 feet from a side wall and the other 4-feet from its side wall, the front wall is 4-feet or 6-feet behind. Speakers are separated by 5 feet and form an equilateral triangle with the sofa listening position. The room plays a role in the sound, though speakers still exhibit stable imaging, a frequency response that is consistent throughout, and wave tracking that is respectably accurate. The loudspeakers are a rear-ported two-way dynamic bookshelf design with a stand designed to complement its characteristics; each contains a 1-inch tweeter and a 7-inch driver (with a phase-coherent crossover at 1.5 kHz); their operating range is 59 Hz to 20 kHz. The speakers are driven by a quality multi-channel receiver/amplifier that can produce 100 watts/channel (the receiver also drives a separate pair of bookshelf speakers in another room and is capable of simultaneously powering a modest home theatre system). A universal disc player (capable of CD, DVD, DVD-A and SACD playback) rounds out the system.

The headphones I use are an open-back design with sealing ear pads. Though some acoustic energy escapes (and releases some air pressure), they produce stable, detailed resolution. I have learned how they transform the spatial and timbral aspects of tracks; I know how far I can trust what I hear.[19]

Some Questions for Recording Analysis

What is the goal of my analysis? What analytical question do I wish to answer?

What kind of listening system(s) will I need to meet the goals of my analysis?

How well does my listening situation and playback system function in relation to these analysis goals?

What are the specific ways my system changes the sound of the track? Is there a way I can compensate for these distortions? How might I take these distortions into account when making observations of recording elements? Of music and lyrics?

Is the record in a compressed format? A lossy compression format?

Might an element be best characterized in a typology table?

Are section duration(s), element(s), and variables clear in typology tables?

Which recording elements display activity or characteristics that warrant close observation with appropriate X-Y graphs?

Of the listening activities described, which might be outside my previous experience?

How might I guide my attention to focus on any single recording element?

What biases do I bring to the listening experience, and my interpretations?

How might I begin to neutralize them to bring my analysis to be relevant to the broader audience of my cultural group?

What are the macrorhythmic durations of sections that reflect a cohesive sound from element activity?

Am I able to direct my attention to the described microrhythmic activities?

NOTES

1 See Brøvig-Hanssen & Danielsen (2013) "The Naturalised and the Surreal: Changes in the Perception of Popular Music Sound" and Moore & Dockwray (2008) "The Establishment of the Virtual Performance Space in Rock," among others.

2 Carol Krumhansl (1989, 44) observes the difficulties in separating timbre from pitch: "Can we really assume the differences in spectral energy distributions are completely uncoupled from pitch perception mechanisms in hearing?"

3 The reader will find examining Allan Moore's chapter useful, and much to gain throughout *Song Means*.

4 *Klangfarbenmelodie*, German for "tone-color-melody" is a technique that splits a musical line (melody) between various, different instrumental colors in order to introduce timbre and texture into the line. Arnold Schoenberg (1966, 503) called these "timbre structures."

5 Albin Zak "Electronic mediation as musical style" (2010, 316–319) provides more detail.

6 Pattern perception and grouping are complex processes worthy of study and significant coverage—both outside the scope of these pages. This section offers the briefest of introductions, to flag their significance in analysis. Pattern perception is important to rhythm—and typically underlies patterning within any element. Meter is a pattern, as a recurring sequence comprised of stressed and unstressed pulses; hypermeter is a pattern; we commonly conceive of rhythms as surface patterns; etc. As the metric relationships become less prominent and the further activities diverge from the surface level, patterning of rhythms takes on different temporal characteristics.

7 These insights are from a letter Pierre Schaeffer wrote to me on 3 April 1983. Recently Serge Lacasse and Sophie Stévance provided me with this fresh and articulate interpretative translation of Schaeffer's letter. They gave generously of their time and talents, and I am most grateful to them.

8 My correspondence with Pierre Schaeffer was related to my dissertation research, and directed to inform my *An Analytical System for Electronic Music* (Moylan 1983). My initial thoughts and approaches to transcribing and notating the elements of recording grew out of this study. These methods of notation have had significant development and broadening since, owing much to on-going research and scholarship by many others as well as my own.

9 'Typology' is a term that has been used in many ways, some quite different from how it will be used here. For example, Simon Zagorski-Thomas (2014) approaches 'typology' in a broader context of examining the qualities related to an entire discipline; over the span of "eight chapters that . . . constitute a functional typology of the key issues that need to be addressed if recorded music and record production are to be integrated into musicology" (*ibid.*, 3).

10 This 'typology' approach to examining elements brings together aspects of "type 1 categorization" and "prototype effects" as explained by Lawrence Zbikowski (2002, 36–49), with various programming techniques for data collection and analysis, and the spectrum of characteristics and spectrum of contrasts employed by LaRue (2011).

11 These terms for stereo location are defined in Chapter 8.

12 Snyder (2000, 196) refers to 'secondary parameters' "that cannot easily be divided up into very many clearly recognizable categories . . . there is no way of establishing standardized scales of proportional values for them that are *repeatedly identifiable across different experiences*" within music elements such as dynamics. This same concept is applicable to recording elements.

13 Some background information can prove invaluable for recognizing the qualities of recording's elements; it can provide a resource to understand the concepts, bringing greater ease toward ultimately being able to experience the sound qualities of recordings. One will benefit from some background in acoustics, psychoacoustics and an understanding of the dimensions of recordings. This background establishes the knowledge necessary to talk universally about the physical characteristics of sound, while acknowledging the sounds are transformed within our perception. Sounds will be conceptualized by what is physically present in the sound, as one perceives them. Different levels of accomplishment in the above areas will bring different levels of detail

to communication, and commensurate precision to the descriptions. Throughout this writing, some acoustics and psychoacoustics are woven into explanations. Though more detailed information is beyond the scope of this title, engaging acoustics and psychoacoustics more deeply will benefit the reader in navigating the sounds of recording's elements.

14 Tagg 2013 and A.F. Moore 2012a offer important and contrasting studies on interpretation and meaning that are valuable here.

15 Thomas Porcello (2004) approaches this problem by examining language devices such as onomatopoeia. He also finds metaphorical descriptions of sounds to be inherently vague.

16 This term is a word play on 'point of view,' transferring reference of sight and connotations of thought into hearing and sound; this concept is fully developed later, in Chapter 8.

17 Resources to explore assembling a system with an eye on cost are available. Books such as *The Complete Guide to High-End Audio* (Harley 2015) and others provide detailed information; it is important the resource is reasonably current. Sources related to audio production can provide some welcome guidance to technologies, components, interconnections, and so forth; one such source is *Sound and Recording: Applications and Theory* (Rumsey & McCormick 2014). Audiophile, high fidelity, home theatre, and consumer electronics publications will offer more topical, current and commercial information.

18 This disappointed me, as I recognized and respect the significant expertise of that author, and that an otherwise strong methodology was applied to important issues and topics. Convention is to name the scholar who made this error in judgement and reference and engage the study. I believe this convention was once constructive in debating and exploring materials, focused on the study itself; the current approach to this convention often contributes to eroding the civility of academic discourse, and has limited value in scholarship. I will not diminish another's work—especially work that is other-wise commendable and representative of a laudable scholar—in order to further my own study; I strive to be civil and supportive, and will be silent rather than damage another's reputation. In its place, I identify this matter in confidence readers will recognize the value of these observations and the acute nature of this problem, that they will be careful in examining the methodologies of others and will approach their studies seeking to understand how listening to the track impacts one's observations.

19 A number of years ago I used these headphones for an extensive recording project on location. I learned their characteristics well; within known limits, I can transpose sounds I experience on them to what I expect to hear over loudspeakers.

Timbre and Pitch in the Recording Domain

Chapter 7 is divided into four parts: (1) the linkage of pitch/frequency and timbre as recording elements; (2) pitch/frequency as a recording element; (3) timbre as a recording element, and its significance and contributions to the record; and (4) observation and analysis of timbral content and character.

Before beginning these sections, it will be informative to explore the linkage of pitch/frequency and timbre, in order to set the context for the individual elements. While this chapter divides pitch and timbre, it does so to emphasize their unique qualities and approaches for observations. In perception and in function, timbre and pitch are closely linked in recording, and are in some fundamental ways inseparable.

Pitch in recording differs significantly from pitch in music. The psychological phenomenon of pitch and the sensations of frequency levels are entwined, with pitch and frequency levels both used in observing pitch as a recording element. Pitch areas and pitch density establish both functional and conceptual linkages between pitch and timbre. This linkage is made clear in examining timbral balance and in observing the content of timbre.

Timbre's role as a recording element comprises part three. Timbre is one of recording's two dominant elements. Its functions and characteristics at all hierarchical levels embody the track. Timbre immediately reaches listeners through the 'sound' of the track, and the nuance of timbre manifests throughout the track's structure, and is ever present within all sounds. The concepts of sound object and of acousmatic listening establish points of reference for observing timbral content. Common challenges of timbre perception are explored to prepare the reader for listening within sounds.

Observing timbral content and character fills part four, including approaches to listening and uses of sound analysis software. Typology tables and graphing notation of timbral components lead to recognizing physical content and to describing timbre's overall character as a recording element. These form the foundation of the approaches to timbre analysis and description.

During the course of this chapter and these discussions important terms will be defined. It is important to take note: some term definitions may not align with those one may have encountered from other sources, especially within ancillary disciplines; the usage here is sometimes more precise and nuanced, and definitions and usage are deliberately directed toward facilitating timbre analysis.

THE LINKAGE OF PITCH/FREQUENCY AND TIMBRE AS RECORDING ELEMENTS

The word pitch refers to the mental representation an organism has of the fundamental frequency of sound. That is, pitch is a purely psychological phenomenon related to the frequency of vibrating air molecules. By "psychological," I mean that it is entirely in our heads, not in the world-out-there;

it is the end product of a chain of mental events that gives rise to an entirely subjective, internal mental representation or quality. Sound waves—molecules of air vibrating at various frequencies—do not themselves have pitch. Their motion and oscillations can be measured, but it takes a human (or animal) brain to map them to that internal quality we call pitch.

(Levitin 2006, 22)

'Pitch' is a tone or note we perceive as a result of sound frequency. Pitch might be conceived as an interpretation of sound waves, though it is actually a psychological phenomenon; we perceive pitches similarly to the way we perceive colors within visual perception—as something identifiably discrete that is observed to exist or occur within our (visual and aural) perception.

While a rough analog is often drawn between the pitch (and also often timbre) of a sound wave and the color of a light wave (since both are based on frequencies of a waveform), it can provide some insight into our mental representation of pitch; importantly, though, our perceptual mechanisms for each are profoundly different.[1] The vast majority of people are able to identify the 'pitch' of optical frequencies—in the sense that when we are shown a red object we identify it as being the color red, without needing to compare it to a reference color; we experience each color as an absolutely identifiable sensation from the light stimulus—in other words, we have absolute pitch within the optical frequency band (this is simplified here and not strictly accurate). Most humans do not possess the ability to similarly identify pitch; absolute pitch—the ability to "immediately, unthinkingly tell the pitch of any note, without reflection or comparison to an external standard" (Sacks 2008, 129)—is rare in humans.[2] Most of us utilize 'relative pitch' in our listening process—we identify pitch levels (and appreciate the movement of melody and the content of chords and the movement of their successions) by comparing them to others. Pitch, for most of us, is simply a sound that is heard with some sensation of placement in the hearing range; it is abstract and without bias until in context with other pitches. Relative listening has advantage; it allows a performance tuned to 'A' 445 Hz to still sound like the same performance if tuned to 'A' 430 Hz—as long as everyone retains the same pitch relationships. Relative pitch also allows transposition of pitch patterns or entire tracks. Discrete relationships of pitch appear differently between various cultures world-wide (establishing different scales, sonorities, etc.); the commonality appears to be the octave, though its divisions and its emphasis vary significantly.

The octave is the duplication of the pitch sensation at a level that is higher or lower than the original; this duplication may result from doubling (octave higher) or halving (octave lower) of the pitch and its fundamental frequency, and it extends further by successive octaves in either direction to two (or more) octaves above or below. The percept of the octave is critical to our sense of pitch, as it relates to frequency, wavelength and the simplest harmonic relationship within spectrum.

Pitch is related to the repetition rate of the waveform; it is a subjective attribute with an assigned value that specifies the frequency of a tone with the same subjective pitch as the sound. For a pure tone the repetition rate of the waveform corresponds to its frequency and for a complex tone the repetition rate is (with some exceptions) its fundamental frequency (B. Moore 2013, 203). The fundamental frequency is the frequency at which a body (such as a guitar string) vibrates; this vibration along its entire length is its resonant frequency (see Figure 7.1).

Pitch Becomes Timbre, Timbre Becomes Pitch

Complex tones contain frequencies additional to the fundamental frequency. These additional frequencies are 'partials' which in aggregate comprise the 'spectrum' of the sound—spectrum is one of the three components of timbre, the other two being dynamic envelope and spectral envelope (explored below). The many frequencies that might be contained within a spectrum can be understood as being either integer multiples of the fundamental frequency (called 'harmonics') or being at frequencies that are more distantly related (these are 'overtones').[3] This is an important distinction for our purposes, and one that is not acknowledged by certain other disciplines. Harmonics are at frequencies in a ratio above the fundamental frequency that reinforces its presence and the sensation of pitch (for our purposes this ratio is 1:17). Overtones are frequencies in an inharmonic ratio above the fundamental frequency (any ratio more distant than 1:17, excluding those ratios that are multiples of harmonics); these frequencies reduce the dominance of the fundamental; they diminish the level of 'pitch definition,' and can distort the 'pitch clarity' of the sound. This distinction allows partials to be categorized by their relationship to the fundamental and by their contributions to the content and character of the spectrum.

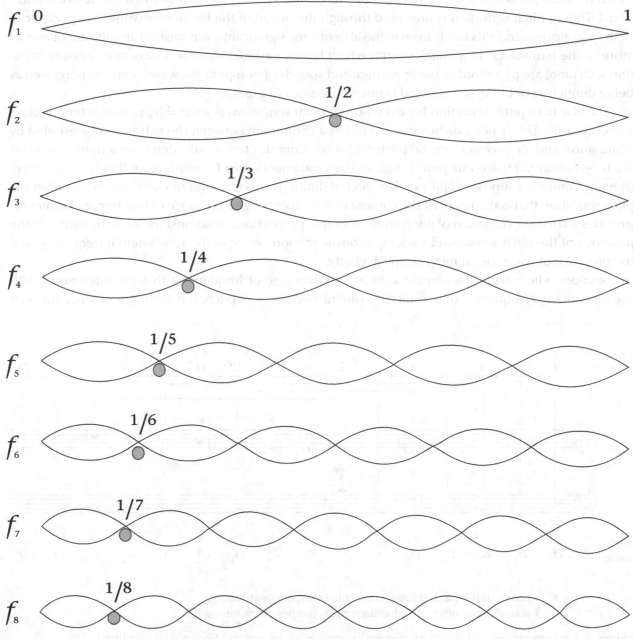

Figure 7.1 Harmonic vibrations of a string through the 8th harmonic.

Figure 7.1 illustrates the vibration of a string. The vibration (and standing wave) along its entire length is the fundamental frequency (the first harmonic). The string also vibrates in halves (2nd harmonic), in quarters (4th harmonic), in eights (8th harmonic), and in sixteenths (16th harmonic) which are all in octave relationships to the fundamental frequency and reinforce its perceptual presence. The third harmonic represents a perfect fifth relationship (plus an octave) to the fundamental; it, along with its octave doublings (the 6th and 12th harmonics), also reinforce the dominance of the fundamental frequency and strengthen it as the pitch sensation. Notice 8 of the first 16 harmonics illustrated in Figure 7.2 have been listed here.

This reinforcing of the fundamental frequency by the harmonic series is so strong that the harmonic series alone can establish the sensation of pitch even in the absence of a fundamental frequency; this is termed a 'missing fundamental.' We perceive the pitch of the tone not only by its fundamental frequency, but also by the periodicity of the waveform that is generated from strong relationships between its harmonics—particularly those that are low-integer multiples of the fundamental (or octave replications). Thus, a pitch sensation is produced through the fusion of the harmonics within the periodicity of the waveform—and this holds even in the absence (or significantly diminished amplitude) of one or more of the harmonics. In general, sounds which have a periodic waveform (acoustic pressure variation with time) are perceived as being pitched, and sounds of non-periodic waveforms are perceived as being unpitched or being less defined in pitch, to various degrees.

The clarity of pitch sensation (or definition of pitch sensation) is a variable, to which timbre plays a decisive role. This is not a dichotomy, but rather a continuum between the extremes represented by white noise and an uncompromised pitch sensation. Complex tones can often have a more dominant pitch sensation compared to pure tones, as the fundamental can be reinforced; this is not a given, though. Complex sounds exhibit variable 'pitch definition'; this is a level of clarity or focus within the pitch sensation that is the result of the content of their spectra. The presence of low-integer harmonics provides a stronger sensation of pitch (as their simple proportional relationships serve to reinforce the presence of the pitch sensation by adding acoustic reinforcement of the fundamental frequency) and the presence of overtones diminishes pitch clarity.

Non-periodic waveforms contain a decisive percentage of frequencies that are inharmonic with the fundamental frequency; this results in a diminished sense of pitch. It is common practice for such

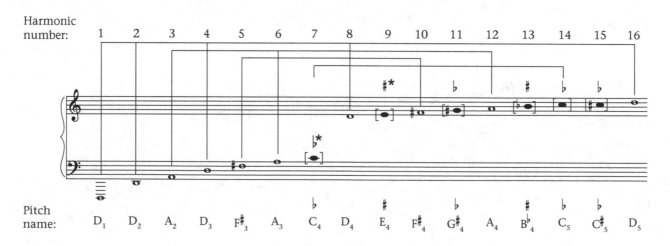

Figure 7.2 Harmonic series through 16 harmonics, with pitch names and frequencies identified.

sounds to be described as unpitched, though there is not a complete absence of pitch sensation. Elsewhere I have identified sounds of this type as occupying a 'pitch area' (Moylan 1985; 2015, 153–161), in contrast to being either pitched or unpitched. The sensation of a pitch area is the result of a dominant frequency band or bands within such sounds; these frequency bands are defined by their upper and lower boundaries and by the density of frequency information they contain (the number of partials and their spacing, which produces density). Among such sounds, drums and cymbals have timbres comprised of pitch areas; the pitch areas these instruments produce are often wide enough to span intervals of a second or a third, and sometimes more. This allows these instruments to appear to conform to the tonality, harmonic progression and or other pitch-organizing principle incorporated into a track. The pitch areas can appear to reflect any pitches they contain; with a different pitch dominating the percept depending on context (chord, tonality, etc.). Thus, even for sounds with non-periodic waveforms and dense in-harmonic partials, spectrum plays a defining role in pitch sensation and the clarity (or definition) of any pitch impression.

Timbre Defined

We have often noted the significance of timbre to the track and to popular music. Yet for all this acknowledged significance timbre is not easily defined. The American National Standards Institute formally defines timbre by identifying what it is not: "Timbre is that attribute of auditory sensation in terms of which a listener can judge two sounds similarly presented having the same perceived loudness and pitch as being dissimilar" (ANSI 1994). Timbre is often framed as 'tone color,' and in doing so it can easily substitute for 'pitch' within the discussion of sound waves and light waves that began this section. Timbre is a gestalt percept; it is a singular impression that coalesces from its many parts. Complicating matters, the acoustic component parts within timbre contain dimensions of pitch/frequency and loudness/amplitude.

Timbre is an overall quality of a sound, or a collection of overall qualities exhibited by a sound source, or an overall quality of a combination of sound sources (as examined in a moment or observed as fluid over a period of time), or of a track or genre of music—it functions on all levels of perspective, and can represent the percept present within a moment, or extended to any duration. Timbre is an overall quality that is comprised of physical attributes that together establish an auditory sensation and also a psychological phenomenon of a singular whole that is different from the sum of its parts. Thus, timbre has often been called 'tone color' because it brings sounds to be readily identifiable objects in a unique representation—not dissimilar to visual color. Timbres can be representative of their source (instrument or voice) and of what the source is communicating (language, urgency, emotion, etc.); a timbre can span a generic type (an acoustic guitar) or be specific to a sound (a friend's laugh); timbre itself can express and communicate energy, expression, emotion, and more. The approach to observing and examining timbre we will use acknowledges timbre has (1) an overall quality (interpreted within its character, representation, affective quality, etc.) and timbre has (2) acoustic content (measurable within its spectrum, spectral envelope, and dynamic envelope). A brief summary of the components of timbre (that incorporates some of the information just presented) follows.

Components of Timbre and Clarifying Related Definitions

Timbre is immensely complex in its acoustic content. If reduced to simplest terms, timbre is comprised of three component parts: its overall dynamic shape (dynamic envelope), its frequency content (spectrum or spectral content), and the dynamic shapes within its spectral content (spectral envelope). These components themselves are complex, with each holding the potential for different attributes and significant variation of states and/or activities within them. Each of these components typically will change throughout the duration of the sound, and as one component changes it is likely to initiate a change in the others; their interaction can be intricate.

'Dynamic envelope' can be the most easily observed component of a timbre; it is the overall dynamic/loudness level of a sound as it changes throughout its duration. Dynamic envelopes of vary greatly in contour—this is especially important when observing voices and acoustic instruments. Dynamic envelopes are often considered as divided into time segment parts; a four-segment concept of ADSR (attack, decay, sustain, release) emerged from early synthesizers, and is greatly generalized. Dynamic contours are typically much more varied—especially when contours reflect the details generated by attack types, changing sustain levels, secondary increases of loudness, tremolo and other types of performance expression. Figure 7.3 provides a more detailed contour, though it, too, is overly generalized. Many musical instruments have more complicated contours, and a few are simpler; most sources have some dynamic control over the sustain portion of the sound.

The 'prefix' is the initial portion of the sound's duration; spanning the time segment from the attack through initial decay, the prefix displays markedly different characteristics of dynamic contour, spectral content and spectral envelope than the remainder of the sound—which is sometimes called the 'body' of a timbre. The body of the sound typically has a much longer duration than the prefix. Many of the unique attributes of the prefix are the result of the way the sound is produced—bowed, blown, plucked, struck, etc. The intensity level of the performance can bring significant changes to the components during the prefix.

Recalling from above: spectrum is a collection of all the frequency information present within the sound. A partial is an individual contributor to the spectrum; it may remain stable or vary in frequency; typically, we conceive partials as single frequencies, but they can appear as narrow bands of frequencies such as is present within the attack of many piano sounds. A host of partials are present within timbre's spectrum that combine into a single percept that is the spectrum. We can understand the values

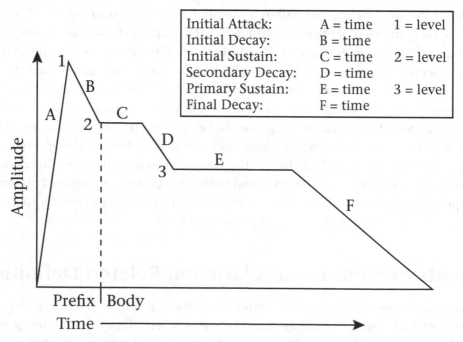

Initial Attack:	A = time	1 = level
Initial Decay:	B = time	
Initial Sustain:	C = time	2 = level
Secondary Decay:	D = time	
Primary Sustain:	E = time	3 = level
Final Decay:	F = time	

Figure 7.3 Dynamic envelope general shape and parts.

(frequencies) and relationships (intervals between frequencies) of partials and the overall qualities they establish for the spectrum, by relating them to the harmonic series (from above).

Spectrum is much more complex than multiplying harmonics, though. A string not only vibrates in multiples of the fundamental, it makes other excursions as well. The start of the sound brings the string it to vibrate in other ways, adding other partials—ones that are not proportional to the string-length (overtones). Further, consider the string residing on an instrument: the body of the instrument will resonate and will generate another set of frequencies, called 'formants,' formant frequencies or formant regions. Formant frequencies will not shift when the fundamental frequency changes, but will be excited by some fundamental frequencies more than others; this brings all pitches on an instrument to have some frequency information in common. Formants are potentially always present at some level, and are independent of the source spectrum. Thus, frequencies other-than harmonics are nearly always present in the spectrum—and many of these might be present, perhaps more prominently than harmonics.

Revisiting, these frequencies that are not harmonically related to the fundamental are considered overtones. Once a proportional relationship of a frequency to the fundamental exceeds 1:17 (except multiples of the fundamental, and 3rd and 5th harmonics) it is more appropriate to consider those partials as overtones. Partials lower in frequency than the fundamental may be present; these are called 'sub-harmonics' (if in a harmonic relationship to the fundamental) or 'sub-tones' (if inharmonic). The acoustics and geometry of a string vibrating is very straightforward; many sound sources—including vocals—are considerably more complex in how they establish vibrations. Add to this frequencies and frequency bands created by the mechanical or physical action of starting the sound, breath and body sounds, sounds created within the performance (such as delicate or pronounced bow noise); many other qualities of sound may be embedded that create 'partials' of frequencies or of pitch areas.

It is becoming clear a great complexity is common within the frequency content of the spectrum—adding to this complexity is the presence of a 'spectral envelope.' The spectral envelope is the aggregate established by the presence all partials as they exist at their individual dynamic levels and each partial's dynamic envelope, as each partial potentially changes loudness throughout the duration of the sound (Moylan 2015, 6–10).

Consider, each partial typically has a unique dynamic contour (envelope); each partial exhibits changes in amplitude (loudness, dynamics) over the course of the sound's duration, and typically changes independently from all other partials. In aggregate, these dynamic envelopes of each partial comprise the timbre's spectral envelope. This aspect of timbre also appears in literature as spectral centroid and spectral flux. The spectral envelope contributes substantially to the intrinsic quality of timbre, and is perhaps its most defining characteristic. The spectrum is in constant flux; it changes over time related to the presence and amplitudes of its partials. The number of partials and their loudness levels, contours and inter-relationships establish a complex and changing sonority that is the timbre's spectrum. Spectral envelope changes as partials enter and exit during the sound's duration—and as partials change in amplitude. Observing the evolving spectral content within timbre is central to defining its content and understanding the origins of its character; indeed, spectral envelope functions at the level of greatest nuance and shapes the overall character of timbre intrinsically.

This explanation of the physical content of timbre is rudimentary, but should prove adequate for our immediate purpose of engaging timbre and pitch/frequency in recording analysis.[4]

Acousmatic Listening and Timbre as Sound Object

To learn about the sound itself and to identify its content, we can directly engage its substance by conceiving of the sound out of context, as a sound object. Pulling a sound out of context isolates it in our perception, and removes the influences of other sounds and its influence upon other sounds. It can also remove it from thoughts of its origin. This is a different way to approach sound, to conceive sound, to hear sound—one that facilitates the understanding of timbre as a recording element, and also pitch density and pitch area (below).

In our life experiences we engage individual sounds 'in context' within analytical listening. Sound sources and sounds carry connotations and meanings, bring associations and affects to individuals, these can contribute to or detract from the analysis of the content of timbres. We have discussed before the challenges related to describing sounds and talking about sound; language typically moves into personal perception, interpretation and impressions, and even culturally accepted terms are not well defined when applied to sound.

What might we accomplish if we were able to remove the notion of a source of the sound—a sound independent of any causal reference? Some core concepts of Pierre Schaeffer's research in *musique concrète* might open a portal into finding at least a partial solution to this important matter. Schaeffer's writings were directed towards creating a generalized music theory,[5] and partially intended to inform and guide his compositions. The core concepts of his work that are important to recording analysis relate to the sound object, and to reduced and acousmatic listening. While these concepts may find a different application in recording analysis, their central thrust of isolating sounds from their context is critical here. The approach to timbre analysis incorporated here does not contain direct traits of Schaeffer's *Programme de la recherché musicale* (1952, 2012) however. The approach offered below is intended to be productive in observing and evaluating the timbres of popular music records, rather than Schaeffer's process which is "inextricably linked to the actual process of composition" (Dack 2010, 278).

Acousmatic Listening

Beginning in his earliest experiences in radio, Schaeffer's work engaged and revealed a different way of listening: acousmatics. Radio presented sounds without the aid of visuals, as did the telephone, and these provided direction to his separation of the sound and the originator of the sound;[6] of course, the record functions similarly—as performers are invisible to the listener. Acousmatics is a state where one does not see what caused the noise that one hears. Schaeffer was drawn to create sounds that were ambiguous in their origin—unknown sounds of unimaginable origins—so timbre itself could be the substance of the music, and could function as an abstract musical element. While this is not our purpose in recording analysis, this concept brings us to access 'pure sound.' Approaching listening as an experience of 'pure sound,' we might acquire the ability to observe the inner workings and qualities of timbres without identifying and interpretative layers (whether personal, visual, cultural, or other) (Schaeffer 2012). We might accomplish capturing details on the sounds *themselves*—the sound decoupled from its "physical-causal source" (Kane 2014, 16), and all that accompanies that source.

Sound becomes an object, disassociated from its source and its contexts; it is also suspended outside the passage of time (Moylan 1992, 64–65). The *objet sonore*—or sound object (sometimes translated as sonorous object)—leads us to take interest in the sound itself.

> The dissociation of seeing and hearing here encourages another way of listening: we listen to the sonorous forms, without any aim other than that of hearing them better, in order to be able to describe them through an analysis of the content of our perceptions.
>
> (Schaeffer 2004, 78)

Pierre Schaeffer's work also introduced 'reduced listening' "whereby the listener perceives the sound for its own sake, as an object isolated from its source" (Bayley 2010, 9). Schaeffer further explains:

The concealment of the causes . . . becomes a precondition, a deliberate placing-in-condition of the subject. It is *toward it*, then that the question turns around: "What am I hearing? . . . What exactly are you hearing" — in the sense that one asks the subject to describe not the external references to the sound it perceives but the perception itself.

(Schaeffer 2004, 77)

The sound object does not reach outside itself to make connections, meanings, references, etc. It is self-contained, self-defining. Sound itself is the object to be observed, in contrast with the sound as a vehicle for something else (Chion 2012).

In this way, we can seek to define or describe the perception itself, not external references (its origins, causes, effects or affects). We might come as close as possible to a common experience; the experience of the sound itself, void of its source and void of its meanings and associations; the phenomenological query "am I hearing what you are hearing?" might have some relevance when 'hearing' is directed to the pure sound and decoupled from its physical origin.

Approaching sounds through acousmatics and as sound objects, we might find ourselves able to engage the content of sounds "independent of its source and any semantic content" (Cox and Warner 2004, 413). In so doing, we might be able to establish some way of navigating the physical information that characterizes a sound. We might describe sounds by their content, and talk about sound in terms of the perception of the sound object itself. With these concepts, we might exchange meaningful observations about sound(s) that are shared by others.

The sound object is clearly applicable to observing and analyzing timbre. It is also central to observing pitch areas and the 'scenes' of pitch density that we will engage next.

PITCH/FREQUENCY AS AN ELEMENT OF RECORDING

Pitch as an element of recording contrasts substantially with pitch in music. 'Pitch' in recording is connected to 'frequency'; it is rarely considered in relation to the pitch levels and the organizational systems of music.

Pitch is a mental representation of certain frequency activity—a "perceptual experience related to the frequency of vibrations" (Tan, Pfordresher & Harré 2018, 30)—relayed to (through) the listener's perception into cognition; "it requires a human listener to make a perceptual judgement" (Howard and Angus 2017, 131). Frequency levels, in contrast, can be scientifically measured, as periodic oscillations of a waveform create regular displacements of air pressure. Periodic oscillations are the result of regular displacements of air pressure, created by a regularly vibrating body. A waveform of repeating cycles results; frequency is a measure of the number of these cycles occurring within a second. Frequency is not systematically organized, such as we find with pitch in music—neither in recording nor in other ways that the measurement is used scientifically or in recording production applications.

Contrasting frequency, pitch is a psychological phenomenon that results in the perception of a single, dominant impression of frequency as reinforced by the spectral content of the dynamic waveform. The frequency level of a fundamental frequency forms relationships with other frequencies (partials of the spectrum); the frequencies are understood in relation to the harmonic series. This is a complex process: the sensation that a pitch quality results from frequency relationships of the harmonics of the waveform weds timbre and pitch; it also leads to understanding how pitch is perceived as replicated at octaves (Butler 1992, 50–55). Pitch perception is certainly more complicated than this brief explanation, and various conditions allow a variety of frequency levels to be accepted as the same pitch level.[7] The point of all this is to illustrate the intrinsic connection of pitch to frequency (and how the two are interconnected to timbre). In recording analysis, it can be useful to understand the relationships between frequency and pitch, and equivalent nomenclatures. Connections between pitch and frequency are evident within Figure 7.4.

Praxis study 7.1 promotes establishing a personal sense of pitch that can bridge pitch and frequency, and can also develop a sense of pitch that can assist music analysis as much as recording analysis. Most of us have not developed absolute pitch, and cannot recognize specific pitches with culturally defined labels—as identifiable, measured and precisely labelled percepts that conform to our cultural definition of divisions of the octave. All of us can, however, recall pitch levels when associated with the beginning pitches of specific melodies within particular tracks that are well known to us (Levitin 1994); we usually start singing melodies on pitch (or within one or two half-steps) of the original melody—this holds for untrained musicians as well. Pitch memory without pitch labelling appears to be common among the musically trained and the general population (Tan, *et al.*, 2018, 72–75). By developing and using the skill of recalling the starting pitch level of a known melody, one can acquire a well-honed sense of pitch recognition that is different and distinct from absolute pitch, in that it is *learnable* by adults of all ages and that it does not rely on the labelling of pitch levels. This skill can ultimately lead to being able to move between pitch recognition and identifying frequency levels—an important step towards identifying the subtle frequency levels and relationships within timbre's spectra (below).

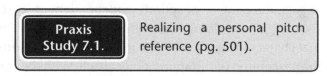

Praxis Study 7.1.   Realizing a personal pitch reference (pg. 501).

As approached as a recording element, pitch units and frequency levels might appear in the same sentence, describing the same sound or material. Pitch levels allow analyses to compare the recording element to musical contexts. Frequency levels allow access to how the element is measured within the processes of the recording, allow the material to be related directly to various listening processes, and also allow for more detailed examinations. At times it may be appropriate to an analysis for pitch and frequency to be addressed simultaneously, allowing an interpretation to be more widely applied. For example, the pitch areas of unpitched timbres might be identified by the frequencies of their boundaries, as pitch area boundaries will rarely align with equal temperament pitch levels, and the pitch of the boundary is not as significant as the registral placement of the pitch area.

Commonly used language identifies frequencies of a greater number of vibrations per second (Hz) to be 'higher' than frequencies with fewer. References to 'higher pitches' and 'lower frequencies' are widely-accepted conventions of conceiving and discussing pitch and frequency; in fact, our vocabulary seems to have no others. This practice and limitation do not present a problem until we engage the elements of recording; then confusion can emerge between physical height and the range between 'highest and lowest' pitches. In other words, is 'higher' greater elevation or faster vibration? Language confuses placement of a frequency/pitch within the pitch/frequency range of the track (or within the range of audible frequencies of the listener) with physical height, or the location of sound at an upward or downward trajectory from the listener location; this discrepancy is important within the context of spatial relationships of sounds, and will be discussed later.

The language of higher and lower pitches is part of our vocabulary for engaging music. Talking about 'higher pitches and lower frequencies' in most contexts communicates clearly. Clarity about which percept is engaged is needed in recording, though. One needs to recognize that the 'vertical space of music' (and sometimes 'vertical space of recording' reflected in timbral balance's connection to music's arrangement) is a metaphor and pitch related, and the 'vertical space of recording' may be spatial (as generated by most virtual reality and some surround sound formats). Denis Smalley (2007, 45) describes this vertical distribution of pitch/frequency-related concept: "[S]pectral space is concerned with space and spaciousness in the vertical dimension." This distinction can help clarify the vertical aspect of the soundbox as clearly metaphoric related to pitch, where the "frequency of a sound determines its placement on the vertical plane, with higher frequencies perceived to be placed in the upper zone of the soundbox and lower frequencies occupying the lower section" (Dockwray and Moore 2010, 183).[8] Space in recording is distinct from any vertical aspect of frequency, though conceiving of sound as distributed throughout the pitch/frequency range vertically is central to our understanding of timbral balance and pitch density (below). Lastly, this same vertical distribution is inherent in frequency/pitch-related X-Y graphs used here—though without intention related to 'height'; with low frequencies at the bottom and high frequencies at the top, it relates to familiar musical pitch notation.

Individual pitches impact the element of recording only when recording functions in tandem with melody; such an alignment is unusual, though. An example appears in "Here Comes the Sun" (1969) with the Moog glissando that ends the introduction—and within pitch density of lines with moving melodic materials. In essence, pitch/frequency as a recording element is a link between music's pitch elements and the distribution of sounds within the recording's timbral balance that is distributed among the pitch/frequency registers of the track. The following sections explore this, as unpitched sounds are recognized as occupying pitch areas (bandwidths) within the track's range, and as timbres of sound sources and their materials are recognized as occupying an area (a bandwidth) of pitch/frequency and form pitch density 'scenes.' This leads to conceiving the track's overall pitch/frequency dimension as the record's timbre (timbral balance).

Pitch/Frequency Registers

A sequence of frequency (pitch) bands has been devised to divide the hearing range into pitch/frequency registers that might be readily identified. This sequence will facilitate observing and evaluating sounds for precise pitch/frequency content; it also aids in sound placement (from general to precise observations) within the range of the track. The concept of pitch/frequency range divided into registers is a central component to accessing the pitch/frequency/spectral content of the track offered here. Noting "rock has strands at different vertical locations, where this represents register" Allan Moore (2001, 121) promotes the placement of sounds on the vertical axis of the soundbox; conceptually this is similar to the continuum that is proposed here for transcribing pitch/frequency content. The information the two systems collect is reflected in the strata of pitch density and the textures of timbral balance (discussed below).

Figure 7.4 illustrates the division of the audible pitch/frequency range into pitch registers. These pitch registers will appear in numerous graphs: pitch area, pitch density, sound quality/timbre, timbral balance, environmental characteristics, and others.

This arrangement of pitch registers has been specifically devised[9] for dividing the hearing range into recognizable pitch/frequency bands. These registers establish important points of reference—both for ease of use in recognizing pitch/frequency levels, but also in developing this skill. Learning to recognize the sound of the boundaries between registers establishes an important resource for many processes that follow. These boundaries are areas of transition and not thresholds that separate registers; they are generally defined for ease of entry to the process.

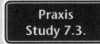
Praxis Study 7.2. Recognition of sounds in relation to pitch/frequency registers (pg. 503).

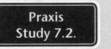
Praxis Study 7.3. Conversion of pitch levels into frequency levels and the reverse (pg. 504).

Praxis Study 7.4. Developing facility at frequency estimation through transcribing melody into pitch/frequency registers and mapping melody against a timeline (pg. 505).

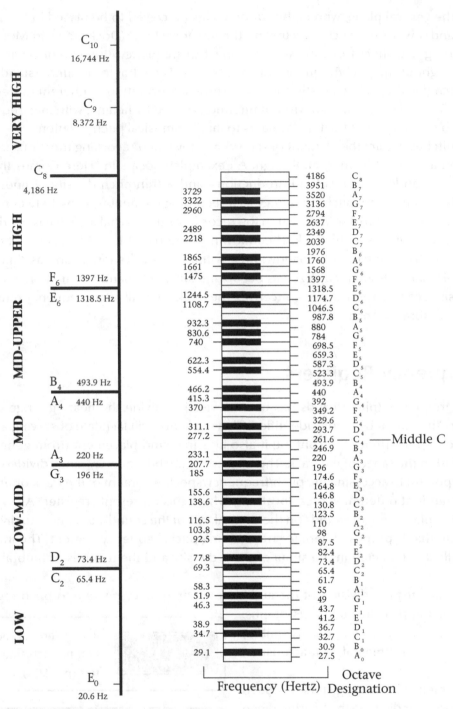

Figure 7.4 Pitch/frequency registers in relation to keyboard with pitches, octave designations and equivalent frequencies.

This sequence of pitch/frequency registers was carefully devised to allow the listener/analyst (and also recordist) to most readily recognize what they hear. The distribution of registers has been refined over time, with the hearing range divided into pitch/frequency bands (registers) based on physiological and psychoacoustic factors.[10] Hearing mechanisms process frequencies within each register consistently, and the bands that separate the registers are significant as transition areas. The registers 'appear' to have something consistent throughout, and to be different from the others; this helps them to be recognizable and learnable. This range of registers forms the Y-axis of many graphs, providing readers a way of writing the pitch/frequency information they hear out of the context of musical pitch notation and temperament.

Pitch Areas

Sounds that occupy a pitch area, rather than a specific pitch level, offer a gateway into the inner qualities of timbre/sound quality. Many percussion-related sounds have noise-like qualities with waveforms that are non-periodic in many respects; they do exhibit a pitch quality, though. We recognize these sounds as being 'higher' or 'lower' than other similar sounds—some drums appear to be 'higher pitched' than others. These sounds often contribute prominently to the textures of timbral balance; recognizing their pitch areas opens one to recognizing a number of important concepts.

Unpitched percussion sounds occupy pitch regions. Their imprecise pitch quality comes from an area of dominant frequency content within their spectrum—this is the sound's primary pitch area. These sounds appear to occupy a bandwidth between two pitch/frequency boundaries (these define the primary pitch area); within this bandwidth will be a certain amount of frequency/pitch activity that creates a density of partials. These sounds can be perceived as aligning with (or representing) any pitch within their primary pitch area; this allows some drums and cymbals to be perceived as approximating pitches within different keys, and perhaps having harmonic functions.

Typically, unpitched sounds will contain several pitch areas—a primary pitch area that is most prevalent, and secondary pitch areas that are less pronounced. Any number of secondary pitch areas will be present; the secondary pitch areas are bandwidths of frequencies/pitches that are distinct from the primary area. Secondary pitch areas are located at frequency levels higher than the primary pitch area, and sometimes also below; often they are in inexact harmonic relationships with the primary pitch area. Secondary pitch areas will have a dynamic relationship to the primary pitch area, as well as their own characteristic density; an interval void of frequency/pitch activity may be present between any two pitch areas, though some areas can be contiguous. A secondary pitch area will be present in nearly all unpitched sounds; the number of secondary pitch areas can vary significantly, from a single area to a considerable number. Further, when such instruments are performed at different loudness levels (intensities), different secondary pitch areas might appear, and the densities of existing pitch areas typically change. With changes of performance intensity, the dynamic relationships of secondary pitch areas to the primary pitch area will change. Performance intensity plays an important role in determining the timbre of all acoustic sounds, whether these are unpitched instruments or any pitched sound source.

Representing Pitch Areas of Unpitched Sounds

Unpitched sounds may be graphed, defined or described by identifying these variables:

- Width and register of the primary pitch area (established by the two boundary frequencies/pitches of the pitch area)
- Density of pitch/frequency activity within the pitch area
- Presence and locations of secondary pitch areas (identified by the frequencies/pitches of their upper and lower boundaries)
- Density of the pitch/frequency activity within each secondary pitch area
- Dynamic relationships of secondary pitch areas to the performance intensity of the primary pitch area

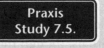 **Praxis Study 7.5.** Recognizing pitch areas, pitch area characteristics, and the relationships of pitch areas within percussion sounds (pg. 506).

The bandwidth of pitch areas is the span established by the lowest frequency/pitch (lower boundary) and the highest frequency/pitch (upper boundary). It is a region of activity perceived as exhibiting a unified quality. It will have a mostly uniform density and loudness level. Pitch areas exist at the perspective of spectral components within sounds; they are the rough equivalent of partials within a spectrum, discussed below under timbre/sound quality.

In this approach to pitch areas, the sound (instrument) is pulled out of context—observed as a sound object. The content of pitch areas is deliberately generalized in a number of important areas:

- time/temporal characteristics: the entire duration of the sound is brought together into a single impression
- frequency/pitch content and characteristics: pitch areas are defined between two frequency/pitch boundaries, and a general impression of density, or the number of frequencies within the pitch area is recognized
- loudness relationships and levels: the relative loudness levels of pitch areas are recognized, allowing loudness levels of areas to be compared

These three characteristics of the sound are summarized temporally into an overall impression. Changes over time are not included in these observations. Instead, pitch area observations collect data on all of the frequency/pitch information over the duration of the sound, and formulate an overall impression of the sound. The instrument's timbre is examined for the pitch areas that are formed, and that are present at any time throughout its duration—though the pitch areas will certainly change in density and loudness during that duration. The component parts of the unpitched sound's timbre are observed as an overall impression. This simplifies much during these initial observations.

These generalized observations may be adequate for a given analysis. Pedagogically, this generalized approach forms a portal into hearing the inner workings of timbres; pitch areas are more easily distinguished than individual frequencies, and learning to bring one's attention to these more prevalent portions of spectrum will bring the reader/listener experience in hearing 'under the surface' of timbres. Learning to hear the overall quality of pitch areas leads to identifying their lower and upper boundaries to bring clarity to its width and registral placement; this will ultimately allow one to more readily distinguish individual frequencies of partials as will be examined within timbre analysis, below. Practically, this approach provides a condensation of considerable frequency, loudness and register information into a single object, and as a summary of all activity that took place over the sound's duration; this is often adequate for pitch density and timbral balance observations (below).

At this stage of observation, the sound is being engaged as a conception that summarizes all of the activities of the sound throughout its duration into a single, non-temporal sound object. The changes that take place in the sound at each moment are not engaged here; rather this process makes observations about the sound as a singular impression, observations of the sum of all moments of the sound event. There is no left-to-right progression of time in the pitch area graph in Figure 7.5; this is unusual. This adds some inherent subjectivity caused by generalization, as well as an absence of detail to these observations.

The amount of density and the loudness levels within pitch areas are purposefully generalized. This allows faster observations that are accurate within limits, though still informative; these observations can later be used to generate more detailed information on the sounds, if that is desired. Density of activity within the pitch area is calculated in general terms on a rough scale between "very dense" (conceptually similar to a cluster chord of quartertones or perhaps pink noise) to "sparse" (a density with a relatively small number of partials, though the fundamental remains veiled or vague). Table 7.1 lists five levels for this general scale of density: moderately dense, sparse, dense, very dense and moderately

Table 7.1 General scale for the relative density of pitch areas.

Relative Density Level	General Impression of Density
Very Dense	Conceptually similar to a cluster chord of quartertones or perhaps pink noise
Dense	Contains a significant number of partials throughout the area
Moderately Dense	Contains an opaque quality, where its presence can bring another sound to be partially covered or masked; density may shift slight throughout the area
Moderately Sparse	Contains a translucent quality, where it does not mask other sounds; density may shift slight throughout the area
Sparse	A density with a relatively small number of partials; though the fundamental remains veiled or vague, one begins to sense its presence

sparse. These terms carry no small degree of subjectivity and are inherently imprecise, but represent a roughly accurate assessment that can be learned and accomplished reasonably quickly (with some listening skills and control of these concepts). Even this generalized approach can communicate significant information about the substance of the sound. This can lead to a sense (for example) that an unpitched sound morphs into being pitched, as density diminishes and as the spectrum begins to emphasize harmonics.

Pitch Area Graph

Notice Figure 7.5 identifies a time for each of the four drums; this locates the sound in the track and represents a specific strike of the drum. This single timbre is the object of our attention, wherein we observe its pitch areas as broadband partials within its spectrum. Small changes of loudness and density within pitch areas may be heard with changes in the level of intensity of hitting the drum that translates into a level of performance intensity. An instrument produces a characteristic timbre when performed at a given level of intensity; its spectrum changes when played more forcefully, or with less exertion. As performance intensity changes, so does the sound's timbre; densities of pitch areas change, as do the dynamic relationships between pitch areas. On some instruments, boundary frequencies of pitch areas will also shift with pronounced changes of performance intensity. Loudness levels of pitch areas are addressed by identifying the general performance intensity of striking the instrument, and using that level to represent the primary pitch area's loudness. That level then serves as a point of reference for that particular strike of the instrument.

The primary pitch area is designated by a 'P' in the first position. It is replaced by a number designating loudness levels of each secondary pitch areas; loudness levels of secondary areas are gauged by degrees of louder or softer in relation to the primary pitch area. Note the numbers and defined levels of 0 through 6 in the Figure 7.5 key. As with density, these are general observations intended to pull one into the sound, rather than provide a detailed and precise representation of the sound. Here we are observing the generalized dynamic/loudness levels of and relationships between pitch areas; later we will be observing the more specific dynamic/loudness levels, contours and relationships of partials as individual frequencies within a spectrum.

Figure 7.5 is a pitch area graph of four drum sounds from the Beatles' "The End" (1969, 1987). The bandwidth of each pitch area is defined clearly by the frequency of its boundaries; these frequencies can be translated into pitch (with varying degrees of conformity to equal temperament) if needed by using Figure 7.4. An instrument produces a characteristic timbre when performed at a given level of intensity; this timbre is what is being explored by pitch areas.

Examining the three tom drums, we can identify the primary pitch areas occupy a similar register; their dominant energy is within the 'low-mid' range. All three drums contain a lower, sub-harmonic

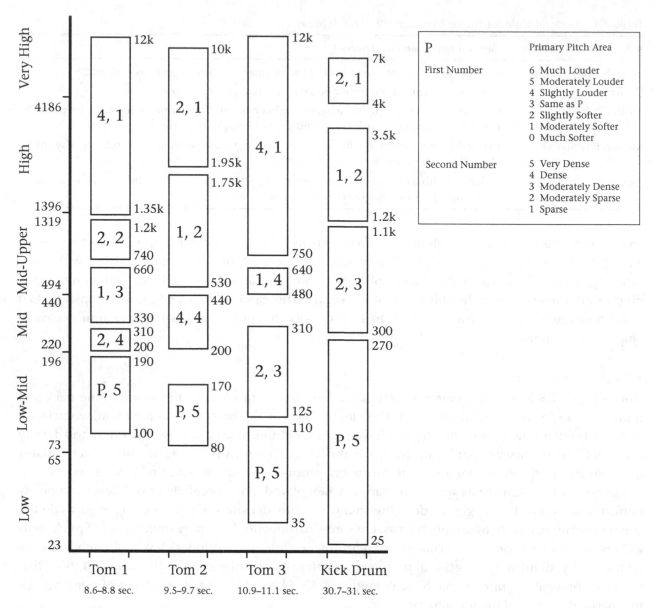

Figure 7.5 Pitch area graph of individual drum sounds from "The End" (1969, 1987).

pitch area of lower loudness that emerges when performed at lower intensity levels; these are within the 'low' range, where the primary pitch area of the bass drum resides. Notice the secondary pitch areas between the three tom drums; a significant difference in number of secondary areas, and frequency distribution exists between them. These drum sounds have been examined for their own qualities, out of the context of the track; they are approached as sound objects. Graphing pitch areas in this way, before engaging the more complex and subtle qualities of timbre (below), allows one an opportunity to make pitch/frequency observations in isolation, and independent of time. Further, it allows one to engage accessing the inner workings of spectrum at a slightly higher level of perspective and with 'spectral components' (secondary pitch areas) that are more easily distinguished from individual frequencies. Accessing this information as a generalized impression outside time simplifies the task of understanding the content of these sounds, yet reveals pertinent information. Without tracking the changes that take place over time, those without experience of listening within sounds have the opportunity to acquire skills in calculated stages of development. Our next observations, and most of those that follow, will observe elements (or attributes of elements) as they change over time in the context of the track.

Other general observations might be made, should it best suit the analysis or the skill of the analyst. For example, one may be less specific about the lower and upper thresholds of pitch areas by not identifying specific frequency/pitch levels; a more-or-less clear sense of their register and their placement within that register (such as 'at the bottom' or 'in the upper third') might be deemed adequate, and would be more generally located on the graph as a boundary line of the pitch area. This approach might also be used as an initial step (perhaps as a study) that precedes defining pitch area boundaries more precisely. The detail-level of information gathered should be adequate for the goals of the analysis—detailed enough to discover traits about the track to support analysis goals or perhaps to lead the analysis to be slightly broadened; detailed pitch area information is not always needed in an analysis, though it is central to timbral balance, and thus pertinent within many recording analyses.

Other instruments and sound effects are unpitched, not only drums and cymbals. Pitch area graphs can bring those sounds as well as drum and cymbal sounds to be located within the musical texture of the arrangement. This can serve to form a substantive link between music's arrangement and recording's timbral balance. Unpitched sounds—often central to popular music—can be observed for the placement in and contributions to the track by approaching them as occupying pitch area and having pitch density placement.

Pitch Density

Pitch density is an element that may be discussed or analyzed from the perspective of either the recording and music domains, and it is significant to the track. Pitch density collects and presents information that can lead to understanding the track's arrangement and timbral balance. This section is intended to augment—reframe, broaden and add detail—the discussion on pitch density and the arrangement within Chapter 5. Pitch density allows access to understanding certain qualities of an arrangement—both as a dimension of the recording and as a dimension of the music. It will be addressed here for its significance in building timbral balance, and in doing so contributing to both the arrangement and the 'sound' of the record.

Pitch density fuses the individual sound source's timbre plus any variation in pitch over the time period that represents its duration. It localizes the musical material and its timbre relative to pitch/frequency and bandwidth—these may be static or changing. Pitch density defines a sound source in relation to its placement in the pitch/frequency range of the track. It includes any changes of the bandwidth; these are both changes in pitch material reflected in the lower boundary and in the source's timbre (that is reflected in the upper boundary and density). Pitch density is how the sound source resides in the arrangement (as a music element) and how it is situated in timbral balance (as a recording element).

To reframe, the pitch density of a sound source is identified by its pitch/frequency content and location within the overall frequency structure (timbral balance) of the track. The musical materials a sound source presents are fused into a single idea (sound event), that includes the substance of the musical idea with the dominant timbral characteristics of the source; this brings a sense of the source occupying a variable pitch/frequency bandwidth (pitch space) for a certain segment of time. This is pitch density.

Pitch density unfolds as 'scenes' of the source's materials. As the material unfolds, it coalesces into a singular impression of what happens/is happening in the time window of the present; this is the 'now sound' of syncrisis (Tagg 2013, 417–484). Pitch density is mapped out as a series of snapshots of time, passing successively from one to the next—often (though not necessarily) aligned with phrases or with the prevailing time unit, it forms a syncrisis unit. Note the unfolding piano line of "Let It Be" in Figure 7.6, where the range and contour of the line establish the lower boundary and the piano's timbre establish the upper boundary and the time unit demarcates the 'scenes.'

Structurally, pitch density appears at the perspective of the individual sound source and contributes separate lines to the perspective of groupings of sound sources (where each is perceived as equal, such as background vocals). This situates it in the middle dimensions, where the pitch ranges of materials and the sound sources that deliver them function. Pitch densities, being of individual sound sources, may be conceived as 'partials' within the composite texture that forms the timbral balance of the track.

To collect observations on pitch density, we recognize and engage the following attributes:

- Beginning and ending points for the 'scenes' (sound events) of the lines
- Pitch/frequency bandwidth(s), boundaries of lowest pitch and the highest significant partial of its timbre
- Relative amount and distribution of pitch/frequency information within bandwidths, the sense of density of pitch/frequency information and activity
- Distribution of spectral energy within the pitch density; this may vary by register and may change over time
- The pitch bandwidth unfolds over time as an experience, or an event
- Pitch materials within the individual scenes may be active and changing (such as a distinct melodic contour), or static (fused into a single impression)
- The lowest pitch of this pitch-band might (1) unfold and reflect the material's melodic motion, (2) remain steady if pitch does not change, or (3) group (fuse) all changing pitches of a line together into a single impression

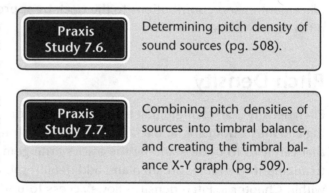

Praxis Study 7.6.   Determining pitch density of sound sources (pg. 508).

Praxis Study 7.7.   Combining pitch densities of sources into timbral balance, and creating the timbral balance X-Y graph (pg. 509).

Pitch Density and Timbral Balance Graph

Observations on pitch density in recordings are readily notated on an X-Y graph, with pitch/frequency registers as the Y-axis and time (utilizing the metric grid) on the X-axis.

All instruments and voices can be represented in their pitch density on a single graph—including unpitched percussion such as drums, cymbals, and other unpitched percussion that occupy pitch areas rather-than a specific pitch level. This reveals the entire timbral balance of the track. Graphs displaying the pitch densities of select sources can often display important traits of a track.

The time unit that defines this material (a gesture described above as 'scene' or 'sound event') might take the form of a number of durations; it might reflect the prevailing time unit, hypermeter or be defined by the phrasing of the musical idea. These units may be readily recognized as the syncrisis unit brings a sense of what has coalesced in the scene of the present time. The prevailing time unit and/or the syncrisis unit can each represent a strong underlying rhythmic and phrase structuring that can provide appropriate structural divisions of pitch density scenes, and their durations can often align. The concepts of both pitch density and syncrisis are based on the window of the extended present, and to perceiving sound gestures as a single, simultaneous sound object—one that is formed by simultaneous and successive sounds clumped into a single aural image within the middle dimension (Tagg 2013, 19–20). In this way, the appropriate time division for pitch density will reflect the unique qualities of the material within the context of the individual track, and will likely vary between sources and perhaps from gesture to gesture. The pitch density gesture might also be identified as being present within a specified *time period*, that is defined by its number of beats or number of measures, or by its start and stop points.

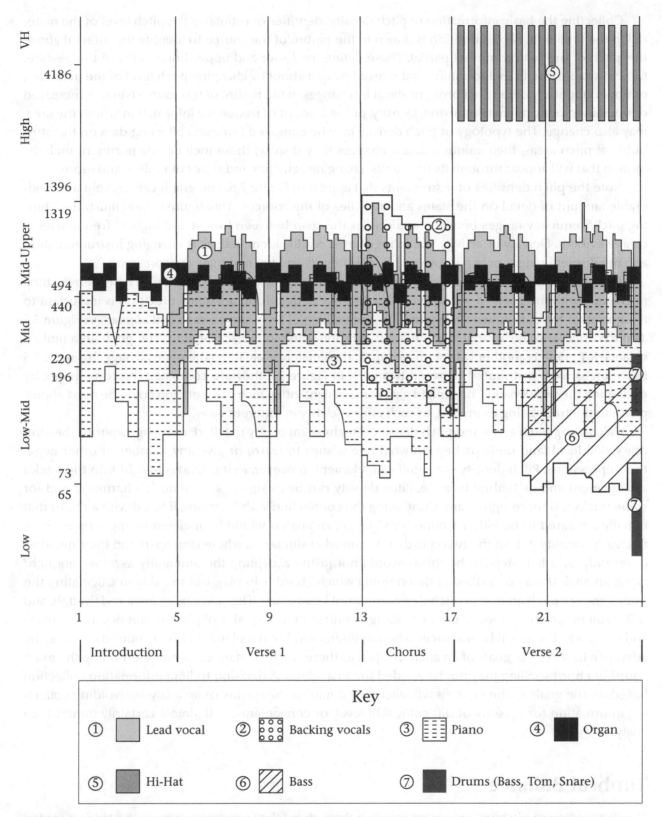

Figure 7.6 Pitch density and timbral balance (including all instruments in the texture) from the beginning through the second verse of "Let It Be" (*1*, 2000).

Collecting the basic information of pitch density identifies or estimates the pitch level of the material in the time unit. Next, attention is drawn to the timbre of the source to identify the interval above this pitch of the last prominent partial. These define the lower and upper boundaries of the gesture. Other variables might include shifts of this area brought about by changing pitch level of the materials, or upper boundary of the area brought about by changes in the timbre of the source (such as increased or decreased performance intensity). Density, or the amount of frequency information within the area, may also change. The typology of pitch density may be compiled through collecting data on the attributes of pitch areas, their values and any changes they display; these include any pertinent melodic motion that will appear through its time units, changing activities and states of timbre, and more.

Note the pitch densities of instruments that appear in Figure 7.6. The graph can present a considerable amount of detail on the states and activities of the sources. This figure clearly illustrates shifting pitch/frequency ranges between sections, as the span between lowest and highest frequencies is clearly evident. Density of activity within registers is readily identified, as is changing instrumentation and the discrete ranges occupies by various instruments and the materials they present.

The level of detail in graphing sound source boundaries, densities and their changes over time might vary with the goals of the analysis. Consistency of detail and adequate precision is important to the quality of the data, however. Figure 7.6 is a reworking of the same passage that appears in Figure 5.5 as pitch areas and the arrangement, there it is represented in music notation. The pitch area/timbral balance X-Y graph is another way of observing (and ultimately evaluating) the same passage. In Figure 5.5 the sources were distributed between two tiers, in Figure 7.6 they are combined into one graph; by examining the two versions, one might gain some insight into which presentation could be most appropriate to the track being examined, as each affords clarity in different ways.

Calculating pitch areas and pitch densities might seem overly detailed, exacting, onerous, beyond one's skill level (and perhaps beyond what one wishes to learn) or give any number of other negative impressions. Pitch density is a significant element, however, as it provides insight into the track's arrangement and its timbral balance. Pitch density can be as significant as music's harmony, and for some tracks it is more significant. Continuing the connection with harmony, if faced with a chord that initially appeared to be either F minor or G major, an analyst would be inclined to apply themselves to clearly identify it. That the two chords are somewhat similar—a whole step apart and their qualities differ only by a half-step in the third—would not justify accepting the ambiguity as 'close enough.' Here, an analyst may be skilled at determining which chord is in play, but the skill in calculating the boundaries of pitch areas and pitch density may not be as acute. The skill can be acquired though, and will result in sensitivity toward understanding this fundamental quality of pitch/frequency in recording and the track. There will be instances when a general level of detail in identifying boundaries may be adequate to serve the goals of an analysis—just as there will be instances when determining the exact chord or chord spelling may not be needed for an analysis. A decision to limit information collection based on the goals of the analysis will likely not diminish the quality of an analysis; avoiding collecting information for reasons of difficulty, skill level, or convenience will almost certainly diminish an analysis.

Timbral Balance

The distribution of pitch/frequency information throughout the frequency range of the track is central to its overall sound. This has been a significant characteristic of records dating back to earliest popular music tracks. The interconnectedness of pitch and timbre in recording is clearly evident within the element of timbral balance.

Timbral balance represents the track's overall frequency spectrum; it is the combination of all of the pitch densities of all of the record's sound sources. Timbral balance is one of the dimensions of the overall texture of the track and is situated in the large dimension. It is characterized by the distribution and density of the pitch/frequency of all sources that coalesce to establish its overall sound.[11]

The pitch density/timbral balance graph (Figure 7.6) can be viewed at the perspective of the individual sound source or at the perspective of the composite sound, comparing one to another. When viewed at the perspective of the overall sound, the graph represents the distribution of dominant frequency/pitch information for the entire track and is the content and sonic quality of all sound sources combined, and as distributed throughout the pitch/frequency range of the record. A spectrograph will represent an image of this aggregate texture of timbral balance, and will show great detail of the loudness levels of specific frequency areas; covered in detail under timbre, these images have proven helpful to some analysts for data collection of certain materials.[12] The timbral balance X-Y graph has different advantages; it allows one to see individual sound sources, and observe how their individual pitch information and timbres contribute to timbral balance. A spectrogram of the same passage that appears in Figure 7.6 and Figure 5.5 is included on the webpage.

Timbral balance is rarely static; in most records it exhibits changes spanning the entire continuum from the most subtle change to the most profound. The pitch density/timbral balance graph allows these changes to be notated and observed; changes over time and against structure can be noted as well as the content of those activities and states of timbral balance before, during and after changes. The graph can be an important analytical tool; much data on individual sources and on the texture of timbral balance is observed in creating it, and is present within it.

Table 7.2   Some primary variables of pitch density scenes/gestures and of the composite texture of pitch density and the overall texture of timbral balance.

Individual Pitch Density Scenes or Gestures

Range of pitch-area (interval between lowest pitch and highest significant partial) of its timbre

Density of pitch/frequency information (relative amount and distribution of pitch/frequency information within its pitch-area)

Changing levels of lower boundary (melodic motion)

Changing interval of range

Shifting density within pitch-area

Duration (time span) of the gesture (either time unit or time period)

Pitch Density/Timbral Balance Graph

Range of pitch-area of each source

Number of sources (voices)

Relationships of sources (voicing)

Density of voices (relationships between voices)

Density (distribution of frequency energy)

Changes in any pitch area (interval, density, level)

Degree of changes (subtle to very large)

Speed of changes

Sudden or gradual changes

Range of aggregate texture

Highest and lowest pitches of each section; structural placements

Widest and narrowest ranges of each section; structural placements of extremes

Ranges of little or no activity

Ranges of peak activity, least activity

Time spans of peak density, least density

As noted before, the vertical dimension of the soundbox places sounds in frequency registers and presents the overall texture in similar ways to timbral balance; the soundbox has distinct value in its general increments and multidimensionality, and might be a more suitable choice for some analyses. For many analyses, though, the pitch density/timbral balance graph may be a more useful tool; its precision and detail in representing data will allow timbral balance to be explored in a more exacting manner.

The pitch densities of sound sources weave an intricate tapestry of timbral balance that is unique to each record. While producers may have traits that make similarities between tracks, the individual record will be distinguishable within those traits. Pitch densities of sounds tend to be located 'in their place' within the distribution of pitch/frequency information of the track (and within human hearing); the patterns created by the pitch/frequency areas (bandwidth locations) of sources is a primary factor that determines timbral balance, and production style. The variables of Table 7.2 can serve as a basis for observations of pitch density and timbral balance, and the building of a typology for how they appear within a given track (or a section thereof).

The natural world provides an interesting validation to this approach to timbral balance, and the pitch densities that comprise it. Recordings of natural soundscapes from all over the world have been analyzed for many purposes, and for many factors and qualities. Within the acoustic analysis of each of these recordings it was revealed species keep to their own bandwidth (pitch area) of the audio spectrum: insects typically in the highest frequency bands, birds in the middle registers, mammals in the lowest frequencies, and families within species further localized (Krause 2015, 34–43). While some overlap of ranges is present, most ranges contain only the sounds of birds, insects, or mammals; "[T]he natural world soundscape necessarily pinpoints 'acoustic niches,' the special ways different species in a single soundscape jostle for sonic territory" (*ibid.*, 41). In *How Music Works* David Byrne (2012, 33) also found interest in this phenomenon, observing: "[N]ot only have the calls of songbirds evolved to work best in the acoustic environment in which those birds live, they also have evolved to stay out of the way of the other critters that live there. . . . everyone always stays in their place." The record's emphasis on the distribution of sound sources in strata and by frequency levels is a practice that stretches farther back in time than we can determine—in many ways the domains of music, lyrics (language and speech), and the recording have evolved similarly.

Timbral balance connects and establishes interrelationships between the overall sound quality of the recording and music's arrangement. It is as substantial a dimension of the arrangement of the record as it is a significant dimension of the 'sound' of a record. Timbral balance is a measure of the sound quality of the overall texture, and as such it is a dimension of the timbre of the track, and of crystallized form.

TIMBRE IN THE RECORDING DOMAIN

This next part will explore timbre as a recording element—its qualities, its manifestations and its roles in the track. In this initial section, some pertinent information on timbre perception is presented; this section also serves as an introduction to the sections that follow.

The element of timbre permeates the record's fabric in all its layers. At the highest perspective, it is the 'sound' of the aggregate timbre of the track. Timbre is also a strong and defining presence in all other levels of perspective, and plays a dominant role in shaping the recording.

The sound of a track immediately captures the listener's attention. Robert Walser (1993, 41) observed: "Before any lyrics can be comprehended, before harmonic and rhythmic patterns are established, timbre instantly signals genre and affect. . . . its significance can hardly be overstated." It can cause one to instantly recognize the genre of music and its affect, and can spur immediate recall of a specific track or an artist. It may also manifest as 'a sound' that distinguishes a track's instruments and voices, especially its performers and vocalists as unique timbres. Within the track it is common for these timbres to be like no others, unique sonic personalities and qualities. While the impact of timbre can be substantial, it is comprised of a complex interaction of subtle qualities. In discussing Brian Eno's work with U2, Albin Zak (2001, 69) summarized: "the most subtle sonic nuance may hold a profound significance."

Everyday Timbre

In our everyday world we interact with others, with objects and other creatures. We function at a basic-level where we categorize others and other things in a way that is similar to ourselves. We relate to the objects of the world on a certain basic-level where things are similar, and where we can recognize them by their sounds; whether another person, a vehicle or a bird (etc.) in our everyday life, we continually identify sounds and their origins.

Timbre's basic-level categorization is perhaps the clearest of any element, and it leads to the formulation of auditory streams (Bregman 1990). These streams delineate the activities of sources, and substantially guide observations in all other elements.[13] Timbre's basic-level is the perspective of the individual sound source (single instrument or voice—as reflected in the materials they present; nested into this level are individual presentations of the sound—appearances ('notes' or 'soundings') of a sound source. Timbre is "the primary perceptual vehicle for the recognition, identification and tracking over time of a sound source" (McAdams and Giordano 2011, 72). Basic-level categorization is that level where sounds and their materials "have similar and recognizable shapes . . . [and represent] a single mental image" (Zbikowski 2002, 33).

The depth and strength of our basic-level categorization toward seeking the source of origins for sounds are ingrained. This presents a significant challenge for acousmatic listening and conceptualizing the sound object. To separate a sound itself from its source goes against (perhaps all of) our previous behavior—and is perhaps contrary to an innate trait, as we continually seek the source of food, danger and other things we need to survive or thrive. This acknowledges that learned intention is needed to engage listening to sound as an object independent from its cause and source.

In our everyday way of listening—practiced since birth or likely before—the subtleties within timbre's component parts are not the focus of attention, and are rarely observed. The inner workings of timbres are rarely (if ever) engaged by typical listeners; in fact, these nuances are only noticed by their impact on the whole, by the way their subtle details transform meanings and characteristics of the timbre. While the inner workings of timbre are the substance of its overall sound, listeners are not drawn to directly engage them; the identity and/or meaning of timbre are reflected in its overall quality, and the listener has no reason to direct attention 'inside' the sound.

Our ability to differentiate timbres, to recognize and to remember timbres is acute—perhaps more so than any other element. This ability functions similarly within all of the domains—timbre in music, and the sound qualities of language and vocal performance. We know many categories of timbres (voices, types of instruments, sounds of daily life and of animals, bird and vehicle sounds, etc.), and many distinct timbres within categories. Consider, we have the ability to recognize vocal timbres of a great many individuals, and even subtle changes of timbres (and the meanings they represent) of the voices of people we have never met or do not recognize. Not surprisingly, "everybody is an expert to some degree" in differentiating timbre, though some listeners are prone to bring attention to spectral properties most often, while others are drawn to focus on temporal aspects (McAdams and Giordano 2011, 73). We will rely on this innate ability and sensitivity in observing timbres and timbral characteristics.

With differentiating timbres, patterning of timbre functions similarly to those of other recording elements. While sequences of different timbres struggle to establish identifiable patterns (like pitch does in melody) (Dowling and Harwood 1986, 158–159), timbral sequences do exist in records—but differently. Timbral sequences function strongly on microrhythmic and on macrorhythmic levels. At macrorhythmic levels timbral shifts between sections (at any middle dimension level or higher) become a significant determining factor of structure and shape. Microrhythmic timbral changes are found between soundings by a single source (Handel 1993) and are also present within a source's individual sounds (B. Moore 2013); they may provide nuance of expression or direction to sounds and sound source materials. These patterns incorporate characteristics that are most readily approached using pitch density (or pitch areas) and pitch/frequency registers. McAdams and Giordano (2011, 75) concluded:

> [O]ur predisposition to identify the sound source and follow it through time would impede a more relative perception in which the timbral differences were perceived as a movement through timbre space [as a sequence of values bonded through syntax] rather than as a simple change of sound source.

Focus remains on simple changes to the overall sound, rather than on sequences of individuated timbres. This works against patterning of different timbres (between instruments or sources) or of timbral qualities (changes of timbre within a single source). When such patterns are established, those patterns are the result of much repetition or of established convention—such as those that occur through interactions of various instruments (timbres) within a groove, and those within drum set beat patterns of drum timbres; note, in both instances these timbral patterns are supported by rhythmic groupings.

In its simplest definition, timbre is the overall quality and character of a sound—a sound that has coalesced from numerous measurable multidimensional characteristics. The term 'gestalt' is often applied to timbre, supporting this concept of a unified whole comprised of many variables; the term also communicates timbre is a perceptual pattern possessing qualities as a whole that cannot be derived from the summation of its component parts. This duality is important for recording analysis, though in everyday listening it is the character and origins of timbre that dominate. For all its physical complexities of content, timbre readily appears as a singular psychological phenomenon—or aural image, quality, character or representation—that is easily recognized; indeed, we naturally perceive timbre in this way.

The Duality of Timbre

Timbre can be conceived in two ways: as a function of component parts, and as an aural image or an overall (all-inclusive and global) quality. Character is the impression of the sound's global quality that results from listener interpretation of its content (and often other factors). A sound's timbral content and character reside at different levels of perspective; content is the inner workings of timbre, and character of the sounds as a whole. The duality of timbre is this contrast between acoustic content (that which occurs in nature) and the psychological phenomenon of character (that which is resident in culture, and in personal interpretation).

This duality approach allows timbre to be conceived in two ways that taken together provide a more complete and detailed observation and understanding of a sound's unique timbre.

Timbral content is the activity of numerous acoustic components. Details of timbral content were discussed earlier. Timbral content also manifests at other perspectives, as we engaged in discussing timbral balance as an aggregate of timbres, each functioning as a partial in the timbre of the track. This is timbre as a complex, multidimensional set of sonic attributes, a collection of internal physical properties. This conception of timbre often removes it from context (appearing as a sound object) where its parts can be perceived as attributes with variables without concern of how they relate to other sounds within the track. While these physical dimensions are scientifically measurable, they challenge aural analysis. Such data is observed by focusing one's attention inside timbre, to its physical components and their attributes; this is often aided by acousmatic listening.

Character is timbre as aural image—timbre as a singular character, or an overall characteristic—and an auditory event, or sound event. Timbre's character is an interpretation; our interpretation will seek to address the impression created by its content, and also (potentially) the influence of the track's context, associations of the sound's origin, associated meanings, and the emotion, energy and expression it portrays.

Timbre as an overall quality is used to identify sound sources, to recognize the origin of sounds; we navigate life recognizing the source and causes of sounds. Further, timbre might be a sonic representation of that source "that we experience phenomenally" and also linked "to the external world" (Fales 2002, 91) outside the listener as well as outside the track. This conception of timbre often leads one toward descriptions of those overall qualities; descriptions of character and characteristics often result. The sonic representation is a timbral signature. It often communicates a "rhetorical aspect" that "involves the conventional associations that sounds have, which allow them to stand as symbols suggesting dialogues and resonances beyond the boundaries of the track" (Zak 2001, 62). Timbre as an overall character can be either isolated from context (observed for its own inherent qualities) or observed as situated within the context of the track (with qualities that reflect its role in the track and the sounds of the track in addition to—in reinterpreting—its inherent qualities). Clearly observing and defining timbral character can incorporate many factors.

These two conceptions of timbre—character and content—are woven throughout much of this discussion, and the basis for our approach to understanding timbre. The substance of content establishes qualities that generate character, aided by the interpretation of the listener and the context of the track.[14]

Various manifestations of timbre function at all structural levels. Timbre at the basic-level plays a significant role in the record. This is the most common of human experiences with timbre. At the basic-level sound sources are identified (causal origins) and their streams of information (music, lyrics or recording) are differentiated. In addition, subtle nuances of timbre establish inherent sound qualities of specific instruments and individual performers, qualities of artistic expression and performance intensity, the degree of detail within a timbre's component parts, and much more.

Timbre as a Recording Element

Allan Moore (2012a, 45) articulated the intrinsic significance of timbre in tracks when stating: ". . . the difference between styles such as 'rock' and 'pop' are . . . frequently timbrally defined." Yet timbre can be elusive, and it is so closely related to pitch as recording elements, at times they are not entirely separate. As covered above, a percept of pitch can be established by the spectrum of timbre, and "certain kinds of timbre changes can actually change the pitch content of a sound" (Snyder 2000, 199). The approaches to pitch/frequency registers and pitch areas (discussed above) not only open access to the pitch/frequency-related inner workings of timbre, they also reflect the blending of timbre and pitch percepts. The roles of pitch and timbre in music merge with those of the recording element's pitch/frequency and timbre; recording's timbral balance is fused with the song's arrangement.

As with other elements, a hierarchy of timbre exists; a "hierarchy of embedded distinctions" (Krumhansl 1989, 45). Table 7.3 provides a general summary of the many manifestations of timbre within tracks, at the three dimensions, and delineating timbre as sound object (out of context and unrelated to the origin of the sound) and as sound event (sound as it appears in context of the track, connected with its cause and its origin, and as it unfolds over time).

Table 7.3   Perspectives of timbre in recording, including timbral qualities of other recording elements and related to content of the sound object and character of the sound in the context of the track.

Analytical Listening	Within Context of the Track; Related to Source; Sound Event
Large Dimension	Timbral balance of the track's overall texture established by strata of sound sources Timbral balance as a quality of crystallized form
Middle Dimension	Pitch density qualities of groupings of sound sources within overall program; often relevant for a rhythm section, may be any grouping of sources; timbre related to orchestration, arranging Pitch density interrelationships of individual instruments/voices Overall characteristics of an individual sound source as it relates to the track; includes inherent qualities of sound sources, sound qualities of performance expression and intensity, sound qualities of language and paralanguage as contributors to the track
Small Dimension	Timbral detail within the individual sound as it contributes to perceived qualities such as distance positioning, expressive qualities, etc.

Critical Listening	Disassociated from Source; Removed from Context; Sound Object
Large Dimension	The content and character of the overall 'timbre of the track,' observed by acousmatic listening as an aggregate of sound occupying frequency bandwidths characterized by their densities and amplitude from loudness balance (Chapter 9)
Middle Dimension	Sound quality of sound sources detached from context; overall characteristics of the sound source as an isolated entity; the objective sound itself, attributes include inherent qualities of sound sources, sound qualities of performance expression and intensity, sound qualities of language and paralanguage in themselves
Small Dimension	Timbral detail within an individual sound of a sound source; defined by the activities and states of its physical dimensions (dynamic envelope, spectral content, and spectral envelope) Timbral detail that contributes to qualities of environments, distance, performance expression, etc. Timbre of environments Timbre of time durations (microtiming)

Large Dimension

In the largest dimension, there is the overall texture and quality of the track. This concept has some similarity to what Allan Moore (2012a, 29) calls "the 'tactility' of a sound recording. . . . it will also carry a 'feel.' It is this feel that is frequently the first aspect to attract (or repel) a listener, but it is also the hardest to discuss." Pitch/frequency and timbre are just two of the contributing factors; they function related to timbral balance's contribution to the 'timbre of the track.'

While more a texture than a timbre, at the highest dimension is the 'timbre of the track' where all elements of all domains interact as a rich tapestry; it is a multidimensional, multi-domain gestalt, similarly to the concept of timbre. The 'timbre of the track' is a composite timbre comprised of the sonic and expressive qualities of sound sources and their musical materials, any language or paralanguage sounds and sonic characteristics of the performance, and qualities imparted by the elements of recording. It may even contain external associations and meanings elicited by the sound source, and the persona of the performer. The timbre of the track is one of the dimensions of the track's crystallized form. This will be explored in great detail in Chapter 9.

Situated within the composite texture (of interacting sound sources), the track's timbral balance is at a perspective on the plane immediately below the overall sound. Timbral balance observes the interactions of sources and their pitch/frequency materials, and perhaps their dynamic relationships.

Timbral balance is comprised of all sound sources and their materials that shape them, the unfolding overall characteristics established by the qualities and dynamic relationships of the timbres of all sound sources. Timbral balance is entwined with the music's arrangement at this large/high middle dimension level of perspective; the two work synergistically to establish and articulate structure. Denis Smalley (2007, 44) is describing timbral balance with:

> Sounds occupy areas of spectral space. Each piece of music will have its upper and lower boundaries within which spectromorphologies act—in narrow bands, concentrated knots, masses, layers, extended spreads, diverse clouds; they may remain stable or evolve over time, moving through ranges and registers with greater or less energy and alacrity, smoothly or by step or in leaps, in an orderly or erratic manner.

Small Dimensions

Central to timbre as a recording element is its level (amount or degree) of detail—subtle, low-level attributes and values—contained within the sound source's timbre. Subtle variations of timbral qualities at this level of perspective should be expected; these are the nuance of expression and performance technique, and the recording typically contains significantly more of this information than is heard from audience seats at acoustic performances. The amount of detail present within timbres enhances the expression of the source, serves to define its distance or degree of detachment between the listener and the source, establishes the identity of individual sources, and much more. While this detail is within a timbre's components, its impact is in shaping the overall quality of the sound source, and therefore contributes to timbre at the basic-level of individual sources.

At the lowest levels of perspective resides the nuance within timbral components of individual sounds. The qualities within timbres that fundamentally shape sounds, these are the attributes of timbre's physical properties. Observing these subtle qualities reveals the content of timbre. Depending on the goals of an analysis, these observations can be pertinent and significant. These inner details are observed through the process of deep listening.

Middle Dimensions

Timbre's basic-level categorization is situated in the middle dimension. Here sound sources form streams of materials and they forge relationships with other sources that are similar and recognizable as within the same level of perspective. Timbre is used to recognize sources and to identify and track their activities over time. Timbres bring sounds to coalesce into and represent an aural image. The unified whole of timbres clearly establish singular mental images and recognizable shapes. These are deeply engrained in our real life experiences, where we continually identify sounds and their origins. This basic-level categorization is in the middle dimension, and identifies sound sources by their timbres.

In the record, sound source timbres hold some unique qualities as a recording element. The qualities of the recording fuse with those of the other two domains, and form a global quality. This is a composite timbre of the source—one comprised of the inherent qualities of the sound source, the musical materials, any language or paralanguage sounds, characteristics of the performance, and qualities imparted by the recording; it may even contain external associations and meanings elicited by the sound source, and the persona of the performer. In describing the timbral character of the source, it is this composite and global quality that is addressed.

In the middle dimension, the timbres of individual sound sources anchor the 'sound' of the record as they interact as individuals. Timbres might also form groups of sources, such as a string sections or rhythm section; an overall timbre of those sources is established at a slightly higher perspective; the blending of these groupings is part of the mixing process, and timbres of these groups are formed that can be unique to the track. In the small and middle dimensions, individual lines and sounds are shaped to heightened moments, to blend a texture, to provide distinctive qualities to the track—this shaping is the product of the recording process, and provides timbral qualities unique to recordings.

This is a short introduction to sound source timbre to establish context; it will be discussed more thoroughly below.

Linkage with Other Domains and Other Elements

The ubiquitous nature of timbre—as a recording element and as an element of music or of lyrics—carries a sense that timbre in recording cannot readily be separated from timbre within the other domains. We have just encountered several forms: composite timbres of sources, timbral balance and arrangement, and timbre of the track.

This sense of timbre integrating with other elements extends further. Examples above demonstrated timbre is often inseparable from pitch in perception and in its presence within the track. Timbre's connection to dynamics is equally synergetic; the two often work in tandem as timbral content is often reflected in the loudness of performed sources (a level that may or may not be adjusted within the mix). Timbre, especially timbre as a recording element, shapes and represents the expressions, affects and energies within and of the track—at all levels of perspective; these are readily reflected within the pitch and dynamic qualities of recording elements.

As succeeding chapters unfold, timbre's interrelationships with other recording elements will be explored in detail. In the next chapter we will explore how the sound of the track is also fundamentally shaped by space. Timbre plays an important role in the track's spatial qualities; timbre plays a defining role in distance positioning, in the frequency-related characteristics of environments, in the distribution of pitch densities across the lateral axis of the sound stage, and other percepts.

Sound Source Timbre as a Recording Element

Bring to mind the timbre of an instrument you know well, perhaps an acoustic guitar. That timbre has generic characteristic qualities; listening closer, signature qualities emerge that separate it from all others—timbral qualities that are inherent to that particular instrument, as well as qualities that are consistent with others like it of the same model by the same manufacturer, and also more generalized as 'acoustic' as opposed to 'electric.' A level of specificity exists in observing these timbral qualities as they relate to others—a comparison of overall character and of physical properties. These sounds carry a reference, without which the nature of timbral differences cannot be identified. This is the instrument's neutral state, where its unaltered inherent qualities are all that is present; though entirely hypothetical in practice,[15] this state is not altered by the energies and expression of performance, nor is it modified by the recording process.

There is a characteristic timbre of that instrument within its state of normalcy—its natural sound when being played in a neutral manner. While such a state might be impossible in practice, the concept is none-the-less an instinctive point of reference to observe timbre changes. Using the state of normalcy as a reference, we might observe the type of change that occurred within (or to) the source, and the degree of that change. Allan Moore (2012a, 45) frames this idea: "the most important questions to ask of timbre . . . concern deviations from implicit norms." Such deviations are significant in meaning as much as in sound—such as a vocalist shifting to the timbre of falsetto to bring a different context to the line, or a guitarist shifting timbres by applying more exertion, leading to still more in order to heighten urgency or some other expression.

Inherent Sound Quality

This state of normalcy ('implicit norms') is the source's inherent sound quality, or timbre. The inherent timbre (sound quality) of a sound source has unique qualities particular to individual voices and particular instruments—a sonic identity, a timbral signature. These are the qualities that naturally occur within the instrument or voice. This aural image of unaltered, inherent quality establishes a point of reference against which any alterations of timbre might be observed. This presumes some level of familiarity with a source or some contextual reference for 'normalcy' to be established; normalcy can be established by the performance within a track or calculated by what preceded a timbral shift; it may also be a reference from prior experience. This distinction should be noted.

The 'aural image' (as the term is used here)[16] is the real, virtual, created or imagined causal factor of the sound. In records, the causal factor is always invisible and manifests as a mental image conceived from the source's timbre. The source may well be recognized by the listener, though perhaps not. When the source of a sound cannot be identified, the listener's reaction is to imagine its origin; one extrapolates a source that caused the sound. Often this is accompanied by a mental visual image of the source of the sound—instrument, playing technique, performer, other. Listeners attempt to re-unite the sound sources of the record with the physicality that created it. The listener conceptualizes the performers and performance through these aural images. The listener applies their own interpretation (and imagination and experience) to unknown sources.

The aural image of inherent sound quality allows changes in timbre to be recognized.

As previously noted, the significance of source timbral qualities is heightened in recorded song. Timbral qualities—no matter how subtle—become integral to the communication and expression of the track, and fuse with its artistic voice. For the timbres of instruments and voices, every detail is significant to the whole of the sound, and for the experience and expression of the track. These source qualities are heightened in timbre as a recording element—containing more detail and greater clarity than one encounters acoustically from a distance many times greater than a close microphone. Further, the timbres of performances are carefully crafted, performed many times until just right (perhaps technologically manipulated to make them so), and shaped to be the most suitable presentation of the material; "the performance is treated to the same sort of considered deliberation and decision making as musical composition" (Zak 2001, xiii). The platform for the performance is the timbral signature of the source. Recording's timbre is the vehicle that helps establish "musical ideas and performances [shaped] into permanent sounding form" (*ibid.*, xii)—the sonic qualities of recording contribute fundamentally to this. As the recording process imparts an imprint, these same inherent qualities are further transformed.

Recording's Transformation of Sound Source Timbre

We have learned sound source timbre is transformed in performance and within the materials the source presents; when recorded its timbre becomes something entirely different. A myriad of sound qualities can potentially be applied to the original timbre, and at widely varying levels. Accepted practices of making records place significant focus on shaping these performances and the timbres they contain—with attention to artistic results. These practices exploit the sonic palette of the recording process to provide qualities that are often unique to each track. Recording's timbral imprint is woven into the fabric of the sound source's timbral content and character; it transforms the source into something synergistic to the track, and often particular to that record.

[O]ne of the most pervasive practices in rock recording is the search for some distinctive sonic quality—distinctive, that is, in relation to conventional associations. As they go about getting sounds, often in adventurous ways, rock recordists' search for distinctive colors may lead them far from acoustic reality and deeper into a sound world born of the imagination.

(Zak 2001, 65–66)

Here Zak identifies some fundamental concepts of ways recording transforms the sound source timbre. Timbres are often provided with qualities outside acoustic realities, timbres are often provided with qualities that forge new ground and give sounds a unique character, and the process is often deliberate in pursuing a distinctive sound. The source timbres that deliver musical ideas and lyrics infuse them with character—character imparted by recording. For recording analysis, this is fertile ground.

Following are several examples of recording imparting qualities onto tracks. These examples provide readily recognized transformations; certainly all tracks have examples, though some transformations are subtler than others. The degree to which the recording transforms timbre is also a trait recognized (albeit unconsciously) by the listener and reflected in their interpretation of the record; this often relates to the realism of the record's context, and the degree to which the sound's timbre relates to real life situations. Sounds that are highly transformed, or timbres that are invented and highly original, create a context of a transformed sense of reality; sounds and even a record might appear to be from or within a world of their own. The following examples will illustrate timbres captured, shaped and otherwise modified by the recording process; invented timbres—sounds that are synthesized, emerge out of samples, or are otherwise created—will also be introduced.

While this discussion will identify and discuss several sources of timbral transformations, it does so to illustrate particular qualities and identify the origins of those changes. The reason for doing so is that by identifying these traits and dimensions of how recording shapes timbre, the reader will gain some sense of the ways recording's qualities might appear. The goal of recording analysis is typically not to determine 'how' the transformation was accomplished (i.e., specifics of performance, recording process or synthesis technique), but rather to observe 'what is' the resultant timbre, and its qualities. Goals for the recording analysis should be clear on the perspective of the analyst. An analysis is typically at the listener perspective—on 'what' the listener is perceiving. To explain 'how' it was accomplished the analyst assumes a different point of observation: that of the performer, recordist or others. Seeking to understand the processes that transformed the sound from its inherent qualities is generally outside the scope of recording analysis; engaging this can become a distraction from exploring and observing the timbre itself, and (perhaps more importantly) how the timbre is situated within the track, its expressive qualities, and any cultural, semiotic meanings it may elicit.

Shifts in timbral content bring associated changes in character. Over-driven electronic circuitry in amplifiers distorts the signal (timbre of the sound source) passing through; this creation of distortion that has long been widely used to modify electric guitars, and other sources including the voice. Soon after this sound became fashionable, devices such as the fuzz box, among many others, were created to bring more control to the process of distortion.[17] Distortion alters the spectrum of the timbre, bringing the sound to be a more complex waveform, increased energy in higher partials, less-defined in pitch, and to contain a much denser and less stable spectral content than the original sound. When Jimmy Page's heavily distorted guitar announced "Whole Lotta Love" (1969) he initiated the sound of *Led Zeppelin II*—the album that would definitively codify "the sound that would become known as heavy metal" (Walser 1993, 10). Distortion happens naturally within the human voice with shouting and screaming; with this connection (among others) to common occurrence, distortion semiotically links heavy metal with other human experiences of distortion as "a sign of extreme power and intense expression . . . materializing the exceptional effort that produces it" (*ibid.*, 42). Serge Lacasse has noted the expressive affects of distortion as presenting an associated level of 'aggressiveness' (Lacasse 2000a, 158–166; 2000b, 3) in his study of extramusical significations related to vocal settings of recorded rock music. That timbre can denote energy, expression and connotation is exploited in shaping the sound source.

Substantially transforming a timbre into something unique—but with enough original quality remaining to allow the source to remain recognizable—fills popular music styles. A clear example is the iconic snare drum sound devised by Phil Collins (collaborating with recordists Steve Lillywhite and Hugh Padgham) for Peter Gabriel's 1980 self-titled album; its first use appeared in "Intruder" (the album's opening track). The drum timbre became a signature sound for Collins, and prominent within his "In the Air Tonight" (1981) and other solo works. The sound was adopted by other artists throughout the first half of the 1980s. To 'get' the sound, the snare drum was fed into a reverb tuned to the sound characteristics of an exceptionally large space (with a high density of reflections and a substantial decay time), and set to a balance giving the reverb substantially more prominence than the direct (original) sound. This reverb/direct sound mixture is then fed through a noise gate, to alter dynamic envelope. The result is a sound with a trace of natural attack, almost instantly overwhelmed and masked by a dense and unnaturally immediate and pronounced reverberation that does not decay in loudness before the sound abruptly stops. The dynamic envelope is nearly immediately 'on' as the noise gate allows but a few milliseconds of the direct sound to establish the onset attack of the sound before the gated reverb dominates; the dynamic envelope ends the sound abruptly when the noise gate closes to cut off the sound while still near its highest amplitude. The timbre's concentrated pitch area is pronounced in its density with the lower threshold of its bandwidth clearly delineated.

Surreal qualities are applied to John Lennon's lead vocal in "Tomorrow Never Knows" (1966). During the first half of the track his voice goes through the treatment of artificial double tracking (ADT). The process uses a variable tape delay of the original sound; it duplicates the original signal, delays the signal by a determined time, and recombines the delayed signal with the original—this is an early time processing technique. After the solo (1:26) Lennon's voice is run through a Leslie speaker—a cabinet system of combined amplifier and loudspeakers, with the loudspeakers rotating (at controllable speeds) to modify its sound—to add its unusual quality George Martin described "as that strangled sort of cry from the hillside" (Everett 1999, 36). These are two strikingly different timbral modifications, each shifting the content of the sound and its character to suit the musical expression of the line and the artistry of the track.

Recording can also heighten realism. Sources may contain (or even emphasize) timbral qualities in a sound that are not perceived in real life. A microphone's close proximity to a sound source can present the listener with qualities of sound not otherwise audible and with unnaturally detailed timbres; sound qualities of instruments or voices not detectable from normal, social distances can be absorbed into the performance and the timbre of the source. Breath and paralanguage sounds blend with language, expression and melody in performances by Björk. Containing sounds inaudible from normal distances, the qualities of her voice in the track "Cocoon" (2001) contain mostly real-world timbre qualities but at unnatural levels of detail. Included in her performance are subtle uses of language timbres and expression in her vocal technique; breath and paralanguage sounds shape lines and add drama, the presence of sound sometimes wavers, timbres of low performance intensity are juxtaposed against musical balance dynamic levels that place them at a higher loudness level than all other sources, the voice is intimate as it contains sounds only audible when someone leans over your shoulder—these are most complex immediately after 3:00, though they permeate the track. As Burns, Lafrance and Hawley (2008, 57) describe:

> Although she sings very quietly, . . . [we] hear every movement of her mouth from the articulation of the text to the movement of air when she breaths in. . . . these details gives the listener the sense

that she or he is right next to Björk's mouth as she sings . . . that the listener has immediate access to Björk's private thoughts.

The notion of inventing timbres through synthesis or sampling processes (electronic and *musique concrète* related) has been imbedded in popular music for fifty years. While these sounds are not new, they can continue to generate an experience of 'something' created, of something outside real-life experience; these are sound sources left to the listener's imagination. Synthesized and sample-based timbres can bring a sense of performance intensity and drama based on listener interpretation; this perception is based on the timbral characteristics present and what physical actions the listener imagines necessary to produce them; listeners process the sound relative to instruments and timbres within their experience. This interpretation is within the context of the performance they imagine is occurring within the record; it is an interpretation of the timbre's character, as well as their appreciation of sound's content. The listener relates what is heard to other timbres they have previously experienced; the qualities of known sources and the sound qualities created during their performance are used to make comparisons and assessments (here exists the danger that the source being compared is not compatible). Invented timbres have a life of some sort resulting from the sound qualities, their place within the track, and how they are interpreted. Paul Théberge (1997, 159–160) explains this connection:

> [I]t is primarily through their use that technologies become musical instruments, not through their form. . . . musical instruments are not 'completed' at the stage of design and manufacture, but, rather, they are 'made-over' by musicians in the process of making music.

"Here Comes the Sun" (1969) subtly introduces three synthesizer timbres within its introduction; these timbres are generated by an early Moog synthesizer's analog oscillators, which produce simple waveforms (sine, triangle and square waves) and are given rudimentary dynamic shape (attack, sustain, release), with no control of spectral envelope (any audible changes are the result of the sound changing registers, not of the partials having shaped dynamic envelopes). George Harrison's pioneering use of the Moog modular synthesizer might be considered an early example of an invented timbre, but in reality it is more closely aligned with the inherent sound of that instrument, in utilizing what limited sounds and processes it had available. As synthesizers became more complex, the timbres that could be invented grew more varied—some more realistic in resembling traits of acoustic instruments, and others more unusual—and more creative decisions were involved in crafting their sounds. By the early 1980s (less than fifteen years after *Abbey Road*) the entire approach to arrangement and the sonic palette of synthesis had shifted. As one example, Michael Jackson's album *Thriller* (1982) contains tracks that use a myriad of synthesized sounds to comprise its complex sonic tapestry. Sampled sound effects play prominent roles throughout the first minute of "Thriller," the album's title track; timbres (virtual instruments) devised specifically for each musical line substantially dominate the texture; acoustic instruments are mostly relegated to providing punctuation and assuming supportive roles. While an important album, *Thriller* is by no means an isolated example—any from a long list of notable albums might be referenced here. The point is the use of synthesis and samples—inventing timbres or using existing sounds as a resource to generate others (sampling synthesis)—had become common place in popular music, generating not only unique individual sounds and virtual instruments, but a different approach to texture and emerging musical genres. Performances are also 'synthesized' (or 'electronically generated') and part of popular music vocabulary (likewise well established by this time); this was most prominent with drum machines. Prince's use of the Linn drum machine (LM-1 Drum Computer) on the tracks "1999" (1982) and "When Doves Cry" (1984) are prominent examples of looping beat patterns and of drum sounds generated by a specialized synthesis/sequencing device; integral to the drum machine and its mechanized performance is its robotic sound. While the sounds become part of what defines those tracks and the music of the time, the devices themselves are accepted into the act of the performance, with the drama and timbres generated—they participate in the groove. The synthesizer became less defined by any inherent sound qualities, and became more flexible in generating and shaping the musicians' (performer, synthesist, engineer, producer, arranger, composer) aural image of the sound source.

This very cursory history of synthesis is presented to illustrate unknown timbres at their most basic, and allow the reader to anticipate the breadth of creativity that might be encountered. It is also presented to illustrate synthesized sounds have qualities and origins outside previous experience. This brings one to listen exclusively to the timbral qualities of the sound without attending to its cause, a concept central to acousmatic listening.

The guiding principle that every track (recorded song) is unique also embraces the potential for any sound source to be unique in some small way, or in its entirety. Observing and evaluating these created timbres can reveal substantive information on the character of the sound qualities and also their physical timbral content. Depending on the track and the goals of the analysis, this information can provide pertinent information for understanding the track.

The descriptions of these examples presented some contrast between observing timbres for their overall, characteristic qualities and their functions within the track, and observing timbres as sound objects out of the context of the track. As sound events, timbres contribute to the structure of the track; they may be observed for the timbral qualities that create motion and tension, communicate affects, generate associations outside the track, and more; these might be represented as the character of the timbre, and how that timbre is situated into the music, lyrics and recording of the track. As a sound object, a timbre can be observed for the content of its characteristic qualities, and often emphasizing its physical characteristics; the timbre is observed in isolation, without concern for other sources (or even other iterations of the same timbre) or for its contributions to the track, and perhaps without concern for its causal origin.

OBSERVING TIMBRE AND TIMBRE ANALYSIS

> Timbre is difficult to even define, let alone analyze; it is by far the hardest parameter to say something meaningful about, because theorists have yet to develop and adopt a reasonably comprehensive analytical language to describe it.
>
> (Doll 2019, 5)

Approaches to timbre analysis by others are not directly examined in this writing.[18] This decision to limit review of the work of others is based both on length, and as a reflection that the approach to timbre analysis offered here is unique in its focus on analysis of recorded popular song. More recent research is bringing new ideas that may find relevance within certain recording analysis goals; some of this research has been incorporated within these pages: Lacasse (2000a, 2010a and 2010b) and Heidemann (2016) have devised approaches to analyze vocal timbre within the context of recorded popular music, and Wallmark (2014) links music perception and cognition of timbre with ethnography and cultural musicology. Brackett (2000), Cook (2009), Lacasse (2010a), Lavengood (2017), Zagorski-Thomas (2015) and others have incorporated spectrogram images within popular music analyses, with Bjerke (2010), Burns (2019), Danielsen (2010b, 2015, 2019, among others) and Lacasse (2010b) also incorporating waveform images—these are strong examples of using these devices to illustrate concepts and evaluations toward meeting the goals of their analyses. Fink, Latour and Wallmark in *The Relentless Pursuit of Tone* identify a relationship between "tone, a complex quasi-object shaped by cultural networks, and timbre, the 'real' physical and perceptual correlated of that object" (2018, 10) similarly to the character and content duality I propose here, including that tone is built on the psychoacoustics of timbre; their collection of writings assembles chapters from a broad spectrum of scholars, and broadens considerably beyond this notion. The character of timbre we will explore here embraces a number of dimensions outside their collection. Composition-related approaches to timbre analysis (largely by composers of electronic and contemporary art musics) emerged in Pierre Schaeffer's work (1952, 1966) and continued in the work of Wessel (1979), McAdams (1999), and numerous others. Broader examinations of timbre related to art music are undertaken by Erickson (1975), Lerdahl (1987) and others. Several writings appeared in the 1980s that in different ways blended musical timbre with linguistic research (Cogan 1984 and Slawson 1985); though both studies contain little of direct relevance to popular music and records.

The approach to timbre analysis offered here is intended to directly address recording. It embraces timbre's two dualities: two levels of perspective, and the two information streams of character and content (detailed below).

As we have previously learned, timbre analysis and description is challenged by a lack of adequate language; further, approaches and methods for evaluating timbre are few. To engage timbral character we must navigate our tendencies to personalize our listening experiences, and rather consider timbral content and the context of the sound. This requires listening to detail we rarely if ever direct our attention toward, to aspects of sound so subtle they may not be detected during initial listenings. All these challenges might be mitigated—in as much as we are able—by establishing a process of observing timbre's physical properties themselves (for content analysis), and by keeping descriptions of timbres directed towards culturally shared experiences (for character observations).

This part will explore observing the content of timbre, and engaging timbral character. Perhaps an analytical approach and language can emerge from something offered here, to allow us to say something meaningful about this element that is so important to the track; I make no such claims, but I see the potential for the approach presented here to be useful (to some extent) to others and to be developed further by others.

General Issues of Timbre Analysis

Listening within the sound to physical properties, and making observations of the component parts of timbre (dynamic shape, spectral content, and spectral envelope) is integral to the process of content analysis; observations might be detailed or take more general forms. Examining the component parts of timbre brings the sound itself to be observed, removed from its causal factors and approached as a sound object. Observing timbre as an aural image and for its rhetorical aspects leads to examining its character and overall qualities; this may also include causal factors, meanings and associations within the context of the track, and its changeability as a sound event.

Moving away from considering the sonic qualities of timbre, we might conceive timbral content as a principle of organization for sound's structure. Timbre's construction—of dynamic shape, spectral content, and spectral envelope and their relationships to a sound's gestalt—may be used to recognize relationships of or within component parts of certain other recording elements, and elements of other domains. Timbre's concepts can be examined for instruments and voices, language and paralanguage of lyrics, and the content of performance qualities; environmental characteristics can be approached with clarity as the timbre of spaces (Chapter 8); similarly, so can the timbre of the track (Chapter 9). Timbre as a concept allows us to disassemble these elements into component parts, and to observe and recognize the contributions of activities within those component parts to the element's overall quality (gestalt).

Analysis of timbre's qualities has been initially engaged within the domains of music and lyrics. Timbre is not so discrete, though. The gestalt of timbre is a fusion of all traits, in all domains. To separate out traits by domain might assist in observations and assessment of musical materials, language usage, etc., but does not reflect the actual experience of timbre. The experience of timbre is a rich composite of inputs; inputs that are fused most convincingly (and permanently) within a record. The composite timbre is the collection of characteristics, and is the collection of all dimensions of content—separately or jointly, depending on the goals of analysis. "Timbre is a multidimensional phenomenon, a fact obscured by the use of a single term to cover a bundle of characteristics. Its multidimensionality has much to do with why timbre has been so hard to organize" (Lerdahl 1987, 142). Engaging the character of timbre is perhaps more complicated than its content, as timbral character bumps up against one's own human condition and interpretations as well as the unique nature of each track, each performance, each sound and their expression, energy and emotion.

Timbre at the upper dimensions manifesting as timbral balance has been covered under 'pitch.' While timbre manifests on a number of levels of perspective, the perspectives of individual sounds, and of the basic-level of the sound source are the focus of this section. Herein an approach to timbre analysis will be introduced that might address these matters (and generate relevant details), allowing observations and evaluations to examine various levels of detail—from the detailed to the general—as needed.

As a recording element, timbre is typically analyzed at two levels of perspective: the individual sound and the individual sound source. Timbre analysis can be applied to an individual sound, perhaps in the form of a single sound object; the analysis is restricted to the one, isolated sound. Alternatively, timbre analysis may be an observation of a sound source (instrument or voice), covering many individual sounds of that source; these individual sounds may be widely varied in timbral quality, be at different pitch levels, different loudness levels, exhibit different expressive qualities, and so forth. The results of a sound source analysis might be inherent sound qualities and formant areas, or perhaps a typology of its dominant or most representative characteristics—to identify only two.

Other perspectives for timbre analysis might emerge within an analysis, though they will likely originate with examining the timbral content and/or character of an individual sound. The exception being the timbre of the track, which will build on pitch density (temporal movement of musical material plus the dominant spectral energy of the source).

It is important to remember that a basic hermeneutic position acknowledges neutral observation is an unachievable ideal. This applies to timbre analysis, just as it does to analysis in all domains. We will always carry our sense of self when approaching material. Issues of interpretation arise in both timbral character and timbral content analysis. While the ideal is to be unbiased in data collection and evaluation, we need to be aware to avoid being biased—in as much as it might be possible to avoid bias. We each have different experiences with music and sound (as well as a host of other factors) that will influence what we hear (content) and how we interpret what we hear (character). Perception is central to analyzing content, and the cognizant processing of those perceived qualities brings (personal and cultural) understanding of character.

Character and Content

Timbre analysis may seek two distinct and separate streams of information; it may be directed toward physical content or it may be directed toward a listener's interpretation of the timbre's character. The two will be examined separately throughout the remainder of this chapter, after general summary here. Table 7.4 provides a summary outline of contrasts between content and character.

Table 7.4   Timbre as content contrasted with timbre as character.

Content	Character
Physical dimensions can be measured	Perceived impressions are interpreted, defined
Shared experience; the soundwave might be delivered to others relatively unaltered	Interpretation of the sound is personal; seeking cultural common ground in descriptions
Sound object	Aural image, singular character
Isolated, removed from context of the track	Entwined with context (defined within)
Suspends time	Sound event, occurring over time
Function of component parts	Function of overall quality
Unambiguous content, levels of, and changes within component parts	Rhetorical aspects, semiotic associations and global impression

Composite timbre:
  – contains the physical dimensions of all other recording elements; physical content of the recording
 	
  – incorporates the physical dimensions of other domain elements; complete waveform
	Composite timbre:
	  – fusion of sound qualities of recording elements; sonic character of the recording
	
	  – fusion of recording element sound qualities with those of music and lyrics; all data for interpretation

Timbral content analysis will seek to define the sound itself—disassociated from its origin and meanings. It will be an observation of what is present within the sound and its overall qualities—before evaluation and conclusions. It seeks to define what is physically present.

Timbral content analysis removes the individual sound from the context of the track. It suspends time so the timbre can be observed for its unique features, without concern for other sounds within the track. The timbre is considered as a sound object; its function is solely within its physical dimensions, its component parts. Timbral content is measurable; it is unambiguous in the amplitude levels and contour of its dynamic envelope, in the frequencies of harmonics and overtones within its spectrum, in the loudness contours and levels of each partial of the spectrum. Its sonic qualities are the object of study, not 'what it is,' or 'how it contributes' to the track; its defined content can be compared to other sound objects to identify measureable differences and similarities—as in comparing the timbres of George Harrison's J-200 Gibson acoustic guitar sound as it appears (as a recording element) in various releases of "While My Guitar Gently Weeps" and "Here Comes the Sun"—though any meaning, musical function or aesthetics of those differences are not a concern. In considering the goals of an analysis, this data might well be of greater interest and value to recordists, or perhaps to certain performers, than to analysts. In contrast, these observations of the guitar sounds' content might well lead one to evaluate *how* those timbral differences are situated within and contribute to the track—note though, this is a step *into* context, and perhaps toward defining timbral character. In this way timbral content observations can be used to enrich timbral character descriptions (evaluations).

Timbral character is an interpretation; character is defined through an assessment and interpretation of the factors that comprise the timbre as perceived within the context of the track. Therefore, the character of a timbre is the result of evaluating information that has been collected on the sound, and describing the findings (conclusions). Timbral character is an assemblage of pertinent traits and characteristics, and their unique qualities; which traits may be pertinent traits can vary markedly between sounds, and between and within contexts. Many factors can come into play; timbral character might reflect its content traits, its expressive qualities (characteristics), its musical materials, what it communicates, its contributions to the track, its source/identity, associations and much more. Definitions and descriptions seek to articulate experiences common to one's culture (or the projected culture of the track); should the analysis seek greatest value beyond one's self, descriptions will minimize personal interpretation.

Timbral content analysis will be explored in the following section. We will return to timbral character description afterward.

Timbral Content Analysis Process

The process of analyzing timbral content seeks to identify what is occurring within a sound's acoustic components as they are perceived.

The challenge of this process is to recognize the physical qualities of timbre within one's perception, and to calculate them accurately. Learning to hear these relationships can be elusive, and recognizing specifics within the nuance of content even more so. Still, humans are very adept at processing timbre; that ability may be redirected to engage the different levels of detail and to the components of timbre. Guidance for accessing this detailed content follows.

Conceptualizing a single sound as an isolated sound object can provide the most direct entry into timbral content analysis. Observing and evaluating the individual sound source—independent of its source, context, or associations—is one of the most common applications of timbral content analysis.

Timbral content analysis makes observations of spectral content, spectral envelope and dynamic contour. Just like defining pitch areas, timbre analysis will reveal the register(s) where spectral energy is present, and some information on the density and dynamic relationships present. Timbral content analysis provides much more detail on these components of timbre—not only locations, densities and dynamic relationships of spectral energy, but specific frequencies/pitches above the fundamental might reveal themselves. The analyst may also engage time (often in clock time increments) to observe changing values within various component parts—these changing values are most audible as the dynamic contour of the overall sound, and dynamic changes can be recognized and traced within the spectral envelope and spectral content. A timbral content analysis will contain some or all of the following, depending on the goals of the analysis and the content of the sound object:

- Provide a visual representation of its component parts, so an analyst might better observe their characteristics and interrelationships
- Define the duration of the sound and establish a timeline for observations
- Identify the performance intensity of the specific sound
- Trace the dynamic contour of the sound
- Identify specific frequencies of the spectrum, particularly those partials most prominent and those that substantially contribute to a sound's distinguishing character
- Map partials' dynamic levels/contours and dynamic relationships to other partials

As it was not necessary to transcribe each nuance of music for a transcription to be of use, every detail within a timbre does not need to be observed, graphed or described for the content of timbre to be engaged. The analyst can and should be selective of what to explore, and of course be willing to change direction as needed, in reaction to what has been discovered. All partials do not need to be identified; the partials that provide a timbre with characteristic or unique qualities are significant, as are the dominant or prominent partials. The frequency levels of these significant partials and/or their relationships to the fundamental will assist one in identifying the timbre's unique content; they can be important to define precisely, and at times a more general observation of pitch register might suffice. The loudness level changes of these partials over time, and their relationships with the dynamic envelope may also be significant; general contours may suffice, rather than specific levels rigorously calculated against the timeline. The level of detail of timbral content analysis can vary widely, and is (as we are aware) contingent on the goals of the analysis and the nature of the timbre(s). The analyst will acquire skill in determining the appropriate level of detail for the analysis over time and experience, just as one is acquiring a sense of how much detail within a music transcription is needed for it to be of use.

Collecting Observations on Timbral Content

The object of observations will be set by the goals of the analysis. Identifying pertinent sound sources for observation is central to understanding the track's intrinsic qualities. Often the significance of certain sources and sounds is evident, and often significant sources reveal themselves after some immersion in the nuance of the track. Some exploration of the track's sound sources can bring some clarity.

Observing a single iteration of the sound source often represents the most appropriate entry into observing its timbre; identifying and observing an isolated sound will allow its qualities to be most easily perceived and observed. Each appearance of the source will have many unique and identifiable traits: pitch level, performed intensity (performed dynamic) level, method and intensity of articulation, expressive shaping of spectrum and dynamics, etc. By limiting an observation to a single appearance of the sound one is able to access detailed information directly and more deeply, than if one is trying to summarize findings of a number of appearances of the sound.

A more detailed examination across multiple examples of the same source will prove valuable in many analyses. One might seek to explore the inherent sound qualities of a sound source, or establish a typology of a source's varied appearances, as examples. Observations of any number of appearances of the same source allows the analyst to compare them and to chronicle their many appearances. Sounds may be pulled out of the context of the track for examination, then returned to observe them within the track's context. Sounds can be compared between various sections of a track (for example verse against chorus), between sounds within phrases, and much more. Further, the same instrument might be compared for its sound qualities in two or more tracks. The analyst can observe how the sound may have changed or remained consistent, or deeply examine a source of particular interest.

Observations of multiple sound sources may prove valuable in some analyses. Timbral content of prominent or important sources can assist in evaluating timbral balance, the arrangement, structural relationships, and much more.

Timbral content analyses will seek to notate (graph), define and/or describe the states (stable values of attributes) and activities (varying values of attributes) of the sound source's (1) dynamic envelope, (2) spectral content, and (3) spectral envelope. To facilitate listening to these qualities, the listener's perception of (4) pitch definition might be incorporated into the process. These are explored in detail later.

This content analysis process can be accomplished substantively through 'transcribing' the sound into graph form (into a timbre analysis graph). This is a representation of the physical content of the sound notated in a graph form—generated through the interpretation of perception. Reliance on perception alone, though, is difficult especially when first engaging timbral content.

Using Technology Tools

Techniques and technologies might be used for guiding data collection and validating perception. Though none will replace experienced listening, they can provide corrective information and validate observations, and bring some sense of stability to the unknown. Without a means to objectively verify a perception, one might never be certain of the accuracy of observations. Verifying perceptions with empirical data (those that can be measured) from spectrograms and waveform graphics can allow one easier access to listening inside sounds, or even identifying spectra and learning to hear partials and dynamic shapes.

Spectrogram software depicting spectral data[19] and waveform tracings outlining dynamic contour[20] transform sound into graphics, allowing one to visualize sound.[21] Spectrograms represent time (left to right) and frequency (top to bottom) on an X-Y graph, with amplitude of the frequencies shown by color (or black and white shading); they can provide some insight into where sounds start and stop, their pitch density and spectral content, and dynamic characteristics. Different software settings can allow one to focus on certain aspects of sound, and diminish the prominence of others. Some settings show activities over a period of time, others the data of a specific moment. Waveforms are especially useful for timing information and dynamic contours. These visualizations can be used to supplement the ear—in certain instances they can "help transform listening into analytical interpretation" (Cook 2009, 226).

Digital audio workstations (or related sound editing and processing software) and spectrograms assist data collection with various degrees of success, depending on what information is being sought and on the amount of unrelated activity occurring simultaneously. Results of spectrograms and waveforms are often only partial—as visuals often contain information generated by other sounds. In the best of situations, they will provide or allow:

- A sound to be edited out of the track and isolated for closer examination (though simultaneous sounds will remain)
- The isolated sound can be looped to repeat indefinitely, allowing it to be heard without interruption
- Spectrum partials can be visible in spectrograms and identified
- Timelines of a sound's duration can be accurately identified, and divided into relevant time units
- Dynamic envelope to be visible
- Amplitude and frequency activities or changes may be visible

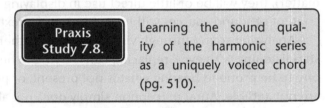

Praxis Study 7.8. Learning the sound quality of the harmonic series as a uniquely voiced chord (pg. 510).

It is important to note here that visualizing tools have limitations. Individual sounds are rarely isolated within the track, and visualization tools transcribe all of the sound presented to it. This creates difficulty in seeing only what is under observation, as individual sounds may not be clearly visible. Therefore, sounds with little else occurring simultaneously need to be identified and observed in order for their content to be clearly visible (and not confused with other sounds). Information on duration and dynamic envelope might be readily visible, though spectral content is often visually blended with the content of other sounds. Spectral content and spectral envelope might be deduced—or largely identified through listening. Constructing the timbre analysis graph to organize data collected from these visuals will present the data in a form that might be readily interpreted and evaluated.

These visualizations of the track require interpretation to yield pertinent information. What is visually present in these forms must be read by the analyst and translated into an understanding of timbral content. Here it is important to remember we are seeking to understand and represent sound as heard—the sound of the track as perceived by the listener, not the visuals of sound analysis software. Cornelia Fales (2002) articulates this paradox clearly:

[A]s the technology of digital sound analysis becomes more sophisticated and accessible, increasing evidence accumulates that what we hear—the source we identify so easily—is often different, especially in timbre, than the sound that digital analysis tells us was actually emitted. The difference is blatant because sound analysis can reveal only the *physical* characteristics of a sound, and if it shows features different than those we perceive, we know that the act of perceiving the sound has changed it. The paradox emerges with the observation that while timbre is a dimension of central importance to identifying sources, it is also the dimension that is most divergent from the sound in the physical world.

(Fales 2002, 57–58)

What is seen in spectrograms and waveforms is not necessarily what is heard—in fact, it rarely is. The visualization of physical, acoustic properties does not directly transfer into the listener's psychoacoustic perception. This is a complex matter, and a field of study in itself. To illustrate, consider loudness and frequency: the sensitivity of our hearing varies as the frequency varies, requiring some frequency regions to have greater amplitude in order for sounds to be 'heard' as equal in loudness.[22] It is entirely "possible for a sound wave with a larger sound pressure amplitude to sound quieter than a soundwave of a lower pressure amplitude" (Howard and Angus 2017, 92); this becomes increasingly prominent as frequencies extend from the top of the bass clef downward in frequency. A bass line may need an amplitude 30 dB SPL (or more) greater than a female voice in the upper-register to be perceived as the same loudness; this is substantial, and separates what we see in a spectrogram's visual representation of acoustic energy from what we hear in our perception. That the spectrogram presents all of the sound present in the texture is often problematic. Most applications of timbral content analysis seek to examine individual sound sources that cannot be clearly visible within a spectrogram's texture—or perhaps not even identifiable. Until human stream segregation (allowing individual sounds or lines of materials to be clearly distinguished) is part of the algorithms of spectrograms (an unthinkably complex matter), they will be of little direct use in displaying and tracking complete information on the timbral content of sound sources within a typical track's texture.

Analyzing sound through visuals brings up other issues. Perceptually, one is likely to be inclined to start listening with eyes, replacing ears—trying to hear what is seen, instead of the reverse. This causes one to be prone to hearing what is not present, or perceiving aspects of sound that may be visible but are not audible. Aural perception simply does not align with the visual data of spectrograms; the processor does not compensate for anomalies that differentiate physical, empirical data (that which can be measured) from the perceptual—what we hear.

Finally, the listener's attention, prior experiences and a myriad of dispositions bring a potential to hear timbral attributes that may not be present, or to transform them. This can be a similar experience to a listener misperceiving a speaker's statement—whether due to misperceived tonal inflection, listener expectations and desires, or some other factor. Sometimes we hear what we believe should be present, or what we wish to hear, instead of what is present.

It is clear transcribing timbres onto the timbre analysis graph is a detailed process, and even with the assistance of sound analysis software it will be time intensive. While a detailed graph will facilitate thorough data collection of sounds, it might also contain less detail to be more suitable (and workable) for generalized observations. Whether or not a detailed examination is sought, the graph allows the sound itself to be notated, and data collected for future evaluation. After some time working with this process, one is brought to think more about the pure sound, and to describe its substance through its components; from these observations descriptions about the sound itself can emerge with fluidity.

Hearing into Timbre

The following sequence is focused on a single appearance of a sound. This process provides opportunity to explore this most common application of timbral content analysis, in perhaps the most straightforward manner.

Table 7.5 illustrates the levels of perspective and levels of dimension contained within timbre. The outlined process utilizes these levels to bring the analyst/listener to focus on specific information within

Table 7.5 Dimensions and levels of perspective of and within the timbre of sound sources.

Level of Perspective	Timbral Component	Object of Attention and/or Point of Reference
Middle Dimension	Sound source timbre	Overall quality or character; Basic-level
Small Dimension (level of the sound's overall timbre)	Individual sound (from sound source timbre)	Overall quality of the sound; PI is reference for levels within timbre
-1 (one level lower)	Dynamic envelope	Performance Intensity as RDL
-2 (two levels beneath the level of overall timbre)	Spectral content in aggregate	Use of harmonic series as template to identify quality of 'sonority'
-3 (three levels below the overall timbre level)	Spectral envelope	Changing quality of the spectrum from dynamic flux of partials
-4 (four levels lower)	Individual partials within spectrum	Use of harmonic series as template to identify individual partials
-5 (five levels lower)	Dynamic envelopes of each individual partial	Performance Intensity as RDL

timbre, and to maintain segregation of percepts. Note how the basic-level recognition of overall quality unfolds in layers, from revealing an overall quality of an individual sound through to the dynamic envelopes of an individual partial within the sound's spectrum.

While a small number of readers may be attuned to hearing some of the inner workings of timbre, we have been conditioned to process the qualities of the overall sound. Since our earliest experiences we have learned to bring our attention to the overall quality of timbre, as that is where the significant information resides. In our daily experiences it is important to know who is talking, and perhaps to identify if they have a cold; we do not consider the spectral content of our friend's voice in the course of daily events. That information is simply not the focus of our attention—though, importantly, we do process that information as it pertains to the whole of the sound. When we recognize the spectral content of our friend's voice changed as reflected within vocal resonance, caused by her stuffy nose and congested lungs, we do so as an alteration of the overall sound—the coalesced sound remains the object of concern, not the content that was altered. Listening inside the sound brings attention to its subtleties, and to the variables within timbres; we clearly detect minute differences, but are not practiced at bringing attention to them. Realistically, most readers will not have had reason to attempt to access the components of timbre. Therefore, to do so will require some deliberate effort, and intentional listening, especially at first (Moylan, 2017).

Repeated hearings of the sound will be needed; listening with intention focused on a specific component will successfully guide this process. At each listening attention will be brought to (1) seek specific new information in one of the component parts, (2) confirm or correct previous observations, or (3) scan the sound for additional qualities within a specific component part. Working from what is known to determine what is unknown, observations can open to the unexpected and to those qualities never before experienced.

Defining a Timeline

Most often timbral content analysis will use clock time to observe and chart changes to the attributes of timbre. The sound is understood as an object that has a characteristic shape unfolding over time. As this is outside the context of the track, the metric grid is counter to the material being observed.

Timbre analyses of single sounds will nearly always use clock time, with the timeline divided into tenths of seconds. The timeline is often determined first in the analysis process. The overall duration of the timeline is followed by subdividing it into time units, and identifying reference points of notable activities within the timeline. Skill may need to be acquired to perform these tasks. Typically, visualization programs can assist this task—sometimes substantially.

Defining the Four Tiers of the Timbre Analysis Graph

Observation of the physical components of timbre may be collected with the timbre analysis graph. The graph incorporates four tiers: pitch definition, dynamic envelope, spectral content, and spectral envelope. Each tier might reflect any appropriate level of detail; the goal is to observe the significant and defining qualities of the sound, as reflected in each component.

Pitch definition provides a tangible way to access timbral details by beginning with a more general observation. The prominence of the fundamental frequency is reflected in the pitch definition tier; this aids in recognizing information within the three components. The loudness level of the fundamental frequency in relation to the remainder of the sound's spectrum, as well as the degree to which harmonic partials dominate are reflected in the amount of pitch definition present at any moment within the sound. These four parts of timbre analysis are observed throughout the sound's duration and are plotted against a single timeline. Pitch definition and dynamic envelope are at the perspective of the overall sound (of the individual sound source), and spectrum and spectral envelope are internal components of that sound, situated at the next lower level of perspective.

Pitch Definition: Clarity of Pitch

Clarity of pitch quality, or pitch definition is the relative strength of pitch sensation within a timbre (discussed above). This is a general observation, and carries some subjectivity—just as did the densities of pitch areas earlier in this chapter. It is useful for getting a general impression of spectral content. As a quality of the source's overall sound, this impression is primarily established by spectrum and spectral energy. Pitch definition is scaled on a continuum that spans from an uncompromised pitch sensation (one where the fundamental frequency has nothing diminishing its presence) through to completely void of pitch (perhaps white noise). A more defined pitch quality is likely to contain a greater dominance of harmonics. As a general rule, as the presence of overtones increases (in either number or loudness level) the prominence of the fundamental frequency (and resultant pitch sensation) diminishes. As overtones increase in prominence, the fundamental and the focused pitch it carries become less noticeable and perhaps distorted; the more periodic the waveform, the greater the pitch definition.

Pitch definition or clarity may be identified immediately after the timeline has been drafted, or started during that process. Observing pitch definition is useful in making preliminary observations that will guide exploring the inner details of spectral content. Indications of the content of the spectrum and the spectral envelope are typically revealed by observing the degree to which the pitch sensation dominates at every moment.

A contour of pitch definition is present within all sounds. The pitch definition tier of the graph will help to characterize a sound. For example, the definition of fundamental frequency might be described verbally as having a certain pitch quality for a certain portion of its duration, then another certain quality for the remainder of its duration. Pitch definition is often mostly stable during the sustain portion of a given sound; changes in pitch definition are common between the onset and the body of the sound.

Dynamic Contour

The dynamic contour of the sound is mostly apparent at first hearings. Details of loudness changes and alignments of changes against the timeline require attention. Perceptual difficulties are common between loudness changes and changes in spectral complexity; this makes tracking loudness challenging to some listeners.

A reference for mapping the dynamic contour is established by the sound's performance intensity. This is the level of energy, force and/or intensity at which the source was performed. This intensity level is represented as a precise dynamic level—it is a specific point within a dynamic area such as 'mezzo-piano' or 'forte'.[23] The same reference dynamic level will be used in the spectral envelope tier, explained below. In this way, the same reference performance intensity level will be used on these two interrelated levels of perspective. Performance intensity is covered in great detail in Chapter 9.

Dynamic contour can be described through its general shape and levels, with or without reference to time; the contour and speed of the dynamic envelope changes, and its dynamic levels at defined points in time, can all be described by states or activities. By discussing dynamic contour by how loudness changes—by defining the levels and speed of those changes—one is describing this physical element as others are experiencing it. This is an important trait of the unique character of the sound, and a mostly objective way of describing it (Moylan 2015, 193).

Spectral Content

In a complex process, harmonics and overtones fuse with the fundamental frequency, and blend within the timbre's unified whole; included are breath, body, and performance sounds embedded in the timbre. To a great extent, to observe spectral content we not only work against our previous listening processes for timbre, we also explore pitches/frequencies that are perceptually blended (Roederer 2008, 133–138). Physiologically, each frequency is analyzed in a specific 'area' location on the basilar membrane (providing the basis for the "place" theory of hearing). These 'places' are critical bands, wherein tones fuse and are indistinguishable (Rossing 1990, 74); this becomes increasingly at issue in higher partials, where several tones can be present within a single critical band. It should not be a surprise, then, for few or no partials to be perceived at first, and that some may never emerge.

The harmonic series itself, though, can be an important tool for identifying spectral components. Envision the harmonic series as a chord above the fundamental frequency—a chord of a specific sequence of intervals, with a specific voicing. The harmonic series as a chord provides a template that can be used to identify partials. Knowing the voicing and the sound quality (timbre) of this 'chord,' it is possible to scan the spectrum above its fundamental, as one might arpeggiate a chord above its root. Individual pitches (or frequencies) present within the harmonic series can, in this way, be identified. Frequencies/pitches other than harmonics will also be recognized and located; they do not conform to this 'chord,' and are non-harmonics, overtones that may be recognized as placed between specific harmonics. One can learn to calculate where any overtones fall in relation to the envisioned harmonic series, as long as one is able to discern those frequencies.

Pitch matching with an instrument or tone generator, emphasizing frequency areas with EQ or filtering the spectrum to remove portions of the spectrum or the fundamental, using the tools of a digital audio workstation, or using spectrograms may all assist in exploring the content of spectrum.

Spectral Envelope

Spectral changes are important traits contained in spectral envelope. The spectral envelope contains the dynamic levels and contours of all partials throughout the sound's duration; this activity is mapped against the same timeline as the other components of timbre. The spectral envelope uses the performance intensity level identified for the dynamic envelope. This level is a reference used to calculate the dynamic contours of each partial. All partials present in spectral content are represented as contours here in the spectral envelope tier.

The shifting prominence of harmonics and overtones is an important trait—though much of this activity is subtle, especially at first hearings. The presence and activities of these partials will be revealed

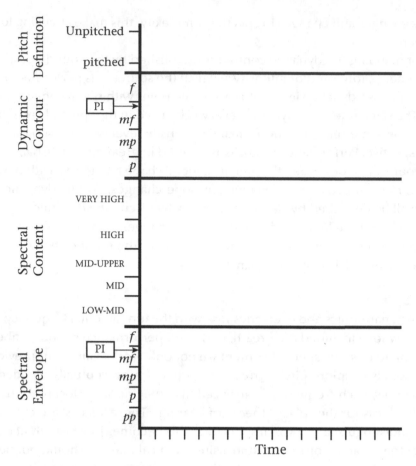

Figure 7.7 Timbre analysis graph.

through focused attention on the inner workings of spectral information. Visualization software may or may not aid this process, depending on circumstances described earlier.

Substantive information about the timbre is revealed by observing the dynamic contours, entrances and exits of partials. Even general observations about changing spectral content will enrich descriptions of the sound's timbre. Guided by the goal of the analysis, the amount of detail observed may vary widely.

Timbre Analysis Graph

The timbre analysis graph is comprised of:

- A four-tiered Y-axis for: (1) pitch definition, with a pitched-to-unpitched scale, (2) dynamic envelope, with relevant dynamic areas and the sound's performance intensity level identified for reference, (3) spectral content, with pitch register designations, and (4) spectral envelope, with relevant dynamic areas (and the PI reference)

- A timeline divided into increments of clock time (typically tenths of seconds, depending on the material), on the X-axis

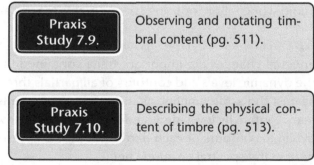

Praxis Study 7.9. Observing and notating timbral content (pg. 511).

Praxis Study 7.10. Describing the physical content of timbre (pg. 513).

- Each partial is plotted as a single line, against the two axes (its pitch/frequency characteristics appear on the spectral content tier, its dynamic contour on the spectral envelope tier)
- Each partial is identified consistently between the two tiers, allowing partials to be observed in the spectrum and spectral envelope tiers simultaneously

The following table on timbre attributes and related values outlines areas for typology tables of timbral content. Typologies of sounds might pursue a separate typology table for each component, generating values of dynamic levels and contours, frequency/pitch content and ranges, etc., as relevant to the attributes and the sound. This presents a point of departure that is malleable to the goals of the analysis and individual sounds. A typology may be initiated, if not generated, by engaging the variables of the attributes listed. Typology tables are of particular value in comparing sounds.

Table 7.6   The primary attributes and their variables for timbre analysis, this comprises an initial listing of topics that may generate a typology of timbre.

Component Parts	Attributes (variable) and their respective Values (by level, location or other unit)
Timeline	Duration of sound Divided into time unit for component detail to be clearly visible Threshold between prefix and body identified
Pitch Definition	General observations concerning dominance of pitch Time locations of most and least pitched qualities, assign values Pitch definition values at beginning and end of sound Detail on values and locations on prominence of fundamental Speed of change between pitch quality values Contour of pitch definition throughout sound's duration
Dynamic Envelope	Performance intensity of the sound: reference level General observations on shape of sound's overall dynamic level Dynamic levels: at beginning, peak of attack, sustain and end Relate levels to performance intensity reference level Time periods of attack, initial decay, internal dynamics, final decay Locate time of attack peak, initial decay end, start of final decay Time location and levels of other important envelope characteristics Detail on speed and amount of changes between identified levels Detail on shape, including speed of level changes, time locations of start/stop of level changes Dynamic contour throughout the sound's duration
Spectrum	Frequency/pitch level(s) of fundamental and audible harmonics Frequency/pitch level(s) of overtones, sub-tones, and sub-harmonics Locate points on timeline partials begin and cease Changes in frequency/pitch of partials: starting, ending, identifiable internal levels Changes in frequency/pitch of partials: speed and amount of change between levels; contour details: including time locations of levels, speeds of changing levels, periods of frequency/pitch stability
Spectral Envelope	Dynamic levels and contours of partials Partials (linked with spectrum) aligned in time with spectrum tier Beginning dynamic levels of partials, related to performance intensity Ending dynamic levels of partials Prominent contour internal levels identified and placed in time Details on shape, including speed of level changes, time locations of start/stop of level changes Dynamic contours of partials throughout the sound's duration

A typology table can be compiled from timbre graphs. It might also be created through observations directly transferred to a table. Ultimately, the analysis goals should drive data collection and organization decisions.

Describing Timbral Content

Timbral content might be described by interpreting or explaining information that has been graphed or assembled in a typology table.

The Moog sound from "Maxwell's Silver Hammer" appears in Figure 7.8, and is followed by a description of timbral content.

Pitch definition begins at unpitched at the attack of the sound, and moves to pitched by 0.1 seconds, where it remains pitched for the remainder of the sound's duration. Its dynamic envelope begins in *mp* and gradually rises to mid-*mf* by 0.8 seconds, where it remains before beginning a gradual decay at 0.9 seconds (note: Praxis Studies 6.1 and 7.9 address hearing in tenths of seconds). All dynamics are relative to the performance intensity (PI) of the sound source, which has been identified as being about 10% from the bottom of the *mf* range. The spectrum of the sound is composed of the lower four harmonics based on a fundamental frequency of F#$_2$. The spectral envelope demonstrates dramatic changes in level of the second harmonic (F#$_3$), as it moves from much softer to louder than the fundamental, and the third and fourth harmonics increase loudness by lesser amounts but at a similar time and speed as the second harmonic; the fundamental frequency is at the top 5% of *mp* at the beginning of the sound and changes loudness in a slight arc over the sound's duration. The components of timbre exhibit limited interaction, which is reflective of the simple construction of the timbre.

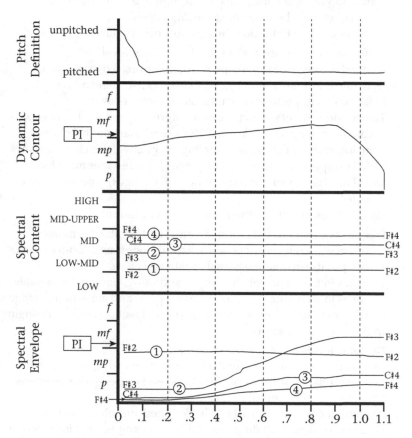

Figure 7.8 Timbre analysis graph of Moog synthesizer sound from "Maxwell's Silver Hammer" (1969, 1987) at 51.1 seconds.

More generalized observations of the components of timbres might be possible. Depending on the goals of the analysis and related significance of timbral content, general descriptions may well be more appropriate than graphing timbral components. These may be clearly assembled in typology tables. A general description might typically address:

- Qualities that are different in the onset than in the body of the sound
- General shape of the dynamic envelope (perhaps with some relation to the timeline),
- Registral placement of the spectrum, with some indication of its density and complexity, perhaps including some information on prominent harmonics or overtones
- General changes in the spectrum may be pertinent, and can be traced to spectral envelope

Though not as common, timbral content analysis may also take place within musical contexts, and the metric grid used for the timeline. These observations are often focused on relationships of the sources and materials; this usually is at the higher perspective of the sound source rather than the individual sound. In these instances, using typology tables for data collection might be most productive, and graphs (or portions) used to focus materials or illustrate points.

Matters of the physical content of the sound can contribute substantively to understanding sound sources. When those components are described, the content of the sound can be more precisely and clearly communicated—it can also be incorporated into the timbral character.

Defining, Observing and Describing Timbral Character

The timbral character of a sound is an impression of overall quality—a quality that contains much more than its role as a recording element. Character is also an interpretation of the physical content of the sound of a recorded performance—this quickly extends to the musical materials and any lyrics, as they quickly become inseparable. Thus, character includes performance, and is situated in the context of the track. Character also embraces outside matters that may arise—timbre's rhetorical forms: semiotics, extramusical connotations, etc. While timbre may be conceptually distinct, it is thoroughly entwined with other elements, other domains, and outside factors. This section will initiate discussion on characterizing timbre as a recording element—though timbre is so intrinsically enmeshed with all other elements and domains, inevitably discussion broadens to mention other matters.

Characterizing timbre as a multidimensional, multi-domain percept will be explored in Chapter 9 under the confluence of domains. There we will return to the matters below in greater breadth and detail.

Description as Interpretation

Timbral character carries subjectivity of the individual—emerging from preferences, knowledge, experience, abilities in listening and evaluating, and much more. Engaging character brings the analyst into the equation, carrying their biases and sensibilities, backgrounds and skills. Creating personal interpretations of timbres and sounds (and entire tracks) might be enjoyable and meaningful for the individual, but rarely adds substance to an interpretation (analysis). Descriptions from a vantage point of cultural norms allows the opportunity for communication between the analyst and others to be successful, and to engage the substance in common with others. Cultural norms for sounds are typically rather vague, poorly defined, rarely articulated, visceral, steeped in convention and usage, etc. Still, we know the associations, meanings, connotations, etc., when we hear certain characteristics of timbre and certain sounds—we simply have trouble talking about them from common ground. Additionally, cultural norms play out very differently between subcultures, age groups, social groups, etc., let alone between individuals.

By keeping descriptions directed towards culturally shared experiences (of timbral character), relevant information might be successfully communicated to many people. In this way the description will identify qualities that may be similar to those experienced by many others. Should the interpretation be guided and at least generally described by qualities adequately defined, vague descriptive terms become more usable (revisit Table 6.6 for examples). Terms have more value and communicate more clearly when at least generally defined and described. This blends content with interpretation of character—a process that relies on one's ability to assimilate and address many qualities and articulate results accurately, and to select an appropriate descriptive term to define its character. This is no small feat, although it is something we likely have already attempted.

Terms chosen to define a timbre's character communicate something about the sound. Some of our previous practices may lean us toward affects for character terms; this can be unproductive—that is unless the term and affect are defined (especially semiotically) by the characteristics and qualities of the timbre. A definition that provides a term with a clear description as to its nature, context, content, and associations (as appropriate) is likely to produce successful characterization.

Words hold various meanings, and they elicit connotations and connections; one hears words through the definitions they privilege. These subtleties of meaning can transform definitions between individuals. The relevance and usefulness of a term in delivering accurate meaning and substance is dependent upon how it is defined. The definition, then, becomes as important as the term itself in providing an analysis with a clear sense of timbral character.

Structural Levels and Time Spans

Description of timbral character often begins with naming the sound, identifying (or imagining) its causal origin. This holds for the individual sound and for the sound source. These are the only structural levels identified here, as timbral character at higher structural levels increasingly acquire traits from other domains into its fundamental character; even here it is difficult to limit observations to the recording.

The time span for timbral character of the individual sound is the sound's duration (unless extended by time processing, below). As a recording element timbre will often bring focus on the sound object, and its physical content. This then shifts for timbral character to address what the sound elicits in its overall characteristics; acoustic content might also include added qualities such as those of mechanics that produced the sound, breath or body sounds, performance noises, etc. embedded into timbral content. Typically, this begins with identifying the cause or origin of the sound (instrument or voice). Qualities that follow might include an indication of its general content (simple, complex, static qualities), its overall shape, affect or expressive quality; these may perhaps allude to performance qualities (energy, effort, tension), and any other relevant influences. Timbral character typically describes the sound in action, in an event that changes over time, and thereby has some type of motion and direction. Timbral character will have a relationship to the context of the track, and perhaps some associations outside. Timbral character can often broaden to articulate other qualities of the recording imparted upon the sound.

These hold not only for the individual sound, but also for the timbral character of a sound source. In contrast, the sound source's timbral character is established by the voice or instrument's overall quality as reflected in its inherent sound quality (perceived as it performs throughout its pitch and expressive ranges). The instrument or voice assumes character based on its acoustic properties, and implied qualities from the physical gestures of performance, the expressive and affect traits that are perceived, and perhaps others. These fuse into a global impression that is the representation of the instrument/voice character. The traits of inherent sound quality are identified through observing many diverse sounds from the source.

The character of a sound source is fluid; a very similar timbral character might be present throughout a track, or it can change at any instant. Timbral character may change and create contrast from section to section or between phrases or prevailing time units, and, as is the case for many lead vocals, it may change within a phrase or possibly within a word (or a note). Often some instruments (especially those in a contextual layer) may not change character substantially throughout a track, and their stable character frames a point of reference for the other sound sources' activities, expression and character. Timbral character might in some way align with the syncrisis unit, as it can be a defining aspect of syncrisis divisions that establish pitch density and timbral balance units.

Timbral character can draw one into recording's transformations of the sound source (instrument or voice) or of an isolated sound. As a recording element, timbral character may broaden to incorporate attributes of additional recording elements (all or an identified few); this forms a composite aural image of a sound or sound source. Timbral character might reflect the content of positional locations and image size, distance cues and characteristics of host environments; it can reflect modifications from time, amplitude and frequency processing, dynamic contours wed to timbral qualities of performance and their associated exertion and energy, and synthesized or unfamiliar sounds.

In summary, characterization of timbre as a recording element may contain any or all of the following:

- Name or origin(s) of the sound or sound source
- Acoustic content
- Breath, body, performance sounds embedded in timbral content
- Breath, body, performance sounds that elicit associations
- Relationship to context of the track
- Emotion and affects
- Expression
- Energy, intensity, exertion, speed
- Performance qualities, technique
- Any external connotations and associations
- Semiotics
- Impressions of qualities within other recording elements
- States or activities of all contributors to global timbral quality

Following this outline will provide some guidance to the task of describing timbral character.

TIMBRAL SIGNATURES AND CONCLUDING REMARKS

The readily recognizable and distinctive qualities of timbre bring it to be well suited to establishing 'timbral signatures.' These are timbral characters that might appear as themes and provide musical substance or assume a defining (as character) role. Sonic signatures are distinctive sound qualities that define the 'sound' of an artist (performer, group, engineer or producer), a track, or a style of music. A timbral signature may be restricted to a track, or span a career.

The Rickenbacker electric twelve-string guitar timbre that opens the Byrds' "Mr. Tambourine Man" (1965) immediately provides a timbre that defines the track—not only the artist and the passage. Its sonic distinctiveness places it in a thematic role in the track; further, that guitar timbre not only defines the instrument and its contributions to the track's texture, it permeates and defines the track's timbral character. Its continued use

> was a central timbral element on most of the Byrds' subsequent recordings. . . . it became associated with the musical style that [the Byrds] exemplified: a ringing, amplified mixture of urban folk

elements with a rock beat often referred to as 'folk rock' . . . any later use of the instrument can hardly avoid calling those records to mind.

(Zak 2001, 61–62)

This timbral character of this single instrument that most distinctively announced their cover of this Bob Dylan song, soon became part of the persona of Roger McGuinn and the timbral character of the group, and grew to represent a significant musical genre of the late 1960s. Timbral signatures are distinctive sonic personalities—timbral characterizations that reflect the persona of an artist, the concept (tangible or abstract) of a record, a producer's production 'sound,' and more.

Timbre's content and character represent one of two dominant elements of the recording. Spatial properties, the other dominant element, is the subject of the next chapter.

We will return to timbre analysis in Chapter 9 to study the impacts and contributions of all domains in shaping the global character of timbre. Formulating descriptions will emerge from evaluating the content of timbres and the observations of character traits to draw conclusions that generate the relevant and detailed descriptions analysis requires; several descriptions of the character of specific timbres will appear in Chapter 10. Further, sound source timbre as the basis of performance intensity will be examined carefully in Chapter 9. Before this, in Chapter 8 we will explore the level of detail within sound source timbres as it functions to position sounds in distance from the listener, and we will encounter the role of timbre in the characteristics of environments. Timbre has a connection to seemingly all other elements in one way or another.

Timbre's complexities and deep connections to all other facets of sound are vexing, and certainly the reason there are few existing methods for engaging in-depth study of timbre. The overall, global quality of timbre "poses a delightful and yet frustrating challenge—it is a facet of musical experience that cannot be denied any easier than it can be explained" (Heidemann 2016, 1.2). In the end, there is no way around timbre's multidimensional and fundamental significance to popular music records; timbre must be engaged if one is to learn about the substance and character of the track.

Some Questions for Recording Analysis

What are the pitch areas of the instruments of the drum set?

Are the pitch areas of the drums and the pitch density of the bass line within similar pitch/frequency registers?

What are the pitch densities of the groove instruments? How are they distributed throughout the pitch registers?

How is the pitch density of the lead vocal situated in register compared to significant accompaniment instruments?

How does the timbral balance of the track fill the audible range? Does it shift between sections?

Does the DAW or spectrogram you are using reinforce or contradict your perception? Are you able to keep the visuals from influencing your perception of sound?

Which sound sources warrant analysis for timbral content and character?

Is the analysis of one isolated sound adequate for the goals of your analysis?

Does the timbre of any central source shift during the track? Between sections of the track?

Does the timbral character of a certain significant source shift with its performance intensity, its range or other aspect of the sound? How does this manifest in its timbral content?

By describing, graphing or through typology tables, do the content and the character of the dominant sound sources gain clarity? What are similarities and differences? Which change between sections? How do these qualities support the track?

What are the significant inherent characteristics of the lead vocal? Of instruments that play a significant role in the texture?

How would you describe the timbral character of the lead vocal in the chorus sections? How does it compare to the timbre in the verses?

NOTES

1 Both sound and light exhibit oscillating wavelike characteristics—various frequencies, amplitudes, wavelengths—and both are detectable to humans within a more-or-less defined bandwidth of frequencies. There are very significant differences between light and sound; sound waves are pressure fluctuations of longitudinal waves propagated in a direction determined by a vibrating source; in contrast light waves are transverse, consisting of electric and magnetic fluctuations perpendicular to the direction of propagation. Frequencies of visible light and audible sound differ markedly; the frequency range of visible optical sensation is approximately 380 trillion Hz to 760 trillion Hz, and the audible range is roughly 20 Hz to 20,000 Hz. The range of audible frequencies covers about 10 octaves, while the range of light spans slightly more than a single octave.

2 Perhaps one person in ten thousand in Western cultures hold the ability of absolute pitch, though within cultures with tonal languages the percentage of individuals with absolute pitch appears substantially higher; there appears a connection between acquiring absolute pitch and early childhood music training. See Deutsch, *et al.* 2004 and 2006, and Sacks (2008, 129–139) for more.

3 The terms 'harmonics' and 'overtones' are often used synonymously in music theory and music technology, and by some in physics and acoustics. Here, the terms are clearly defined as a different set of relationships to the fundamental—they are partials of different types. This is an important distinction in timbre analysis (Moylan 2015, 7–9).

4 Readers might benefit to pursue further readings on the subject, depending on their goals in recording analysis. Among many helpful sources are Howard and Angus (2017), B. Moore (2013) and Roederer (2008).

5 Though identified as *solfège concrete*, it was not intended to be limited to any one musical language (Dack 2010, 271).

6 Schaeffer (2004, 76–77) explains acousmatic is the "Name given to the disciples of Pythagoras who, for five years, listened to his teaching while he was hidden behind a curtain, without seeing him, while observing strict silence. Hidden from their eyes, only the voice of their master reached the disciples."

7 See Roederer (2008, 27–54) for a detailed description.

8 This might also clarify the approach to contemporary art music as containing vertical density as a dimension of or a replacement for harmony—whereby the amount and distribution of pitch information throughout the pitch range of the composition might generate characteristic densities as sonorities or generate the motion of tension and resolution.

9 I formulated the boundaries of these registers to approximate the functioning of the cochlea and its basilar membrane, with its regions appearing 'tuned' to frequency bands.

10 The registers align with sound placement locations within the cochlea and factor in critical bands and resonances; the registers also coincide with important ranges in music, and frequency aspects of the human voice and language sounds (Moylan 1983, 1985). Brian Moore (2013) provides a detailed accounting of the hearing mechanism's processing of sound and frequency, pitch perception, and the entire transference process of acoustic energy to neural impulses. Schnupp, Nelken and King (2012, 64–69) provide a clear explanation of this "transduction from vibration to voltage."

11 Simon Zagorski-Thomas approaches this concept slightly differently. His use of the "term 'timbral staging' refers to choices and treatments that shape the timbre of recorded music. These may be analogous to orchestration and arranging, in the sense that they influence the tone or the gestural shape" (2014, 81).

12 Notable examples incorporating spectrogram images within popular music analyses are Brackett (2000), Burns (2019), Cook (2009), Danielsen (2010b, 2015, 2019), Lacasse (2010a, 2010b), and Zagorski-Thomas (2015), among others.

13 All other elements except all occurrences of rhythm, where cross rhythms between sources delineate those activities.

14 A relationship between "tone, a complex quasi-object shaped by cultural networks, and timbre, the "real" physical and perceptual correlate of that object" is identified by Fink, Latour and Wallmark (2018, 10). It has some similarity to the character and content duality I propose here.

15 The very act of performance imparts characteristics on timbres, and the recording process can never be entirely neutral or transparent.

16 'Aural image' is also a central concept within acousmatics: the state where one hears a noise without seeing what caused it (Schaeffer 2004, 76).

17 It is interesting to note that to recordists, distortion of any type is an undesirable alteration to audio signal paths of the recording. Engineers labor to eliminate distortions in order to create recordings free of undesired sounds and sound qualities.

18 A thorough literature review would be lengthy and detailed, and out of place here. That said, a breadth of literature has certainly been engaged while researching the approaches offered, a process that continued to publication.

19 Basic, free spectrogram programs include Audacity or Sonic Visualizer, and more sophisticated programs such as iZotope's RX4 software are commercially available.

20 These are readily generated from digital audio workstations and even simple, free production programs such as Apple's GarageBand.

21 Nicholas Cook and Daniel Leech-Wilkinson's "A musicologist's guide to Sonic Visualiser" (2009) is a source for detailed guidance on using a spectrograph for analyzing recordings. Also of interest, Megan Lavengood (2017) discusses in much detail her use of iZotope's RX4 software in "A New Approach to the Analysis of Timbre."

22 An equal-loudness contour is a measure of sound pressure (measured in decibel SPL), over the audible frequency range, charting levels against frequency which a listener perceives a constant loudness level when presented with pure steady tones. The use of pure tones is important to note from an ecological perspective.

23 The concepts that define dynamic 'areas' are covered thoroughly in Chapters 3 and 9.

CHAPTER 8

The Illusion of Space as an Element of Recording

Every record re-invents physics—the relationships and dimensions of our natural world. More precisely, space is redefined in every track to serve the recording's artistic intentions. The physical positioning, relationships and the spatial qualities of sound and sound sources are presented in ways that cannot occur in the physical world, in ways that defy the basic principles of acoustics and physics and how we have experienced sound around us in real life. Sometimes these differences are subtle and other times they are pronounced, though their impacts on the track are profound; many of these percepts may not be apparent to the untrained listener. In records, there is an artistic use of space that serves, shapes and contributes substance to the track at all levels of perspective.

Spatial properties of the track play a dominant role in shaping the sound of the track as a whole and the sources it contains—a role shared with timbre. This connects fundamentally with two of the framework's guiding principles: that every record is unique, and that of equivalence (each element has the potential to be significant, or contribute substantively at any time).

The significance of ecological perception in engaging recording elements was introduced in Chapter 6. Perhaps nowhere is it more profoundly in evidence than in the hearing of spatial properties. Research in psychoacoustics has offered much about sensations within the ear and its transformation of acoustic energy into neural impulses, but offers little in the way of information perceived and 'heard' (understood). Facets of ecological psychoacoustics (Neuhoff 2004a) and ecological listening (Clarke 2005) provide the concept of opportunities (affordances) for states of spatial properties and for their contributions to the track—and to listener interpretation. This is especially important for distance and environments, as we will discover. We hear spatial properties in context of the multidimensional layers of information within the track, not in isolation within a controlled laboratory. This distinction allows us to approach the properties of space as aesthetic variables; variables that can shape the track as much as any other. Thus, spatial properties open to the principle of equivalence.

Created (composed, invented) spatial qualities are integral to the individual track and the listener's experience of it, and establish an individualized 'reality' for and 'space' of the record. They provide a sense of 'place' for each sound source, and a 'stage' for the 'performance' that is the track. Spatial properties bring the track to life for the listener; the listener accepts the virtual reality as part of the context and expression of the track—part of what makes it unique. This happens no matter the level of realism of the spatial properties.

This chapter will define the track's spatial properties, and explore how they appear in and shape the record. It will navigate some of the ways we hear and perceive each spatial property, and engage in observing their attributes.

Before progressing with the spatial properties, however, we need to define the listener's point of audition, and we need to examine initial challenges of hearing spatial properties of invisible sounds.

HEARING INVISIBLE SOUNDS IN VIRTUAL SPACE

Spatial properties are aesthetically central to the record's content and its expression. This establishes a demand on our listening that most previous experiences have not prepared us to engage. Recorded popular music's use of space (including amplified popular music performances) sets it apart from nearly all other music-listening experiences. Experiences that include visuals—such as motion pictures and video games—employ spatial properties, too, though rarely are they aesthetically central.[1] In our everyday, casual listening our aural sense of space is not central to our experiences; it is a tangential quality that provides context or enhancement to the central focus—such as the sonic quality of a room around a speaker's voice. When a spatial property arises and captures our attention, it is in the presence of visual experience. The aesthetic roles of spatial properties, and especially that we encounter the properties without the support of sight, further separates the record from real-world sound experiences.

When we engage space in everyday life, we process it as a multimodal experience. We bring sounds into our field of vision when they grab our attention; we open our eyes, and we turn our head or body toward the source. The geometry of a space is seen, in all its dimensions; our orientation to sources comes from looking at them; distance is estimated by sight much more than by sound; we turn our head to bring sources into the center of our vision to localize them. Vision takes over data collection; we quickly process and dispense with what was heard by shifting to fully engage and then define the experience with sight. Sound grabs our attention, but sight confirms the source. Sound may lead sight to discover and verify, as it provides the impetus that confirms physical relationships, though it does not provide the defining information in our daily life—for example, we look to identify a sound perceived as threatening. Sound may continue to contribute to the spatial experience, but it is typically the subordinate sense.

Listening to invisible sounds—to identify them or to localize them—is something we do not often attempt. When listening into the darkness, sounds in the night bring on imagining and often not knowing 'what's happening.' Because we localize sounds and identify sounds so rarely by listening alone, we can feel confused or uncertain about what we hear.

We question ourselves: *What* is that sound? *Where* is that sound? What is it *doing*? Most typically, though, the sound is past tense—as 'What *was* that sound?' brings realization that our attention was elsewhere when the sound began. We wait, listening to hear the sound again to gain more information. We begin to question ourselves even more deeply: 'Did I really hear that?' We often do not trust our ears, our hearing—which is to say, what we perceived at the periphery of our attention and awareness. We experience a sense of uneasiness when relying on listening alone to figure out those things we cannot see—those things that go bump in the night. Acousmatic listening brings one to engage listening differently.

Clearly, real world listening leaves us inherently unskilled in engaging sounds without the aid of sight. Yet, this is exactly the context we find ourselves embracing with records. The track is wholly invisible, yet the track provides the illusion of spatial properties. Further, and very importantly, spatial properties provide significant substance to the record. Observing, recognizing and identifying spatial properties present challenges.

A reorienting of prior listening processes concerning space, a shift of attention toward sonic attributes previously not experienced, unknown or dismissed, and a sensitivity to the attributes that define spatial properties will each be required of the reader. It can present a challenge to hear what is now also unseen, to listen for the information of spatial attributes rather than allow the attributes to trigger sight. Sonic qualities never before experienced may be encountered by some. Finally, sonic qualities may need to be relearned or re-conceived to engage distance and angular orientation to the source; these may require reframing how to listen and what information to seek.

We search for a reference to make sense of the world around us. When listening to spatial properties in records, and hearing into its space (or spaces), the void of a visual reference can be disorienting. We may seek to use the only visual available: the loudspeakers. Those loudspeakers and their locations in the room will offer no guidance. Should loudspeakers be used as visual-to-sound reference they will mislead all percepts, except for the rare source located specifically *at* the speaker. Attempting to visualize sound locations by relationships to loudspeakers will mix two different contexts: the physical sound within the listening room and the virtual world within the track. Any common ground between the two will be by chance, and completely a coincidence of the particular track and the qualities of the physical room plus loudspeakers; observations will be distorted, and it is a distraction of effort to seek common ground. Further, sounds can appear at positions beyond the loudspeaker positions, extending the stereo array (see Figure 8.3).

Fortunately, we have inherent skills at hearing the attributes of direction, distance and environments that are capable of being developed (Blauert 1983, 47). These skills simply have not needed to be developed in our casual listening, and they can be honed to observe spatial properties within the track. Guided by clear understanding of the attributes that define spatial properties, we will use these skills for collecting information and for evaluating the track.

Listener Perspective and Track Playback Format

The listener's perspective is used to calculate and define the spatial locations of sounds, and to understand the qualities of source host environments. An "implied physical perspective" (Williams 1980, 58) for the listener brings the impression of a point in space from which the track is heard; this is a vantage point from which listeners observe the track and its sounds. This conceptual location is the listener's "point of audition";[2] this perspective is their perceived physical relationship to the track, that is defined as (or located at) a specific point in space. This term is adapted from film and television sound; I have transformed that definition to allow the term to identify and locate an unchanging listener's position from which the track is observed.[3]

In everyday life, a point of audition is where one finds themselves at any moment; that position from which the direction and distance of sounds are perceived and unconsciously processed. Of course, as we navigate our lives our point of audition travels with us, and is continually changing with our moves of position. Here within recording analysis, though, we are concerned about this location more specifically; it is fixed throughout a track. The record establishes this position with its mix; the mix holds the assumption that the audience would hear the track from this same virtual position. This listener location needs to be recognized, and it needs to be stable in order to be of use for calculating sources positions and positional changes observed within the track. 'Point of audition' establishes a point of reference for calculating the angle and the span of space existing between listener location and sound source. The analyst will consciously process the qualities of space from one illusory physical location, a single point of audition. This defines the point of reference for all spatial calculations of the individual track; the point of audition is that point of reference, and establishes the listener perspective related to spatial properties.

Two-channel stereo (short for stereophonic sound) is the default format being addressed in all discussions; exceptions will clearly specify a different format such as surround sound (though 'point of audition' is also relevant to surround). This acknowledges the vast majority of music listening takes place using the two-channel version of a track; it is what nearly all consumers purchase and hear regularly.

Indeed, the overwhelming majority of records are only available in two-channel stereo. Therefore, the spatial properties of stereo sound are integral to the vast majority of tracks.

Two other track formats are in common use: mono and surround. Either may be of interest to the analyst, and be examined in one's analysis. All three playback formats will yield a different spatial experience. The three formats are:

- Stereo (with two independent channels)
- Surround sound (typically with 5 or 7 independent channels, with the potential for an additional channel that contains all of the lowest frequency range, which is directed to a subwoofer),
- Mono (a single channel containing all of the sound).

Mono was the only format of early popular music recordings. Stereo (with its two independent channels) established a presence in the mid-1960s, and quickly dominated the market; initially two independent mixes were created: first for mono and an after-thought mix in stereo. Stereo is currently the default commercial format, and mono versions are now typically reductions of the stereo to a single channel. 'Collapsed' or 'folded-down stereo' merge the two channels of stereo into mono; this results in phase cancellations and other anomalies that alter the track, sometimes considerably. Mono records can be reproduced over two channel systems (sending the same information to both speakers producing "mirrored mono") or over a single loudspeaker; these appear as either the center speaker of Figure 8.1 or the combined left plus right speakers. I have chosen to limit coverage of monaural

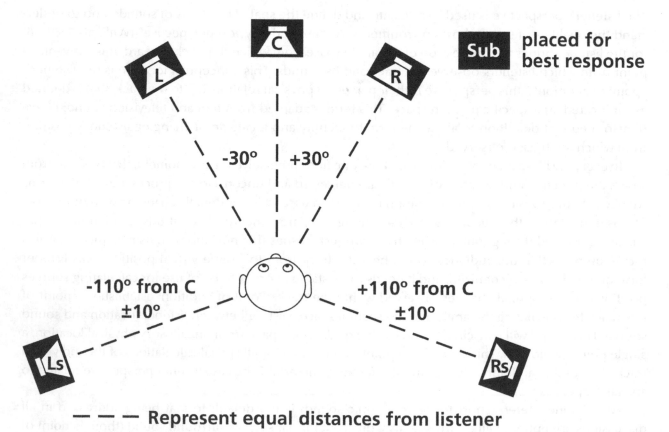

— — — Represent equal distances from listener

Figure 8.1 Left-right stereo loudspeaker configuration imbedded within the 5.1 surround sound layout recommended by the ITU (International Telecommunications Union); mono reproduced by center speaker or L/R stereo speakers.

versions of records primarily because of the overwhelming dominance of stereo in the literature. The spatial properties of mono are restricted to distance and depth, environments, and to a limited extent source size; Peter Doyle (2005) provides an extensive examination of spatial properties in mono tracks.

Stereo and surround sound playback formats locate sources very differently in relation to the point of audition. The listener is presented with sound arriving from different directions, different number of directions, and listening cues differ between the two formats, altering percepts. Each format provides a very different experience, with striking differences to source localizations, width and depth, and artistic treatments of sources; they shape the artistic statement of the track in very different ways.

The surround sound format, and all it brings to the track, will not receive coverage in this book. I have written about surround elsewhere (Moylan 2012, 2015, 2017); these provide a basis for recording analysis considerations that readers can pursue, though there is much left to be written. As surround sound continues to struggle for consumer acceptance and stereo substantially dominates over surround, it was decided not to engage it here. The format, however, has striking, distinguishing attributes not found in stereo; attributes that add substantive dimensions to the track's aesthetics. The listening public has largely not experienced surround sound music production, but has embraced it for motion pictures. Perhaps these qualities may bring surround sound to become a more important part of the public's music listening experience, just as home theatre sound has become widely embraced. For those of us who have experienced (let alone studied) surround recordings on a high quality system properly tuned, it is an experience that makes records new again—even tracks one knows well in stereo are rediscovered from their mix in surround.[4]

SPATIAL PROPERTIES AND ATTRIBUTES

Spatial properties and their attributes establish the sonic world of the track and a spatial identity for its sound sources. They can create "the appearance of a reality that could not actually exist—a pseudo-reality, created in synthetic space" (Moorefield 2005, xv), and they can provide a vivid real-life context for the track.

The track's spatial properties present sounds in space where none are present. Sonic illusions locate sounds at direction and distance positions from the listener, and also provide sounds with size. Sounds may be localized anywhere within or around the listener's listening field, at any conceivable depth and any angle reproducible by the playback format. Virtual spaces bring the experience of instruments and voices emanating from surreal rooms—illusions of rooms of any size, perhaps infinitesimally small or immensely large places, even spaces of impossible dimensions and geometry.

Curiously, perhaps astoundingly, the listener is perfectly willing to accept (albeit unconsciously) sounds emanating from these unknown, strange places. Worldly limitations are ignored, and listeners are willing to experience these illusions of distance and size, and accept them as the unique reality of the record. These qualities are wed to the fabric of the track, are an integral part of its sound and of its context; simply, they are part of the experience and substance of the recorded song. As such, the track may often be conceived as a performance emanating from a place of different sonic realities.[5]

The spatial properties of recording that establish these illusions fall into three categories:

- Angular direction and width
- Span of distance location and depth
- Dimensions of the environment within which an individual sound appears to be located ('host environment'), and dimensions of the overall environment the track occupies ('holistic environment')

This section will define the attributes of these three spatial properties in more detail, and discuss ways they interact, fuse, and work in complement. This is in preparation for a more detailed coverage of each that will fill this chapter.

Spatial Properties and Levels of Perspective

The dimensions of lateral angle (direction), distance, and illusory environments (simulated physical spaces) function most significantly on three (3) levels of perspective—levels we have already engaged:

*   Individual sound sources and their attributes
*   Composite texture of interrelationships of sound sources
*   Overall sound

Table 8.1 adds detail to the three fundamental spatial properties, and how they manifest at levels of perspective.

Listener attention naturally falls onto the perspective of the individual sound source. It is at this level that we interact with other humans, and at which we perform on instruments and sing. The perspective of the individual sound source represents the basic-level categorization; further detail of categorization

Table 8.1   Spatial properties and their attributes or variables at the three levels of perspective.

Level of Perspective	Spatial Properties	Variables, Attributes
Individual Sound Source	Lateral, Horizontal Location	Source position, image size, Angular trajectory from listener, Phantom image
	Distance Location	Source position, depth of image, Distance from listener location, Aural image
	Host Environment of Sound Source	Size of enclosed space, Patterns and timings of reflections, echo, reverberation characteristics (duration, density, dynamic contour), Frequency content of the environment and reverberation, Ambience, spaciousness, Aural image
Composite Texture	Sound Stage	lateral and distance positions in aggregate, Relationships and interrelationships of source positions, Distance may include depth from source host environment
	Simultaneous Environments Space Within Space	Relationships between host environments of all sources, Relationships of each host environment to the holistic environment
Overall Sound	Sound Stage Dimensions	Boundaries of sound source locations, Left-to-right width and front-to-back depth of an all-inclusive, overall staging area
	Holistic Environment Space Within Space	Overall environment (spatial 'place') of the record Relationship of the holistic environment to the aggregate of host environments of the sound stage

brings more specific subordinate types (such as a particular performer or a type of guitar) (Zbikowski 2002, 31–33). Spatial properties shape the basic-level individual sound sources substantively; they add dimensionality. A spatial identity for each instrument, voice, or any other sound within the track results from (1) their left-right lateral placement, (2) their position of distance from the listener, and (3) the attributes of the individual host environment (space) they are perceived as occupying. The spatial identity is a virtual aural image of the source (1) having lateral location and size, (2) having distance from the listener, and (3) a sense of occupying a space that provides it with depth and a host room, space, or environment within which it exists and sounds.

These three properties are observed for each sound source at this perspective. The anomalies that establish environment and distance cues are at times much subtler than those of lateral sound location—both in real life and in the record. Sounds of environments can be pronounced though, and even incomplete sets of cues can establish illusions of physical, enclosed spaces. The perception of distance is fraught with misconceptions; distance attributes are often overlooked and confusion tends to replace distance with what is actually loudness, or reverb (an attribute of environments), or prominence, among others.

In the composite texture, sources are situated at a level of equal significance and are potentially balanced within the listener's attention. This has been discussed earlier, and the concept continues for spatial properties. Here, individual sources are perceived in their interactions and relationships to other sources, and the interrelationships they might establish. The composite texture is where several important spatial traits manifest; these are based on the interrelationships of sound locations that coalesce on the sound stage.

In evaluating the sound stage, placements of sounds may establish groupings; the sounds dispersed across the stereo field and the depth of field can bond in various ways as a result of their timbral content, musical functions, staging placement, etc. Sources coalesce into groups within regions of the sound stage, and some may be isolated or delineated. This provides a connection or separation of sources and also of the materials they present; it also impacts the density of sound, and all that might entail.

Host environments of sources also establish relationships with the host environments of other sources, perhaps also generating percepts of distance and depth. Each instrument or voice might have its own 'host environment' (its own acoustic space, artificial room, reverb, etc.); relationships forming between instruments/voices are rarely akin to occurrences of naturalistic acoustic spaces. The sound stage houses rooms (spaces) that are positional in relation to other rooms—each containing a sound (instrument, voice), a performer, and an aesthetic idea. Each room (source host environment) is of its own geometry, size, sonic properties—real or surreal—and may change at any time. The sound stage has the potential to be active and dynamic, as well as contextual; in aggregate it establishes a context for the track, and much activity can exist within that framework without altering its fundamental context.

At the level of composite texture, the interaction of sources and spatial dimensions is evaluated. This contrasts with the spatial attributes of individual sound sources that are observed and evaluated at the track's basic-level, above. Figure 8.2 illustrates these three perspectives: lateral and distance positioning of individual sound sources, composite staging of source image positions, and the placement of all sources within a holistic environment for the track. The identified sound stage width and depth can vary widely between tracks, as can the distance between the listener and the front edge of the sound stage.

At the highest level of perspective (the overall level of spatial properties), locations of all sounds coalesce into a single aggregate group. The grouping of sources establishes an area that is defined by width and distance—this area is the sound stage. This single all-inclusive 'ensemble of sound sources'

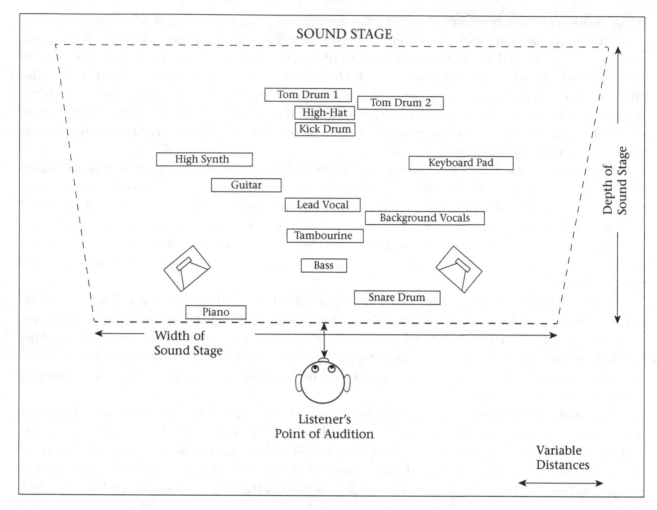

Figure 8.2 Sound sources positioned by lateral (stereo) location and distance from the listener, grouped into a single area or sound stage, which is contained within the track's holistic environment.

resides within a single venue, a single space within which the track as an entirety resides—this is the track's holistic environment. The holistic environment is an all-encompassing, global space or environment for the track; it also contains all of the track's spatial properties that are generated from the individual sources and the sound stage—again, this aggregate is likely to be surreal.

In many genres of popular music or individual records, the listener experiences the illusion of a performance of the track; a performance that emanates from a single area that encompasses and binds all of the performers, and all of their sounds. This represents one conception of a sound stage for the track. The sound stage is positioned within an overall space that may also contain the listener's position—the holistic environment, or the environment of the track, can be conceived as including the location of the listener, or detaching the listener as an observer. The holistic environment is at the highest level of perspective of all spatial properties; it contains all spatial properties. The sound stage and the holistic environment is contextual; they each establish a stable point of reference that may be used to understand lower-perspective activities. Some tracks are not staged performances, and others will not be conceived as performances by listeners; even in these situations, the sound stage can remain a helpful point of reference.

Elevation (the perception of sound located at an upward or downward angle along the listener's median plane) has not become incorporated into stereo records, as the cues cannot be reliably or convincingly produced by two loudspeaker locations, on a common horizontal plane. For sounds at vertical angles to be consistently reproduced, an additional channel or channels of audio directed to a loudspeaker(s) located above and/or below the listener ear-level are required; these are found in the several emerging surround formats, with ceiling channels largely dedicated to environment cues. References to a vertical plane of the track typically refer to frequency or pitch register, or the 'height' frequency or pitch content: "All rock has strands at different vertical locations, where this represents their register" (A.F. Moore 2001, 121); rarely is activity on the vertical plane, situating sounds top to bottom, proposed (Hodgson 2010, 183–185). Neither of these references to the vertical plane is applicable to the discussions of this chapter.

In summary, the spatial properties of lateral placement, distance location and host environment are the basis for all that is space-related in the record. They have the potential to provide each instrument and voice with a unique spatial identity, and working together they establish the spatial identity of the track. These properties rely on the listener's point of audition as a reference, allowing consistency to observations; the point of audition affords the performance of the record with some degree of separation from the listener, with sources at angles from the listening position and at some separation of distance, and with a sense of depth and other attributes from each source situated within its own performance space.

The following sections will present (1) stereo location of individual sources (in aggregate comprising the width of the sound stage) and (2) the distance location of sources (in aggregate comprising the depth of the sound stage). The sound stage (3) will be explored in detail afterward, before moving on to (4) the roles of environments in records, and their sound properties.

STEREO LOCATION: ANGULAR DIRECTION AND IMAGE WIDTH

Lateral (or stereo) location is the topic that arises when many think of a record's spatial properties. It is the perceived lateral position of sound sources; their locations within the boundaries of the stereo array, calculated at an angle left or right from the listener's forward facing center. Sound sources may be perceived at any lateral location within the stereo field, Figure 8.3. Sounds may be situated at either loudspeaker, though the majority of sound sources are located elsewhere, where no loudspeaker exists. "The stereo space acts as a sort of window through which the listener can 'view' the location of sounds. Not only in an overlapping construction but in a complex and *dispersed* structure" (Camilleri 2010, 201).

Illusions of sound placements can be established at positions where a physical source is not present. These illusions are produced by the interaction of the two independent stereo channels that emit from the two loudspeakers—speakers that are correctly positioned in relation to the point of audition—as each channel arrives asynchronously at both ears. Sound sources that appear without a physical presence are *phantom images*. The majority of sound images in nearly all stereo tracks are phantom images. Phantom images (and the sound sources they represent) may appear anywhere between the two loudspeakers, and up to 15° beyond (outside) each loudspeaker position.

Lateral placement of sounds establishes the width of the sound stage. The left-edge of the furthest left sound source image and right-edge of the furthest right sound source image define the sound stage lateral boundaries.

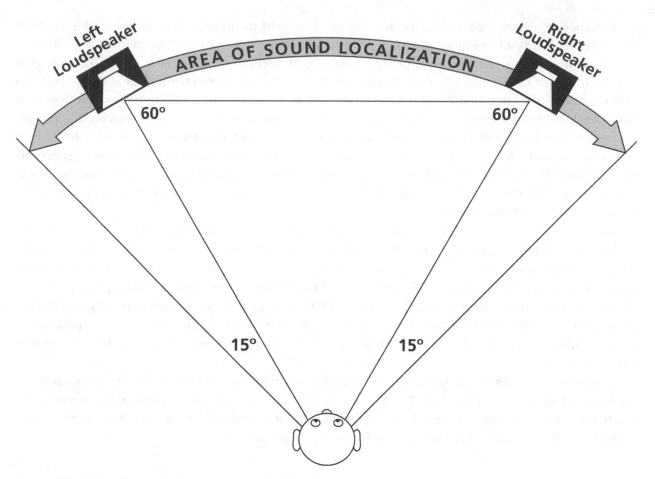

Figure 8.3 Stereo field: area of sound source localization in stereo.

Perception of Direction and Phantom Image Lateral Localization

Locating sound sources on the lateral plane relies on the perception of direction. Understanding a bit of the psychoacoustics for localization might assist the analyst in observing and assessing sources. In listening to the track, as in real life, the sound wave is different at each ear. These waves differ by time/phase, amplitude/intensity, and/or spectral content. These differences are essential to the perception of the direction of sound sources; they also play a role in the perception of environment attributes. Additional cues are required for the perception of distance and the attributes of spaces. Interaural cues provide decisive information for sound source location and angle on the horizontal plane, and for soundstage width.

The head, neck and shoulders act to produce time differences and intensity differences between the sound that arrives at each ear; these two types of interaural differences provide the primary cues used for perceiving direction. Jens Blauert (1983, xi) has noted the significance of sound at the two ears for perceiving spatial properties: "The acoustic signals presented to the two ears are by far the most important physical parameters of spatial hearing. It would be appropriate to discuss spatial hearing in terms of these signals alone. . . ." Handel (1993, 98) frames this a bit differently: "The human body acts to generate the physical cues for object localization. If we were only points in space with central ears, there would be no way to infer the direction of sound." Direction of sound and locations of sounds are perceived largely through interaural differences of very similar waveforms.

Interaural time differences (ITD) are the result of the sound arriving at each ear at a different time; the physical separation of the two ears produce these differences. A sound wave will reach the ear nearest the source before it reaches the far ear. These arrival time differences also generate phase differences, as the wave at each ear has travelled a different distance; sound at each ear may be almost identical except the sound at each ear is at a different point in the waveform's cycle (sound at each ear will also contain minute spectral differences) (ibid., 99).

Interaural amplitude differences (IAD) work in conjunction with ITD in the localization of the direction of the sound source. IAD are also identified as interaural intensity differences (IID). IAD is the result of sound pressure level differences at high frequencies present at the two ears. Reflections established by shadowing of the head, pinnae and upper torso produce interaural intensity (amplitude) differences at frequencies whose wavelength is shorter than the distance between the listeners two ears (frequencies above approximately 1600 Hz) (B. Moore 2013, 247–275). The human head acts as a low-pass filter (of sorts) where frequencies above approximately 2 kHz are attenuated at the ear opposite of the side where the signal originates (Mather 2016, 114). This disparity between 1600 and 2000 Hz as the approximate threshold for dominance of IAD percepts reflects the inconsistency of human physiology—as we each have a uniquely sized and shaped head, and each outer ear, hearing canal, inner ear (etc.) is different from the other and between individuals.

Interaural spectral differences (ISD) occur throughout the frequency range. While they may be subtle, they are important for the localization of objects in frequency ranges where IAD and ITD are ineffective. ISD are produced by the ridges of the pinna (outer ear); as sound reflects into the ear, the ridges introduce small time delays between their reflections and the direct sound that travels directly to the ear canal. Resonances also appear to be excited by, or produced within the outer ear. These also alter the frequency response of the sound source in predictable ways that vary between individuals. Important to recognize for surround sound, distance and location judgments are not as accurate to the sides and the rear. The absence of this spectral information generated from and collected by the outer ear seems to play a central role.

Pinnae serve a critical function in front to back localization. When sound arrives at the head from the rear, ridge reflections are not generated. When sounds are generated beyond 130° from the front center, pinnae block the rear-arriving direct sound from reaching the hearing canal and its ridges (Tan, et al., 2018, 40). The sound source is recognized as being present at our rear because of the absence of pinnae-generated spectral alterations (Mather 2016, 114).

With this information an analyst may direct attention to how specific interaural cues might be acting upon sound source placements. In doing so, one might most accurately identify the location of sounds by considering their prominent frequency content. Table 8.2 provides some guidance of which cue may be most appropriate for initial observations; the physiology of individuals brings frequency ranges

Table 8.2   Interaural sound localization cues by frequency range.

Frequency Range	Interaural Difference	Description
Below 500 Hz		Location accuracy progressively diminishes as frequency decreases
Up to 800 Hz	ITD (Time, Phase)	Cues determine localization
800 Hz to 2 kHz	IAD & ITD	Both cues are used in localization
1250 to 1500 Hz	IAD (Amplitude, Intensity)	Becomes a significant factor
2 kHz to 4 kHz	ITD	Dominates, though localization is poor
Above 4 kHz	IAD	Cues determine localization
Throughout the hearing range	ISD (Spectral)	Present cues for localization

and thresholds to vary slightly. This table is a point of departure for exploration, not a definitive guide. The nature of spectral content can bring localization of sources to manifest in unexpected ways. For example, a lower pitched sound (such as a bass) may localize more clearly than its presence well below 500 Hz might suggest; it might be localized by amplitude differences resulting from high-register frequency content in its attack, resulting in a narrower and more focused image than would result in other bass timbres. Localization typically occurs within the onset of a sound, bringing greater significance to this initial window of time; in real life we quickly determine where a sound is, then shift our attention to process other information (i.e. what it is doing, and whether it is necessary to react or take action).

There is no reason to believe we hear equally well in each ear, or that both ears share the same functional characteristics. We don't see equally well with both eyes; we are comfortable in this knowledge, as the majority of us experience this regularly while being fitted with eye glasses. We do not have our hearing assessed regularly, and nearly all of us have no idea of how well our ears function in relation to 'normal hearing.' Given our physique is not fully symmetrical, and our outer ears are not identical and eye glasses need adjustments to conform to our head and ear location irregularities, it bears that we hear at least slightly differently in each ear. Further, we seem to have a dominant ear—anecdotal observations have proposed most people consistently put a phone up to the ear of their dominant hand, and others have proposed 'creative' people put their phone to their left ear. I offer no validation of these, as few formal studies have engaged ear dominance, or examined acuity imbalance except under trauma and restorative conditions. Anecdotally, though, it does appear many of us have a preferred ear we use to lean into a conversation, to talk on the phone, and so forth—just as we consistently make our first step off the bus with a certain foot, and stumble should we start with the other. Consider, you cup one ear (rather than the other) to hear more clearly. These matters of imbalance of hearing acuity between each ear and the possibility of a dominant ear have potential bearing on interaural perception—a bearing that is largely undefined.

This is a meaningful place to remember the explanation from Chapter 6 of how headphones present the spatial properties of tracks differently from loudspeaker listening—in ways that directly transform the spatial qualities being examined in this chapter. Headphones eliminate interaural cues and thereby establish voids in the stereo field where images cannot be formed (and those contained in the track reproduced); further, this establishes "an unnatural stereo image which does not have the expected sense of space and appears inside the head" (Rumsey 2001, 59); lastly, the closeness of the drivers to the ears exacerbates timbral detail, and alters distance and depth positions of sources and the timbres of environments. The sound stage and localization are wholly different experiences over headphones as compared to what is heard out of loudspeakers. The vast majority of records are created while listening over loudspeakers; the spatial properties generated by loudspeakers are integral to the sound established as the track's finely crafted artistic statement (including recording elements); the sounds emanating from loudspeakers are integral to the track's primary text.

In order to collect observations that accurately reflect the lateral characteristics of sounds, interaural cues need an accurate and consistent point of audition. Listener location at the apex of the equilateral triangle that defines the loudspeaker to listener position is critical; a shift of position brings a shift of angle/location of sources—even a small shift can make a substantial difference. This is a significant concern, as the positioning and sizes of sources are integral to the track; they play decisive roles in the spatial identities of individual sources and the track as a whole.

Image Width

Aural images (whether phantom images or located at a loudspeaker) also have a width dimension. This attribute is significant for the track, and it is often overlooked. Perhaps width does not get noticed because it is a quality rarely encountered in nature and life situations; when width then is present, we are ill-prepared to give it our attention. We are unaware of the presence of width, and do not have experience directing attention to that property. Further, its cues are typically subtle, though size can be perceived as more prominent when sounds are intimately close or when in highly reverberant spaces. Listening for width will be a new experience for many.

Width provides the illusion of a physical size to the sound source. Aural images have edges or boundaries on the left and right sides.[6] They may be of any size width, spanning the extremes from occupying the entire breadth of the stereo field, to a very narrow point. Images may also change in width—at any time and by any amount. Subtle changes are common within instrumental or vocal lines, and more pronounced changes are common between song sections. Interesting examples of shifting image widths and positions give the sparse accompaniment of Phil Collins' "In the Air Tonight" (1981) motion and direction, as well as suspense and tension, beginning with the first electric guitar sound.

Images that are very narrow in width, and clearly distinct as occupying a concentrated spot are point source images. Examples of point source images are not common. Sources in high frequency ranges produce point sources more readily, as these sounds tend to radiate less and be more directional. Lower frequency sources typically have resonant bodies that help the lower frequencies to radiate more, and thus provide the sounds with a sense of width. Paul Simon's *Graceland* (1986) provides some interesting examples of point sources. Unusual point source electric guitar sounds are found in the opening riff to "Gumboots" and a similar guitar sound in the introduction of "Crazy Love, Vol. II." Both sounds are near the center of the stereo field, and both are widened inconspicuously by reverb; each appearance has the guitar in a focused spot of direct sound, situated within a subtle and broader width of its space. More typical point sources appear in the collection of metal percussion sounds within the introduction and coda sections of "Under African Skies"; these sounds remain as focused points while shifting between lateral positions.

A spread image is one perceived to occupy a span of area; phantom images very often cover some expanse of width. The spread image is defined by the locations of its left and right boundaries (edges of the image), and by the area it is perceived to occupy. At times, a spread image may appear to be split, where it might occupy two more-or-less equal areas, one on either side on the stereo field. An example of a split image, polarized to each side of the sound stage, is the tambourine image during the first chorus of the Beatles' "She Came in Through the Bathroom Window" (*Abbey Road* 1969, 1987).

In tracks, images are provided with width by the interaction of the two speakers each with different amplitude, timbre, and/or time-based characteristics of the source. The source is provided size by these differences between each channel. Width may also be the result of the attributes of a source's host environment; the attributes of environment may produce an expansion of the edges of sounds. The sound of spaces may be prominent and distinct as a second presence, or may fuse with the source to create a blended sound.

Size is significant to stereo images, and contributes substance to their spatial identity. Sound source presence may be established and shaped by the amount of space they occupy; their prominence can be impacted by their size. As a result of their widths, images may overlap, occupy the same space, or be delineated in individuated areas; innumerable relationships are possible. When images are expanded by reverb or other environment cues, the edges of images that might otherwise be precise in their boundaries can acquire blurred and indistinct edges. This is significant, as images are defined by their size, as outlined by their edges; images with blurred edges take on different qualities, and may function differently in relation to other images.

These indistinct edges are established when an environment has a width greater than the sound source. This situation will typically also provide a sense of a different level of density (or amount of frequency present) within portions of the image. The central portion that is the source typically contains a greater level of substance than the edges (the environment). An example of these are the exposed drum sounds in the introduction to "The Boy in the Bubble," also from Paul Simon's *Graceland*.

Image size can bring unnatural qualities to a sound source, and unnatural relationships between sources. Imagine a flute occupying the entire breadth of the sound stage, and a piano confined to a single point in space. Width can have a strong impact on the aural image, and also on the realism of the track.

Sound Sources in Motion

Just as image sizes change, image locations are not fixed. It is as common for images to change in location as it is for them to change width. Images often move; they can abruptly change positions or gradually sweep to a new position. Abrupt changes in image location are common. They often shift positions between sections of the track, as mixes for verses and those for choruses can alternate throughout the track. Change can happen at any time, though, and sources may shift location by any amount along the entire stereo field. A change of position may be accompanied by a change of image width. Changes in position tend to be more readily noticed, as this skill is commonly used in daily life.

Actively moving sounds are not typical in tracks, but are certainly found. Moving sound sources may be of any width, travel at any speed, move to or between any locations. Narrow spread images and point sources most closely resemble our real-life experiences of moving objects, but the track is rarely governed by reality. Motion can be gauged by the amount of movement, or the difference between the starting and ending positions. For example, at the end of the introduction to *Abbey Road's* "Here Comes the Sun" (1969) a Moog sound travels from the left loudspeaker to the center of the stereo field. The speed of movement, and the consistency of that speed represent other variables; here the sound moves steadily over the span of 4 beats. The image's width remains stable throughout its motion.

Returning to *Abbey Road*, the lead vocal in "You Never Give Me Your Money" begins the song as a narrow image. It soon begins to gradually grow wider, until it occupies a significant portion of the stereo field; the sound's environment contributes to this change with its gradual addition and varying qualities of cues and its changing proportion with the direct source sound. In the second section of the track, a new lead vocal sound gradually moves from the right to the left side of the sound stage; throughout the movement, the spread image maintains a similar size.

Motion between sounds may be present in a track. In this case a rhythmic structure can be established by interacting sounds located in different locations. The changing locations add spatialization to the rhythms, changes in timbres may also be present. Such occurrences are common in percussion parts, though examples between other instruments (especially instruments of the rhythm section) and voices (between lead and background locals) abound. Peter Gabriel's "In Your Eyes" (1986) binds numerous percussion and drum sounds into a "spatialized rhythmic structure" (Théberge 1989, 104) that presents a strong rhythmic pattern between the instruments spread widely across the sound stage.

Observing Stereo Images

Data collection of source images can be engaged directly using the stereo location graph. Observations of sound source images—positions, size (width), and movements of size and locations—can all be notated on an X-Y graph. This image data can be clearly notated with as much precision as might be needed for the goal of the analysis; images can be located with great precision, or in a more general manner.

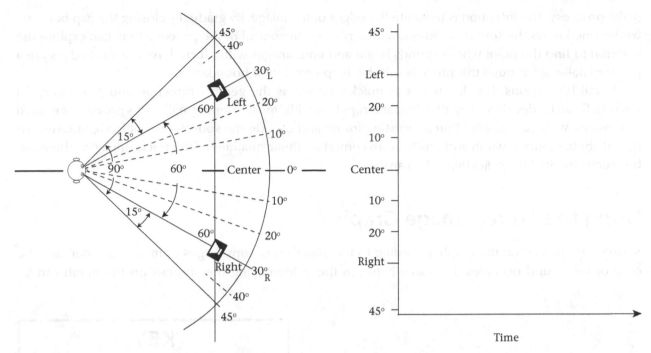

Figure 8.4 Calculating degree increments for stereo image source positions, and the vertical axis of the Stereo Location X-Y Graph.

The Y-axis positions the listener in the center, with the left loudspeaker location above and the right below the point of audition. This allows the listener to turn the graph to orient themselves at the point of audition. The axis is divided into

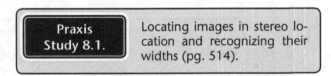

degrees to the left or right of center, locating each speaker at 30° and the furthest left and right boundaries at 45°. Figure 8.4 provides guidance in calculating degree increments on the stereo image graph.

Hearing Images

The challenge of hearing images successfully lies in accurately identifying their boundaries, or edges. As previously noted, in real life we do not directly engage the widths of sounds—and often they are not audible.

Beginning the process of hearing images, one is naturally drawn to the center of the image. This tendency is helpful to establish initial observations. With the eyes closed and head positioned correctly, one can readily point in the direction of the sound, to the core of its position;[7] the skill can be refined with just a bit of practice. This will identify the general position or location of sources in the stereo field; the general distribution of sources might be assembled in this way, but the data would be quite incomplete.

Hearing edges of spread images requires focused attention. Once one has identified the center of the image, its edges can be sought. The point where an image begins can be elusive to start. It is one of the sound experiences of the track many have not previously experienced. Repeated listenings can reveal them through a process of guided exploration between 'knowns.'

Knowing where the sound is not can assist one in identifying where it is. By directing attention outside the image, and gradually closing the point of attention toward the identified location one can make progress. The intention is to locate the edges of the image. By gradually closing the gap between where one knows the sound is and is not, one remains in control of the process. One can explore the material to find the point where sounds begin and end. Images with blurred, reverberant edges are a greater challenge, though this process will also help reveal their boundaries.

It will be obvious this skill is not as quickly honed as the general direction and positioning of sounds. It can be developed by attention and repetition, like most listening skills. This process of guided discovery—with guided attention alternating inside and outside the sound—will help significantly. As the attribute of image width and position are central aesthetic qualities of the track, engaging them can be crucial to meet the objectives of an analysis.

Using the Stereo Image Graph

Sources are placed on the graph according to the area they occupy. Edges of images are defined; the core of the sound occupies the space between them. Identifying the sources on the graph can be

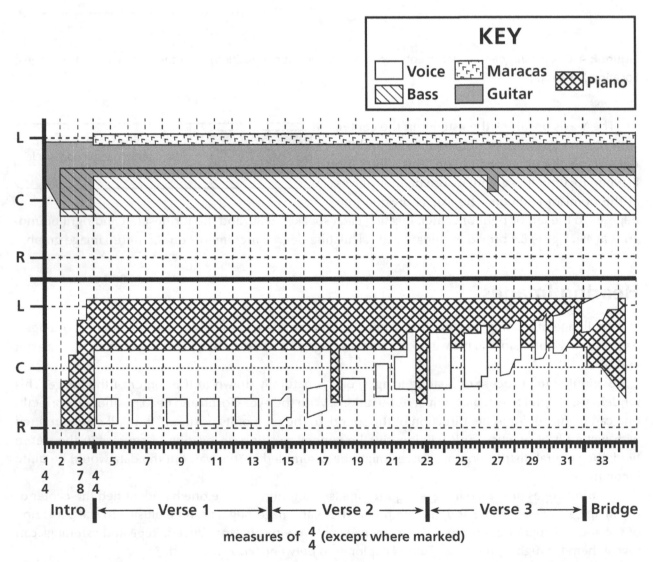

Figure 8.5 Stereo imaging graph of "A Day in the Life" (1967, 1987). Graph contains two tiers of sources against the timeline.

challenging without incorporating color or graphic patterns to fill the space between the edges. The graph can become unclear when numerous images are included, especially with wide spread images. In these situations, tiers can be stacked one above the other; this allows several groupings of sources (one on each tier) to be compared against the same timeline, as in Figure 8.5.

A suitable resolution to the timeline will be identified as observations progress; resolution is determined to clearly show the smallest degree of change the graph needs to clearly present.

The graph is capable of revealing considerable nuance of image positions, widths and their changes. Great detail may not be relevant to the goals of some analyses, though; in these instances, the Y-axis of the graph can be used without the detail of identifying positions by degree increments. The graph may also be dedicated to more general observations by changing the timeline resolution—perhaps to a level representing general positioning of particular sources within a section, rather than detailed positions and changes against a timeline.

Figure 8.5 contains two tiers of sound sources, following their activity throughout the first sections of "A Day in the Life" (1967, 1987). The Y-axis does not contain the detailed scale of degrees around the center, though the center position and left and right loudspeaker locations provide clear guidance of source locations and widths. The graph is the result of reducing a higher resolution version. Changes in image sizes are evident in the piano and acoustic guitar. John Lennon's vocal gradually shifts across the stereo field, from right to left, over the duration of three verses and into the bridge; note several changes in width also appear. Percussion sounds have been omitted from the graph; these could be added to the graph overlaying the existing sounds, or placed on another tier.

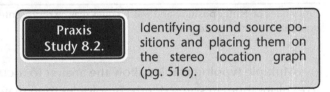

Praxis Study 8.2.   Identifying sound source positions and placing them on the stereo location graph (pg. 516).

Stereo Imaging Typology

Table 8.3 is a general listing of topics that could be collected in a typology table for stereo location. The table can be applied to data collection from X-Y graphs in a variety of ways—related to the number of sounds, the time area, and the variables under observation. For instance, a table might be dedicated to specific sounds within a specific section of the track; others might observe an attribute and its changing values of a single sound, of a collection of sounds, or of all sounds.

Table 8.3 Typology table attributes and values for stereo location images, and for the stereo field.

Variable or Attribute	Values* or Characteristics
Positions of sound sources	Center point of source position Identified for individual sources Collected for all sources (stereo field)
Widths of sound sources	Characteristic: widths individual sources, defined positions of left and right edges Comparisons: largest, smallest, most common widths calculated by degrees
Groupings of sound sources	Sources within identifiable regions of the stereo field (bonded by proximity) Mirroring of sources balancing opposite sides of the stereo field Size of image (like-sized images bonding) Sources moving in gestures with other sources, establishing movement patterns Groups of stationary sources alternating in rhythmic positional patterns

(Continued)

Table 8.3 (Continued)

Variable or Attribute	Values* or Characteristics
Movement: both width and position	Speed of motion (duration sound in motion and distance traveled) Regularity of motion's speed Beginning and ending positions
Rhythms of locations	Patterns of alternating source locations Patterns of reiterating sound in motion
Stereo field width	Location of left edge of left-most sound Location of right edge of right-most sound Amount of area spanned from left to right edge
Stereo field density	Region(s) of source congregation Region(s) of overlapping sources Position(s) of isolated sound sources Region(s) void of activity (silent areas) Amount of space separating sources or groupings of sources
Stereo field profiles by structural division	All source locations and widths during specific section(s) Contrasting sections observed as separate profiles Patterns of alternating profiles (i.e. patterns created by alternating verse and chorus sections)

* All values of center position and width edges may be represented as an angle to the left (L) or to the right (R) of center, precisely identified in degrees.

Multiple typology tables allow the analyst to focus on specific variables, attributes and sources, and to explore them in some depth. Typologies then may be compared and contrasted to gain a more holistic perspective and understanding. Separate tables for various structural sections of the track could be appropriate for many analyses that examine lateral placement of sound sources.

As in many other recording elements, the number of tables and their formatting is determined by the goals of an analysis—the type of information the analysis is seeking to explore and understand.

DISTANCE AND SOURCE POSITIONS IN THE TRACK

Distance is often misunderstood and misconceived, and therefore misperceived. Distance as a recording element is the amount of separation between the listener's position and the position of a sound source. Framed differently: it is the degree of separation between the listener's point of audition and the source's location on the sound stage. We often confuse loudness for distance, amount of reverberation for distance, prominence for distance, and more; qualities that draw our attention can seem closer, something louder can create an impression of being closer than something softer, and an association such as a gentle breath can pull a vocal sound intimately close despite other contradictory distance cues. Distance perception is multidimensional and complicated.

Several important distance concepts shape the track: (1) the distance from the listener to the front of the sound stage, (2) the distance position of each individual sound source away from the listener, and (3) the distance placement of each sound source within its individual host environment.

The first two of these distances rely on the concept that the entire recording emanates from a single, holistic environment. This all-encompassing, global environment establishes a reference space for the track—a space within which the listener is also located. The listener position within the holistic environment is the point of audition—the listener's position from which distance is calculated (and that was also used for calculating angle in lateral localization).

The stage-to-listener distance establishes the location of the front edge of the sound stage with respect to the listener. This is the distance between the closest source within the sound stage and the listener's point of audition. This stage-to-listener distance also localizes the sound stage within the holistic environment of the recording, and provides a location for the listener inside the track's overall environment. This distance plays a significant role in defining the listener's level of connectedness to the track.

Each sound source is located at a more-or-less unique position away from the listener. These distances may be vastly different spans of space, ranging from unnaturally close to the listener to unimaginably far away. Distance positioning can differentiate sources from one another, as can lateral location and image size.

The depth of sound stage is the area occupied by the distances of all sound sources as they appear within their own host environments. The boundaries of sound stage depth are the nearest and the furthest sound sources—fused with the depths created by their environments, discussed below. The source's host environment extends the sound source to have depth; this directly establishes depth to the sound stage. The perceived distances of sound sources within the sound stage may provide the illusion of great depth and a large area, the exact opposite of minimal depth and a minute area, or any state within a continuum between the two extremes.

Understanding Distance in the Track

Distance cues in the track are different from those in real life. This disparity—along with misconceptions about distance perception we may have often heard—is the source of many misunderstandings about distance in recordings, misunderstandings that result in misperceptions of source distances within the track.

In most of life's contexts we first engage our prior perceptual experiences to understand what is newly encountered. In the track, what we have experienced for distance in the real world is often present for individual sources, as they are situated within their own environments—a very natural perception no matter the qualities of the environment. This is not the complete spatial identity of distance, though, and this partial presence serves to further confuse distance perception.

Within the track there are two distance cues: the distance placement of the sound source within its own host environment (just described) and the distance cue of the sound source in relation to the listener's point of audition. It is this second percept—point of audition to source—that positions instruments and voices on the sound stage, and that is a dominant factor in distance perception *in the track*.

Let us briefly look ahead to examine Figure 8.10. Several sources are within their own performance spaces, the host environment within which they are located. These rooms and spaces place the sources within a sonic (virtual physical) presence at a specific distance *within* the sound stage. Note the sources are contained within their spaces; the listener is outside all spaces. The distance cues we have learned and rely upon for distance judgement are only useful within the spaces of the sound source; they are not fully in play in determining the placement of the sound away from the listener.

For distance localization in the track, timbral detail is the overriding attribute that defines a source's position.[8]

In real life we hear the distance of sources within the spaces we occupy—whether enclosed spaces or free field spaces (out of doors, for example). We commonly rely on the cues of direct to reflected sound, of changing loudness and of spectrum changes when attempting to consciously judge distance—bringing mixed results. We process these cues, and sometimes they contribute to a reasonable estimation of distance position, and sometimes we make assumptions based on attributes that are not presenting distance cues. Our distance judgements are relatively inaccurate and our skills unrefined;

we tend to rely on certain cues to make universal judgements about distance (especially loudness and reverberation), when those cues often have limited influence within a given context. Related to this, we have great difficulty judging the distance of sounds we do not know or cannot recognize.

Perception of Distance

Loudness is often considered a determinant of distance. A common notion is that louder sounds are closer sounds. Experiments have appeared to have born this out, but under certain test conditions that examine psychoacoustic perception without also investigating ecological and cognitive psychology (Neuhoff 2004b, 1–4). This is an important distinction. We often hear close, loud sounds in the world, and closer sounds may well be louder—at times. In music, a louder sound does not move toward the listener—neither in real-life acoustic performance nor in the track. A trumpet does not surge toward the listener during a crescendo. Louder is simply louder. In the track, loudness can increase markedly without adding the timbral detail that is gained when sounds move closer in the world.

As sounds move further from us, higher frequencies diminish more rapidly than lower frequencies—being absorbed by the air, attenuated by air friction. Timbral detail is diminished with increasing span of space between the source and the listener, and timbral detail is increasingly heightened with decreasing distance. Timbre is fixed when the source is recorded. Raising the amplitude level of a source in the mix does not change distance, when the changing loudness is not accompanied by a change of low-amplitude detail in timbre—loudness changes without timbral detail changes will not shift distance location. An increase in loudness might allow a sound's timbral detail to be more apparent, in which case the timbral detail that establishes distance was made audible by the increased loudness—the loudness did not establish the shift, it revealed what was already present in the source. This is an important distinction for accurate localization of source distances.

While loudness and reverberation are often identified as "determinants" of distance, they are inconsistent and unreliable gauges of distance location and often not valid indicators. Loudness and reverberation are matters of coincidence and circumstance when they align with actual physical distance; these changes may be present because of changes of distance, and their qualities may reflect change in distance, but these are not causal. They are not directly transferred from one context to the next.

The ratio of direct to reflected sound and the time delay between direct and reflected sound may provide cues to distance within enclosed spaces. In the track, reflected sound (including echo) and reverberation are attributes of the space within which a sound is produced—its host environment. These create some confusion, as they may localize a source within an environment in the real world (this sonic experience suggests sources are in their own spaces contained within the space of the track itself, establishing a space within a space, discussed later). Still, a common misperception is to perceive a sound to be at a considerable distance, when presented with a sound appearing within a large environment containing a high percentage of reverberation. In the track, a sound can be placed intimately close to the listener while it is performing in an unnaturally enormous, overwhelmingly reverberant space—a space that contains the individual sound, a space that is then situated within the hierarchy of the holistic environment. Clearly, distance has several levels of dimension in the track.

Distance is very easily confused with other sound qualities. Perhaps this is because we have such little experience identifying the span that separates us from a source by sound alone. Sound elements are tangential to the visual (Schnupp, et al., 2012 177–189), so those sound elements that are prominent or are easiest to recognize (such as reverb or loudness) take over our perception—we seek to make them fit our experiences. For example, we equate loud with close, and highly reverberant with far, when the real world provides vivid experiences of close sounds in highly reverberant environments (singing in bathrooms?) and distant loud sounds (crack of thunder?). Handel (1993, 183) notes: "Listening is 'making sense,' trying to come up with the simplest and most plausible percept." It 'makes sense' to us: if its loud then its near. It is helpful to remember what seems simplest and plausible may be a misinterpretation, a misperception, or misdirected attention. Schnupp, *et al.* (2012 188) contrasts loudness and visual perceptions related to distance to clarify this matter:

> [S]ound level declines by 6dB for each doubling of distance. Louder sounds are therefore more likely to be from nearby sources, much as the size of the image of an object on the retina provides a cue as to its distance from the observer. But this is reliable only if the object to be localized is familiar, that is, the intensity of the sound at the source or the actual size of the visual object is known.

Obviously, we readily recognize distance positions of the known visual source; we are practiced at judging how visual objects change, decreasing in size proportionally as distance increases. We are not as skilled with recognizing the attributes of the sound that diminish with increasing distance—or that gain in resolution as distance decreases; instead we apply what 'makes sense' and cease searching for 'what is.'

The dichotomy between the distance position of the source within its own space, and the distance of the source within the track can be a confusing one. It might be clarified with a central focus on timbral detail. Listening with attention on the level of subtle detail within a source's timbre provides the cue to sound source position with respect to the point of audition.

Fortunately, we have the capacity to improve distance perception by bringing attention to a sound's physical content. We locate sounds in distance by timbral detail, by observing the content of the sound. Also, we personally and culturally sense into distance of sources as they relate to personal space, bringing us to define our place and relationship to sources as they are situated in their location (more on this below).

A significant study of distance perception performed by Mark Gardner (1969) is often referenced and used to explain the perceptual process.[9] Gardner's study asked listeners to judge distance for shouted, normal and whispered voices; test subjects readily and accurately identified general locations and changes of distance from these sources, aided by instructions. Subjects also identified similar distance changes when presented with these sounds produced from the same location and at the same sound pressure; whispers were identified as closest, shouts as farthest. Examining this from experiential and ecological perspectives we might understand the percept was not established by loudness or perceived loudness differences, but rather by the recognition of timbre, timbre's shaping of the experience, and what the timbres represent to the listeners (especially reflected in the energy required to produce the sound) based on their previous experiences. The percept is a product of interpreted context and connotation; it is not based on sensation and is not based on valid information. Not loudness, but timbre—both generating interpretation and producing associations—brought the percept of changes in distance and established distance positions.

We all engage distance (just as all percepts) from the vantage point of our human condition. Our interpretations of distance can easily be based on inaccurate data, should we not seek information based on its defining attributes.

Personal Space and Proxemics

As we learned in the previous chapter, timbres may be approached as situated within context; through context, timbres have character as well as content. Timbral character elicits interpretation. Augmenting our use of timbral detail to position sounds on the sound stage, we can also incorporate our sense of personal space to localize distance of the sound relative to the point of audition. Perception of content brings location; perception of location generates the context of a sense of physical relationship to the sound; content through context produces (or allows) interpretation. With distance, interpretation relies on our sense of occupying a personal space or territory.[10] From a sense of being safe to an instinctive visceral reaction of being threatened, from intimate connection to the detachment of formality, our sense of distance takes place within this context of personal space.

Humans have a sense of occupying an area of territory—just as do other living creatures: insects, birds, mammals, fish. We unconsciously radiate a bubble around us, a sense of the space we occupy. This bubble is individualized—some people have bigger bubbles than other people—and otherwise variable in its size and qualities. It can change or be redefined from factors that are personal (personality type, inclinations), cultural (national customs of social interaction, those of social groups), environmental (size of space), or situational (number of people or objects in an area, or how one feels about others present). Each of us has a somewhat unique sense of territory, though we share social norms within our own, diverse cultures; further, the sense of territory can expand or contract from our feelings or intentions, such as feeling threatened or attempting to control a situation. Moving about within another culture informs us that others process space differently. We navigate our distance from others through a sense of interpersonal distance.

Interpersonal distance is how we can gauge appropriate action based on the distance of others (or other sounds); it is the basis for our social interactions, and also our sense of place. Personal space (the area we sense ourselves occupying) might serve as a reliable reference for distance location, with knowledge of its defining conditions. Personal space was the basis for the 'area of proximity' I proposed in my earlier writings on distance analysis (1992, 119–122; 2015, 218–220). The area of proximity aligns with the combination of intimate and personal zones of *proxemics*. We have a heightened sensitivity to all auditory differences within and between near sources, including distance cues (Shinn-Cunningham, Santarelli & Kopco, 2000).

Edward Hall (1969) formulated the study of proxemics, which psychology describes as interpersonal distance. He proposed we are surrounded by a series of invisible bubbles, each of measureable dimension. The radiating sequence of bubbles represent four distance zones, each containing two phases; each zone represents a different type or level of social interaction, that might be applied to different cultures. The zones and dimensions he identified are:

- Intimate distance: close phase, touching to six inches; far phase, six to eighteen inches
- Personal distance: close phase, 1.5 to 2.5 feet; far phase 2.5 feet to 4 feet
- Social distance: close phase, 4 to 7 feet; far phase 7 to 12 feet
- Public distance: close phase, 12 to 25 feet; far phase 25 feet or more

His research defined the size of these zones and their attributes from interviews with and observations of test subjects, and anecdotal evidence. The subject pool was narrow and non-inclusive (representing professional, white, upper-middle class individuals in the United States during the early 1960s dominated the pool) and resulted in skewed observations. Reading his descriptions today, many will recognize they emanate from a different culture.[11] However, the zone dimensions, both physical and perceptual, establish several tangible points of reference that can be adapted for recording analysis; these might guide observations that are contextually based on interpersonal distances, as culturally defined. While Hall (*ibid.*, 115) observes "Concepts of [how we respond to distance zones and territory] are not always easy to grasp, because most of the distance-sensing process occurs outside awareness," the basic concepts, and certain core attributes of distance zones, hint toward references for engaging distance in recorded song.

Simon Zagorski-Thomas (2014, 78–79) observes proxemics, along with metaphorical models of embodied cognition and image schema, "provide an interesting avenue of analytical and interpretive potential." He continues by noting a parallel:

> Intimate space, as defined in Hall's proxemic categories, is associated with physical and emotional warmth but intimacy can also be associated with honesty and sincerity. The use of intimate space in recorded music may then create metaphorical meaning that suggests a personal and direct relationship with the performer.

The distance of space brings associations and meanings of many origins; in the track, many will emerge from the singer's persona.

Allan Moore (2012a, 184–207) explores the relationship between singer persona, the 'personic environment,' and the listener through a modification of these proxemic zones; distance zones are adapted to refer to various states of presence of the persona and its 'personic environment' within the track. Persona is "the result of the activity of singing" (ibid., 189) that encompasses lyrics and 'vocality' in addition to melody; the 'environment' of the persona includes accompaniment (texture and harmony) and "formal setting or narrative structure" (ibid., 190). Here, proxemics is used as a set of categories to examine the character and content of the persona and its 'environment' (the lead vocal and its accompaniment), the narrative and form (structural and formal patterning), as well as distance-conceived interpretations based on the voice and its lyrical content. This approach is rich in nuance for examining the relationships between the lead vocal and all else (including lyrics), and how they might be interpreted by the listener. Allan Moore's table on proxemic zones (ibid., 187) identifies some qualities of listener to sound source (lead vocal) distance that are useful in understanding spans of physical or virtual space; those will be incorporated into the approach offered herein.

With these constructs offered by Hall, Allan Moore and Zagorski-Thomas, we find ourselves mixing distance perception with other concepts and percepts. They augment distance observation by connecting it to other concepts that may hold great value for some analyses. Let us recall, now, that we intend to identify the position of sound sources (all sources, including the lead vocal) relative to the listener's point of audition.

Distance Perception in Records

> Hearing usually takes place in the presence of complex stimuli in complex acoustic environment. The neural mechanisms and processes responsible extend beyond the peripheral auditory system and in some cases beyond the auditory system itself. . . . It is understanding the complex interaction of acoustics, physiology, sensation, perception, cognition, and behavior that is the puzzle of audition.
>
> (Neuhoff 2004b, 11–12).

Distance is a complex percept that may be understood more clearly by including ecological perception for perceiving distance positioning. As sensations give way to (or coalesce into) information, that information affords particular possibilities that "cannot be measured as we measure in physics" (Gibson 2015, 128). Physiology, psychoacoustics, perception and ecological psychology contribute to and blend within our perception of distance. The following outline summarizes what we (seem to) know about distance perception pertinent to records, generated from the above discussions and background research:

- Potential to Learn Distance Perception

    ○ Our skills at hearing distance cues are unrefined, perhaps due to our reliance on vision to identify distances (Handel 1993, 108).

    ○ Distance perception, whether in real or virtual environments, is dynamic, and is dependent on the listener's knowledge of the sound and experience in listening within the room; we adapt to and learn the properties of sound sources and the conditions and attributes of spaces (Blauert 1983, 47).

- Relevant Aspects of Distance Perception

    ○ Research points to timbral content (spectrum) as decisive for distance perception (B. Moore 2013, 279). Spectra of sound sources change with distance; high frequencies are attenuated (absorbed by air friction) more than lower frequencies as distance increases.

    ○ The reverberation time and the early reflection timing tells the size of the space and the distance from the source to its surfaces (Rumsey 2001, 35); this is not listener to source distance. The spectrum of the reflected sound may differ from that of the direct sound caused by several influences (Roederer 2008, 80), and may provide some distance cue (B. Moore 2013, 280).

    ○ Loudness changes may parallel distance changes of steady-state moving sources tested in free space, and within certain real world experiences (Blauert 1983, 117); such changes are rarely present in records. In records, a change in loudness typically does not result in a change of distance percepts.

    ○ In enclosed spaces, the ratio of direct to reflected sound, and the time delay between direct and reflected sound, can provide certain cues to distance (Howard & Angus 2017, 46–50); in records, these distance cues represent the sound source within its own host room/environment, not the listener to source distance.

    ○ Timbral detail, and the changes of spectral content that shape it, positions sound sources at a distance from the point of audition. In records, timbral detail (the amount of low intensity information present within the sound source's timbre) is a consistent and reliable distance cue.

- Basis for Analysis of Distance in Records

    ○ Our life experiences of distance are comprised of observing (hearing) sources within acoustic spaces, and with the assistance of sight. In records, we observe distance from outside the space in which a sound emanates, and we observe it acousmatically without the source being visible.

    ○ Distance perception in records blends percepts of listener position, sound stage, and distances of individual sound source positioned within their own environments.

    ○ Loudness levels often do not reveal distance cues in records; they often present information that conflicts with timbral detail or timbre as performance intensity.

    ○ Reflected sound of environments are rarely relative to the point of audition, and rarely provide reliable listener to source distance cues; the relative loudness of the direct sound to reverberant sound can be a reliable percept in some contexts.

    ○ Level of timbral detail is the conclusive percept of distance, also encompassing changes of loudness and reverberation; loss of high frequency content with increasing distance contributes.

    ○ A timbre's state of normalcy represents a reference that is reliable for calculating changes in sound source timbre due to distance. Identifying distance is difficult for unknown sounds.

- o We hear distances most accurately as relative positions between sources, by comparing positions of sources, and by placing sounds away from our bodily position related to degree of timbral detail.
- o Distance may be understood as a sense of territory. We are surrounded by distance zones representing various levels of culturally defined (or influenced) social interaction.
- o Distance zones may help establish tangible points of reference for analysis, affording more-or-less discrete distance positions of sound sources. Placing sounds within these zones/areas relies heavily on timbral detail cues, supplemented by social distance constructs that carry a variable degree of subjectivity.

An approach to distance analysis must function through addressing these factors. Examining this list, we see a familiar pattern emerge. Woven throughout are the physical, sonic content of distance cues—the waveform of the sound source and the components of timbre. Also present throughout that list are the psychological context of personal space and related conceptualizations and perceptions. These will form the basis for observing distance positions, and to analyzing distance.

Devising an Approach for Observing Distance

We will incorporate the above distance factors into an approach to observe distance in records. The approach seeks to draw attention to percepts that can become readily recognizable, are realistically learnable and are pertinent. Further, the approach strives to be readily transferable to different contexts—different musical genres and cultures, to begin.

It is possible to learn, or hone, a skill in perceiving distance of sounds; this can be most directly engaged by recognizing processes we already perform, even if we are unaware of them. Processes of our sense of territory and of our perception of timbre (already engaged in the previous chapter) are used regularly in localizing sounds in distance. We have experience with these tasks—though little awareness of how we engage those processes. We also have experience relating the distance of sounds to each other—as in which conversation taking place behind you is further from another, even within a noisy environment.

These three factors are the basis for our approach:

- Timbral detail
- Territorial zones surrounding the listener
- Positioning sounds in distance relative to one another

Context

Our approach will engage the psychology of our sense of occupying space; this provides a context for our observations. Through it we might recognize sound sources (and their performers) as a virtual presence in a (perceived) physical relationship to the listener. The approach will use some more objective observations from Hall's (1969) research, and will incorporate select concepts of proxemic zones offered by Allan Moore (2012a) and his colleagues Ruth Dockwray (2017) and Patricia Schmidt.[12] Central concepts from my previous work on distance in records (1992, 2012, 2015) will be blended into the approach as well.[13]

We have a sense of occupying territories, radiating in zones of various qualities; this sense may assist our navigating of physical distance, measured by inter-sensory modalities and by metaphor, by potential physical relationships to sources, and by psychological implications. These create references that can be useful, when used with some awareness of what is personal and what might be common to others within a culture. Though our sense of proxemic zones is dynamic—with situational changes, highly varied cultural conventions and individual differences—and the zones themselves are at best vague and inexact, the fundamental concept of proxemic distance zones establishes a reference that can assist distance judgements.

Providing some validation of personal space (and intimate and personal zones of proxemics), there appears to be physiological evidence supporting the theory that "listeners can discriminate between sources that are reachable and those that are not" (Neuhoff 2004c, 94). We are most sensitive to near sounds, especially those within 'arm's length.' Further, it appears listeners are also sensitive to the distance region of bending at the hip to reach an object, versus solely extending an arm; both allow reachable distances to be accurately estimated, when differences of arm and torso lengths are factored (Rosenblum, *et al.*, 1996). These findings are based on experiments implementing an affordance paradigm with listeners judging reachability of a natural, live sound source within a familiar acoustic environment. "Thus, the auditory perception of what is within one's reach appears to be scaled to one's body dimensions" (Neuhoff, *ibid.*). This demonstrates, the context of our personal sense of space has a physical basis, as well as being psychological.

Content

Timbral detail can be used as a spatial cue to place sounds within these territories. We 'recognize' the timbral attributes of known sounds through experience (including life experiences), and can further refine this skill with attentive study. Timbral content guides distance localization, and our ability to hear subtleties of timbre is acute.[14] We recognize subtle differences between instruments and within instruments; our ability to judge timbral details and changes within details in voices is deeply imbedded as part of communication and in many other aspects of everyday life. These and other aspects of timbre were covered thoroughly in Chapter 7. We can use this skill to localize sounds from the listener's point of audition, within distance zones. In time, we will recognize that loudness changes do not create distance changes, but rather *may* add or subtract timbral detail to cause distance to shift; in time, we will recognize that increased reverberation *may* mask timbral detail and decreased reverberation may reveal timbral detail. Loudness and reflected sound do not contribute the defining attribute of a sound source's distance, though they may assist in allowing that information (timbre) to be available— or masked.

Timbral detail is the percept used to establish distance positions:

- From the listener
- Within distance zones
- Among sound sources

Once a source is localized within a certain distance zone, we can recognize the positional relationships of sounds relative to one another. This is a skill we already carry, and it is a skill that can be further refined. Just as we already carry the ability to sense objects within our territorial boundaries, we can place sounds in distance relative to one another within those zones with considerable acuity.

It might be obvious by now: bringing attention to distance positioning will require a learning process. As listening to any recording element, some may have an innate ability to acquire awareness to one property or attribute and less natural inclination for another. This obviously is normal; we have predispositions to listening in certain ways, to having certain elements dominating our attention (McAdams & Giordano 2011, 73). Though

| Praxis Study 8.3. | Developing distance perception (pg. 517). |

localizing distance positions may require some effort to develop, it is learnable. It requires becoming aware of possibilities and of where and how to direct attention.

COLLECTING DISTANCE OBSERVATIONS

A continuum for distance positioning of sources has been generated. Loosely based on proxemics, the continuum assimilates timbral content and the context of personal space to define its zones. These establish references to assist in positioning sound sources on the continuum; the goal is a consistent and relevant process of identifying distance positioning.

Context for identifying distance positioning pertains to the psychoacoustic, psychological and sociological underpinnings of personal space. A territory system provides the contextual reference. It is inherently subjective—personal and culturally variable. It is interpreted from perceptions that are not measureable.

The concept of content in distance positioning reflects timbral detail and the attributes of timbral components. Content is the physical waveform; it can be measured. Content is objective, and can be examined in isolation with the timbres pulled outside the musical, lyrical and sonic contexts of the track. The reference is what the listener perceives as the 'normal' attributes of a source timbre—this adds a level of subjectivity based on the analyst's knowledge, experiences and biases.

The subjective perceptions of personal space are heard through the filter of listener biases, skills and experiences. In all this subjectivity we seek the common ground for communication whereby these perceptions might be more objectively communicated to others, and used to represent some percentage of a shared experience. By identifying timbral content, we might locate the source into a position relative to the context of personal space.

Establishing a Content Reference for Timbral Detail

Detailed and accurate judgement of distance is only possible when the listener knows the 'normal' sound of a source. It relies on the analyst's experience and memory. One's knowledge of a timbre is used as a reference to recognize changes and attributes within the source's timbre. Skill in listening *into* timbre guides this assessment in hearing subtle qualities. Memory of timbral content and of timbral detail are critical to detecting and processing any changes; it is those changes that reflect change of distance, and timbral detail defines positional location.

Within that memory of timbral content is a sense of the source timbre in a state of normalcy—normalcy of timbres and the inherent timbral qualities of sources were discussed in Chapter 7. By knowing the qualities of an instrument or voice in its state of normalcy, we are able to identify changes in its timbre resulting from distance change or repositioning. This 'state of normalcy' is the listener's expected timbral content of the source, one that has not been altered. This 'normal' timbral quality varies by individual; it is what is usual or expected by them under performance and/or listening situations they perceive as normative.

This timbral content the listener 'expects' under 'normal' conditions also establishes a reference. It also exists at some known distance (even if initially the listener is only vaguely aware of it). The timbre is known and remembered, though perhaps currently not articulated in their awareness. It is possible for the listener/analyst to define the distance zone (even a specific location of the source within a zone) of *their* timbral reference for the instrument/voice; this serves as a reference to gauge changes of location, even when the source moves into other zones. This 'normal' or 'expected' timbre contains a specific level of timbral detail, an expected level of detail that is present at an identifiable distance from their point of audition. This distance position might be generally or very specifically defined within a certain zone, depending on listener experience and/or the needs of an analysis.

For many timbres, this content reference will reside in the 'social zone,' where the performer and the instrument is clearly outside their area of proximity. In this zone many of life's activities take place. Should the analyst have experience performing an instrument, the reference might be closer, perhaps in the 'personal zone' or even the 'intimate zone' depending on the instrument and the listener's depth of experience. A reference timbre located within the 'public' or 'distant' zones would be atypical.

Establishing a Context Reference of Personal Space

Contextual traits for each zone are "by no means universal" (Hall 1969, 118); they will change between cultures (even between groups within cultures) and to some extent between individuals. For example, an intimate distance in one culture may be a social distance in another. Considering personal disposition, one person's perception of activity within the intimate distance zone may differ markedly from those of others; the nature of musical materials or ideas, and the qualities of a sound source's attributes will also impact how the sound is received—in the intimate distance zone, or at any other distance. This sense of occupying a space that we can claim as our own is strong, and it is a sense to which one can bring awareness.

Distance zones are, therefore, adaptable references; they can be redefined by individual analysts, according to their personal and cultural norms. The continuum of distance zones is useable, though—it allows for notating the relationships of sound sources. While source positioning might be relative to the individual's interpretation, relationships of sound sources can be expected to be consistent with other, similar groups of listeners and analysts. This will allow communication between individuals to be effective in sharing and discussing information on the experience of tracks—though details within interpretations may well vary.

The notion of distance zones, loosely based on proxemics, represents a useful classification from which a more objective continuum can be formulated—though one that differs from that offered by Hall and modified by Allan Moore and his colleagues. While still far from empirical, I attempt to provide definitions that emphasize timbral qualities (physical content of sound sources) and relationships between the bodies of the listener and of performer/source (a context established from awareness to interpersonal distance). It is intended that jointly the levels of timbral detail and the relative physicality of distances and relationships of and between 'individuals,' will provide a bit of commonality (between analysts and readers) for a workable distance continuum. These are incorporated into the following discussion of distance zones.

The Continuum of Distance Zones

The continuum for distance location used herein is comprised of five zones. A fifth, 'distant' zone has been added to the four proxemic zones. Together, these five zones represent the space spanning 'one molecule away' from the listener to the furthest distance imaginable. A span of innumerable distance positions exists within each zone, the beginning to the end of each distance zone as a continuum from its closest point to the most distant point it contains. Within each zone there is a 'close' region and a 'far' region, representing the bottom and top halves of the zone; the center point of these regions can be identified, making visible each quarter of the zone, and allowing one to calculate a percentage above the lowest position a sound is placed. This makes it possible for a great many sound sources to appear at slightly different distances within the same distance zone—a quality that is common in some production styles.

Table 8.4 outlines the content and context factors that define each of the five distance zones. The ever-widening 'bubbles' of personal space each provide different contexts, and sources located within each exhibit unique levels of timbral detail.

Table 8.4 Physical content and psychological context of sounds within the continuum of five distance zones.

Zone	Context	Content
Intimate Close: to 6 inches Far: 6 to 18 inches	Presence is felt as well as heard Clearly within body space Voice: whispered or very low level Alternants & qualifiers are likely to be present Instrument noises (string squeaks, etc.) Close: physical involvement or touch Far: source is readily touched	Extreme timbral detail Potential distortion of timbre components Modifications to spectrum from extreme formant levels Delicately produced vocal & instrumental sounds are prominent Rarely are environment attributes present
Personal Close: 1.5 to 2.5 feet Far: 2.4 to 4 feet	Kinesthetic sense of closeness Can reach the sound source extremity Voice: level is soft to moderate Less clarity to paralanguage sounds Fewer instrument production sounds Far: source is at arm's length	Timbres are unbalanced throughout Close: appreciable timbral detail, especially in dynamic envelope and spectral flux; Unnaturally prominent spectral components remain but less so; Far: at rear edge, slight modifications of spectrum and dynamics remain; Moderate level of timbral detail Some early time field reflections may be present at low amplitudes

Threshold ending listener's area of 'personal space'

Zone	Context	Content
Social Close: 4 to 7 feet Far: 7 to 12 feet	Mid-zone: an object can be handed to another with outstretched arms Noticeable shift between close and far Far: clear separation from others Few (if any) paralanguage and instrument sounds are present Voice: level is moderately loud	Close: some unnaturally heightened timbral components remain Mid-zone to Far: timbral detail and timbral components are in "normal" balance; this is the reference timbre for most listeners Far: slight changes to timbral content and detail for most sources Early reflections become slightly more prominent in medium-sized enclosed spaces; low level reverberation is present
Public Close: 12 to 20 feet Far: 20 to 35+ feet	Substantial listener to source distance No paralanguage or instrument sounds are present Voice: loud volume (close); full volume, semi-shout and shout (far)	Close: slight changes in timbral detail, loss of high frequencies may begin Far: moderate changes to source timbres in both detail and content; noticeable loss of low amplitude partials and subtle changes in dynamics Far: early reflections become prominent in large-sized enclosed spaces; level of reverberation to direct sound is noticeable

Horizon of detailed distance perception; localizing sources in relation to one another becomes difficult

Zone	Context	Content
Distant Close: 35 to 60 feet Far: 60 feet to ∞	Close: this amount of distance is often aided by sound reinforcement in real life (i.e. stadium performances) Far: this amount of distance is rarely encountered in real life (for example, distant thunder); extraordinary, unnatural distances; little sense of position	Close: moderate changes to timbral content and detail; sounds begin to lack definition; distance positions ill defined; reverberation and reflections match or surpass level of direct sound Far: sounds difficult to recognize; few low-level partials and subtle loudness changes; considerable changes in timbral content; positions are vague; pronounced influence of reverberation, little/no direct sound

The intimate distance zone envelops us closely. Sounds in this zone are unnaturally close, clearly within our body's space; their presence may be felt as much as heard. An extreme level of timbral detail is present, and there is strong potential for some attributes (especially dynamic levels and contours of partials within the spectral envelops) to be grossly out of proportion. Modifications to timbres through over-exaggerated formants are common, as are selective emphasized frequency bands within attack transients; these bring the characteristic sounds of instruments and voices to be overemphasized, and out of balance from what we expect of the sound. In the farthest end of the zone the over-exaggerated qualities begin to diminish, though still noticeably present. Here voices in real life speak in whispers or at a very low level; communication at this distance in real-life is rare and with people with whom we are intimately connected. Alternants and qualifiers (see Chapter 4, and Lacasse 2010a, 228–230) are present in the voice, and may be the only voiced sounds in an exchange; instrument noises such as the squeak of fingers on guitar strings are present. These vocal and instrument sounds that are only noticeable in real-life at hyper-close distances can quickly transport a source into the intimate zone.

The personal zone is from about 1.5 feet to approximately 4 feet. Some kinesthetic sense of close-ness between the listener and the sound source may be retained; a sense of the motion of making the sound can be present. One can easily touch sources at the front of this zone, and stretch to touch the ones at its far boundary (where they appear 'at arm's length'). Voice is at soft to moderate levels, with alternants and qualifiers at a reduced presence, and diminished clarity. Sound production noises from instruments are reduced. At the closest points significant timbral detail remains, especially in subtle changes of dynamic envelope and spectral flux; unnaturally prominent spectral components (notably those from formants) remain, though at reduced levels. At the farthest position, slight exaggeration of spectrum and dynamics remains; a moderate level of timbral detail is present. Some early time field reflections may begin to appear at the middle positions; these would be at low amplitudes, increasing as a source appears further from the listener.

The intimate zone and the personal zone together comprise our area of personal space. As we move into the space beyond the personal zone, we no longer consider it our own. Rather we might regard it as a shared space, as a public space, or perhaps as a space belonging to others.

The social distance zone begins at about 4 feet; its far phase extends from about 7 feet to 12 feet. At mid-zone an object can be passed from one to another with outstretched arms. The distance is more detached and formal; the space is shared with others. A pronounced shift occurs between the closest and furthest locations in this zone. Sounds at the closest locations retain some unnaturally heightened timbral components. Those in the middle portion of the zone exhibit timbral detail and timbral components that are in 'normal' balance; this is the reference timbre for many listeners. The farthest points exhibit slight diminishing of timbral content and detail is present in most sources. Early reflections become slightly more prominent in medium-sized enclosed spaces, and low-level reverberation can be evident; these may change depending on room geometry and source location relative to reflective surfaces. Few (if any) paralanguage sounds are present. Spoken voice is typically moderately loud.

The public distance zone is at a substantial distance from the listener; this is often the distance of a public speaker, of little personal connectedness or direct interaction. Close is from 12 feet to about 20 feet, and far extends to around 35 feet. Real-life voice levels are loud volume (close) and full volume, semi-shouting and shouting progressing into the far phase. Close sounds exhibit some loss in high frequencies and slightly diminished timbral detail. Far sounds show noticeable changes in both timbral content and detail; there is a distinct loss of low amplitude partials and the subtleties within dynamic contours. Early reflections become more prominent in large-sized enclosed spaces; level of reverberation is apparent in relation to direct sound.

Between the public zone and the distant zone is the horizon of detailed distance perception. Beyond this threshold localizing the position of sources relative to one another is confusing. Progressing deeper into the zone localization quickly becomes increasingly difficult, and soon it is impossible.

In real life, we rarely encounter sound sources emanating from the distant zone. Often when we encounter them, these sounds are not the focus of attention, but background—distant airplane, traffic noise, a rumble of unknown origin, a dog barking from the other side of the neighborhood. The closest sounds in this zone have moderate changes in timbral detail and content; they begin to lack definition. Even in the nearest third of this zone, timbres are ill-defined, and reverberation and reflections match or surpass the level of the direct sound. From the mid-portion of the zone onward, sounds exhibit a pronounced influence of reverberation and reflections, with little or no discernible direct sound present. Sounds become difficult to recognize; moderate-level amplitude contours and spectral partials are no longer present, resulting in considerable changes of timbral content and no timbral detail. Positions of sounds are vague; comparing positions between sounds is rarely possible.

Distance positions are perhaps most readily apparent in lead vocals, because of their strong presence in the track. This distance position is also a significant factor in the character of most tracks. Lead vocal lines will be examined here to illustrate distance positions; in Chapter 10 we will explore evaluation of distance positions.

From the very beginning of the track, Björk's vocal in "Cocoon" (2001) provides a clear example of a sound in the 'close region' of the Intimate zone. As breath and mouth sounds mingle with paralanguage sounds and exaggerated timbral definition and formats, Björk's voice is eerily close; if not for the gentle performance intensity it might be uncomfortably close. In contrast to Björk, George Harrison's vocal in the *LOVE* version of "While My Guitar Gently Weeps" (2006) is positioned in the 'far region,' near the rear of the Intimate zone; while clearly containing the detail required for intimate placement, the vocal timbre does not contain the breath and mouth sounds and exaggerated timbre of extreme closeness. The vocal's timbral detail and distance position is fixed from the start and does not change positions even as the track progresses into bridge sections.

In "Valentine's Day" (2007) by Linkin Park, the lead vocal of Chester Bennington is positioned in the center area of the Intimate zone during the first sections. Paralanguage and vocal noises, heightened timbral detail and a sense of restraint in performance intensity bring expressive qualities as well as establishing the vocal's distance position. As the track progresses, the vocal slowly recedes into the Personal zone; then, as the refrain finally and suddenly arrives, the lead vocal shifts position radically to the Public zone's mid-area. Natalie Maines' vocal during the initial verses of "Not Ready to Make Nice" (2006) by Dixie Chicks is positioned a bit further back within the Intimate zone, about 75% away from the closest position of the zone; it is centered within the 'far region'; the vocal timbre contains heightened detail and breath sounds despite a low performance intensity. As the track moves into the first chorus, the vocal moves into the Personal zone, positioned in the middle of the far region in that zone as well. These two tracks (both produced by Rick Rubin) present the lead vocal placed clearly within arm's reach of the listener in the verses (especially in the opening verses), and shift the vocal to a farther distance position during choruses/refrains; this relationship has become a convention in the records of the past few decades.

George Harrison's vocal in "Here Comes the Sun" (1969, 1987) alternates distance positions between chorus and verse, but in an opposite manner from the two tracks just cited. The vocal resides in the rear of the Personal zone in the first chorus, and shifts to the front of the far region in the Social zone for the first verse; as the track unfolds this alternation continues, though the precise placements of Harrison's lead vocal differ subtly from these first appearances. It is unusual for the lead vocal to be placed closer to the listener in choruses; typically, the lead vocal is positioned closer to the listener in verses, allowing the lyrics to speak more directly to the audience.

Paul Simon's vocal in "The Boy in the Bubble" (1986) reflects a unique approach to distance positioning. The lead vocal begins in the front-third of the Social zone, and gradually creeps closer to the listener position. By the last sections of the track, Simon's vocal is positioned in the middle of the Personal zone.

Table 8.5   Distance positions of John Lennon's lead vocal in "Strawberry Fields Forever" (1967, 1987), listed by structural section.

Structural Section	Position of John Lennon's Lead Vocal
Chorus 1	Social zone, mid-area
Verse 1	Social zone, 10% from the front threshold
Chorus 2 (location of the splice joining the track's first and second versions)	Personal zone, mid-area
Verse 2	Social zone, 10% from the front threshold
Chorus 3	Personal zone, mid-area
Verse 3	Personal zone, mid-area
Chorus 4	Personal zone, mid-area

John Lennon's vocal in "Strawberry Fields Forever" (1967, 1987) changes distance positions with the changes of structural divisions as well, but in a quite complex set of relationships. The lead vocal is placed in the front of the Social zone in Verses 1 and 2, and this establishes a temporary distance position reference; Lennon's vocal is positioned in the mid-area of the Social zone in Chorus 1 (the first vocal section of the track) and in the mid-area of the Personal zone in Chorus 2. Table 8.5 lists the vocal's distance position by song section; there we can observe any stable reference of distance position moves from the verses to the choruses, as the mid-area of the Personal zone becomes established as the place where the vocal settles. In Chapter 10 we will explore how "Strawberry Fields Forever" is the result of combining two separate, and very different versions of the track; these shifting distance positions of the lead vocal are the result of the qualities of those versions.

John Lennon's vocal in "Come Together" (1969, 1987) provides an example of a lead vocal in the Public zone. As the vocal enters its diminished timbral detail and absence of high frequencies are evident; the delay on Lennon's voice aids in masking timbral detail and also contributes early reflections of a substantial distance between it and the listener. Lennon's vocal is located at the front of the near region in the Public zone. It remains at that position throughout the track, with the exception of a few vocal gestures where the vocal recedes noticeably but momentarily to the mid-area of the Public zone.

Collecting Observations Using the Distance Location Graph

Figure 8.6 introduces the format of the distance location graph. Data collection of distance can be engaged directly using the distance location graph. A suitable resolution to the timeline and to the distance location continuum will be identified and refined as observations progress. The graph is capable of revealing any nuance of distance change that is present in the track's sources, to the degree it is relevant to the goals of the analysis. The graph may also be dedicated to more general observations—such as a level representative of the source's general placement within verses (for example).

Sound sources may change distance positions at any time, and by any amount. Position changes will be found more abundantly in some tracks, genres or artists than in others. Changes are often subtle when they occur within sections of a track (i.e. within a verse). More substantive changes of distance positions are common between verses and choruses, where a shift of materials, singer persona, lyric content and arrangement also shifts relationships with the listener; though here, too, this generalization will apply to some tracks and not others.

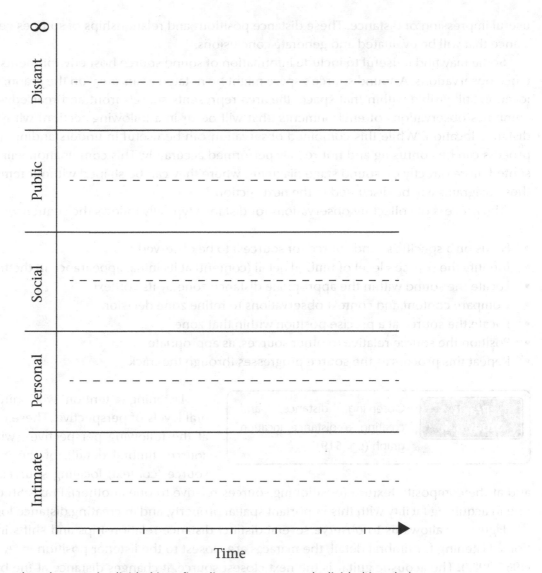

Figure 8.6 Distance location graph, divided into five distance zones, each divided into halves.

The Y axis is divided into the five distance zones. Zone sizes in the graph do not reflect their disparities of physical distance. Zones are of equal size here, to acknowledge equivalence—that all zones hold equal potential to contain sources. All five zones occupy a similar amount of conceptual space, but represent significantly different amounts of physical area.

The size of the zones can be adjusted to most clearly present the positions of sources. Zones with considerable activity are enlarged, and those with no activity are contracted. Public and distant zones may be omitted from the graph entirely when they contain no sources, though the public zone should be present (though perhaps contracted in size) when a sound appears in the distant zone. The intimate zone should be present in all graphs (whether or not it contains a sound) as the empty zone provides the sense of separation from the listener position to the other zones.

In practice, it is not unusual for tracks to locate more (or most) sounds within one zone than in others, and even to exclude some zones. This distribution and grouping of sources within distance zones represents an important spatial characteristic of tracks.

Sound sources are placed on the graph as thin lines, marking a precise location of the source from the listener. Observing the source at a discrete distance from the listener yields the most pertinent and useful impression of distance. These distance positions and relationships of sources generate the substance that will be evaluated and generate conclusions.

Some may find it useful to include information of sound source host environments (spaces) to distance observations. A sound source's space might be notated as an 'area' on the graph, with the source location still visible within that space; the area represents space's front and rear edges. This process combines observations of environments (that will occur in a following section) with observations of distance location. While this combined observation can be useful in understanding many tracks, the process can be confusing and not readily performed accurately. This combination will often be represented more directly on sound stage diagrams, where they can be shaped without temporal concerns; these diagrams will be discussed in the next section.

The process of collecting observations for distance typically follows the sequence:

- Focus on a specific sound source (or sources) to be observed
- Identify the source's level of timbral detail (content) at its initial appearance in the track
- Locate the sound within the appropriate distance zone by its context
- Compare content and context observations to refine zone decision
- Locate the source at a precise position within that zone
- Position the source relative to other sources, as appropriate
- Repeat this process as the source progresses through the track

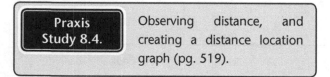

Praxis Study 8.4. Observing distance, and creating a distance location graph (pg. 519).

Listening attention will shift between several levels of perspective. These observations are at the following perspectives: within the sound source (timbral detail), of the individual sound source (context, locating sources within zones), and at the composite texture (positioning sources relative to one another). Praxis Study 8.4 can assist one in acquiring facility with this important spatial property, and in creating distance location graphs.

Figure 8.7 allows us to observe several distinct distance relationships and shifts in distance positions. Listening for timbral detail, the maracas are closest to the listener position in "A Day in the Life" (1967, 1987). The acoustic guitar is the next closest source, it changes distance at the beginning of the passage, and again during measure 26. John Lennon's vocal is slightly farther from the listener than the guitar; while the line contains a significant amount of reflected sound, its timbral detail establishes its presence within the listener's personal zone. The lead vocal is embedded within a small area shared by the bass and the acoustic guitar; they are near to each other yet distinctly separate. The graph does not contain all of the nuance of distance changes that occur within each of these sources; their presence on the graph is somewhat generalized, though representative of their location and general activity. The remaining percussion sounds are omitted; it will be informative to notice how the percussion sounds extend the sound stage depth by their distance positions.

Typology of Distance

Table 8.6 is a general listing of topics that might be collected for distance positions, assembled as a typology table. The table can extract data that has been collected from X-Y distance graphs in various ways depending on the goals of the analysis and the qualities of the track. A typology table will often be defined by the sources it contains and the time space of the observations. It might encompass all sources, or be limited to a specific few; it may observe the content of verses, or of any other section; many other options exist.

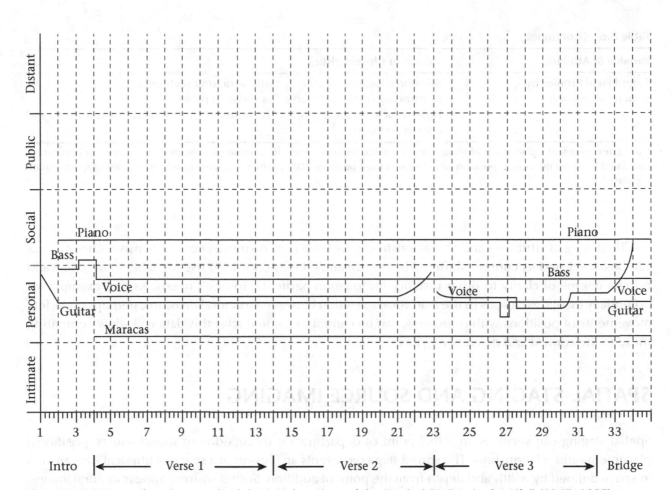

Figure 8.7 Distance location graph of the initial sections of the Beatles' "A Day in the Life" (1967, 1987).

Table 8.6 Typology table of general attributes and values for distance position of source images, and for depth of sound field.

Variable or Attribute	Values* or Characteristics
Positions of Sound Sources	Direct sound, or front edge of distance images
	Identified for individual sources
	Collected for all sources (sound field)
	<u>Comparisons</u>: nearest, farthest, most common positions identified by zones, or regions within zones
Groupings of Sound Sources	Sources within identifiable regions of the sound field (bonded by proximity to one another)
	Sources bonded by appearance within the same distance zone
Movement of Distance Position	Speed of motion (duration sound in motion and distance traveled)
	Regularity of motion's speed
	Beginning and ending positions
Depth of Sound Field	Location of nearest sound source
	Location of farthest sound source
	Amount of area spanned from nearest to farthest sound source
Density of Sound Field	Region(s) of source congregation
	Position(s) of isolated sound sources
	Region(s) void of activity
	Amount of space separating sources or groupings of sources

(Continued)

Table 8.6   (Continued)

Variable or Attribute	Values* or Characteristics
Sound Field Profiles by Structural Division	All source locations during specific structural section(s) Contrasting sections observed as separate profiles Patterns of alternating profiles (i.e. patterns created by alternating verse and chorus sections)

*All values represent a placement of image within specific distance zones; for precise identifications of position, a percentage calculated between the closest position (0%) to the farthest position (100%) within a specific zone's continuum of space can be stipulated.

The listing of attributes and characteristics in the table is by no means comprehensive. Other topics can be added, and its content shifted. Thereby, an analyst may choose to use the table to organize and bring a degree of clarity to other distance information particular to one's needs. As one example (of potentially many), variations in the lead vocal's position between structural divisions, perhaps including movements of positions within sections, can be central to some tracks; this data will allow comparing these to the context and delivery of lyrics.

SPATIAL STAGING AND SOURCE IMAGING

Spatial staging can serve as an initial point of departure for discussions of sound source positional placements and relationships. The sound stage represents an illusion of a space of physical size, an area in space defined by width and depth from the point of audition. Sound sources appear as aural images within the hollow space between the furthest possible left and right lateral positions and the nearest intimate and farthest distant locations. 'Staging' recognizes the positions and interrelations of sources.

The sound stage encompasses the area within which all phantom sound sources (invisible aural images of instruments and performers) are perceived as being located. It is a two-dimensional area of width (stereo field) and distance (the illusion of depth of field). The sound stage is the area from which all sources of the track emanate, or where they are staged in a single grouping. This concept of 'staging' may be a metaphorical or a functional representation of a performance—depending on the track.

'Sound stage' is a convenient metaphor, connecting the virtual area within which source images appear to be congregated with the concept of the staged performance, with all sources congregated in ensemble. As such, the term does not imply staged performances are always present in popular music records—or even that they are the norm. The metaphor of staging is helpful in conceptualizing (and ultimately recognizing) the boundaries of the sound stage. When taken a step further, the metaphor establishes the front edge of the sound stage; the relationship of this front edge to the point of audition allows the degree of listener connectedness to the track to be observed.

The sound stage does not impose a fixed model for understanding positions of recorded sounds; nor does it impose a fixed area within which sources may emanate (except, of course, those imposed by principles of acoustics). Rather, it is an open void within which sounds of the track appear; it is a hollow space where sources are positioned without predisposition other than conventions of the track's style. The sound stage provides a platform where the positions of sources can be observed, and the spatial uniqueness of the track recognized.

In 'Listen to My Voice' Serge Lacasse (2000a) has offered an extensive study of vocal staging, engaging many significant matters of the voice in recorded popular music; while his study has its focus on the single vocalist, it can help us focus some broad issues engaged here. He provides a useful distinction of process and object: "in that 'vocal staging' refers to the practice taken *as a whole* on an abstract level, while 'vocal setting' refers to a specific 'embodiment' of the (general) practice of staging" (Lacasse 2000a, 5). In applying these definitions to applications of sound stage, the term 'staging' is here defined as the process of establishing or observing the sound stage; it is the overall act and conception of recognizing stage dimensions and attributes in the abstract. The term 'setting' is a bit problematic here; the sound stage is at a higher level of perspective (see Table 8.1) and more complex than the setting of a single source.

The term 'scene' would be more appropriate for an individual sound stage diagram, representing that which results from the act of staging—the 'embodiment' of the sound stage content. 'Scene' brings to mind a portion of a dramatic work, a section unified by some identifiable context, and that exists within a specific period of time within a larger whole comprised of additional scenes. Further, there is also a connection between the sound stage and 'auditory scene analysis' (Bregman 1990). Auditory scene analysis is a psychoacoustic phenomenon that involves dividing up the components of a complex wave (a complicated auditory experience) into auditory objects on the basis of grouping cues; groupings may be based on spatial location, spectral content, time or onset. The individual sound stage might be considered a 'scene,' as a complex auditory image based on spatial groupings of successive and simultaneous sound sources. Considering each sound stage as representing a span of time allows it to coalesce into a singular complex auditory image or scene; this image or scene contains the arrangement of the positioning of all sources as a single group (sources may be either stable or exhibit change). The notion of 'auditory scene' provides a useful way to consider the particular arrangement of the myriad of qualities within an individual sound stage (Schnupp, *et al.*, 2012 223–267).

Track as Performance

The track is its own performance (A.F. Moore 2010, 264). No matter if its performance appears to happen like a live event, or if the performance is utterly impossible to have occurred given human and worldly limits, the sounds of the track unfold in time and are linked to the origins and expressions of its sounds. Zak (2001, 43) frames this matter: "[R]ecord making represents "out loud" musical thinking. Ideas are not merely *expressed* in sound; rather, ideas *become* sound. Thus, concept and performance enter into an integral relationship that we perceive as a whole."

The track is "a schematic representation of some real or constructed performance" (Zagorski-Thomas 2014, 6); a performance that is directly experienced, often deeply and viscerally. We hear it, feel it, react to it, sense its physical gestures, interpret the effort and expression of the performances. Further, we seek to identify the location where the performance is happening, our physical relationship to it, and so forth. There is a natural human need (a survival trait with inherent skill, that is more visual than aural) to identify where sources are located—and attend to them accordingly. Spatial staging provides the analyst (or listener) access to observing the positioning of performers and the performance.

Our "natural tendency to relate sounds to supposed sources and causes, and to relate sounds to each other because they appear to have shared or associated origins" (Smalley 1997, 110) is an important part of experiencing the performance; the performer is source-bonded to the instrument (or voice) which generated its sound. Sound sources are bonded with our sense of their origins—a human or technological player, generic or specific instruments/voices, timbres, aesthetic materials, and also spatial environments and spatial identities. The sounds of sources evoke the presence of their performers, and the gestures and interactions of performances. The spatial positioning of the virtual performers adds dimension and richness to the experience, as sources bond by association or sharing perceived origin or cause.

Spatial staging locates the track's performance and its sounds—and all that they carry. It provides a sense of where the performing ensemble is, in relation to the listener. The sound stage not only localizes the ensemble, but places each sound source (its aesthetic materials and sound qualities, its instrument and performer) in space. Sources are in a place; they are at angles from the listener's center, and at a distance from their point of audition. And, sources can move. Zagorski-Thomas (2010, 252) relates the staging of tracks to those of staging a play:

> Transferring these ideas to the staging of recorded music provides certain analogous forms of mediation: those relating to the size and nature of the space in which the performances are being situated, those relating to the positioning of performers in that space; those that change the perceived form and character of the performance; and those that direct our attention towards, highlight of obscure a particular feature of the performance.

Earliest stereo recordings placed sounds in relationships that performers typically formed on stage. This was a rather simple matter for orchestral and choral recordings, with but a few choices of locating sections (King 2017); jazz groups could get a bit more complex, but not significantly. For popular music performances, a group's live stage layout is often casual and varied between performances; staged positions of players are quite arbitrary, not rigidly established. On stage, drums might be more or less centered, with instruments to either side, and vocalist front and center, though even this loose connection is meaningless to the sound and performance, except for performer interaction. Amplification and mixing the sounds of all instruments and voices determines the perceived positions of sources; the stage locations of performers is not reflected in what is heard through sound reinforcement systems. In the track, convention established—for technical reasons of LP record grooves as much as musical—the perspective of a listener facing the bandstand; vocalist at center stage, drums behind (kick drum centered, snare drums centered or slightly left, high hat perhaps about 10° to the right, toms and cymbals balanced left to right), bass centered, other instruments spread out to balance the stereo field and to be clearly distinguished (or blended). It did not take long for these positional relationships of rock performers on records to break their connection with a virtual stage. While some tracks retained a connection to the stage, others used spatial positioning of sources as variables to shape the track. Staging was used to create unique spatial relationships between aesthetic ideas and the sources (performers and instruments) that produced them. As early as 1966, with the Beatles' "Tomorrow Never Knows" we witness a dramatic shift away from common staged relationships; the track fully embraces an aesthetic that instruments and voices (and source-bonded representations of the lads that played them) are untethered by the laws of physics. The positioning, motion and environments of the sources on the sound stage also shaped the character of the track, both from drawing listener attention and from being absorbed into the track's texture.

Popular music recordings treat space as an aesthetic canvas. Locations of sources may be at any location (lateral, distance or room) an artist believes best suited to their expression and aesthetic. Spatial properties blend into the essence of sounds and materials, become part of their character and content. Much artistry is involved in inventing worlds where sounds have unique or surreal spatial qualities, relationships that defy physics. "The more unique the space is, the less it represents experiences of sound in the natural world and the more the record takes on the quality of a dramatic stage" (Zak 2001, 145).

Staging and Listener Interpretation

The concept of a sound stage *may* be predicated on the listener conceptualizing the track as a staged performance, one occurring in front of them (Moylan 1992, 2012, 2015). While this may not hold between individuals—or especially between cultural segments—it is *one* tangible way to conceptualize, organize and interpret sound location. The concept of a performance in front of the listener, within some virtual place, is not a stretch; in fact, it is a common perception to those with considerable exposure to staged performances. Staged performances (on stages humble and grand, from concert halls to pubs, churches to amphitheaters) are cultural norms to a great many listeners. With personalized listening experiences, "listeners often conceptualize acousmatic sound by comparing it to previous experiences with sound" (Brøvig-Hanssen & Danielsen 2017, 194); listeners are prone to interpret the staging of the track as an association to some listening event of the past. Whether tracks are 'heard' as emanating from a stage or as staged in some other manner is a matter of listener interpretation.

Production shapes the spatial qualities of the track—and the listener interprets them. Recent production practice for popular musics has not aligned spatially with concert listening. It is rarely relevant to carry expectations of concert listening spatial properties and relationships into an analysis. Individuals have internal spatial references, shaped by past experiences of live music listening in various venues and by listening to certain production styles (Brøvig-Hanssen & Danielsen, 2013).

The contexts of live performance vary by culture, and by styles of music; formal staged performances will be foreign to some, common to others. While some listeners—for example those who regularly hear classic rock records—might recognize relationships of performers that might realistically be staged in performance, others' perceptions might transport them to the corner of a local pub, others to a street scene, etc. Dance music, and other genres (those existing and those yet to be devised) may well be heard differently.

Some tracks cannot be (or will not be) interpreted as staged performances. The materials, character and content of sound sources can bring spatial positioning of performers (sound sources) within the track that have no relation to prior live listening experiences. A 'performance' may not emerge from the track at all for some listeners; further, the track may not be *presented* as a 'performance' by an artist. This is among the aesthetic statements of the track, and amidst the factors interpreted by listeners. For example, it is unlikely numerous listeners would imagine a staging of Peter Gabriel's "Intruder" (1980) as a humanly possible, live performance taking place in front of them. If the listener conceives "Intruder" as a gestural performance at all, it might tend toward the cinematic, or perhaps some other, more personal manifestation.

The listener may locate their point of audition at center stage, sound coming from a performance in front of them—though perhaps not. Even the point of audition can be a variable for some listeners, along with the acoustic variables of automobile seating and ear buds, there are conceptual matters where some listeners are prone to projecting or placing themselves *within* the contexts and activities of songs. Listeners will have diverse experiences, and will bring diverse backgrounds and expectations.

A sound stage cannot represent a singular cultural experience—contexts are not universal among many listeners, and many diverse cultures are represented within the expanse of popular music genres. Popular music is mostly experienced *as* records, not in live performance. When popular music is experienced (performed live), sound reinforcement systems very typically render the physical locations of performers meaningless, as their sounds are blended and presented to the audience through the same loudspeakers. In the end, listener interpretation of the 'performance' cannot be anticipated; the listener's conception of the sound stage may be highly personal at some level, and hold cultural norms in other ways. The notion of a sound 'stage' may simply be a location for the 'staging' of sounds—the notion of an area in which the sources are congregated.

Sound Stage and Analysis

Based on the previous discussion, the concept of sound stage may be defined to serve the analyst's purposes and conceptions. The concept is sufficiently malleable to allow the record's representation of spatial properties to be organized in a way meaningful to the listener or the analyst—and pertinent to the unique attributes of the track. For the analyst, the sound stage provides a reference; it can aid in explaining the track's experience of spatial positioning of sources to others, as it provides a conceptual common ground based on the analyst's perception of the spatial properties produced by their playback system.

This approach to sound stage as a conceptual common ground, can allow positioning of sources and the size of the overall area they occupy to be identified and observed. The dimensions of the sound stage and its content can transfer to the listening experiences of others, and provide common points of reference. While some variation will appear between playback systems, a basis for common experience (of some significant substance) of the sound stage will emerge. As a single space, its size is defined by its boundaries left to right, front to back; it can be precisely calculated in degrees left and right, and by nearest and farthest placements in percentage of specific distance zones.

The listener groups all sources to occupy a single area—an area from which the track is staged, and its 'performance' is heard. Even sound sources occupying significantly different locations within the sound stage may bond and group into the illusion of a single ensemble; an aesthetic alternative, sources may be caused to segregate into subgroups, or some individual sources appearing isolated. For example, it is common for the lead singer to be detached from the group of the accompanying instruments—the textures established by the functions and relationships between the lead vocal and accompaniment take many forms, and may at times segregate into separate streams. Other permutations are possible as the individual track asserts its own aesthetic—for example, lead singer separated from background voices, separated from accompanying instruments. Still, some rock styles position the singer very close to the instruments, as if with the band.

Sound Stage Diagrams: Notating Image Positions

An early writing on the sound stage articulated its relationship to the performance: ". . . a two-dimensional area (horizontal plane and distance), where the performance is occurring . . . The sound stage is the location . . . where the sound sources are perceived to be collectively located as a single ensemble" (Moylan 1992, 207–208). The dynamic nature of some sound stages in rock music could not be reflected in a single sound stage diagram; these diagrams were intended to synthesize observations from stereo and distance graphs. Early approaches to charting stereo location and distance placement of sounds made clear the need to engage sounds in surreal positions and relationships (Moylan 1986). Spatial staging became a way to account for perceived locations of sources found within tracks, whether or not they conform to real life experiences; it allowed for the collective location of the ensemble to be conceived in relation to the overall environment of the track and to the listener position. In this way, the sound stage could be used for production planning; as an analytical tool it could also be used for evaluating the aesthetic of the track in relation to its simulation of real-life experience(s) or representing the surreal (Moylan 1992, 79–89).

The sound stage can function as an analysis tool (Moylan 2012, 180). In illustrating the locations of sources, the following sound stage variables might become visible:

- boundaries of area containing all source placements
- densities of sound source distribution throughout the sound stage area
- the relationship of the collected sources to the listener
- the relationship of sound stage and the holistic environment

In the physical world, lateral cues and proximity discrimination work together for positional localization. This two-dimensionality is transferred into the record. Sound stage diagrams reflect these real life percepts and interrelationships; the physical objects that are heard and localized with the significant assistance of sight in everyday experiences are represented as phantom, aural images in the track. The placement of aural images on the sound stage establishes an illusion of positional relationships between sources *and* between sources and the listener; "[T]he apparent location in space of sound sources, near or far, left or right, does refer to the simulation of actual spatial dispositions" (Doyle 2005, 27). It may be obvious, but is easily overlooked, these relationships also extend to encompass the musical materials presented by the sound sources, as well as the location of the narrative and persona of the lyrics.

Staging diagrams do not include non-spatial elements. They are dedicated to the two spatial properties, and are capable of significant detail and accuracy. Other elements, such as the frequency registers of timbral balance may be compared to spatial staging at later steps in the analysis process;[15] this will be covered in Chapter 9. The sound stage is intentionally limited to representing two-dimensional space in order to provide focus on those percepts and to make possible a high degree of detail and precision in observations. Staging brings the percepts and observations of stereo location and distance positioning into a single two-dimensional perspective.

Two sound stage formats can be of use for different purposes: (1) proximate sound stage and (2) scaled sound stage.

The (1) proximate sound stage is visually uncomplicated; it is an empty staging area, with loudspeakers placed as a reference for angle only. This sound stage format is useful for showing approximate (or generalized) sound source placements and their size; it can be refined to be fairly accurate, and to locate images very near their actual positions. The (2) scaled sound stage format is capable of illustrating precise locations—it incorporates a grid of increments specifying degrees left and right from center and distance zones (adapted in size to suit the example), scaled for precise placement of images.

Importantly, all staging diagrams are snapshots of time; they represent a specific time period (with a beginning and an end point) or a specific moment in time. The time period might be any suitable length: a phrase, a song section, or in some cases, the diagram might represent the entire track. The longer the time period, the more generalized the observations. Moving sources are difficult to notate on the diagrams, as there is no temporal axis. The X-Y graphs for lateral and distance location can clearly notate any source exhibiting motion.

The proximate sound stage for general positioning of sources is valuable for sketching initial observations of source positions. This format does not position the listener and the loudspeakers in an

Figure 8.8 Proximate sound stage, for generalized image placements.

equilateral angle, but rather allows a conceptual placement of sources in relation to relative positions of loudspeakers and the listener. Placements have the potential to be more or less precise in relation to the loudspeakers, and the listener, as determined by the analyst (their needs and abilities); the diagram does not contain the divisions necessary for exact placements. Instead the diagram serves two functions that might be important to analysis.

First, the proximate diagram is useful for sketching image locations during initial listening sessions. One's first impressions of stereo imaging and distance locations can be noted, and developed upon repeated hearings. The analyst can obtain a sense of the positioning of sources and the extent to which more detailed analysis might yield significant information. Listener position and distance zones can be sketched along with the aural images; the analyst can adjust the diagram to best suit the analysis and the material being observed. From these sketches one is able to obtain a sense of the important attributes and levels of activities within the track's sound stage; and the need for further detail in the observations. One can decide an appropriate course for the analysis from the available options to:

- Create an X-Y stereo location graph to explore locations in detail, with a sense of temporal progression of the track, or of sources that change widths or locations; this may be for all or for select sources
- Create an X-Y distance location graph to explore locations in detail, with a sense of temporal progression of the track, or of sources that change locations; this may be for all or for select sources
- Refine a proximate sound stage diagram for all or for select sources
- Establish a scaled sound stage diagram for all or for select sources (perhaps with the remainders on a separate general sound stage diagram)

Second, the proximate sound stage diagram can provide proportional and relational placements of sources with enough detail to be useful in many contexts. This generalized diagram may often be all that is needed to notate the positions of sources, and to adequately observe these properties in sections of the track.

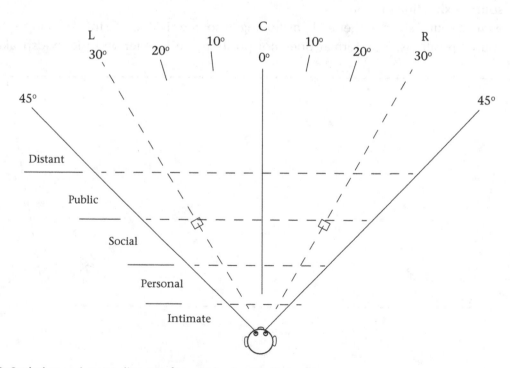

Figure 8.9 Scaled sound stage diagram for precise localization of images.

A scaled version of the sound stage diagram locates the listener and loudspeakers in an equilateral triangle relationship, and with a sense of distance zones. This scaled sound stage is capable of illustrating precise locations of stereo placement, and the listener's considered interpretation of distance placement within specific zones. This data can be sketched first on a proximate graph, and refined in final form here. Perhaps most directly, the scaled staging diagrams can organize and reframe data collected in stereo and distance X-Y graphs; those graphs bring focused attention to the percepts and their placements, and are directly transferable to these diagrams.

Figure 8.9 illustrates the increments of degrees for lateral positioning of sources and distance zones; these are scaled for precise placement of images. The area for image localization projects from the point of audition, in a 90° cone that stretches to the horizon of the sound stage. The positioning of the loudspeakers pertains to lateral imaging only; extra reference lines guide image calculation and positioning.

The visible loudspeaker placements on the diagram are not relative to the distance zones, and are irrelevant to distance positioning; they are included to assist placing sounds in the stereo field and should be ignored in establishing distance locations of sources. It will be helpful to identify distance of the source by timbral detail (perhaps placing the sound on a general diagram or a distance X-Y graph), and then proceeding to determine the position and width of the image at that distance position on the sound stage. Distance zones incorporated into the scaled sound stage may be size-adjusted to most accurately or clearly illustrate the track being examined. The subjectivity inherent to some distance calculation can be noted and minimized by describing one's observations and processes of deduction during analysis discussions.

The point of audition perspective of the track will often bring closer images to be wider sources; this is clearly evident in this approach to staging. Here again this approach brings sound stage imaging to mirror our perception of how lateral images appear to change size (occupy more of our perceptual field) as they move closer and farther from our point of audition.

Praxis Study 8.5.

Making detailed and general observations of sound source images, and placing images of sources on proximate and scaled sound stage diagrams (pg. 520).

Typology of Staging

Typology tables will assist in organizing the many variables of the sound stage, and can reveal their characteristics and interrelationships. The typology table can also make clear the way the sound stage functions on all three levels of perspective.

Functioning at the overall perspective, the sound stage establishes a sense of physical boundaries for the track. The dimensions of the sound stage and its containment area establish important contextual attributes of the track's spatial identity. These can be clearly identified or defined in the typology table.

These sound stage dimensions and attributes characterize sections of the track; with movement from one section to another staging changes are common. The characteristics of the verse sound stage typically return with the verses, perhaps with some modification; choruses, bridges and middle eights typically have a contrasting overall size of the sound stage, and internal attributes. Progressing between sections of the track can be accompanied by a sense of movement from one 'auditory scene' (Rumsey 2001, 43–44) or set of spatial relationships to another. Separate sections of the track are most readily observed and evaluated using individual typology tables, and individual sound stage diagrams, for various sections of the track (as appropriate).

Table 8.7 Typology table of proximate and scaled sound stage attributes and values.

Variable	Values* or Characteristics
Sound stage boundaries (dimensions): • Front edge of nearest source • Rear wall of depth of farthest source • Left edge of left-most sound • Right edge of right-most source	<u>Proximate sound stage</u>: image placements are approximate, though fairly accurate; generalized locations of left-most, right-most, nearest, and farthest sources <u>Scaled sound stage</u>: values are specific degrees left and right of center, nearest and farthest distant zone locations as a percentage within a zone
Area of sound stage containment	Size of area covered: degrees left + right, by nearest source and the depth of the farthest source
Localization of sources: lateral and distance	Positions of selected sources (such as lead vocal) or of all sources
Sizes of source images: width and depth (depth is a source host environment attribute)	Image size of selected sources (such as lead vocal, drums) or plotted for all sources to represent the entire staging area
Location of sources by their functions	Primary parts: lead vocal and instrumental lines Groove elements: dominant rhythmic parts and often bass line Secondary parts (such as backing vocals) Supportive and ornamental parts (often keyboards, guitars, string sections, etc.) Contextual and ornamental rhythmic, accompaniment and timbral gestures and parts Thematic, riff, or other defining gestures of sounds: melodic, harmonic, rhythmic, or timbral
Density of lateral location distribution	Region(s) of source congregation Region(s) of overlapping sources Position(s) of isolated sound sources Region(s) void of activity (silent areas) Amount of space separating sources or groupings of sources
Density of distance location distribution	Region(s) of source congregation Position(s) of isolated sound sources Region(s) void of activity Amount of space separating sources or groupings of sources
Groupings of sound sources (calculated in proximate or in scaled locations)	Sources within identifiable regions of the stereo field (bonded by proximity) Sources within identifiable regions of the sound stage (bonded by proximity to one another) Sources bonded by appearance within the same distance zone Sources bonded by appearance within the same portion of the sound stage (center, mid-left, left, mid-right or right side)
Areas (of space) void of activity	Points and areas in space defined by boundaries of distance zones and lateral area regions
Listener connection to the track	Distance from nearest source and to lead vocal
Depth and horizon of sound stage	Back edge (wall) of farthest sound source
Expanse of sound stage	Outer edges of farthest right and left sounds
Time frame of diagram	Structural section, time segment, etc.

*Lateral location values may be approximate and generalized, or precisely defined in degrees of angle to the left (L) or to the right (R) of center; distance location values may identify distance zones and generalized placement, or precise identifications of position as a percentage calculated between the closest position (0%) to the farthest position (100%) within a specific zone's continuum of space.

Because stereo (lateral) localization and distance location are part of the sound stage, their typological concerns (just discussed) are imbedded into those of the sound stage. The perspectives of activities of individual sound sources and of composite texture come together in these localizations. Individual sources reveal their positions and widths or depths in the sound stage's two dimensions; this positioning and size information is then available for comparisons at the composite texture's perspective. Sources can be compared to one another; from this process, patterns and groupings of locations and attributes might be recognized, areas of density and voids within the stage become apparent. The stage can be fluid or fixed; source locations in flux or those that are unchanging mark the extremes within which much gradation can occur. The left-right and front-back boundaries and the dimensionality of the sound stage have the potential to change, just as each source has the potential to change; the sound stage as the overall space that contains all of the sources may change, and in so doing shapes the character of the track (by what its size and other attributes mean to the listener) and its sonic content.

For example, the great spatial complexities of the sound stage of Kate Bush's "Get Out of My House" (1982) combine lateral image locations and sizes, distance positions and depths, and attributes of sound source environments into a complex spatial tapestry of great and continual flux. Studying such a sound stage thoroughly is facilitated by examining all of these elements in significant detail; this cannot be accomplished by any single set of data. In contrast, the sound stages of numerous tracks depict scenes that unfold, often moving alternately between positions for choruses, and continuing to return to a place of the verse's story, and a similar sonic quality of relationships of sounds; the result is a sense of stability within a changing landscape. Examples of tracks that alternate scenes are many; the primary variables are the types of changes of source positions (lateral or distance) and the amount of change. For instance, "Something" (1969) by the Beatles is typical in that the chorus retains several distinguishing features from the verse while adding several new spatial qualities and shifting the distance position and lateral size of the lead vocal; Adele's "Hello" (2015) transports the listener to a very different world between the verse and chorus—and a very different character of the sound stage and the track.

Notating the significant activity (on one extreme) and the stable spatial context (on the other) of the sound stage can present challenges. X-Y graphs are able to clearly illustrate these activities and changes for either stereo location or distance positions as they change over time, but comparing the two dimensions is difficult with these graphs. Using these graphs in tandem with sound stage diagrams may best illustrate these essential qualities of many tracks; the diagrams allow the numerous qualities to be visualized, and the X-Y graphs allow for details of size, position and movement. Distance and lateral positioning of sources and size of their collective area are not the only spatial dimensions, though.

The aural image holds the *spatial identity* of the sound source. This is more than its width and lateral placement plus its distance from the listener and depth. The characteristics of sources' host environments contribute additional depth cues to aural images, and may also provide added width. Images appear in virtual space with qualities of direction and distance from the point of audition; they also contain breadth and depth from their individual environment's cues (including those generated by reverb and discrete reflections). The spatial identity of each source is an amalgam of these three attributes, or spatial properties. These will be explored in the following section.

A kaleidoscope of spatial activity can churn within the sound stage—activity that is not only of spatial properties, as the sound stage is the platform on which the sources and all of their activities and qualities are heard and interpreted. The sound stage is the venue for the track's mix. Within it music elements and lyrics fuse with the spatial qualities; placement of aesthetic ideas in space and their relationships to the listener and the overall sound stage can be observed and evaluated. We will return to staging in Chapter 9 within the mix. At that time additional elements, such as timbral balance, will be observed simultaneously with spatial positioning—other approaches to examining positional, environmental and frequency spaces of the track will also be introduced there.

Peter Doyle offers the following words. They are a pertinent way to transition from positional staging to the spaces within which sounds emanate—and point to the significance of each in communicating the track's message.

A circumstantial case—I believe a strong one—can be made for the existence of a coherent spatial semiotics operating in popular music recordings. So the issue is not simply, I would argue, that recorded voices are "staged" (they clearly are) but rather, what might be the nature of that staging? What characterizes these ancillary, sonically rendered but implicitly visual representations? And how might the two domains, the music and the effects (the actors and the scenery perhaps) interact?

(Doyle 2005, 31)

Discussed next, the holistic environment (which might also be considered the host environment of the sound stage) coupled with the attributes of the sound stage establishes the spatial identity of the track.

THE SOUND OF PLACE

In nature, the space in which an event occurs has a sonic quality, a sonic character—indoor spaces and spaces in the open air. From caves to cathedrals, closets to concert halls, all places produce a sonic sense of their proportions and the materials of their surfaces. Places have a sonic presence. Instruments, voices and ensembles acquire sound qualities from the spaces in which they perform. In much of our daily activity, this 'presence' is a backdrop for our life experiences in the physical world. The places (spaces and physical environments) of our life events have sonic character that provide context for our activities and experiences. Within all discussions of space, the term 'environment' will be defined as the perceived or virtual space within which a sound source seems to be sounding (emanating or performing); an 'environment' can be any enclosed space or the open air.

Sounds excite the acoustic attributes of spaces from the actions of their sound waves; sounds are bonded with the spaces within which they were produced. We expect sounds to emanate from within spaces—to be linked with the physical world. In records the perceptions of sound sources are wed with a sense of the space from which they are *perceived* to emanate; a source is perceived as fused with the space within which it is performing—the fusion is both sonic and conceptual. "The idea of source-bonded space is never entirely absent" (Smalley 2007, 38). "Sounds . . . carry their space with them—they are space-bearers" (*ibid.*). The sound of a space is bonded to the track's instruments, voices, performers, performances, materials, etc. Bonded to the source, too, is all that the space represents culturally and to the listener, and its cultural meaning (Doyle 2005). This blended sound informs us of the space of the source, and the connotations carried by the space; the sonic environment of a cathedral might conjure some sense within a listener, or metaphorically represent it within a track.

We know the sounds of spaces; we remember the sound of a place. The listener remembers or 'knows' the sounds of spaces, as overall qualities comprised of many factors that are mostly perceived unconsciously, similar to timbres; this perception and prior experiences are used to recognize and understand the spaces encountered in records. "When people engage with acousmatic musical sound, which has no visible source, their experiences with these sorts of different acoustical reflection patterns allow them to imagine specific actual spaces" (Brøvig-Hanssen and Danielsen 2017, 197); reflection patterns and other attributes of environments that characterize spaces are recognizable.

In records, the sound attributes of spaces become part of each sound source ('host environments'), becomes a dimension of the overall quality of the track ('holistic environment'), and becomes integral to the track's content and expression. Each sound source will appear within a 'host environment'; these environments may be unique for each sound or shared with one or more other sources—each instrument and voice in a track has the potential to be in a different space. These 'host environments' are situated at the basic-level of the sound source. The bonded sound sources plus their host environments are localized on the sound stage, where all sources within the track form a singular grouping (that is the 'performance' of the track). The 'holistic environment' is the space of the sound stage—the place where the track is perceived as existing. All instruments and their spaces are amalgamated into a shared space, and a holistic character forms, if not a recognizable space. The holistic environment is resident in the large dimension level, and is a dimension of the overall sound. Thus, a hierarchy of space exists, where spaces can be imbedded within other spaces—spaces of sources are housed within the overall space that contains the sound stage.

It may now be getting clear, tracks present environments in unnatural ways; they change space and our relationships to spaces. They bring us spatial variables and attributes that are well outside our worldly experiences.

The spaces within tracks might be polarized ranging from natural spaces (captured while the sources are recorded), to fully fabricated, with their attributes crafted and invented—though most fall somewhere between. Such invented virtual spaces are often sonically and experientially unnatural, having qualities that nature cannot generate. Physics and acoustics are no longer limitations; sounds of environments in records are created technologically (from many options, each with its own sonic imprint), captured (within the recording process) or manipulated (through various mixing and processing techniques); their attributes need not align with the real-world. The listener accepts unreal or surreal qualities as being suitable places for sounds and tracks with little hesitation—we seem willing to readily accept as plausible a physical environment that is not of this earth, that was fabricated to support the qualities of the sound source, and that is integral to their experience of the track. Brøvig-Hanssen and Danielsen (2016, 27) observe: "[M]usical spatiality has a tendency to point the listener toward a real-world physical phenomenon even as it acts to undermine that reality." We might recognize a continuum, of sorts, spanning from two extremes. On one end are real sounding spaces; environments that embody sonic attributes of the natural world (and that listeners may have actually experienced). On the other extreme are those that are utterly liberated from physics—void of Earthly sonic attributes. Environments may take a myriad of forms stemming from how they are established, the values of their attributes, and their relationship to known spaces.

Even those environments that are 'natural sounding' hold traces of technology, however; a microphone imparts some degree of its unique, inherent sound quality onto the space that is recorded along with the source (Moylan 2015, 360–363). "Any sound recorded by a microphone bears some degree of patina imparted by the space in which the sound was produced" (Zak 2010, 313). All recorded sounds bear the sound of the place of origin; even the subtlest of room sound situates a sound in space. All sounds come from places, whether natural spaces or created environments; within the track all environments are to one extent or another simulated spaces, holding the imprints of the technologies that captured it. It is natural for the listener to perceive the source within a space, no matter the source's spatial content. When sounds display no noticeable environment attributes, "the most likely interpretation . . . would be to hear the voice as sharing, or as sounding inside, the listener's own environment" (Lacasse 2000a, 193).

As this section unfolds, we will engage the content of environments and their character; this linkage with timbre is intentional, and will become clear. Discussion of physical content of spatial environments, and of echo and reverberation, will allow observation of the subtle qualities of spaces, and of their appearance in tracks. Such appearances give opportunity to engage the character of spaces. Spatial environments organize into hierarchy; bonding of sources with their environments, the rich potential of relationships of spaces, and the holistic environment bring potential new richness to spatial character. Entwined in the microscopic details within the attributes of environments, is the deep listening required to hear their values; this is another connection to timbre—this time for analyzing and recognizing portions within sounds that we previously ignored, as they coalesced into the whole.

Sound Source Host Environments

All sound sources in the track emanate from a conceptual performance space, or environment. This space is a 'host environment' for the sound source; it 'hosts' or contains the sound source and the sounds it produces. In real life, sound sources will most often share a host environment with other sources—we talk amongst friends in the same room, sounds are around us on the street, performers share the same stage, etc. Often in tracks, though, individual sound sources are situated within their own space, contained in a separate room or environment.

The many sound sources in tracks may each be located within a different host environment; each source may be bonded to and hosted within a uniquely different space. Sources may share environments in the track, with several sounds appearing to emanate from within the same space, though perhaps at different locations within the space. Further, it is common for sources to change spaces as a track progresses, and/or for them to change distance locations within their host environment(s). When this happens, it is typically between structural sections, such as between verse and chorus.

The source 'host environment' is situated at the basic-level of the individual sound source. As such, it provides the same tangible point of reference of activities of individual sources that we have observed for positional locations of sources.

The attributes of the host environment become part of the 'sound' of the instrument/voice; the two fuse into a single percept. They are entwined, as the sound of a space draws out the qualities of an instrument or voice within it, just as a sound source excites the qualities of its host environment. Environment qualities contribute to and shape the sound's content and character, and add other dimensionality to its spatial identity. The host environment also reinforces the source's performance, situates the source and performance in a physical setting, and can influence stream segregation's delineation or blending of sources and the materials. Host environments, and their environmental characteristics, have equal potential to shape the track, along with all other elements.

In surround sound, the environment can appear disassociated from the sound source (direct sound), appearing as a source in itself. Mirroring the original sound to some degree, such an environment is separated spatially from the source, in a clearly different position. The tendency of the source to bond with its space makes this segregation very difficult to accomplish in stereo. This separation is often found in surround sound, though, and it can take many forms (Moylan 2017, 44–49). For instance, in surround sound sources may often be located in front of the listener with their host environments behind, providing some dispersion qualities similar to the way the sound of live performers might fill a space (the *LOVE* version of "Strawberry Fields Forever" is an example of this separation); localization of sources (in lateral location and size and distance positioning and depth) and their host environments (either bonded or separated) is complicated in surround and carries distinct differences from stereo (see Moylan 2012, 2015, 2017).

The hi-hat sound in two contrasting versions of "Let It Be" allows us to recognize the unique qualities of their host environments; as they are rather isolated in the second verse (beginning at 0:52), this contrast is starkly apparent. Diffusion is important to the host environment of the source in both versions. Diffusion of a reverb is the lateral dispersion of the reflections and reverberation (in playback) and the density of those reflections.

George Martin's version, remastered for the *1* (2000) release, provides the hi-hat with a host environment with sparse reverberation diffused over a significant area. The reverberant sound is quite uniform in the amplitude of its reflections and is only subtly different in spectrum from the direct sound. The pre-delay is very short (in the range of 3–4 ms) with the reverberation established quickly; the content of the host environment does not mask or disguise the subsequent attacks of the hi-hat. This host environment provides some significant sonic interest, as the instrument's reverberation disperses to one side of the sound stage away from where the attack of the sound occurred. The hi-hat is located in the area between 17–25° right of center. After the attack, the reverberation disperses to the left, ending at 25° left of center; it takes place in less than a half-beat for the sound's reverb to travel across the sound stage, and the reverb tail is a bit less than the duration of one beat. The reverberation is increasingly sparse as it spreads across the sound stage; there is an impression that something causes the reverberation to stop at that point in space, perhaps interpreted as a reflective wall on the left that is the source of the subtle echo that is established as the reverberation stops spreading.

The host environment of Phil Spector's hi-hat has width of approximately 19°, extending from 7° left of center to about 12° right. The direct sound of the hi-hat image occupies the space from 3° left of center to 3° right; this direct sound width is confirmed by the closed hi-hat strikes in the chorus. The sound pulsates within that environment and the echo iterations spread the image around the direct sound to fill the area. The pronounced tape echo on Ringo's hi-hat generates a prominent 16th note pre-delay that provides the impression of a large space with highly reflective surfaces (and that also reinforces the rhythm of the music); the echoes continue at the 16th note (with subtle 32nd note echoes appearing over time) as the length of the string of echoes coalesces into an aural image of reverberation. As with Martin's, this host environment has a different reverberation density in various positions. A large cavernous space is implied by all these reflections, and the high frequency area is attenuated as the iterations progress and the low and low-mid range frequency area gradually builds in amplitude, supporting a sense of distant reflective surfaces. As the verse progresses, the attacks of the hi-hat begin to blend into the host environment's reverberation, the instrument begins to mask itself (reminiscent of Spector's wall of sound techniques). Spector's host environment is comprised of rhythmic echoes that become reverberant and become perceived as reverberation; this is explored in detail within the next section.

The host environments of both versions provide the hi-hat with a sense of space and place, and also an impression that an area is spanned with a changing density and varying timbral content. Spector's environment has more unusual and pronounced qualities; Martin's space is surreal in other ways, though, with its sparse density and motion of dispersion. The character of both spaces has some commonality in their sizeable spaces (though Spector's is slightly larger) and sense of motion. Here we can witness the host environment contributing to, and supporting, the track's sentiment and lyric content — though the sound source it contains is presenting a simple back beat.

Echo and Reverberation

Echo and reverberation are reflected sound; each can manifest in various forms. Most of their forms carry a content relationship and/or association with a host environment, though some of their appearances are independent. While they may be conceived as incomplete forms of environments, echo and reverb are acoustic properties of their own. Echo is the result of an extended early time field;[16] it results when a sound is distinctly reproduced as it bounces off some distant surface. The time between direct sound and reflections is sufficient for the reflection(s) to be heard as a copy of the original; echo may appear as a single repetition or can multiply, depending on how many times a sound bounces. In contrast, reverberation occurs when the sound is reflected so many times, and in very close time successions, that all reflections fuse into a single sonic impression. As echoes' discrete reflections move closer in time (time intervals smaller than 100 ms) they cease to be perceived as separate sounds; as this time interval shortens, discrete reflections ultimately blend into a diffuse sound, what we experience as reverberation (or reverb). As such, "echo is perceived as the repetition of an original sound event, while reverberation is perceived as a prolongation of that sound event" (Lacasse 2000b).

The use of these spatial techniques in records began by recording and manipulating acoustic rooms and chambers. Artificial reverberation techniques were soon developed to generate the many reflections—by passing the sound through springs and through metal plates—and ultimately to also control the dynamics, reflection speed and density, and the spectrum of the reverb. Reverberation is an integral attribute of environments; reverb alone (as it commonly appears in tracks) is without the early time field reflections found in real-life spaces. Echo was created artificially through various tape delay techniques, until the digital delays of the early 1980s allowed regenerating sounds to create any echo delay.

Echoes occur naturally in certain real-world environments, as repetitions of sound are generated as the original is reflected. Echoes are generated within very large spaces, containing walls, ceilings, floors and other objects or boundaries of hard surfaces, located at considerable distances apart and from the listener.[17] They can also be produced in open air, with sound reflected off distant surfaces and objects before returning to the source location as a repetition or a partial repetition of the original. While in records echoes can appear with qualities nature could never produce, echoes all evoke a sense of extraordinary space. R. Murray Schafer (1977, 218) notes: "[E]cho suggests a still deeper mystery . . . every reflection implies a doubling of the sound by its own ghost, hidden on the other side of the reflecting surface." Echoes can elicit images, interpretations, meanings and representations of many sorts from within its unique timbral and time qualities.

Further, echoes, and especially a string of echoes, can establish the illusion of a peculiar environment. Echo can provide the experience of extraordinarily large spaces, the outdoors, etc. At shorter durations, a single or small number of echoes represents an incomplete space; certain qualities of real, enclosed spaces are missing. This is neither an issue for the listener nor for the track, it is an assessment of the potential content of reiterated sound. The variables of echoes are the time difference between the original sound and its repetition(s), dynamic relationship between the original sound and each repetition, number of repetitions, dynamic shape of repetitions, and more. Reiterated sounds of echoes may be altered in frequency content, given dynamic shape, and other characteristics to provide greater alignment with the attributes of real-life spaces.

Echoes may appear in other forms, notably (1) echoes bonded to a single source (as in the above hi-hat example from "Let It Be") and (2) as a separate source. As connected to the original sound, it is bonded with and refers back to its source. The sound source is reiterated, fully or partially. ADT (artificial double tracking) is an example of the echo bonded with its source for timbral effects as well as spaciousness; "I'm Only Sleeping," "Eleanor Rigby" and other tracks off *Revolver* (1967) are clear examples of the Beatles' first uses of this technique. Bonded with the source, echo is used as an effect in many ways, such as the tape echoes appearing at the end refrains within the Beatles' "Paperback Writer" (1966). As echoes are temporal, they can become part of the rhythmic fabric of the track, bridging the recording and music domains. Echoes can be tuned precisely by recordists, to integrate with the track's rhythmic elements, or those of sources. Echo is used as part of the musical and rhythmic textures in countless records of the 1980s; two differing examples are David Bowie's "Let's Dance" (1983) and "Don't Come Around Here No More" (1985) by Tom Petty.

An echo may be a separate entity, representing a second sound source that is a duplicate of the original but "separated from the sound in space and time," with "a presence of its own" (Zak 2001, 77). It is often a single echo. An organ chord inside the left speaker location and its echo inside the right speaker location can be heard distinctly in the introduction through first verse of Bob Dylan's "Love Sick" (1997);

the echo ceases as the texture thickens, and the organ retains its original position without the motion of echo. Elvis Presley's "Blue Moon" (1954) provides another example, as "the voice's echo casts a languid shadow clearly audible in the spaces between phrases. Here, the echo takes on a ghostly character that, again, enhances the track's overall stylistic effect as it reflects the dreamy character of Presley's performance" (Zak 2010, 317). An echo's form will have some defining quality that distinguishes it from the source—perhaps timbre, or position, or some other characteristic.

Reverberation is one of the potential attributes of natural environments. It is comprised of innumerable reiterations of the original sound. Among its variables are: duration of the reverb, time density created by the spacing of reflections, dynamic contour of the reflections in aggregate, dynamic relationship between the original sound and the reverb, frequency content of the reverb, and others. Reverb is also an incomplete representation of the sonic character of space, just as is echo. Reverb in itself does not contain the early time field qualities of spaces, though it has often been used successfully to represent environments—environments of unnatural qualities. Reverberation can fuse with the original sound, or it (just as echo) can have a presence in the track that is separate from the original sound. Simple reverb, generated solely from reiterations of the original sound was common in early popular music recordings. From the earliest records, reverb devices (and processes) carried attributes not found in natural reverb. Since the beginning of popular music records, reverb devices were used to create unnatural spaces, spaces that in some way were original and complemented the sound and artistic intentions of the sound sources.[18]

Reverberation can be used for dramatic effect, to support "the message of a song," and as a metaphor (Lacasse 2000a, 179–180). It can function to bind and provide tension, as in the opening measures of the Beatles' "Nowhere Man" (1965). Appropriate observations of host environment attributes, values and applications can be relevant to reverb when it is encountered. Environments (containing an early time field) and reverberation (that does not) can be difficult for a listener to distinguish under many contexts commonly found in tracks.

Echo and reverberation can emulate or simulate spaces in themselves; for listeners, they provide an acceptable and believable environment. Each alone does not contain the characteristics of an acoustic space. Some qualities of naturally occurring host environments will be missing, though this does not detract from it simulating space. This does not hinder their representation of space, or of a sound source seemingly inhabiting their sense of place. Enough is present within these qualities to allow the listener to situate a sound source—or themselves—in that environment.

Holistic Environment of the Track

The holistic environment is the space in which the track, in its entirety, resides. It is the environment of the sound stage; it contains all sound sources fused with their host environments. The holistic environment provides a sense of place for the track.

A track projects an aural image (a sonic imprint) of the space within which it is located. This is the place where the illusory 'performance' of the track appears to occur, perhaps where a listener might 'hear' or 'conceive' the track as existing. This is the holistic environment; a singular space that binds all of the sounds of the track into one place (no matter its size). This allows for the listener to experience the space of a performance within their living room (or other listening space), or conversely transporting the listener into a studio or concert venue—as examples of innumerable possible experiences. It also presents other significant attributes and connections of space and place; it establishes an aesthetic context for the track, and provides its character (including its ambience and spaciousness). The holistic environment contributes substantively to the unique world of the track.

Elsewhere I have often referred to this space as a 'perceived performance environment'—identifying it as the environment in which 'the performance that is the track' takes place, and that is intended to be perceived by the listener.[19] 'Perceived performance environment' is a term inherently directed toward understanding and crafting production processes; within recording analysis this 'space of the track' might be more appropriately identified as 'holistic environment.' The term 'holistic environment' is intended to reflect the all-inclusive nature of that environment, and also that it represents a sense of size, contains attributes of content that together are more than the sum of its parts, and projects a complex character that can pull the listener out of the track to external connections. This overall environment manifests 'holistically'; while it represents a whole with interdependent parts, its character and content represents a complex percept that is an amalgamation of its parts into something greater.[20] Of course, the reader may choose between the two terms, as they are synonymous.

The holistic environment is a conceptual place or space within which the record exists. It can be conceptualized as the environment of the sound stage.

The overall environment can appear to shift between different structural divisions of a track, when various sections exhibit unique qualities. As those sections progress, they crystallize into a higher-order, global impression of the track as a whole (the holistic environment). Thus, sections might have an overall character and content, but contribute to a larger whole. It is how those sections interact to shape the track that bring the dominant features to ultimately find balance that defines the track's holistic environment.

A hierarchy of environment relationships is established in popular music records. It is the result of the track's holistic environment and the potential for a multitude of host environments containing individual sound sources, or groups of sound sources; these will be simultaneously present at respective levels of perspective: the overall sound, the basic-level of individual sources, the composite texture, and perhaps others.

Multiple Spaces and a Hierarchy of Environments

The holistic environment is situated at the highest level of perspective of the overall sound. As discovered for other elements, a composite texture is present between the highest level of perspective and the basic-level of individual sound sources. At this composite level, the bonded host environments and sound sources ('sources+spaces') establish relationships and interrelationships with others.

Host environments may blend and create relationships between sources; multiple sound sources might be situated within a single host environment, though at various locations (of lateral and distance positions). Host environments may also establish relationships and interrelationships with others—for instance, to support connections or delineation between sources or materials, perhaps to mark similarities with other spaces, contrast with other environments, and more.

Table 8.8   Potential levels of pyramiding relations of environments in tracks.

Perspective	Environment or Relationships
Overall Texture	Holistic environment
Interrelationships of Overall Texture and Composite Texture	Space within space
Composite Texture	Sound stage positions and interrelationships of host environments (sources+ spaces)
Groups if sources sharing a host environment	Shared host environment
Basic-level: individual sound sources	Host environments

Any number of source host environments (sources+spaces) may occur simultaneously, and coexist within the same sound stage. All host environments (with their defining characteristics of width and depth, and source distance locations) are conceptually bound by the spatial impression of the holistic environment. As we recognize the sources plus their environments situated on the sound stage, we might begin to recognize spaces related to other spaces—just as the sources they host establish relationships, these spaces may have relationships to other spaces. On the sound stage sources+spaces are situated as parallel presences with other sources+spaces. Multiple environments are experienced simultaneously, just as multiple musical ideas and sound sources stream the materials of the track; in this way "the listener is aware of different types of space which cannot be resolved into a single setting" (Smalley 1997, 124). These simultaneous environments each carry their unique traits, supporting or enhancing their source and aesthetic ideas. The interrelationships of sources+spaces appear simultaneously with all of the other interactions of sources. A collage of host environments overlapping, sharing area, delineating, separating, etc. can readily established. A sense of function between host environments (sources+spaces) might emerge; as examples, a host environment might enhance its source to support the manner in which the source functions to cause tension and motion, other host environments might contribute to a sense of resolution or arrival, a shared environment might establish a context and character for groups of sources, and of course there are innumerable other possibilities.

Further, each host environment (being situated on the sound stage) is contained within the environment of the sound stage (the holistic environment). This creates an illusion of a *space* existing *within* another *space* (Moylan 2015, 209). The holistic environment contains all host environments; the host environments are nested within the holistic environment. This situation establishes relationships of host environments to the track's overall space (Blaukopf 1971, 162).

This places the individual sound sources with their individual environments within the overall, perceived performance environment of the recording; this brings many spaces (sources+spaces) to 'exist' within the track's holistic environment. Space within space and the hierarchy of environments allows for some unusual percepts—unusual, in that they defy what is possible in the physical world. Among these are:

- Simultaneous spaces, host environments existing side-by-side on the sound stage, with attributes distinguishing different types of spaces
- Overlapping spaces, where two spaces occupy the same area at their edges
- Embedded spaces, host environment emanating from within another physical space (perhaps the holistic environment)
- Multiple sound sources positioned at various distances within the same host environment (their common performance space)

Somehow the containment of all the track's sources+spaces within its single overall environment is believable to listeners; one overarching place for the track can contain the many spaces (or worlds) from which diverse sources deliver and contribute their voice. Large spaces have no trouble fitting within small spaces in the world of a track. Again we find spatial qualities and relationships that do not align with life experiences, but that are readily accepted by the listener. Just as we allow ourselves to accept and believe the impossible within the invented worlds, storylines and actions of motion pictures, we accept the created and crafted worlds of records as the reality of a track. Each track re-invents physics and rewrites the laws of acoustics; any set of spatial qualities, dimensions and relationships can be found within the platform for track. The holistic environment often contains attributes of an environment that is significantly smaller than the spaces it contains (let alone the aggregate of all of those host environments). Individual host environments of sources may seem of enormous size, but fit into a tiny holistic space; this is a convention of space-within-space relationships.

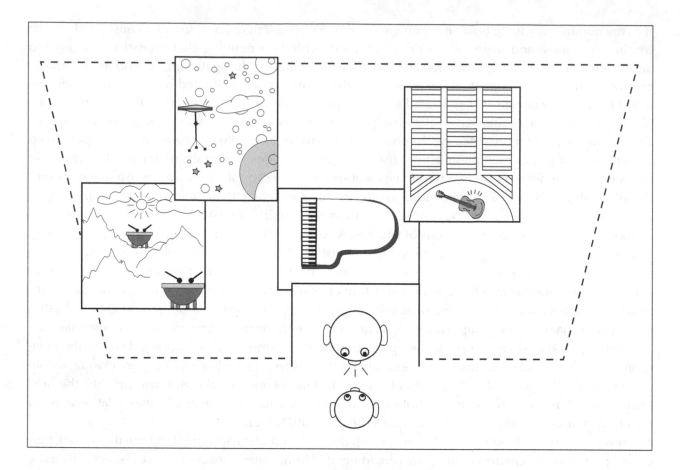

HOLISTIC ENVIRONMENT

Figure 8.10 Host Environments (sources and their spaces) on a sound stage; simultaneous spaces, overlapping spaces, spaces within space (host environments embedded within the holistic environment).

There is one additional level of environment, though this one is external to the track.

The listening environment and the listening process can enter the spatial hierarchy outside the track itself. The track is 'heard' by the listener from their own point of audition, and with their own sound system and listening sensibilities. They hear the track from within their own listening space. This space may add a layer to the experience, where "A listener might . . . apprehend a recording and experience a sense of a physical space, other than the one he was actually occupying" (Doyle 2005, 57). The sense of one physical space for the track and the listener in another may enter one's perception. This separation might result in some detachment, perceiving the track at some point removed from the listener—a slightly removed position to one of significant separation. The environment produced by the playback system—or of headphones or earbuds—may become integral to the perception of the track, and bring attention to the listener's point of audition. To a great extent, the impact of this playback environment will be heard similarly in other tracks played at similar loudness levels. A carefully located listening position at the peak of an equilateral triangle with the two loudspeakers placed to minimize room sound, with a balanced playback system will minimize the sound of their room and system, and may well make it negligible.

The Duality of Content and Character of Spaces

The contrast of content and character returns here; the attributes of spatial environments have physical properties (content) that establish a quality of character. Like timbre, environments can be conceived in two broad categories:

- a collection of measurable, physical, component parts or attributes that combine into a single, identifiable percept
- an overall impression (that at times acknowledges inner details) that results from the perception and interpretation of those physical attributes

Like timbre, environments can be approached to observe their content, or to observe their character. Indeed, to an extent, spaces may be conceived as timbres; the acoustic content of environments (spaces) combine in a gestalt, in conceptually the same way as the physical components of timbres. This might be conceived as a 'timbre of the environment.' In full, environments have a timbral quality comprised of component parts, plus integral spatial attributes, described later.

Environments are a collection of numerous physical components—with time, frequency, amplitude, and timbre dimensions—that represent its content. Environments are comprised of a complex, multidimensional set of sonic attributes, a collection of internal physical properties. Spaces can be largely observed through measureable parts—as just described. While these physical dimensions are scientifically measurable, they challenge aural analysis in many of the same ways of timbre analysis, but also in new ways. Spaces contain timbral information, but also discrete reflections and reverberation, and the time and dynamics matters they bring. Records contain spaces of various content; some of these spaces have a full set of attributes, others only one or a few. To observe environments, attention is focused inside to its component parts, to observe their attributes:

> Focusing on space . . . requires a reorientation of listening priorities and attentions: in my experience we are not that used to listening out for spatial attributes . . . because there is so much else to listen out for. . . . or that we think it tangential rather than central.
>
> (Smalley 2007, 35)

Environments can be approached as sound objects to evaluate content, just as we approached timbre—pulled out of context to unravel its subtle content.

We recognize spaces, and the character of spaces; environments carry qualities of ambience, spaciousness, size, and bear significance, connotations, associations, and more.

Environments can represent an aural image—as a singular character, or an overall characteristic. Spaces form an overall impression; environments are fused with the sound source and are also situated in an identifiable place of origin. Space assimilates into the source's timbre, and contributes to any sonic representation of that source; we can experience space as links to places external to the track, and outside the listener. This conception of spatial environments often leads one toward descriptions of those overall qualities; descriptions of character and characteristics often result. The sonic representation is part of the sound's spatial identity. Environments can often communicate conventional associations, and stand as symbols resonating outside the track. Environments "made it seem as though the music was coming from somewhere—from inside an enclosed architectural or natural space or 'out of' a specific geographical location—and this 'somewhere' was often semiotically highly volatile" (Doyle 2005, 5).

The character of the overall quality that emerges from the perception and interpretation of the physical attributes establishes spaces and places that can produce subjective effects, can signify and represent, and can communicate to the listener.

The Sonic Content of Environments

Sounds in natural spaces interact with those spaces. The sound that results contains the timbral attributes of original sound source fused with a transformation of the sound brought about by the sonic imprint of the space. Rooms react to sound energy produced within; the results are a frequency response of the room, timings of reflections of the sound produced within, and dynamic contours of the room frequency response and the reflections. The interaction of the sound source and the environment produces a new, unique sound—one that has been transformed in timbre and provided with additional spatial properties. Throughout this explanation of the components of environments, one will recognize how timbre works with spatial properties (and loudness and time) to provide spaces with a unique profile. Engaging the sound of environments in deep listening, our attention is brought within the sound to hear these diverse and subtle dimensions.

The sound of a natural environment can be divided into three parts: direct sound, early reflections and reverberation. Environments within tracks may contain all of these parts and their attributes, or fewer—as they are often created, their content cannot be assumed. These three components (or whichever may be present) can be observed to identify the content of spaces.

Direct Sound and Pre-Delay

Direct sound travels on the shortest distance between the source and the listener, and is therefore the first thing the listener hears. Direct sound delivers the information of the signal (sound source) to the

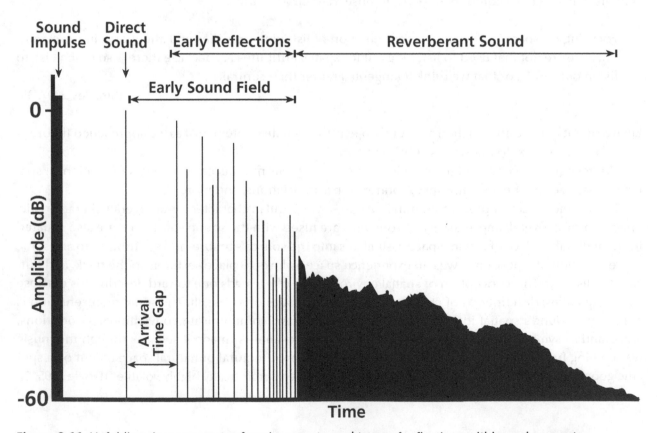

Figure 8.11 Unfolding time segments of environments, and types of reflections within environments.

listener in an uncontaminated form; it is unchanged by the environment, as it has yet to interact with it. A high proportion of direct sound provides clarity of the signal, and shares the attributes of sound in free space, because it has yet to interact with any boundaries. "Sound emanates from the source radially in all directions" (Everest & Pohlmann 2015, 97). In a free field, a source's direct sound moves past the listener, never to return. With the same source in a room, the direct sound moves past the listener once in a direct path, then strikes a room boundary. As the source radiates in many (if not all) directions, these sounds reflect off one or many surfaces (walls, floor, ceiling, etc.) before arriving at the listener. The first of these reflections will arrive at the listener very shortly after the direct sound; these are early reflections.

The arrival time gap (also called pre-delay) is the time that separates the arrival of the direct sound and its first reflection. This time-length communicates important information about the size and dimensions of the space; it is determined by the distance the direct sound travels from the source to the reflective surface nearest the listener and then on to the listener. The time units of this distance play significant roles in our sense of the size of the space, and we have 'learned' the sounds of these time units, as blended into the overall quality of the source. For example, if conditions were right to allow sound to travel at 1000 feet per second (a rough approximation of what one might expect) the perception of a 50 ms delay places the length of distance the sound traveled at (very) roughly 50 feet (source to listener, inclusive of reflection path). Pre-delay alone is capable of adding depth to a sound source, simulating an environment.

Early Reflections and Early Time Field

Early reflections arrive at the listener within a window of time up to about 80 ms after the direct sound; another set of discrete reflections may arrive after this initial 35 to 80 ms, depending on the characteristics of the environment. The individual reflections will each have a potentially unique amplitude that is different from other reflections and that of the direct sound. The amplitude levels of early reflections are the result of distance travelled and the type of reflective surfaces; surfaces absorb some of the sound energy (varying by type of materials) and diminish the intensity of the reflection. The amount of energy that is removed when a sound strikes a surface material is its absorption coefficient; absorption coefficient varies with frequency. Early reflections also arrive from different directions (potentially from all directions); these different angles of arrival provide critical information on the location of the source within the space, the location of the listener within the space and also the size and geometry of the room.

Patterns of reflections are generated by spaces, providing a significant quality of its sonic imprint; these patterns combine spacing in time and of amplitude levels; they are 'rhythms of reflections.' Through this the early time field provides significant cues on the geometry of rooms, the source's location within the room, and much more. The early time field exists in micro-timing, and often in rhythmic patterns and dynamic patterns; its short durations are less than 100 ms and as short as approximately 2 ms. Early time field reflections can simulate spaces (host environments) in themselves. Alone the reflections of an early time field can establish an illusion of a small space, or of unnatural spaces. Figure 8.12 illustrates the simplest reflection paths between the source and the listener; sound will continue to reflect around the space in more complex trajectories, establishing reverberant energy.

To summarize, early reflections bring the following traits; some will also appear within reverberation:

- Time delays from the arrival of the direct sound
- Amplitudes of reflections will differ from one another
- Frequency content of reflections is altered, from absorption by reflective materials
- Reflections arrive from different directions
- Patterns of reflections are established by repeated time delays and perhaps amplitude levels

Reverberation and Frequency Response

As sound continues to reflect in the enclosed space, it is reflected many times and arrives at the listener from all directions. The early time field dissolves into reverberation; the time this takes is a function of the size of the room, with smaller rooms taking a shorter amount of time to produce reverberation. Reverberant sound is a composite of the many reflections (from different reflection paths) arriving at the listener in close succession. The many reflections that comprise the reverberant sound are spaced very close in time to other reflections, and fuse into a single sonic impression. These many reflections are therefore perceived as a single entity. The time it takes for reverberant sound to die away is reverberation time, and is dependent upon both the size of the space and reflective materials.

Reverberation can be considered in three stages. The (1) initial portion of reverberant sound is the rate at which sound builds, and is determined by the time between reflections and absorption by materials and air; as more reflections and stronger reflections are generated by the space, the loudness level of the reverberation increases. This build-up will reach a (2) sustaining or steady-state level for the reverberant sound; this constant level is a balance between the amount of sound being produced in the environment and the amount being absorbed. This brings rooms with little absorption to have a higher steady-state level than rooms with more absorptive surfaces or longer distances between surfaces. Reverberant sound continues and decays after sound within the space stops. (3) Reverberation time is the rate (speed) of this decay; this is an important behavior that strongly reflects the character and content of a room. This rate is determined by the amount of reverberant energy continuing in the space, and is a product of the amount of absorption and the distance between reflective surfaces. Each time sound reflects in a room it loses some energy, absorbed during travel between surfaces and by contact with surfaces; between reflective surfaces, higher frequencies are absorbed faster than low frequencies. Surface materials absorb frequencies in different and unique proportions; some absorbing low frequencies more than high, etc. Longer reverberation times tend to be generated by larger spaces; sounds produced in spaces containing highly reflective materials of little absorptive capacity will take longer to decay.

With this we understand that reverberation will have different frequency content from the direct sound, brought about by loss of energy due to air absorption and also from uneven frequency absorption (and deflection) by reflective surfaces. Spaces typically have additional frequency content based on the geometry of reflective surfaces. Not all reflections travel in random directions; some reflections

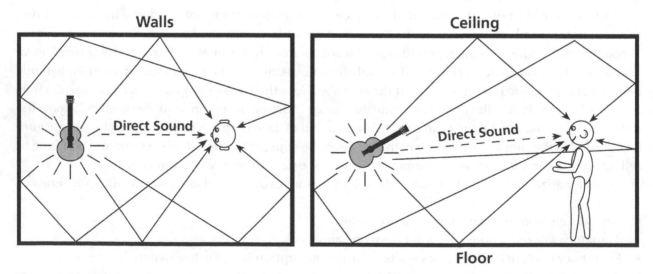

Figure 8.12 Simple paths of reflected sound within an enclosed space.

travel cyclically around a room, other reflections occur between opposing surfaces. These 'standing waves' are resonant modes that are spatially static; their pressure and velocity distributions stay within a certain subset of surfaces, are fixed in trajectory angle and wave length, and bring energy back to an original surface in a cyclic path. These paths exist for discrete frequencies that are determined by the geometry of the room, and bring significant variations of sound pressure levels around a room at those 'modal frequencies.' These are axial modes (between two surfaces), tangential modes (between four surfaces) and oblique modes (between all six surfaces). These modes might be conceived as 'formant frequencies' within the reverberant energy of an individual enclosed space.

The balance of the amount of direct sound to reverberant sound will shift with listener position relative to the sound source. In any space, there will be some distance between the listener and the source at which the reverberation will start to dominate. This distance (called 'critical distance') is highly variable depending on the size of the room and the materials of construction. This variability is precisely the reason reverberation (or more specifically ratio of direct to reverberant sound) is an ineffective measure for the perception distance location.

To summarize, reverberation contains the following traits. These traits can be extreme in created environments, and might appear in unnatural ways:

- Reverberation exhibits a dynamic shape that results from its build-up, sustain and decay; each stage has the potential for a dynamic profile, as each may contain changes in its dynamic level
- A density of reflections exists in reverberant sound; densities (number of reflections in a time unit) vary from one space to another
- Reverberant energy contains a different frequency spectrum (frequency response) from the direct sound, as a result of air and reflective surface absorption and of room modes
- Frequency response within reverberation is not static, and can exhibit considerable shifts throughout its duration; some frequency areas may get louder while others get softer, and at different speeds

The acoustic content of environments fuses together into a single percept. This overall quality of the environment is a gestalt; a complex overall experience that is different than the sum of these individual parts. These experiences—the inner workings of all environment attributes, or the overall quality of the environment—might be readily engaged by listeners in a general manner; for the goal of a recording analysis, heightened attention to details and relationships may reveal pertinent data. Denis Smalley (1997, 122) identified: "most listeners cannot easily appreciate space as an experience in itself. Spatial appreciation can be acquired by consciously listening to the spaces in works as distinct from regarding space as only . . . enhancement." This manner of assessment and appreciating environmental content is a skill one can develop; one that can also be extended to the character of environments.

Observing Acoustic Content of Environments

> We never hear sound without ambience [without its space], and it is therefore impossible to disassociate the first, immediate reflections in a space, especially a small room, from the timbre of the sound source itself—another feature of ecological perception.
>
> (Zagorski-Thomas 2014, 65)

Hearing the qualities of spaces requires directing attention to listening to subtle details of time, of activities in pitch/frequency areas, and and of dynamics—at a microscopic level—and also the resulting gestalt of the source+space sound. This is possible with some knowledge of what to listen for, and some guidance in how to hear it. The attributes of environments are subtle, and some are not perceivable as they are measured; in these instances, we perceive the *results* of the variable's appearance or activity rather than the variable itself. Here we will engage the process and challenges of hearing the attributes of environments.

In observing these attributes, we are concerned with identifying (ultimately observing and understanding) the room's sound separately from that of the sound source. We are attempting to understand the space in which the sound is resident—its sonic attributes, so that in the end we can examine how the source's host environment contributes to its aesthetics.

The sonic signatures of environments are present and perceived, but difficult to segregate from the timbre that is exciting it. Environment characteristics are most clearly heard after the source stops abruptly; at that time a certain percentage of its reverb is audible—how much depends entirely on the attributes of the unique space. Within the context of the track, sources do not always stop abruptly; it is likely that in any track most sounds will not stop abruptly, as many naturally sustain until they are muted or decay below audible levels. Should a source stop abruptly in a track, it is very common for reverb information to be concealed by other sound sources.

This brings the analyst to need to engage the sound of the environment within context. A knowledge of the sound source, and the listener's memory of that timbre are used to calculate 'what is the direct sound (and its qualities)?' and 'how has it been altered?'. This extrapolates (estimates, applies an educated guess) toward identifying the physical content of the environment. This process is part past experience, and is in part an acquired skill; some detail is impossible to hear, so one directs attention to the result. One's past experience of the timbre of the source allows one to recognize changes to spectrum (frequency response), changes to timbral detail (which is likely to diminish after the early time field); one's past experiences of spaces allows a sense of size based on timbre within the early time field, as well as the duration of the reverberation and the ratio of direct to reverberant sound once the reverb cycle has reached saturation. In essence, we are trying to hear (recognize, determine) the dimensionality of the space and the content of the space's timing and timbral signatures.

This process might extract the sound source from context—as a sound object—and observing the above qualities. Focusing on the perspective inside the sound, and out of the context of the track, allows isolation, and reduces distraction.

This is an appropriate time to remember that a significant percentage of environments encountered in tracks are artificial and/or incomplete. Observing the attributes of a host environment situated within a track, one often confronts contradictory percepts; further, many environments will not contain the full set or sequence of attributes found in natural environments. To observe content of room sounds in the detail they can be measured is impossible; it is unrealistic to attempt to identify many percepts with precision. To observe the basic traits of dominant content is possible, though, and can yield information significant to a track. Table 8.9 presents a typology table for the content of environments; it can be modified to complement individual environments, and the values it recognizes can be generalized when specifics cannot be identified (or when they are not needed for an analysis).

The content of the environment as an overall percept allows most direct and simplest access to observing an environment. This can be approached in three parts. First, one can approach the reverberation plus early time field as a single gesture—one with a dynamic contour, a duration, and a sense of density. This data would in itself be significant in defining the environment. Second is the frequency response (or spectrum) of the environment; remembering that the space reacts to sound energy that is produced within it, certain frequencies are emphasized and some attenuated. These can be identified related to pitch register, to obtain data on, for example, whether the environment boosts low frequency information and by how much (as in 'the Low pitch register is moderately emphasized'). Third is the dynamic relationship between the direct sound and the reverberant/

Table 8.9 Content typology table for observation of individual host environments; may also be applied (in whole or in part) to observing the content of holistic environments.

Component	Attribute	Value
REVERBERATION		
Reverb contour as gesture	Duration	Related to pulse, or to clock time; RT60 – duration (time) needed for reverb to diminish 60dB
	Density of reflections	Relative value (sparse – very dense)
	Loudness contour	Overall shape of reverberant sound
Reverb contour segments		
Rise or build-up	Duration	Length of time required for reverb to build to saturation level
	Density	Number of reflections; amount of density; shape of varying density
	Contour	Shape of loudness gesture over time
Saturation or steady state	Duration, Density, Contour (as listed in rise/build)	Duration of sustained saturation; amount of reflections; shape
Decay or reverb tail	Duration, Density, Contour (as listed in rise/build)	Duration of reverb decay after direct sound has stopped; number of reflections; loudness shape
Frequency response	(the spectrum of the space)	
Low register	Altered discrete frequencies or frequency bands	Identify frequency information accentuated or attenuated
	Loudness contours or levels	Relative to nominal level of environ.
Low-Mid register	Altered freq's or freq bands	accentuated/atten freq levels/bands
	Loudness contours or levels	Relative to nominal level of environ
Include all other registers as needed	Altered bands or discrete pitch/ frequencies;	Accentuated and/or attenuated frequencies or freq. bands;
	Loudness contours or levels	Relative to environ's nominal level
Spatial image of reverb	(connected to sound stage)	
(depth & width of the environment)	Sense of depth of environment	Density of reflections; lengthy timing of reflections
	Width of reverberation with some detachment from sources	Stereo image that supplies additional width to source, extending beyond the sound source stereo image
LOUDNESS LEVELS (of reverberation and of initial reflections)	Ratio or balance of direct sound to reflected sound	Loudness level of direct sound relative to levels of initial reflections and reverberation
EARLY REFLECTIONS (prior to reverberation)	Microtiming spacing of reflections	Echoes if spaced by more than 30ms (a wave path of 33 feet)
Arrival time gap	Time between direct sound & arrival of first reflection	Timed in milliseconds, or relative to context
Early reflections	Duration	Timbral content/quality
	Density of reflections	Relative value
	Loudness of aggregate grouping of reflections	Level relative to direct sound

environment sound; this is a relative proportion ('direct sound is slightly louder than the reverb'). The higher the proportion of reflected sound to direct sound, the more the level of distinguishability of the sound source diminishes—resulting in diminishing the level of timbral detail of the source (indication of distance as well as space). Time-related observations can be calculated against the track's metric pulse (for some environments, but certainly not all); durations of reverberation can be calculated against clock time as well. If necessary, general time comments such as 'short' or 'long' might be a starting point, and 'shorter-than' might begin a comparison with another environment—either within or outside the track.

Observations of greater detail might be accomplished, as might be appropriate for the analysis—or the skill level of the analyst. Explanations of these more detailed observations and processes follow.

Already noted above, the arrival of the first reflection provides important indication of the size of the environment. This time unit is very often smaller than the threshold of pitch fusion (50 ms) and typically within the threshold of order (3 to 25 ms).[21] Therefore, the reflection is not perceived individually, but rather as part of the timbre of the source. This percept manifests as an alteration of timbre brought about by time; we can learn to recognize this subtle alteration *as* timbre should we wish, as we already process this information to identify room size. A single short-duration reflection is commonly used to add substance to sound sources, particularly vocals. Further, this single reflection alone is enough to provide the impression of an environment size and character.

The arrival of multiple reflections in the early time field (ETF) has three important factors for observations. The first two are reflected in the overall timbre of the source, though their origins might be recognized as related to density and loudness. First (1), it can bring a sense of density, or an amount of activity within its very short time window; a relative density ('sparse' to 'very dense') can be noted as part of the observation. Second (2), the loudness levels of all reflections in the ETF can provide an impression of proportion of direct sound to reflected sound; this ratio can be observed on a general scale, to help define content. The ETF will be primarily recognized as a shift in the timbre of the sound source; this timbre shift might occupy a significant portion of the sound's onset. These reflections can provide further information on the size of the space, and the position of the listener and the positioning of the sound source within the space. The timbral alterations brought about by these ETF reflections are subtle, and difficult to define. Third (3), of significance here is the duration of the ETF, as it separates the arrival of the first reflection from the beginning of reverberation. Should it not be possible to calculate a duration of the ETF, one might acquire a sense of how quickly the reverberant energy begins, and provide a general comparison to a known room of similar size (for example, "the ETF appears to be similar to that of my church").

The three stages of reverberation (build-up, sustain and decay) might be identifiable, but may be obscured or simply not present in others. When present, each might be observed for their duration, density of reflections and dynamic relationship to the direct sound. The overall reverberation time is an important element of the environment, and can be calculated in clock time or against the pulse of the track. The frequency ranges emphasized and attenuated are apparent within the reverberant energy; some ranges will have different reverberation times than others. With practice one might recognize these when they are present (and prominent).

Summarizing all of this: the process of observing the content of spaces (environments) acknowledges that the attributes of environments only exist when they are excited by the sound of the source(s) performing within it. The source's different performed pitches and the variety of frequency content they contain reveal certain characteristics of environments slowly over time and many different soundings. Other data is more consistent over the source's frequency range.

The typology table of Table 8.9 lists the attributes with their value types of spaces. Hearing these broader attributes of spaces will be readily accessible to many readers; it can start with the broadest and most apparent qualities of reverberation, especially the residue after the direct sound has concluded.

This process can then progress to the most detailed and thorough observations, such as microtiming (millisecond delays heard as timbres of time increments) of the early reflections.

These have been consolidated into four stages, each containing several attributes:

1. Reverberation: (a) duration (RT60), (b) density of reflections, (c) loudness and density contour of the reverberation divided into three segments (Praxis Study 8.6)

2. Reverberation: (a) frequency response (frequency bands or discrete frequencies that are accentuated or attenuated by the environment), (b) loudness levels (the ratio or balance of the direct sound to reflected sound) (Praxis Study 8.7)

3. Early reflections: (a) arrival time gap (time before the arrival of the first reflection), (b) early time field (duration before beginning of reverb, density of reflections, loudness of ETF reflections in aggregate compared to direct sound) (Praxis Study 8.8)

4. Environments: (a) cues that at times establish a sense of depth to the source+space image, and (b) potential of reverberation sound to extend the width of an image with the reverberation heard as separated from and *around* the direct sound of the source (Praxis Study 8.9)

Praxis Study 8.6.	Hearing surface attributes of environments: reverberation duration, density and contour (pg. 522).
Praxis Study 8.7.	Listening inside environments for frequency response and ratio of direct to reverberant sound (pg. 524).
Praxis Study 8.8.	Hearing microtiming within the early time field and predelay; experiencing the timbre of time (pg. 525).
Praxis Study 8.9.	Observing the attributes of depth as created by environments and reverberation, and of the potential of added width to images (created by extending the environment around the image of the direct sound) (pg. 527).

As with timbre, the level of detail within observations can shift with the goals of the analysis; all attributes of an environment do not need to be revealed, indeed that would be great effort that would often produce little relevant data for many tracks (or sources). Selective observations will hold value when they reflect those components that provide the environment its uniqueness and that recognize its prominent traits.

Character of Environments

Spaces (including echo and reverberation) produce "powerful subjective effects in listeners" (Doyle 2005, 31). These effects are embedded in the character of environments, and linked to semiotics and interpretation. Relations exist between the sounds of spaces within tracks and "what those spaces signify to those producing and hearing the sounds in specific sociocultural contexts."[22] Engaging a study of spatial semiotics, or developing a vocabulary for the character of environments are well beyond the limitations of this writing, though some approach to describing character must be attempted here, in order to incorporate this important spatial trait into the recording analysis.

Character is inherently a description and an interpretation; just as encountered with timbre in Chapter 7. Character can be defined to address the affects and effects of, and the activities within, its content. As character can engage the subjectivities and quasi-subjectivity of what spaces elicit within individuals, describing character pulls the analyst into the equation—along with their biases, knowledge, experience, abilities and more. Spaces carry cultural meanings, associations, connotations, etc.; approaching descriptions from cultural norms allows the opportunity for communication with a wider audience. The potentials for individual interpretation and for differing meanings within social groups and subcultures, let alone between cultures, ensure any approach will be fraught with difficulties. Still, there is some notion of common ground—the sound itself.

Descriptions that emerge from the examination of an environment's physical content carry a common ground. The interpretation of the content may differ between individuals, but the sound itself with be largely the same (assuming similar playback variables). Describing the attributes of the sound provides a useful starting point, and can allow the description to unfold with relevant information. One can then extend observations to other areas. Just as with timbre, a description of an environment that provides a clear indication of the nature, context, content and associations (as appropriate) is likely to be most successful.

The term 'ambience' has been used to describe varying qualities of the tracks, and sometimes used synonymously with reverberation.[23] Ambience is the sonic presence of a place, and as such is interpreted. In some areas of production (such as classical music recording, television productions and films), ambience is room sound (often called 'room tone'); it is the sound of the space where a recording is being made. 'Ambience' can describe the sound of the recording space itself—alone and void of any sonic activity. This concept can be transferred to spaces—real or virtual—within the track. The sound of the room, separate from the sound of the source sounding within the room, may bring environments to be sound objects that can be frozen in time for examination out of context as content; environments can also be examined as events changing in real time and observed in context of the track as character. In this writing, 'ambience' is used to denote an affective response to the overall character of an environment, within a broad cultural context (in as much as it might be possible). This is an extension of the common use of 'ambience' to represent the mood, feeling, aura or atmosphere of a place.

Our approach to the character of environments will (1) seek to recognize their overall qualities and to describe them relative to the context of the track; this might include addressing some portion of the physical attributes of a space. Character descriptions will also be (2) relative to how an environment might be situated culturally—and perhaps how the listener interprets them.

Qualities that elicit impressions of character are innumerable, including the most ethereal described under ambience. Other qualities might directly engage the space itself, often involving past experiences to interpret the physical dimensions one perceives. A listener may compare the environment they are hearing with what they have previously experienced (and remember); they make connections and comparisons, perhaps associations. A comparison might engage the size of the heard space ('this space is larger than space A and similar in size to space D') or the types of spaces they have experienced (lecture halls, hallways, concert halls, rooms, etc.).

Spaciousness is a spatial impression of an environment. It describes the sense of open space around a sound source, a sense of expanse. Spaciousness is an impression of the size of a room, and also a sense of the amount and type of reflection within the space (Rumsey 2001, 38). Rooms with longer times between direct sound and early reflections, especially those arriving from the sides, have a greater sense of openness or spaciousness. The sense of spaciousness is commonly a dominant characteristic, giving the host environment a sense of depth extending behind the location of the sound source and a sense of breadth or width to the environment, as it can extend to either side of the lateral location of the source image.

Observing the Character of Environments

The process of observing environment character pulls together observations of the physical content of the space with what those dimensions elicit in the listener. Interpretations, as we have noted, are potential grounds for overly personalized observations; here, again, it is encouraged that cultural norms be emphasized in observations. Personalized observations, if used, are best identified as such, so as not to imply they are those the reader might share.

Environment character situates and colors the performance within a space, and we recognize its impact on the source within the context of the track. Character also embraces outside matters that may arise—it has rhetorical forms (just as does timbre): semiotics, extramusical connotations, connections to real spaces within one's experiences, and more. The size of a space carries connotation and meaning, as well as sonic properties. The affects of spaces set a scene of the track, or of each of its sources—ambience, mood, aura, level of tension, and more. The character (and content) of environments can generate new meanings for musical materials and sources. Brøvig-Hanssen & Danielsen (2016, 134) observe that

> the track's variety of digital spatialities might affect the listener's interpretation of the musical meaning of "Get Out of My House." [Kate Bush, 1982] Its many spatial environments both support the meanings already communicated by other musical aspects of the track and generate new meanings of their own.

Placing a sound within a space provides it with added meaning (enhance or support the meaning that is present), and the space itself can transform a source and its materials into something different. Spaces can minimize and can expand a source, or they can dominate a source. Still, the sound of a space does not exist without the source that sounds within it, and the two have a significant bond between them.

Table 8.10 presents a general typology of attributes (with related values) of the character of an environment. This is presented as a point of departure. Certainly some tracks will contain spaces that

Table 8.10 Potential typology attributes (with values) of the character of individual host environments; may be adapted for observing character of holistic environments.

Variable or Attribute	Values or Characteristics
Size of environment, space	*Descriptors of size*: large, medium, tiny, immense, intimate
Spaciousness	*Amount of space around the source (all sides)*: open and airy, source occupies the space
Recognizable type of space	Bedroom, cave, closet, chamber
Comparison to known spaces	Like a bathroom because . . . Spatial presence of a tunnel
Realism	*Realistic bond or conceptual connection*: association between a known acoustic space and the track's virtual space
Surrealism	*Qualities outside reality*: pattern of echo, dynamic contour of reverberation, changing impression of size, *etc.*
Spectral qualities	*Examples*: Emphasis of high frequencies; attenuated low register
Source clarity	*Degree to which room sound (reverberation) is masking timbral detail*
Reverb decay time	*Examples*: compared to rhythmic pulse: 'three beats'; relative to another space: 'like a cathedral'
Direct to reverberant sound	Prominence of source within its environment *Amount the direct sound dominates*: slightly, significantly, moderately *Amount the reverberant sound dominates*: substantially, about balanced
Defining traits	*Examples*: Hard surfaces; slap back echo;
Impact on source or track	*Description of energy, motion, character*: suspended motion, relaxed energy, blurred texture, widened image, added depth, *etc.*
Ambience (affect)	Mood, feeling, atmosphere, aura, *etc.*
Meaning	Connotations, implied dramatic interpretation, cultural associations, external connections to specific spaces

require observation of additional attributes; those will be identified as data collection unfolds. This table does present an approach to organizing a description of character based on content and utilizing listener experience.

These all have an element of subjectivity stemming from interpretation, experience and exposure. Still, this typology could communicate a certain degree of objectivity around some commonality. Different analysts will engage these attributes in varying ways depending on skill and experience; the reader will interpret the analyst's prose through their own set of experiences and level of knowledge and listening skill levels. As an interpretation, character carries the inherent subjectivity of the analyst's biases; as with the character of timbre, the most useful interpretations minimize personal biases and emphasize those that are cultural, and that engage pertinent qualities of the track itself.

The aesthetic and affect-related descriptions of character include those that the physical content elicit, and the analyst's perception of the ambience's qualities as it is situated within the context of the track. Importantly, character is the sound of the environment as it contributes to the sound source's gestalt, as it contributes to the sound stage and the track, and more. In this way, describing the character of an environment is an evaluation; a step undertaken after observations of content, by further observing and evaluating the space within the context of the track.

Observing Holistic Environments

Observing the holistic environment may be directed toward an environment that has been applied to the track—accomplished by routing the mix through a reverb/environment simulation processor or plug-in. Should an applied environment be absent (which is most common), the holistic environment is approached 'holistically' (as a whole with interdependent parts), by observing the spaces of all sources, perhaps emphasizing those that provide the track with its strongest qualities. Observing the holistic environment involves both content and character. As in host environments, observing content will provide some degree of context toward interpreting a definition of its character.

Remembering a track's holistic environment is a single impression of space—the track's space—we recognize it contains all host environments. It encapsulates the many potentially disparate spatial impressions, which interact within the unique mix of the track, and it may also reflect them.

In records, a holistic environment is rarely a real space; it is also rarely even a single, fabricated virtual space. Rather, the holistic space of the track is the result of a blending of numerous factors, and is to varying degrees influenced by all sources and materials. The holistic environment is most likely to manifest its sonic impression with some of the following content attributes:

- Frequency levels or regions that are consistently accentuated or attenuated, thus altering the timbral balance of the track, along with the manner(s) in which these alterations unfold over time (dynamic contours)
- Reverberation time and density of reflections
- Ratio of levels comparing direct sound to reverberant sound
- Early time field arrival time gap and spacing of early reflections
- Dimensions of the sound stage (depth and lateral boundaries)

Attributes of the source host environments contribute to the impression of the holistic environment; for the majority of records the holistic environment is an overall impression generated by the content/character and the interactions of all host environments and their sources. There can be no formula to engage the process of identifying the quality of the holistic environment—each track is unique, and to some degree will generate its holistic environment in its own way. Some portion is an interpretation, an impression of character and surface features; some portion is the dominant or prominent or unique spatial attributes within the overall sound. The source host environments will not all contribute equally; the track itself will offer up what defines its holistic environment. The most significant or prominent sources may serve a dominant role, or they may be overshadowed by a unique environmental quality in a secondary source; the contextual layer of a track might establish a sonic space against which all other activity is calculated. Every track will be at least somewhat unique in how its holistic environment is formed—as well as its content and character.

The defining questions are: What is the environment in which this unique track is housed? What are its sonic attributes? To what extent is it an audible presence or a conceptual context?

The typology tables for content of environments and character of environments can be used to observe holistic environments. The attributes of each table will be somewhat unique for each track. The process of exploring the holistic environment to engage those tables' attributes will provide the analyst guidance.

In talking about the environment of the soundbox representing Prince's "Kiss" (1986), Brøvig-Hanssen & Danielsen (2013, 76) identify qualities and dimensions of the track's holistic environment, as having the width and resonance characteristics of a large hall, while the depth reflects a small, "'dry' or dampened space." Their assessment continues: "this single space comes forth as surreal . . . the hyper-presence and lack of depth imply a small space with almost no reverberation, but the high intensity and voluminous sound imply a larger, resonant one." Allan Moore (2001, 163) is discussing common traits of holistic environments of U2 when he identifies:

> Aside from the registral gaps at the openings to many songs . . . the great use of digital delay on the guitar (giving very immediate echoes) and the high degree of apparent reverb (the "atmospheric" background, wherein any textural holes are felt to be only temporary) seem to contribute greatly to this effect, as if the sound were bounding around in a great amphitheatre. This, combined with the use of long sustained organ chords/notes and sometimes high sustained guitar pitches, becomes most apparent on *The Unforgettable Fire*.

Both examples acknowledge and describe the content of this overall space of the track; note, they examine frequency content along with time delay and reverberation as attributes of the experience.

The character of a holistic environment represents the listener's (analyst's) impressions of the overall spatial identity of the track, as a global character trait. This impression is elicited through the experience of the track—an impression that can shift somewhat with re-hearings, further observation and evaluation. Like all thoughtful interpretations, the interpretation of the holistic environment is a work in progress, with new observations and discoveries building on the former. This process is on-going, and continual for the attentive listener.

CONCLUSION

Spatial properties provide the track with a sense of place and space, at multiple levels of perspective.

Each sound source in the track contains and projects a spatial identity. Spatial identities of sources can be defined by their lateral and distance positioning, and the attributes of its host environment, though this would be incomplete. The spatial identity of sources contains a semiotic level, a functioning role within the track's materials, and unique, multidimensional presence within the track. A dimensionality of the spatial identity is also generated by the character of its environment, and by how its distance, size and lateral position can elicit powerful subjective associations and effects within listeners.

The spatial identity of a sound source provides it with a place—a place that coexists with the spaces of every other sound source (overlapping, situated side-by-side, or separated by some variable space), as all sources+spaces are resident in the sound stage.

The track has a spatial identity as well. It is one of the dimensions of the track's crystallized form—which will be revisited in greater detail in Chapter 9.

Some Questions for Recording Analysis

Are you able to determine the edges of sound source images? Do you have a sense of some being wider than others?

Do the sources move rhythmically, or between sections?

Do any sources have images that change between structural divisions?

Are sources of similar size, or are there some that are significantly larger or smaller than others? What are these sources? Is there a correlation with the materials they present?

Are you able to distinguish distance placements relative to the listener's point of audition?

Do sound sources form groups, related by distance position?

In which zones are these groups located?

Do source distance positions within groups shift during a section or between sections?

Does the lead vocal shift proxemic zones between verse and chorus (or middle eight, or bridge)?

What proxemic zones are in use throughout the track? What are the nearest and furthest sources, their positions?

How does the use of proxemic zones shape the listener's connection to the track, to the lead vocal and to other sources?

What is the nature of the staging of sounds on the sound stage? The triangulated position of sources in distance and horizontal angel from the listener's point of audition?

What is the breadth of the sound stage—its left to right area—and its depth—closest to furthest source. What is the size of the area that is spanned?

Is there some relationship between the dimensions of the sound stage and the aesthetic of the music or lyric qualities of the text? Does this area shift in size and character? Do changes occur between sections?

To what degree are sources stable or in flux? How do shifting sound sources—sources that change lateral locations, width, or distance positions—affect the sound stage?

Do the boundaries of the stage shift?

Are there areas void of sources, or of activity? Are sources isolated, or coalesced into subgroups, or a single group?

Where is the lead vocal, and how does its position relate to the others?

What are the dominant content attributes of each host environment that contains the most significant sources of the track? Examining them in typology, are their recurring or dominant qualities?

Which source has the most pronounced host environment, and the least noticeable environment?

Which source has what appears to be the largest host environment, and which the smallest?

Does any source change host environments during the course of the track?

What sources carry an element of depths from their host environment?

> What size is the holistic environment of the track?
>
> What are its qualities?
>
> How does the holistic environment establish a context for the expression and concept of the track?
>
> Does the holistic environment complement the lead vocal, or present a space that contrasts with it?

NOTES

1 Electroacoustic art music also embraces space as a central aesthetic concern, see Smalley 2007.

2 This word play on 'point of view' transfers reference of sight and connotations of thought into hearing and sound.

3 Høier (2012) explores the conceptual difficulties the many perspectives of the "point of audition" present. Of interest to recording analysis are two spatial concepts. First, Kassabian (2008) explains 'point of audition' used in a spatial sense; it refers to 'the spatial position,' within the world of the film, that the audience will 'hear from'; or from what position it will hear the story and action of the film *The Cell*. Second, Michel Chion (1994, 91) notes that in film scenes listeners located throughout a room hear mostly the same qualities, while what they each see might vary considerably; "So it is not often possible to speak of a point of audition in the sense of a precise position in space, but rather a space of audition, or even a zone of audition."

4 Some might perceive surround as irrelevant, given its small listenership and the costs and efforts of establishing up an accurate surround playback system and environment. The reasons to be open to studying surround are simple, though that surround has the potential to profoundly enhance the listening experience, surround recordings continue to be made by some of the most talented recordists and acclaimed artists, delivery and storage technologies are improving to the point where streaming may soon be practical to the average consumer, and—not the least—the listening experience can be extraordinary in its sound qualities of the recording, dramatic impact of the lyrics, and musical expression. I regret space limitation prohibits coverage in this writing.

5 Brøvig-Hanssen and Danielsen (2013) explore this perception of the 'naturalized and the surreal' use of space in records.

6 These images are common and integral to popular music records. In classical music recording, though, convention is different; microphone techniques commonly "have the effect of adding 'air' or 'space' around sources so that they do not appear to emanate from a single point" (Rumsey 2001, 37).

7 Neuhoff (2004c, 92) cites several studies where listeners had better localization performance using point or orienting responses to sound location stimuli. The elimination of visual distraction by closing eyes is helpful to many, to aid in recognizing positions of phantom images.

8 The timbral qualities of sounds being performed at diverse distances can be experienced in audio files on the supporting website to the author's *Understanding and Crafting the Mix* (2015). Within these tracks you can also experience slight changes in reverberation between certain examples, and no loudness changes throughout (save those of the performer within lines).

9 See Blauert 1983, 45–46 and Handel 1993, 108.

10 I use these terms synonymously for our purposes in recording analysis. Robert Sommer, in *Personal Space* (1969) articulates a clear distinction between territory which is stationary (such as our home) and personal space as something we carry around with us; territory may be visible to others, but personal space is invisible. This is an important distinction in other contexts.

11 Some of this work actually resounds more loudly today, read between the lines. In particular, I read Hall's (1969, 129) observation: "[D]ealing with proxemic patterns for people of different cultures, . . . show[s] the great need for improved cultural understanding. Proxemic patterns point up in the sharp contrast some of the basic differences between people—differences which can be ignored only at great risk."

12 See also Dockwray & Moore (2010); Dockwray (2017); Moore & Dockwray (2008); Moore, Schmidt & Dockwray (2009).

13 This previous work includes several decades of observing timbre at various distances and contexts from recording projects and productions, listening deeply in many environments and in analyzing sounds in records, and in teaching listening for timbral content and detail pertaining to distance location.

14 The concept of image schema—so clearly articulated by Simon Zagorski-Thomas (2014, 9–14)—may hold some future promise for engaging timbral detail with some manner of organization.

15 Identified previously, the soundbox devised by Allan Moore (see Dockwray & Moore 2010) is a diagraming approach that has some similarity to sound staging; it is distinctly different, though, in that in addition to the stereo image and distance (which he identifies as "perceived proximity of aspects of the image to . . . a listener") are "the perceived frequency characteristics of sound-sources" (A.F. Moore 2012a, 31). A similar conceptualization of a three-dimensional space of tracks was independently devised by Anne Danielsen (1998) as *lydrom*, meaning 'sound room' in Norwegian.

16 The early time field is the duration between the arrival of direct sound and the onset of reverberation.

17 A delay of more than 0.1 seconds (100 ms) between the arrival of the original sound and the echo is required, this represents a distance of about 17.2 meters (56.43 feet) between the source and the reflecting object, for an echo to be perceived at the source (given the velocity of sound in air of low humidity with a temperature of 25° C). This time varies with a number of factors, and is somewhat indefinite. Some researchers set this threshold at 62 ms (1/16 sec), and some at 80 ms; 100 ms represents an absolute threshold beyond which an echo is clearly distinct.

18 See Brøvig-Hanssen and Danielsen 2016, Doyle 2005, and Zak 2010 for more in-depth information.

19 See Moylan 1986; 1992, 233; 2012, 164; 2015, 206.

20 The term 'holistic environment' emerged while considering 'crystallized form' as a Gestalt and a product of holistic hearing. The term 'holistic environment' appealed to me as an apt substitute for 'perceived performance environment' which was unwieldy and can be misleading (as not all tracks simulate performances, or elicit a sense of a 'performance' within the listener). Denis Smalley's 'holistic space' (Smalley 2007, 37) contrasts sharply with my definition of the 'holistic environment.' Smalley's 'holistic space' represents an analytical stance that brings together "an array of spatial forms into a unified spatial view" (*ibid.*, 55); his 'spatial forms' also include percepts of frequency, as applied to acousmatic art music. I see Smalley's 'holistic space' as having much in common to the notion of 'crystallized form' (especially with its disregard of temporal evolution) that is presented elsewhere here, while also including some of the qualities of 'holistic environment.' I will return to his concept in Chapter 9.

21 A 3 ms delay represents a distance of about 1 meter.

22 While this phrase is from Philip Tagg (2013, 145), appearing within his definition of the 'semiotics of music,' it is equally applicable in this context.

23 Albin Zak (2001, 76–85) uses the term 'ambience' in place of reverberation; his discussion does often include qualities of what are called 'host environments' here, though he does not differentiate between reverberation and environments with more complex content. In films, 'ambience' is the collection of background sounds (not room sounds) present at a location, such as wind, crowds, traffic and so on.

CHAPTER 9

Loudness, the Confluence of Domains and Deep Listening

This chapter explores three subject areas: the recording element of loudness, the confluence of all the track's contents and resulting timbre percepts, and a more thorough coverage of deep listening (a subject briefly introduced in Chapter 2). These are presented by dividing the chapter into four parts. Together these topics (1) conclude our exploration of recording elements, (2) situate recording elements within the confluence of all domains, (3) acknowledge timbre as a confluence, and (4) introduce the reader to new ways of listening to tracks.

In the first section, loudness as a recording element is the last of the recording elements to be explored. It is discussed last because loudness determines if and when all other elements are audible. It is also, perhaps, the most misunderstood of the elements. Loudness is examined in its appearances at all levels of perspective, ranging from the individual loudness contours of partials inside the spectral envelope up to the overall loudness contour of the track (program dynamic contour). Loudness balance, loudness levels and contours of sources, and source performance intensity (a timbral quality) are each in turn presented in detail; each are central concerns of records. These percepts commonly bring challenges to the listener's sense of prominence, and largely shape expression leading to listener interpretations.

The second section presents the 'confluence' that leads to the perception that all elements blend; all elements of the three domains, all contributions of performances within the track, along with the outside semiotic associations, affective states, and so forth entwine within listener interpretation. The mix stage of the recording process is used to illustrate confluence concepts. Discussion begins with the confluence of recording elements, shedding light on how they may combine to establish larger concepts of how recording shapes the song; this is often associated with the three percepts of lateral image and sound stage width, distance and depth, and timbral balance. Prominence is further explored in this context. The exploration of confluence leads to the third part, 'timbre as confluence,' and a presentation of timbre as a multi-domain and multi-faceted percept. Timbres of sound sources and the timbre of the track are investigated as a confluence of domains; this allows the analyst to engage the contexts, character and content of sound source timbres and the timbre of the track, and leads to engaging crystallized form. A detailed account of crystallized form concludes this section.

The last section of this chapter expands coverage of deep listening. Deep listening concepts have been introduced throughout this book, though not explicitly. It is the basis for numerous approaches offered for hearing many of the qualities of the track, especially recording elements. This chapter concludes by contrasting open listening and directed listening. The deep listening principle to allow whatever comes along to hold equal potential is a significant aspect of our framework.

LOUDNESS AS A RECORDING ELEMENT

Loudness is the perceived magnitude of sound; it is the sensation of the amplitude of the waveform. Loudness is the percept resulting from the psychological impression of the physical intensity of pressure (sound pressure) (Handel 1993, 63). Loudness as a recording element utilizes our sense of this sensation of amplitude.

Loudness, though, can encompass more than sensation. Loudness "is a concept that has implicit meaning for nearly everyone" (Schlauch 2004, 317) yet its sense of magnitude is subjective to the individual and to context (both cultural and environmental). All of us might 'know' when a sound is soft and when it is loud, but we all carry our own sense of these concepts, and this sense of 'loud and soft' shifts. These ratings of loud and soft are largely meaningless out of context—they carry no association to any actual and measureable sound pressure level. Environment context impacts 'loud and soft' so what is perceived as loud in a small room might be moderately soft from the back of a lecture hall. Cultural and social context are even more complicated, adding norms, notions of acceptable levels, and more; a softly spoken word in the midst of a noisy crowd is far different than the same level of speech within a quiet theatre. Loudness percepts are subjective between individuals, are subject to social and cultural conventions, and are perceived differently within various contexts. In addition, perceived level of loudness can vary under influence of other elements—whether or not the actual amplitude of the signal differs.

Perceived loudness can be significantly transformed without an actual change in sound level. We learned, beginning with earliest psychoacoustic studies, that functional dependencies exist between loudness and all other elements, and that there are underlying psychological and physiological factors that may impact loudness perception. Spectral content, duration, time relationships, bandwidth, frequency/pitch level/range, and even visual cues (experienced or imaged) are among percepts that can transform the experience of perceived loudness. Even surprise, shifting attention, pleasure, interest, and discomfort impact loudness perception—psychological influences on perceived loudness can be highly influential, and highly personal. Loudness is a subjective impression we can "assume to be influenced by different nonsensory factors and biases" (*ibid.*, 318), and by selective attention and focus on perspective level, as much as the impression of loudness is influenced by other elements within the sound and by the sound qualities and aesthetic activities of other sound sources. Loudness perception has many layers of subjective influences, laid upon the subjective impression of the sensation of the magnitude of sound.

Loudness manipulation is an important function of recording. Loudness levels of sources and sounds can be profoundly shaped by the recording process. When many of us consider the recording process, our thoughts turn to changing loudness levels of sources and mixing them together in different proportions from what occurs live (or what might have occurred live, given appropriate circumstances). The subtle qualities of loudness, then, can provide qualities and relationships that distinguish tracks as well as the sources they contain. Loudness can create surreal relationships of sounds, where gentle whispers can be incorporated at very loud levels, and instruments performed with great exertion are altered so the timbre generated by such a performance appears at a subdued loudness level in the mix; the balanced mixture of sound sources that comprise the track's aggregate texture need not reflect acoustic realism.

Loudness as a recording element is focused on the sensation of amplitude, separated (in as much as it is possible) from context of the performance (performance intensity, timbre, etc.). This sensation is loudness as loudness alone; it distinguishes loudness as a recording element from dynamics as a music element, and the dynamics coupled with loudness of performance. Here, the loudness percept and observation is separated from all of the influences of interpretation and impression; it is the perception of actual physical sensation. Loudness can be perceived as sensation, but typically we process loudness quite differently. In our daily experiences, loudness is rarely experienced solely as the sensation of the magnitude of the sound. It is counter to our natural listening tendencies to process (hear) loudness as sensation alone (without influences outside the sensation of magnitude). To separate listening to loudness from the above mentioned factors requires intention, and a control of attention and focus; this listening process seeks to isolate sensation from other influences within the sound. In this way, listening for loudness alone pulls the sound or sound materials out of context and the subjective factors they contribute to the interpretation of the sound's loudness. This attention to loudness is a critical listening process; it examines the experience as a sound object, void of causal factors and contextual implications. Attention to dynamics, as stated above, incorporates context and character of sounds and materials; dynamics is as much (and often more) about timbre, expression, energy, intensity, and nonsensory factors as it is about the sensation of loudness.

This separation makes it possible to approximate actual loudness levels against a reference, and to calculate loudness contours over time. The sensation of loudness is observed within the context of the track, a context established by its reference dynamic level. Loudness as an element of recording shares an equal role in shaping the track. Hearing the percept of loudness as a sensation of the magnitude of sound can unveil sonic characteristics and gestures inherent to the track that might otherwise go unobserved. The recording element of loudness/dynamics can add dimension to a recording analysis at all levels of perspective. Loudness establishes the presence of sounds and sound sources, but does not in itself establish prominence.

Prominence and Loudness

Attention itself can play a role in loudness perception. Attention is the act of bringing active awareness to the listening process. It can also be holding something within the center of that awareness—an act that by its nature diminishes the prominence of all other sounds, impressions, thoughts, aesthetic ideas, materials, etc. What is held in the center of one's attention is most prominent to the listener; this prominence from focus of attention is often mistaken to also be the loudest aspect of the track.

Prominence is what is most noticeable or conspicuous at a particular moment in time; it is what has grabbed the listener's attention. It is not *necessarily* the most important or most significant sound or material in the track, but only what the listener—at that moment, or in this listening—finds most interesting (Moylan 2015, 452). What is most prominent is also not necessarily what is loudest; prominent sounds are often not the loudest. A sound can be most prominent in the listener's consciousness, while being at a lower loudness level than all other sounds.

A sound, sound source, aesthetic idea, lyric or musical idea can dominate the listener's attention for a myriad of reasons. All elements of the track have an equal potential to provide qualities that cause sounds or materials to standout and be noticed. The entry of the lead vocal very often captures the listener's attention, and at least for a moment can seem to dominate by loudness; when language is present, or pronounced affects of a voice, our life experiences direct us to give it our attention. Prominence can be personal, and influenced by the listener's prior experience; what stands out to listeners on a personal level can be highly unique.

Unexpected events can attract attention and thus bring prominence, as can those that are unusual in some way. There is a prominent hi-hat entrance in Phil Spector's mix of "Let It Be" (*Let It Be*, 1970) at 0:53. The instrument is not the loudest in the mix. It is prominent because it is the first appearance of a percussion sound in the track, and it is a new addition to the texture. The listener's attention is likely immediately captured by the new sound that is unlike anything that has preceded it—though they may also remain engaged in the lead vocal melody or the content of the lyrics. Immediately the instrument's unusual spatial identity becomes pronounced as its delayed iterations provide movement on the sound stage. Never is it the loudest sound, or the track's most significant aesthetic voice. Whether or not it is more prominent than the lead vocal rests in the attention of the listener. Shifting one's focus of attention between the lead vocal, piano and hi-hat, one might experience a shift in loudness accompanying a shift in prominence, or awareness. Recognizing this shift in prominence can lead toward experiencing the sensation of loudness decoupled from other aspects of sound.

Prominence can be influenced by interpretive factors, and our perception of it may be detached from the actual context of the track. Prominence can be brought into context, though. This is accomplished through intention to perceive all sounds and activities in the track as being equivalent to all others. A balance of prominence (an equality of all that is occurring) will allow one to 'hear' all sounds as equivalent. This can be accomplished by directing attention to a higher level of perspective, where the sources (or whatever is being compared) can be heard as equals; this can reveal their differences and unique states most accurately. In this way we can perceive the balance of loudness levels without the influence of prominence. This chapter's praxis studies can assist the reader in acquiring skill in these areas.

Measuring Loudness Perception

Measuring loudness and establishing identifiable (or perceptible) loudness increments is highly problematic, and not entirely possible.

First is the matter of equal loudness throughout the hearing range of pitch/frequency. As we have learned, two sounds of different frequencies will very likely require different physical amplitudes to establish the sensation of equal loudness. The sensation of loudness can change throughout the hearing range, while levels of acoustic energy might remain consistent—and these inconsistencies of loudness sensitivity are non-linear, and vary significantly at different pressure and frequency levels (as the equal-loudness contour previously informed us). Loudness perception of the track is linked with the loudness level of playback, and its ability to reproduce the frequency range with relatively even response. Identifying equal loudness levels between diverse pitches and frequencies (and the timbres presenting them) is one significant problem of measuring loudness.

Identifying increments of loudness levels is the second significant problem we encounter.

Loudness (a subjective measure) is often confused with sound pressure (Mather 2016, 128–131). Sound pressure is a physical characteristic of how tightly air molecules are compressed together; as the displacement and compacting of air molecules increases (as a sound body moves farther) the greater the pressure increase in the waveform. Sound pressure is measured in decibels as sound pressure level. When we measure the physical amplitude of a sound, we identify its sound pressure level and establish its relationship to a reference level; we arbitrarily choose one value of sound pressure as a reference, and then measure all sounds as relative multiples of that reference level. Thus, the decibel is a comparative measure (a ratio) relating a reference value of the threshold for human hearing[1] to the current sound's measured sound pressure level. This measure of amplitude does not transfer into perception for numerous reasons. The most obvious might be the range of loudness we can perceive. If we accept the threshold of pain (for most listeners) to be 160 dB SPL, the loudest sounds are about 100,000,000 times more intense than the slightest perceivable sound (threshold of hearing for young ears); workable numbers are derived by converting the ratios to the logarithms of the ratios as decibels (dB). The sensations of loudness and sound energy are not proportional, but are calculated as a logarithmic function. Decibel's logarithmic scaling is not helpful for establishing perceptual increments, as an increase of 3 dB is a doubling of power, but this is unrelated to an increase in perceived loudness; a sound 10 times the intensity as another is 1 Bel greater in loudness. (B. Moore 2013, 133–167) We "can only infer loudness from objective measures" (Schlauch 2004, 318); the measure of sound intensity does not transfer into a measurable *perception* of loudness.

The methods of measuring loudness explored in psychoacoustics, music psychology and recent ecological studies have little to offer recording analysis. The methodologies used to study loudness sensation have ranged widely. They have included paired comparisons, loudness matching, magnitude scaling, category ranking, cross-modal matching and psychological scaling; included also are studies in ecological loudness—the relation between loudness and the naturally occurring events that they represent. Much about loudness has been examined—both as simple stimuli in laboratories and as environmental sounds.[2] While these studies have further informed our understanding of loudness perception, they have not opened a path toward devising a way to identify and compare loudness levels against some scaling or objective measure (like we do pitch and frequency). Our means to engage loudness levels—identifying levels and differences (intervals) between levels—has not advanced in a way that can be incorporated into studying the content of tracks. Loudness levels cannot be doubled to establish an identifiable clone of itself at another level of sound pressure, as pitch can be recognized as being an octave higher; the range of loudness cannot be divided into equal increments that are perceivable and recognizable, such as the half-steps of pitch.

Loudness in the Context of the Track

The category ratings of dynamic markings remain the most readily useable scale available to compare loudness levels and loudness relationships; dynamic markings are range areas of dynamics/loudness, not discrete levels. Discussed at length in Chapter 3 and the following: dynamics differ from loudness in that they reflect timbral characteristics, energy, expression, intensity; dynamics carry subjectivity of interpretation and intermodal connections (physical exertion, connotations of expression, etc.); dynamics are relative to the context of the track. Further, a range of loudness levels exist within each dynamic area—and the actual loudness levels of areas can overlap (also a reflection of dynamics privileging timbre over loudness).

Loudness as a sensation of sound intensity may be situated within this context, with recognition of its sensation being associated with, though distinguishable from those of dynamics. As a recording element, the sensation of loudness is understood within the context of the track. Its level may be defined within the continuum of dynamic ranges.

The reference dynamic level is used to relate dynamic levels, and has been adapted and adopted here (from what was presented earlier) to calculate loudness. It is a holistic impression of all of the sounds of the track and their musical materials and performances; this impression also embraces the content of the lyrics and the meanings and affects it generates. Reference dynamic level was discussed at length in Chapter 3, and will be discussed in more detail here.

The RDL is a specific level within a specific dynamic area; it can be placed at a precise point on the continuum of dynamic ranges, representing a clearly defined level of intensity. The RDL is a stable, unchanging reference within a track; no matter what occurs within the track, the RDL remains a consistent reference.

The track's reference dynamic level (RDL) provides a stable level of intensity against which levels of sources and sounds can be calculated. Loudness will be identified on the basis of sensation alone, and placed within the context of the track against the 'conceived' loudness level that is implied by the reference dynamic level; this carries modest subjectivity from (1) the analyst's determination of the RDL and also of (2) the transference of the RDL's 'dynamic' level into a loudness level sensation. The RDL implies a 'loudness level' that reflects the energy, intensity and expression of its 'dynamic level' related to music contexts (**pppp**, **mp**, etc.). To allow this transfer, the thresholds of hearing and pain need to have corresponding dynamic levels. The reference levels of the threshold of hearing and the threshold of pain have been assigned the dynamic markings:

> **pppp** for the threshold of hearing
> **ffff** for the threshold of pain

The selection of these markings is somewhat arbitrary; five **p**'s and/or five **f**'s might seem appropriate to some analysts. Some analysts may find tracks that require more than five increments to reflect its dynamic continuum. These were chosen because they are rarely encountered extremes in musical score markings, and the markings represent rarely encountered extremes of loud and soft within musical contexts. These markings will typically allow for a clear observation and presentation of data. The analyst may adjust this scale if it seems appropriate to the track; for instance, some metal tracks may use only a range from **f–ffff**, while a folk track might use a range of **pp–mf**.

Using the RDL and these references for extreme levels, loudness levels encountered might be understood and calculated within the context of the individual track. Calculating and comparing loudness levels utilizes numerous processes. They can differ somewhat at various levels of perspective and with various types of observation. Nearly all can be related to one or several of the following steps, however:

- Identifying the level of the sensation as compared with that of the RDL; this determines its loudness level within the context of the track
- Matching or comparing loudness levels within the same gesture or percept stream (a musical line, for instance) at different points in time; this establishes points within a gesture, and allows loudness contour to be mapped against time
- Matching the loudness sensations of two or more different sounds/sources occurring simultaneously; this allows separate parts to be observed for their relationships and interactions, as well as their individual characteristics
- Placing loudness within dynamic ranges from an interpretive calculation of sustainable physical exertion (see Chapter 3)

Interpreting Reference Dynamic Level and Crystallized Form

Crystallized form is the highest dimension of form; it is the track (in its entirety) existing out of time, 'heard' non-temporally within the experiential present. Perceived in an instant of realization, it is a single, multidimensional shape, and a large-scale nonverbal conception and experience. Crystallized form is an aural image and a large-dimension sound object—a unified presence of all qualities present at once. It establishes a sense of knowing the track's fundamental substance (manifesting as a core essence, inherent nature, a unique presence) and its individual form (multidimensional shape) and character (energy, expression, affects, meanings).

In the silence after the track, allow the listening experience of the track to dissipate into a single awareness and reflect on the impression that remains. Reflect on this presence of the track that lingers in your memory and psyche, in your consciousness and awareness. The goal is to not 'make sense of it' but rather to 'recognize it' for what it is, perhaps to 'feel' its character. This overall presence is—at least in part—crystallized form.

Within this impression and conception is a sense of the amount of energy and the level of exertion of the performance, the tempo and the speed of motion, and the magnitude of intensity within its expression; these are supplemented with the drama and meanings of the lyrics and the spatial attributes and other characteristics of the recording. These all coalesce into a manifestation of the performance intensity of the track.

This performance intensity of the track is the reference dynamic level (RDL). The goal, here, is to experience the RDL as a sense of the intensity of the track. Perceived holistically and considered without interference from verbalization, it is a single level of intensity that embodies the track.

The reference dynamic level is the part or quality of crystallized form that embodies the intensity of the track—in all the elements and materials and outside associations that shape and establish it. While Table 9.1 presents factors that potentially influence RDL, it is important to remember this it is the result of an experience, not a calculation. Listening from the position of accessing musical expression and an appreciation of the singular, coherent whole will open attention and awareness; attention is directed toward nonverbal qualitative reflection and a recognition of its expression; one that is based on *musical thought* and aural imagery. This contrasts markedly with the analytic reasoning required of analysis, which engages the *rational thought* processes of calculating, deducing, problem solving and otherwise attempting to assemble a result for the RDL—a process that quickly and prematurely brings verbalization, and does not access the inherent character of crystallized form and the intensity of the RDL. This will be addressed in more detail later under crystallized form.

Revealing an RDL involves listening and finding the level at which the track's energy and expression reflect its intrinsic character. Table 9.1 is presented to identify sources that *may* contribute to the RDL of a track. They are not intended to bring the reader to divert attention to these sources; attention should remain at the highest level of perspective, to experience the intensity dimension of the track. Each track, every individual track, will have a different combination of factors, and different proportions of factors that formulate a sensation of RDL that is unique. There is no formula for RDL, as it is an interpretation. One can expect one's sense of the RDL to evolve—especially while one is becoming more acquainted with a track. As an interpretation, the RDL will (to some unknown extent) reflect the analyst's biases, though this interpretation should (as much as is possible) reflect an objective interpretation of culturally shared perceptions.

> **Praxis Study 9.1**   Recognizing reference dynamic level (pg. 529).

Table 9.1   Potential factors that can influence reference dynamic level. These factors potentially provide a variable level of influence toward establishing a singular, overall impression of intensity for the individual track.

Tempo, energy and directed motion	Dramatic expression of lyrics
Density of information	Meanings and associations of lyrics
Levels of exertion of performances (reflected in intensities of timbres): all sound sources, potential emphasis of dominant or significant sources, potential emphasis of contextual layer, with respect to the overall texture's performance intensity	Recording element influences: pitch density, timbral balance, program dynamic contour, loudness balance, holistic environment, sound stage qualities, and other
Loudness levels and contours	Affects and emotions of the whole
Content and tension of expression	Musical expression

Overall Loudness Level of the Track

A hierarchy of loudness-level strata exists similar to hierarchies of other elements. The above comparison types for loudness levels will occur at all levels of perspective, except one sensation of loudness exists at the highest level of perspective and comparisons will be made only within that single contour. Table 9.2 illustrates the hierarchy of loudness levels.

Attention to the sensation of loudness can be directed to any of these levels of perspective. Simultaneous sounds or successive sounds can be compared in all of these perspectives—allowing a single stream of sounds to be followed, or various streams to be compared to one another. These levels of perspective differ greatly, ranging from the singular loudness of the aggregate texture to the strata within a sound's components and its reverberation (loudness levels within timbres were encountered in Chapter 7, and Chapter 8 introduced loudness levels within environments).

At the highest level of perspective, the track is distilled to a single sensation of loudness; we perceive this level to change continually and to form a contour across the entire track. In earlier writings (Moylan 1992 and 2015) I have referred to this as 'program dynamic contour.' In these I was writing from the vantage point of an engineer/producer (I prefer the term 'recordist')[3] where the term 'program' has been commonly used to describe the overall track or its singular sound; 'dynamic' was used synonymously with the sensation of 'loudness' to connect the two concepts (though this connection was rarely explicitly articulated) in order to connect the role of loudness sensation to the musicality of the track. This overall loudness level could be re-named as 'track loudness contour'[4] should one wish to be more accurate and perhaps less confusing. I will use both terms synonymously as this discussion unfolds.

The track loudness contour is the single loudness-level of the track's aggregate sound; it is the result of the combination of all source loudness levels. It helps some to envision this sensation and concept by thinking of a single VU (voltage unit) meter that displays a representation of the signal level, following the loudness level of the program as it potentially changes at every moment. The contour is the shape of the track's changing loudness that is revealed as it progresses from beginning to end. This loudness shape often has structural significance to tracks, and can support its drama; this sensation in itself it is capable of generating movement and tension within the track. It is unusual to listen for overall loudness of many sources in our everyday lives; many practicing musicians, including conductors, are not aware of or seek to engage this sensation of combined loudness levels, though some certainly do. Recordists, in contrast, often bring their attention to this level of detail while shaping various stages of the recording process. This overall loudness contour may be deliberately shaped in the recording

Table 9.2   Hierarchy of loudness-level strata as a recording element in relationship to levels of perspective.

Perspective	Loudness/Dynamic Levels and Relationships
Overall Texture	Reference dynamic level Loudness/dynamic contour of overall, aggregate texture
Composite Texture	Loudness/dynamic balance relationships of sound sources (voices and instruments)
Basic-level: individual sound sources	Loudness shapes/contours of materials or lines Loudness levels of individual sound sources
Individual sounds: overall loudness level and shape/contour	Dynamic/loudness contours of individual sounds Contour/shape of dynamic envelopes
Individual sounds: internal loudness levels and contours	Contour/shape of spectral envelopes Dynamic/loudness contours within reverberant energy

process; its subtle changes are often the result of the track's arrangement or of its mix; these are significant contributions to the character of the track.[5]

This loudness shape often has structural significance to tracks, and it can support its expression. Jada Watson and Lori Burns (2010) use the loudness-wave shapes of a track's two channels to illustrate this overall loudness as amplitude; while this is not precisely aligned with program dynamic contour (it depicts a physical measure and not the perceived dynamic shape, and it treats channels independently instead of bringing attention to a single impression), such a diagram can guide perception and observation, especially when acquiring this skill. They note their amplitude diagram of the Dixie Chicks song "Not Ready to Make Nice" (2006) "reveals that the song has an overarching increase and decrease in dynamic amplitude (< >), a design that reflects the growing intensity of the vocal gestures and instrumentation and complements the intensification of anger and resistance in the lyrics" (Watson & Burns 2010, 345). Here they connect structure, performance intensity, dramatic expression and the content of the lyrics into a statement that provides much important information to the track, and to its overall loudness shape.

A program dynamic contour graph allows the loudness contour of the track to be observed and notated. Listening to this overall loudness sensation will be a new experience to some readers, though it can be developed with directed attention; praxis study 9.2 can guide this experience. Engaging this level of perspective will lead to new observations, even for tracks one already knows well. When engaging the track's loudness contour care should be taken to remain focused on the sensation of loudness alone, and to remain uninfluenced by the timbres and intensity levels of the ensemble or the drama of the music. Timbre and expression characteristics (and other percepts) can bring the impression of increased or diminished loudness, without an actual change in acoustic energy; likewise, loudness can change without a change in timbre (or another element of recording or another domain). Remember to monitor playback level; consistent playback level is needed for accurate observations between listening sessions.

| Praxis Study 9.2 | Hearing overall loudness and creating the track loudness contour graph (pg. 530). |

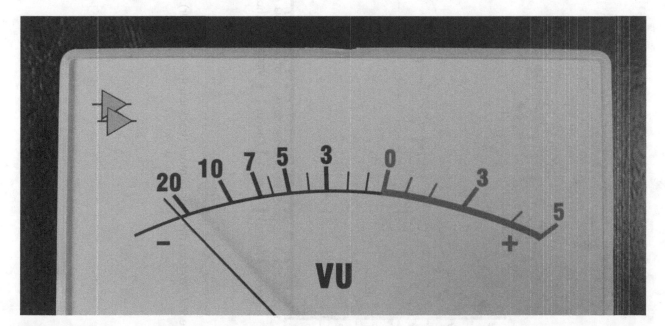

Figure 9.1 VU (volume unit) meter. Image courtesy of API (Automated Processes, Inc.).

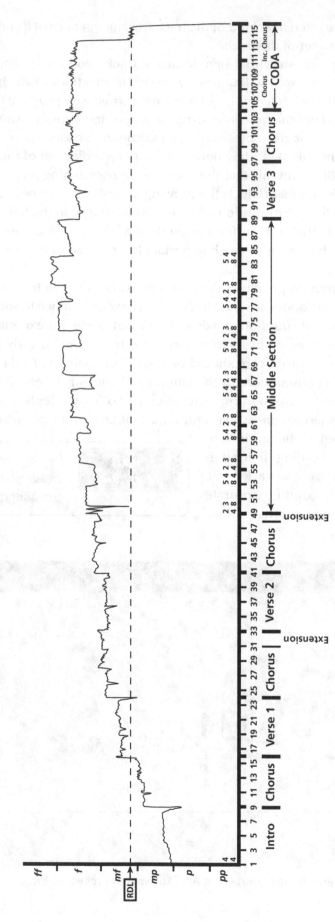

Figure 9.2 Program dynamic contour (track loudness contour) graph of the Beatles' "Here Comes the Sun," *Abbey Road* (1969, 1987).

Figure 9.2 illustrates the changes in overall loudness level throughout the track "Here Comes the Sun" (1969, 1987). The graph contains the reference dynamic level of the track, against which the contour can be heard. Imbedded in the contour are shapes of loudness that correspond to structural divisions; as the shapes emerge within one's hearing of the track, their role in defining sections through their repetition becomes apparent. The loudness shape of the track is clearly evident from its beginning at the lower portion of *mp* to its peak within *ff*. The wide dynamic range of the track contains subtle changes of loudness as well as large and sudden shifts.

"Here Comes the Sun" is among the uncommon tracks in which the reference dynamic level is prominently experienced. During the final moments of the coda, the level of the track loudness contour matches the track's RDL; the reference dynamic level is audible as the track's overall loudness arrives at the track's overall sense of energy, exertion, and expression (that is the RDL). At this moment, the low *mf* RDL delivers a sense of arrival and a settling in the place of the conception and expression in which the track exists. It is common for a track to arrive at its RDL as an important occurrence, but it is not common for it to be a point of arrival that provides aesthetic closure to a track.

Loudness Balance, Musical Balance

The balance of sound source loudness levels established by recording exerts significant influence on the track. The level of loudness of sources and their resulting interrelationships represents an important element of recording—this is loudness balance. Though loudness balance is but one of the recording elements applying influence on the track, it has the potential to prominently shape the content and character of the track.

The recording's role shaping relationships of sounds is often reflected—to some degree—within loudness balance. Loudness balance situates each source in the mix in terms of loudness; as loudness brings all sound into perception, loudness brings it to be audible and establishes its presence. Observing the actual loudness levels of sources can bring an understanding of the loudness contours of each source, of their loudness relationships, and of the composite texture's loudness balance. In music settings, loudness balance of instruments and voices (sound sources) is commonly framed as 'musical balance.' In earlier writings (Moylan 1992 and 2015) I have used 'musical balance' synonymously with what is referred to as 'loudness balance' here. 'Loudness balance' will be used here to more clearly differentiate this recording element from the elements of music.

Loudness contours of individual sound sources can be notated on an individual or a collective 'loudness contour graph.' This can provide a clear way to notate even the subtlest loudness/dynamic changes of individual sources and their gestures, or musical lines. It can also reveal loudness relationships and groupings of sources where they combine to create particular contours. The loudness balance X-Y graph will illustrate the actual loudness levels of sources; the graph can be used to make general loudness observations, or it may be calculated in detail against the reference dynamic level. Either approach may be most appropriate, depending on the goals of the individual analysis. Sound sources are represented by a separate line of the graph, allowing their contours to be mapped as it changes over time. The graph displays loudness as loudness—it does not factor in the influences of timbre, register, or prominence of any other origin. We remember that loudness is the perception of the amplitude of sound as we bring our attention to this element. It is important to remember to hold all sounds in equal prominence, as loudness can easily be distorted by listener perspective and focus. A source in the center

> **Praxis Study 9.3**
>
> Perceiving loudness balance: the contrasting percepts of loudness relationships and the levels and contours of individual sources; relating source levels to the RDL and formatting loudness balance against a timeline (pg. 531).

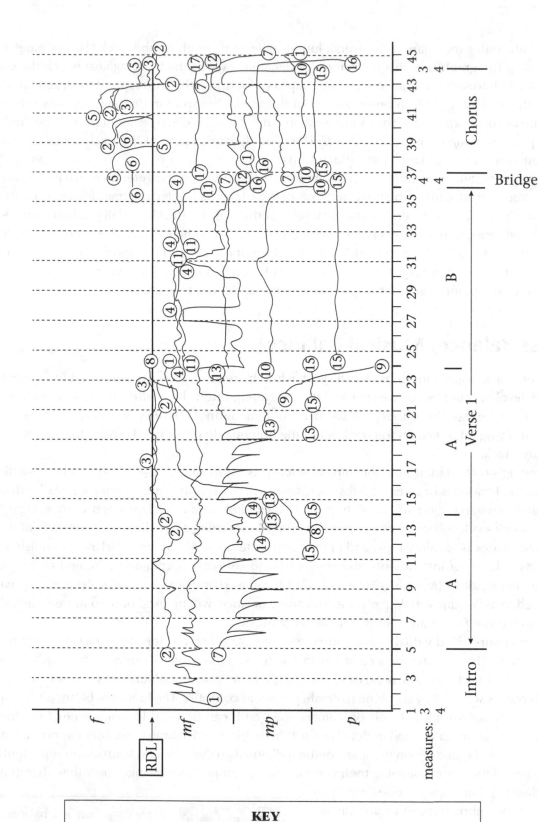

Figure 9.3 Loudness balance graph of the Beatles' "Lucy in the Sky with Diamonds" from *Sgt. Pepper's Lonely Hearts Club Band* (1967, 1987).

KEY

1 Lowrey Organ	7 Bass	13 Open Hi-Hat
2 John 1	8 Tamboura	14 Hi-Hat Closed
3 John 2	9 Piano	15 Kick Drum
4 John 3	10 Guitar Acoustic	16 Ride Cymbal
5 Paul 1	11 Guitar Melody	17 Snare Drum
6 Tom Drum	12 Guitar Bassline Chorus	

of one's attention will be emphasized in one's awareness, and cause loudness judgement to be skewed. Listening for loudness contours, and judging relationships to the RDL does, however, bring one to listen at the perspective of the individual source. Praxis study 9.3 can guide these listening experiences.

Figure 9.3 illustrates the loudness balance of "Lucy in the Sky with Diamonds" from *Sgt. Pepper's Lonely Hearts Club Band* (1967, 1987). This graph provides great detail on the loudness shapes and relationships of all sounds against the track's RDL. Note the designation of the RDL's placement related to the dynamic areas; the loudness levels of individual sounds are interpreted against that level just as occurred with the program dynamic contour (track loudness contour). The RDL is the same in each graph (observing the same track).

This graph brings the activities of all sources into keen focus, and closely observes their loudness interrelationships.

Not all analyses require this highly detailed data collection, though. A more general assessment of loudness levels might provide adequate insight into tracks. Figure 9.4 illustrates the loudness levels of sound sources in three versions of the Beatles' "Let It Be." The sources are identified within the more generally-defined (though closely consistent with traditional dynamic areas) loudness areas ranging from very soft to very loud. These graphs contain general contours of loudness levels of little detail; they provide an impression of the overall loudness level of sources during a section or passage. A reference dynamic level is absent from each version as well; this also contributes to the general nature of these loudness level and relationship observations. A quick glance at these graphs identifies fundamental loudness differences between the three versions, even with observations quite generalized. Much detail could be extracted from these examples, adding to what has been provided here.

The original, single-release version produced by George Martin has a narrow range of loudness levels. All sounds are in the lower half of 'moderately loud.' The lead vocal is loudest, except for a short phrase at the end of Verse 1. The background vocals have a similar loudness, to the lead vocal and piano, and is situated between them until the piano increases in loudness at the end of the chorus. The piano's rise in level at the end of sections is obvious from the graph.

Phil Spector's version from the *Let It Be* (1970, 1987) album has clear contrasts. Loudness varies between high 'moderately loud' to the upper portion of 'soft.' The background vocals passage changes loudness level markedly. The piano retains its general loudness contour, though the chorus level is some magnitude lower in this version. Billy Preston's Hammond organ resides at a soft loudness in the chorus, removed from all other sources until its level is approached by the backing vocals at the end of the chorus.

The graph for the *Let It Be . . . Naked* (2003) version makes clear it has the widest range of loudness levels of the three versions. The lead vocal is loudest in this version; just as in the other two versions, it is the loudest within the track. The background vocals are louder in this version than in the others, and the general loudness contour of the piano is slightly modified in the chorus. The Hammond organ is present throughout all sections graphed; though it is extremely soft and at times barely present, it has a distinct loudness contour.

Performance Intensity

Contrasting performance intensity with loudness balance allows the analyst to observe the impact of the mix on the performances of sources. Here, loudness contrasts with timbre instead of working in synergy. Loudness changes not accompanied by timbral changes emerge, and the reverse occurs as well, as timbres remain stable while loudness is altered. This contrast will often provide information on the relationships of sound sources to the overall loudness level of the track (program dynamic contour).

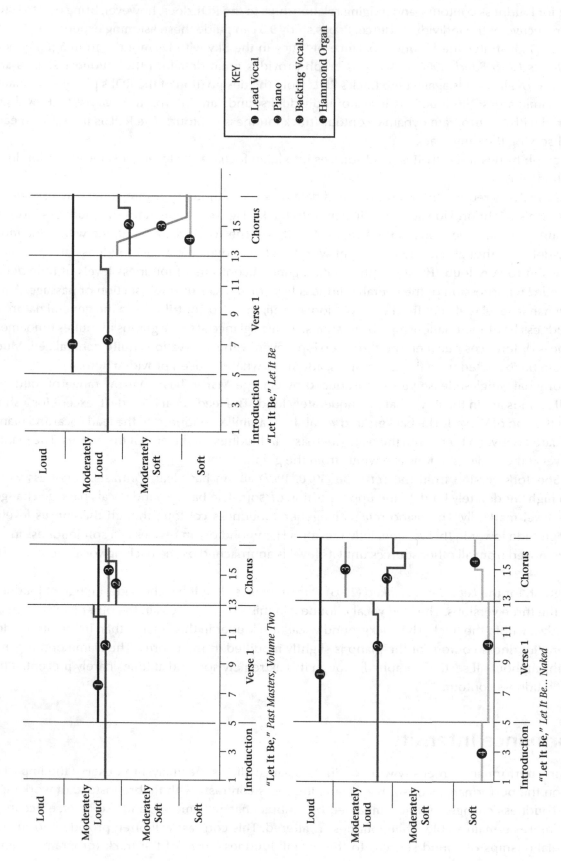

Figure 9.4 Generalized loudness balance graphs of three versions of the Beatles' "Let It Be."

Performance intensity reflects the qualities present when a voice or an instrument is recorded; it is reflected in the timbre of the sound source. Performance intensity is the timbre of the instrument resulting from the levels of physical exertion, energy, expression, performance techniques, and any other timbral qualities related to performance present in the sound and the performance when it was captured (recorded) (Moylan 2015, 450).

The loudness level that established performance intensity when the source was recorded is transformed within the mix process. This allows performance intensity—and its expressive content—to be a separate percept in the track. This dichotomy separates reality from the crafted world within the record; this is an inherent trait of popular music recordings. The timbre of performance intensity is often used aesthetically to carefully shape the performance itself. In this way it is used to enhance the drama and expressions of the track, and for many other purposes. For example, sounds of low performance intensity often appear at higher loudness levels in the mix than their timbres indicate; this is especially common within vocal performances, where a whispered word can appear loudly in the track. In expressing lyrics and the musical line, a lead vocal can vary in performance intensity considerably and often—sometimes within a word or a syllable. The voice—"full of concrete meaning that is not conveyed through lyrics is to be found in all forms of musical expression, but in recorded music, precisely because it is recorded . . . the effect is especially discernable" (Lefford 2014, 44)—clearly illustrates the significance of performance intensity to shaping interpretation by performers and listeners. Performance intensity is often the primary carrier of musical expression, and also of the drama delivery and the shaping of meanings within the lyrics. It can create and carry the level of urgency in the track and establish a sense of tension or of ease.

The levels of performance intensity are interpreted by listeners by observing timbral qualities as they relate to physical exertion and expression. Denis Smalley (2007, 39) notes: "[O]ur experience of the physical act of sound making involves both touch and proprioception—the tensing and relaxation of muscles in relation to all types of body movement." This experience can be one of observation as well as participation. A listener's prior experience with and knowledge of the particular instrument's timbre, as well as their abilities to remember and match the timbre in a state of normalcy, play central roles in this interpretation, and its relative accuracy. The deeper the experience and prior knowledge, the more likely the listener will successfully and accurately recognize the instrument's performance intensity through perceived timbral qualities. In absence of experience, this interrelationship of timbral quality and performance intensity may have its "content . . . simply assumed, or even invented, by the listener" (A.F. Moore 2010, 259). No matter the accuracy of the percept to the actual act, performance intensity contributes substantively to a listener's interpretation of important aspects of the track.

The threshold level that separates expending and restraining energy—the transitions from force to resistance as discussed in Chapter 3—is a reference for calculating performance intensity. At this level, resting between *mp* and *mf*, performance intensity (and source timbre) is in a state of normalcy for timbral characteristics, where the timbre is not altered by the energies of performance. This level of effort can theoretically be sustained indefinitely. In "Burnt Norton" from *Four Quartets*, T.S. Eliot (1943, 15–16) called this idea "the still point"; his poem describes this "still point of the turning world" as being "neither from nor towards . . . neither arrest nor movement . . . neither ascent nor decline . . . except for . . . the still point, there would be no dance, and there is only the dance."

This "still point" provides a reference to allow loudness related to the listener's perception of body movement's role in sound making to be calculated. Embodied music cognition tends to recognize music perception as based on action—those of the listener body movements, and perception of kinesthetic properties and musical gestures involved in creating the sounds of the performance (Zbikowski 2011, 181–190).

Performance intensity in itself can produce directed motion in a line. The urgency of performance intensity's expression and perceived actions of its physicality can create motion and tension in music, and in subdued expression it can create calm, ease, and repose, and thus mitigate all but the slightest tension; this is not a duality, but rather a continuum that may establish many states between the greatest tension and most urgency imaginable, and the slightest motion of ease and a most minimal sense of tension. Performance intensity can communicate urgency and energy, drama and expression, directed motion and stillness, tension and release, exertion and restraint, and more. Changes can be extreme and fast, subtle and gradual, but all contribute character and substance to sound sources and their materials—all are embodied in the timbre of the sound source and its invisible performance gestures. Performance intensity does much to shape the character of timbres and • the affects of their delivery, and define the qualities of individual performances and the delivery of their musical idea.

A confluence of all domains and many of their elements can be found within performance intensity—as timbre, musical lines, and perhaps lyrics blend into one gesture. Performance intensity—and all that it carries—is an important aspect of performances in the track (Moylan 2015, 323).

In the context of the track, performance intensity may be accompanied by loudness changes, but not necessarily. Timbre can change without loudness changes, and performance intensity is reflected in timbre much more than loudness. These are controlled separately in the mix process.

Performance Intensity and Loudness Balance Graph

Figure 9.5 is a performance intensity and loudness contour graph. This graph is uniquely suited to the analysis of recorded performances, and can reveal a wealth of pertinent information.

Performance intensity is notated on the upper tier of the X-Y graph, charting the dynamic levels and contours of each source's original performance. The timbres of sources are not related to the reference dynamic level; rather their intensity is calculated based on their timbre and expressive character. The lower tier of the graph provides the loudness balance of sources; the loudness levels of sources are calculated relative to the track's RDL. Contrasting performance intensity with loudness balance, one is comparing two separate elements against the same timeline—the interaction of implied loudness of timbre and actual loudness in the track illustrates a confluence of the recording elements as well as their independence. The graph allows direct comparison of two different states of the same sound sources, as they change over the duration of the example.

Performance intensity and loudness contour graphs may contain all sources, or a selection of sources of significance to an analysis. When needed, a single source might be observed through this approach. The level of detail of data from both performance intensity and loudness balance can also be adjusted to reflect the needs of an analysis, from a significant attention to detail to general impressions.

Praxis Study 9.4	Observing and notating performance intensity; graphing performance intensity and loudness balance as two tiers of an X-Y graph (pg. 533).

Figure 9.5 provides much detail of contours and levels in both tiers. The graph could have greater detail, though. Subtle changes of timbre in the vocals and in several instruments are not contained here. In examining the two tiers of Figure 9.5, one can immediately recognize the disparity in levels of the Lowrey organ, as it is much louder in the mix that its performance intensity suggests. Comparing the two tiers, one can determine how sources deviate from their original levels, and just how far the loudness balance has strayed from the original

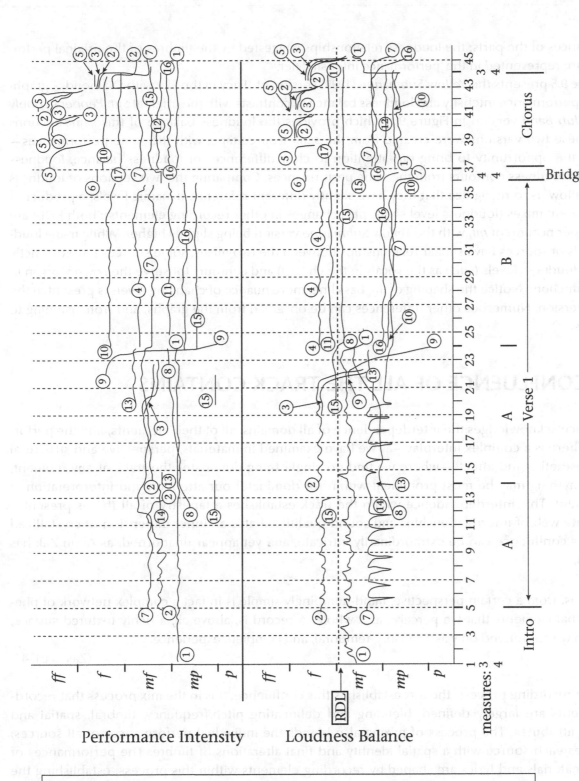

Figure 9.5 Performance intensity and loudness balance graph of the Beatles' "Lucy in the Sky with Diamonds," *Yellow Submarine* (1999).

KEY

1 Lowrey Organ	7 Bass	13 Open Hi-Hat
2 John 1	8 Tamboura	14 Hi-Hat Closed
3 John 2	9 Piano	15 Kick Drum
4 John 3	10 Guitar Acoustic	16 Ride Cymbal
5 Paul 1	11 Guitar Melody	17 Snare Drum
6 Tom Drum	12 Guitar Bassline Chorus	

performances of the parts; the loudness relationships suggested by the timbres of the original performances are represented in the 'performance intensity' tier.

Figure 9.5 presents the *Yellow Submarine* (1999) version of "Lucy in the Sky with Diamonds," graphing both performance intensity and loudness balance. It contrasts with the original *Sgt. Pepper's Lonely Hearts Club Band* version in Figure 9.3, which provides the loudness balance of this same sections as 9.5. These two versions were created from the same source tape, with the same performances—allowing the opportunity to bring our attention to clear differences of loudness balance, loudness levels, and loudness contours from the same performances. Comparing the two versions of loudness balance allow us to recognize that they have slightly different reference dynamic levels, a product of their different mixes (loudness level shifts, plus changes in other recording elements); both RDLs are in the upper portion of *mf*, with the *Yellow Submarine* version being slightly higher. While many loudness levels of sources have similar relationships between the two mixes, some sources have distinctly different loudness levels (such as the vocals in the chorus) and contours (observe the Lowrey organ in the introduction). Notice the shaping of the bass line; more nuance of changing levels is present in the original version. Numerous other differences can be observed from the graphs, and from listening to the tracks.

THE CONFLUENCE OF ALL THE TRACK CONTAINS

Confluence acknowledges the interdependence of all domains, all of their elements, and the performance. There is a complex interplay—as we have examined in materials, perspective and structural levels, aesthetics and affects—where we find no single thing dominates the track at any moment, though anything may be most prominent within (or dominate) our attention and interpretation at any moment. This interdependence within the track establishes a tapestry of all that is present—an intricate web of interrelationships and sound qualities that is an inner dimension of crystallized form. This confluence can be extraordinarily intricate, and yet appear unadorned, as Albin Zak has observed:

> What is, from a certain perspective, mind-numbingly simple is in fact a complex network of phenomenal elements that we perceive as a whole. A record is, above all, a richly textured surface, which we apprehend only as a sensory, temporal, and complete experience.
>
> (Zak 2001, 43)

In the recording process, the mix establishes this confluence. It is in the mix process that recording elements are largely defined, blending and delineating pitch/frequency, timbral, spatial and loudness attributes. The process of mixing also blends the individual performances of all sources; it provides each source with a spatial identity and final alterations of timbre. The performances of musical materials and lyrics are shaped by recording elements within this process, establishing the fundamental qualities of the final track—qualities that provide the track with a level of distinction. The mix process occurs at the basic-level of sound source and at the composite level that compares them as equivalent and equals, though its confluence influences relationships of all elements at all levels of perspective. The mix shapes and blends even the subtlest aspects of each element of each domain into a sonic tapestry that ultimately manifests as the overall timbre of the track and the track's spatial identity.

The mix manifests the confluence of the track. It is also a metaphor for the confluence of all the qualities, proportions, character and expression the track contains. Elements lose their individuality as they blend into gestures and materials, aural images and aural events, and a rich and complex texture. The acoustic wave (the result of all of the sounds of the track) that arrives to us (slightly differently at each ear)

> is a smear, because of the additive superposition of the sound waves generated by each event [sound source]. Each event contributes a time-varying frequency and amplitude pattern, but the integrity and connectedness of each pattern is lost physically among the other patterns of the overall wave. The acoustic wave is thus inherently ambiguous, because each event loses its identity when it is woven into the acoustic wave. . . .
>
> (Handel 1993, 185)

There is good reason we often have trouble making sense of what we hear in the record; confluence establishes a rich and multidimensional texture that can blend sounds so they are no longer distinguishable.

Prominence Emerging from Confluence

The mix establishes a balance of all parts within that complex texture. From that balance, any sound or element may emerge to be more prominent than others. As we learned above, prominence is established by listener attention. Attention is drawn to what stands out from all else—this exists at all levels of perspective, comparing sounds of sources, observing the elements within sounds or musical lines, observing the interactions of musical materials, hearing a text emerge from a musical fabric, and more. Albin Zak (2001, 157) significantly notes that "prominence is perceptible only in relationship. That is, to assess prominence we need a frame of reference." A frame of reference exists from the materials and elements within the track, whereby some ideas emerge as more significant, and others fall into other roles; sounds can also emerge because they are interesting in some way, or simply discovered and brought into the center of one's attention. From the latter, we might begin to understand how prominence can be personal—what emerges from the texture for one person (and their listening interests, skills, sensibilities, experiences, etc.) may not emerge for others, and certainly will carry some level of individual interpretation.

Allan Moore and Albin Zak both use the term 'prominence' in a manner that contrasts with this writing (see earlier this chapter). Zak states: "[R]elations of prominence are analogous to 'depth,' among Massenburg's four dimensions of the mix. They impart impressions of proximity and emphasis along with whatever associations these may have" (*ibid.*, 155); this points to Massenburg's blending of the terms proximity (depth) and prominence. Allan Moore (2012a, 31) uses the term 'prominence' to represent "sounds . . . more (or less) distant than each other" referring to perceived proximity or distance, the second dimension of his soundbox.

I use the term 'prominence' as a perceived emphasis of one material (element, domain, etc.) over another that is determined from a manner of attention, and from a direction of focus. I use the terms and concepts of 'depth' and 'distance' as dimensions of physical space—dimensions one might physically measure, or perceive, and/or interpret depending on context. The percepts of depth and distance are distinctly separate, and both are removed from prominence, which I approach as a manner of interpretation that may emerge as evoked from any element, sound or material.

When Albin Zak (2001) discusses 'prominence,' he approaches the concept more broadly, thereby touching upon several key concepts that are relevant here. First, he clearly identifies depth as existing at many levels of perspective, from the overall texture to the individual event—a central consideration related to confluence and the mix that applies to all elements. Next, he also extends his use of the term 'prominence' toward engaging the ways tracks reveal and emphasize elements and materials;

especially significant is his recognition of the "multifaceted nature of prominence perception" (2001, 156). He makes it clear 'prominence' is distinct from 'loudness;' though prominence might be established by loudness, it is only one facet that might influence the impression. Prominence may also be established by timbre, its level of diffusion in the mix (environment quality), ambience, sense of distance or location in the stereo field. This acknowledgement of the multifaceted nature of prominence is significant. Sounds will emerge from the confluence of the track at all levels of perspective, and within all domains—including the confluence of recording elements and the confluence of all that the track contains. These concepts support the roles of equivalence as a factor in the potentials of any element to have significance, and in bringing attention to prominence perception (Moylan 2015, 320–321).

A sense of shifting perspective—intentionally shifting the focus of attention from one level of detail to another—allows prominence to be recognized as a matter of context, relative to its surrounding materials and elements. Eric Clarke (2005, 188) describes:

> One of the remarkable characteristics of our perceptual systems, and of the adaptability of human consciousness, is the ability to change focus, and what might be called a "scale of focus," of attention—from the great breadth of diversity of awareness to the sense of being absorbed in a singularity.

A sense of control develops as attention is deliberately and clearly focused to various perspectives, various domains, various elements, and so forth. With this sense of control, a perception of prominence that is the result of context rather than bias has the potential to emerge. The analyst might then be able to choose whether to seek "what is most prominent within the texture" or "what appears to them the most prominent based on their own sensibilities"; the deliberate choice is what is important here. One choice allows the analysis to be based on (or at least emphasize) content within context of the track and within a culturally bonded interpretation, and the other emphasizes personal interpretation—of course, a continuum of shadings exist between these two poles.

Confluence of Recording Elements

The interrelationships and interdependence of recording elements manifest within the confluence of the mix, as each individual recorded performance (or track) is combined and mixed with others. Richard Middleton (1990) has framed this process as 'polyvocality':

> . . . the 'polyvocality' created by multi-mike or multi-channel recording. . . . within 'polyvocality' there is a *range* of effects, stretching, at one theoretical extreme, from montages of totally separate voices or sounds to, at the other, a completely blended mix-down; and different positions within this range, embracing different 'balances' of 'foregrounds' and 'backgrounds', changing 'perspectives' within stereophonic 'panoramas', different 'layerings' and 'dissections' of the musical 'space', are connected to different aesthetics.
>
> (Middleton 1990, 88–89)

This section examines ways data collected from several recording elements can be displayed to allow their individual traits and their interactions to be observed. What is offered is far from exhaustive, but can lead the analyst to determine how to most suitably explore elements within individual tracks. The most significant difference in these approaches to comparing elements lies in the factoring of time into observation methods. Some recording elements in some tracks are substantially fluid and temporal, changing over time. Other elements may be largely or completely static or stationary, their qualities fixed throughout a track or within sections; non-temporal graphs or diagrams bring a visual representation of the data of these elements. The temporal nature of any recording element within any track may establish surface rhythms aligned with the metric grid—though this is uncommon, especially an element like environments; elements establish their own pacing and morphology (changes of quality) within individual tracks, and also at each level of perspective.

The number of sources examined in diagrams, graphs or within any process might range from all sources present to a smaller number of select sound sources; a single source could also be graphed in all of these forms, allowing it to later be examined in great depth. Timelines for graphs might range from the entire track, to major sections, or perhaps single measures or more extended phrases. The span of time represented by illustrations and diagrams might be defined similarly.

Temporal Graphs for Comparing Recording Elements

Interrelationships of recording elements can be observed by comparing the observations of elements. The juxtaposition of loudness balance and performance intensity X-Y graphs discussed above allowed those elements to be observed simultaneously, as they evolved temporally over the duration of the example; changes of levels (either general or in detail) could be displayed in their magnitude and at the time of their change(s). Time marks the place (or moment) of change, and comparing these places of change represents rhythm.

Graphing two or more elements against a common timeline might display the elements' data so as to allow their interrelationships to emerge more visibly—in other tracks this effort might yield less richness.

This approach can be used similarly for any other combination of two elements to be compared at the same level of perspective. At the perspective level of the composite texture where the interdependence and interrelationships of elements manifest, we have identified five qualities of this texture:

- Pitch density
- Loudness balance
- Performance intensity
- Stereo imaging (image positions and sizes, transferable to surround sound)
- Distance positions

The reader may notice host environments and timbre are omitted from this listing. Timbre and host environments do not function directly at the composite level, interacting with others; they function most strongly at the basic-level of defining the character and content of sources into an identity ("an acoustic guitar in a small hall with vaulted ceiling") or at the overall texture (as timbral balance and holistic environment). These more complex percepts are comprised of several elements from this list functioning at a lower perspective.

These five qualities may be coupled into ten (10) X-Y graph pairs—ten ways the qualities of the composite texture might be observed in groups of two. Among the possible permutations, interesting evaluations might emerge from observing the following pairings at the perspective of the composite level, plotted against the same timeline:

- Distance positions and stereo imaging X-Y graphs
- Stereo imaging and pitch density X-Y graphs
- Performance intensity and pitch density X-Y graphs
- Loudness balance and distance positions X-Y graphs
- Loudness balance and pitch density X-Y graphs

In addition to the coupling of performance intensity and loudness balance offered before, other dual combinations of elements may be desirable, as they hold potential to generate pertinent observations

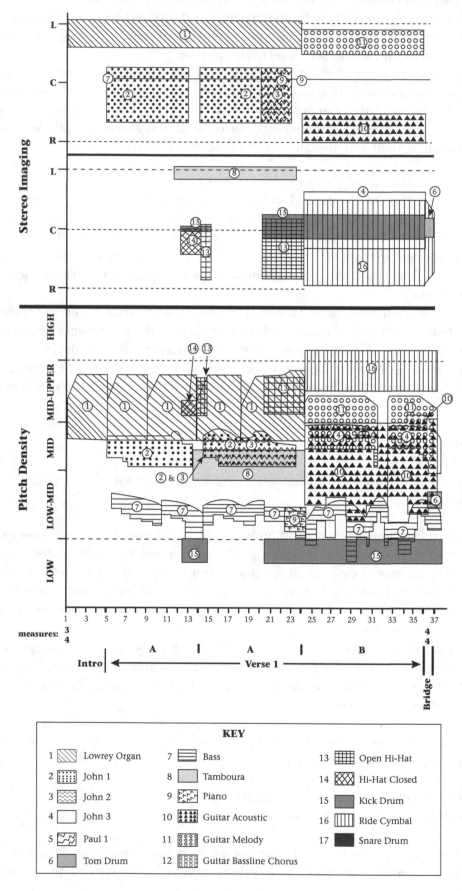

Figure 9.6 Temporal X-Y graph comparing pitch density and stereo imaging against a common timeline; from the Beatles' "Lucy in the Sky with Diamonds," *Yellow Submarine* (1999).

for an individual track (or section thereof). These other combinations are (1) pitch density and distance position, (2) loudness balance and stereo imaging, (3) performance intensity and stereo imaging, and (4) distance position and performance intensity. Some graphs are more workable than others, and the value of any graph is related its usefulness or appropriateness for an individual track; a graph's usefulness is based on the content of an individual track, the goals of an analysis and the intentions of the analysist. All temporal graphs can allow subtle changes to be illustrated and observations can be detailed, or they may be approached observing more general values.

Examining combinations of three or more elements might progress similarly at the composite texture, illustrating the various elements of basic-level sources. Figure 6.2 illustrated loudness balance, stereo imaging and distance positions for four sources against a common timeline. Any combination of elements may be examined in this way, so long as observations remain at the same level of perspective. Returning to the five composite texture qualities listed above, there are ten (10) possible combinations of three qualities, and five (5) possible combinations of four qualities that could appear on a single graph, on separate tiers and against a common timeline.

The combination of pitch density, stereo image width and location, and of distance position is one of the ten possible combinations that could comprise a three-tier X-Y graph against the same timeline. This combination is the same as the soundbox (explored later).

Non-Temporal Graphs and Diagrams for Comparing Recording Elements

Certain combinations of elements may also be charted as opposing axes on the same X-Y graph. Some combinations of elements are better suited than others for illustrating data. Figure 9.7 plots the stereo image size and position and the pitch density of several sound sources, positioning the sounds

Table 9.3 Possible recording-element combinations of three (3) and of four (4) elements that may interact and/or establish inter-dependence within the composite texture.

Combinations of Three (3) Qualities	Combinations of Four (4) Qualities
Pitch Density, Loudness Balance and Performance Intensity	Pitch Density, Loudness Balance, Performance Intensity and Stereo Imaging
Pitch Density, Loudness Balance and Stereo Imaging	Pitch Density, Loudness Balance, Performance Intensity and Distance Position
Pitch Density, Loudness Balance and Distance Position	Pitch Density, Loudness Balance, Stereo Imaging and Distance Position
Pitch Density, Performance Intensity and Stereo Imaging	Pitch Density, Loudness Balance, Stereo Imaging and Distance Position
Pitch Density, Performance Intensity and Distance Position	Loudness Balance, Performance Intensity, Stereo Imaging and Distance Position
Pitch Density, Stereo Imaging and Distance Position	
Loudness Balance, Performance Intensity and Stereo Imaging	
Loudness Balance, Performance Intensity and Distance Position	
Loudness Balance, Stereo Imaging and Distance Position	
Performance Intensity, Stereo Imaging and Distance Position	

in perceived lateral space and by frequency/pitch content—what some refer to as "spectral space" (Smalley 2007) or "pitch space" (A.F. Moore 2012a, 31). This juxtaposition works visually, whereas graphing other combinations from the ten pairings described above may not be as successful—for instance, materials in a graph of performance intensity on the X-axis and loudness level on the Y-axis may be confusing. These two axes are capable of presenting these two source attributes with as much precision and accuracy the analyst wishes to seek; this graph allows considerable detail to be observed for these two dimensions of the "soundbox," described in the next section.

Subtle changes or aspects such as small gradations of size, location, etc. are often significant to recording elements, and are often temporal in some way. All subtle changes and qualities have the potential to be significant to the track. An analyst choosing to use non-temporal graphs or diagrams

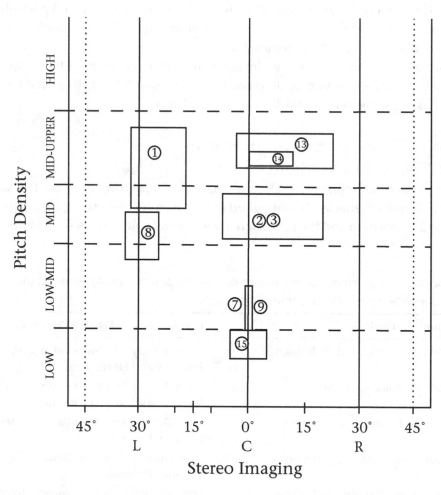

KEY		
1 Lowrey Organ	7 Bass	13 Open Hi-Hat
2 John 1	8 Tamboura	14 Hi-Hat Closed
3 John 2	9 Piano	15 Kick Drum
4 John 3	10 Guitar Acoustic	16 Ride Cymbal
5 Paul 1	11 Guitar Melody	17 Snare Drum
6 Tom Drum	12 Guitar Bassline Chorus	

Figure 9.7 Non-temporal X-Y graph comparing pitch density and stereo imaging; the Beatles' "Lucy in the Sky with Diamonds," *Yellow Submarine* (1999), 0:00–0:31.

may be forced to omit or condense details that cannot be incorporated into the format; changing formats for displaying data may be required if the data of the track cannot be clearly represented.

As these graphs are non-temporal (do not change over time), materials or elements that change over time are typically generalized into a single image. Elements that exhibit changes are difficult to notate, or illustrate, and might be generalized similarly. It follows that these graphs are inherently less detailed, imprecise to some degree (depending on the track and materials), and represent some span of time.

As time is not incorporated into these illustrations, the time period represented needs to be identified. These graphs (or illustrations) represent snap shots of time, or defined durations within which the graph's content is present. This time period can represent syncrisis time units (Tagg 2013, 385) of an extended present, an integrated auditory scene (Bregman 1990) of some duration less than a syncrisis unit or extending beyond the window of 'now sound,' a structural song section (an appropriate division for numerous elements of many tracks), or a generalization of an entire track. Any time span appropriate to the element(s) and to the track may be represented here. Typically, the longer the time span the more likely the illustration contained missed details, as the track's subtle information is increasingly absorbed into an overall impression.

The sound stage diagrams introduced in Chapter 8 are examples of non-temporal illustrations. Those diagrams represent defined periods of time; these may be considered as 'scenes' (Chapter 8) that take place over defined periods of time. Sources placed on those diagrams can be precisely located on the scaled sound stage (see Figure 8.9), for image location and width boundaries as well as distance position; those diagrams incorporate scales allowing detail to accurately place sounds.

Sources are located with less precision on the proximate sound stage (see Figure 8.8), where lateral and distance axes are not scaled. Image boundaries and positions are more generalized, though their relationships with other sources and the listener position are apparent. Figure 9.8 uses this proximate sound stage to localize sources in lateral size and distance from the listener point of audition, and also to provide an illustration of the size of source host environments, and their relationships.

Thus, the perceived depth of the sound source plus their individual host environment might be incorporated into the proximate sound stage, providing additional detail to the depth of the sources and of the sound stage. This allows source and environment placements to be conceptualized on the sound stage, though not placed in exact positions. Sources can be localized with proxemics (with as much detail as needed) though environment size is an interpretive approximation of the size of the space and the distance of the source from the front of the space and its rear wall. Some of these qualities are not present in artificial spaces. Sizes and locations of host environments are determined largely by comparing one sound to another, one space to others, etc. The non-scaled placements of sources within this format is more appropriate for these percepts which cannot be placed with proxemics or against a scale. The increments of space used to divide the scaled sound stage do not conform to these percepts; while one can identify precise widths and depths of spaces, we do have a sense of the distance of the source from the boundaries of the space, and can use that sense to assess the relationship of the source to the geometry of its host environment. Figure 9.8 demonstrates placements of sources plus their host environments on a sound stage—stated differently, this non-temporal figure allows three elements (stereo imaging, distance location, host environment size) to be illustrated in one place, and for comparisons to emerge. The track's multiple spaces, the interactions of spatial simultaneity (Smalley 1997, 124), and the interrelationships of host spaces to the holistic space might be observed aided by this illustration.

The Soundbox and Other Approaches to Observing Multiple Recording Elements

The approaches discussed here—including the soundbox and various sound stages—mark a transition from data collection and display, toward and engaging the potentials of the elements in shaping the

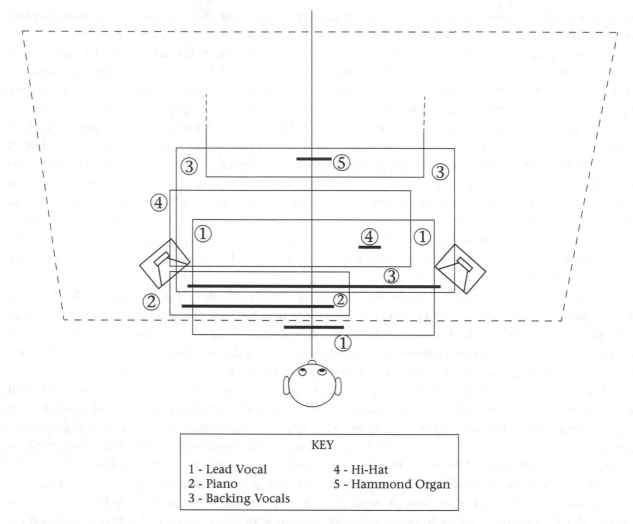

KEY

1 - Lead Vocal   4 - Hi-Hat
2 - Piano   5 - Hammond Organ
3 - Backing Vocals

Figure 9.8 Proximate sound stage of "Let It Be" (*Past Masters, Volume Two*, 1988) 0:00–1:01. Diagram illustrates sound source image positions and size, outlined by the widths and depths of their host environments.

record (we will seek to retain clarity between the acts of collecting or observing data, and of evaluating it). Each offer guidance to access and recognize the contributions of recording elements to the track, and the interdependence of elements as each contributes to confluence. The acts of examining and of recognizing contributions of elements are the evaluation and conclusion processes—processes that will be explored in detail within Chapter 10. Only a few of these approaches below offer ways to illustrate or notate elements. While illustration and notation (of all sorts) is rife with issues we have discussed before, it holds benefits of collecting, refining, visualizing and holding data; as observation progresses, it is simply impossible to accurately hold all information in one's mind.

The general qualities of the proximate sound stage have some similarity to the soundbox (briefly discussed in previous chapters). The soundbox contains the dimensions that numerous scholars engage when discussing tracks: stereo field, depth of sound stage and frequency range.

Allan Moore (1992) offered the soundbox as an approach to illustrating some of the primary elements of records; while in a way that has some similarity to sound staging it was devised quite separately. It is also distinctly different. The soundbox "is a heuristic model of the way sound-source location works in recordings, acting as a virtual spatial 'enclosure' for the mapping of sources . . . locations can be described in terms of four dimensions. The first, time, is obvious" (A.F. Moore 2012a, 31). The other three are the stereo image, distance (which he identifies as "perceived proximity of aspects of the image to . . . a listener") and "the perceived frequency characteristics of sound-sources" (*ibid.*). The soundbox is "almost like an abstract, three dimensional television screen" (Moore & Martin 2019, 149), positioning sound sources in frequency/pitch range (as in pitch density, above), in stereo positioning and image size, and in perceived proximity to (distance from) the listener position; using terminology offered within this writing, it combines stereo imaging, distance positioning, and pitch density. Like sound stage diagrams, the soundbox is at the perspective of the composite sound; it illustrates strands of instrumental timbre "conceived with reference to a 'virtual textural space,' envisaged as an empty cube" (*ibid.*). Fourth dimension of time represents a span of time, much like sound stages; illustrating changes to source positions in any of the three dimensions requires a new soundbox. It is challenging to make motion of images (changes of positions) clear in any illustration that does not incorporate a timeline — including the soundbox.

Allan Moore and colleagues have applied the soundbox to numerous tracks pursuing a variety of goals,[6] including a taxonomy study (Dockwray & Moore 2010). The soundbox can convincingly illustrate the relative placement of a moderate number of sources (adequate for many tracks) within a conceptual three dimensionality of space. It combines percepts of two perceived physical dimensions and one metaphorical conception of the "'highness' or 'lowness'" of pitch/frequency (Doyle 2005, 27). What the soundbox loses by way of precision of displaying data, it often gains in establishing a readily identifiable three-dimensional visual representation of sources. The similarity of the soundbox to the visual approach of representing sound sources as circles used by David Gibson (2005) has been acknowledged (Dockwray & Moore 2010, 224–225). Soundbox diagrams use simplified representations of specific sound sources — images of instruments and voices — morphed to occupy the three-dimensional space of the sound.

It should be clear the soundbox examines individual sound sources at the basic-level, as does the sound stage; this perspective allows comparison of sources at the level of the composite texture as well. These levels of perspective are the basis of approaches offered by the following scholars as well. As we have engaged many times to this point, the same percept can be defined differently on different levels of perspective. The audible pitch/frequency range (divided into registers) that I use to chart "pitch density" on the composite level (and timbral balance in overall texture) is defined as 'register' or the "height" of a sound by Allan Moore (2012a, 31) for the vertical axis in the soundbox. Lelio Camilleri (2010, 202) conceives the audible pitch/frequency range as "spectral space"; it is "height" to Anne Danielsen (2006, 52) and "frequency spectrum (height)" to Albin Zak (2001, 144); Jay Hodgson (2017, 220) approaches the audible pitch/frequency range as a "vertical plane." Considering a percept from a slightly different conceptual angle — perhaps as an object experienced in crystallized form — can change how one perceives the concept (element, percept, or confluence) without altering its substance; if solely for this reason, each of these approaches (and those of others) holds value for our analyses. There are other reasons to be sure; each considers similar aspects in unique ways, and some explore other dimensions of tracks. These approaches have some inherent differences, but largely the same percepts engaged from different angles.

Lelio Camilleri (2010, 201) approaches the interaction of recording elements as a three-dimensional "sonic space" "to indicate the space in which the piece unfolds in recorded format." The three dimensions are localized space, spectral space and morphological space. Localised space is the area wherein sounds are placed in stereo and mono, and includes depth, position and motion; this reflects two axes of the soundbox and also of the sound stage. Camilleri offers: "[T]he spectral content (timbre) of sound plays a relevant role in the overall perception of space. . . . the notion of spectral space . . . is metaphorical since there is no such physical space" (*ibid.*, 202). Within the spectral space [the pitch/frequency range of the track], the spectral content of the sounds used can establish experiences of saturation or emptiness within the space; in addition, Camilleri acknowledges the perspective of spectral space at the perspective of overall sound (timbral balance): "[T]he combination of the spectral content of sounds and their disposition can accentuate the various sensory experiences to be had from listening to the overall sound structure" (*ibid.*). The third dimension is morphological space, as sound unfolds temporally to develop the shapes of sounds, and perhaps evoke motion and a sense of direction; this can be at the perspective of sound source timbre, though its implications can manifest at all structural levels if one remembers the equivalence of elements at all structural levels, and timbre's central role as a recording element.

Albin Zak (2001) views the confluence of recording elements as a four-dimensional space, supported by incorporating concepts of mixing music offered by George Massenburg. The approach "highlights the interactive nature of the relationships among individual elements and larger composites—artifacts and gestures—and points to the ongoing shifts in perspective that a record makes available through its manipulation of 'four-dimensional space'" (Zak 2001, 144). Three of the dimensions are familiar: the stereo soundstage (width), the frequency spectrum (height), and "the combination of elements that account for relations of prominence (depth)" and the "fourth dimension is the progression of events, the narrative or montage" (*ibid.*). The fourth is temporal (as also identified by A.F. Moore and Camilleri), though it seeks information on all levels of perspective and acknowledges the unfolding of drama, structure and simultaneous, perhaps unrelated materials, elements or sounds. His use of 'prominence' was discussed earlier.

The soundbox is reframed with six components by Jay Hodgson (2017, 218–221). The components are auditory horizon, horizontal plane, horizontal span, proximity plane, vertical plane and vertical span. The auditory horizon "constitutes the total reach of the mix's 'earshot'"; horizontal plane "describes where a sound is heard in relation to center, and we call the total horizontal expanse of a mix its 'Horizontal Span'; the proximity plane "describes the position of sounds . . . vis-à-vis its Auditory Horizon . . . [and] represents a mix's ability to hear depth, with the Auditory Horizon comprising its far limit" (*ibid.*, 220). Hodgson identifies the proximity plane as perhaps the most significant component of a soundbox. The vertical plane and vertical span describe the mix's "capacity to hear vertically" (*ibid.*) and work together to identify the span (highest to lowest) of frequency content (vertical plane) in the mix. The "width, height, depth and temporal change" dimensions of the soundbox offered by Allan Moore (2010, 258) to discuss the textures established by interacting recording elements are engaged by Hodgson as dimensions of horizontal plane, vertical plane, and proximity plane; each 'plane' is then observed for its activity, with spans of farthest left and right image locations for horizontal span, lowest and highest pitch/frequencies for vertical span, and an auditory horizon used as a contextual reference for recognizing proximity (distance) of sources. At the perspective of the overall texture, horizontal span and auditory horizon align with three of the boundaries of the sound stage and vertical span represents the range of timbral balance for a track. Hodgson's work offers the defining dimensions of the soundbox, incorporating some additional concepts to present an approach that has the potential to open readily to typology and application to analysis.

Anne Danielsen (1998) has provided a unique conception of a soundbox—originally described by the term "*lydrom*" (meaning 'sound room' in Norwegian)—with some similar dimensionality, though defined with a sense of functionality. "Her conceptualization of the sound box . . . was an attempt to capture processes *within* the sound—for example, radical change and the lack of continuity in time and/or space caused by the montage-like aesthetics" (Brøvig-Hanssen & Danielsen 2013, 72) found within Prince's *Diamonds and Pearls* album (1991). This approach recognizes the recording elements that comprise the sound room (arranged similarly to the soundbox) as potentially acting in confluence and interdependence with the materials and elements of music. In discussing the interactions of pitch, timbre, dynamics, rhythm and melody merging into a heterogeneous groove sound of funk, Danielsen (2006, 51–52) identifies their interaction with the track's space and the spatial differentiation of sounds:

> [I]t seems useful to think of the music not only as a process in time but as a virtual *room* of sound. The organization of the many layers in a multilinear rhythmic structure may be imagined as a set of different positions in this three-dimensional sound box. Left to right might refer to a stereo mix and constitutes one axis (x), while two other axes indicate height (y) and depth (z), respectively. High to low (y) is linked to pitch and frequencies, while front to rear (z) depends on timbre and dynamics, or, in other words, close/distant and strong/weak.

The sound room—as applied here—illustrates a means to recognize the spatial delineation of musical materials, as they function to provide rhythmic propulsion to the mix of the groove sound.

With Danielsen's sound room, we transition toward the confluence of the three domains and the many performances within tracks. The temporal change dimension is the most problematic within the approaches cited. Except for collecting and observing rhythmic patterning, temporal change leads directly to evaluation, to structural hierarchies, and much more. We will return to these approaches to recording elements, and add others, as we engage data evaluation and formulating conclusions in Chapter 10.

The Mix as Metaphor: Confluence and Interdependence of Domains

> The analyst will . . . have to take this generic relativism [of the more or less equal importance of composition, performance, and phonography that varies by culture and genre] into account, as well as take into consideration parameters related to composition, performance, and phonography. In other words, not only the analyst will have to study, say, the piece's structure and relationships between melody and harmony, but also the ways in which vocal and instrumental timbres are exploited, or the relationships between vocal performance and lyrics. Furthermore, it will be necessary to examine how this ensemble is phonographically staged: [stereo imaging, host and holistic environments, distance positions] . . . this whole set of parameters could be considered as constitutive of what I would call the *phonographic discourse*.
>
> (Lacasse 2002, 9; emphasis in the original)

I read in Serge Lacasse's offering of "phonographic discourse" a recognition of the confluence and interdependence of the domains of recording, music, and lyrics and their performance. As examined in Chapter 1, the act of 'composition' reaches not only through the domains, but encompasses their performances as well; initiating a cyclic process, performances (their interpretations, expression, elaborations and improvisation) provide compositional elements, additional ideas and considerable nuance. The domains are the raw materials that the performance (with gestures both predetermined or spontaneous) is shaping—composition is realized in performance, and performance adding substance becomes the artifact that is the composition. The concept of 'mixing' all these becomes a metaphor for their interdependence and their confluence. Confluence acknowledges (as the domains combine and their elements lose their independent characters) that materials and elements blend to become something else; the performance (recording included) transports materials into other gestures and shapes by means of absorbing qualities from any element in all domains. The composition and the performance are one—as are the track's recording, music, lyrics, expression, and so forth. Each performance "is never exactly re-executable" even by the same artist(s) (*ibid.*, 8); only the captured moments of performance and the resultant crafted confluence that is the track remain unchangeable, and unique to all others.

Figure 9.9 illustrates a conceptualization of the interdependence of domains and the performance, with the resultant interplay and enmeshed texture established. Our activities thus far have been to explore the individual elements of each domain, and how they manifest in performance. We have also discussed their interactions in a higher dimension, and that they ultimately blend into a unique and coherent whole. To begin engaging this complexity of their confluence, we embraced all aspects of the track as equivalent, and we seek to observe all elements/domains at all perspectives with that intention.

The diagram of Figure 9.9 may be more properly examined from a polar opposite position: that the domains and their elements (and all else in the track) are *extractions* from the whole. Analysis deconstructs what is present into smaller parts, it does not add the observed parts to establish an overall texture. The parts we identify and observe cannot be summed to make the whole—the sonic experience and aural image of the track are different from any assemblage. What is present within

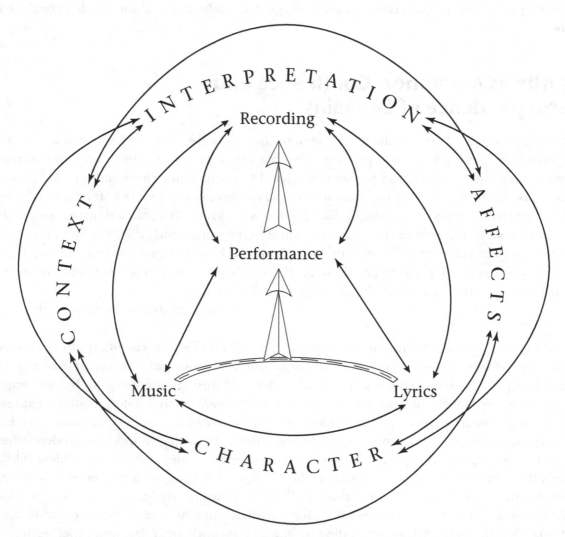

Figure 9.9 Confluence of domains and performance, within the context, character, affects and interpretation of the track.

our observations is what we have chosen—chosen to include, and chosen to omit. Our observations will not include all aspects of the track, whether by choice or from some eluding our attention. Further, and importantly, there are materials and meanings that are generated by the interdependence and interrelationships of domains and performance, as well as attributes of the track emerging from outside the domains and from outside the track. The track is not self-contained and isolated—it is situated in context, it establishes a unique character, and it produces affects within (and/or from) listener interpretation. The confluence of the track contains much, but it is the whole that is its essence, and its unique voice.

Observing the Confluence of the Track

Peering into this confluence is facilitated by observing the domains (and what they contain) simultaneously, at least to some degree. Observing these materials allows observing their interactions—in the best of situations, these observations would allow evaluations to follow, and lead to conclusions from those evaluations. We may thus identify covariances among dimensions/domains/elements, or the lack thereof, and other patterns or characteristics to connect domains. This is less straight-forward when comparing across domains (though challenges within domains can at times be considerable). It is through all of this that our characterization of the track—based on content and context—emerges.

A considerable amount of data in each domain has been assembled (over the course of the previous chapters). Displaying the data of all these observations in one place—in such a way as to allow evaluation—is challenging. In whatever manner one approaches information display, at this stage (with so much data) some editing will take place. This has already happened (to some degree) in the processes of examining materials/elements and notating or noting their qualities and activities. Seemingly significant features of the track guided the analyst to collect more detailed information on some elements than others; it is likely one doubled back later and collected more detailed information on elements or materials that were at first deemed less significant or went unnoticed. The process of displaying data will also acknowledge functions of materials; this allows contextual and supportive materials/elements to be included and their significance observed alongside the primary and secondary ideas. This begins the evaluations stage of the process.

Figure 9.10 offers a format to display organized and summarized data in a way that might facilitate comparison. Information collected from each domain can be referenced in this timeline chart without the actual data appearing, or with data appearing in little detail. This is a summary of observations that can be used to reference more detailed observations retained elsewhere.

Reviewing Figure 9.10, each domain has an area in which observations may be listed. The three domains are located separately; performance observations are woven within those domains. Domain areas are distributed around the track's timeline (which is divided into sections, but could have more detail if desired). Recording elements are located closest to the timeline, and in the most prominent location, because our goal is *recording* analysis—not music analysis, performance analysis, or lyrics analysis of popular records. Should one undertake an analysis that would emphasize music—or lyrics or performance—the arrangement of domains could justly be transformed to be appropriate for the track being studied.

Observations might be generalized for entire structural sections. Alternatively, element activity or materials may be placed against the timeline to illustrate their placement in time.

Element data displayed within any domain might represent specific material, salient features, features that *may* be significant, or some type of generalization. In placing information on this chart the analyst is engaging the evaluation process by the selection of what to include. Leaving space to add features to the chart can be important; further, at some point information on the timeline might be removed or condensed. Acknowledging primary, secondary, supportive or contextual roles will aid in

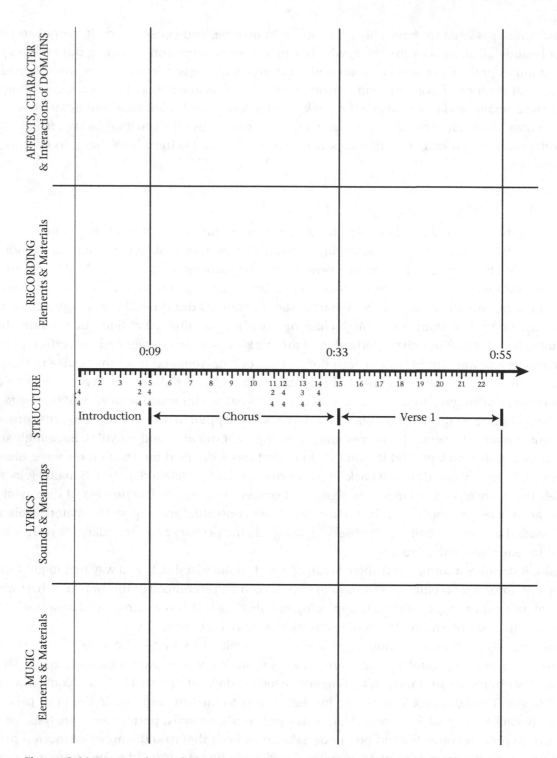

Figure 9.10 Master chart of element activities and of materials within the three domains, qualities added by performance, and of interpretation, context, affects, character and characteristics of the track.

organizing and delineating materials—in as much as it does not prematurely cause one to evaluate materials before examining all (or at least the most relevant) conditions. That which is prominent to the analyst could be appropriately noted—knowing what captures focus and attention allows one to determine attributes that bring interest, and allows one to willfully ignore what is prominent in order to discover subtle characteristics of attributes and nuance.

Interactions (i.e. complementary, parallel, delineated, or blended activity, etc.) of domains have potential to become visible here, and recorded in a location that sets them apart. Among interactions that may be explored are:

- Music and lyrics
- Lyrics and recording
- Recording and music
- Performances of individual sources, reflected in various elements/domains
- Performance of lyrics
- Performances of bonded groups of sources

An area of the chart is reserved for observing these interactions. There the affects generated within the track might also be noted. Aspects of interpretation, character, context and other observations can be included in this area. This chart serves as a way to track all of the information collected in a central location.

TIMBRE AS CONFLUENCE

Timbral percepts are not delineated by the domains we have been so carefully dissecting; no clear line separates it from other elements, or from one domain to the others. Timbre binds and infiltrates all other percepts into a coherent whole, a single aural image. This is—at least in large part—what makes talking about timbres, and observing and describing them so vexing.

Timbres are gestalts, are overall qualities; beneath this surface they are comprised of many dimensions containing subtle attributes of intricate nuance that are beyond our capacity to readily engage and perceive. Timbre is independent of its parts, however—it is a whole that is different from what the sum of its parts establish. Timbre does not belong to any single domain; a timbre blends domains through the soundings of performance. We know timbres by experiencing them; they defy notation and visual representations as readily as they defy description.

The relationships of and between timbre's acoustic properties (across elements and domains) and its physical dimensions, between its affects formed from interpreting musical expression and projecting performer physical exertion, between its sonic character, and between all these and listener perception and interpretation remain blurred. We cannot easily access timbral content or define its character because it is not simply an element of music, or of vocal production and language, or of recording; it is not simply a product of performance, or of instrument selection; it fundamentally relates to frequency (and pitch) and to amplitude (and loudness), but is much more complex; finally, it elicits symbols, images and associations from outside the track.

Here we acknowledge that the sound's content and character span all domains, as timbre functions and manifests in significantly different ways within the dimensions of tracks. Timbre represents the confluence of sound source, of musical materials, of performance, of recording's imprint, perhaps of language—and more.

Reaching back to Chapter 7, recall that to describe timbre is to address the sound's context and character as much as its content—this holds for whatever level of perspective we seek to understand. Here we will reframe 'talking about sound' by broadening our conception of timbral content, and thereby also broadening the scope of its character and its context. We will not seek to invent a vocabulary for sound, rather we will strive to define and describe timbres by the attributes and interactions of elements articulated earlier.

Two significant levels of perspective for timbre will be explored: (1) the timbre of sound sources and (2) the timbre of the track (the timbre of the track's overall sound).

The Confluence Within Sound Source Timbres

Timbre demands we engage interpretation to identify a source's sound, and we typically cease inquiry once we know (or can imagine) the timbre's source and the qualities it is expressing (message, degree of force, level of urgency, nature of expression, etc.). Timbre brings meanings and qualities that frustrate description and explanation and that are an important facet of the track.

Timbres are associated with sound sources, with their origin. Identifying timbres is typically naming a sound source; when timbres are heard we seek to recognize 'what it is,' and judge the timbre related to the inherent qualities of instrument types (their acoustic content). Source timbre represents a blending of causal factors and modifying influences (from the performance) with the materials performed (music and lyrics).

The sound source itself can carry cultural and stylistic associations within the musical context of the track, and can also summon personal meanings related to the source within the listener's interpretation. Further, the source may present drama and language, and a performer might contribute persona.

The performer adds dimension to the sound source, with their personal interpretative style and performance technique—this is also content, but blends into physicality. The listener's interpretation acquires a sense of the physical gestures of the performance; the level of physical exertion and expression (performance intensity) blends into the content and context of the timbre, and attention might be shifted to timbral character.

As this progresses, we often seek to define the character of a timbre to identify it by some analogy or cross-modal metaphor or other associative reference. This is all linked to interpretations, as Kate Heidemann (2016, 1.2) offers: ". . . describing timbre in the context of an interpretation motivated by visceral experience . . . [it is] difficult to find satisfying words or representations, and misunderstandings abound." To provide an over-simplified example: when we hear the voice of a friend, we identify the person (by the gestalt of vocal timbre, comprised of content) and immediately attend to the character of the voice (how her expression and mood are manifest within the timbral content and the context of the communication); this attending to character applies interpretation, which is rife with opportunities for mistakes (misinterpretation).

Within the sound source timbre is complex acoustic content, physical activities and gestures, and some sense of emotion or feeling and association or meaning. Our perception of timbre is different, though. We interpret an overall quality that is independent of these parts and other intangibles (affects, etc.). The sonic signature (see Chapter 7) that is timbre is not an addition of its parts, but a reality of its own; it is a coherent whole, or a gestalt.

As we seek to communicate with others about this coherent whole, about 'sound,' we attempt to share our subjective experiences and our resulting interpretations with others—we continue our attempts to describe sound with some clarity.

The Continuing Quest to Describe Sound with Shared Meaning

It is important to recognize descriptions of sounds are meant to inform others. We do not need to verbalize the qualities of timbre to make sense of it for ourselves. Our personal and even contextual understanding of timbres is nonverbal (including those expression and mood cues of our friend, above).

Timbres of sources (as well as the timbre of the track and crystallized form) are auditory images; as such they are sensory memories (Baars & Gage 2013, 30) and are not time-dependent and are nonverbal (Snyder 2000, 216). Auditory images are sound objects in content, but they carry much about context external to their source, and the confluence of all these establish its character. That tracks are listened to as containing auditory images might partially explain why so many aspects of music's aesthetic meaning, moods, emotion and expression defy verbal explanation (Clarke 2011, 197–202), and that their richness and clarity diminish as they are forced to conform to language.

With timbres (at all levels of perspective), we are engaging an overall quality that defies verbal description. Should we use language, when no single word can explain its complexity? Should we attempt to verbalize about them when even lengthy descriptions in language diminish, and do not adequately reflect, their multi-dimensional character and content?

Lawrence Zbikowski (2011, 186–187) offers a contrast that might help:

> I take the position that the primary function of language within human culture is to direct the attention of another person to objects or concepts within a shared referential frame (Tomasello 1999, Ch. 5). The primary function of music, by contrast, is to represent through patterned sound various dynamic processes that are common to the human experience. Chief among these dynamic processes are those associated with emotions . . . and the movements of bodies—including our own—through space.

Language externalizes a perception of the experience as an offering to others, and phenomenological musical consciousness is introspective and private, nonverbal, and infused with affects and the abstract.

To use language to describe the nonverbal is clearly incongruous—yet simultaneously it appears utterly necessary if we wish to communicate our interpretation of our experience of the track (phenomenologically or otherwise framed) to others.

For at least part of what we seek to accomplish in an analysis, some description of timbre appears necessary—regardless of the difficulty, or perhaps impossibility. The central role of timbre to the track's sound—and all the sounds it contains—is overwhelmingly obvious. As we engage the inherent multi-domain nature of timbral content and perception, this difficulty becomes clearer.

Chapter 7 examined the challenges of talking about sound at great length. In the end, the matter was advanced, but remained unresolved. Here there might be a bit more resolution, as the notion of timbre is broadened, the richness of its multi-domain gestalt is acknowledged, and confluent elements across domains plus external factors might form a more viable approach to timbre analysis and description.

To make this shift will require discipline and attention, though. Our natural inclination is to seek to describe any sound—any timbre—with a single word or a few descriptive words; often with words utterly unrelated to sound.[7] We articulate an interpretation to represent a complex gestalt, reaching for language to describe what we interpret as core qualities and character of the 'sound' plus all it brings forth in us, but rarely do we address its content, context or nuances of character. When we use ecological terminology (Clarke 2005, 197), we can be prone to overly simplify observations into highly personalized interpretations, though this need not be the result—should we decide to approach timbres differently.

As our data set of attributes increases with the confluence of all aspects of the track, it becomes clear no single word can represent all that timbre contains. We can transition to a definition of timbre that includes a description of its attributes and their interactions; terms such as those in Table 6.6 might have adequate meaning when the sound is defined, should the analyst (listener) wish to continue this practice of using them. Such terms could just as readily be abandoned.

We can expect no direct vocabulary to emerge to address timbres, though. In effect, the complexities of timbre ensure the unique qualities of each to be a multivariable calculus formula in itself; the relationships of its parts as important as their content, their interactions establishing further depth and breadth of content and meaning the formula could not predict. Each sound is different from others—note to note, between and within instruments, one vocal sound or voice to the next; the timbre of each track is different. Engaging these relationships and interactions, as well as content and context, might provide some tangible timbral information, albeit an incomplete approximation. Some shared meaning and understanding of timbres might emerge from descriptions based on observations of their content, character and context.

To summarize and recall what has been covered elsewhere: a hierarchy relationship exists between character and content. Character describes the overall quality (gestalt) of the timbre; content defines the attributes and traits of the component parts within the timbre. Context is external to the timbre, to the sound or the track—timbre's associations and meanings connect it to matters outside the track, and also how these may situate the timbre within the track or establish a conceptual frame of reference for the sound.

Typology of Timbre in Confluence

As domains become blended, examining the confluence with a goal to define or describe its timbre gestalt appears overwhelming, and provides no real access point. It seems more appropriate to shift to an articulation (observation and examination) of physical attributes, of perceptual impressions and interpretations, and of the perceived physicality of sound production. Describing timbre might then turn to the attributes and their values (or variations) within three views of timbre: (1) those that are interpretation-related, (2) those that engage the content of the waveform, and (3) those that relate to the physiological. Information pulled from these categories might function toward defining a timbre by contributing to (1) its character, (2) its acoustic content, and/or (3) the context in which it is situated or which it establishes; some attributes may clearly be associated with content, character or context, while other attributes may apply to several, though differently to each.

Table 9.4 is a listing of attributes that might be included within a timbre typology table at the perspective of the sound source. The collection of variables (attributes) selected will be most effective when it conforms to the salient features of the sound source being studied. This listing will provide some guidance in assembling a suitable timbre typology table; it is not intended to be all-inclusive.

Rarely will one attribute (or element) dominate or dictate the character of a timbre. Timbre's components are always interacting and interfering with one another—like micro-auditory streams. Timbral content is essential to the gestalt context; it emerges from all three domains and is the basis of what happens physiologically as it elicits interpretation from perception and other factors. Descriptions of timbre attributes are incomplete without content.

Using these categories and functions as references, the inner workings of timbre might be described with some consistency and substance. This approach might establish a framework of sorts for others to understand what one is identifying and to communicate something of substance. Clearly a single word for a sound will not emerge from this process. Describing sounds will be far more involved than offering the first descriptive term that comes to mind; this will be a decidedly positive step toward discussing a timbre's character and content within the contexts of tracks, or as sound objects independent of context.

It is possible to discuss timbre by describing elements relative to one another. Each timbre has its own formula or algorithm of how the domains and elements (in their content, character and context) blend in confluence to establish its unique nature. An open process of description based on

Table 9.4 Interpretation-related, acoustic and physiological attributes that might be included within a timbre typology table at the perspective of the individual sound source.

Interpretation-Related Psychological, Perceptual	Physical Content Acoustic	Physiological Visceral, Implied
Source identification or recognition	Inherent acoustic content of sound source	Visceral connection with performance
Expression: musical, dramatic	Dynamic envelope	Implied physical gestures
Levels of energy, force or exertion	Spectrum	Levels of energy, force or exertion
Strain and ease of performer	Spectral envelope	Strain and ease of performer
Clarity or distortion of performed sound	Definition of fundamental frequency	Visceral feelings of affects
Affects, moods, emotions	Space: width and location	Idiomatic modes of performance
External connotations and associations (such as cause)	Space: distance and depth	Deviations from idiomatic playing
Tension level of sound changes and musical movement	Space: echo, reverb, environment	Tension driven motion of performance technique
Semiotic meanings attached to sound source	Space: spectral content of reverberation	Performance techniques
Language communications	Timbre of time	Implied meanings of imagined physical gestures
Symbolism	Modifications to physical dimensions	Level of difficulty of materials
Perceived meanings of paralanguage sounds	Noise elements within spectrum	Athleticism of performance
Realism and surrealism	Inherent timbral traits of performance style	Energy, intensity, exertion, speed
External associations elicited by breath, body, and performance sounds	Content of breath, body, and performance sounds	Persona
Drama, persona	Language and paralanguage sounds	
Listener connection with performer		

observation and evaluation of all pertinent timbral components and of its gestalt qualities is proposed here—typology might facilitate this.

Describing sound can become an act of addressing the values of its attributes. The process of investigating and observing attributes will allow one to identify the features that are defining features of the sound, to guide further observation of other attributes. The attributes of Table 9.4 can be observed, and their values collected; when collected, observations can be recognized as relating to content, character and context and categorized on the typology table. With acquired facility, these two steps might be combined. Formulating a description requires some evaluation of data; entering evaluation now, prominent attributes that provide the timbre with distinctive traits are identified and described. Attributes that provide distinctive ornamentation to the timbre may be pertinent, and certainly

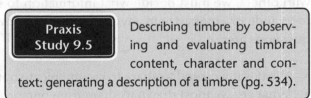

Praxis Study 9.5   Describing timbre by observing and evaluating timbral content, character and context: generating a description of a timbre (pg. 534).

Table 9.5   A general typology table for timbre, with attributes spanning timbral content, character and context.

	Variable (Attribute, Dimension)	Value (Traits)
Content	Dynamic envelope	
	Spectral content	
	Spectral envelope	
	Definition of fundamental frequency	
	Noise components	
	Formant frequencies and characteristics	
	OTHER	
	OTHER	
Character	Overall quality or distinguishing nature	
	Emotions or sub-emotions	
	Expressive qualities	
	Aesthetics	
	External associations	
	Physiological connections	
	Energy, intensity	
	OTHER	
	OTHER	
	OTHER	
Context	Source or origin of the timbre	
	Semiotics	
	External associations	
	Cultural meanings and connections	
	Conformity to the texture (blend or revealed)	
	Functional relationship to other sources	
	Aesthetic	
	OTHER	
	OTHER	
	OTHER	

other observations will bring further connections and interrelationships between elements. A detailed 'definition' replaces the single word description; the definition identifies the distinctive features within its content, the nature of its character, and its relationships to contexts in which it appears. It will be helpful for this definition to acknowledge that summing these parts will not adequately reflect the whole of the sound. The whole of the sound is something other than what these parts put together might represent.

Given the purpose of language—and the description of timbre is verbal—is to share information with others, we must decide what information to share. The questions that arise, then, relate to what to include or emphasize in a typology table, or how to interpret what has been observed. Some useful questions are:

* What is it we need to share to communicate our observations?
* What are we most drawn to share from personal bias?
* What is needed to represent the fundamental traits of the timbre?

- What traits provide the timbre with its unique character?
- What physical components provide it with its unique sonic quality?
- What within the timbre is important to context, to expression or to its function?
- What is required to achieve the goals of the analysis?

The Timbre of the Track

This timbre of the track is the 'sound' that is a significant trait of any record. The confluence of all sound sources—and all that they contain and represent—brings all percepts within the track to fuse into a single distinct timbre; it is a global, aggregate texture, and also a single impression.

The timbre of the track has a duality of content and character. It also establishes and reflects the context of the record—its overall affects and aesthetics, its energy and expression, its sense of directed motion and level of intensity, its atmosphere and sense of space, its drama and more, all coalesced into a singular and complete intrinsic character of the track—into a coherent whole. The aspects of content, character and context are the core 'sound' of the track—they establish and embody its sonic signature (sometimes called signature sound) that can often be recognized from just a brief exposure to a track.

The timbre of the track exists simultaneously as one dimension of crystallized form, and also as the fabric of all the track's content at the highest level of structure and of perspective. This extends the duality of the track (1) as 'character' conceived as a single sound object or aural image (where all is present at once and exists without constraint of time), and (2) as 'content' conceived as an event experienced as the confluence of all materials and elements as they unfold over time within context of the track's activities and structure, and as they coalesce into a single overall, changing texture or fabric (or timbre).

When manifest within the structure of the track, this overall timbre of the track is at the highest structural level, and is conceived at the highest level of perspective. Within structure, the 'timbre of the track' is temporal, changing dynamically over time, and it contains all of the materials and elements of all domains—included are the nuance and subtleties of all elements and materials at all dimensions, wherein the qualities of all can be observed and 'appreciated' for their contributions to the whole.

The timbre of the track's aggregate texture is comprised of superimposed strands of materials and activities of musical materials, lyrics, and recording elements. The features of this texture are unique for all genres of music and to some degree for each track. A typology of the aggregate texture may be related to density and range, to the number of strands and the placement of the strands within the range of a particular element, to the relationships of elements and domains, etc.

In considering recording elements, insight into this confluence can be obtained by contrasting program dynamic contour, timbral balance (pitch density) and loudness balance as if they were (respectively) the dynamic envelope, spectrum and spectral envelope of the track. The interdependence of these establish a gestalt that is at the core of the timbre of the track.

It may be helpful to pause for a moment to remember these concepts, and to reframe them here. Program dynamic contour is how the overall loudness level established by all sounds and the materials they present evolve over the duration of the track. Pitch density is the pitch/frequency range that each sound source (and their musical materials) occupies within the spectral space (frequency range) of the track's timbral balance. Loudness balance is how each sound source of the track (each that is also represented in pitch density) change and interact over the duration of the track. These three dimensions change fluidly, unfold temporally, and shape the sound structure of the track—they embody the timbre of the track just as dynamic envelope, spectral content, and spectral envelope are the content of an individual sound's timbre. As timbre reaches across all elements and domains to establish character, the qualities of music and lyrics (etc.) contribute to the timbre of the track. The spatial identity of the track (highest dimension) that includes the traits and interrelationships of host environments and the holistic environment also provides components within the timbre of the track.

Timbre of the Track and the Dimensions and Domains of the Track

Confluence (of all domains/elements, performance and outside influences) permeates the timbre of the track, as well as crystallized form. Confluence resides within the content of its texture, and character. Confluence includes the affects of emotion, energy and expression, along with the semiotics of meanings emerging from each domain. These—along with listener bias—are all included within the listener interpretation of the character of the timbre of the track. The reader will recognize these have all been explored previously in great detail.

Here we will consider how the timbre of the track and crystallized form are reflected in and relate to the character, context and content of the track.

Crystallized form—which will be explored more deeply in the next section—might be conceptualized as a stationary physical object; it can be identified as a stable 'multidimensional outer layer' with 'a myriad of activity on a host of perspectives' occurring internally. This conceptual 'outer layer' is the context of the track and the overall shape of the track; the conceptual activity that is internal to the object is the timbre of the track. The timbre of the track is the myriad of activity of internal content (structural hierarchy, and the domains/elements that comprise it) of crystallized form; crystallized form also carries additional qualities beyond the timbre of the track. The totality of the inner activity and the outer layer is crystallized form, as an all-encompassing character of the track.

The 'outer layer' of crystallized form is the highest level of perspective of elements and confluence within each domain, and interdependencies of domains. These contribute directly to the overall character and context of the track. The timbre of the track is at the structural level just below. The timbre of the track has an overall quality of character—just as do timbres of sources—that is established by the content of all that is within the track and the contexts that they carry or establish. The timbre of the track is structural, temporal and changes of over time; crystallized form is an aural image, non-temporal and conceptualized outside time.

Recalling what was explained in Chapter 2, recording elements can serve a primary, supportive, ornamental or contextual function in shaping the track at the highest level of perspective. Of these functions, the contextual function creates references for the activities of materials (and the elements that create them) to assume primary, supportive or ornamental functions.

As recording elements manifest at the highest perspective, contextual recording elements are present in all tracks. Their quality provides a consistency throughout the track, and also establishes a reference against which other elements can be observed consistently. Contextual elements are stationary or static; their values or qualities are unwavering and do not change throughout the course of the track. While some values of these elements may be revealed slowly over the duration of the track, in total they establish a context and frame of reference against which all activities may be related (and evaluated).

Table 9.6 lists contextual dimensions, beginning with those that are established by the interdependence of all the track contains. Contextual elements for music and lyric domains are more fluid than those of recording. Tempo (for example) will always be present and contribute to context, though (conversely) tempo can be quite fluid; other stylistic traits such as tonality, groove, beats and ostinatos, etc., may or may not provide an individual track with context. Context represents a backdrop

Table 9.6   Elements that may function to establish a sonic context within the timbre of the track.

Domain	Element
Confluence of Domains	Reference dynamic level
	Timbral Balance
	Overall shape or form of the track
	Overall emotion, energy, expression
	Embedded degree of tension and motion
	Degree and qualities of any final resolution
Elements of Recording	Holistic environment
	Sound Stage boundaries
	(stage left-to-right width, front edge of sound stage, & rear wall from depth)
Music Elements	Tempo, beat, groove, ostinato patterns, tonality, hooks, *other*
Elements of Lyrics	Singular impression of drama, mood, tension (etc.) of overall conception
	Subject, meanings of lyric, story or drama, language style, *other*

against which all other like attributes can be gauged, and their essential value does not change. The elements of music are highly variable in function, and those that are contextual in some tracks may well differ in others; the key or tonality of a track is a prominent exception. Lyrics' contributions to context are also quite variable, and often to some degree unique to individual tracks; lyrics' content is often linked to subject matter or message, and what they communicate can vary (even markedly) between individuals.

The temporal qualities of domains at the highest dimension contribute the substance of timbre of the track's character and content; these present materials that are fluid—melodic relationships, harmonic motion, morphing timbral balance, evolving storyline. These, as all structural components and relations, occur over time and cannot be instantaneous (Handel 1993, 186). This content is embedded within all structural levels (levels of dimension or perspective), and can function as primary, secondary or support materials/elements in shaping the track. Table 9.7 lists the structural materials, organization and relationships within the domains at the timbre of the track's structural level.

The timbre of the track manifests into one of the dimensions of crystallized form. It is a large dimension concept of aggregate texture of all domains that result in a timbral quality of the track—one of internal content, overall character and connections with external contexts. Crystallized form provides a different angle on the character and context of the timbre of the track.

As will be explored next, crystallized form contrasts starkly with the timbre of the track; the timbre of the track is an unfolding gestalt-sound at the highest level of structure and of perspective, whereas crystallized form exists outside temporal experience and coalesces in memory. The timbre of the track represents the internal activity within crystallized form's shell—a shell that is frozen outside time as an instantaneous manifestation of the track's presence.

Crystallized Form

The character of the timbre of the track leads to crystallized form. Crystallized form will conclude this discussion of confluence. Our discussion will broaden, though, as when we engage crystallized form we will progress into the connected topics of deep listening.

Crystallized form is (1) a quality inherent to the experiencing of a track and (2) a sense of deeply knowing and comprehending one's interpretation of the track as an all-encompassing presence and impression. It is the combination of perceptual attributes, abstractions, social significances, aesthetic

Table 9.7   An incomplete listing of elements that are temporal and variable within the content of timbre of the track; these represent the structural materials, organization and relationships at the timbre of the track's highest level of perspective.

Domain	Element
Highest Structural Dimension of Domains	Hierarchies of music and lyric materials
	Hierarchies of recording elements
	Variable affects: emotion, energy, expression
	Semiotics within each domain
Elements of Recording	Program dynamic contour
	Timbral balance and pitch density
	Loudness balance
	Timbral qualities and performance intensity
	Sound stage positions of sources (lateral and distance positions, including image width & depth)
	Host environments of sources
Music Elements	Musical syntax creating motion and tension
	other
Elements of Lyrics	Unfolding story or drama
	Word usage and meanings
	Sounds and rhythms of text
	Language syntax

gestures and embodied experiences that for the listener/analyst personally, from one's own vantage, characterize the track. Crystallized form is what is most memorable about the track—to a particular listener, or listening analyst—as it is formulated in reflection. What crystallizes in this quality of form may be highly personal, or one might acquire the ability to step outside their subjective vantage into a position of greater connection with others or that may allow substantive academic discourse.

The defining and holistic qualities and concepts of crystallized form are:

- Cognized as a whole, is apprehended in an instant as a singular manifestation
- An impression, atmosphere, ambiance, aura that constitutes the track's presence
- Realized through the temporal experiencing of the track
- Coalesced retrospectively through introspection
- Multidimensional sound object, aural image (or auditory image)
- An inherent, singular identity with multidimensional features that results from the convergence of the content and character of all elements and materials of all domains (including affects, energy, message, drama, etc.), and all that they illicit from within the listener's biases and experiences, and their cultural context
- A sense of awareness and of knowing the core, essential nature of the track
- Coalesces within listener interpretation
- Establishes and embodies the context of the record
- Represents the character of the track

Inherent within crystallized form is (1) what is unique to the track and what constitutes its content, character, and context, and also (2) what is most salient and meaningful to the listener. Listener interpretation and their subjective vantage, and the (perhaps) more objective position of the analyst are explored and contrasted below.

Intrinsic Nature of Crystallized Form

Crystallized form is the manifestation of the entire track in an instant of realization; the entirety perceived at once. It is a single large-scale aural image, that is a nonverbal percept, an interpretation of the track, and a memory of the listening experience and reflections. It is the highest dimension of form. It is experienced as a sense of knowing or understanding.

Crystallized form may be framed as the essential nature of the track, as its singular intrinsic character as a whole. Crystallized form may be considered as the presence of the track—equivalent to that 'feeling' or 'understanding' (or something other) we experience after a motion picture, as we nonverbally reconcile the story, drama, characters, plot, ending, etc. into a single 'sensation' or 'mood' or 'spirit' or 'impression' or 'whatever it is that we experience.'

It is, perhaps, the 'higher essence' any work of art contains that allows it to reach beyond the human condition to transcend its combined materials, reason, imagination and emotion; as the quality that allows the track to communicate similarly to many; perhaps at times some may experience it as a 'higher consciousness.' Crystallized form can be "something felt to be greater than oneself, yet somehow within oneself" (Burnham 2001, 195). The phenomenological within experience and consciousness are inherent to crystallized form; as such crystallized form is (1) a quality inherent to the experience of an art object—whether visual, aural, dramatic, etc.—and (2) a state of knowing and comprehending the track as the memory of its experience.

As an aural image and a large-dimension sound object—a unified and multidimensional entity or conception—it might be understood "as equivalent to an image schema: a cognitive construction that represents the abstract qualities of sound rather than a single perception of it . . . [with] invariant properties that can be examined by 'looking' from a variety of perspectives . . ." (Bourbon & Zagorski-Thomas 2017, 3). Its substance reflects not only its formal shape and domain content, but also the affects, meanings and aesthetic expressions of the track.

Origin of the Term

The term 'crystallized form' aligns with the principle that every track is unique. Each track is multidimensional, reflecting different dimensions, shapes, qualities—just as the surfaces, levels of transparency and angles of a geological crystal. Each track may be conceived as a physical, sounding object, all qualities present at once—that can be viewed from a variety of perspectives. Approached in this way, as crystallized form the track can be turned (like a crystal) to appreciate it from another different vantage points, though its content has not changed; each hearing of the track represents "an incomplete view of the 'object' from a single perspective" (Bourbon & Zagorski-Thomas 2017, 3). The track's richness will be further revealed from observation at each successive perspective, of every new vantage point, from considering character and confluence from a variety of angles (and so forth); as implied above, crystallized form contains all strata of perspective and structure, providing access to all levels of detail (from the all-encompassing whole to the microscopic) without altering content.

The term 'crystallized form' also emerged in analogy with ice. Water crystallizes when it freezes; while all motion of the liquid is stilled, it remains comprised fully of its original substance. Further, the ice crystal captures the moment of its formation; time is frozen at the moment of its realization. Within each ice crystal is a unique light patterning with many rarefactions, a unique size and shape, colorations and more; each track brings its unique patterning and sense of motion, its size and shape, the qualities of its materials and elements, and an inherent, singular identity that emerges from the convergence of its many parts into a holistic whole.

Chapter 2 compared visually observing a sculpture to crystallized form; this analogy could apply to any complex physical object, such as a crystal. This analogy is also helpful in recognizing that the object can be observed as a coherent whole, independent of its component parts—connecting with the concepts of acousmatic listening and sound object. At this highest level of perspective, crystallized form is most readily approached as an object and holistic perception; "holistic perception implies that objects are not broken into their component parts but simply perceived as wholes" (Neuhoff 2004d, 250). When shifting perspective to begin to notice the track's component parts at their highest perspective, the relationships of components, as much as the components themselves, establish dimensionality to crystallized form. As listener perspective draws closer to the object, detail is added to dimensions and analytical perception and the parts can be separated from the whole to be observed (or analyzed), and then "glued together to form the whole" (*ibid.*)—as with timbre in Chapter 7.

Continuing this physical object analogy, we might accept the track in crystallized form as stationary and without external motion. Internally is the myriad of activity of the timbre of the track—with its independent motion on a host of perspectives—that is contained by the stable, multidimensional outer layer of crystallized form. In this way, the impression of crystallized form establishes a context for all that happens within the track. The crystallized form is complete at every moment in time; in essence, it exists out of time. Crystallized form disregards temporal evolution[8] because all that it is—its entirety—is present at every moment. Denis Smalley (2007, 37–38) notes: "I can collapse the whole experience into a present moment, and that is largely how it rests in my memory."

Aural Image and Memory

Crystallized form reflects the notion of music as memory; the track existing out of time, 'heard' simultaneously in an instant. Its totality[9] is held in long-term memory and accessed with retrospection and introspection. The experience of crystallized form relies on memory, and each person's memory of an experience will differ. Our memory

> is dynamic and mutable and interacts with other processes. Thus two people experiencing the same event may have different memories of it. It is not simply that one person is right and the other wrong, each person's outlook, knowledge, motivation, and retentive abilities may alter what is retrieved.
>
> (Baars & Gage 2013, 285–286)

Each person's sense of a track's crystallized form will differ; within the same cultural group the differences may be slight, but not necessarily. When one accounts for personal interpretation of lyrics, of performance intensity gestures, and the like, subtle differences can make for substantial meanings.

Crystallized form is manifest within the listener's awareness after the track is experienced. It comes to be known (perhaps understood) on a nonverbal, intrinsic level. This is an interpretation that may result from an informed accumulated process, or at some level it may be noticed immediately. Formulating an impression of crystallized form likely begins the first moments the listener hears a track; the non-reflective, casual listener will obtain their interpretation of this overall impression, just as will the seasoned analyst. During the stages of collecting data for domains/elements/materials observations, a sense of crystallized form might gain richness. The impression of crystallized form will become more nuanced during the process of hearing, assimilating, recognizing, discovering, and experiencing the track on many levels.

It is entirely possible that crystallized form is what attracts one to a record. To learn about crystallized form is to learn how the track comes together, and also what is memorable and significant within the track—for us personally, or as an object for analysis. It is the opportunity to perceive its complexities as they fit into its grand scheme. Alva Noë (2004, 118) has observed "thought and experience are, in important ways, continuous," recognizing that between perceptual awareness and thought awareness there can be no clear distinction. Denis Smalley (2007, 40) recognizes this overlap allows for understanding the condition of his concept of space-form, "which although gathered in time, can be contemplated outside the time of listening. . . . think about the . . . soundscape now, without perceiving it." Perceptual awareness informing memory, brought to awareness through reflection.

Recognizing Crystallized Form

Crystallized form is approached through holistic listening, and also through reflection.

Crystallized form might be engaged early in the analysis process—as the analyst becomes aware of salient qualities during sessions of open listening. In a way, it is an impression that is continually evolving for the analyst, though for the lay listener its impression might become rather fixed. All listeners—even those with little experience and knowledge to inform their impressions—will formulate an interpretation of crystallized form, and that impression, mood, expression, understanding may be quite personal. This impression of crystallized form can seem to arise instinctively.

As the analyst engages the content of the track, focus shifts away from crystallized form in many directions. Crystallized form is a central topic within the analysis process when establishing the context and character of the track. In the end, the last task of the analyst might be to reflect on what they remember. What is it that remains? Think not just of the track, but also of your memories of it. Bring emphasis to your memories of the experience of it, as opposed to your analysis of its parts.

A few questions that may guide thoughts are:

- What is it that remains and establishes a nonverbal aural image of the whole?
- What is most memorable? Can you conceive this without labelling it with language?
- What in it has stuck with you, personally? (This is awareness of subjective vantage.)
- What is most meaningful? Can you recognize this without naming it?
- What appeals to your sense of taste, your listening biases?
- What reflects the cultural norms relevant to the track?
- Can you settle your memories of the track into a single, nonverbal impression?
- How much of this impression might be shared by others, and how much is personal?

Reflection is on the presence that remains after the experience, allowing a memory of the whole, as one impression, to coalesce. Crystallized form is not approached through considering the specific content, or deductive thought of prominence or significance; these higher cognitive functions of analysis elucidate the inner workings of structure (content), but are counterproductive for recognizing the singular presence (character) that is crystallized form. Crystallized form is a complex aural image in memory; it is not a temporal experience. Its qualities are those that are most memorable and meaningful to the listener; this inherent nature is based on listener interpretation of the nonverbal aural image, along with any context that arises.

Holistic listening—a type of deep listening—is the act of perceiving the complete track as a single object (aural image) that is not broken down into its "component parts but simply perceived as a whole" (Neuhoff 2004d, 250). Crystallized form is encountered through holistic listening, through opening to the experience of all at once—the work as a coherent whole, one single impression within one's awareness and as a conscious experience. The qualities of crystallized form evade verbalization just as do the timbres of sound sources, and the timbre of the track—for all the same reasons.

An interpretation emerges in memory, in the silence after the track has ceased. In quiet reflection on the impression (perhaps a feeling or mood) or the presence (perhaps an air, atmosphere, aura or ambience) that comes forth—any specific materials recalled in memory are 'heard' within this context. Many of the factors for approaching the interpretation of reference dynamic level apply to recognizing crystallized form. These will be explored next, as deep listening and open listening lead us to a position of being aware and open to all that arrives, while also being fully passive toward seeking and processing sounds; one holds a position of only observing.

DEEP LISTENING

Deep listening and open listening can offer guidance for experiencing and recognizing crystallized form:

- Allow one's self to be open to sense its presence; searching to try to analyze or to make sense of crystallized form may often divert attention from the overall presence
- Attempt to limit the tendency of rational processing, calculating, comparing and searching for a deduced answer; these pull attention to other perspectives, introduce prominence and utilize different memory functions
- With intention, be receptive to the track's global auditory presence as an elevated experience, transcending the confluence of sounds, accessed through holistic listening and nonverbal reflection
- Experience the silence after the track as transcendent of the sounds that preceded it; hold the opportunity to experience the track's crystallized form within that silence
- Awareness is nonverbal, and directed to character that defies language, not to qualities that can be defined

Deep listening guides the experience of the track, and also can guide the process of recording analysis. It opens the analyst to experience the track differently—such as the presence of crystallized form can be revealed or the holistic listening of the track experienced. Deep listening can allow discoveries of what was not previously apparent or perceivable—for example, dimensions of sound not previously experienced, such as the subtle sonic dimensions of space.

Woven throughout the previous sections and chapters have been encouragements to listen deeply. To listen deeply can take many forms, with all based on or partially reflecting one or more of the following:

- Being fully attentive to what is present at this moment
- Listening without processing significance or imposing function or structure
- Listening that minimizes memory's influences
- Listening without expectations based on prior materials or activities within the track
- Listening that arrests anticipation
- Listening without judgement, prejudice, personal bias
- Listening intentionally: without an agenda or with a specific agenda
- Listening with focused attention to a specific level of perspective and element
- Listening with attention to all that arrives: adapting equivalence
- Listening with open awareness: attention that accepts all that arrives
- Listening without language: nonverbal listening to aural images and unfolding events
- Listening that permits deeply knowing any object at the focus of attention
- Listening that facilitates deeply knowing the singular impression of the track
- Listening that opens awareness and retention for introspection and reflection to follow

Some of these forms have already been thoroughly introduced in Chapter 2. Many of these listed will be explored as this discussion unfolds. In this writing, deep listening functions as a common thread that interconnects observations, evaluations and conclusions. Deep listening's aware attention and its openness to all that arrives are resident throughout the framework for analysis, in all its steps and concerns.

Records are created by deep listeners, for deep listening. Yes, they speak immediately, viscerally and profoundly and this is what grabs our sensibilities and shapes our memories—they also present great richness and detail. There is much more present on, under and over the surface, and what is there is often what shapes tracks fundamentally, and what is there can be of great relevance and interest in studying the track. Albin Zak shared a pertinent personal experience of deep listening:

> The aural images that records place before us have a detailed sonic intricacy about them, a sensory richness that invites us to listen to them again and again. I have often found careful listening can be repaid with delight even when the song and the performances are not to my taste. I also find that the harder I listen, the more I am aware of how much there is to hear and how much the record artfully withholds from me. For on many records textural depth is provided by instruments that are barely heard, or often not consciously heard at all. Like the invisible undercoats that build up the surfaces of paintings, sounds and musical parts may lose their individual identities and become inaudible distinct characters, yet their presence is felt in the overall sound and affective sense of the texture.
>
> (Zak 2001, 86)

A Tradition of Deep Listening

Deep listening has a long tradition that has come forward from experimental music throughout the twentieth century. It is evident in music composition, performance and the listening practices to engage the new ideas of the experimental, of what has not previously been encountered. A connection with listening to nature is evident in many approaches: many emphasize open listening and listening within the present; the notion of intermingling real-life listening and music listening is commonly used to contrast open listening and directed listening.

The latter is evident in the 1920s compositions of Erik Satie and Darius Milhaud that produced "furniture music" and brought surrounding noises into the experiences of music performances, and incorporated into music listening. Referring to Thoreau, John Cage has stated "Music is sounds, sounds around us whether we're in or out of concert halls" (cited in Schafer 1971, 1).

John Cage represents an early key figure in which deep listening was substantive in his musical thinking and practice. In his book *Silence* he advocates open listening and non-directed deep listening by his typical abstract inference:

> A technique to be useful (skillful, that is) must be such that it fails to control the elements subjected to it. Otherwise it is apt to become unclear. And listening is best in a state of mental emptiness. Composers are spoken of as having ears for music which generally means that nothing presented to their ears can be heard by them. Their ears are walled in with sounds of their own imagination. . . . cause and effect is not emphasized but instead one makes an identification with the here and now.
>
> (Cage 1961, 154–155)

The listening within the present 'here and now,' not hearing because the composers' imagination and expectations distort the experience, listening with mental emptiness (open attention) are all integral to the deep listening concepts we have been engaging.

Pierre Schaeffer—already covered within acousmatic listening, with its approach of pure sound disassociated from its origins and associates, void of external context—could not have engaged sound objects without deep listening. Since its beginnings in 1948, his approach to sound analysis had been based on directed deep listening within timbres (Schaeffer 2012). Occurring at the same time as to John Cage's open and non-directed use of deep listening, Pierre Schaeffer utilized a deep listening technique that was directed inward, toward exploring the nuance within sounds; interestingly, the two hold in common a position that the origin of a sound is not pertinent to the experience of it.

R. Murray Schafer (2004, 34) identified a "blurring of the edges between music and environmental sounds is the most striking feature of twentieth century music." The alternative he put forward:

> My approach . . . has been to treat the world soundscape as a huge macrocosmic composition which deserves to be listened to as attentively as a Mozart symphony. Only when we have learned how to listen can we make effective judgements about the world soundscape.
>
> (*ibid.*, 37)

He goes on to identify the significance of silence to listening and to acoustic ecology, and notes with the deep listening techniques he teaches: "the whole body becomes an ear . . . and [students] have heard music as never before" (*ibid.*, 38). Schafer relied heavily on deep listening concepts in his research and compositions related to acoustic ecology and soundscapes.

Pauline Oliveros shared a pertinent personal experience of deep listening:

> The canyon and creatures joined us as we played, and we played until our awareness became imbedded in the canyon and summoned a ghostly, floating train, an apparition of metal meeting metal, reflected doubly, triply, endlessly in the canyon, from the mind, from the flickering passenger windows, the rumbling ties. OUR EARS FELT LIKE CANYONS. We didn't speak until morning.
>
> (Oliveros 2004, 105)

The listening experience of open awareness widened enough to include how the sound was interacting with the large environment, sounds from creatures, as well as their performance; these all intermingled and became one; using open listening, this experience included all that arrived without judgement or prejudice.

These different approaches to deep listening can all be useful in recording analysis.

Deep Listening for Recording Analysis

Deep listening provides new listening opportunities for recording analysis, and also in record production and engineering. These opportunities relate to the listener's intention for how they will use their attention. Attention and awareness may be either *directed* by deliberate focus or it may be *non-directed* by holding an open awareness. Both forms have purpose, function and value to the analyst. Each is most effectively engaged by the exclusion of the other. Both directed and non-directed deep listening require concentration, disciplined attention and an ability to remain focused on awareness of sound.

Deep listening may be utilized to bring the analyst a sense of focus and attention to sound that can be particularly effective for engaging the sonic worlds within records. This is the 'deep listening' that has been mentioned previously. Deep listening can be *directed* with intention to any singular aspect of the track. This approach was engaged often in discussions of recording elements—such as with listening *inside* the gestalts of timbres and environments, identifying distance positions and relationships, recognizing the edges and size of stereo images, and so forth. Deep listening such as this can be directed to any element in any domain, and at any dimension. The focus of deep listening can be directed to nuance in the overall impression of an element (such as program dynamic contour) progressing through a continuum of levels of dimension to the nuance of activities within timbral components. Listening for this nuance—as well as hearing and recognizing the substance and contributions of all that is within the track—can be facilitated by adapting and broadening the openness to sound framed as equivalence.

Equally important, deep listening can be *non-directed*—listening without guiding intention. With this approach to deep listening attention allows all that arrives to be an equal presence within one's awareness. This allows discovery, allows experiencing the unexpected, of what could never have been anticipated, of the utterly unique, or the experiencing of what has never before been experienced, of being prepared to hear the unexpected.[10] Listening with open awareness allows all sounds to be experienced without consideration of their origins or functions; sounds can just be, they can simply exist, and their emerging and diminishing presences experienced.

Deep listening with open awareness is the experience within the present. When in practice, it inherently arrests the distortions of anticipation and minimizes the effects of the prejudices and preferences within our bias and subjective vantage. Practicing non-directed, open awareness is not so simple, though. It is dependent upon one's ability to concentrate without fixating, to not be distracted by what occurs and yet to be fully aware of it, to withhold judgement and yet perceive significant detail and also broad perspective.

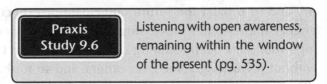

Praxis Study 9.6 — Listening with open awareness, remaining within the window of the present (pg. 535).

Deep Listening and the Present

The psychological states of 'being present' and being aware within 'the present moment' are integral to deep listening. Since the beginnings of the practice of deep listening these factors have been central to it. Being present and of being aware of the present moment are abilities that can be developed—developed to the benefit of performing recording analysis. They are mutually supportive: deep listening itself cultivates being present, being present allows deep listening, the awareness of the present moment is the window of attention that allows being present, that opens deep listening.[11]

The practice of listening deeply is the act of listening only. 'Being present' is having one's attention consumed with what is happening. For us, it is a concentration on the listening process to the exclusion of all other thoughts and distractions. One's attention is dedicated solely to listening. One acquires the ability (skill) to remain focused on what is being heard, and being unaware of all other thoughts and sensations. This cultivation of concentration and attention is central to deep listening; it is also relevant to the other types of deep listening that involve memory (discussed below).

The present moment is what is now; it is the moving time window of our existence. The present moment appears resident within the 2 to 3 second time "window of consciousness" (Snyder 2000, 9)—a duration curiously similar to the 'buffer' of echoic memory. Echoic memory functions in early processing[12] where "information persists as an echoic memory, which usually decays in less than a second like an echo" (Neisser 1967, 189–194); indications are that this storage and processing may extend from about 2 seconds to perhaps as much as 4 seconds depending on context. Conscious awareness relies on perceptual processing of sensory systems (including audition) and as a storage system (echoic memory) (Snyder 2000, 4; Zbikowski 2011, 185). Echoic memory holds unprocessed auditory stimuli for a short period of time until the following sound(s) are heard, and the sound made meaningful. This is the 'specious present' that we experience as the present moment, our 'right now.' Within this space of time the experience is pre-reflective "with a certain breadth of its own on which we sit perched, and from which we look in two directions into time [to the past and the future]" (James 1890, 609).

From this position that we experience as now, our natural tendency is to shift our concerns and attention to higher-order mental functions where we can process sound in memory and where we can anticipate what is most likely to follow (based on our knowledge and experience of context). These are activities that open awareness seeks to mitigate (discussed below). To be present is to remain in the moment, to not reflect on and make connections with the past or to anticipate the future. Deep listening establishes a vantage that the past is irrelevant (what has happened is over and thoughts of it are distractions), and any notions of the future (expectations, anticipation) are conjecture (we cannot predict what will happen and any thoughts of it are distractions).

Memory and Reflection

Deep listening is keen awareness. This keen awareness can aid in establishing memories that are rich in detail, accurate and objective. All aspects of the track—such as auditory images, musical materials, structural relationships, and so forth—can be retained through deep listening. Much in the track can be perceived nonverbally and without cognitive processing (Damasio 2003, Edelman 2006) and made resident in memory through deep listening.

Deep listening's goal when non-directed is to shift the natural tendency to process sound in memory and to project the future (anticipate or expect something to follow). Open awareness does not engage higher-order processing for anticipation or expectation, yet can allow memory without imposing order. This deep listening can experience the track's inherent qualities without verbalizing them, without privileging some over others, and so forth. Auditory images may be retained in the abstract, appearing in nonverbal form, but nonetheless establish a fixed impression or presence; they appear resident in long-term memory as "an auditory image at the center of musical thought" (Butler 1992, 188), where the ordering of events are not fixed (Baars & Gage 2013, 286; Neisser 1976, 112). This approach to deep listening might be considered holistic listening.

Directed deep listening involves higher-order processing when it engages memory; it can also enhance and develop the ability to retain information. We hear backwards in time, and all sounds are resident in memory—including those within the window of the present. Directed deep listening can utilize memory to retain information and experiences of the track. Memories that can form the basis of reflection and introspection are important to the character of timbres and crystallized form—among other interpretations of tracks. Deep listening can aid introspection, allowing it to be employed without extreme subjectivism; the overly personal (and even added levels of fantasy) can be avoided with refined attention that cultivates a stable and vivid awareness. As sounds are perceived without judgment and prejudice, the influence of one's subjective vantage can be minimized.

Deep listening's retention of aural images and events that allows reflection is integral to evaluation processes. One's experiences of the track—its materials and ideas—can be evaluated in memory and related to other experiences within the track. This allows the analyst to more easily perform analysis steps to reach insightful conclusions, and also to hold the experiences needed to interpret the character of crystallized form. Long-term memory is enhanced by deep listening, and it is vital to numerous recording analysis processes. With deep listening a perception of the track that is not based on working memory during listening, and not of recalling episodic memory events, can be realized. Working memory is highly influenced by our predispositions and attention, and recall of episodic memories has a lower level of accuracy than recognition of remembered episodes (Baars & Gage 2013, 253–288). This state might be observed with intention to be unencumbered (as much as this is possible) by personal inputs of interpretation, listening or aesthetic preferences, cultural influences, past experiences with the track or artists, personalized outside associations, and more. Intention and choice are what is crucial here, as they shape the experience and interpretation.

Arresting Anticipation and Minimizing Bias

Deep listening engages what is present; it allows us to hear what is there. This contrasts with our natural tendency for anticipation, an expectation that something related to the past will follow. We can listen hoping to hear what we believe will happen, or what we want to happen. Expectations and anticipation can bring one to hear what one wishes, whether or not it occurred.

Deep listening can arrest anticipation by listening in the moment only—by listening to what is present, when it is present. It does not process what has happened to project what will follow, rather it is concerned only with what is happening now.

Deep listening can bring listening without predicting. Being open to all that arrives, and not privileging some over others, mitigates the memories of what has happened. Without holding those memories there is no basis to project the future, no basis for anticipation or expectation.

An equal awareness and accepting all that arrives can in itself minimize bias. Deep listening allows being present without judgment of what arrives—no matter the qualities. What happens is the object of observation, not one's preferences or personal reactions to those objects. Should the analyst desire the experience, deep listening can be used to identify prejudices, expectations, likes/dislike, preferences, what one is drawn to, and so forth. Deep listening can allow awareness to shift to these topics; by mitigating the influences of memory and expectation, an awareness of one's biases can become more acute.

In these ways, holding an open awareness allows the track to unfold without imposed distortions of listener bias. Listener bias can be directed toward the content of the track, aspects of imposing order on what is heard and identifying substance and functions; listener bias can also take the form of the subjective vantage of the listener.

Subjective Vantage of the Analyst Listener

Each of us carry our own vantage from which we engage the world—including the worlds within records. Our listening experiences are unique as they carry all of our sensibilities and histories, our hearing mechanisms are unique, and how we make sense of what we hear will be different for each of us. We all are attracted to what we are attracted to, and for our unique (typically nonconscious) reasons; we remember what is prominent to our attention, whether significant to the track or superfluous. The analyst, as well, is human and holds these qualities.

In framing an analysis, analyst/listeners will benefit from perceiving their own subjective vantages. This allows one to become aware that their memories are their own—and may not be shared by others. The analyst may focus on that which is most memorable to that analyst, on what is most meaningful to themselves, on that which seizes their attention. This is useful information, and deep listening can facilitate revealing it.

In deep listening from the vantage of the subjective and personal, one can learn much about one's self. One can be aware of their biases—their preferences and prejudices, the ways they are distracted and attracted by materials or elements, and so forth. Remaining inside the subjective vantage, an analyst might learn why a track speaks to *them* (if that is desired), they might develop their personal practice and skill levels, and they may learn how their subjective vantage is impacting the analyses they wish to be objective. To listen from the position of the subjective is to be deeply within the personal. This is not always desirable or functional in recording analysis, though; one's personal perception of the track will likely not reflect what is actually present.

The analyst may choose to remain within the subjective vantage or to move out of it—in much the same manner as the analyst shifts attention from one level of perspective to another.

Stepping outside the subjective vantage allows the analyst to partake in objective analytic discourse. A neutral vantage can be established (as much as this might be possible) from which the track can be analyzed with minimal bias. A dispassionate position, objective in its assessments, and centered on the objects of study is the ideal basis of many assessments and analyses—these lay in stark contrast to the subjective vantage.

With intention, the subject of the listening experience can be shifted just as we have previously shifted between objects of focus or between levels of dimension and perspective. Learning to recognize one's listening position is central to this. Learning to be aware one is listening from within one's personal subjective vantage is possible; through deep listening with attention to how one is reacting to what is heard—to what is speaking to the individual's affective senses, to what *they* are finding meaningful or moving, and so forth—the analyst can become aware of the nature of their subjective vantage, and how it manifests. Aware of our own biases, we might develop an awareness and ability to slip in and out of the subjective vantage.

It is equally important to learn to listen objectively and dispassionately, putting aside the personal and subjective. This is easier when listening to content than it is when listening for character, and the subjective vantage often privileges aural images over their inner workings. We will benefit from the skill to shift between these two vantages, because they clarify each other, as well as supply different—often pertinent—information about the hearing of the track.

Once aware of the two states, and once skillful in distinctly assuming one vantage over the other (listening from within each position), the listener/analyst can choose which vantage to assume. One vantage can then become the basis for the analysis, privileging one over the other. Even in an analysis keenly focused toward neutral observation and objective assessments, the analyst might learn to recognize their subjective vantage in relation to the analyses they perform; the subjective vantage can inform how they are inherently prone to distort data, and also the subjective vantage may be broadened toward the culturally inclusive, beyond the personal.

> **Praxis Study 9.7** Cultivating awareness of being within your subjective vantage and of listening from a more objective position; shifting between these positions with intention and with attention to what each offers (pg. 537).

A Knowing of the Track

Deep listening and the recording analysis process affords a deep learning of the track, and establishes a deep sense of 'knowing' it. With every new hearing, any amount of richness, in detail and breadth, might be added to our sense of the track. Greater clarity of crystallized form's presence and character may emerge as our memories and experiences of it accumulate—and also the materials and the details down through all structural levels. The analyst/listener's interpretation gains insight and encounters new attributes and nuance. The deeper the understanding of the details of the track (while maintaining a sense of perceptually balanced prominence), the more aware one may become of its overall substance and presence. The result is a sense of deeply 'knowing' the track as a unique and singular presence— a sense of awareness of its fundamental nature.

One comes to a place of awareness in 'knowing' the record for its core substance through immersion in the track's broad-reaching concepts and the unique affects that define its context. Included are small details and large-dimension gestures. Through deep listening, this 'knowing' can encounter the track's holistic character, and an experience of its essential nature. Included are all that the track contains in its confluence, and of course all that the track's content can elicit—'knowing' can ultimately extend to each of any of the individual elements and ideas, aural images and structural elements, etc. within the track.

This 'knowing' awareness is important functionally as well. It allows recognition of the largest-dimension characteristics of the recording that are essential aspects of crystallized form—allowing a recognition of the full dimensionality of crystallized form will likewise impact understanding of the music, of the track's performances, of the lyrics, and of recording's attributes. It also acknowledges and includes all of the functioning of elements give rise to the motion and movement of all domain elements. It is a sense of the kaleidoscope of the content and character of the track, in all its nuance, and also the complex web of their interrelationships.

Deep listening provides the portal for acquiring this sense of knowing. Deep listening allows one to encounter the track from a position that privileges none, a neutral position that holds all that occurs as intrinsic to the track, a vantage that does not allow the future to be distorted by the past. Deep listening allows the listener/analyst to develop an interpretation that is considered on the basis of what is experienced, with minimal prejudice and bias, with minimal memory-related expectation, and with a sense of the passing window the present, that ultimately broadens to encompass the track's entirety.

CONCLUSION

As nine chapters have unfolded, we have encountered many forms of listening; they are listed in Table 9.8. Within that list, many are related to or interrelated with deep listening. Throughout this book you have been encouraged to listen deeply, to search for nuance and dimensionality, and to be open to the notion that everything in the track may be an important defining feature. Nearly all of these listening techniques can function within the processes of recording analysis.

As we progress further into the analysis process in Chapter 10, let us remember here that it is only through listening that we engage tracks. At the core of successful listening is attention and intention, discipline and concentration, and nonjudgement and awareness.

There is much to discover within tracks. The challenge of recording analysis is to discover what is within the track, and to accurately perceive its content and character through listening alone. Our ability to listen deeply and accurately will be rewarded by what we can unveil. Our ability to engage deep listening in all its facets, and an open awareness to many potential qualities within tracks, will uncover qualities within records that may well otherwise go unnoticed.

Table 9.8 Forms and approaches to listening.

Acousmatic listening	Listening inside sounds
Active (engaged) listening	Listening with attention (attentive listening)
Analytical listening	Listening with intention (intentional listening)
Aural analysis	Music listening
Casual listening	Non-directed listening
Critical listening	Open listening
Deep listening (directed & non-directed)	Passive listening
Detailed listening	Pharmaceutical (mood modulating) listening
Directed listening (deep listening)	Real-world listening
Ecological listening	Recreational listening
Holistic listening	Reduced listening

Some Questions for Recording Analysis

Are you able to focus your attention on the sensation of loudness—separating it from timbre, performance intensity and other percepts?

Are you able to condense all sounds within the track into a single sensation of loudness that changes moment to moment (program dynamic contour)?

Can you get a sense of the experience of a reference dynamic level from reflecting on a track you know well? Reflecting on a track you have recently encountered?

Are you able to distinguish between listening to the loudness dimensions of an individual sound source and, shifting perspectives, listen to the loudness balance between two or more sources?

Are you able to identify a track's reference dynamic level?

What are the loudness relationships of sound sources in the first few major sections of the track you are studying? Are you able to relate them to a clearly defined reference dynamic level?

Are you able to distinguish between prominence and loudness? Prominence and distance?

Considering the track you are analyzing, how might recording element X-Y graphs support your observations of loudness at various hierarchical levels?

Which recording elements are displaying the greatest activity within a track? Which are establishing the context of the track?

How might recording element X-Y graphs support your observations of several elements simultaneously for the individual track you are studying?

Are you able to list 2 or 3 attributes with values of a lead vocal's timbral content, character and context?

How will you describe (what attributes will you address, and how will you describe their values) the timbre of a sound source? Of one specific sound?

Are you able to define the timbre of the track of your current analysis project?

Are you able to experience crystallized form by reflecting on a track?

When you reflect on a track, and especially on what it means to you, does a single impression emerge?

What is it that you experience when you listen to a record? Where is your attention? What is meaningful to you? Interesting?

Can you describe your subjective vantage?

How is your subjective vantage playing a role in your current recording analysis?

Are you proficient at shifting into and out of your subjective vantage, and into and out of a more objective, dispassionate analytical position?

NOTES

1 The threshold of hearing (audibility) is often commonly used as this reference. This physical reference pressure is 0.0002 dyne/cm^2.

2 See Handel 1993; B. Moore 2013; Roederer 2008; Schlauch 2004; and Schnupp, Nelken & King 2012.

3 Those who hold functions in the recording process: producers and a wide collection of engineers such as tracking/recording, mix, mastering, sound design, editing, sound reinforcement, broadcast, audio, etc. One person may serve a single function or many.

4 'Aggregate loudness contour' or 'overall loudness contour' are viable terms as well.

5 Compression applied to this overall sound is responsible for the increased loudness of many current and recent tracks. Legendary mastering engineer Bob Ludwig (2013) provides a clear explanation of this matter of "Loudness Wars."

6 See A.F. Moore 2001, 121–126; 2012a, 29–44; 2012b; Dockwray & Moore 2010; Moore, Schmidt & Dockwray 2009; among others.

7 I often notice that I, too, can feel an innate compulsion to describe a sound with a single word or two—this tendency to distill timbre—though I am deeply aware of the consequences of doing so and refrain. This appears to be a strong, natural tendency.

8 Similarly, Denis Smalley (2007, 37–38) recognizes an important aspect of a holistic view of his space-form as an "experience [in which] I disregard temporal evolution."

9 Or, more precisely, those portions that arose as significant enough to consciously remember, along with all that was acquired and resident in non-conscious and unconscious awareness.

10 Still relevant here, in *The Principles of Psychology* (1890, 402) William James wrote of attention: "My experience is what I agree to attend to. Only those items I *notice* shape my mind—without selective interest, experience is an utter chaos. Interest alone gives accent and emphasis, light and shade, background and foreground—intelligible perspective, in a word."

11 These concepts can be cultivated from a variety of approaches—such as mindfulness meditation, meditation in other traditions including contemplative and spiritual practices, and a myriad of other, more secular ways of improving attention, concentration and intention. If the right fit for the person, any one of these practices can successfully aid in developing the skills of concentration and awareness that are sought here; the reader may find any of these worthy of their personal exploration.

12 In early processing "the inner ear converts sound into trains of nerve impulses that represent frequency and amplitude of individual acoustic vibrations" (Buser and Imbert 1992, 156–171).

CHAPTER 10

Analyzing Recording Elements: Their Contributions to the Record

This final chapter will focus on analyzing recording elements and bringing the analytical framework into practice.

Throughout the majority of this chapter each recording element will be examined for ways it has contributed to specific tracks. Numerous and varied examples will be incorporated into these analyses. These examples not only illustrate how the concepts presented previously apply to different types of tracks and different approaches to analysis, they also will illustrate interrelations of elements and connections between structural and analytic perspectives.

The goal for recording analysis within the contexts of this book is to explore and begin to unravel some understanding of how the elements of recording contribute to the track—to the sonic qualities of the track, to its artistic expression, to what it communicates. To this end, the functions of recording elements within the confluence of domains will be examined, as will their contributions to the track. When examining recording elements within the context of the track, the contributions of the music and lyrics (Chapters 3, 4 and 5) to that context cannot be ignored. Music and lyrics, though, will not be given substantive coverage in the evaluations and conclusions presented—as these are not the focus of this book and in-depth coverage is offered by other sources.

In preparation for analyses of recording elements, the framework and its four guiding principles will be briefly revisited. Then the three-step process will be re-introduced in greater detail. Evaluations and conclusions (the second and third steps) will be explored in depth here, as observations (the first step in the sequence) have been discussed throughout Chapters 3 through 9. These steps then form the basis for analyses and discussions of recording elements that will take place throughout the remainder of this chapter.

Before we engage these topics, let us revisit briefly the broader goals and uses that analyses may serve, and a general set of variables for analysis. This will prepare the reader for approaching the analysis of recording elements.

A GENERAL TYPOLOGY OF ANALYSIS

Breadth and depth of coverage are central to defining the goals of an analysis. An analysis might seek to be comprehensive, blending what one unveils in the three domains with pertinent outside connections into a single study; this would be quite an undertaking, and would produce a chapter of great length, if not a stand-alone book. In contrast, a broad study such as this could be undertaken without significant depth of detail, and still yield significant findings for some analyses. Shifting to other approaches, an analysis may, instead, seek to focus on one domain above all others (such as a track's music) or it might explore the cultural significance of a track; this brings a sense of narrowing the analysis by its focus on subject areas. Within a narrower focus, one may seek to study some single aspect of the track (such as its rhythmic content) or to study a single element over all others (such as its use of spatial environments). Ultimately, scope and focus must match purpose.

The reasons for conducting an analysis will mean different things to individual analysts—and individual listeners. To some it will be music analysis that comes forward as the subject of interest, and to others it will be lyrics. There will be those that are fascinated by artists and their intentions, and cultural studies will draw many to social concerns. Subject areas within tracks are numerous; many disciplines can be brought into an analysis. There are a myriad of goals and reasons for analyzing records, and as many angles; many of these have various sorts of well-established methodologies from related disciplines that may be brought to bear for a successful evaluation.

An examination of recording elements may be performed within any analysis process, and should occur concurrently with any musical analysis or examination of lyrics. The techniques for data collection are of course different, and the manner and approaches to evaluate that data, and to draw conclusions of course differ. Still, recording elements should be regarded as equivalent to the other domains, and subjected to the same rigor of evaluation within the process. Music analysis methods and approaches that might be pursued within a recording analysis may include reductionist practices, psychological approaches, formal approaches, comparative analysis techniques (Cook 1987), kinetic-syntactic, referential (Meyer 1967, 42), among others. Music analysis methods were devised for art music, and their relevance to the analysis of recorded song will vary with the track being studied and the approach and operational method of the analyst; music analysis was addressed in detail in Part I. In concept, the notion of emphasizing an approach toward the substance of the track and a method of operating on that substance can be very generally transferred to the recording domain; this will be illustrated in certain areas as the chapter unfolds (for example the different approaches to structure, below). It will also be evident that divergent results can be identified from the same collection of data; any single track can be analyzed and interpreted in numerous ways.

Table 10.1 has been loosely adapted from Ian Bent's *Analysis* (1987, 80) and has been augmented with topics from our framework. This table covers general approaches to setting analysis goals, accounting for the biases of analytic approach and methods, media for representing observations and conclusions, and other matters. Many of these areas have already been discussed, and others will appear in the following pages; generally accepted analysis techniques may be found in a variety of sources (follow this note for sources).[1] From this collection of topics, the analyst may explore and choose what is most appropriate for them and their analysis practice—or the specific analysis being undertaken. This is offered to provide some guidance in an analyst's self-reflection on their approach and how it may be privileging some results over others. It is by no means comprehensive, and should be modified to suit one's needs and preferences.

All concepts within Table 10.1 embrace the three domains equally. Although the rhetoric in the table aligns most closely with music analysis, the concepts should be connected to recording elements and to lyrics as appropriate. There is great diversity in this compilation, and no method is being privileged in it—nor are any being promoted. On the contrary, all that is being proposed is an approach whereby the elements of recording may be recognized in equivalence with the domains of music and lyrics, and each is worthy of analysis in their own right.

Intention and Recording Analysis

Exploring the track for *recording* analysis assumes the elements of recording will be central objects of study, that recording elements will dominate attention. An analyst will often be drawn to topics outside

Table 10.1 Recording analysis process areas and their variables.

Analyst's individual traits	Biases, preconceptions, preferences, tastes, etc.
Goal of the analysis	Breadth and depth of coverage
	Subject areas covered and emphasized
	To determine how the track works, what traits make it unique, or what brings it to stand out above others
	Determine how the track speaks to the audience and what is says
Analyst's approach to data collection (Observations)	Selection of domains, elements and materials to be examined
	Derived or existing techniques and approaches for each element
Analyst's view and approach to the substance of the tracks	Form as shape, and a closed network of relationships of parts
	A concatenation of interconnected structural units
	A field of data in which patterns may be sought
	A linear process of tensional motion
	A string of symbols and emotional values
	These categories are not exclusive; two approaches may exist at different levels (large, middle, small dimensions)
Analyst's operating methods (Evaluations & Conclusions)	Reduction techniques, separating out non-essential information, foreground and background
	Comparisons; recognition of unique identifying traits, similarities, or common properties
	Segmentation of structural units
	Search for rules and syntax
	Counting of features
	Reading-off and interpretation of expressive elements, imagery, symbolism
	Cyclic retracing of analysis steps to confirm evaluation results or broaden observation and evaluation steps
	Methods and topics of inquiry
Media or formats for presentation of data and findings	Score-related: annotated transcription of materials, compilation of related materials for comparisons
	List of musical units, probably accompanied by some kind of syntax describing their deployment
	Reduction graph showing hidden structural relationships
	Formulaic restatement of structure in terms of letter- and number-symbols
	Graphic drawings: contour shapes, diagrams, graphs, and visual symbols for specific [elements of each domain]
	Waveform and spectrograph images
	Statistical tables or graphs [typology tables]
	Verbal description – using strict formal terminology, imaginative poetic metaphor, suggested program or symbolic interpretation
	Such media can be used together, or combined
Subsidiary matters	Purpose for which the analysis was carried out
	Context in which the analysis is presented
	Audience or recipient the analysis is addressing

the track—for example, to the socio-historical context surrounding a record, or to the technical issues related to its realization. Such matters can enrich one's study, when approached with the intention and purpose of understanding how these matters function within a record.

In exploring the track, popular music analysis may bring its focus to lyrics, music or their interplay. It may also incorporate or be directed toward its sociological context, and a vast array of cultural studies-, humanities-, and psychology-related observations. Analysis need not focus solely on the track, and may serve to examine larger questions or a research agenda—comparing tracks, examining a producer's sonic signature, exploring common practices, among many others.

Popular music is a cultural phenomenon, and as such it has significance, meaning or value beyond itself. The track has meaning and communicates; it does so in many ways. The sounds themselves communicate semiotically, bridging musical syntax and social meanings (Tagg 2013, 143–193; 2000, 77–84). A social dimension plays a role in popular music, sometimes a crucial role, and it is possible "to analyse a popular song in a manner sensitive to both its social message and its specific musical content" (Burns 2005, 136) and to also include elements of the record as 'expressive strategies' (Burns 2010). However, pulling the social sciences and other humanities into recording analysis is often problematic; aside from analyst expertise (or lack thereof) Allan Moore (2003, 7) notes reconciling music analysis with these social disciplines leads to "trying to sort out at what point objectivity intersects with intersubjectivity. . . ." John Covach (2001, 466) offers one can undertake engaging these "without necessarily becoming a student of sociology; while an undertaking of social context is crucial for understanding of popular music, it need not be the central, fundamental consideration." Social and humanities streams of examination can represent challenges for those of us who lack training in these disciplines; resulting observations might (unknowingly) be overly personalized. The extent to which the situating of the track within social constructs might be engaged depends on defining what the analysis seeks to accomplish, and what it seeks to explore.

The track contains much within its *sounds*. Tension exists between its *sounds* on the many streaming levels of domains and also within the outside associations, connections and connotations those *sounds* elicit. That sounds 'elicit,' evoke or arouse reminds us analysis is interpretation; the italics reminds us the track is nothing if not for its sound. Sound is experienced over time, and interpretations coalesce over time from that experience (and its repetitions); "analysis is also performance" (Clarke 2005, 201), as knowledge is accumulated of the track with the subtle shaping of the track by an analyst.

The goals of an analysis emerge from the interests of the analyst, the intended audience for the analyst (to whom the analyst is speaking), and the fundamental nature of the track. At some level, we explore the track pursuing what we find interesting—with some general or specific guiding interest or purpose (Clarke 2005, 202). Ideally, though, the analysis begins from the neutral position of wishing to observe and discover what it presents. Personalized interpretation is not analysis; it is a personal and subjective impression. We have a marked tendency to privilege our subjective, emotional experiences of records—to make sense of them as we make sense of our place in the world. Analysts, especially new analysts, need to make conscious effort to maintain an objective stance, detached from their own personalized interpretations.

Recording analysis and the analysis of recording elements are the core of this book. Exploring the track for characteristics and activities of the elements of recording will reveal 'how the record shapes the song'—how the recording elements contribute to the track, how they provide substance to the record, how they might provide context and depth of expression and character to the recorded song. We will now briefly revisit the analysis framework before we begin the analysis of recording elements.

ANALYSIS PROCESS OVERVIEW AND ANALYZING RECORDING ELEMENTS

The framework provides a generalized analysis process that does not stipulate technique, method or subject matter. The process and the framework's guiding principles are intended to guide the analyst—dispassionately and without adding bias of direction or focus—to the content, character and contexts of tracks that are pertinent to their inquiry and interests. The framework can be functional no matter one's individual analytic tools and agenda from many origins and traditions, as well as those that have yet to be devised.

This process will be especially useful for engaging the elements of recording, and examining how the record shapes the song. While the analysis of recording elements is emphasized in this chapter, music and lyrics may also be examined within this process.

Providing a foundation for this process are the framework's guiding principles. They bring the track itself to be the starting point and center of attention, and not analytical methods. The guiding principles allow the recording elements to be approached as of equal significance and potential to the other domains, and that they require a shift of attention and listening skill to be successfully engaged. To revisit, the guiding principles are:

- Every record is unique
- Equivalence
- Perspective
- Listening with intention and attention

The framework is an analytic tool for identifying and observing elements of recording (elements that are held as being potentially "equivalent" to all others), for 'listening with attention and intention' at specific levels of "perspective" towards drawing conclusions (through evaluations) about the interrelations of elements and their materials, in such a way that an analysis can reveal and address how records are comparable and also how "every record is unique."

The analytical process contains three steps—observation, evaluation, and conclusions—that are largely sequential. We have explored the first stage throughout the past seven chapters, as we have been observing and collecting information for each element of all three domains. All of what we have learned about observing each element—deep intentional listening, graphs, typologies, transcriptions and otherwise making a record of their qualities and activities—now comes together as a multidimensional set of skills and knowledge available for our use. Just as we sought to observe the elements of all domains without distorting their values by our biases or collection practices, we will continue this practice for the next two steps of the process.

While this three-step process is useful for music analysis and for some aspects of lyrics, it is especially valuable in the analysis of recording elements, where no (or few) methods of analysis exist. This process provides a working mechanism and a sense of control and purpose to analysing recording elements. It will be the basis for our approach to analyzing recording elements.

Once information is collected, it is then evaluated (step 2); data is studied and organized, and analyzed in a way that conforms to selected analytical practices. Results of those evaluations are next studied and analyzed, allowing conclusions to emerge or to be formulated (step 3).

Some cycling between the three stages can be expected to occur. An evaluation might reveal a need for additional observations of an element; it is not unusual for one to determine more detail is required, or that information that was neglected now requires coverage. Further, conclusions might require more detailed or broadened evaluation of observations in order to confirm or validate findings. Conclusions are often the result of inquiry; ideas are tested, some get reframed, others get discarded, before arriving at supportable positions. This is especially important when examining recording elements; as there is minimal syntax for recording elements, the basis for evaluation, and the contextual basis for conclusions is often the track itself.

Syntax of Recording Elements

A track's unique sounds and ideas, properties and relationships, are often what is most interesting to us. This idea of uniqueness implies a deviation from the norm, from a convention. In recording elements, this 'norm' can be as much the vocabulary (language or syntax) of the individual track (or producer's or artist's signature sound) as it is a convention of standardized practice. Conventions of how each recording element appears in the track have become loosely established by production practices of recordists; just like the melody, harmony and rhythm of music elements, within the contexts of individual tracks, these recording element conventions will bend and unique qualities and relationships can be established. While these conventions and practices vary between genres of music and producers as part of their style of production, certain relationships and qualities remain a general standard, to which each track may be examined to identify its unique qualities. Again, these relationships and qualities are generalized; no explicit or unified syntax is present, though some loose norms of convention have evolved to establish some sense of syntax (Moylan 2015).

Syntax is a reference for evaluations and conclusions. While each element has its own qualities that contribute to syntax, recording element syntax is most strongly grounded in how the elements work together to establish gestalt relationships at the basic–level.

Perhaps the central example of syntax for recording elements is a sense that records establish 'scenes' of relationships among sound sources based on recording element attributes (see Chapter 8); further, each sound source is uniquely shaped by recording elements (by timbre, for example) in each scene. Recording elements define positions and qualities of sound sources that remain mostly stable within song sections; these may include lateral and distance positioning of sources, loudness relationships of sources, timbral qualities, host environments, among others. A series of changing and recurring 'scenes' becomes established as a track progresses, as scenes flip between song sections. These scenes often provide important character or context for major sections (such as verse or chorus), and at times can provide substantive materials. Scenes represent a set of qualities and relationships of sound sources unique to those sections, and are created in the mix stage of recording production where distinct qualities are added to each type of section. This set of qualities and relationships recur as the track unfolds in changing and returning sections; this sequencing can often add motion and movement to a track, and contribute to its drama and expression.

Recording elements that shape these scenes typically function at the basic-level of sound sources. Element characteristics and qualities may have already been observed in X-Y graphs. This information is now evaluated to identify how sources are provided their character; this can then lead to examining their relation to other sources for the content and character of individual recording elements. It is at this point in the process that X-Y graphs and sound stage diagrams might be given added detail, if those elements are functioning strongly within the track. Using this basic-level as a reference, one is able to begin to evaluate the interrelationships of sources (at the aggregate or composite level of perspective) or shift perspective to the large dimension; one will also be able to look more deeply inside the sounds of the basic-level, and shift perspective to perceive inner qualities of timbres and environments (as examples). Timbral and spatial qualities and relationships contribute most strongly to establishing scenes and an overall syntax, with loudness relationships at times being prominent.

To conclude, it will be helpful to remember that a generalized syntax for each recording element, and within each level of perspective, is present in productions. These conventions will be incorporated into the analyses of elements presented later. As a track is intended to communicate to an audience (in some way, no matter how vague), recording elements contribute to this communication.

The Evaluation Stage

> To 'write about' something is to analyze it, to adopt a distanced approach, to observe rather than participate.
>
> (Brackett 2000, 158)

A blurring of the observation and evaluation stages can easily occur. We may be tempted to evaluate data before we have finished compiling it; this is our inherent tendency (and real life need) to make sense of information, to seek out relationships and patterns, and to recognize them within even incomplete information. We often engage in anticipation when we begin to recognize traits and patterns we expect certain things to follow. It is important to recognize when one is collecting information and when one is evaluating it, though.

During the evaluation stage discoveries are made that bring one to return to observations, and to collect more information on elements or materials. This is a stage that can bring skewed observations, and distort an analysis. Care must be taken to listen dispassionately and objectively for data collection, and to listen without bias to what is present during evaluations. Too easily we can imagine we hear qualities or connections that do not exist—imposing our imagination and wishes onto the material. This is particularly a problem when the primary text is a record rather than a written score.

The individual imprint of an analyst can become strongly evident beginning with this evaluation stage, as analytical methods and techniques begin to influence the content of an analysis, and as personal bias begins to accumulate. Analysts make their imprint by choosing their vantage and perspective(s), the elements for observation, and mode of analysis.

At this point where evaluation and observation intersect and interact, one can notice how elements interact, and how domains might interact. The interdependence of all elements can be explored through collecting data on the interactions of specific elements; these can be groupings of two (such as loudness balance and performance intensity, or distance and lateral positioning), groupings of three (stereo positioning, timbral balance, and loudness balance, for instance), or larger groups, and may also include elements from other domains. The interactions of these elements can be perceived as forming aural images, and evaluated as gestalt qualities. Observations of these interactions might be collected in typology tables or X-Y graphs.

In the process of evaluations, an analyst may find their analysis goals shifting. As evaluations reveal insights into a track and its unique traits, new matters of interest often arise. Goals can shift and broaden, and one's personal curiosity or interests can be triggered as one learns more about a track. As analysis goals begin to transform, the focus and purpose of an analysis may shift as well. The focus in an analysis "may come down to what people think is interesting—which is a quite different matter, and a function of people's broader aims and agenda" (Clarke 2005, 202), and this may vary greatly from an academic or empirical study. With awareness of one's objective for the analysis, this shift of goals and purpose might remain controlled.

Evaluating Observations of Recording Elements

Deep listening practices continue to be used when verifying observations. The analyst returns to listening with intention, bring attention to specific matters or widening perspective for open listening.

It is a process that embraces equivalence of all that is present, and seeks to hear what is there with minimal impact of expectations and biases. Evaluation also requires interpretation of what is heard, and interpretation's inherent subjectivity. Multiple listenings, repeated over time, can be expected to never appear quite the same; our experience of the track continues to grow, and also our frame of mind and attention shifts. Materials and relationships within the track that we recognize as significant or interesting also change with time, attention and further exposure to the track.

During the evaluation stage observations are studied and organized; we begin to recognize how the materials of each element are shaped, and how they might function. Equivalence is important in guiding decisions of elements that will receive attention in evaluation. As all elements are capable of shaping and contributing to the unique, distinguishing qualities of the track, each element is to be considered carefully (within the context of the goals of the analysis), and elements then chosen for more in-depth study. This may require returning to the observations stage to collect more information.

Evaluations make sense of observations; observations are investigated and analysed to identify the characteristics of each element and the content of the materials they shape and their relations to others. During evaluation, we may begin to recognize the organization of materials, and how they interrelate ('materials' here refers to what we traditionally call musical materials or ideas, such as vocal melody, accompaniment parts, groove, and so forth). Patterns may emerge from within observations of each element, and an organization of material may be recognized. Characteristic traits of sounds and interrelationships of sounds and elements might also emerge.

Initial decisions on structural divisions take place throughout evaluations. For many analysts, structural divisions are most easily recognized by the activity in the music and lyrics domains; structural divisions are often most readily engaged by identifying major sections such as verse and chorus divisions. After identifying the sections, one could next determine their length and divide them into phrase lengths, including the prevailing time unit. This level of structural detail will assist evaluation at this stage.

Sample Topics of Evaluations

As evaluations seek to make sense of observations, certain topics are examined. Topics that will provide insights into one track may be less effective on others. Evaluations of recording elements might seek:

- Compiling the collection of basic-level sound sources
- Establishing the basic structure of the track at the large and middle dimensions
- Focus on middle dimension (basic-level) activities, and magnify into more detail (smaller dimension) or withdraw to less detail and broader gestures (large dimension)
- Identifying the attributes and their values or states of each recording element
- Identifying characteristic traits of materials, and how recording elements contribute to them
- Identifying functional roles of recording elements and/or of materials they form (primary, secondary, ornamental, contextual)
- Identifying functional roles of the materials recording elements help form (primary, secondary, ornamental, contextual)
- Roles of recording elements within the characteristics and interplay between the lead vocal and accompaniment
- Rhythmic patterning and relationships of recording elements
- Structural placements, relationships, and patterning
- Aggregate scenes established by recording elements and their interrelationships
- Structural (essential) versus ornamental (elaborative) states or changes related to the prevailing time unit

The evaluation process seeks to distill what is significant from observations. Observations can produce a large amount of information, though often they contain repetitions, patterns of attributes, variations, and contrasts. Evaluation makes sense of observations, and organizes those findings.

The Conclusions Stage

Analyzing a piece of music is an act of faith. We pose questions we assume we can answer; we use techniques that should result in what we seek. There are times, however, when what we expect does not occur, and our efforts to find specific musical norms . . . are frustrated.

(Stein 2005, 77)

Conclusions pull together analyses. They link disparate information from evaluations and establish some sense of direction and stream of thought, in the pursuit of the analyst's goals. In recording analyses, recording elements hold equivalence to the musical norms Deborah Stein describes, and within their specific attributes and actions they establish order and coherence to the track.

The whole of the analysis is focused and formed at this conclusions stage, including any concerns from outside the track.

In the broadest sense, conclusions may seek to recognize (within some predefined limits): (1) how the track's music works, (2) how the sonic qualities of the record shape and enhance the track, and/or (3) how the lyrics add meaning and structural and sonic dimensionality and communication to the track. Within these areas can be discovered the qualities, materials and relationships that comprise the track, that make it unique, and that make it work.

The process of evaluation may be revisited and refined during this conclusions stage; impressions that were delineated during evaluations may be modified here. All observations and evaluations may be refined at this stage.

Conclusions are what has been discovered and learned; they are the results of our analysis and study of evaluations. Analysis and study are guided by a sense of purpose, a sense of what *questions* one wishes to answer, what one wants to know about the track. Pursuing conclusions will lead to dead ends and unexpected discoveries, to new ways of hearing, and deeper understandings, and into unknown areas. Inquiry guides this quest for discovery.

An interrogative approach to recording elements will be pursued for examining their qualities, content, functions and contributions to tracks—in other words, to arrive at 'conclusions.' Analysis is guided by a series of questions that are "invaluable in getting to grips with what a production is actually achieving" (Moore 2012a, 38). Strategic questions for specific elements will be found woven throughout this chapter's examination of ways recording elements contribute to tracks; many of these questions will have previously appeared in the 'Questions for Analysis' sections at the ends of previous chapters. Through such investigation, analysis might discover, reveal, or identify the recording dimensions that make a track unique; the salient features and inherent qualities—established by or reflected within the recording elements—that contribute to how the track 'works' or the roles of elements within that larger context.

Conclusions are drawn at all levels of perspective, and for all elements and materials. It is a pyramiding examination and exploration (with evaluations and conclusions that are perhaps branching, hierarchical, or fractal in their interrelations) that builds an analysis towards the bigger questions of analysis goals. Lower perspective information builds into basic-level findings; basic-level materials mix into confluence of interrelationships; all matters unfold in temporal relationships and establish movement; large dimension shape and relationships are established. Much is possible within that rough continuum of elements, materials, and dimensions.

Streams of Inquiry

Inquiries guide the exploration of recording elements and the materials they shape. Important among the questions to consider are:

- In what way(s) do each individual element impact individual musical materials? As examples, in what way:

  ○ Does the source timbre impact the vocal's melodic line?
  ○ Does the distance placement and/or lateral positioning impact the guitar accompaniment's presence and prominence?

- In what way(s) do each individual element enhance (or contribute to) the tensional effects of music movement? To musical expression? To the affects of the track? Its communication? Its characteristic sound qualities?

- Does and/or how does a particular instance of an individual element:

  ○ Represent substantive material or ornamental embellishment through its own activity and presence?
  ○ Shape the musical idea(s) or material?
  ○ Impart character onto specific musical ideas and materials?
  ○ Function as a primary shaper of the material, in a supportive role, or to provide ornamentation?
  ○ Establish a contextual reference for a basic-level instrument or voice?
  ○ Establish a contextual reference at the large dimension?
  ○ Impact the track directly?
  ○ Shape the experience of the track for the listener?

- Is the activity or the character of the individual element functioning, through its own recognizable identity, as a musical idea, gesture or statement?

Elements also add detail to understanding structure and motion, and form and shape. To these ends, queries can be made such as:

- How do elements interrelate between themselves and with music elements and lyrics, within and between all levels of dimension?
- What are the elements (within each domain) that serve to control and shape the primary ideas?
- At which dimension do these primary ideas exist (though often at the basic-level, this can deviate)?
- How is synergy between domains and their web of interrelationships unraveled and revealed?
- How do the activities and characteristics of pertinent recording elements contribute to structure and movement?
- How do the activities and characteristics of pertinent recording elements contribute to form and shape?
- How does the tension continuum that establishes structural movement, and/or the ebb and flow of the track's momentum (musical, dramatic, kinetic-syntactic, and other types of movement) manifest in this track? At each level of perspective?
- What is the shape of sequencing of structural units, and large dimension activities? How is the singular shape established?

Conclusions also seek to answer larger questions. Often such questions are directed towards the goals of the analysis, either directly or to investigate matters toward that end. In addition to the questions posed in the 'Questions for Analysis' section of Chapter 1, questions such as the following may be pursued:

- How does this track work? What does 'work' mean within the context of *this* track?
- What makes this record unique? What makes this track similar to others?
- What provides this track with overall traits such as: variety, level of complexity, unity and consistency, coherence and balance?
- Are there particularly interesting, unique or unusual sounds related to recording elements?
- How do the unique qualities of this track relate to standard syntax of recording elements?
- What are the expected qualities of sounds or of the overall sound related to recording elements? How are expectations established?
- What are the expected relationships of sounds related to recording elements? What establishes these expectations?
- What is it that brings this track to communicate its message and meaning?
- Is there something that makes this track significant (to the artist, the genre, the time, place, or other)?
- How does this track conform to the norm of the genre's style?
- How does this track reflect the artist's or producer's sonic signature?
- What brings this track to speak to a wide audience? To a specific group?

There will be analysts who may be inclined to seek to identify reasons or justifications for 'what makes this track great' or 'superior to others.' It is important to recognize the inherent subjectivity of such undertakings—and the relevance of such questions to one's goals for an analysis.

Final Consideration Before Beginning Analysis of Recording Elements

'Conclusions' coalesce from our interpretations from evaluations; as such they are subjective in two stages—from the biases of evaluations and from that which we choose to extract from the evaluations. It is easy to overly personalize interpretations from even relatively objective data. As we have learned, there are also subjective aspects to tracks that coalesce to form our impressions of character and other qualities in tracks and sources.

"[A]s a listener, you participate fundamentally in the meanings that songs have" (Moore 2012a, 1); only you know how a track stirs you, what it moves within you, and the ways it speaks to you—and what it is communicating (especially nonverbally) to you. Tracks speak to you just as genuinely as they speak to others (no matter their level of technical or academic expertise), especially within a track's subjective traits. Meaning making is also shaped by knowledge, experience, culture, and so forth—as already discussed. Recognizing what is deeply and inherently personal and what is a culturally shared experience and response is central to analysis.

When interpretation turns to character and affects, an understanding of identifying mood and emotion with awareness and intention can guide one through these inherent challenges.

Identifying Emotion and Affects

To assess the character of sounds and of tracks, an analyst identifies affective cues and infers conventional or likely audience responses to those cues. It is this inference of cultural meanings and affect that can guide a grounded analysis.

Identifying emotional content or one's affective response to a track or the track's elements can be difficult for some; distinguishing between the two may also be challenging. Certain people are more aware of their feelings, moods and emotions than others; those individuals are more likely to identify mood and character in greater detail or clarity. Then there is the dichotomy of whether the mood or emotion is observed *within the track* and (approximately) consistent between listeners of the same culture, or if it is an emotional response *within the listener*. This relates to the difference between *perceived* and *induced* emotions, as tracks both express or convey emotions and arouse or evoke emotions (Juslin 2019). The underlying psychological mechanisms of musical emotion are extraordinarily complex; in a rough summary, perceived emotion is intellectually processed as a perception of an intended or an expressed emotion, and, in contrast, induced emotions reflect an introspective perception of psychophysical changes that may involve emotional self-regulation (Song, *et al.* 2016, 473).

Emotional response can be highly personal and variable; there is an "inherent variability of emotional response, from listener to listener, and within the same listener on different occasions" (Sloboda 2005, 214). Some listeners will be prone to privilege the personal over the cultural, others the reverse; with attention and intention, one may learn to recognize what is present within the track, and what was elicited within themselves from what is within the track.

Affects and emotions must be approached to address and describe (1) the character of timbres, (2) the expressive materials within the track, and (3) the surface-level qualities of the track itself. We are again faced with a need to approach a decision from either a cultural or a personal position. As one engages the analysis processes, bearing this distinction in mind will provide clarity to any evaluation or description that is generated.

Table 10.2 is a list of emotions devised by W. Gerrod Parrott (2001). It identified six primary emotional categories then branches them out into two levels of related emotions. Other models[2] have defined similar sets of primary emotions; it is important to note, the more complex the emotion (the further it extends from the basic or primary emotions) the more subjective the appraisal. This particular listing has been chosen for its breadth and its clarity. Its use here is to serve as a source of identifiers and terminology to guide the reader in considering affective and emotional characteristics of timbres and tracks.

Table 10.2   Tree-structured list of emotions (Parrott 2001).

Primary Emotion	Secondary Emotion	Tertiary Emotion
Love	Affection	Adoration, affection, love, fondness, liking, attraction, caring, tenderness, compassion, sentimentality
	Lust	Arousal, desire, lust, passion, infatuation
	Longing	Longing
Joy	Cheerfulness	Amusement, bliss, cheerfulness, gaiety, glee, jolliness, joviality, joy, delight, enjoyment, gladness, happiness, jubilation, elation, satisfaction, ecstasy, euphoria
	Zest	Enthusiasm, zeal, zest, excitement, thrill, exhilaration
	Contentment	Contentment, pleasure
	Pride	Pride, triumph
	Optimism	Eagerness, hope, optimism
	Enthrallment	Enthrallment, rapture
	Relief	Relief
Surprise	Surprise	Amazement, surprise, astonishment

(Continued)

Table 10.2 (Continued)

Primary Emotion	Secondary Emotion	Tertiary Emotion
Anger	Irritation	Aggravation, irritation, agitation, annoyance, grouchiness, grumpiness
	Exasperation	Exasperation, frustration
	Rage	Anger, rage, outrage, fury, wrath, hostility, ferocity, bitterness, hate, loathing, scorn, spite, vengefulness, dislike, resentment
	Disgust	Disgust, revulsion, contempt
	Envy	Envy, jealousy
	Torment	Torment
Sadness	Suffering	Agony, suffering, hurt, anguish
	Sadness	Depression, despair, hopelessness, gloom, glumness, sadness, unhappiness, grief, sorrow, woe, misery, melancholy
	Disappointment	Dismay, disappointment, displeasure
	Shame	Guilt, shame, regret, remorse
	Neglect	Alienation, isolation, neglect, loneliness, rejection, homesickness, defeat, dejection, insecurity, embarrassment, humiliation, insult
	Sympathy	Pity, sympathy
Fear	Horror	Alarm, shock, fear, fright, horror, terror, panic, hysteria, mortification
	Nervousness	Anxiety, nervousness, tenseness, uneasiness, apprehension, worry, distress, dread

Talking About Tracks

It may have been noticed, the above discussion clearly implies evaluations and conclusions will result in verbal description—whether oral or written, externalized or an internal conversation. The act of description itself implies interpretation and analysis, and bringing ideas (often nonverbal and abstract) into words.

To no small extent, a recording analysis is a description of some aspect of the track. The content and activities of recording elements, and of all else within the analysis will be brought into words as description. We have struggled with this before. Our previous discussions of describing sounds and the activities of sounds within tracks remain pertinent here—as does our need to describe the track and what it contains in order to perform the analysis we seek.

In describing tracks—just as discussed with timbre—we can focus our discussion towards its context, its character, or context. Context may be qualities internal to the track or those that the track establishes for its communication/expression; context may also be external to the track. The track's character emerges from interpretation and psychological tensions, and content from quasi-objective measures, assessments and calculations—quasi-objective because interpretation is always dependent upon context, culture and genre, upon listener experience, knowledge, attentiveness, and so forth. These can lead our discussions with intention of what is being described and to which type of experience the description is addressing.

We have previously recognized that little language or vocabulary exists for much within music, sound and recording elements; while this might be seen as limiting and can cause frustration and miscommunication, there are some positive aspects to this situation. Considering that every track is unique, every track—in some way(s)—establishes its own set of representations and relationships; every track establishes its own (nonverbal) vocabulary of how elements appear and work. A common 'vocabulary' or 'lexicon' to describe tracks (or any element within them) would have a normalizing effect; common terminology would often simplify content and concepts, and certainly would push against the framework's notion that every track is inherently unique.[3] In our current state, our ideas are not distorted by many predefined terms, and our descriptions are not directed by existing language and its meanings or implications. Unencumbered by the bias of language, this brings a certain openness to this position from which we need to describe the track to share our perceptions and interpretations. Whether the description be for academic discourse, the processes of production, to reflect informally with ourselves, or to share with others in our societal group (as examples), our position can be an opportunity to speak objectively about the track from a detached and dispassionate vantage. With this intentional choice of language and content, one also remembers to be wary that language diminishes what it seeks to describe, and to approach description with an intention for clarity and accuracy.

The remainder of this chapter will focus on analyzing each of the recording elements—the core topic of this book. Within these examinations will be many opportunities for interpretation; each will be an opportunity to hone our skill to objectively engage topics dispassionately, and with some detachment. We will now examine how the record (recording elements) generates some of the substance that speaks to us.

BEGINNING STAGES FOR ANALYZING RECORDING ELEMENTS

Initial observations and some evaluation of structure at the start of an analysis—especially in the large dimension—can establish some context and provide organization to an analysis process centered around recording elements. Allowing structure to broaden into form, and establish more detail of middle dimensions will follow. Using timelines to notate the qualities and activities of elements will assist keeping information organized, accessible, and clear for performing evaluations and for conclusions to be formulated.

Structure and Form

Structure is shaped and determined within the confluence of domains. Music, lyrics and recording elements all contribute, though music typically dominates our perception of large dimension structure—and overall shape, or form.

Structure represents architecture; it is the qualities and relationships of the track's dimensional parts. Form is the overall shape established by structure's large-dimension parts. This contrast (explored in Chapter 2) allows recognition of the track unfolding over time as a linear experience and process to be contrasted with the large-dimension aural image of the track's singular shape; a contrast between a dynamic process and a (mostly) static impression.

Evaluating structure identifies points of conclusion and of beginnings; places where sections conclude, and where a new section begins. Sections often have differing character and sonic relationships/content; a different sense of tension and motion may be present between sections.[4] Typically, lyrics and musical materials most clearly define song sections. Obviously, the lyrics of verses work in tandem with music, and the recurrent choruses or refrains present the same (or nearly the same) lyrics and music at each repetition. Recording elements often also shift between sections. This shift will often function synergistically with the expressive nature of musical materials or the dramatic qualities of lyrics; an example are the changes in distance and host environment as Alanis Morissette's vocal in "You Oughta Know" (1995) moves between each verse, prechorus and chorus sections of the track.

Section identities are established by their materials, qualities and functions. Three processes relate sections to one another, and build structural relationships: repetition, contrast and variation; respectively, these are typically expressed as AA, AB, and AA' (Bent 1987, 88). Recognizing distinction between variation (modification or elaboration of what has come before) and contrast (the arrival of new material) can be challenging with recording elements at first listen. Clarity may come with recognizing the recording element attributes within defined sections, and comparing those attributes to other sections.

Dividing the track into sections allows further, more detailed evaluation to recognize middle dimension relationships between phrases; the motion of the track's unfolding can be evident as the prevailing time unit establishes a recurrent meter of measures that provides cohesion and reference. The materials of the middle dimension often define the identity (verse, chorus, bridge, etc.) of the large-dimension section.

Evaluation of section qualities and their interrelations will reveal important information on the track; some of the most pertinent areas to explore are:

- Characteristic traits established by each domain within sections
- Functional qualities of sections
- Weighting of sections by function, significance or dramatic impact
- Durations of sections and interrelations of section durations
- Sequencing of sections and groupings of sections within sequences
- Contexts of sections (sense of place and time)
- Unfolding experience and drama of the track

Conclusions identify the qualities, patterns and relationships that emerge from evaluating structural sections. Those examinations may reveal symmetry created by the sequencing of sections, how individual sections balance against one another and in relation to the whole, and sectional proportions. Shape emerges from sequencing of sections, proportional relationships of sections, and any groupings of sections. Factors that provide unity and variety, or that provide the track with its unique traits and character might be identified, for how they contribute to sectional identities—and therefore to structural sections. At this large dimension level, structure might be conceived as a dynamic process as well as sectional successions; tension and repose, instability and stability, directional motion or stasis, and other factors might be considered for their contribution to the unfolding experience of the track.

Figure 5.1 presented the timeline of "While My Guitar Gently Weeps" (1968), and illustrates its structure. The figure is formatted to make visible the repetition of the verse+bridge+verse sequence that is framed by introduction and coda; from this sequence, the track's form can be visualized. Figure 10.1 depicts the shape of a palindrome that is created by the order of those sections. The form of the track is shaped in the mirror, as the beginning half reversed for the concluding half.

The structure pivots after the second verse, reversing the order of sections—second verse becomes instrumental solo and Verse 3 is essentially Verse 1 repeated—establishing a distinctive ordering. Figure 5.1 makes clear the symmetrical shape of Harrison's "While My Guitar Gently Weeps," despite the introduction and coda sections not being proportional or balanced in length. The aural image is one of a balance of regularly succeeding sections; there is a recurring proportion of a regular tide (PTU) of eight-bar gestures, and a symmetry of retracing steps, returning to where one started.

Harrison's "Here Comes the Sun" (1969) is quite different structurally. It is based on a recurring verse+chorus grouping—an eight measure verse and a seven- (or eight- or nine-) measure chorus. The four iterations of this pairing are interrupted by a prolonged middle section after the third group. The format of Figure 10.2 makes this unique structure clear. The individual extensions of the second and third choruses contribute to a sense of unpredictable phrasing; the choruses are each of different

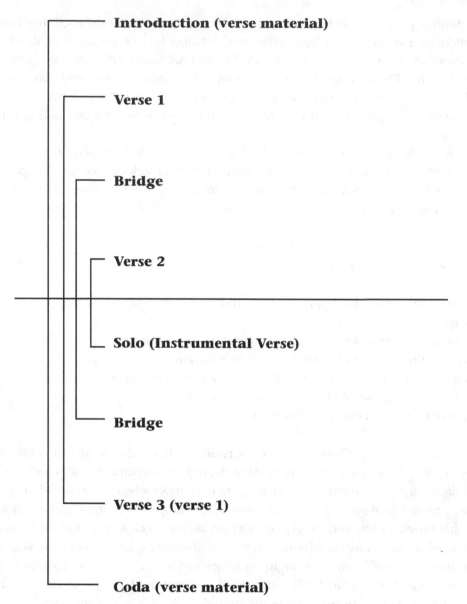

Figure 10.1 Palindromic shape of "While My Guitar Gently Weeps" (1968, 1987).

lengths, until the last chorus repeating the first and mirroring itself during the coda (the last chorus being incomplete). These shifting chorus lengths anticipate the middle section and its clear, recurring pattern of shifting meters (one measure 2/4, three measures of 3/8, one measure 5/8, one measure of 4/4). The proportions of sections are unusual, as outlined in Table 10.3.

In describing this track, we might engage the structure and form of "Here Comes the Sun" speaking from a kinetic-syntactic position (Meyer 1967, 43) and hermeneutics' extra-musical experience (Bent 1987, 79). The unfolding dynamic process of the track might be described as three verse+chorus groupings of gradually accumulating momentum and tension; they combine into a single prolonged gesture that culminates at the middle section. The middle section continues this building of tensions and heightened anticipation; its directed motion continually gains energy through waves of swelling loudness (see Figure 9.2) to the middle section's end—the dramatic and dynamic apex of the track. Added instrumentation, broadening range and density of timbral balance, and rhythmic pattern repetition contribute to the loudness contour and building tension. With the arrival of the fourth verse+chorus

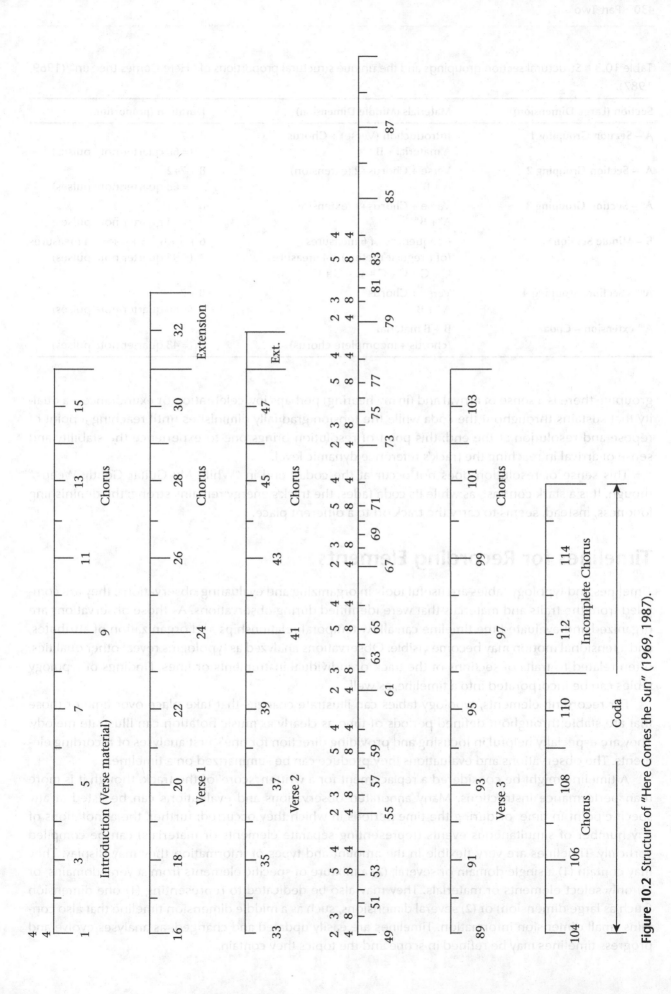

Figure 10.2 Structure of "Here Comes the Sun" (1969, 1987).

Table 10.3   Structural section groupings and the unique structural proportions of "Here Comes the Sun" (1969, 1987).

Section (Large Dimension)	Materials (Middle Dimension)	Duration (proportions)
A – Section Grouping 1	Introduction (Verse) + Chorus A material + B	8 + 7 (= 60 quarter note pulses)
A' – Section Grouping 2	Verse + Chorus+2 (extension) A + B'	8 + 7+2 (= 68 quarter note pulses)
A" – Section Grouping 3	Verse + Chorus+1 (extension) A' + B"	8 + 7+1 (= 64 quarter note pulses)
B – Middle Section	6 sequences of 6 measures (of irregular meters) + 4 measures C + C + C + C + C + C + C'	6 + 6 + 6 + 6 + 6 + 6 + 4 measures (= 94 quarter note pulses)
A"' – Section Grouping 4	Verse + Chorus A" + B	8 + 7 (= 60 quarter note pulses)
A"' extension – Coda	B + B material (chorus + incomplete chorus)	7 + 5 (= 48 quarter note pulses)

grouping there is a sense of arrival and (in my hearing) perhaps joy, celebration or exuberance—a quality that sustains throughout the coda while the tension gradually diminishes until reaching a point of repose and resolution at the end; this point of resolution brings one to experience the stability and sense of arrival in reaching the track's reference dynamic level.

This sense of resolution does not occur at the coda's end in "While My Guitar Gently Weeps," though. It is a stark contrast, as while its coda fades, the track's energy remains strong; the diminishing loudness, instead, seems to carry the track off to a different place.

Timelines for Recording Elements

Timelines and typology tables are useful tools in organizing and evaluating observations; they are compiled from the traits and materials that were identified during observations. As those observations are organized and evaluated the timeline can allow temporal relationships and organization of attributes, and a tensional motion may become visible. Observations analyzed as typologies reveal other qualities, often related to traits of sections of the track or individual instruments or lines. Findings of typology tables can be incorporated into a timeline, as well.

For recording elements, typology tables can illustrate changes that take place over time or those that are stable throughout defined periods of time as clearly as music notation can illustrate melody. They are especially helpful in focusing and providing direction for one's first analyses of recording elements. The observations and evaluations they produce can be summarized on a timeline.

A timeline might be considered a replacement for a written 'score' of the track, though it is more than performance instructions. Many annotated observations and evaluations can be listed at any specific point in time, or during the time periods in which they occurred; further, the annotations of any number of simultaneous events (representing separate elements or materials) can be compiled vertically. Timelines are very flexible in the amount and types of information they may display. They may contain (1) a single domain or several, (2) a mixture of specific elements from several domains, or (3) only select elements or materials. They may also be dedicated to representing (1) one dimension (such as large dimension) or (2) several dimensions, such as a middle dimension timeline that also contains small dimension information. Timelines are easily updated and changed, as analyses evolve and progress timelines may be refined in scope and the topics they contain.

Stan Hawkins (2000, 59–64) uses several timelines for music elements and structure at several dimensions to examine Prince's "Anna Stesia"; Jada Watson and Lori Burns (2010, 342) use a timeline (divided by clock time into sections) to illustrate amplitude form, vocal/instrumental narrative and harmonic structure of the Dixie Chicks' "Not Ready to Make Nice"; Figure 5.7 incorporated select observations of lyrics into a timeline to illustrate "While My Guitar Gently Weeps." Each of these timelines frames different perspectives and levels of detail, and each allows the reader to recognize the unfolding development of tracks. Both Hawkins and Watson/Burns place their times near the top of their illustrations, with most elements and objects of analysis placed below.

Hawkins' timeline (2000, 59) is divided into eight-bar segments against which the elements of pitch and tonal centers, chord sequences, bass and rhythm lines, voicings/textures, and overall dynamic shape are compiled; phrase structure and song sections are also illustrated. Some materials change within sections and others occupy entire sections; Hawkins' figure allows this activity to be clearly visible, and also allows the dynamic shape of the track to be clear. Watson and Burns' figure (2010, 342) divides the timeline by song sections, the starting time of each section is noted by elapsed clock time; against these sectional divisions the harmonic structure and 'vocal/instrumental narrative' are laid out. The harmonic structure tier of the figure displays tonal centers and modulations at one level of perspective, and chord progressions at a smaller-level dimension; further detail is added by referencing scale degrees of important vocal cadences and by drawing attention to the elision that occurs at the transition from the bridge to the instrumental bridge at 2:13.3. The 'narrative' tier also presents information on several levels of perspective; matters of instrumentation, timbral qualities of sources, affects of materials, content of vocal melodies, musical texture, some generalized sound stage information, and other topics are included in the explanatory boxes for each section. Against these tiers the amplitudes of the two channels provide a sense of the acoustic energy of the track throughout its duration. These figures successfully illustrate the unfolding of these two tracks and the overall dynamic (Hawkins) or loudness (Watson/Burns) shapes, and display the co-development of individual elements (mostly, but not exclusively musical elements) in such a way that the selected content, motion and shape of the track are evident—though they do so in different ways.

Figure 5.7 plots elements of lyrics against a timeline that is divided by two-measure units, and broken into song structure sections. The points in time where element activities occur are illustrated in this figure. The Hawkins and Watson/Burns examples devote activities within sections, not at specific points in time; this is a marked difference in level of detail and function of the timeline, though (importantly) this positioning of activity at precise moments was not relevant for these analyses. Figure 5.7 brackets off verbal space and illustrates word spacing against the timeline, and also identifies the positions in time of stressed words, repeated words and rhymes. This type of precision will be found in nearly all X-Y graphs in this writing.

Figure 9.10 illustrated how domains, structure and affects might be incorporated into a single X-Y graph against a timeline. That figure can be adapted to suit the needs of any analysis—tiers may be omitted and others provided more detail, and the entire vertical axis might be reformatted. The timeline of this graph is centered along the vertical axis, with domains and their elements situated above and below it; the timeline is divided into measures and elapsed time is shown for the beginnings of each section and structure is incorporated beneath it.

Timelines at the basic-level of sound sources will be common; this holds even for recording elements. The X-Y graphs of recording elements often display detail within individual sources as well as the composite texture of interrelationships and group combinations of sources. If desired, large dimension qualities may be noted as well.

Figure 10.3 is an annotated timeline of "Here Comes the Sun." The timeline presents a cursory summary of recording element qualities and activities. The elements are distributed so one may compare them within the perspective of the individual sound source; elements at the perspective of the overall

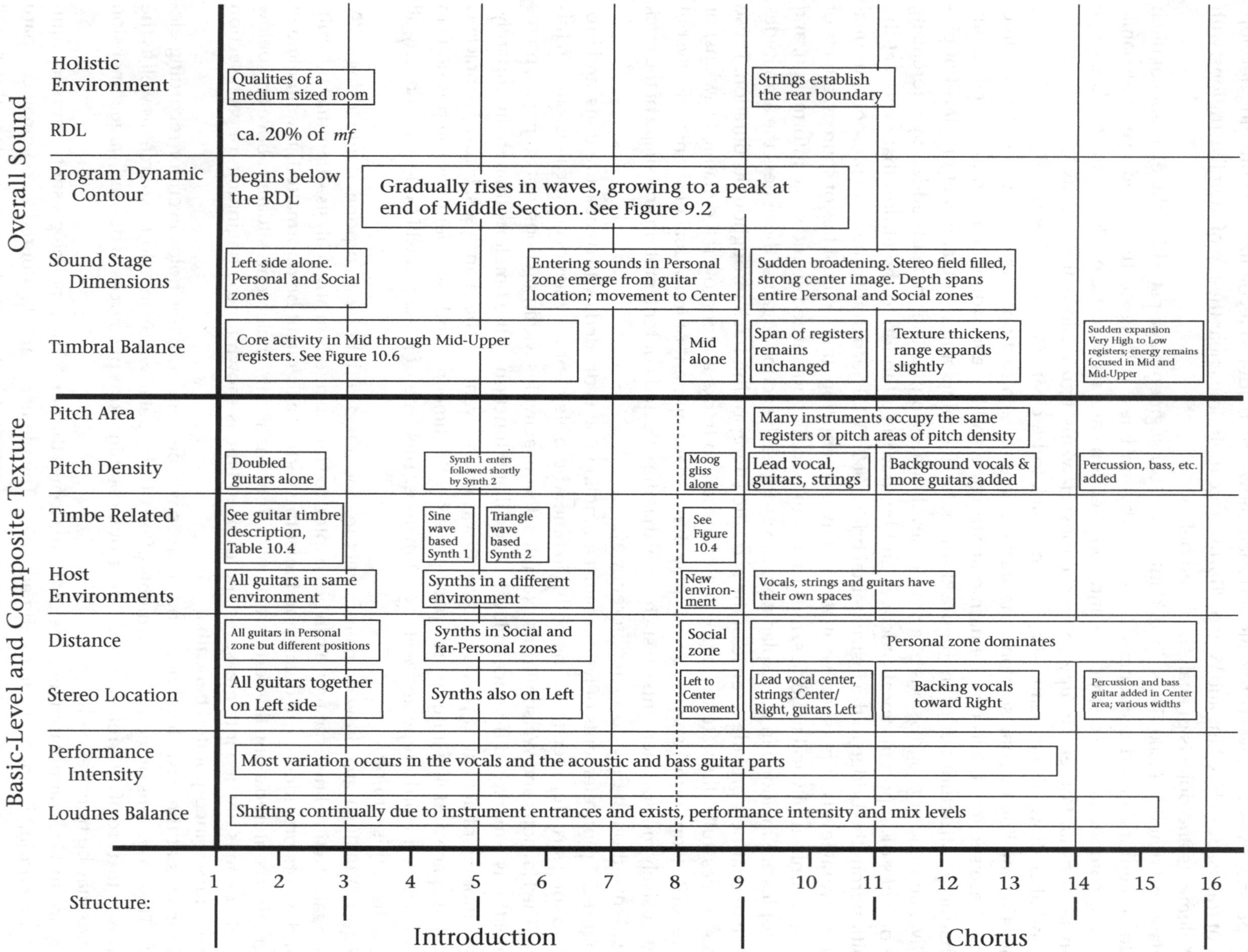

Overall Sound

Parameter	Annotations (measures 1–16)
Holistic Environment	Qualities of a medium sized room (m. 2–3); Strings establish the rear boundary (m. 11)
RDL	ca. 20% of *mf* (m. 2)
Program Dynamic Contour	begins below the RDL (m. 3); Gradually rises in waves, growing to a peak at end of Middle Section. See Figure 9.2 (m. 3–11)
Sound Stage Dimensions	Left side alone. Personal and Social zones (m. 3); Entering sounds in Personal zone emerge from guitar location; movement to Center (m. 6–9); Sudden broadening. Stereo field filled, strong center image. Depth spans entire Personal and Social zones (m. 11–13)
Timbral Balance	Core activity in Mid through Mid-Upper registers. See Figure 10.6 (m. 2–6); Mid alone (m. 9); Span of registers remains unchanged (m. 11); Texture thickens, range expands slightly (m. 11–13); Sudden expansion. Very High to Low registers; energy remains focused in Mid and Mid-Upper (m. 14–15)

Basic-Level and Composite Texture

Parameter	Annotations (measures 1–16)
Pitch Area	Many instruments occupy the same registers or pitch areas of pitch density (m. 9–13)
Pitch Density	Doubled guitars alone (m. 2); Synth 1 enters followed shortly by Synth 2 (m. 5); Moog gliss alone (m. 8); Lead vocal, guitars, strings (m. 9–11); Background vocals & more guitars added (m. 11–13); Percussion, bass, etc. added (m. 14)
Timbre Related	See guitar timbre description, Table 10.4 (m. 2); Sine wave based Synth 1 (m. 5); Triangle wave based Synth 2 (m. 5); See Figure 10.4 (m. 8)
Host Environments	All guitars in same environment (m. 2); Synths in a different environment (m. 6); New environment (m. 8); Vocals, strings and guitars have their own spaces (m. 9–12)
Distance	All guitars in Personal zone but different positions (m. 2); Synths in Social and far-Personal zones (m. 6); Social zone (m. 8); Personal zone dominates (m. 9–13)
Stereo Location	All guitars together on Left side (m. 2); Synths also on Left (m. 6); Left to Center movement (m. 8); Lead vocal center, strings Center/Right, guitars Left (m. 9–11); Backing vocals toward Right (m. 11–13); Percussion and bass guitar added in Center area; various widths (m. 14)
Performance Intensity	Most variation occurs in the vocals and the acoustic and bass guitar parts (m. 2–13)
Loudness Balance	Shifting continually due to instrument entrances and exists, performance intensity and mix levels (m. 2–14)

Structure: Introduction (1–11) · Chorus (12–16)

Figure 10.3 Annotated timeline of "Here Comes the Sun" (1969, 1987).

sound are also represented, and are divided into tiers of those which are stable throughout the track and those that hold the potential for change ('variables'). Several of this track's recording elements will be examined in later sections; these annotations and comments may be more deeply understood as this chapter unfolds.

This illustrates some of the ways timelines can bring the results of evaluations to be visible and to make them accessible and functional for further study. In this way, evaluations might open most readily to the investigations that bring conclusions.

The Context of the Timbre of the Track

The timbre of the track is the fabric of all of the track's content and of the character of all it contains—the domains, affects, outside associations, etc. It unfolds over time, like the timbre of a single sound, and it contains a multitude of information, like the timbre of a sound source. The above timelines that present multiple recording elements (Figure 10.3) or that contain multiple domains (Figure 9.10) provide opportunity to simultaneously observe some of the qualities that comprise the timbre of the track.

The timbre of the track carries interactions of all elements and domains. In this chapter we are concerned about the activities and the interactions of the recording elements. During the following discussions it will periodically be noted how certain elements and interactions establish themselves within the larger context and texture of the timbre of the track.

All recording elements contribute to the timbre of the track. At the highest level, the elements that directly shape the timbre of the track are:

- program dynamic contour
- loudness balance
- timbral balance, and
- spatial identity (sound stage and space within space)

The interrelations among individual elements link to establish larger percepts—such as the sound stage. Elements will continue to hold their own influence on the overall sound of the track—such as image width or distance from the listener. All activities and qualities have a place in the content, character and context of the timbre of the track.

The following sections present 'sample' analyses of recording elements as they are functioning within select tracks. Numerous analyses will interrelate, establishing connections between elements and identifying the larger constructs that are established; these are designed to bring awareness of the richness within those tracks, and their overall sound. Certain interactions of recording elements with lyrics and with music will also arise, though these discussions will not receive significant coverage.

The reader is encouraged to remain aware that each analysis of each element occurs within this overall context of the track.

ANALYZING AND DESCRIBING TIMBRES

As we have discussed, in records timbres exist within a confluence of the three domains (when lyrics are present) and they blend acoustic content, character from affects and interpretation, context, performance matters, and outside associations into a single percept—one that is not the addition of its parts, but rather represents something else. This 'something else' is an overall quality that can often be immediately identifiable and potentially a unique characteristic; this occurs on a number of levels of perspective.

Throughout this section, several basic-level timbres are explored, and several descriptions offered. These descriptions cannot be definitive, as they are at least partially based on interpretation while defining timbral character and other matters.

The process of collecting observations and evaluating them follows an interleaving of the following steps:

- Identify the basic-level sound source, its cause or origin and related matters
- Observe the context of the timbre as it is situated with the track, how it contributes to that context
- Perform content-related observations, collected and evaluated through acousmatic listening.
- Interpret these observations, performance matters, affects and moods, semiotics and outside associations as appropriate to identify its character

Evaluating the Timbre of an Individual Sound

Individual sounds may be the object of concern in an analysis. Describing its timbre may illuminate an analysis, or may be desirable for any number of reasons. Evaluating the sound's timbral content, character and context can lead to a description that can be tailored to the needs of the analyst. An individual Moog synthesizer sound concludes the introduction of "Here Comes the Sun" (1969, 1987), and will serve as the object for analysis to explore this activity.

Figure 10.4 displays the physical content of the sound, as it is perceived. The sound has a clearly defined fundamental frequency is a strong pitched quality (plotted in the "Pitch Definition" tier). The spectral content is comprised solely of harmonics that reinforce the fundamental; this tier also illustrates the glissando of the sound (reflected in the pitch level of the fundamental frequency). The spectral envelope is mostly linear and parallel to the dynamic envelope, though the interaction of harmonics creates some uneven dynamics of upper partials. The dynamic envelope does not contain a rise time; instead it is 'instant on' and no immediate decay is present. This dynamic envelope and the harmonic content of the spectrum are reflective of acoustic sounds performed at a gentle intensity.

> A spectrogram of the Moog glissando sound of Figure 10.4 appears on the webpage, though that image was not used to generate this description. All observations and the X-Y graph were performed unaided by acoustic analysis methods, and reflect the qualities of the timbre *as heard*.

Performance intensity of synthesized sounds is gauged by association with acoustic sounds that approximate their physical characteristics. This glissando has a moderately-low performance intensity and character of ease of performance, as the sound contains no distortion or fast transients that result from quick attacks that would suggest an assertive level of energy. The character is relaxed, and this is reinforced by the descending glissando that provides a sense of release of energy—a sense of falling or floating without effort. This sense of movement is reinforced by the sound gradually shifting locations from 20° left to the center of the stereo field. A sense of quiet anticipation is established, as the sound appears to be moving from where we have been to what will arrive next.

Context situates the sound concluding the introduction, and accompanied only by hand claps in the background. The sound provides a segue between the guitar passages of the introduction and the

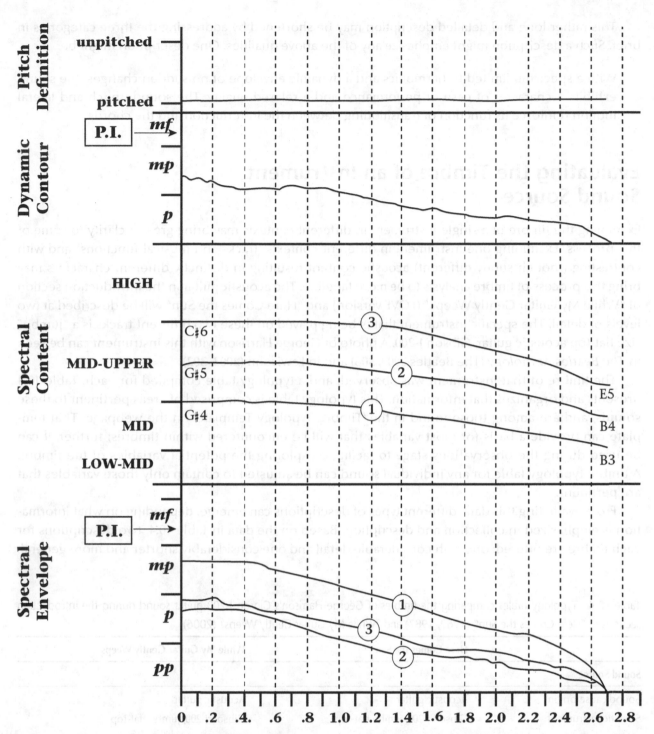

Figure 10.4 Moog synthesizer glissando at 0:12 in "Here Comes the Sun" (1969, 1987).

chorus' arrival of strings and the lead vocal. The sound may have a surreal association to some listeners because of its synthesized qualities; this surreal association is not likely now, as synthesizers have been common place for 50 years, but this association could almost be expected when the track was released, as *Abbey Road* (1969) was one of the first albums to include a Moog synthesizer. The sound's timbre is largely harmonic and contains no sudden loudness shifts; these bring it to have qualities of gently played strings and flutes, and allow it to blend with the track's other sounds.

This rather long and detailed description may be shortened by addressing the three categories in brief. Such a description might emphasize any of the above qualities. One description might be:

> With a spectrum limited to harmonics and a dynamic envelope of no sudden changes, the sound exhibits a character of ease of performance and a relaxed quality. The sound's pitch and lateral motion reinforce its function of transitioning between the introduction and the chorus.

Evaluating the Timbre of an Instrument Sound Source

Examining the timbre of a single instrument in different contexts may bring greater clarity to some of this process. Examining one instrument in different contexts (tracks and musical functions) and with contrasting (though subtly different) acoustic content resulting in distinctly different characters may bring this process of timbre analysis to be more tangible. The acoustic guitar in the introduction section of "While My Guitar Gently Weeps" (LOVE version) and "Here Comes the Sun" will be described at two levels of detail. The specific instrument that is being played on these two different tracks is a "jumbo" size flat-top acoustic guitar, Gibson J-200. A photo of George Harrison with this instrument can be seen in *The Beatles Anthology* (The Beatles 2000, 306) and in Lewisohn (2003, 302).

The timbre of that instrument was observed, and a typology table compiled for each. Table 10.4 presents and organizes that information. This typology table is comprised of areas pertinent to these sounds, and are among those found in the 'Timbre Typology Template' on the webpage. That template can provide a basis for most variables that will be encountered within timbres; further, it can be used during the observations stage to facilitate exploring the potential variables of the timbre. A timbre typology table for any individual sound can be adjusted to contain only those variables that are pertinent.

From evaluating this data, different types of descriptions can emerge, depending on what information is emphasized in collection and description. Based on the data in Table 10.4, two descriptions for each timbre are offered: one with considerable detail and one considerably shorter and more general.

Table 10.4   Typology tables comparing the timbres of George Harrison's Gibson J-200 guitar sound during the introduction sections of "Here Comes the Sun" (1969, 1987) and "While My Guitar Gently Weeps" (2006).

	Here Comes the Sun	While My Guitar Gently Weeps
Sound Source		
Sound cause/origin	Acoustic guitar	Acoustic guitar
Specific instrument	Gibson J-200, jumbo flat-top	Gibson J-200, jumbo flat-top
Modifications/processing	Capo on 7th fret	Capo on 5th fret
Performance technique	Flat pick	Flat pick
Context		
Other sources present	Doubled guitars, Moog sounds	None. Solo passage.
Part/function of material	Verse melody within accompaniment	Layers of voices: chords (strummed and arpeggiated), moving bass line, and treble motives and ornaments.
Mood it establishes in track	Light, sunny, joyful	Somber, reflective, pensive

Character

Performance intensity	Low *mf*, stable	Fluctuates slightly: high *mp* to low *mf*
Exertion level	Ease, relaxed, low exertion	Moderate level of exertion, complex part appears to take moderate effort
Energy level	Quick passage, though barely expending energy, little resistance	Energy is through technique and motion of bass line; shifting loudness between voices
Expression	Timbre and material blend	Significant expressive shaping (OTHER)
Clarity/distortion	Timbre is clear, though not detailed	Clarity/distortion varies by layer
Tension of sound changes	Tension is slight and occurs from tempo and source's interaction with other guitar lines and Moog sounds	Tension through interplay of layers, directed motion, and dynamic shaping that create performance intensity shifts
Tension from performance		Moderate level of exertion and difficulty of execution
Affects, mood	Light, joyous, playful, airy	Wistful, tender, plaintive
OTHER	Bonded with singer by distance position and performance intensity	Performance places dynamic stress and expressive shaping of harmonic motion

Content

Inherent acoustic content	Little alteration of inherent sound	Expression creates shifting timbre; heightened timbral detail
Dynamic envelope	Narrow dynamic range between start of sound and peak loudness; pick provides quick rise time of attack; low peak loudness level, little sustain	Upper strings have fastest attack times, bass line also fast attacks; bass pitches and chords allowed to sustain and decay; chords slower attack times
Spectrum	Narrow bandwidth; little resonance from body; upper partials barely audible; only moderate timbral detail	Significant resonance of jumbo body is present, at times pronounced; distinct spectral differences between strings, upper partials prominent in high strings
Spectral envelope	Greatest density at attack	Bass-line pitches exhibit pronounced shifts in spectral envelope
Definition of fundamental	Pitch is clearly defined	Varies: Upper-Mid pitches clearly defined, some distortion in Low-Mid
Noise elements in spectrum	Noise present during attack for very short duration, at low level	Noise and distortion present in bass line from forceful picking
Space influences	Separation from listener, though linked by position with lead vocal	High degree of timbral detail places instrument near to listener

Should it serve the analyst, a description could focus more on character than on content, or the reverse; descriptions serve the analysis.

A detailed description of the primary acoustic guitar line of the introduction to "Here Comes the Sun" might read:

This is the sound that dominates the introduction, presenting melodic motives from the verse and establishing the context and character of the track; other guitar lines support it, and several Moog patches provide ornamental and secondary materials. There is little alteration to the inherent acoustic content of the instrument beyond a capo on the 7th fret. This brings the instrument to have a spectrum with narrow bandwidth, as the resonance of the instrument is not excited; only moderate timbral detail is present bringing spectral envelope to be largely concealed except during the onset; a narrow dynamic range from initiation of the sound to its peak level at the attack, the dynamic envelope exhibits a quick rise time from attacks with a pick and steady succession of pitches allow little sustain of sounds. The character of the sound is defined by the ease at which its materials are played, the low level of exertion required; performance intensity is expending energy, though minimally and with little resisting its output; the low *mf* intensity is stable, as the tension and directed motion of the passage also changes little throughout. The character is light and playful, and reflects the overall joyful quality of the track.

This description clearly illustrates how performance technique and intensity connect with character, and are also linked with affect. A shorter description of this acoustic guitar line could summarize these central points and conclusions, or it could approach description from a different angle:

The content of the sound occupies a narrow band of frequencies and has a dynamic envelope of fast attack that is moderate amplitude and little sustain. This content provides much space or 'air' around the source that complements the light, playful character of the timbre and its musical materials.

The solo guitar part in the introduction to "While My Guitar Gently Weeps" presents a different timbral quality from the same instrument:

This guitar part is the only source during the entire introduction, though it presents three different timbral qualities; strummed and arpeggiated chords, a moving bass line and treble motives and ornaments establish three musical and timbral layers within the passage. The instrument establishes the reflective mood of the track that George Martin described as "wistful" and "tender" (The Cirque Apple Creation Partnership 2016). The spectrum contains a resonance from the jumbo body that is at times pronounced; there is a distinct spectral difference between strings, with higher partials prominent in the upper strings. Bass-line pitches contain clear shifts in spectral envelope and at times distortion. Timbres of the upper strings and bass part have the fast attack times, with the higher pitches a bit faster; the bass pitches and strummed chords are allowed to decay; chords have slower attack times. A low peak dynamic level to the attack reflects a moderate performance intensity—that fluctuates expressively between high *mp* and low *mf*. The passage has some complexity that appears to require a moderate level of exertion and a moderate effort; energy is present through performance techniques, motion within the bass line, and shifting loudness of the layers. Tension is created by the interplay of layers, dynamic shaping, performance intensity shifts, and expressive shaping of harmonic motion; these combine to establish directed motion. The level of clarity and distortion varies between layers, with the bass containing distortion and the highest strings the greatest clarity. The high level of timbral detail places the sound near the listener, and prepares the listener to be directly addressed in the first line ("I look at you all . . ."). The character is rather pensive, plaintive and somber; this instrument reflects and establishes the character of the track as a whole.

A shorter description and set of conclusions might be:

The content of this instrument and line is multi-timbral, with the lowest notes containing some distortion and the highest pitches clear and highly detailed. The resonance of the large body balances the prominent partials of the highest strings. A moderate performance intensity is reflected in the attacks' low dynamic peaks (except a few strategic bass notes) and in the tender and plaintive character of the timbre.

These shorter versions could readily emphasize other qualities of the timbres and explain different conclusions. Here they condense and focus what I interpret as the primary attributes of these timbres. As lengthy as they appear, these longer descriptions might hold even greater detail in both character and content; an exploration of timbre can be much more substantive than these, though for recording analysis purposes this would rarely generate information worthy of the effort. This said, every recording process will generate minute differences in timbre (as well as those that are pronounced), and every performance introduces unique qualities (some subtle, others quite apparent); these are defining qualities of sound sources and often of tracks themselves, and may well be worthy of examination (and the requisite time and effort) according to one's analysis goals.

At this point, these two timbres of the same instrument could be compared, and similarities and differences identified. The typology table illustrates these two timbres side-by-side, allowing for this to be explored. Taking this further, the reader may wish to examine other tracks where this instrument is prevalent. One could compare the two tracks just examined to the version of "While My Guitar Gently Weeps" on *The Beatles* (the "White Album"); the 2018 remix of the track situates the timbre distinctly differently and more prominently in the context of the track and establishes modest alterations of content from the first CD release of 1987, and the 2009 remaster (both mono and stereo versions) provide further opportunities to hear subtle changes of content (especially noticeable in the Mid-Upper register) and character (with attacks that appear more forceful).

Evaluating Timbre and the Voice: Vocal Timbre, Lyrics and Performance

Lead vocals are the most complex and complicated of timbres. Within these timbres is a confluence of all domains—including the sounds of language—plus associations from outside the track such as language, paralanguage sounds, moods/emotions, and performance factors. Any or all of these factors may shift at any instant. A description of vocal timbre may address any or all of these factors, at any suitable level of detail. For example, the timbre of the lyrics is embedded within the timbre of the voice, but may also be separated out for study—as, for example, is possible in examining the recurring lines between choruses. Much timbral detail is present and could be explored in each note, in each syllable or word, and within each nuance of expression. This section will explore more global timbral qualities, any of which could be followed in greater depth for additional detail.

The single line "Nothing is real" from each chorus of "Strawberry Fields Forever" (1967, 1987) will be examined, and descriptions offered. John Lennon's lead vocal is transformed at several stages during the track. By evaluating timbre as it sequences through the four choruses, these different qualities may provide a clearer sense to the timbre of a sung vocal. The line will be evaluated for context, vocal timbre content, timbre of lyrics (content), performance, and the resultant character.

Context

Considering context, Lennon's vocal is central to the track, as are nearly all lead vocals. Its timbre represents primary musical materials and the lyrics; the timbre is also altered in performance offering added expression and meaning. This line is one of the primary themes of the lyrics, and much in the track other-than the vocal also affirms that "Nothing is real." The vocal is prominent in each chorus, though in the first chorus there is less competition from the accompanying parts; the accompaniment gains in complexity and density beginning in the second chorus, when a significant shift occurs. Two versions of "Strawberry Fields Forever" were recorded: as Lennon was not pleased with the first version that began with McCartney's Mellotron line, a second version at a faster tempo, a higher key and added cellos and trumpets was recorded. In the end Lennon decided he preferred the beginning of the first version and the end of the second, and requested they be joined together. In the world of analog tape, this was nearly impossible; fortunately, as George Martin often stated, "the gods smiled down upon us" (Emerick & Massey 2006, 139) and it was made to work. The tape speed was changed in both versions so they could align in pitch and speed at a defined moment during the beginning of the second chorus. Geoff Emerick (who engineered the track) explains:

> George and I decided to allow the second half [second version] to play all the way through at the slower speed; doing so gave John's voice a smoky, thick quality that seemed to complement the psychedelic lyric and swirling instrumentation. Things were a bit trickier with the beginning section [first version]; it started out at such a perfect, laconic tempo that we didn't want to speed it up all the way through.... With a bit of practice, I was able to gradually increase the speed of the first take and get it to a certain precise point, right up to the moment where we were going to do the edit. The change is so subtle as to be virtually unnoticeable.
>
> (Emerick & Massey 2006, 140)

Listening carefully to the track, one can be aware of pitch and timbres shifting where the two versions were spliced together at the 0:55 and 0:59 marks—the most significant change being at the word 'going' within the first moments of Chorus 2.[5] This slowed second version ensures an unreal vocal timbre during the longest portion of the track.

Content: Voice Timbre

Lennon's vocal contains greater timbral detail and clarity in the first chorus than in the others. While the double tracking adds substance and depth to the voice it does not diminish the line's precision. This line is at the speed of the original performance, the voice's spectrum and formants and its dynamic envelopes occur naturally. The dynamic envelope attacks that begin each word contain only a moderate rise time and reach a moderate loudness level, indicative of the line's restrained, low-*mf* performance intensity. The voice resonates freely; its energy is mostly relaxed though it has focus. Little distortion from processing or performance is present.

The timbre of Lennon's voice in the other three choruses is substantially different. These recorded performances were double tracked, and slowed down; the result is a dense spectrum with some distortion within the spectral envelope. These timbres contain double attacks about 10ms apart; the dynamic envelope profile is blurred and a pronounced unnatural quality results—this may be what Emerick describes as "a smoky, thick quality" (*ibid*.). These speed shifts and double tracking timbre modifications enhance the distortion within the voice and support greater intensity. Each of these three choruses has a slightly unique timbre profile: Choruses 2 and 3 share the most prominent and assertive quality of the four and the high-*mf* performance intensity, and Chorus 4 a slightly less intense quality. Choruses 2 and 4 have a higher level of slurring of the vocal line, with 3 containing more clarity but substantially less than Chorus 1. The timbre of the voice no longer has the impression of ease, and though its resonance remains it appears more agitated.

/ ˘ ˘ /
Nothing is real

nəθɪŋ ɪz ril

(N)-o-(th)-in-(g) i-(s) (r)-e-a-(l)

Figure 10.5 Strong and weak stresses, phonetic sounds, and the lyrics as sung by John Lennon, "Strawberry Fields Forever" (1967, 1987) (See Figure 5.10).

Content: Timbre of Lyrics

Figure 4.1 is a scansion of the entire chorus text, and illustrates strong and weak stresses. Figure 10.5 isolates the line "Nothing is real" and illustrates its poetic stresses, phonetics and a rendering of how the line is divided into sounds as sung by Lennon.

The text is clearly articulated in Chorus 1, though not exaggerated; the voice timbre assists this clarity. "Nothing" begins with a natural 'n' sound and the 'g' is clear by not out of proportion with the line, and the word "real" has a speech-like quality; the unfolding of the text is one of ease, and has a natural, spoken rhythm. The word "real" becomes emphasized in the other choruses, with Chorus 2 extending and blending the "(r)-e-a" sounds more than the others; this word is prominent in each of Chorus 2, 3 and 4 with each of them providing a slightly different timbre of this combination of word sounds. Of all the choruses, Chorus 3 places the strongest articulations on the 'n' and the 'g' of "Nothing"; Choruses 2 and 4 also strongly articulate these 'n' and 'g' sounds, though less than Chorus 3. Each chorus contains a different interpretation of the same lyrics (and the same music)—different word sounds are emphasized or receive unique articulation, and different words given emphasis in each of the four choruses.

Performance

In Choruses 2, 3 and 4 this line is shaped expressively, and does not have the natural, spoken pacing of Chorus 1. Internal dynamics, rhythmic pacing and pitch inflection converge to shape each iteration uniquely. No paralanguage sounds are contained in the performance; though the hard 'g' sound in "Nothing" can appear extreme, it is clearly situated within the lyric.

Turning back to Chapter 5 for a moment, Figure 5.10 is a transcription of the line "Nothing is real" in its four performances. A slightly different rendering of the line is performed in each chorus, with rhythm and pitch varied for expression and the lyric articulated and shaped differently in each. The rhythm of those phrases reflects the unfolding of the line within each Chorus. Note that the word "real" anticipates the downbeat in Chorus 1, and it lags behind the beat in the other three choruses; while this may well be the result of altered tape speed, it is situated in the track as performance technique.

Character

This line is often quoted as a description of the overall atmosphere of "Strawberry Fields Forever." The timbre of the voice and performance of the lyrics within the context of the track produce a unique timbre of Lennon's lead vocal; the shift of pitch and tempo provide "Lennon's vocals an unreal, dream-like timbre, especially in the second, slowed-down portion of the song" (Everett 1986, 369). The mood or emotion of this specific line seems one of disorientation and describing the surreal nature of Strawberry Fields; though the lines that surround it assert action and direction.

Here are two examples of short descriptions that were generated from information extracted from the above observations and evaluations:

Lennon's lead vocal in Chorus 1 contains a resonant quality that is reinforced by double tracking; its character is relaxed with an intensity of low-*mf*. Lyrics are clear, though presented without stress or urgency.

Lennon's lead vocal in Chorus 2 contains distortion in its spectral envelope that provides a sense of assertiveness, with a high-*mf* intensity. The dynamic contour has a wide range as the first syllable is heavily stressed and exhibits a fast and strong attack that is then punctuated by a strong 'g' sound word's end. Timbre of the lyrics is blurred from slowing the line and double tracking, and the surreal quality of the voice reflects the line's sentiment that "nothing is real."

The reader will benefit from listening carefully to this line in its four forms to appreciate its development as the track unfolds; one could also benefit from adding detail or other observations to the information and conclusions presented. One's own descriptions might follow. An interested reader might audition different reissues of "Strawberry Fields Forever"; the 2009 remaster of *Magical Mystery Tour* (in both mono and stereo versions) and the original mono mix contained in the 50th anniversary *Sgt. Pepper's Lonely Hearts Club Band* collection offer subtly different interpretations of the line—and the track. Those differences can be compelling, though they may appear subtle.

It is significant to notice here: timbral descriptions make choices. A description will emphasize voice content, character, context, performance or the lyric's timbre content over other factors, by some degree. Intention of the description's purpose should guide its formation.

TIMBRAL BALANCE AND PITCH DENSITY

Timbral balance represents a confluence of all domains, where the individual sources and elements remain distinguishable, though considered as a group where each are of equal significance. It is comprised of the pitch densities of basic-level sound sources—combining source timbre, the materials they present, and the content of their performance—within the syncrisis unit of their materials. As it is a confluence of pitch densities of all sources, it represents the track's pitch/frequency spectrum—each source acting as a partial within the track's frequency fabric.

Timbral balance does not factor loudness levels and relationships of the basic-level sources it contains. When evaluated side-by-side with a loudness balance graph, the two illustrate how each source/part is situated in the mix by both loudness and dominant frequency band. A spectrogram can illustrate the loudness and frequency content of the overall texture, though details of individual sources are often largely concealed; further, the spectrogram illustrates acoustic content which differs from our perception. Engaging timbral balance allows observation and evaluation of the track as heard—as experienced by the analyst and by listeners.

> A version of this graph illustrating the complete timbral balance of "Here Comes the Sun" is on this book's website. Spectrograms of the track and of the first 24 measures are also found there, for purposes of study and comparison.

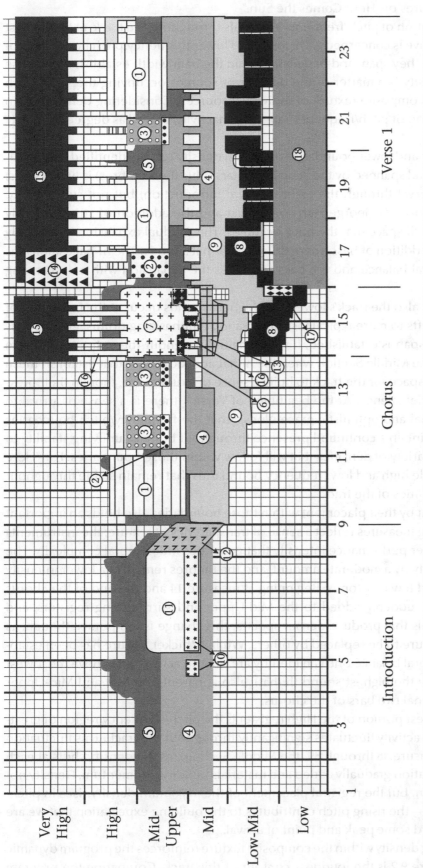

Figure 10.6 Pitch density and timbral balance graph of the beginning through Verse 1 of "Here Comes the Sun" (1969, 1987).

Key

1 - Lead Vocal	13 - Harmonium
2 - Backing Vocal 1	14 - Crash Cymbal
3 - Backing Vocal 2	15 - Hi-Hat
4 - Acoustic Guitar 1	16 - Snare Drum
5 - Doubled Guitars	17 - Tom-tom
6 - Acoustic Guitar 2	18 - Kick Drum
7 - Guitar 3	
8 - Bass	
9 - Strings	
10 - Moog Synth 1	
11 - Moog Synth 2	
12 - Moog Synth 3	

Figure 10.6 plots the pitch densities of sources and the timbral balance of the composite texture and overall sound of the first 24 measures of "Here Comes the Sun."

Timbral balance is the distribution of pitch/frequency materials throughout the track, at the highest level of perspective. This perspective is concerned with upper and lower frequency/pitch boundaries of the overall sound and the interval they span, and of density within the bandwidth established by those boundaries. Timbral balance density is a matter of the number of sources occupying the same pitch region; it is evaluated within the composite texture of individual sources. Considering density within timbral balance allows identification of pitch/frequency ranges of inactivity, as well as degree of activity within ranges.

Through Figure 10.6 the upper and lower boundaries of timbral density can be identified and evaluated. In the introduction the interval spanned by the acoustic guitars (plus their timbres) is two octaves; there is some growth of this interval through the eight-measure introduction, but not substantially. In measures 9 through 13 (of Chorus 1) George Harrison's vocal and the addition of strings slightly widens the range of frequency/pitch space that the track occupies. This gradual expansion is suddenly increased in measure 14 with the addition of bass, snare drum and hi-hat sounds. The hi-hat establishes a stable upper boundary of timbral balance and the bass continues the expansion with a descending line, leading into the first verse.

The widest span of range, and also the track's lowest and highest sounds occur at the beginning of Verse 1. Here the arrangement shifts to more typical instrumentation emphasizing drums, guitars, bass and vocals. Now that this widest span is established, it will remain throughout the entire track; there will be phrases and moments in the Middle Section when the hi-hat and kick drum that establish these boundaries trade rests, and allow space for the track to build a sense of exuberance, though their presence remains felt even in their brief silence. At the beginning of Verse 1 there is a sense of arrival to the story of the song and its casual and optimistic gracefulness; that the frequency/pitch boundaries have been set at this time and maintain a continued presence throughout brings a unifying stability to the track that will frame the rich variety of densities that will follow within those boundaries. For "Here Comes the Sun," establishing stable high and low frequency boundaries that remain stable throughout is one of the unique, defining qualities of the track.

The density of sources brought by their placement within those boundaries also holds an important function for the track. The opening measures reflect a sense of transparency of texture; the instruments are few in number and have a lower performance intensity that generates little timbral complexity. The arrival of the chorus adds to density by a moderate amount, and the timbres remain at a lower intensity and contain mostly harmonics and few overtones. Beginning at measure 14 and the end of the chorus density thickens; more parts are suddenly added to the Mid-Upper and surrounding registers, and some sounds are at loudness levels that produce denser spectra; this change is temporary, though, as the instruments will leave the texture to be replaced by others. While a thicker texture has been established, glancing at the Verse 1 timbral balance one can identify areas of no activity just above the lowest sound (kick drum) and just below the highest sound (hi-hat); also apparent, the Mid and Mid-Upper registers are less dense than the final two bars of the chorus.

The Mid register up to the lowest portion of the High register is the pitch/frequency region of greatest activity. The amount of density activity fluctuates significantly throughout the track. This fluctuation is most clearly observed from measure 48 through 89 (the Middle Section), as the density shifts at each phrase repetition with instrumentation gradually added and materials building momentum. Density not only increases with each repetition, but the registral placement of the maximum density rises in pitch/frequency throughout the section—the rising pitch contributes to the building expectation that we are moving towards something, toward some peak and point of arrival.

This ebb and flow of increasing density within the composite texture reinforces the program dynamic contour's building loudness. Figure 9.2 is the loudness contour of this track. Comparing the program dynamic contour and timbral balance graphs one can trace their complementary motion throughout the Middle Section to the loudness and dramatic apex of the track.

It is a general convention for timbral density to have some shift between sections—roughly akin to lighting changes that shift between 'scenes' of a play. Sections often share a timbral-balance identity with like sections—where verses are similar in timbral balance to other verses, and choruses to other choruses. It is not as common for timbral balance to so closely mirror program dynamic contour (track loudness contour) and phrase-level structural divisions as occurring in "Here Comes the Sun." In this track, timbral balance functions strongly in providing movement and tension, and also to establish unity throughout the track.

On the perspective of the individual sound sources, the pitch densities of sources are determined by timbral content as well as materials performed; this results in the range of the source and also the density within its sound. Mentioned earlier, a relative absence of overtones and dominance of harmonics result from the moderately-low performance intensity of the introduction's guitar lines; the spectrum is less dense than if the guitar was played at a higher intensity level. Greater contrasts of performance intensity and spectral complexity within timbres are found in the backing vocal lines, especially prominent between Chorus 1 and the middle section. Such contrasts are more significant in many tracks. In "Here Comes the Sun" most sources hold a largely consistent performance intensity, and thus a consistent spectral complexity and content. It should be clear, pitch density relates strongly to timbres of sources, and is shaped by performance intensity as well as the varying pitch levels of the materials they present.

ANALYZING LOUDNESS OF THE TRACK AND OF SOUND SOURCES

Program Dynamic Contour, Track Loudness Contour

An overall loudness contour is established by the singular loudness level that results from the combination of all sources. This is a perceived loudness level that may change at any or every moment during the unfolding of the track. The 'track loudness contour' or synonymously 'program dynamic contour' is a contour mapping of the integrated loudness of all sources over the entire track duration.

There may be temptation to use the waveform of the mix buss within any digital audio workstation as a replacement for transcribing or aurally observing program dynamic contour; the two are not equivalent, though. Human perception of loudness differs from the measurement of acoustic energy, as humans are not uniformly sensitive to all frequencies. A DAW's wave shape is acoustic energy registering as amplitude; in most software it appears as two independent channels. Program dynamic contour (track loudness contour) is our perception of the changing loudness of the track (and all that influence this overall loudness), and a singular impression of both channels combined—it is what we actually hear when we listen to the track as primary text. These are significant differences. To illustrate this, Watson & Burns (2010, 342) use a track's two-channel wave shape—accurately referred to as an amplitude diagram—to successfully illustrate several key points within their analysis of "Not Ready to Make Nice" (2006) by the Dixie Chicks; a close examination of the waveform reveals significant acoustic energy in the introduction, though in listening one quickly realizes the passage is quite soft—especially in comparing the introduction to the following sections. The bass and piano parts display distinctly low loudness levels; the low frequencies of these instruments require more acoustic energy to enter our perception than the acoustic guitar that is also playing. That the waveform successfully illustrates Watson & Burn's analysis is a product of what they were studying; the waveform may be adequate for a given analysis depending on how it is used. The program dynamic contour will be the appropriate approach if examining the loudness of the track's overall sound—as it is heard.

The track loudness contour (or program dynamic contour) is at times a significant element to a track; it can establish a stable context, and conversely it may provide movement and shape to the track as it changes over time. Referring again to Figure 9.2, the program dynamic contour of "Here Comes the Sun" reinforces the structure of a track; this is reasonably common, but not typically so clear. Loudness contours are unique to each individual track. Shapes may vary widely in the amount of changing levels and the degree to which levels change, and one may well encounter tracks of near steady-state overall loudness levels.

Figure 10.7 is the program dynamic contour of "Strawberry Fields Forever" (1967, 1987). Shapes emerge from within this track loudness contour, such as the beginnings of each chorus exhibiting a similar profile. The overall contour of the track is perhaps most immediately evident, though; its gradual building spans the entirety of the track well into the coda. The general contour evident in the graph is a large dimension gesture; in the middle dimension are the changes that occur that outline each verse and chorus.

Table 10.5 is a typology table for the contour that organizes and collects some information on the contour. Within the contour are subtle shapes, reflected in the speed of changing loudness, increase or decrease of loudness, and the amount of this change of loudness—both moment to moment or over the duration of the gesture. This typology table addresses some of the broader aspects of the contour, and can pull one into and illuminate an evaluation. These two collections of data (typology table and X-Y graph) on overall loudness allow discovery of attributes of the contour from different approaches; each may more readily reveal something not as evident in the other.

The overarching, building gesture of the contour is punctuated by regular drops in loudness that reside in the middle dimension. These sudden drops are listed in the table, and occur at the end of the first three verses and choruses. This idea of ending of sections with an abrupt decrease in level continues throughout the track. The pulling back of loudness is a reset that allows each new section to proceed from its own dynamic position; it separates sections from one another yet provides a gesture that is (in a generalized way) repeated in each section with variation. The drop in loudness allows the building to begin again while maintaining a controlled loudness range, the motion of building tension continuing while the aggregate loudness remains contained within the *mf* area. Finally, these drops in loudness bring the contour to articulate the sections of the track.

An expectation is established that each section will diminish in loudness at its end. This sets off each section and allows their unique contours to be more apparent; each section has its own contour and build and release of motion and tension brought about by dynamic levels and shape. This should become apparent as we evaluate these contours and compare their levels, contour shapes and durations of changes of the sections. Within the Chorus 1 and Verse 1 sections, loudness plateaus of little change are present; these are the simplest contours of all sections. Chorus 2 continues the gesture of increasing loudness and establishes the largest increase between any two sections; further it presents a more complex contour (compared to the previous two sections) in its increasing level followed by a more extreme and terraced diminishing level. Here the only break from the steady increase of level over several sections occurs, as Chorus 2 and Chorus 3 occupy the same dynamic range; this frames Verse 2 at a slightly lower dynamic level and accentuates the differences in its contour compared to Verse 1. After Chorus 3, Verse 3 resumes the building of level by resuming the level of Verse 2 and immediately increasing loudness level; all three verses have a unique contour, with the contours becoming more complex as the track progresses. The most erratic changes to contour patterns occur at the end of Chorus 3 and within Chorus 4; these participate in the increasing complexity and unpredictability of the contours as tension and directed motion begin to speed and build. This

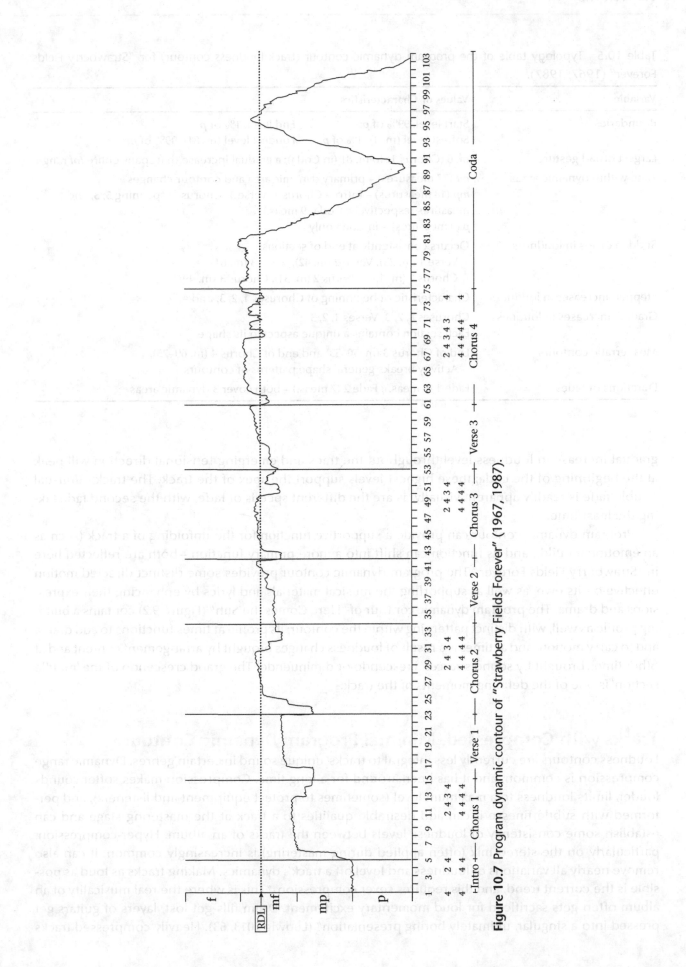

Figure 10.7 Program dynamic contour of "Strawberry Fields Forever" (1967, 1987).

Table 10.5   Typology table of the program dynamic contour (track loudness contour) for "Strawberry Fields Forever" (1967, 1987).

Variable	Values or Characteristics	
Boundaries	Start level: 99% of *p*;	End level: 1% of *p*
	Softest level (m. 1): 1% of *p*;	Loudest level (m. 81): 99% of *mf*
Largest broad gesture	m. 6 (Chorus 1) to m. 81 (in Coda); a gradual increase that spans entire *mf* range	
Time within dynamic areas	*mf* (77 measures) – primary dynamic area and contour changes	
	mp (18 measures) – Intro + Chorus 1, Verse 1, Chorus 3; spanning 5, 3, and 1 measures respectively; Coda (9 meas.)	
	p (9 measures) – in Coda only	
Sudden drops in loudness	Occurs consistently at end of sections:	
	Verse 1 (m. 23), Verse 2 (m. 42), Verse 3 (m. 61)	
	Chorus 1 (m. 13), Chorus 2 (m. 31), Chorus 3 (m. 49)	
Stepped increases in loudness	Characteristic of beginning of Choruses 1, 2, 3, and 4	
Gradual increases in loudness	Choruses 1, 2, 3; Verses 1, 2, 3	
	Each section contains a unique aspect to its shape	
Most erratic contours	End of Chorus 3 (m. 50–53) and end of Chorus 4 (m. 69–75);	
	Activity breaks general shape patterns of contours	
Durations of fades	Fade 1 (9 meas.); Fade 2 (7 meas.) – both cover 3 dynamic areas	

gradual increase in loudness level throughout the track and emerging tensional direction will peak at the beginning of the coda; these highest levels support the apex of the track. The track's unusual double fade is readily apparent visually, as are the different speeds of fade, with the second fade taking the least time.

Program dynamic contour can provide a supportive function for the unfolding of a track (such as an emotional build), and its function can shift into a more primary function—both are reflected here in "Strawberry Fields Forever." The program dynamic contour provides some distinct directed motion effective on its own, as well as supporting the musical materials and lyrics by enhancing their expressions and drama. The program dynamic contour of "Here Comes the Sun" (Figure 9.2) contains a building profile as well, with distinct patterning within the contour; its profile at times functions to add drama and to carry motion, and at times is a result of loudness changes brought by arrangement content and at other times brought by sound sources' crescendo or diminuendo. The grand crescendo of the 'middle section' is one of the defining moments of the track.

Tracks with Compressed, Minimal Program Dynamic Contour

Loudness contours are currently less integral to tracks' unique sound in certain genres. Dynamic range compression is common, and it has been around for a long time. Compression makes softer sounds louder, limits loudness to a maximum level (sometimes to protect equipment and listeners), and performed with subtle finesse it can add desirable qualities to a track at the mastering stage and can establish some consistency of loudness levels between the tracks of an album. Hyper-compression, particularly on the stereo mix (often applied during mastering) is increasingly common; it can also remove nearly all variations of loudness and level off a track's dynamics. Making tracks as loud as possible is the current trend, and this requires sever compression. "This is where the real musicality of an album often gets sacrificed for loud momentary excitement. Drum fills get lost, layers of guitars get pressed into a singular, ultimately boring presentation" (Ludwig 2013, 63). Heavily compressed tracks have a characteristic sound—a sound that may be attractive to some (and not to others), that is inherent to certain genres and production styles, and that provides an impression of extreme loudness and exaggerated presence.

Program dynamic contours of wide dynamic ranges remain prominent in many tracks, and still provide shape to many albums. An analyst must be aware this element may be in play in a track—and that, conversely, the dynamics of the overall track may be minimized into insignificance. In such tracks, another recording element will be prominent in shaping the overall sound—often timbral balance.

Loudness Levels, Loudness Relationships and Performance Intensity

Loudness levels and relationships are altered in the mix process. The result is a balancing of loudness relationships that is akin to the musical balance within an ensemble. Figure 9.3 presents a loudness balance (musical balance) X-Y graph of "Lucy in the Sky with Diamonds" that illustrates the loudness levels and relationships of all sources.

A great deal can be learned about a track by contrasting loudness of sources with that of other sources (both as levels and as gestures over time), and also by contrasting loudness of sources with their performance intensities. The mix process combines recorded performances of individual sound sources that are typically isolated from others; this allows the mix process to apply a loudness level to each individual performance separately. Any instrument can appear at any loudness level in the track completely independent of the loudness of its performance during the tracking process. The loudness of this initial performance is reflected in the timbre that was captured during tracking; the captured timbre reflects the amount of exertion of the performance, as well as its energy, expression, performance technique, and so forth—this is its performance intensity. We understand how the loudness balance alters the levels of the performed tracks by recognizing performance intensity, and comparing that perceived intensity against the actual loudness of the sound in the mix. This dichotomy is one of the ways recording alters reality—and establishes the realism of the track.

Applying a loudness level in the context of a mix is an artistic decision, and is one of the core ways the recording shapes the track. This applied loudness level can approximate that of the performance, with the loudness of the instrument as it appears in the track being similar to the level at which it was originally performed. It is most common, though, for there to be some discrepancy; typically, the loudness level of a source in the mix contradicts the information provided by its performance intensity. It is most common for the performance intensity of a source (its timbres) to not match the loudness levels at which it appears in the mix—then the question is one of the magnitude. At the extremes are the surreal experiences of the timbre of a loud performance appearing at a low loudness level or the timbre produced by a low-loudness performance appearing at a high volume, and the experiences that very subtly alter the loudness to intensity balance.

Loudness and Performance Intensity as Insight into the Mix

Figure 10.8 illustrates this disparity between loudness and performance intensity for a select few instruments and the lead vocal in the opening sections of "Strawberry Fields Forever" (1967, 1987). The figure makes clear the performance intensity of John Lennon's vocal varies considerably in the chorus and less so in Verse 1; it also shows a reduction in intensity between the verse and chorus. Within the loudness/musical balance tier it is apparent there is less variation of loudness level, and in general the actual loudness of the lead vocal exceeds its performance intensity in the chorus and in the verse this disparity is greater as the lead vocal is at the top of *mf* in loudness and in performance intensity it is circling the

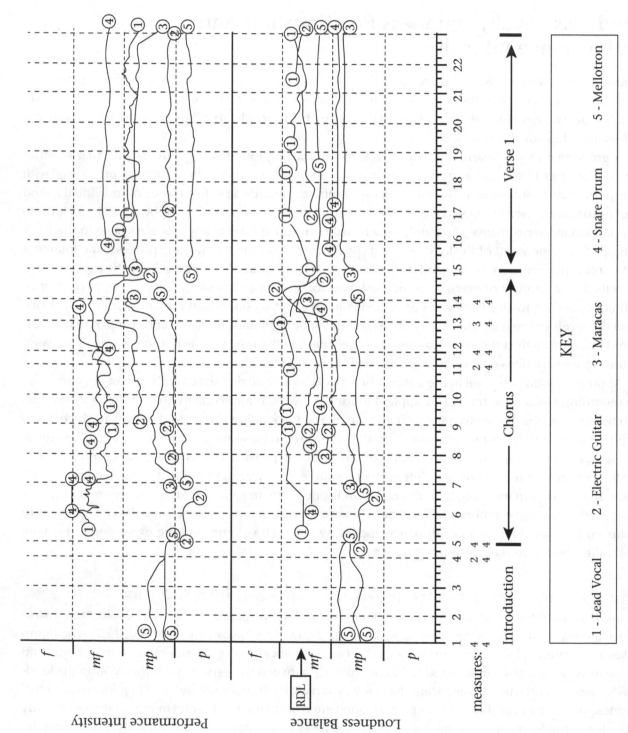

Figure 10.8 Performance intensity and loudness balance of "Strawbery Fields Forever" (1967, 1987).

threshold between *mp* and *mf*. The result of this lower performance intensity is a more relaxed, less assertive lead vocal in the verse compared to the chorus; loudness is not sacrificed, though, as the line remains at the upper edge of *mf*. This verse to chorus relationship of varying performance intensity is common enough to be considered a convention.

The tape loops of flute sounds on the Mellotron vary little in performance intensity but the lines change loudness considerably between the chorus and Verse 1; conversely, the snare drum remains at a similar loudness level throughout these sections, while its performance intensity has a wider range, especially in the chorus. The electric guitar exhibits the widest loudness range, spanning two entire dynamic areas, and its range of performance intensity is wider still; in the chorus the electric guitar's loudness levels are altered rather significantly from its performance intensity. The maracas change of loudness level is exaggerated in the chorus, and is reduced significantly between the chorus and the verse; the performance intensity of the maracas is at a lower level than its actual loudness in musical/loudness balance.

By evaluating these tiers, comparing loudness balance against performance intensity, we are able to recognize the mix—the impact of changing loudness on the balance of sounds and the presentation of their performance intensity (and the musical materials, etc., they contain). Each source in the track has its own changing relationship between loudness and performance intensity; the reader may wish to explore this track in greater detail. While the loudness alterations of performances tell us a great deal about the mix process, they also reveal a great deal about the track's interpretation of those performances. When Lennon sings in a more relaxed voice and his loudness level does not change, a distinctive quality of his presence shifts; the line speaks differently when performed at a different intensity and when incorporated into the mix at a different loudness. When high energy is not required to be heard above all others (especially other sources playing with more force), the intensity of the track itself is altered, and its sense of realism is in some way diminished. This brings "Strawberry Fields Forever" to possess a duality of energies—one for the verses and one for the choruses—that are supported by the performance intensity of Lennon's vocal.

Conflicting Impressions Between Loudness and Timbre of Performance Intensity

"Cocoon" (2001) by Bjork contains a clear example of a source performed at a low performance intensity that appears at a high dynamic level in the track. Her vocal line at 0:29–1:04 contains a performance intensity in the *p* area and the actual loudness could place it in the upper portion of *mf*, if not higher. The backing vocals in the track also present a similar conflict but more subtly. They are at low performance intensity (though a bit more intense than the lead vocal) and they are in the *mp* area of loudness balance; their first appearances are at 0:14, 1:06 and 1:11.

Because performance intensity is embodied within timbre—and timbre is altered by energy, expression, and exertion irrespective of actual loudness level—an artist can manipulate performance intensity without changing the loudness of the performance. Timbre also represents and exemplifies performance intensity; the character of timbre provides an impression of effort, from strained to effortless. This is accomplished by shaping timbre to have qualities of intensity; one can change the expression of a line without shifting loudness. The manipulation of timbre and performance intensity shapes performance style, and can be highly personalized. Rod Stewart's vocal in "Every Picture Tells A Story" (1971) is a clear example of this. The grain of his voice adds a sense of strain; at times it appears he is making an extreme effort, perhaps approaching the maximum intensity he is capable of producing. All this intensity is a variable in his performance, but in this track's first half the tension and intensity barely relax though dynamics provide considerable shape. Still, the actual loudness of his vocal line is noticeably less loud than the acoustic guitar in the left channel, and the drum kit filling the right side of the stereo field; both accompanying parts are distinctly louder than Stewart's vocal. While the acoustic guitar and drum set are louder than the lead vocal, their performance intensity levels are often less; they occasionally rise enough in intensity to nearly match Stewart, though only at critical moments.

Even with live music and speech, performance intensity can be manipulated by the performer irrespective of the amplitude of sound generated. Perceived performance intensity need not be matched by loudness level—in real life as well as in records. This dichotomy is exploited in records though, and at times to great dramatic effect, and at times simply as a characteristic timbral quality (as in Rod Stewart's vocals).

Performance Intensity and Distance Position: A Different Reading of the Same Perception

Timbral content defines performance intensity—it also defines distance position. Timbre is approached differently for these two percepts; the same information is processed to observe different qualities.

Performance intensity is defined by timbre alterations of a sound source, by recognizing the changes to its timbral content in a state of normalcy as related to the alterations that occur as it performs at a particular level of energy, expression, exertion, and so forth.

Distance position is defined by a sound's timbral content also. It is calculated by timbre alterations of a sound source that occur between distance positions. Distance position is recognized by relating the level of timbral detail present within the sound to that sound source's timbral content when it is in a state of normalcy. We will engage analyzing the contributions of of distance positioning next.

ANALYSIS OF SPATIAL ELEMENTS

Distance Position

Identifying distance positioning of sources relative to the listener's point of audition can be challenging initially (and for some time afterwards). It is common to confuse distance positions with loudness level and with amount of reverberation; neither determine distance location. Distance is a percept of timbral detail—the amount of low energy spectral information present—and of timbral content; as sound moves further from its source, high frequency information is absorbed by air more than lower frequencies, and the most-subtle changes in the wave form become lost. Listening for distance, we listen into sounds for subtle timbral content and performance sounds.

Conventions for the positioning of sounds in distance vary somewhat by genre, with certain styles emphasizing closer positioning than others—country songs tend to locate vocal within the listener's personal space. Some tracks will fix distance positions from the outset, not changing them throughout; others establish a set of certain distance positions in verses, and to change some of those distances in choruses or other sections; this structural use of distance positioning—establishing alternating scenes of position—is common. Overt changes in distance positions within sections are not common; when they do occur, it is often a dramatic effect. Subtle changes are more common; these can often be a result of a changing texture that reveals or masks timbral detail of a source. Changes of distance position can occur during fades at the end of a track, though diminishing loudness may not result in diminished timbral detail.

The following sections will examine these, and other uses of distance positioning. Typically, lead vocals in tracks reveal distance cues most readily—as they are often the most prominent sounds in tracks. For this reason, this discussion will initially focus on lead vocals. Further, the lead vocal typically sets the level of intimacy and connectedness of the listener to the track; it serves in the central and referential role in distance positioning and establishing relationships in most tracks.

Structural and Dramatic Use of Distance Positioning

"Not Ready to Make Nice" (2006) by the Dixie Chicks is an example of a track that uses distance positioning as a structural element; it also is an example of a striking change of distance placement for dramatic effect in this this track. Rick Rubin produced this album, *Taking the Long Way*. Figure 10.9 illustrates the distance positions and movements of the track's vocal lines: Natalie Maines' lead vocal, two backing vocals, and layers of descant vocals.

The structural use of distance positions is reflected in the lead vocal, which changes positions between the verses and choruses. During each verse it is toward the rear of the Intimate zone, and in each chorus it is the central portion of the Personal zone. With this change of position comes a changing scene where the lead vocal joins the two backing vocals (one closer and one further) in the same zone and within close proximity to one another. The two backing vocal lines are fixed in place throughout the track. Additional layers of vocals appear in chorus 3; these new lines of descant vocal layers that "soar above the texture with emotional intensity" (Watson and Burns 2010, 342) are the vocal lines farthest from the listener, though among the most intense of all the track's parts.

Maines' lead vocal changes distance positions markedly during the bridge. As the lyrics' content and the performance intensifies the vocal becomes more distant; building tension with the increasing anger of the lyrics is (perhaps counter-intuitively) reflected in the line being stepped away from the listener's position. The bridge vocal begins at the Intimate position of the verse; it moves away from the listener with each passing phrase until the text intensifies, and it moves further away with each measure. The bridge's final explosive lines (beginning "Sayin' that I better . . .") are accompanied by a shift from the middle of the Personal zone into the Social zone before very quickly returning to the nearest region of the Personal zone, with the greatest motion away then suddenly toward the listener occurring on the held and intensely sung word "over." This movement is quite unusual and it is effective in its support of this dramatic moment, in its quickly building tension and in its faster and more directed motion. There is movement in the lead vocal again near the end of the track. During Chorus 3, which has the thickest texture of the track (and also contains the layers of descant vocals), just after the point of highest tension the texture thins and the lead vocal suddenly returns to the Intimate zone position of the verses and a softer tone—again to dramatic effect, though this time to enhance the release of tension. During the last section, which is a slightly modified Verse 1, the lead vocal leans closer still within the Intimate zone—ending the track with the lead vocal at its closest position to the listener.

Dramatic use of distance positioning that is not always aligned with structural divisions permeates "Get Out of My House" (1982) by Kate Bush. This is a track with multiple vocals, many with their own, individualized persona. The track presents sources entering and exiting the texture, often at unexpected times to create a sense of shifting distance (and often lateral positioning). That cast of persona are distributed in the depth of the track to their own distance position; each source also has a unique host environment that impacts the track substantively (Brøvig-Hanssen & Danielsen 2016, 28–34). Kate Bush's lead vocal alternates between the close-Personal zone and the close-Social zone; it and an incessant backing vocal in the near-Public zone establish some semblance of a distance reference for the track's depth, and a way to gauge the myriad of other vocals. The many layers of ancillary vocal parts contain some sung lines, some heavily processed vocal timbres, and male spoken lines; they appear from the mid- and far-Intimate zone to the far-Personal zone and mid-Social zone. The result is a level of disorientation, an uneasiness brought about by the track's shifting landscape of physical relationships.

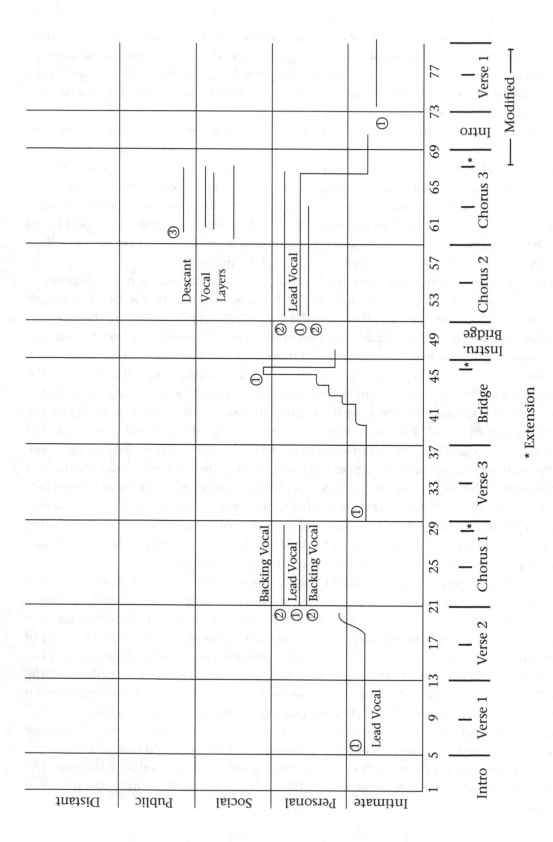

Figure 10.9 Distance location graph of vocal lines from "Not Ready to Make Nice" (2006) by the Dixie Chicks.

Interplay of Distance Position and Performance Intensity

Table 10.6   Distance positions and performance intensity levels of lead vocal from "Not Ready to Make Nice" (2006) by the Dixie Chicks.

Section	Distance Position/Activity	Performance Intensity Level/Activity
Verse 1	Intimate zone (far region)	*p* dominates, some *mp*; intensity is gently controlled, withheld
Verse 2	Intimate moves into Personal zone	*mp* dominates, some *p*; expression intensifies and is no longer withheld, reaches *mf* at end
Chorus 1	Personal zone (mid area)	*mf* to *f*, energy is being expended with strength and conviction
Verse 3	Intimate zone (far region)	*mp* to just over *mf* back and forth; assertion is present, though mostly restrained
Bridge	Begins Intimate (far region); moves in steps through entire Personal zone and into near Social zone; return to Personal zone (near region)	Begins high *mp*, builds steadily in waves up to *ff*; struggles against restraint, releasing anger in steps until last phrase is at full emotional intensity
Chorus 2	Personal zone (mid area)	*f*, high energy exertion is being expended with strength and conviction, some in reserve
Chorus 3	Personal zone (mid area); layered in Social zone; returns to Intimate zone (far region) at end	*ff*, near peak energy exertion, high intensity, some in reserve; descant vocal layers at *ff*; last phrase sudden shift to *mf* ending *mp*
Coda (modified Verse 1)	Intimate zone (mid area)	*p* dominates, some *pp*, and some *mp*

There is often an intricate coupling and interaction between distance placement and performance intensity. This can be integral to the sound and expression of a track. Table 10.6 outlines these elements of the Maines' lead vocal. It illustrates in Verses 1 and 2 performance intensity is reserved and energy withheld; Verse 1 resides mostly in *p* and some *mp* and Verse 2 *mp* dominates, with some *p* present, as expression intensifies and it reaches *mf* at the end of the phrases (transitioning into slight exertion of energy). Distance position in the verses is in the Intimate zone, and these performance intensities are realistic for that zone and the more relaxed vocal phrases are comfortable up close; toward the end of Verse 2 the close vocal becomes more agitated and once it is too assertive for an Intimate relationship to the listener it retracts into the front of the Personal zone—this brings change in distance position and performance intensity to occur in parallel. Chorus 1 brings a performance intensity of energy being expended with strength and conviction, with levels fluctuating between high *mf* and into *f*; the vocal is pushed back into the mid-Personal zone, giving more space between the vocal and the listener for this level of exertion. Following the first chorus, the vocal in Verse 3 returns to some restraint of energy, though its assertion brings it to revolve around the *mp* and *mf* threshold.

The peak of the track builds throughout the successive bridge and Choruses 2 and 3. Performance intensity in the bridge starts high *mp* and struggles to withhold its emotional intensity as it is released bit by bit until the last phrase is unrestrained *ff*; this also parallels changes in distance position: as the line becomes more intense it becomes further from the listener. Space of distance is provided for the high performance intensities (*f* and *ff*) and the charged emotions of the lyrics, singer's persona, and the track's dramatic peak here and throughout Choruses 2 and 3. The return to the Intimate zone at the end of Chorus 3 and in the coda is reflected in the lower performance intensity; at the end is the closest proximity of the vocal and the lowest performance intensity anywhere in the track.

Performance intensity and distance positions are used structurally in "Not Ready to Make Nice." Alternative functions and other alignments will be found in many tracks, and may often be significant to the track's expression and content.

Distance Positions Within a Texture

Figure 10.10 positions the distance location of sound sources within the opening sections of "Here Comes the Sun" (1969, 1987). From this distance location graph we can evaluate the positions of sources and their interrelations. It is immediately apparent no sources are positioned in the Intimate zone, and no sources are placed in the Distant zone and the farthest 80% of the Public zone; all sources are located within the Personal and the Social zones with the exception of the very faint hand claps in the introduction and first chorus. The hand claps are subtle, and can easily pass unheard; for this reason, the strings are listed as the most distant source within the distance position typology of Table 10.7, though the hand claps certainly contribute to establishing the rear of the sound stage.

Table 10.7 organizes some of the data within the Figure 10.10 graph, and outlines an initial evaluation of the graph. Distance positions are used structurally in this track, and subtle changes within phrases or between successive phrases also appear.

George Harrison's lead vocal contains some subtle shaping of distance position. The first shift occurs within its very first phrase, with the motion beginning in measure 10. Throughout the first chorus and Verse 1, all but one lead vocal phrase contains a gesture that shifts its distance position. The degree or amount of this shift varies between 6 and 12% of a distance zone. The speed of change (amount of change per beat) varies, and can be calculated from the graph; the duration of the motion between points varies from one to three beats. These shifts add expression, depth and motion to these vocal lines.

The lead vocal shifts its general position between the chorus (at the rear of the Personal zone) and Verse 1 (to the middle of the Social zone). This relationship is not conventional practice; it is much more common for the lead vocal position in verses to be closer to the listener than in the chorus. In this track this position shift between sections is not substantial, but it is noticeable; throughout the track we sense the closest lead vocals in the chorus. This shifting position is concealed somewhat by another unusual treatment of distance positioning. The two backing vocal lines exchange places with the lead vocal between sections. The backing vocal 1 position in measures 11 and 13 becomes the lead vocal position in measures 16 to 21, and the lead vocal position in measures 9 through 13 becomes the position of backing vocal 2 in measures 16 and 20. This shifting of distance positioning within the same boundaries creates some unity between the sections, as well as the change of relationship with the listener that accompanies any positional change of the lead vocal.

Distance positioning of sound sources is not changed substantially between song sections. The sources that do exhibit change between the chorus and Verse 1 are the lead and backing vocals, acoustic guitar 2, and tom-tom. The amount of change within this mix is not as substantial, as noted, but it is significant. Of these instruments, the acoustic guitar 2 changes the most, and its supportive line does not make a significant impact on the texture.

Worthy of note is the distance area that spans the three Moog synthesizer lines of the introduction. This region establishes an important distance layer to the track. When it becomes occupied by the lead vocal and the second backing vocal in the chorus, the result is a commonality of space between these two source-bonded groups of sources (synthesizer lines and vocals). Progressing into Verse 1, this distance area is occupied by the backing vocals, carrying forward this connection of the area being dedicated to vocals—but this time they are joined by the bass, and a crash cymbal and tom-tom. While the texture continues to thicken, this distance area from the farthest tenth of Personal to the lower third of Social zones is a layer of consistent activity that helps ground the track; the

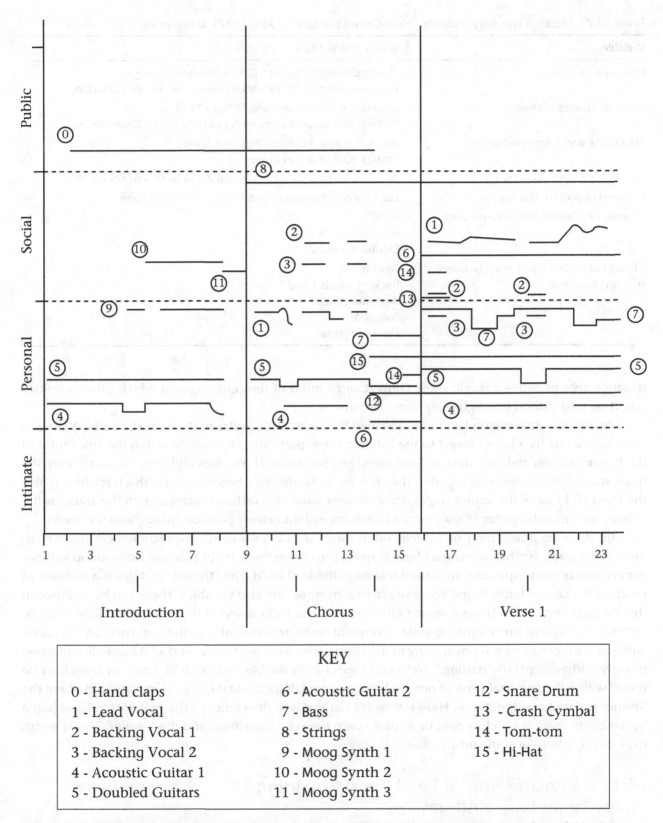

Figure 10.10 Distance location graph of "Here Comes the Sun" (1969, 1987), beginning through Verse 1.

Table 10.7 Distance typology table for "Here Comes the Sun" (1969, 1987), through Verse 1.

Variable	Source and/or Value
Positions of sources	Nearest: Acoustic guitar 2 (5% of Personal Zone) Farthest: Strings (95% of Social Zone) [Hand claps excluded]
Areas of greatest density	Chorus: mid-Personal Zone (25% to 75%) Verse 1: far-Personal Zone (90%) to near-Social Zone (30%)
Stabilizing and referential sources	Acoustic guitar 1 (20% of Personal Zone) Strings (95% of Social Zone)
Greatest amount of change	Acoustic guitar 2 – 5% of Personal Zone to 35% of Social Zone
Gestural motion of position	Lead vocal – changes of 6 to 12% of a distance zone
Change of position between phrases	Bass Acoustic guitar 1 Doubled guitars
Change of distance positions between sections (amount)	Lead vocal Backing vocals 1 and 2 Acoustic guitar 2 Tom-tom Doubled guitars

distance area provides a stable sonic context and a point of reference against which other activities might be understood through their interrelations.

Another distance region layer is established by the acoustic guitar parts. Acoustic guitar 1 and the doubled guitars fix a layer closest to the listener; these parts are consistently within the front third of the Personal zone, and they deviate little from their positions. The strings and acoustic guitar 1 are the nearest and farthest sources; together they represent the distance boundaries for this track (this is also the front and rear of the sound stage). They are also stabilizing factors for distance in this track, as the strings (like acoustic guitar 1) also remain largely consistent in their position throughout the track.

The distance positioning of sources often establishes a level of socio-psychological connectedness to the track. Further, a sense a physical space separating the listener position from sound sources occurs; this is most especially important related to the lead vocal. The strength of these is a variable, as much as the connectedness and sense of physical nearness are also variables. These can be explored at this stage. Doing so may bring a sense of the Personal space closeness of the relaxed and light acoustic guitar lines. Perhaps one might recognize a sense of some detachment with the lead vocal and its music and lyrics; a more universal positioning of the chorus with its closest vocal, and a bit more distant when directly addressing "little darling." We might expect a 'darling' (whether child, lover, or friend) to be found within the Personal zone (if not the Intimate zone); logical distance positioning would orient the listener point of audition there. Here Harrison's 'darling' is in the center of the more formal and polite Social zone; there is no closeness or a small space between the singer and the listener. This is worth exploring further—at some future time.

Distance Positioning of Lead Vocal Providing Shape to an Entire Album

Many albums are crafted with some unifying element that rewards listening to it in its entirety, and considering it as a broader artistic statement. Admittedly, listening to an album from its beginning to end is getting rarer in the 21st century, though this practice was common for many years of important activity in popular music. With changing media delivery systems, listening practices, the economics of buying single tracks—and a great deal more—listening practices have shifted. While today it may not be the norm for listeners to sit for an hour (more or less) and listen to an album without interruption, albums have maintained a sense of an overall singular impression that is sonic and/or conceptual well beyond its package. A high percentage of albums continue to be created as a singular experience comprised of numerous tracks—or conversely, as an overarching experience established by assembling tracks in a certain order, with attention to how their sounds and expressions interrelated. Many albums still reflect some manner of organization, artistry, aesthetic, manner of speaking and message; certainly in albums spanning the 1960s well across the threshold of this century, some unifying factor (whether sonic or conceptual) was central to the artistry of the album. A certain something linked and bonded the songs together. Among a myriad of possibilities, this 'something' could be a concept like *Pet Sounds* by the Beach Boys (1966) or Pink Floyd's *The Dark Side of the Moon* (1973). Albums are often interpreted by critics and listeners as having underlying themes, often without validation of the artist(s). Of course what bonds an album may also be related to sound. Despite its stylistically diverse tracks, Peter Gabriel's *So* (1986) contains a palette of recurring timbres of synthesizers and vocal treatments, and an array of spatial qualities and relationships of percussion and voices that develop as tracks progress. Further, producer Daniel Lanois brought *So* to contain textures of musical and recording elements that are recognizable; as they are developed and recur they become cohesive in interrelationships between tracks—even between tracks that are well separated in time. Situating a track within the context and aesthetics of an album can be an interesting examination; this is a substantial topic, and I will introduce just a small portion here.

Any recording element might function as a cohesive element to provide a link between tracks throughout an album. The role of distance in situating the listener's point of audition influences the experience of an entire album, not only the single track. In some albums this level of connection may unfold throughout its course, changing to support individual tracks and to bind the album into one experience. At extremes, the listener's level of connectedness may shift markedly throughout an album, as each song establishes a contrasting presence; as an opposite, there can be a sameness from track to track that does not shift and does not challenge the listener's connection or level of intimacy with the artist. Here we will explore an important album wherein the distance positioning of the lead vocal is significant to the album's unfolding experience and its content. While distance positioning can be influential in shaping the track, it may not be readily apparent as our attention rarely is drawn to this element.

Bob Dylan's *Time Out of Mind* (1997) is intrinsically shaped by varying distance positions of the lead vocal; this album was also produced by Daniel Lanois. Throughout the album Lanois situated Bob Dylan's lead vocal at a different distance position for each track; while no two tracks are in the same position, tracks tend to group by their presence in similar distance areas. With one exception, each track fixes the distance position of Dylan's vocal from its entrance; in all tracks the vocal remains at that position, not changing throughout the track—the track "Not Dark Yet" is the only exception, as it shifts slightly in gestures related to expression between the far region of the Intimate zone and the front area of the Personal zone. This album's distinctive positioning of individual tracks establishes a sense of motion and shape to the album, supports its expression and the content of individual tracks, and reveals the ways the distance positions relate to the character of tracks and also the listener's point of audition with respect to the singer's persona and the content and delivery of the lyrics. Figure 10.11 illustrates these distance positions, and opens our evaluation of their sequencing and interrelationships.

The first track is "Love Sick"; in it Dylan's vocal is in the Intimate zone, and is the closest of all the album's tracks. In the second track, "Dirt Road Blues," the listener is immediately taken to the album's furthest distance placement of Dylan's vocal in the rear of the Social zone. Thus the first two tracks establish the boundaries of the closest connection and the furthest detachment. Juxtaposing these well separated distance zones sets up an expectation that the lead vocal may not be in the same place in each track, and that shifts can be rather substantial.

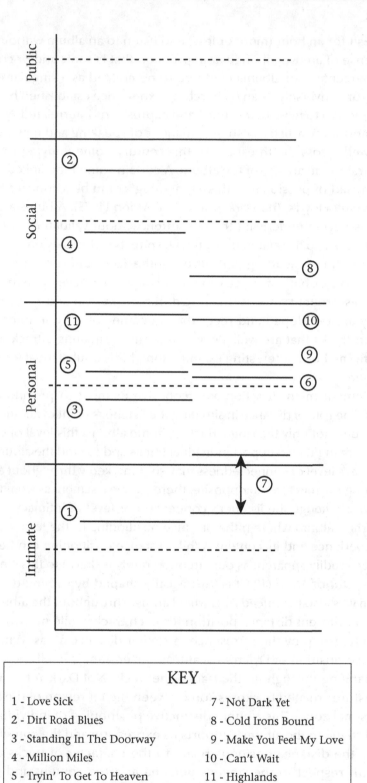

Figure 10.11 Lead vocal distance positions by track throughout *Time Out of Mind* (1997) by Bob Dylan.

The third track ("Standing in the Doorway") returns Dylan's vocal to be near the first track; rather than being in the Intimate zone, though, it is in the closest third of the Personal zone. It is nonetheless close and connected to the listener; it is among the slowest tempos of any track, which is in stark contrast with the second track's fastest tempo of the album. Tempo is noted here because it too is provided a significant change of state, juxtaposed between two tracks at the outset of the album. The sense of the track's tempo and energy, of mood and character, and of subject or story can interweave with distance as listeners form their interpretive connection with a track.

A tendency of how distance positions will change between tracks has been established already. Track 2 moved away from Track 1, Track 3 reverses the motion in the direction of Track 1, Track 4 next reverses motion in the direction of Track 2. Tracks 3 and 4 are noticeably different in distance positions, and have common traits to Tracks 1 and 2, respectively. Track 5 again reverses direction, but is now within the far half of the Personal zone. This concludes the back-and-forth distance patterning.

Dylan's vocal in Track 6 is positioned very similarly to Track 5. The two tracks sit near the mid-point of the Personal zone. This prolonged position establishes a sense of distance that will be recognized (at least unconsciously) when it returns in "Make You Feel My Love" on Track 9. A distance layer is established that links the tracks by their similar placement. This adds a sense of familiarity when the position returns, which helps open the listener to the immediate and personal connections of those lyrics and their delivery.

"Not Dark Yet" (Track 7) is isolated from the other tracks, as it is unique in its distance positioning. Dylan's vocal is again in the Intimate zone, at times very near the position of Track 1, but it does not remain there. Its distance shifts during the courses of this track; this is the only track on the album that does not remain fixed in distance position. This can help direct one to explore the nature of the lyrics and (equally important) their delivery that separate this track from the others.

A significant change of distance position away from the listener occurs for Track 8 ("Cold Irons Bound"). Dylan's vocal returns to the Social zone, though not as extreme as Tracks 2 and 4. When Track 9 follows it joins the layer established previously by Tracks 5 and 6. Track 10 is not a great move from Track 9, but it is about half the distance that was travelled between Tracks 8 and 9. Of significance is Track 10; it establishes a position that is continued (with little change) in Track 11. Tracks 10 and 11 establish another distance layer—this one with the longest duration, as "Highlands" (Track 11) is Dylan's longest track on any of his albums.

This sequencing of distance positioning supports the unfolding motion and tensions between tracks; it is part of the experience, embedded within the album's character and expression, and clearly part of its structural organization. Tracks are brought to contrast with some and to form connections with others through the distance positioning of the lead vocal, as well as through other elements. As implied above, many elements join in such interrelations—such as tempo, mood, expression, instrumentation, lyrics to name a few—for this album, and distance positioning plays an important role.

Table 10.8   Layers of lead vocal distance positions formed throughout *Time Out of Mind* (1997) by Bob Dylan.

Distance Layers	Tracks
Intimate Zone Layer	1 – Love Sick
	7 – Not Dark Yet
Mid-Personal Zone Layer	5 – Tryin' To Get To Heaven
	6 – 'Til I Fell In Love With You
	9 – Make You Feel My Love
	3 – Standing In The Doorway
Far-Personal Zone Layer	10 – Can't Wait
	11 – Highlands
Social Zone Layer	2 – Dirt Road Blues
	4 – Million Miles
	8 – Cold Irons Bound

This concludes our examination of distance position analysis. Distance position will be revisited along with the lateral location and size of stereo images, when these are brought together in examining sound stages. The analysis of sound stage positioning, of the interrelationships of individual sounds on the sound stage, and of sound stage dimensions are explored later.

Stereo Location

The lateral placements of sounds are crafted in making records. This occurs at the basic-level of individual sound sources. Placements of sounds and the widths of those sounds are the two variables for evaluating stereo imaging and their impacts on tracks. Determining the nature of this impact can be relevant for many analyses.

Recordists became creative in their use of stereo location quickly after its development. The main streams of this creativity might be reduced to the following:

- Placement of sound sources
- Widths of images
- Sounds in motion
- Motion and rhythms between locations
- Antiphonal relationships of sources

Rhythms between locations occurring as surface rhythms are not uncommon in tracks. Peter Gabriel establishes surface rhythms between image locations in "Intruder" (1980) and "In Your Eyes" (1986). In "Intruder" (2:19 to 2:40) a xylophone sound establishes an impression of a single instrument placing different pitches between various locations, establishing rhythmic patterns as well as an overall gestural movement as a result. Drum solos such as in the Beatles' "The End" (1969) establish a similar sense of one performer spread in many locations. A texture of motion with embedded rhythms appears throughout "In Your Eyes." It starts at the introduction and continues throughout—it is clearly prominent in sections from 2:05 to 3:05); Paul Théberge (1989, 104; 2018, 335–336) has identified this as a "spatialized rhythmic structure" that binds numerous percussion and drum sounds into groups based on their participation in the rhythm. These rhythms impact the musical textures differently, as the line in "Intruder" fuses into a grouping by timbre and a single performance gesture is established. "In Your Eyes" has instruments playing their own parts and rhythmic passages, additional layers of rhythms emerge as the sounds—in their distinct locations—interact. This is somewhat akin to a groove, where it is quite common for some type of "spatial differentiation" (Danielsen 2006, 51) between instruments to exist and set them apart, and most importantly for a sense of rhythmic interaction to emerge; in some genres—such as funk and afropop—this interaction between parts that are positioned apart in the track generates rhythmic gestures between them. Recognizing these surface rhythms between locations might be pertinent to an analysis; these rhythmic patterns and gestures can be transcribed as one would any performed rhythm.

Rhythms of positions also can take place at a higher structural level, such as between phrases. In the *LOVE* version of "While My Guitar Gently Weeps," string lines containing different materials appear to shift locations with changing phrases between 2:05 and 2:40. In Chapter 4 we discussed how stereo image of the lead vocal travels through space in David Bowie's "Space Oddity" (1969); it finds a different location between phrases, sometimes mirroring itself, and at times appearing antiphonally.

These examples are somewhat unique. Not all tracks will contain surface rhythms between positions, especially ones that shape the musical materials of the track. In contrast, the width size and positional placements of source images shape all tracks substantively; we will explore these next.

Image Widths and Positioning

Two different mixes, by two producers with contrasting production styles, using the same source tracks (here, the term 'track' is meant as an isolated recorded performance on a multitrack recorder, whether analog or a digital audio workstation)[6] provide an unusually rich opportunity to explore how stereo location can impact a track. George Martin produced the version of "Let It Be" (*Past Masters, Volume Two*) that was released as a single in March of 1970; for the *Let It Be* album, Phil Spector produced a different version from mostly the same source tracks that was released only two months later. The mixes are striking in how they reflect the producers' contrasting production styles and their individualized use of stereo imaging. Important qualities of each track are largely shaped by the images and locations of sound sources.

The *Past Masters* version uses image stereo location to delineate sources in their own positions. Imaging establishes a breadth of space for the track. Immediately in the introduction the piano image is to the left, off-center, and occupies over 25% of the stereo field (26°); the barely audible and centered organ (a narrow 8° image) appears to function more as a portion of the piano host environment than an independent source. The lead vocal is centered, and while smaller than the piano it is still substantial at 18°. These two images have a proportional relationship similar to how they might sound together live (though their host environments are distinctly different). With the chorus, the backing vocals make use of the entire stereo field. Their line moves in steps every two beats from the far left position to the far right. The actual movement occurs over three measures, with the seven backing vocals image widths vary between 14°, 16° and 18° during their gesture. This is an example of a sound (or musical line) in motion. The width of the background vocals gradually broadens over its phrase; it reaches the same width as the lead vocal at measure 16, its final measure. Figure 10.12 illustrates these image sizes and locations.

With the second verse, the hi-hat enters, performing on beats 2 and 4 throughout; it is the narrowest of all the track's images at 6°. The hi-hat is located to the right of the lead vocal, but its host environment is located to the left of the lead vocal (overlapping slightly). The result is an impression of motion; the strike happens in one location and the reverb brings the sound to be swiped to the left. The motion is distinct, though subtle. The boundaries of the combined hi-hat sound and its reverb span 48° and occupy much of the stereo field, though a large area (18°) is present between the two images. This void allows the lateral dimension of the sound stage to remain transparent at this time; this type of image is recognized as having a hole in its middle. The entry of the bass at the second half of Verse 2 changes in width; it begins as a narrow image that quickly settles into a width that is the size of the lead vocal. In this version, with the exceptions of the hi-hat and piano, all images are either the same width as the lead vocal, or within a few degrees. Table 10.9 is a typology table of image widths and positions.

Two sources create a sense of motion in this *Past Masters* version. First, the seven backing vocals images are distributed sequentially in increments across the stereo field; this creates a sense the material is moving, and the sound source is changing position. Second, the hi-hat and its host environment (reverb) provide a sense that the source is close to one wall within a rather large host environment, and the sound is heard as moving from where it sounded, to where the room sends it back as diffused reverb to the point of audition—a distinct impression of right to left motion is established. It is unusual in stereo for a sound source and its host environment to be separated, and for the two to be discrete, individual images; conventional practice in stereo is to fuse the two, just as they are in nature. This type of separation is a common treatment in surround sound, however.

The *Let It Be* version grounds most sounds to the center, and most sounds have unique widths. The track's image widths are proportional, though. The piano of the introduction and the lead vocal of Verse 1 are references against which the other widths can be gauged. The typology table of Table 10.9 lists sound source image widths and positions.

The width of the piano is: the same as the background vocals (when they arrive), twice the width of the lead vocal and of the hi-hat, and half the width of the organ. The lead vocal of Verse 2 widens from the first verse, though it remains within the boundaries of the piano; the bass is twice the width as

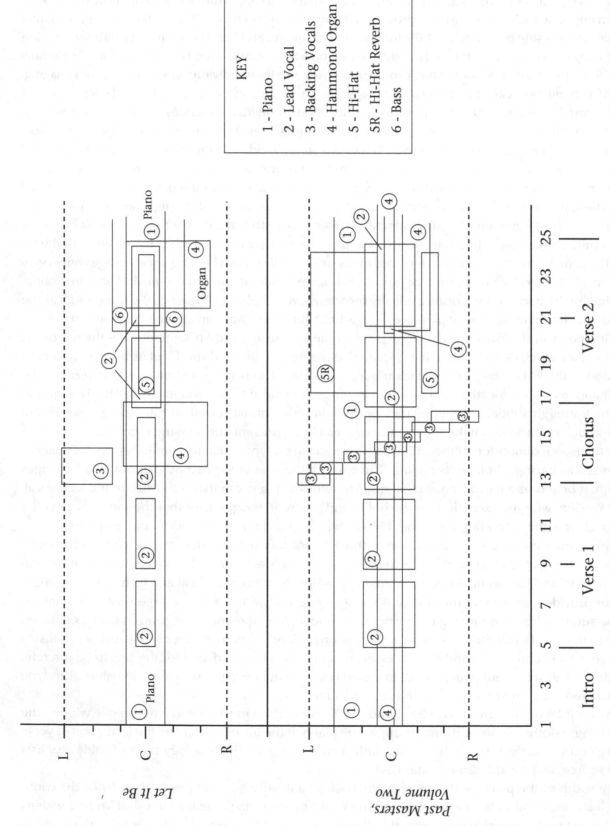

Figure 10.12 Stereo location graph of two versions of "Let It Be," the *Past Masters, Volume Two* (original single release) and *Let It Be* (1970); through Verse 2 (the first 24 measures).

Table 10.9 Typology table of stereo images through Verse 2 of "Let It Be," the *Past Masters, Volume Two* and *Let It Be* versions.

	Past Masters, Volume Two		Let It Be	
	Width	**Position**	**Width**	**Position**
Piano	26°	Left 20° to 6° Right	16°	Center
Lead Vocal	18°	Center	8°	Center
Backing Vocals	14° 16° 18°	Left 35° to 32° Right	16°	Left 32° to 16° Left
Organ	8°	Center	32°	Left 10° to 22° Right
Lead Vocal (V2)	No change	No change	12°	Center
Hi-Hat	6°	Right 12° to 18° Right	8°	Center
Hi-Hat Reverb	24°	Left 30° to 6° Left		
Bass	8° then 18°	Left 4° to 14° Right	24°	Center

this slightly enlarged lead vocal. The hi-hat fills the area the lead vocal occupied in Verse 1; the image remains focused despite the sixteenth-note tape echo that could easily have increased it size. The proportional sizes provide a sense of diversity and also order between the images. Most images are centered, and these different widths allow them their own place. The only sounds that are not centered are the organ and the background vocals; the soft background vocals only appeared for little over one measure, though, and the organ is also soft, and in Verse 2 is barely audible (its left boundary vanishes); this version has many of the characteristics of a mono recording—a format extensively used to creative advantage by Spector. There is little to bring the listener a sense of stereo imaging in this version, yet its broadened images in these sections could not occur in mono.

This rather sparse texture is temporary. At the fourth beat of measure 24 in Figure 10.12 drums and other instruments immediately fill up the texture, and a building process continues. Once the entire instrumentation arrives, this difference between a stereo conceived mix and one that is a hybrid mono mix is exacerbated. Spector's version erupts into his signature wall of sound; source images are indistinguishable within its sound mass. Martin's version retains a sense of balancing the stereo field, and a sense of placing sounds in space.

Recognizing the furthest left and right edges of images, the lateral boundaries of the sound stage are audible. From this, we can recognize how often Martin's sound stage is wider than Spector's—or, rather, that Spector's mix is largely center-concentrated and Martin's is more widely spread. One can observe each sound stage's flexing of width from Figure 10.12, as well as the density of number of sounds in each location or area.

These two production approaches to stereo imaging present the sources and their musical materials differently. Between these two versions are different sizes of lead vocal and contrasting sense of its relationship to the accompanying instruments. It is appropriate here for the reader to consider the manner in which these impart character on McCartney's performance and musical line, and their impact on the track and the musical experience of the listener (Moylan 2012, 168). One image is broader and surrounded by more space and air, perhaps with a sense of reverence. The other is more focused and compact, the accompanying instruments are mostly in the same place and there is more of a sense of connection between the voice and the ensemble. An analysis might choose to evaluate these characteristics and their interrelations with lyrics' message and dramatic themes or the correspondence to expression within performances.

The reader could well find value in extending this examination of stereo imaging throughout these two tracks to their end; the addition of instrumentation and the resulting image placements would prove informative. In addition, several other versions of this track have been released—notably *1* (2000) and *Let It Be . . . Naked* (2003) and the 2009 remaster of *Let It Be*—these each have distinctive qualities; following the approach just presented while examining another version could enrich one's understanding of the impacts of stereo imaging.

Next we will expand this examination of placement to a comparison of the distribution of frequency/pitch content of sources against their distribution in the stereo field.

Interactions and Interrelations of Stereo Location and Timbral Balance

Figure 10.13 illustrates the positioning of sound sources as they are distributed across the stereo field by their frequency ranges. Here we can evaluate where activity is present, the ranges and locations of significant amount of activity, where there are voids, and areas of little activity. As this figure illustrates the *Let It Be* and *Past Masters, Volume Two* versions of "Let It Be" (just as the previous figure) we can observe how this offers a different angle. The previous figure allowed evaluation of how sound source size and positioning within stereo location unfolds over time, moment by moment; both subtle nuances and gestural changes are exhibited in that format. Figure 10.13 provides an opportunity to examine sound source size and positioning within stereo location is distributed across the frequency spectrum—but within defined sections of time. Subtle changes of image sizes and placement are not contained in this figure, and it also omits the subtle and more distinct changes of frequency/pitch range of sources as their musical materials evolve and also (and importantly) as their timbres (performance intensities) might shift. Figures that illustrate shorter time units will yield greater detail, and those of larger sections less.

The sound source timbres remain largely consistent between the two versions, especially within these opening sections. Figure 10.13 reflects this, as it is an aggregate representation of each source's pitch/frequency/timbral content and its stereo imaging over the one-minute duration of the beginning through Verse 2. Greater detail of pitch density and timbral balance for "Let It Be" (*1* version) was presented in Figure 7.6. The *1* version is a remastering of the *Past Masters, Volume 2* version; the timbral balance presented in Figure 7.6 holds for both versions, and would be very similar to the *Let It Be* version through Verse 2 (where significant changes between the versions would quickly appear). The differences between these versions through Verse 2 are changes of timbral balance that are at a different level of perspective, and are not reflected in the pitch density of sources. The difference between these versions is a reshaping of the frequency response of the track is at a higher dimension, and would be observed and evaluated from that level of perspective.

From Figure 10.13 both versions clearly portray the lead vocal is closely connected in frequency/pitch range/register with the piano, and it is well separated from the hi-hat. The organ and backing vocals are also within the lead vocal's frequency/pitch range, and the bass largely below it. This is the distribution of sound sources' presence, activity and energy by location in implied vertical space and lateral space; there is similarity here to the soundbox, though placement of sources is more precise and measured in this format. Frequency/pitch distribution in the vertical space of the figure provides a sense of density within timbral balance that can acquire a different significance when it is examined as spread across the stereo field.

Figure 10.13 illustrates the lead vocal shares its vertical and lateral space with most other sources within the *Let It Be* mix. The piano, organ and bass all joining the lead vocal in the center of the stereo

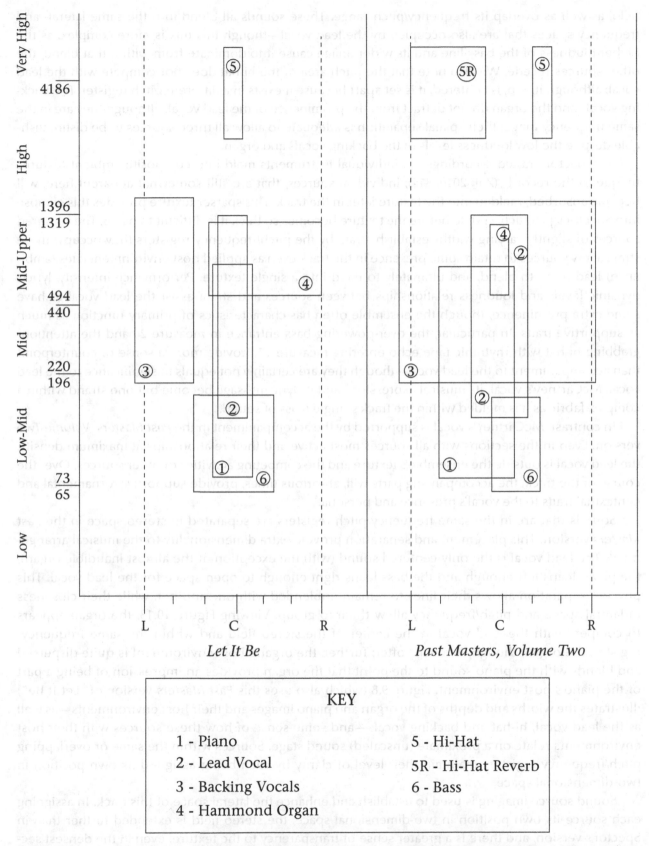

Figure 10.13 Stereo location and timbral balance graph of two versions of "Let It Be," the *Past Masters, Volume Two* (original single release) and *Let It Be* (1970); through Verse 2 (the first 24 measures).

field, as well as overlap its frequency/pitch range. These sounds all blend into the same lateral- and frequency-spaces that are also occupied by the lead vocal—though this mix is more complex, as the higher loudness of the bass line and its wider image cause it to dominate from within that blend, the other sources recede. We also note that the pitch area of the hi-hat does not compete with the lead vocal; although it, too, is centered, it is set apart because it exists in a different pitch register. The backing vocals and the organ do not detract from the prominence of the lead vocal, although they are in the same frequency range; their spatial separation is adequate to allow all three sources to be distinguishable despite the low loudness levels of the backing vocals and organ.

In "Spector crafted recordings . . . individual instruments meld into composite ambient textures unique to the record" (Zak 2010, 315); individual sources, that are still somewhat apparent here, will become absorbed, melded into the texture later in the track. This sparser texture provides this glimpse into Spector's production style before the texture becomes so thick it is difficult to parse. The centered sources of slightly varying widths establish strata by the pitch/frequency registers they occupy; these strata allow sources to retain some presence in the track even as applied host environment cues (ambience) lead them to blend, and ultimately to meld into a single texture. Performance intensity, lyrics, dynamic levels and loudness relationships between sources and strata assist the lead vocal to have some extra prominence, though the ensemble often has characteristics of primary function as much as supportive traits. In particular, the over-powering bass entrance in measure 20 and the attention-grabbing hi-hat with rhythmic tape echo entering measure 17 provide more a sense of counterpoint than accompaniment to the lead vocal—though they are certainly not equals in significance to the lead vocal. McCartney's vocal, its musical expressions and its lyric message become but one strand within a complex fabric as it is melded within the track's single mass of sound.

In contrast, McCartney's vocal is supported by the accompaniment in the *Past Masters, Volume Two* version. Even in the sections with all sources most active and their relationships at maximum density, the lead vocal is outside the ensemble's texture and the competing activities of other sources. Over the course of the track, the accompanying parts will, at various times, provide supportive, ornamental and contextual traits to the vocal's presence and persona.

Sounds that are in the same frequency/pitch registers are separated in stereo space in the *Past Masters* version. This placement and separation provide extra dimensionality to the musical arrangement. The lead vocal is the only centered sound (with the exception of the almost inaudible organ); the piano leans left enough and the bass leans right enough to open space for the lead vocal. This physical separation allows their lines to remain unblended with one another, while their closeness in lateral space and pitch/frequency allow them to group. Viewing Figure 10.13, the organ appears to compete with the lead vocal in the center of the stereo field and within the same frequency-register; the organ, though, is much softer; further, the organ's host environment is quite dispersed and blends with the piano sound to the point that the organ provides an impression of being a part of the piano's host environment. Figure 9.8 (which also uses this *Past Masters* version of "Let It Be") illustrates the widths and depths of the organ and piano images and their host environments—as well as the lead vocal, hi-hat and backing vocals—and some sense of how these sources with their host environments relate on a proximate (unscaled) sound stage. Sources within the same or overlapping pitch/frequency registers maintain their level of clarity by each being assigned its own position in two-dimensional space.

Sound source imaging is used to establish and enhance the lateral space of this track. In assigning each source its own position in two-dimensional space, the stereo field is extended further than in Spector's version, and there is a greater sense of transparency to the texture; even in the densest sections all sources can be heard, their materials are clear, and their widths/positions are evident. Further, the backing vocals and the hi-hat provide this *Past Masters* version with a sense of motion and of host environments that extend to contain these images.

Confluence and Recording Elements

It may now be evident, much can be learned about the track from compiling and comparing these two elements—just as we saw this above, with examinations comparing distance position and performance intensity, performance intensity and musical/loudness levels. What cannot be addressed from these dual views, though, are the other elements. Woven within the above discussion of "Let It Be" are references to loudness, performance intensity, distance and host environments. Recording elements interact to form larger relationships and qualities that ultimately are evaluated as a collective. This becomes clearly evident examining sources on the sound stage.

Analyzing the Sound Stage

Evaluating sound stages occurs at two levels of perspective: (1) the positions of sounds within the sound stage and the interrelationships created by that positioning, and (2) the overall size and dimensions of the sound stage (established by its boundaries). These outer boundaries of the sound stage establish a reference and a context through which the listener can sense the size of the ensemble, and also the amount of space occupied by performers within the track. The positions of sources within the sound stage establish the physical relationships of performers to other performers, and the positions of sources on the sound stage. Very importantly, the front edge of the sound stage and the nearest sound source establishes the listener's distance position relative to the sound stage; this is important for their point of audition.

Two versions of the track "While My Guitar Gently Weeps" will be examined here for sound stage concerns—the *LOVE* (2006) version and *The Beatles* (1968, 1987) "White Album" version. Together they allow us to explore contrasting qualities of sound stages and of imaging.

Sound Stage Boundaries

Sound stage sizes vary widely between tracks. This size is limited by the 90° maximum spread of the stereo field; the depth established by the nearest and furthest sources. Together, these represent the boundaries of the sound stage.

The boundaries of the sound stage are fluid; they have the *potential* to change at any time. Conversely, in some tracks these boundaries can be fully or largely fixed, establishing an unwavering frame around the sound stage. Quite often a significant change within the sound stage will happen at or around section changes; this may be reflected by a change in the locations of sound stage boundaries, though not necessarily. The size of the sound stage and its stability are variables that often contribute to the track's substance and expression.

In the "White Album" version, the sound stage boundaries are established in the first moments. The sound stage spans from the piano with its left edge at 35°L to the hi-hat with its right edge 35°R; the hi-hat sets the front edge of the sound stage at the front of the Personal zone, and the ambience around and behind the acoustic guitar provide a sense of significant space and place its rear wall (and the back of the sound stage) well into the Public zone. These boundaries create a context within which all entering instruments and voices are placed and their locations positioned. While the listener is presented with some sounds in their personal space, those sounds are located at the periphery of their point of audition; the acoustic guitar sets a point of reference within the Social zone, and provides the listener with the impression of observing the track from a greater distance.

The *LOVE* version contrasts starkly. The overall boundaries of lateral width and front-back depth of the sound stage blossoms as the introduction, Verse 1 and first bridge unfold. There is a marked difference as the width of the sound stage moves from little more than 10° on each side of center to more than 80° (nearly spanning the entire stereo field). The front edge of the sound stage begins with the guitar in the Personal zone, and shifts into the intimate zone when Harrison's vocal enters and leans in close to the listener (the front edge becomes fixed in this Personal zone, though during the track it can retreat between vocal phrases); during the verse the rear of the sound stage moves from the guitar's front-Personal zone to the solo cello's rear of the Personal zone, until the string sections enter in the bridge when the rear wall is pushed back further into the Public zone. With these string parts comes a sense that the rear of the sound stage and its left and right walls are in a constant state of change; while not unsettling or disruptive, this undulation of the space has a pulsation that interacts with the track's phrasing. Harrison's vocal, though, is what establishes the listener's point of audition and their sense of intimate connection with the expression and message of the track. We will return to *LOVE*'s sound stage below to discuss space within space.

Individual Sound Source Characteristics and Basic-Level Relationships

This unfolding of sound stage boundaries draws listener attention to the instruments and lead vocal images and their positions. The sources of this sound stage form groupings by distance zones: the lead vocal is rather isolated, and at times grouped with the guitar; during the first verse the guitar and the cello are independent parts thought they are linked by their proximity; the string parts establish a backdrop or a landscape for the voice and (to a lesser extent guitar). The strings are divided, and positioned high strings on the left and low strings on the right—this is an imaging distribution very similar to a typical orchestra layout. The string parts are clearly delineated, and positioned in distinct locations of varying distance positions; while these lines shift in their lateral and distance positions they remain linked, and their performances providing contextual, ornamental and at a few times substantive support functions. The lead vocal retains the primary materials of the track, and establishes a close relationship with the listener.

At 10° the narrowest sound source is the lead vocal, and it is also the closest and the only sound in the Intimate zone; the lead vocal will almost double in width during the bridge passages. The acoustic

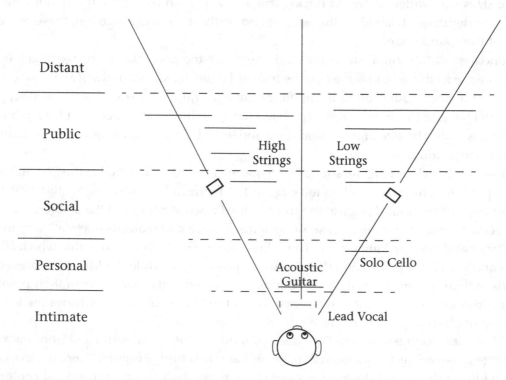

Figure 10.14 Scaled sound stage for "While My Guitar Gently Weeps," *LOVE* (2006) version.

guitar is the next widest source at about 20° and the next furthest. The string section parts are all wider, and vary between 25° and 40°with the exception of one high string line; that line is in the front of the Public zone, and has a narrow 12° spread. In the rear of the Social zone a high string part presents several brief 'weeping' glissando lines, and an intermittent double bass line. Nearly all string section activity is in the Public zone, at a considerable distance from the listener.

From the perspective of today's conventional practices, the sound stage of the version from the "White Album" is quite unique—especially pertaining to the lead vocal and groove instruments (bass, drum set, percussion, piano and organ).

The lead vocal is a large image at a bit more than 30° and it spreads and contracts slightly when it is doubled; it is centered and near the rear of the Social zone. This makes it more detached from the listener, further away; the listener now an observer from an appreciable distance—in contrast to the close connection it had with Harrison in the *LOVE* version. Its more forceful performance and intensity support this more distant placement. The wide vocal image allows it to hold a certain level of prominence though it is at a greater distance; this is an example of stereo imaging (image width) impacting its prominence that is noted by Albin Zak (2001, 156). Harrison's J-200 acoustic guitar is also centered; its width is approximately 10° and it is located in the center of the Social zone; it functions to bridge the groove instruments on the left and the right, but sonically it appears isolated—not connected with the lead vocal because of its different environment qualities or with the groove instruments because it is distinctly more distant and isolated within the stereo array.

The lead vocal and solo guitar are both similarly placed near the Social and Public zone threshold. This positioning groups and bonds their presence, reinforces their exchanging passages, and their alternating hold on the musical center. The solo guitar is the second persona of this track, and contributes greatly to the track's expression. Eric Clapton's Les Paul is located on the left side of the stereo field and its image changes width slightly with register; it is between 13°L and 33°L at its largest. Its distance position is stable near the front of the Public zone; timbral detail clearly places it here. The solo guitar

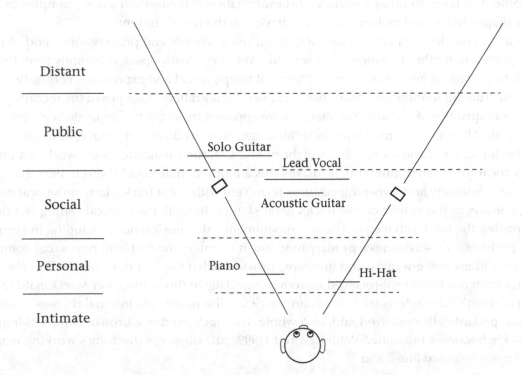

Figure 10.15 Scaled sound stage for "While My Guitar Gently Weeps," *The Beatles* (1968, 1987) "White Album" version.

varies widely in loudness level and in intensity resulting from shifting timbral complexity; its loudest passages will bring some readers to consider the sound has come closer to the listener position—perhaps much closer. It is a natural reaction to assume this, but in close listening it is clear Clapton does not move forward or back.

Other instruments that are grouped are Ringo's drum set on the right, where the snare drum often replaces the hi-hat in establishing the front of the sound stage, and the castanets that are replaced by tambourine on the left side of the sound stage. Here on the left and on the right are the other sources—piano, organ, bass; it is unusual to have the groove at the edges of the sound stage, and for the bass not to be in the middle. Nearly alone in the center are Harrison's vocal(s) and J-200; very little happens at the middle of the sound stage—at any distance—to compete with them.

The largest images are Harrison's lead vocal and Clapton's solo guitar. All other sources are notably smaller. Of significance, the acoustic guitar has the strongest presence of a host environment. Harrison's vocal image has a host environment that obtains greater depth and a sense of open breadth (though the percept does not enlarge) when the vocal is double tracked. The solo guitar's timbre changes—especially when approaching feedback or distortion—and the added ADT (artificial double tracking) provides its image with added breadth. The large number of smaller images allows them to remain distinct and clear within an active and thick texture.

Sound Stage and Lyrics

This section will examine how the positioning of the lead vocal in the two sound stages of these versions of "While My Guitar Gently Weeps" can impact lyrics; this is especially evident, given these two tracks share the same lyrics, but not the same arrangements, vocal performances and recorded vocal qualities, and usage of recording elements. The lead vocal's expression and its connectedness to the listener, and what and how the lyrics communicate differ between these versions. This book has largely left the content of lyrics to other literature; this examination is focused on a few examples of how the recording shapes lyrics, and includes some discussion of the lyrics' content.

There are some close similarities of lyrics, of harmony (chords and progressions), and of melody's contours, pitch and rhythm between these versions. Yet, they clearly speak differently from their vastly different interpretations and performances. Different tempi, mood and expression, contrasting arrangements and instrument timbres and intensities, and the radically different sounds of the records generated from divergent qualities of recording elements make apparent these are two separate versions.

The "White Album" version's iconic character surges from its drive and groove, its abundant energy propels the track. The prominence and weight of Eric Clapton's majestic guitar work not only rivals Harrison's vocal for listener attention, it "makes this a monumental track" (Everett 1999, 202). A sense of grand scale, intensity and sweeping gestures is present within the track. Harrison's vocal and its lyrics are not always at the center of this track's great stature, though, the musical setting is a dominant force in shaping the track's character. The accompaniment—the musical parts including the groove and Clapton's performance—assumes a primary function, it counters the text with nonverbal commentary, it offers other ideas and energies, and (perhaps) provides what the text does not (within *this* setting). These lyrics have not been well-received by some, especially in this setting. Ian MacDonald (2005, 301) refers to Harrison's "characteristically accusatory lyric . . . the quadruple internal rhymes of his middle sixteens are pedantically contrived and, as a whole, the track exudes a browbeating self-importance which quickly becomes tiresome." Walter Everett (1999, 201) observes the song's working manuscript contained many rejected lines, and

> one wonders why Harrison kept such filler as "I look at the floor and I see it needs sweeping." Some of the rhymes are embarrassing, particularly in the second bridge. . . . Still, the imagery is evocative,

and the musical setting is expert and riveting. . . . As they stand, the lyrics express regret at unrealized potential. . . . seem to express a strong dismay that love is not to be unfolded, that the object of the song is not to be put right.

The lyrics' evocative imagery is shadowed by the expert musical setting; powerful, riveting and larger than life, the musical setting competes strongly and at times overwhelms the lyrics. The sound stage plays a significant role in this, as Harrison's vocal is at times partially buried, and its distance position is in the middle of the group, he is detached from the listener and any connection is well beyond conversational distance; this supports the rather assertive presentation of the lyrics that MacDonald hears as 'pedantic' and 'accusatory.' The width of the vocal and its overlaying doubling that comes in and out allows it to occupy a significant area of the sound stage and to compete for attention; though in certain phrases it acquires unrivalled prominence, it is soon matched by Clapton's soloing and by the drive of the groove.

The *LOVE* version could hardly contrast more; the lead vocal dominates the texture throughout, no matter that the accompaniment becomes very active later on. When the *LOVE* version presents the opening lines "I look at you all/see the love there that's sleeping/While my guitar gently weeps" we receive a starkly different performance, interpretation and communication. Harrison reaches out, and makes connection with the listener. An almost opposite tone from the "White Album" is established; the *LOVE* track has a character, intensity and emotion that may retain some dismay but also has concern and perhaps sadness, and appears to be extending support. There is Harrison as a nurturing and caring presence, and he sings with some concern of the sleeping love in all humankind—and his guitar mirrors this. A guitar that 'gently weeps' rings true in this context, much more so than in the "White Album" version. For me, I hear all the lyrics that follow from this reframed position of caring–even those that could be heard as caustic in the previous setting. In this *LOVE* version there is a poignancy, a plaintive quality and a spiritual demeanor—what Mellers (1973, 149) called Harrison's 'lyrical-religious vein'—that could never emerge from the "raucous electric" version (Hertsgaard 1995, 252). There is also an anthemic presence that emerges as an interaction between the lyrics and the other elements—especially those of distance and lateral positioning; this creates an impression that does not align with what lyric analysis offers. The recording elements here shape the meaning, delivery and expression of the lyrics, and thereby the entire track.

It is interesting to recognize, the *LOVE* version vocal along with its simple guitar accompaniment are from the first take of the first EMI recording session (July 25, 1968) of this track. In but a few weeks "While My Guitar Gently Weeps" would undergo a remarkable transformation from this introspective acoustic statement to the powerful energy that erupts from the "White Album" track (which received its final mix on October 7) (Lewisohn 1988, 145–159).

Host Environments, Holistic Environment and Space Within Space

The *LOVE* (2006) version of "While My Guitar Gently Weeps" is a completely different version from the iconic track on the "White Album" (1968). This last version is a reworking of the guitar and vocal from the first session for the song, with orchestral strings added by George Martin 38 years after that first session—for the Cirque du Soleil show "Love." The sparse, stark introduction and first verse reveal the subtle qualities of the acoustic guitar and lead vocal host environments. These environments provide a context for those sources, and enhance their timbres and musical materials; they also set a context for the track and establish a sense of expectation of intimacy and personal space within the track. As the track unfolds, the host environments work synergistically with and within the sound stage to establish

Table 10.10 Typology of host environments for acoustic guitar, verse lead vocal, solo cello, and bridge lead vocal, and the holistic environment of "While My Guitar Gently Weeps" from *LOVE* (2006).

Attribute	Acoustic Guitar	Lead Vocal (Verse)	Lead Vocal (Bridge)	Solo Cello	Holistic Environment
CONTENT					
Reverberation					
Contour segments					
Time of initial rise	No audible rise	No audible rise	No audible rise	No audible rise	Circa 10 ms rise time
Steady state – Density	Sparse density	Moderate density	Less dense, nearly transparent	Moderately sparse	Moderately dense sustain; density increases behind lead vocal position
Decay duration	Less than 10 ms	Shorter, less than 8 ms	Slightly longer, 10–12 ms	10–12 ms	Tail duration approx. 3 beats
Frequency response	No audible alteration to sound source	Distinct loss of high frequencies; Mid register is increased slightly (<1dB) around 200Hz	High frequencies present; Mid register boost no longer present	No audible alteration to sound source	Mid and Low-Mid registers are increased in level slightly
Early Reflections and Arrival time gap	No audible early time field	Pre-delay (ca. 5 ms) No early reflections	Pre-delay shortened or eliminated; not audible	No pre-delay or audible early time field; slow attacks conceal early time field information	No pre-delay
CHARACTER					
Size of environment/ space	Small room	Slightly larger, small room	Larger room than Verse, remains a small room	Moderate size room, slightly larger than voice	Medium-size concert hall
Notable Environment Width and Depth Qualities			Reverb extends width of Voice image		Reverb surrounds sound stage within holistic environment

Level of Realism, and realistic qualities	Realistic spatial balance and quality	Realistic space, includes early time field	Realistic space; audible room reflections		Qualities mostly parallel acoustic spaces
Surrealism	No early time field	Less realistic but remains convincing	Early time field is unnaturally masked		Unnatural density regions reverb; no sense of side walls
Spectral quality	Accentuated instrument timbre in Low-Mid register, not in reverb	High frequencies of reverb attenuated, accentuated Mid register	Reverb mirrors spectrum of sound source	Reverb mirrors spectrum of sound source	Accentuates Mid and Low-Mid registers
Sound source clarity	No masking is present; very clear timbral detail; personal positioning	Clear timbral detail for intimate positioning; some Mid register blurring of detail	Clear timbral detail for intimate positioning; no diminishing of detail in Mid register	Considerable timbral detail; slow attacks establish few low-level partials in onset	Removes some timbral detail by smoothing string attacks
Ratio of direct to reverberant sound	Very high percentage of direct sound	High percentage of direct sound (slightly more reverb than guitar)	High percentage of direct sound (slightly more reverb than voice)	Percentage does not change noticeably from Verse	Direct sound dominates; Direct sound is about 80% of ratio to reverb

a clear sense of a substantial holistic environment; the host environments also provide the track with movement—as some shift between sections, others are present only for a phrase or two before being replaced by another space or by a different line with its unique host environment.

Host Environments

The track opens with Harrison's Gibson J-200 acoustic guitar, in a host environment with characteristics of a small room, no audible early time field, and a reverberation of rather sparse density; the reverberation time is less than one beat, and the direct sound is substantially louder than the reverberant sound. The guitar's Low-Mid register is slightly accentuated, though that is the direct sound of the instrument itself, not the environment. This host environment provides the acoustic guitar with some depth and width—a sense of environment walls containing the sound can be perceived. The host environment provides the guitar sound with unobtrusive support; its timbre remains clear and detailed, while its image has been extended slightly.

George Harrison's vocal in the verses has a host environment with denser reverberation than that of the guitar; though its reverb tail is shorter at less than 8 ms. The host environment contains a frequency response that contrasts the direct sound, with a distinct loss of high frequencies and a Mid register is slightly boosted (<1 dB) around 200 Hz; it also contains a distinct pre-delay that provides the environment with a more natural character. The lead vocal image has more depth than the guitar's host environment, but it is not wider.

When the solo cello enters during the second half of the first verse, the qualities of the voice and guitar host environments do not change, though they become partially concealed. The solo cello adds a different, third host environment to the track. This environment is not substantially different from the others, though it has its own character—a moderately sized room with a higher percentage of reverberation. It features a longer decay, no perceivable change in frequency response, a sense of depth, and no added spread to the direct sound. It is similar to the guitar's environment in that no early reflections are present. The similarity of these environments and their similar distance positions causes the cello to be grouped with the other two sources; though presenting a supportive line, the cello immediately is one of a trio, and this grouping is established largely by recording elements.

Table 10.10 contains a typology of these host environments, addressing both content and character matters. A more thorough listing of attributes may be sought (see Tables 8.8 and 8.9), should it serve the analysis.

With the bridge, the lead vocal's host environment changes. The change is subtle, but important. It alters the character of the line, but the host environment does not call attention to itself—in fact it is easy to miss. This is often the case with recording elements. The lead vocal's host environment broadens the image around the direct sound; the reverberation becomes less dense and almost transparent, and is slightly longer in duration. High frequencies are now present in the direct sound and reverberation, and the Mid register is no longer accentuated by the host environment—this allows the vocal line's change to a higher register to be reinforced with greater clarity in the line and the vocal timbre. The new host environment has the character of a larger and more spacious room (not substantially, but noticeably); it provides the vocal with greater presence, clarity, and sonic substance. It is common for vocals in a chorus or bridge to be provided with some type of larger-than life identity—and to place them in an appropriate host environment. Notice also that the distance position of the lead vocal does not change between the verse and the bridge, but the image is wider in the bridge; the change is subtle but significant (Moylan 2012, 173–174).

The backing strings enter with the bridge and shift the size of the sound stage dramatically. What was once a focused and small space, with an intimate Harrison singing while seated next to the listener is now expansive, a much larger area. The presence of string instrumentation quickly fills the sound stage—both in width and in depth. The listener is immersed in a texture of strings that stretch beyond the confines of the loudspeakers, enveloping as much as stereo can, and in a space that suddenly also has much greater depth.

The strings quickly establish an accompaniment texture of shifting layers. The host environments of the string parts contribute to their separation from the guitar and vocal. The string arrangement contains numerous parts and groupings of violins, violas, celli and double basses. These establish layers that are each in a distinct host environment; most of these environments have qualities ranging from medium-sized rooms to a large hall, nearly all accentuate Low-Mid and Mid register frequencies, and all contain a very short or no pre-delay. Except for a few violin passages, the lines do not have abrupt attacks; this aids in establishing distance positions distinctly farther away than the voice and guitar. Some string parts extend in depth significantly, and place the rear of their image clearly at the farthest points of the Public zone; one line extends into the Distant zone. The interactions of these host environments (in their many different content and character types) and their locations on the sound stage are complex and detailed. Some patterns of environment sequencing appear to be present. Additional host environment and sound stage information could be pulled from this texture, should it provide important detail for one's analysis.

Assessing the typology of these host environments, a common trait is that all environments have a sense of realism; one has a sense that all spaces could exist, and perhaps have been previously experienced. Host environment sizes varied between small rooms and large spaces; there was a progression throughout the track: the spaces encountered were small rooms that gradually grew from the guitar to the lead vocal, then a slightly larger space appears for the solo cello. In the bridge the lead vocal remains in a small room, though (retaining the progression of spaces increasing in size) it is a slightly larger room than in the verse; importantly, the guitar remains in the same host environment, and anchors the listener; its host environment can then serve as a point of stability from which the other host environments with their shifting sizes can be gauged. The string lines in the bridge continued this trend of spaces getting larger, as successive entries presented host environments ranging from larger small rooms (similar to the vocal) to medium, then to larger-sized rooms. None of the string host environments are out of scale, though; they are all appropriately sized, and contain similar-enough reverberation qualities, to allow all sources to remain bonded within an ensemble. This grouping of the string lines is retained despite varied host environment sizes; their interconnection is tested as the track progresses—with smaller host environments for some violin gestures and the track's largest host environment for the double bass's most prominent passages—still their shifting layers maintain a unified presence.

Space Within Space

Recall now, the sound stage of *LOVE*'s "While My Guitar Gently Weeps" expands and contracts in width and in depth beginning in the bridge and continuing throughout the remainder of the track. These sound sources that establish this are bonded to their host environments. The dimensions of the sound stage flex between and within the overlapping string phrases, and the host environments contribute to this imaging; an undulating set of relationships between the sources and their host environments is the result. The largest host environments appear at the edges of the sound stage in the last section; the sound stage settles at its largest width at the very end of the track. While it had approximately this width earlier in the second bridge the host environments play more of a role at the end. The greatest depth of the sound stage is established by the double bass lines in the second half of the track. All of this activity, though, is perceived as taking place within a context that allows all these spaces and lines to have coherence and unity, to have some interrelation and connection. The holistic environment is this context, within which all of this is contained.

This track provides a clear example of space within space. Space within space acknowledges this simultaneity of host environments functioning side by side—as each host environment is bonded to its sound source, and is independent from other environments—and also a sense of spatial layering. The holistic environment is the context within which this happens, as it is at the highest level of perspective and the apex of the pyramid of layered perspectives of sources+spaces; it contains all the host environments as well as the sound stage, and is the environment of the sound stage. The holistic environment is the impressions of the track residing within a single venue, an all-encompassing place and space that is the world of the record.

In this track, all source host environments happen to be smaller in size than the holistic environment—this is unusual. The small rooms of the lead vocal, acoustic guitar and solo cello as well as their positions on the sound stage seem readily contained by the holistic environment; the holistic environment also contains all of the host environments of the string lines, with all their movements and their various sizes (including the double bass host environment that is only slightly smaller than the holistic environment). A layering of host environments at various lateral lateral and distance positions is evident within the holistic environment; string host environments frame the vocal and guitar on either side and also create a layer of greater depth than the lead vocal's space.

Holistic Environment

The holistic environment reveals itself completely once the bridge is fully established. The listener is presented with a size, content and character of an overall environment that is unexpected based on the opening; this space is an entirely different landscape from the small intimate space of the beginning sections. The space that shifts the entire context of the track to be much larger and more universal, and much less intimate and personal; it no longer is speaking to the listener alone, but to a larger collective. Despite this, the listener's point of audition in relation to the sound stage and holistic environment has not changed substantially, as the guitar and voice anchor the texture and the listener's sense of location to the track. Importantly here, the host environment has not changed; its entire size lay latent until brought into play.

The illusory physical context of the holistic environment is presented a bit at a time. It unfolds with growth throughout the introduction and first verse, until it expands significantly as the string lines become established in the bridge. By 1:05 the width and depth of the holistic environment are mostly defined. The boundaries of the holistic environment at this time establish a new reference that will be carried throughout the remainder of the track.

The holistic environment of this track has the sonic character and content of a medium-sized concert hall, and a rather unique spatial imprint. In the region behind the distance placement of the voice and guitar, reverberation gains in density (due to the host environments of the string parts); this helps to maintain a clear presence of the voice even when the texture thickens. The holistic environment's reverberation qualities are convincing as realistic, though this type of tuned density cannot occur in the natural world. Further, the absence of early time field information brings it to appear the track's sound sources are not contained by reflective walls; this establishes an openness, an impression that the track reaches farther to the left and to the right than can be heard within the stereo field sound.

A slight (approximately 10 ms) rise time in the reverb contour leads to a moderately dense steady state, and a reverb tail of nearly three beats. The environment provides an impression that it absorbs the timbral detail of the string attacks, and it slightly accentuates Mid and Low-Mid registers. As sources move increasingly farther from the listener's point of audition the proportion of reverb increases with respect to direct sound—as in nature, sounds contain increasingly more reverberant energy as they move farther away within the same space. Worthy of note, this does not always happen in tracks, in fact it typically does not. In general, the holistic environment has a higher percentage of reverberant sound than do the lead vocal and the guitar. This holistic environment has a distinct character presence and a tangible sonic imprint; these also are not typical of many tracks.

The holistic environment establishes an expanse that complements the expression of the track. It provides is a stable reference and a unifying presence; though it is not revealed in full at the beginning of the track, once its expanse arrives it grounds the track within a space large-enough for its expression. The action of being gradually unveiled provides movement, direction and drama within that unfolding of sections. The holistic environment holds the somber, contemplative nature of the track in a hall that can be intimate and yet have the room for reflection. The size, content and character of the holistic environment enhances the track's message and artistic statement, and its character supports the affects and mood of the track. Of course, the content and character of the holistic environment of other tracks may be very different; the holistic environment is an element that holds potential of great variety of dimension and interrelations with the sources+spaces it contains.

Spatial Identity

The spatial identity of a track is the place and space from which the track appears to emanate. This is a spatial profile that has content and character, and that establishes a spatial context for the track. Its component parts of its content are:

* Sound stage (boundaries, physical size, distribution of sources)
* Holistic environment (content and character of the 'host environment' of the sound stage)
* Space within space (the relationships of the holistic environment to the environments of the sources it contains; acknowledging the influences of those host environments to the content and character of the holistic environment)

The track's spatial identity projects a character, that can potentially be quite pronounced as is experienced in Kate Bush's "Get Out of My House" (as previously described), with its sense of a collage of quickly changing source host environments pushing at the boundaries of the holistic environment. Conversely, spatial identity can be subtle, a gentle binding presence that does not call attention to itself yet functions to fuse the track into place and space. Tracy Chapman's "Fast Car" (1988) provides such an example; the host environment of her lead vocal contrasts with that of all instruments and their variety of spaces, but all sources reside in place and contribute to a characteristic spatial identity that fuses all their qualities into a single venue that is largely transparent. Bob Dylan's "All Along the Watchtower" (1967) contains no obvious manipulations of instrument environments or hint of an applied overall reverb; its spatial identity rests in the interrelations of the natural sounding host environments of the vocal, harmonica, guitar, bass and simple drum kit that establish a sense of blending into a singular impression of room size of width and depth, with room sound qualities established by similar reflection densities in their early time fields, slight timbral alterations of those reflections, and shared reverberation qualities.

A track's spatial identity may also carry associations to listeners, as the spaces of host environments can each provide a listener with impressions of places that can carry meaning. For example, David Bowie's distant vocal and the accompanying crowd noise at the beginning of "Diamond Dogs" (1974) clearly elicit a holistic environment of an amphitheatre, a sudden shift of space then occurs, as the band is performing in a smaller environment and Bowie's vocal shifts to a much nearer position to the listener. Crowd sounds of audience milling about and sounds of an ensemble tuning open the Beatles' track

"Sgt. Pepper's Lonely Hearts Club Band" (1967) establish an impression of a large space (auditorium or circus tent) filled with people and of a stage; the distance between them is established, an impression of holistic environment is formed, and clear associations are elicited. Then the performance begins and the listener is presented with sounds that are close; they are located much nearer the ensemble than the previous impressions indicated, and in a much smaller holistic environment. This position is somewhat closer than the listener position related to the ensemble of "Diamond Dogs" (which had a similar relationship at its beginning); the greatest contrast, however, is that in "Sgt. Pepper's" the crowd returns—this time cheering—as the group Sgt. Pepper's Lonely Hearts Club Band is introduced. The impression of holistic environment now seems unstable—it seems to have shifted again—then the band re-establishes the previous environment with another shift. The crowd returns often as the album progresses; ultimately the crowd establishes itself as an independent character and carries associations, and as such the crowd and the holistic environment which it inhabits contribute to the context of the *Sgt. Pepper's Lonely Hearts Club Band* (1967) album.

While an actual shift of holistic environment is not common, it permeates *Sgt. Pepper's*. Another example that more clearly explores this is "A Day in the Life" that concludes the album. In it we find John Lennon's verses and Paul McCartney's vaudevillian bridge section are in different places and spaces, and could perhaps be heard as from different times. Linking these are the great orchestral build up (beginning at 1:45) that serves to transition from Verse 1 to the bridge; the orchestra also appears in a different space as it emerges from Lennon's intoning ("turn you on"); this same space and orchestral build-up would return to end the track (at 3:50). The wordless retransition from bridge to Verse 2 (Lennon's "ahhh" vocalizations) begins in a new environment that has characteristics of both the bridge and the verse, and morphs into a return to the verse holistic environment. The listener is transported between places to different spaces by design.

As diverse spaces accumulate, the listener's imagination can be activated; those results cannot be anticipated, though a few examples might be a track acquiring a surreal nature, or a sense of shifting places. This can be extended to a sense of parallel places or worlds; one might imagine such a state in the Trent Reznor and Atticus Ross cover of Led Zeppelin's "Immigrant Song" (2011) where Karen O's vocal is disassociated from all others to the point of appearing to be in a different place, separated by significant space.

Many of the observations needed for evaluating the spatial identity of "While My Guitar Gently Weeps" (*LOVE*) have been collected over the preceding sections. Those attributes are brought together here to establish a sense of this larger dimension element. Table 10.11 is a typology table of the track's spatial identity.

Table 10.11   Spatial identity typology table of "While My Guitar Gently Weeps" (*LOVE*).

Holistic Environment	See also Table 10.10	
CONTENT	Reverb density	Moderately dense during sustain; density increases as distance increases behind lead vocal position
	Reverb tail	Tail duration approx. 3 beats
	Frequency response	Mid and low-mid registers are increased in level slightly
	Early time field	No pre-delay
CHARACTER	Size of environment	Medium-size concert hall
	Level of realism	Qualities mostly parallel acoustic spaces; unnatural density regions in reverb; no sense of side walls
	Ratio: direct to reflected sound	Direct sound dominates; direct sound is about 80% of ratio compared to reverb; significant timbral clarity results

Host Environments	See also Table 10.10	
	Smallest environment	Acoustic guitar
	Largest environment	Low strings
	Unusual environment(s)	Solo cello is the most unusual, though quite natural
	Environments of most significant sources	Lead vocal: small rooms (verse and bridge have different environments)
	OTHER	Shifting frequency response between high string environments
Sound Stage	See also Figure 10.14	
	Composite dimensions: overall boundaries	Nearly the entire stereo field Within reach of listener position to the rear of Public zone
	Position of Front Edge	In the rear of the Intimate zone
	Position of Rear Wall	Back of the Public zone
	Number of Scenes:	5 scenes
	Structural sections:	Introduction, Verse 1, Bridge 1, Verse 2 through Bridge 2, Coda
	Areas of significant activity:	Personal zone: acoustic guitar and solo cello Public zone: all orchestral string section parts Intimate zone: lead vocal
	Areas of no activity:	Social zone: no sources present at any time
	Sound stage size and positions of boundaries	Gradually widens from 10° either side of center to nearly the entire stereo field Deepens in steps from: (1) front of Personal zone (Intro), (2) Intimate to front of Personal zone (Verse 1), (3) Intimate to rear of Personal zone, and (4) Intimate to rear of Public zone
	Moving sources:	All sources are stationary
	Lateral motion:	String lines shift positions with changing phrases

The spatial identity of this track is reflected in a sense of significant width and depth that is slowly revealed as the track moves through its first sections. The space is substantial, perhaps equivalent to a medium-sized concert hall. The hall itself has a sonic imprint that appears to transform the timbres of all the sources with similar qualities of few or no early time field reflections and a slight emphasis of Mid and Low-Mid register frequencies.

As sounds are further to the back of the sound stage their reverberation time increases naturally, providing a realistic character to the sound stage and to the track. This sense of realism is also reflected by the host environments of the individual sources; no noticeably unnatural traits arise within any of the host environments or the holistic environment. The sounds furthest from the listener are the string parts. The string lines establish the sound stage left and right boundaries; importantly, the lines establish an undulating character as they provide a sense of spatial motion. The sound stage boundaries and image placements contribute significantly to the spatial identity of the track, and also to the track's overall character and expression.

The spatial identity of this track presents the lead vocal in an intimate setting; it is close and yet its presence is comfortable. Its placement on the sound stage brings it a strong presence by its closest proximity to the listener. At times it has a linkage with the acoustic guitar that provides some impression they may be in different locations within the same space, though clear environment differences are present that provide them independence. The string lines provide a backdrop of context for the vocal (and to a lesser extent the guitar); this is conceptual in the function and also actual as reflected in their placement on the sound stage and in the types of host environments from which they emerge.

The spatial identity of "While My Guitar Gently Weeps" (*LOVE*) might be summarized as representing a moderate-size performance venue (of moderately dense reverb, a 3-beat sustain time, and a slight emphasis of Mid register frequencies). The space is filled with sources that are evenly distributed laterally and are placed in groupings by distance position (as there is separation between the strings and other parts, with no sources in the Social zone). The prominent lead vocal is in an Intimate position and provides a point of reference for the track's spatial identity.

RECORDING ELEMENTS AND THE TIMBRE OF THE TRACK

We have arrived at the end of analyzing recording elements. It should be obvious these analysis examples present a minute fraction of their myriad possibilities for variations of qualities and functions these elements may reflect. The processes of analysis were the focus of the preceding analyses, as much as the content of the tracks. Those examples demonstrate how the elements *might* be analyzed for other tracks; it also usually presented only one *possible* interpretation of the observations and evaluations that were shared.

A final analysis will be a cursory examination of recording elements within a 'timbre of the track.'

Figure 10.3 of "Here Comes the Sun" outlined certain basic qualities of recording elements during its initial stages. Examining that figure one might obtain some insight into how the elements interact—particularly when the comments on that chart are supported by more detailed observations and evaluations. Within that figure we can observe the shifting sound stage by the changes in distance and stereo location, and other qualities that interrelate between perspectives. Further, we can trace the unfolding development of elements and compare them to other elements, such as pitch density and performance intensity. This makes it clear that the extent this chart is useful relates to its detail and organization. Remembering the content of that figure represents a deconstruction of the track into component parts, we might remain aware that it is inherently incomplete; not all attributes can be contained therein, nor are all of their content or characters.

Most importantly, our analyses can contain observations and evaluations of the interactions of elements and their materials. It is within these interactions that the richness of the experience of tracks begins to reveal. When we listen we connect disparate materials into our own understandings. The whole of the track is thus different and more intricate than the sum of its component parts. The more we engage a track, the more these connections become revealed within our perceptions and understandings. The preceding analyses contained several examinations of the interactions of elements; as listed in Table 9.3 and discussed then, many more combinations are present and their associations will emerge from further listening and study. All have the potential to yield insight into tracks—different specific combinations will be most pertinent for each track. These inter-associations and interrelations bring about an interaction that can bring confusion with texture.

The timbre of the track differs markedly from texture. Texture is an aspect of structure, and much of popular music "employ[s] layered textures, which manifest varying degrees of textural stratification" (Covach 2018, 71); texture is the surface fabric of materials and elements, their functions and content. Texture addresses "the structural relationship of parts or layers in the music to one another" (*ibid.*, 53); in recording analysis these relationships of parts also embrace the recording elements. Texture manifests in the middle dimensions, often at the basic-level of individual sound sources, or of groups of sound sources; the timbre of the track is in the large dimension. Timbre of the track engages these layers of texture as multi-domain, complex sound events that are woven (along with their interrelations) within its content to establish a gestalt; these stratified layers contribute to establishing the track's overall timbral character as a single sound object or aural image, wherein all is present at once.

The recording elements that function at the highest level of perspective exhibit the most immediate influence on the timbre of the track. Program dynamic contour, timbral balance, loudness balance and spatial identity strongly shape the track's overall sound, or highest-level timbre. Figure 9.2 presented the program dynamic contour of "Here Comes the Sun," and Figure 10.6 illustrated the timbral balance of its first sections. To complete this set of observations, a study of the track's loudness balance and spatial identity (sound stage and environments) would follow. This track's own, unique sonic signature—its timbre of the track—is reflected within these large-dimension elements, along with their interactions and confluence with all else at that structural level.

The timbre of the track represents the track's aggregate 'sound,' both in content and in character. It is the confluence of all the track contains—all domains, at all dimensions, for each element. Music and lyrics of course are central, and as this writing has illustrated, recording elements are no less integral to recorded song. The timbre of the track also carries affects, outside associations, and the meaning and expression of the lyrics. The timbre of the track reflects the interrelationships, interactions and larger percepts formed by all of these. The listener's overall interpretation of the track often relies heavily on the timbre of the track.

Timbre of the Track

The timbre of the track of "This Woman's Work" (1989) by Kate Bush should illustrate this matter clearly, both related to the recording elements and to musical parts and lyrics. What follows are generalized conclusions related to recording elements that are based on a significant amount of observation and evaluation; all the background collection and examination of information that went into these conclusions cannot be included in these descriptions. A reader's closer examination of "This Woman's Work" will yield much information about how the timbre of the track was built, and would inform these following statements, and others.

Character of the track's timbre embodies an overall mood of "understanding, yearning and regret" (Moy 2007, 111) that is being communicated by the lyrics, the arrangement and the recording. Kate Bush's use of space establishes a strong spatial identity for the track (and also for its individual sources and their materials), and the loudness balance, timbral balance and track loudness contour coalesce into an impression that supports this character. To this end, reverb tails bring vocals to linger after they cease, and the sizes of the vocal host environments contribute strongly to a sense of breadth for the track's holistic environment. Also significant to this overall character, the spaces, timbres and registral placement of sources coalesce to provide a characteristic spacing, or a general sonority for the track; this establishes a reference, or a state of normalcy inherent to the track. Against this normalcy are areas of more activity and of less, of different levels density, dynamic, registral shifts, and so forth; when this state is disrupted the track's tensional motion is quickened, and its drama intensifies towards the track's most significant points of arrival (1:37 and 3:11) that are followed by quick release in silence.

The content of "This Woman's Work" is distinctly evident within its clear layers; they provide a sense of stratified polyphony, and also a sense that all parts coalesce into one. The individual lines enter and recede in distinct registers as if partials of an overall spectrum—each sound source in a confined pitch/frequency space, and also with a loudness contour as if part of a spectral envelope.

The track unfolds in distinct layers of parts, each containing musical materials, and the vocal lines containing both lyrics and non-language vocalizations. The musical lines unfold starting with piano block chords punctuated with delicate melodic gestures. These lines each carry identifying imprints of recording elements, as do the vocals. The layers of musical ideas occupy distinct registers—the piano block chords in one, piano melodic gestures in another, and lead vocal(s) shift by phrase between registers—that build into timbral balance. When backing vocals, synthesized sounds and strings enter, new layers become present which bring timbral balance to widen and create an increase in the overall loudness contour.

The piano accompaniment anchors the track throughout by its register, its performance intensity and its spatial identity; the lead vocal in the verses also provides a strong sense of a normal state, against which its significant excursions of intensity can be understood. These are significant as reference for character and content of this track's level of intensity and energy, its connection to the listener and its holistic environment. This reference allows understanding of the interrelations of the recording elements of all sources, and how they shape the track. All sources each have unique and strong spatial identities, with a unique but pronounced environment, a distinctive distance position and stereo image size and placement that play integral roles in shaping the timbre of the track. In this track, space plays out in the density of host environment reverbs which contribute to timbral balance (pitch density of sources) as much as they play a role in the shaping of the holistic environment, and the performance intensities of sources provide important shifts within timbral balance and loudness balance.

The timbre of the track content of "This Woman's Work" is a tapestry, a woven set of distinctive layers of which the recording elements play important defining roles. This is present in nearly all tracks, though this track reveals it clearly; the clarity of strata is one of the defining stylistic factors of this track. This approach is present in much of Kate Bush's work—sometimes as strata, other times as collage—in the way disparate elements and musical parts are intricately juxtaposed and interwoven.

CONCLUDING REMARKS: LISTENING IN THE FIRST PERSON

Arriving at the end of this journey, the reader is likely well aware that the common thread woven throughout has been listening. Deep listening, listening with attention, listening with intention, open listening, listening at various levels of details, and other ways of listening comprise one's sole access to the sounds that is the track.

Listening is our primary analytic tool. It can become more refined and more accurate, and honed to extract great detail during critical listening. Listening can also become more flexible, while retaining a clear intention; as one gains control of processes, one can shift between modes or objects of analysis with dexterity. The skill itself is malleable, ever-changing.

Listening skill develops over time, and can continue over a lifetime. I am a different listener now than I was when I began writing this book. The processes of evaluating sound and making observations of recording elements bring on-going refinements to my abilities to perceive subtle changes in an element. Most notably, my abilities to identify the content of environments improved, so that all of the observations contained here were performed 'by ear'; this refinement is most vivid related to holistic environments. I directed considerable effort towards timbre of the track, reference dynamic level and holistic environments in order to gain clarity on crystallized form and its manifestation within awareness; my attention to these areas cultivated my ability to perceive (and conceive) character at the highest dimension and to more clearly experience sound holistically in memory. Finally, my ability to define character in each of the many manifestations of timbre has become much more fluid. I spent considerable effort re-evaluating my personal biases; I recognized they had evolved in recent years to be more open in some areas, and also that I was aware that I have become prone to direct attention differently than I had in past years. Further, I noticed my sense of focus in holding attention has refined in recent years, and I tend to be much less distracted by the unexpected, the interesting or the delightful that arise in tracks. I continue to strive to improve, though I do not often engage in exercises for the purpose of improvement; rather, I have found that thoughtfully analyzing tracks with intention can develop one's listening skills. I continue to delight in discovering the hidden sounds and interrelations within tracks—that only reveal themselves through alternating deep and open listening sessions. Interestingly, additional sounds and sound relationships were revealed to me within tracks I had known for decades—they were always there, but my attention was not.

Of course I am a vastly different listener now than I was during my early-adult years as a performer and composer. I have also continued to improve even after achieving some important accomplishments in the recording industry. I was introduced to deep listening in my early twenties, when I was a composer of electroacoustic music, a conductor, and a beginning-recordist—though I sense I was drawn deeply into sounds early in my performance career as I was often focused on shaping timbre and expression. The practice of listening deeply was soon supplemented with open listening, which I encountered performing and recording the experimental music of the 1960s and 1970s. Those listening practices (as well as the acousmatic listening I soon began exploring in the late-1970s) shaped how I heard and what I heard, and how I thought about sound, musical form, musical materials and recording. I cultivated my listening skills in personal ways, too, as I found no tangible guidance to hear the dimensions of recordings. I studied ways we are all unique in our hearing, and learned to embrace my own place with sound; this brought me to understand how to use relative pitch and musical memory to substitute for absolute pitch, and led me to bring attention to frequency/pitch registers. This benefited my musicianship as well as my ability as a recordist (and now provides insight into recording analysis). Along with this self-learning and its challenges, I was keenly aware that the very dimensions of recorded songs were ill-defined, and our understanding of them quite nebulous at the time I was beginning making records—at least from the vantage of a music theorist/musicologist. I spent many hours poring over tracks of all genres (including many classical music records) trying to make sense of their unusual sounds, dimensions (elements) and their relations—matters that still intrigue me, as I continue to reframe some of these qualities, as additional dimensionality and nuance becomes evident to me. The praxis studies I offer emerged from my experiences of learning to listen to the elements of recording, and were refined over years of watching the trials of others (especially my students).

If you are reading this it is likely you, too, want to improve your listening skills. The results of acquiring the listening skills described here will bring you to better understand records; they will also reveal to you a new soundscape.

It is important to embrace that our innate skills are being developed within the ecology around us, developed by the experiences and knowledge of real life. But, this listening to tracks is different. We have to *learn* to hear the qualities of tracks. There is little natural about many of their traits. We need to be willing to experience something new, to be open to the unknown. My listening skills for records have developed because I have a passionate curiosity for what is happening to sounds in all the ways they appear and are transformed by performances, recording and records; this curiosity continues, and is rewarded each time I listen to a new track.

You are able to do this, too. My knowledge and experiences have prepared me well for listening to tracks only because I have learned to make use of them for that purpose. Doing this requires attention and intention—and concerted effort and repetition over time. You already know this.

I continue to exercise the muscle of attention, as well as practice listening deeply and practice listening openly (by choice of intention). Opportunities to practice are always present. I don't decide to practice; rather, I practice whenever I am listening. For instance, as I write this I am hearing several species of birds and can recognize the number of individuals and their locations around me; they are not distractions, they are simply present and I can choose how much and what type of attention to direct to them, or the degree to which I allow them to enter my awareness. Listening to tracks can be approached like that, too. I can bring intention to the process and seek to focus in certain ways; I can listen with an open agenda for discovery or to engage something specific; I can calculate what I am hearing and commit what I hear to memory; I can attend to performances or to the interrelations of sources; the choices continue, and of course I can choose to not choose. The myriad of possibilities will quickly overwhelm one who does not approach listening intentionally. You already know this, too.

The basic skills of listening are paying attention and remembering. Fortunately, both can be developed. These skills may be guided by knowing what to listen for and how to process the qualities of sound in those areas with intention. This is where knowledge and experiences meet skill, and where ability and accuracy get cultivated.

Your musical skills and knowledge, technological skills and knowledge, and life experiences and knowledge have converged to provide you a unique vantage. It is only from this vantage that you can engage the world, and cultivate your listening skills and abilities to understand tracks. One needs to work diligently to perceive and to understand without imposing expectations or desires—this is central to hearing what is present. This is not easy, but it is important in order to control the influences of your vantage so you can experience the track in the form in which it was created. Listening is always work, always takes effort; even open listening is alert while striving to be passive to mitigate interpretation.

I have been teaching critical listening for over three decades, and have been researching and testing pedagogical methods all this time. The experience I have gained has produced the refined set of praxis studies that are included in Appendix A.

As a set, these studies will allow the would-be analyst to engage nearly all aspects of recording elements, and many of these skills can be applied to music and to lyrics. The reader is encouraged to dedicate time and effort to developing listening skill—whether or not they choose to engage these praxis studies. No doubt some of these studies will appear difficult. Some readers may question the relevance of some. There is nothing superfluous here, though. There will almost certainly be topics nearly everyone has not yet sufficiently engaged or understood.

These praxis studies will allow you to develop skills in hearing time, frequency, loudness, distance, direction, image dimensionalities, environments and other topics rarely cultivated by listening studies. As we have learned over these ten chapters, these topics are central to how the record shapes the song, and to the processes of recording analysis.

As listening provides us access to the primary text of the track, we are then faced with the prospect of formulating, describing or defining our discoveries. In my personal experiences, I have found that no matter how clear, detailed or lucid the aural images, tracings of sound events or conceptions of aggregate textures may be as they coalesce nonverbally, some disconnect arrives. I remain acutely aware of the thresholds between the stages of perception, holistic understanding, intellectual processing and language—and the distortion at each transference. The process of explaining and describing what I hear—what I have discovered and perceive, and what I recognize and understand—remains daunting and often insufficient. Still, as Allan Moore (Moore 2012b, 101), has articulated well, while much is "always beyond precise verbalization," it is exactly this that an analysis must seek to accomplish as "to avoid the attempt is to render our evaluations worthless." This is our challenge should we wish to talk about what we hear in tracks. And, we all seem to have an innate desire to share about what we find fascinating or special.

I am taken with how very difficult it is to write about this entire endeavour of recording analysis, and still how very rewarding it is to study tracks deeply. I continue to be fascinated by what I discover and frustrated by attempts to explain it—an experience I first encountered in my youth listening to and trying to make sense of the newly released "Strawberry Fields Forever." This duality pushes me onward, and perhaps forward toward some bit of resolution.

We know recorded songs (records) reach beyond the human condition, and communicate very differently than language. To understand them we need to recognize and learn the nonverbal language that each track speaks uniquely. Tracks (and all they contain) are not only 'always beyond precise verbalization,' they are also a different type of experience, and involve a different way of hearing and of knowing. There is a vast richness within this unique experience that is the record. It is how this richness can be revealed through unique ways of listening, that I have attempted to lead the reader to discover during the journey that is this book.

It is my wish for you that your listening experiences have deepened, and that aspects of the track that were once hidden from you are now being revealed and enriching your understanding and enjoyment. My central goal of this writing is to open you to the possibilities that may be present in tracks, and to deliver you to a place where you can recognize and impartially convey what you have encountered.

May you acquire a deep knowing of the tracks that are meaningful to you, and vividly experience how the recording has shaped them.

Some Questions for Recording Analysis

What establishes a productive and accurate characterization of the sound quality or qualities under analysis?

What within this track elicits the qualities that draw my attention?

What appears significant and worthy of observation and then evaluation?

What, as an overall quality or artistic statement, makes this track unique?

What appears to be unique in the track, and worthy of study?

Which elements appear to be shaping the track most significantly in verses? In choruses? Within other sections?

How does each element (at each perspective) contribute to the timbre of the track?

How does each element manifest within and shape the basic-level sources and materials of the track (reflected in each sound source and their interrelations)?

Working through all potential pairs of elements and elements in groups of three, which interrelations appear worthy of examination at a deeper, more detailed level?

How is the holistic environment of *this* track characterized?

What is the character and the content of the timbre of the track?

How is this track's spatial identity formed through the content, character and interrelations of *this* track's holistic environment and those of all of the individual sources+spaces?

How do the unique qualities reflected within each recording element, within each level of perspective:

- Contribute to the timbre of the track?
- Contribute to the motion, the structure, and the shape of the track?
- Contribute to the surface qualities and character of the track?
- Shape the qualities of significant, individual sources?
- Contribute to the expression and aesthetics of the track?
- Other?

What is the emotional impact of the track, and how is it created?

How does the recording shape this track?

NOTES

1 Here I provide a sampling of sources to encourage the reader's exploration; all have been referenced in previous chapters, and appear in the comprehensive bibliography at book's end. These are but a few of the many sources available, and the list is certain to misalign with the preferences of many scholars; I admit this selection is very incomplete and is not meant to cover these topics thoroughly. This reading list is to start exploration, not to provide a complete background; embarking on this list, the reader may well branch off to other sources, in directions of their own interests and choosing. For a broad cross-section of concepts related to traditional music analysis concepts, see Bent (1987), Cook (1987), Cogan & Escot (1984), LaRue (2011), Lerdahl & Jackendoff (1983), and Zbikowski (2002). Chapters 3 through 5 contained many references to popular music analysis (with some including elements of recording); of all these I select a few (and could have selected them all) see Bayley (2010), Burns (2005), Covach & Boone (1997), Everett (1999, 2008, 2009a, and 2009b), Middleton (1990 and 2000a), Moore (2007a, 2012a, and 2012b), Moore & Martin (2019), Spicer & Covach (2010), Tagg (2013 and 2016), Walser (1993), Zagorski-Thomas (2014).

2 Of the other models is Robert Plutchik's "wheel of emotions" (Plutchik & Kellerman 1980). It is represented as a spinning top of various colors, concentric circles, and multiple layers; this may be the most elaborate, informative and perhaps comprehensive of emotional categorization models.

3 M. Nyssim Lefford offered this insightful perspective on vocabulary and lexicon, and their inherent normalizing effect on analyses. I am grateful for her input here, and in numerous other portions of this manuscript.

4 Common in analyses of art music, sectional divisions can be defined as having expository, transitional, developmental or terminative functions (Spencer and Temko 1994). These may (at times) be pertinent considerations in recorded songs, especially in instrumental sections or sections without lyrics. In popular music recordings these functions will not hold precisely as in art music, though; in recorded song, lyrics typically play a significant role in unfolding tensions and content, and the function of sections.

5 Additional, detailed explanations can be found in Ryan & Kehew (2006, 439) and Everett (1999, 75–83).

6 On analog multitrack recorders that were used for the Beatles' sessions, magnetic tape is divided into segments containing a single channel of audio; a 4-track recorder divided up the tape into four paths or channels (these were also called 'tracks'). The Beatles maximized the use of each channel by placing multiple performances on each, arrived at through submixes, or reduction mixes; near the end of the "White Album" sessions they began using an 8-track recorder, though reduction mixes continued in order to free-up tracks so additional performances/instruments could be added. With the resources of digital, each track usually contains a performance by one instrument or vocalist. As a verb, 'track' can mean 'to record,' as in 'to track a vocal.'

APPENDIX A

Praxis Studies

These praxis studies have been devised to guide the listener to experience, study, engage and develop listening skills and abilities to accurately hear recording elements and the various dimensions of records. The studies are designed to allow each topic to be thoroughly explored in a singular focus of attention; bringing a deeper understanding and higher level of listening acuity than would otherwise be possible. Praxis studies are intended to refine the reader/listener's awareness of their own unique qualities of hearing, and their personal relationships to sound and to the act of listening. Engaging these studies, the listener will learn a great deal about themselves as listeners and will learn to hear attributes that are at the core of tracks, of recorded sound and of the recording elements

In aggregate, this collection of praxis studies/exercises brings key portions of the book's framework into practice. Further, they will serve to develop or maintain the listener's ability to engage all of the dimensions and domains of records.

Though these studies may be undertaken in any order, if studied sequentially as they appear, they not only coincide with the text, but also build on one another. They are designed to lead the reader through the material efficiently and in an effective sequence; when new materials are presented the requisite background will be in place.

Praxis studies/exercises are intended to pull a first-time reader, or a listener first encountering a skill or topic, into the listening process at the time topics are mentioned. The studies allow the reader to be diverted to developing their skill and experiencing a concept without interrupting the flow of the text. Readers and listeners with some prior exposure to the topics will also find value in the studies, as they will broaden their experiences and refine their skills. The goal of these guided listening practice studies and listening exercises is also to inform and open the hearing of the reader/listener—even those with considerable experience might find they acquire some refinement of their abilities and approaches to listening they had not previously considered that might enhance their skills or understanding.

These studies intentionally emphasize recording elements. Recording elements are rarely addressed within the context of developing the conceptual and listening skills needed to recognize their contributions to the artistry of records. There are some praxis studies/exercises that are directed to skills applicable to all domains—such as listening at various levels of perspective; these were devised to support the framework, and are primarily related to structure and form concepts that are found in few other sources. Finally, this collection includes some studies directed towards bringing greater awareness to one's personal listening predispositions, past experiences and latent, innate abilities.

Some of these skills will take significant time to acquire and adequately refine—the amount of time (of course) depends on the individual's abilities, dedication and background. Readers can expect to learn some skills faster than others. The reader may move forward in exploring new material before a challenging task is fully understood or listening skills adequately developed. One should return to pursue challenging studies periodically, however, to continue developing those skills. One will want to learn all skills well enough to allow a thorough examination and understanding of tracks.

It is important to recognize, for those who have not studied certain topics, being able to perceive and arriving at an understanding of certain unique qualities of records—qualities that are rarely experienced otherwise—requires a willingness to open one's listening to the unknown. Those with some background in these topics may find the listening practice studies engaging enough to draw them into discovering qualities not previously experienced in certain records; this is common, and often surprising when it occurs. Many listeners embrace discovering new qualities in records with enthusiasm, some with wonder—and this development of skills and openness to discovery of nuance is an on-going process that can continue to intrigue even the most experienced of listeners. Indeed, even this author continues to find new richness in records I have known for years, and to be fascinated by the sounds of tracks newly encountered. A commitment to listening practice will benefit those wishing to learn to hear the qualities of the record (and recording elements), to discover the full sonic panorama of qualities within records they find meaningful, and to engage important and central topics of recording analysis. It would be the unusual reader that does not find some benefit (and perhaps satisfaction) from working through at least a few of the praxis guided listening practice studies and listening development exercises.

This all begins with learning what to hear within the track, and how to listen in order to recognize its unique attributes and features. Allow the following studies to guide you.

PRAXIS STUDY 5.1 DEVELOPING AUDITORY MEMORY

Goals and Desired Outcomes

The goal of this exercise is to develop the ability to remember what is heard—whether musical materials or aesthetic ideas within lyrics or recording, structural relationships over time or between levels of dimension, or the nuances of recording elements.

This ability can bring a sense of control and confidence to the process of data collection, and may save many re-hearings during analysis processes. The study can be adapted to bring attention to all aspects of tracks, including specific elements in all three domains.

Preparation and Background

Select a record you know well. Next, determine what information you wish to recall about the track; this could be its lyrics, a specific music element or any particular recording element. These and more can be explored within the following process. Exploring the track's structure will be used to illustrate developing auditory memory.

Process and Sequence

1. First, prepare a timeline with measures numbered, up to perhaps 100.
2. Before listening to the recording, sit quietly and try to consciously hear the track from your memory. Bring up as much structural detail as you can of its opening sections.
3. Now, write down the song's meter and tempo. In your mind, listen to the piece unfold from beginning to end from your memory—note the length of the introduction, the ordering and lengths of the succeeding sections, and so forth.

4. On the timeline, write down where the major sections begin and end. If you cannot come up with those divisions easily, you might well be able to deduce that information by thinking about the patterns of phrases in the introduction, verses, choruses, etc. Write down as much information as you can.

5. Think carefully about your observations, and identify aspects you have yet to clearly determine — things that need to be identified when you listen to the recording.

6. At this point you can listen to the recording. Listen intently for the information you identified in step 5. Do not follow your graph. You may wish to listen with your eyes closed. Listen to remember what you hear.

7. Do not write while you are listening and do not correct your graph while you are listening. Listen to aspects of structure only. Do not allow other aspects of the track to divert your attention.

8. Once the track has stopped, write down what you heard from this one hearing, and simultaneously correct what you previously wrote. Then repeat these steps until you have completed the timeline and structure in the large and middle dimensions. Seek to accomplish this in as few hearings as possible.

9. Finally, check your information one last time. While listening to the track follow your graph. Check its accuracy; make corrections.

PRAXIS STUDY 5.2 DEVELOPING FOCUS OF ATTENTION AND INTENTIONAL SHIFTS OF FOCUS AND ATTENTION

Goals and Desired Outcomes

The goal of this study is to develop focus of attention toward a specific aspect of the track, and then to develop the ability to intentionally shift attention in some specific and predetermined manner—such as to another element and/or another level of perspective. It is important to remember not to multitask; we are able to hold only one thing in the center of our attention at one time. We can learn to choose with intention what to hold in our attention, and develop the skill to direct and maintain the focus of attention.

This study promotes a focus of attention solely dedicated on a specific element or other aspect of the track, and learning the ability to intentionally shift to another. This is an active and controlled shifting between elements or materials. These skills may be extended to become an active and controlled shifting between any level of perspective and between any element within any domain.

Preparation and Background

Select a track with five to eight instruments throughout its introduction and only the addition of a lead vocal in first main structural section (such as a verse), and then added instruments in the second section (perhaps this is the chorus). Note the lengths of these sections—using number of measures, phrase groupings of materials, prevailing time unit, etc. This can be a track you know well, or one you wish to learn.

Process and Sequence

1. Select any instrument playing in the introduction. Focus attention on that one sound source solely as you listen from the beginning to the end of the second section; in doing this listen for any single, particular aspect of that instrument—such as its rhythm, dynamic shape, performance intensity.

2. Next listen to the instrument again, allowing your curious, open listening to explore the sound for prominent characteristics—such as those above or melodic materials, a recording element (distance, host environment, etc.), timbral qualities, etc.

3. Next bring your attention to the loudness contour of the instrument—how it changes in loudness over its performance. Stay focused on loudness only, and on that instrument only. Do not allow your attention to roam when another instrument enters; do not allow your attention to switch to another aspect of the instrument's sound.

4. Bring your attention to the same sound source, this time focused on timbre. Notice how the timbre of the source changes or does not change with loudness by alternating focus quickly between timbre and loudness within that instrument.

5. Listen again to that instrument's loudness, but this time shift your attention to the lead vocal when it enters. Notice the loudness and timbre of the lead vocal; alternate listening to the original instrument and listening to the lead vocal.

6. Shift your listening to the level of perspective (the composite texture) where both sounds can be held equally in attention. Note their loudness relationships during one hearing; note performance intensity timbral relationships during the next hearing.

7. Shift listening perspective to the overall sound, and notice how the lead vocal relates in range to the other sounds. Are there sounds higher in frequency, lower, in the same range?

8. Note the span of range between the lowest sound and the spectrum of the highest sound at the beginning of the track; remain focused on this level of perspective and on this span of the range of the track as it unfolds from the beginning through that second section.

9. Now allow yourself to focus on any aspect of the track that is of interest, but maintain a keen focus on that aspect; next, very intentionally shift that focus to some other aspect. The central skill here is being intentional in the shift of attention.

Continue this practice of changing focus and perspective in a controlled and deliberate way by listening for other specific elements and materials.

PRAXIS STUDY 5.3 BRINGING THE FOCUS OF ATTENTION TO THE SMALLEST DIMENSIONS: MICRORHYTHMS, MICROTONES AND SUBTLE SHIFTS OF DYNAMICS AND TIMBRE

Goals and Desired Outcomes

The goal of this study is to bring attention to the lowest dimension details of rhythm (microrhythms), of pitch (microtones), of slight changes of dynamics and possibly to subtle shifts of timbre.

This skill allows the nuance of performances and productions to be observed, recognized, and (if desired) notated. It is within this nuance that the distinguishing features of tracks often emerge. This skill can ultimately be applied to any element within each of the three domains.

Preparation and Background

Select a track you know well. Select a specific instrumental performance within it—a specific line or phrase—with a duration of approximately 8 to 16 measures. The instrument in this passage should have the ability to vary its pitch and loudness. It is best if there is little else going on simultaneously to distract listening attention or to cover (mask) the performance.

Process and Sequence

1. Transcribe the phrase into musical notation. Deliberately generalize the rhythms and the pitches to conform to notation; make note of the tempo, meter, pitch organization, and general dynamic marking. Do this from memory as much as possible, and check your final notation.

2. Now bring into attention the pulse and the meter as a steady, underlying grid, and listen carefully to the passage. Note how the performance aligns with the grid, and how it does not. Begin the process of determining the microrhythmic deviations that are present—the amount of time before or after the notated rhythm the sound actually happened.

3. Using a spectrograph or a waveform representation to find the location of attacks against the metric grid can be instructional in early studies. Also effective is auralizing (internally hearing) subdivisions of the beats—whether sixteenth or thirty-second note divisions, or some other appropriate division—to determine the actual rhythm.

4. Explore the possibilities of patterns of microrhythmic deviations, or of notes receiving unique treatments; the focus is solely on rhythmic placement and hearing the lowest dimension time/rhythm placements of notes against the pulse of the metric grid.

5. Next bring attention to the pitches of the passage. Subtle variations of pitch are present in most (though not all) instruments, and these abound in voices. Notice deviations of pitch during the first moments of the note (for example, the note may start a bit flat or sharp), during its sustain (as vibrato), or as it ends (sustained notes can go flat toward the end).

6. Incorporate these microtones into the notation, first in a general way (noting sharp or flat) and then by adding further detail (calculating relationships to quartertone or other division).

7. The amount of detail here is important. The goal is to bring focused attention to pitch deviation, to recognize the amount of deviation at the smallest level of dimension.

8. This process can now be shifted to the slightest changes in dynamics and these steps repeated. Bring attention to the smallest changes in loudness—as the passage unfolds over its entire span, and also as individual notes begin, sustain, and terminate. Notice the levels and contours, the shaping of phrases and of individual notes—and how this eludes notation and challenges description.

9. This process may be repeated with attention focused on the micro-changes to timbral quality, should the reader feel prepared at this time.

The central concern here is cultivating attention to the slightest changes of rhythm, of pitch, and of loudness, and perhaps of timbre. Establishing a detailed and accurate musical notation of the passage is not the goal; hearing and recognizing nuance of the smallest dimension within pitch, rhythm and loudness is the goal.

PRAXIS STUDY 5.4 ESTABLISHING TIMELINES WITH STRUCTURAL DIVISIONS

Goals and Desired Outcomes

The purpose of this exercise is to create a timeline of a track, divided into major structural divisions and phrases within those divisions. The resulting timeline will be similar to Figure 2.1. The space above the timeline may become the vertical axis of many graphs, as this process provides the horizontal axis for organizational timelines and for X-Y graphs of recording elements.

This exercise will also improve memory recall as well as build skill in directly recognizing and remembering structure at various dimensions. This skill can establish a 'backdrop' or context for one's focused listening, where one can be aware of where activities occur structurally though their attention is focused elsewhere.

Preparation and Background

Identify a track for this study; one you do not know well, but in a familiar genre and from an artist you know well. To be of most benefit, the following exercise should be repeated on a variety of tracks of various musical genres. This exercise can be repeated at the beginning of any new recording analysis.

Process and Sequence

1. Prepare a timeline with measures numbered, up to perhaps 100—knowing it may need to be extended, or that all measures may not be used.
2. Listen to the recording to identify where the major sections fall against the timeline. Try following the timeline while tapping the pulse of the song. When a major section begins/ends, mark the location on the timeline.
3. After its conclusion, write down the names of those divisions. Then, try filling in additional information, such as other verse or chorus beginning/ending points, and phrase lengths. Work from what you remember, and take note of those things you cannot determine.
4. Listen to the recording again, and follow this by writing down any additional information you recognized. Do not write while you are listening; remember what you hear and write later. Incorporate the prevailing time unit into the structural divisions of the timeline.
5. The timeline and graph is completed when it includes all of the major structural divisions, the mid-level structural divisions, prevailing time unit, and the smallest uniform phrase length. Incorporating text information is often helpful.

Remember to wait until the record has stopped before writing observations. Clearly separating the listening and writing activities will assist you in improving listening skills, retaining aspects of the track in memory, and in learning to evaluate sound. These will become increasingly important as this book progresses.

PRAXIS STUDY 5.5 DETERMINING PITCH DENSITY OF SOUND SOURCES, AND TRANSCRIBING THE ARRANGEMENT'S PITCH DENSITY INFORMATION INTO MUSIC NOTATION

Goals and Desired Outcomes

The goal of this study is to recognize the pitch ranges of the musical materials presented by sound sources. These ranges result from the fusion of the source's timbre and the breadth of the pitch range of materials within the syncrisis unit—or prevailing time unit, or some other structural unit that the listener/analyst might recognize as a consistent, cohesive unit appropriate for the track.

This study allows one to recognize how sources occupy a sense of their own pitch space (range), and how that range compares to those of other sources. One will also acquire a sense of how all sources relate to the overall range of the track's arrangement.

Preparation and Background

Identify a track with very sparse instrumentation over the first several sections of the record; it would be best to limit sources to a maximum of 5. The voice/instruments' musical materials should show some noticeable and considerable change in pitch levels. This exercise will use music notation to illustrate the pitch density of these sources over these first sections.

Pitch density will be notated to show the range of prominent aspects of an instrument's spectrum as a bandwidth above the musical material it is performing, as in Figure 5.5.

Process and Sequence

1. During the initial hearing(s), establish the length of the timeline. Make a list of sources, and note their entrances and exits of the instrument against the timeline.
2. Next work to identify the musical ideas that each instrument is presenting, and note their presence against the timeline.
3. Transcribe each voice or instrument's pitch material as a general melodic contour onto the grand staff. Subtle details of melodic contour are not required here, only the gesture of the shape of the line and a few prominent pitches placed accurately against the metric grid. These contours represent the sound source's fundamental frequency.
4. Now notice how the single melodic line falls into phrases that generate distinct recurring time units, and perhaps individual musical ideas. Refine this observation to identify the prevailing time unit or the syncrisis unit. Mark the beginning and ending points of those sections; these are the sections for units of pitch density, where the materials are heard as a singular gesture or event within a sense of the present.
5. Shift your focus to the timbre of one source's first sound. Determine the bandwidth of each pitch area by identifying where the source's spectrum becomes about one-third the loudness of the lowest frequency; this is the upper boundary of the pitch area.
6. Determine the upper boundary of the pitch area of this source throughout these first several sections (prevailing time unit or syncrisis unit); this interval above the fundamental will vary with performance intensity. Incorporate this into the musical notation representation of this these sections.
7. Once these time and register boundaries are finalized, make observations on the general density of pitch and spectral activity (amount of pitch and harmonic/overtone information) within the defined pitch areas. Make note of this density on a scale from sparse to very dense—as explained in detail within Chapter 7.
8. Repeat the processes of steps 5, 6 and 7 for each remaining source.
9. A music notation of the pitch areas of all sources has now been assembled. This allows one to observe the pitch-related content of the arrangement, as it unfolds throughout these initial sections. Should one wish, this process could be repeated throughout the duration of the track—to include all sound sources, or select sources.

PRAXIS STUDIES 5.6.A AND 5.6.B APPROACHING OBSERVATIONS FROM A NEUTRAL, UNBIASED POSITION FOR: A. DATA COLLECTION OF ELEMENTS AND MATERIALS B. INTERPRETING AND EVALUATING MATERIALS

Goals and Desired Outcomes

The goal of this two-part exercise (sections 'A' and 'B') is to cultivate a neutral listening position—one that minimizes personal biases in the collection and evaluation of data or information about the track. In 'A,' we will seek to collect information on the track, to identify data without distorting it; in-as-much as this is possible, we will attempt data collection with a focus on the objective, physical content of the sounds within and of the track. In 'B,' we will seek a neutral position for the interpretation and evaluation of the information (collected data) within the track; we will attempt to temper our predispositions to seek certain information and to cognize and evaluate information in certain ways.

While the very notion of a neutral or purely unbiased observation is utopian, and we approach any material within any track from our unique position of pre-understanding, we can understand how this impacts our listening and what our predispositions might be. With this understanding and with 'listening with intention' to minimize these influences, some semblance of neutrality might be obtained. With concerted focus and effort, one can improve toward assuming and holding a neutral position—though it is important to recognize one's biases will always be present to some extent.

Preparation and Background

Select six tracks from different artists and music genres. These should represent a mixture of two tracks you know well, two tracks you have heard a few times, and two which are new to you (in one you may know the genre or artist, the other should be completely unknown and from another culture).

In 'A,' we will cultivate accurate listening; this requires not making assessments or reaching conclusions until having listened completely, and considered fully all that has happened; accurate listening occurs from a position of neutrality, engaging equivalence.

In 'B,' self-reflection and consideration of personal biases, cultural conditioning, musical conventions, and personal inclinations are explored to open to the possibility of a more neutral, non-personalized, objective, and unbiased observation of elements any domain; the premise is one must learn about one's tendencies in order to mitigate their impact, so one can hear 'what is present' with the least distortion, as opposed to allowing our predispositions and preconceptions to transform the track into something mostly personal.

Process and Sequence

A. Approaching observations from a neutral, unbiased position for data collection of elements and materials

1. Examine Table 5.2 for its summary of approaching music elements as (mostly) objective content. Select one of the tracks you have heard only a few times, and define the timeline for the first two contrasting sections.
2. Work through the Table 5.2 general approaches for unbiased observations for each category of music elements listed. Place collected data against the timeline if appropriate, or transcribe materials into musical notation; some materials such as the shapes of melodic contours might be drawn.

Explore each of these elements to engage whatever comes along from a neutral, unbiased position and to discover unknown qualities.

3. In collecting this information avoid evaluating the materials, or otherwise making assessments or drawing conclusions. Collecting information focuses on raw materials of "what is present" or the raw sounds void of much organization (consider, even meter and tonality are interpretations); avoid interpretations and evaluations of these materials related to "how it works," "what it means," "how it creates motion," "how it moves the listener," for example.

   Review the information you have collected to identify any implied functions or other interpretations/assessments; consider how these might have been avoided.

4. Throughout this process seek to cultivate a willingness to accept and to engage as relevant all that comes along—without expectations or preconceived ideas of what should follow what has already been heard.

5. Note your level of success in collecting material without interpreting it, without assessing its significance or relationship with other material, or privileging one idea over another. Note your predisposition toward assessing information (this can manifest differently in each element) as you collect it; your awareness of this tendency will help you minimize it.

B. Approaching observations from a neutral, unbiased position for interpreting and evaluating materials

1. We need to be self-reflective in order to avoid being biased—it is easy to be biased without being aware of it. Consider your predispositions: your different experiences with music, your listening skill strengths and weaknesses, your knowledge of musical content, musical experiences; how your personality, preferences to certain musical styles, experiences, knowledge, culture, social group, and so on, all unconsciously influence you and what we perceive.

   Write an outline of who you are as a listener with your skill levels and music preferences, as a musician and an analyst, as a person with experiences and unique place in society (see above). Let the list be honest and factual—you need not share the list with others, but these topics are important for your self-knowledge.

2. Listen to one of the tracks that is new to you. Observe how these personal tendencies play out in your comfort level with the track, and how they can steer your listening to it.

3. Now consider what grabbed your attention in that track.

   Were you drawn to lyrics, aesthetics, the beat, groove, affects, certain instruments, overall sound, visceral energy, message, persona of the vocal, or another attribute?

4. Repeat these two steps as you listen to several diverse tracks.

   In considering these tracks, are you predisposed to affective language or an analytical discourse? Does your internal narrative describe what is happening and how the music unfolds? The structural content of the materials or their emotional charge? Do you hear across all domains and elements, or fixate on one or a few?

   Write another outline of your predispositions toward the listening process from 2 and 3.

   Finally, consider carefully how these change between tracks you know well and tracks that are new to you—there will be differences, though they may seem subtle.

5. Reconsider that all within the track has value, and that your willingness to accept and to engage all that comes along as relevant—without expectations or personal desires, preconceived ideas—is an important vantage for exploring the track from a (mostly) neutral position.

6. Listen now to one of the tracks you know less well. Seek to assume a neutral position from this knowledge of your predispositions. Cultivate a sense of exploring the track from a position that neutralizes your preferences and biases, a position that does *not* seek whether something is appealing or interesting, but rather what is central to the unique voice and sound of the track.

7. To end, listen intently to a track you know well from this new awareness of your predispositions and openness to all that is within the track. Be willing to scan the record's entire texture for new information without assessing it and without determining content and function. Has your experience of the track and understanding of its content shifted? If so, how?

PRAXIS STUDY 6.1 ACQUIRING A CLOCK TIME REFERENCE FROM TEMPO MEMORY

Goals and Desired Outcomes

The goal of this study is to bring the listener to readily and accurately conceptualize clock time; to accurately perceive clock time increments from their memory of tempo.

One can learn to recognize their personal, spontaneous tempo—a tempo one carries as a preferred rate of periodic movements, such as foot tapping. One may also access their absolute memory of tempo for familiar songs. You may initially find one of these easier to access than the other for recalling tempos; this may shift as you work with both types of memory, as more deeply engaged engrained percepts may not arise to consciousness as readily, but may be more reliable once accessed.

In this process, tempo can enable one to accurately recall (or arrive at) a recurring pulse at the duration of a second, a half-second, and tenths of seconds—or other useful time units. These can facilitate establishing and using a timeline based on clock time (as for timbral content in Chapter 7), and are useful for tracking the passage of time and of hearing millisecond increments related to environments (as in Chapter 8). Ultimately, one will obtain easy recall of 'known' relevant tempos, and acquire facility in transposing those specific tempos into seconds and fractions of seconds.

Preparation and Background

You will find both tempos of records and spontaneous tempo have the potential to function as useful references for accurately calculating clock time. Through the process of transferring tempo into clock time—seconds or fractions of seconds—the recurring pulse of tempo can be reinterpreted as time with an appropriate and functional formula. A simple formula relating beats per minute to the second or millisecond needs to emerge from the tempo for the reference to be useful.

This study is broken into three parts: (A) recognizing and learning to recall one's spontaneous tempo, (B) identifying and learning to recall the tempo of relevant songs, and (C) identifying a simple formula and learning the process to transform a song tempo into relevant clock time units (second, half-second, and tenths of seconds, to start).

Identify two tracks at each tempo of 60 bpm and of 120 bpm. If a track is not clearly at either tempo, select another track. Next identify two tracks that are at a tempo of 150 bpm; tracks at 75 bpm may be useable, but are not ideal as the process will have an extra step. Having a workable memory of more than two tracks at each tempo can enhance confidence and reinforce accurate recall.

Process and Sequence

A. Spontaneous Tempo

1. Recognize your spontaneous tempo by bringing attention to your body. Begin tapping your foot, walking, bobbing your head or swinging your arms at a speed of repetition that feels natural, 'right' or comfortable. Allow this pulse to settle into a repetitive pulse or tempo at a speed that feels preferable.

We all have a preferred rate of repetition, pulse or beat—which in repetition is tempo. We are surprisingly consistent when we seek to tap at a comfortable rate—which is our spontaneous tempo. "The dominant preferred rate across individuals for beat perception and tapping appears to be around 500ms" (Tan, *et al*., 2018, 97). 500ms is .5 seconds, or 120 beats per minute (bpm).

2. Practice relaxing into your spontaneous tempo by noticing the rate at which you naturally walk, or tap your foot without controlling the speed. Next, tap your hand without thinking of tempo, but rather relax into what feels natural. Do this periodically (every few hours for short 5 minute stretches) over several days; falling into your tempo should emerge without effort or thought. You will be able to compare this rate to tempo, and calculate it in beats per minute.

3. The final step is to be able to call up this rate of repetition conceptually, without first manifesting it physically.

Relax into remembering the feeling of the speed of the pulse, then conceptualize that speed as tempo in your consciousness. It will take practice and repetition for this process of transferring the visceral to the conceptual to be accurate; our tendency will be to try to control the speeds of repetition, which will distort the process and the tempo.

B. Relevant records

1. Identify at least two records at the tempos: 60 bpm, 120 bpm, 150 bpm, 75 bpm. These tracks should be ones you know deeply and that are meaningful to you.

    It is imperative these be specific tracks or records that you recognize as definitive performances—the song's 'correct' version. In this way the tempos are always the same, and they are deeply embedded in your memory.

2. Recall the tempo of one track within your memory; 'hear' it internally. It may help to 'feel' its pulse in your body, but do not let it stray into or to be distorted by your spontaneous tempo.

    You must be able to clearly call up these records in your conscious thoughts, and be able to be consistently accurate in recalling their tempos.

3. When this tempo is clear in your awareness, identify the tempo in terms of beats per minute or a metronome marking. You may wish to use a metronome or some other device or software to verify this exact tempo; you may need to practice recall of the tempo.

4. Repeat steps 2 and 3 for each other track, until you have more than one reference for each tempo.

C. Transposing tempo into clock time

1. To transpose tempo into clock time, a pulse at beats per minute is re-interpreted as pulse per second.

    Tracks at 60 beats per minute readily translate into one pulse per second, bringing a reference of passing seconds. The tempo of 120 bpm translates directly into two pulses per second, which is also a pulse every half second (0.5 seconds). This is a simple transformation of information, and readily perceived.

2. Arriving at tenths of seconds is not this direct. The selected tracks at 150 bpm are important for this.

    150 beats per minute generates a pulse every 0.4 seconds; sixteenth note divisions will divide these 0.4 second iterations into 0.1 second units. Tenths of seconds can be perceived from a tempo of repeated sixteenth notes at 150 bpm. The process is (1) to become aware of the 150 bpm pulse, (2) to establish an awareness of dividing that pulse into four divisions of repeating sixteenth notes, and (3) to reconceive the sixteenth note divisions as tenths of seconds.

    75 beats per minute represents a pulse every 0.8 seconds. 32nd note divisions of 75 bpm will represent 0.1 second.

3. Practice calling up each tempo and reframing its pulse as a clock increment.

Most significantly, practice arriving at a stable sense of tenths of seconds. Then reinforce the process of arriving at the tenths of seconds divisions by gaining facility and confidence in quickly finding and dividing 150 bpm and 75 bpm.

You will now be able to carry with you a way to access clock time in seconds, half-seconds and tenths of seconds. This is critical for recognizing the content of timbre and of environments.

PRAXIS STUDY 6.2 CULTIVATING SENSITIVITY TO LOUDNESS LEVELS

Goals and Desired Outcomes

The goal of this study is to acquire some sensitivity to sound pressure level. Humans experience a physical sensation from the amplitude of a sound wave that is transferred into loudness. Loudness is subjective and its perception can be distorted by context, but the physical presence of sound pressure level (SPL) is a measurable dimension of sound. The SPL remains constant, though the perception of loudness might shift.

Excessive SPL can cause pain in humans; prolonged exposure to excessive levels can damage one's hearing, perhaps permanently. Becoming sensitive to the physical sensation of listening at an appropriate loudness level is equally possible, with practice, attention and diligence.

Preparation and Background

Obtain an inexpensive sound-level meter. If possible, keep it in near of your listening position so you can reach for it as needed to verify playback level.

Set your monitoring level (amplifier output) so that the average SPL registered is between 80 and 85 dB SPL. It is acceptable for occasional peaks to reach 90 dB or slightly above for short periods, and for soft passages to dip below 80 dB by several dB. Monitor levels for tracks with a dynamic range broader than 10 dB should be set so as much of the track as possible sounds in the 80–85 dB window. Keep a record of our playback system settings for tracks you are evaluating; recall those settings each time you listen to that track to replicate playback levels.

Process and Sequence

1. Begin your practice by listening to a record you know well. Set a playback loudness level in the 80–85 dB window, checking the meter frequently to verify listening levels remain within or slightly above or below this window.
2. Over several days, use the meter during all your listening sessions. Maintain equipment settings and a consistent loudness level. Become aware of the physical sensation within your hearing that is created by listening at that level. You should begin to notice energy impacting the hearing mechanism (inner ear) with your focused attention, over time.
3. Be consistent in listening at this level. Check your meter regularly. Over a few weeks (or a bit less or longer) you will begin to develop an increased sensitivity to the physical sensation of this loudness level region. Ultimately you will develop some memory of that sensation.
4. After a week, begin to check your memory. Listen to a recording you know well on your monitor system by starting with the monitor level turned down, and without the SPL meter. Bring up the

level gradually until you believe you have reached this average level of between 80 and 85 dB SPL. Listen at that level for another pass through the track to validate your selection of a monitor level.

Now, check yourself with the SPL meter. Make note of your accuracy, and consider reasons for any inaccuracy. Should you find your level accurate, know this may well be by chance. Repeat this exercise regularly (several times per day if possible, with a minimum of once per day) until you begin to have consistent success in recognizing this level, and confidence in your ability to recognize SPL without the distortion of perceived loudness.

5. You will notice you are not as accurate at first, but this accuracy will increase as you become more aware of focusing on energy impacting your hearing mechanism. In time this skill will carry over between various tracks, between playback systems, and from listening environment to listening environment. As you continue to try to establish a suitable listening level in a new location, check your accuracy with your meter.

This skill will improve your listening skill considerably and serve as an important point of reference, even if a general one. In time you will gain an awareness of when you are listening at a level near 80 and 85 dB SPL. It may be that you come to prefer an average level toward either the high or low side of this range. Arriving at a reliable sense of average loudness level will greatly assist one in maintaining stable and accurate monitoring. Most importantly, you will become acutely aware when SPL extends well above 90 dB SPL. That higher level brings increased pressure on the hearing mechanism that will be very apparent and quickly become uncomfortable. This will trigger your awareness that hearing fatigue will happen quickly and a sense of urgency that if this level extends much higher above 90 dB, you are in danger of damaging your hearing.

PRAXIS STUDY 7.1 REALIZING A PERSONAL PITCH REFERENCE

Goals and Desired Outcomes

The goal of this study is to develop a reliable sense of pitch recognition based on the recall of pitches within melodies that hold significance to the individual. When we focus this skill to consciously recall a melody at its original pitch level, we can rely on conceptualizing the starting pitch, and perhaps other prominent pitches. When we confidently know these pitches, we have a strong pitch reference or two. Using the musical ear training many readers will have completed, any single pitch reference can be used to determine any other pitch that may follow.

Thus, by using the skill of recalling the specific pitch level(s) of a known melody, one has a well- honed sense of pitch recognition that is different and distinct from absolute pitch. It is largely already present in adults of all ages, it can be honed and improved, and it does not rely on the labelling of pitch levels.

Through this process you will establish a connection to your personal relationship with pitch—one that is based on your experiences and your person.

Preparation and Background

The process begins with self-reflection to identify one or several tracks that hold deep significance to you, and that can be readily recalled. The track should be known deeply, and have a place within your sense of musicianship or person. Identifying several tracks that hold this type of significance will be most helpful; as we explore this process you will want to identify which track melody might be most readily recalled consistently for its first pitch to become your pitch reference—especially under conditions of distraction or urgency.

Development of a personal pitch reference is possible with daily attention over several weeks. Daily work will involve any number of 5-, 10-, or 15-minute work sessions throughout the day. With focused effort the following exercise will yield significant results within a few weeks.

Process and Sequence

1. The first steps toward realizing your personal pitch reference is to learn to be aware of your pitch memory as it manifests from the track(s) you know intimately. This may most likely be accomplished as follows:

   a. Know the starting pitch of the original melody, and perhaps some of its other prominent pitches (such as a dramatic high pitch in the line). Learn this by roughly transcribing the melody of the original track.

   b. Bring up the melody into your consciousness relying on your memory. Consider the melody and the track enough to settle into a sense of confidence of the starting pitch, but do not over-think it. Do not rush this process at first, though ultimately you want this step to occur almost in an instant.

   c. Vocalize the pitch to externalize it—where it can be useful in the future. Ultimately this step will be omitted, but in the beginning it may be critical to learning.

   d. Check your pitch to determine accuracy.

   Try not to get frustrated by wrong pitches. Everyone will often sing incorrect pitches at this stage. By evaluating your mistakes, you can learn from them. Keep a journal of the pitches you sing (noting time and day); identify whether you were high or low, and the interval.

   Look for patterns over time, and try to identify what might be causing consistent misperceptions. Random errors are normal and will diminish over time. Know that some days your pitch will vary because of distraction, high or low energy, frustration, or another cause.

   e. Repeat b through d every 2 to 3 hours, for no more than 10 minutes, and assess your progress and your confidence.

2. Consider, only now after you have worked with this for a few days, how comfortable you are with your selected track. You may wish to work with another track at this time—either an additional track, or a replacement. If you select a new track, work through 1 again.

3. Settling on a specific track and making progress in your pitch memory, listen to your reference pitch often throughout four or five days; use your reference song less often to call up its starting pitch (your pitch reference).

   A few times per day, stop your normal activity to sing, play and listen to that pitch level. Use a piano, pitch pipe, tuning fork or another instrument to play your pitch. Sing it frequently and become accustomed to the placement of that pitch in your voice. Work to bring yourself to hear the pitch in your mind before singing it. With this, you are bringing your sense of pitch into your consciousness.

4. After this, try to consciously carry your reference pitch with you throughout the day. Take a few moments at four or five set times throughout the day to sing your pitch-level and check it for accuracy. Before singing, quickly quiet your thoughts and bring your attention to your voice or to your memory of the pitch. Do not sing a pitch unless you are confident you have a memory of it. If you do not have a memory of the pitch, return to the above steps for more practice.

Practice (as we all know) is the only way to create consistency. Begin each day recalling your pitch reference. Check it, and adjust for accuracy as needed. Carry the pitch with you throughout the day. Remember, this exercise will greatly enhance a skill you will use throughout your studies and career.

5. Once you have acquired some confidence in your pitch reference, begin to determine other pitches—first isolated pitches, then pitches within the context of tracks. Using your pitch reference, you can calculate the interval that separates the unknown pitch from your reference and use this interval to identify the other pitch.

Once learned, this entire process can become very quickly accomplished and might function as absolute pitch, but it is clearly not. It relies on skill and self-awareness—and its consistency may well vary with the day. Daily checks and adjustments can be necessary, but the skill remains valid and valuable. Having a pitch reference is as useful for musical elements as it is for pitch as a recording element; it provides direct access to all pitch content within the track and can ultimately lead to being able to move between pitch recognition and identifying frequency levels.

PRAXIS STUDY 7.2 RECOGNITION OF SOUNDS IN RELATION TO PITCH/FREQUENCY REGISTERS

Goals and Desired Outcomes

The goal of this study is to identify the octave placement of pitches and/or the placement of frequencies within the hearing range. To facilitate this skill, pitch/frequency registers have been defined. By learning sound of the boundaries of these registers, one can apply that knowledge to acquire skill in identifying pitches and frequencies at their specific levels in the hearing range.

The pitch/frequency registers divide up the hearing range into regions that roughly correspond to regions on the basilar membrane; each register has qualities that distinguish them from others. Pitches/frequencies within these regions contain a sensation similar to others in the region, and that differ from those in other regions. The boundaries that separate the registers are purposefully a sub-region spanning an interval rather than a discrete threshold; this is how the ear functions.

Preparation and Background

Learn the pitches and the frequency levels that establish the boundary regions that separate pitch/frequency registers. Committing these to memory will be useful for numerous tasks and contexts in the future. Keeping the scale of pitch/frequency registers on hand (see Figure 7.4 and also the webpage) will assist the reader during this learning process.

Process and Sequence

1. Working with a keyboard instrument, some type of tone synthesis device, or other suitable sound source, practice listening to the boundaries of the registers. Seek to remember the sound-sensations of those boundaries, and continually remind yourself of the pitch names and frequency levels that comprise those boundary regions. Audio files of these boundaries are also available on the webpage.

2. It would be helpful to download the audio files of boundaries, or to record each of the boundaries and then randomize their playback. This would allow one to develop recognition of the boundaries from a position of not knowing which will be heard.

     Maintain a record of your errors; evaluate your mistakes and make adjustments.

3. Once confidence has been established in recognizing the general areas encompassed by the registers, begin playing individual pitches against the pitch registers. Identify the pitch register in which the sound is located.

4. Continue to refine this skill by seeking to identify the location of pitches within the registers (i.e., the upper third of Low, or the lower quarter of Mid). Continually seek increased precision in localizing frequencies and pitches within the registers.

5. To further distinguish pitch/frequency placements, add in pitches within the regions to your practice sequence; continually identify the frequency levels of these pitches.

Throughout these steps, you should rely on your memory of the pitch/frequency registers in making these judgments. Use an instrument to check yourself, to validate success or to make adjustments; work towards basing your estimations solely on your growing ability to recognize pitch/frequency unaided.

Once you have established significant skill here, use your pitch reference to begin labelling the pitches that you hear within specific pitch registers. Next, your acuity can be sharpened by engaging frequency levels related to perceived pitch.

PRAXIS STUDY 7.3 CONVERSION OF PITCH LEVELS INTO FREQUENCY LEVELS AND THE REVERSE

Goals and Desired Outcomes

The goal of this study is to gain facility in converting pitch levels into frequency levels, and (to reverse) frequency levels into pitch levels. Pitch as a recording element is closely linked to frequency, and frequency is often an appropriate way to consider this percept as a recording element.

Fluidity in moving between these two percepts and sets of nomenclature—and between the physical and the psychological—is important to observing the content of recording elements, and the contexts of how recording elements impact and play a role in shaping tracks.

Preparation and Background

Reflect on pitches that are meaningful to you. Begin with your pitch reference, then note the pitch or pitches you use to tune your instrument. Consider the lowest and highest pitches in your comfortable vocal range, and other pitches that have meaning to you as a performer, a listener, an analyst, etc.

Figure 7.4 will be used as a reference to transfer pitch levels into frequency evels, and the reverse.

Process and Sequence

1. List pitches that are important to you or to your musical experiences—for example middle-C (C_4) or A_4. List reference pitches around the grand staff, for instance two leger lines above and below each staff, and the highest and lowest pitches on any instruments you play. Consider adding other pitches to this list, making note of why each pitch is present; make certain each pitch contains both its letter name and the octave in which is resides.

2. Add the frequency level of these pitches to your list, noting the frequency is equivalent to the listed pitch:

   $A_4 = 440$ Hz

3. Study this list to gain an appreciation of the placement of these pitches within the frequency range of hearing, and the frequency range within which much music occurs (and where the fundamental frequencies of musical sounds reside).

4. Now consider frequency levels that might have significance to you, perhaps ones discussed earlier or later in this book. List those frequencies and their specific pitch level equivalents. You may quickly find that many frequencies do not align with the pitch levels of equal temperament; for these you can to identify the closest pitch level and how the pitch does not align to the frequency.

   Let us take 60 Hz for instance; it is the nominal frequency of the oscillations of alternating current in the US power grid. In certain conditions, it can become audible—such as is common with florescent lights—and can find its way into playback systems and (occasionally) records.

   60 Hz = neither B_1 (61.7 Hz) nor B-flat$_1$ (58.3 Hz), it is only slightly closer to B_1, and is very close to the quartertone that separates these two pitch levels

5. Seek to commit to working memory these pitch and frequency levels, and seek to gain fluency in shifting between frequency and pitch, and between pitch and frequency.

PRAXIS STUDY 7.4 DEVELOPING FACILITY AT FREQUENCY ESTIMATION THROUGH TRANSCRIBING MELODY INTO PITCH/FREQUENCY REGISTERS AND MAPPING MELODY AGAINST A TIMELINE

Goals and Desired Outcomes

The goal of this study is to gain fluency in using pitch/frequency registers and using timelines. To facilitate this, we will perform a transcription of a melodic line on to an X-Y graph; the graph will contain pitch/frequency registers on one axis and a timeline on the other.

Preparation and Background

Find a melodic line you know well. It may be desirable to use the melody that generates your pitch reference though this is not necessary; any melody can be used. This study functions best with melodic lines that contain some rapid changes of pitch levels, a silence or two, a range of over an octave, and a duration of a minimum of six to eight measures. You will find it beneficial to practice your pitch reference from Praxis Study 7.1 while you identify the pitches of the melody.

Quadrille paper with ¼-inch squares tend to work best for our purposes in nearly all recording element graphs.

Process and Sequence

1. Determine the length of the melody and establish a timeline of the example—noting measures, meter, and beats within measures. Allow at least two squares per beat for the timeline—more if your melody needs a higher resolution to clearly display the rhythms and pitches/frequencies of the line.

Next divide the vertical axis into pitch/frequency registers, as in Figure 7.4. Quickly identify the highest, lowest, beginning and ending pitch/frequency levels; extend this axis a half-register below the lowest pitch/frequency and half-register above the highest pitch/frequency of the melodic line.

2. Place these beginning and ending pitch/frequency levels carefully against the timeline and against the registers; label them with pitch level and frequency number. Seek to be accurate in your placements. The template on the webpage will help formatting the Y-axis.

    Begin plotting the melodic contour against the timeline by identifying highest, lowest, and other prominent pitch levels, and placing them at precise locations on the timeline.

3. Check your timeline again carefully to verify accuracy of length, and the placement of these notes. Identify the fastest change of pitch level in the melody; this will become the smallest time unit the graph needs to clearly present and will determine the appropriate number of squares-per-beat division of the X-axis.

    If necessary, change the two-square-per-beat division of the timeline to clearly depict the line. Keep the number of squares per beat in the timeline consistent so the amount of space occupied by the line reflects the passage of time; this may require making a new graph and starting over.

4. Next, identify other prominent notes of the melody—such as notes before and after silences. Locate these on the graph, and label them.

5. Draw the entire contour of the melody now against the X and Y axes. A single line represents notes when then are present; during silences there is no line representing the melody. Pitches that change abruptly one to another (which is most common) will create a contour that is stepped; angular lines are present for glides and glissandos only.

6. Fill in the remaining pitch information, making certain to check observations regularly. Align pitches carefully in time and against the pitch/frequency registers.

    The graph can illustrate the smallest noticeable pitch/frequency and rhythmic changes accurately; seek to incorporate as much of the nuance as you are able into the graph.

PRAXIS STUDY 7.5 RECOGNIZING PITCH AREAS, PITCH AREA CHARACTERISTICS AND THE RELATIONSHIPS OF PITCH AREAS WITHIN PERCUSSION SOUNDS

Goals and Desired Outcomes

The goal of this study is to learn to identify the pitch areas within percussion sounds. This leads to identifying the density characteristics of pitch areas, and the dynamic relationships of pitch areas.

Identifying and defining pitch areas is an important first step of listening inside timbres. Here you will listen for more general information; these same concepts will be more precisely defined in timbral content analysis later. There are some aspects of pitch areas that are general in nature, yet can still lead to objective assessments. In total, this study will bring the reader to experience timbral content of drum and cymbal sounds in broader and more general terms than will be sought in other studies. This approach may be adequate for many recording analyses.

Preparation and Background

Identify a specific drum sound to evaluate—this should be an isolated sound with little or no competing sounds; it may be from a record, a sound you have recorded, etc. It will be helpful to have the sound in a DAW or other software where it can be easily repeated, and perhaps processed. Repeating the sound quickly and with little distraction of searching for the sound will greatly assist this process. To assist this process, some means for pitch-matching might be helpful. This could be a tone generator, a keyboard, or some other sound source.

Create an X-Y graph with pitch/frequency registers as the vertical axis. The horizontal axis will not be time in this graph; instead a section of 4 squares will place the drum sound against the vertical axis and allow space within for density and dynamic relationships of areas to be noted. This is one of the few X-Y graphs that does not incorporate time—another aspect of the general nature of this graph.

Process and Sequence

1. Identify the dominant or most prominent pitch area of the sound. There will be one area (a band of frequencies) within the sound that stands out from others. This is the primary pitch area. The primary pitch area (frequency band) will present itself as a single percept that coalesces from the frequency content of the sound. At first this may be difficult to identify; this is natural and common, as you will not be experienced at listening inside the sound for areas of activity. Explore the spectrum of the sound by bringing the focus of your attention to specific frequency levels and make note of the level of amplitude and amount of frequency information in that area; areas of most significant activity are likely to be the primary pitch area.

2. Next, define the lower boundary of the primary pitch area. Take the focus of your attention inside the sound in this area of greatest activity. Shift your focus clearly outside that area. Knowing what is inside and what is outside the pitch area brings one a sense of control within the process of determining boundaries. Gradually close this gap (a point of focus within the area and outside the area) until you arrive at the lowest boundary.

   Then shift your attention to determine the upper boundary of the pitch area. Take the focus of your attention inside the sound again, a seek to identify a place within the sound that has a different amount of frequency activity or a different loudness level.

   With the sound in a DAW a steep filter can be helpful in determining these boundaries if pitch matching is not working. Identify the precise frequency of these boundaries as nearly as possible.

3. Shifting from the primary pitch area, seek the secondary area of concentrated activity (these will be identified by either width, density, or dynamic prominence of the pitch area). This is the next prominent area of coalesced activity; identify its lowest and then its highest boundary. Get a sense of its dynamic relationship to the primary pitch area, and number this relationship as in Figure 7.5.

4. Repeat the process for any other pitch areas present. Specific frequencies/pitches of the boundaries are often audible, though at times areas seem to blend one to another—especially in beginning studies. These boundaries of pitches/frequencies can be identified and noted on the graph in similar ways as the secondary area.

5. Note the specific frequencies (and perhaps pitches) of all pitch areas (upper and lower boundaries) and place horizontal a line on the graph at that level. The upper and lower boundaries should be joined to make a box representing the pitch area.

   Now, observe the dynamic relationships between these areas and the primary pitch area (as above). Number these dynamic relationships of areas as in Figure 7.5.

6. We will turn now to the densities of pitch areas. Observe the amount of frequency information present within the primary pitch area; make a general assessment of the amount of this activity from very dense to sparse (this is the relative amount of pitch/frequency activity noted on a numbering scale from very dense to sparse) as explained in Chapter 7. Incorporate that information into the graph as in Figure 7.5.

7. To end, identify and finalize the dynamic relationships between all the primary pitch area and all other areas. Describe this information as part of the analysis (these are generalized dynamic relationships of the area and are noted on a number scale identifying the relative loudness of the pitch areas).
8. Repeat this entire process with a different drum sound; avoid snare drum sounds at first.
9. Now repeat this entire process with a crash cymbal sound; avoid hi-hat sounds at first.
10. Once some skill has been acquired, observe the more complex sounds of the hi-hat and snare drum.

PRAXIS STUDY 7.6 DETERMINING PITCH DENSITY OF SOUND SOURCES

Goals and Desired Outcomes

The goal of this study is to recognize the pitch/frequency ranges of sound sources as they are situated in the pitch/frequency registers. Just as in Praxis Study 5.5, these ranges result from the fusion of the source's timbre and the breadth of the pitch range of materials within the syncrisis unit—or prevailing time unit, or some other structural unit that the listener/analyst might identify as appropriate for the particular track and analysis.

The study allows one to recognize the pitch areas of a sound source. How the ranges of all sources relate to the overall texture of the track is timbral balance, the following study.

Preparation and Background

Identify a track with very sparse instrumentation over its first several sections; it would be best to limit sources to a maximum of 5. You may select the same track as used in Praxis Study 5.5, or a different track. This exercise will graph the pitch density of a single instrument from the track. Select the instrument in the track that plays throughout the time period and that also exhibits the most variation of range and material.

Pitch density will be graphed to show the musical material the instrument is performing and the prominent aspects of its spectrum. The difference between this study and Praxis Study 5.5 is that here a single source will be plotted on an X-Y graph of time against the pitch/frequency registers, as in Figure 7.6.

Process and Sequence

1. During the initial hearing, establish the length of the timeline through the first two major sections.
2. Note the general time locations of the instrument's entrance and exit for each phrase; identify those points with a mark on the timeline.
3. Transcribe the instrument's pitch material onto the graph, with its general melodic contour represented as a single line on the graph.
4. Clearly identify the instrument's phrasing or syncrisis units that generate distinct, individual musical gestures that occupy a sense of a complete statement in the present. Mark the beginning and ending points of those ideas, and smooth out the melodic contour to show a more general outline of the line, eliminating small and fast variations of pitch level.
5. Turn focus to the spectrum of the instrument's first sound. Determine the bandwidth of the pitch area by identifying the frequency area where the spectrum becomes noticeably softer than the loudness of the lowest frequency. This will determine the upper boundary of the pitch area.
6. Listen again to the musical material to identify any significant changes in performance intensity. These changes will often bring about changes in the instrument's spectrum that may expand

or contract the bandwidth of this pitch area. Examine the source carefully and note any such changes on your graph.

7. Map out the upper boundary of the pitch area by observing to the spectral information against the lowest frequency.

8. Once these boundaries are finalized, observe the density of spectral activity within the defined pitch area. Make note of significant shifts in spectral density in any notes pertaining to the instrument's pitch area.

PRAXIS STUDY 7.7 COMBINING PITCH DENSITIES OF SOURCES INTO TIMBRAL BALANCE, AND CREATING THE TIMBRAL BALANCE X-Y GRAPH

Goals and Desired Outcomes

The goal of this study is to recognize the pitch range of the overall texture of timbral balance, and to plot the contributions of individual sound source pitch densities to the track's timbral balance. The individual sources represent the distribution of pitch/frequency activity within timbral balance.

This study allows one to observe the bandwidth created by the lowest and highest pitches/ frequencies present (establishing the range of the overall texture as it changes over the duration of the track, or a portion thereof) and the distribution of frequency information throughout the range of the track (establishing the density of frequency information by register).

Preparation and Background

This timbral balance graph is comprised of the pitch densities of all sources present during the first several major sections of the selected track. Revisit the previous praxis study for any clarification of pitch density that might be needed.

Identify a track with sparse instrumentation over the first four major sections of the work; it would be best to limit sources to a maximum of six, with one being a percussion instrument and one a vocal. Some sound sources should exhibit changes of pitch density within and between sections.

The timbral balance graph will be formatted as an X-Y graph of time against pitch/frequency registers, just as Figure 7.6. The pitch density of individual sound sources will outline the pitch-space occupied from its lowest (fundamental) frequency/pitch up through its prominent spectral components.

Process and Sequence

1. Begin the process with an initial hearing to establish the length of the timeline from the beginning through the first four major sections of the track.

2. Identify and list all sound sources present in the track during the next several hearings.

    Next, determine the presence of each instrument/voice against the timeline. Noting when each source enters and exits will identify the time period when each instrument's pitch area will be present in the timbral balance.

3. Graph the pitch densities of each sound source (as in Praxis Study 7.6).

4. Graphing the pitch density of percussion sounds is in contrast to the pitch area study above. Pitch density depicts a single range from its lowest sounding frequency to its highest dominant spectral

component. Its pitch density is not divided into areas that are internal to its timbre (as above) but rather represents its overall sound. Sources are present in the graph whenever its sound (from attack through decay) is heard.

5. The process of determining the timbral balance graph will use the same skills developed in Praxis Studies 7.6 and 5.5. This graph will plot the pitch density of all sound sources (and their materials) to clearly represent the complete frequency content of the recording.

    The pitch density of all sound sources should be included in this study. While exacting detail of the melodic activity is not warranted, each source should be observed and placed on the graph carefully for placement against time and for accuracy of the defining pitch/frequency levels of each line.

6. Once all pitch density observations are completed, they are combined into a single timbral balance graph. Here the pitch densities of sources establish areas of frequency activity in the timbral balance of the overall texture.

PRAXIS STUDY 7.8 LEARNING THE SOUND QUALITY OF THE HARMONIC SERIES AS A UNIQUELY VOICED CHORD

Goals and Desired Outcomes

The goal of this study is to learn to hear the harmonic series as a uniquely spaced sonority or chord. In hearing the harmonic series as a chord one can conceptualize its partials in a way that is more directly related to one's prior experience. This can make it easier to parse out the individual frequencies within a spectrum.

This approach has proven successful in pulling listeners into the content of spectra, a process that can be illusive and frustrating. It is important to remember, listening inside sounds is contrary to the way we have heard timbres since our first experiences—and yet, it is the only way we can engage and observe the content of timbre.

Learning the 'sound quality' of the harmonic series will be valuable in learning to identify the spectral content of sounds. The 'chord' of the harmonic series will become a template against which timbres can be compared and their spectral content identified; through this comparison, the reader will (ultimately, after practice) be able to recognize harmonics present within timbres as well as frequencies that are not aligned with the harmonic series (overtones within the timbre).

With further development, even the dynamic relationships of harmonics and overtones will emerge.

Preparation and Background

This exercise should be practiced in 5 to 15 minute sessions daily or several times per day over a period of several weeks. It should be supplemented with writing out the harmonic series at a variety of pitch/ frequency levels; it is important to use both frequencies and pitches of partials, as pitches provide an incomplete and inaccurate approximation of timbral content. Speed at conceptualizing and generating the harmonic series is central to being able to use it as a tool to recognizing how the frequencies/ pitches of timbres are distributed throughout the hearing range.

The companion webpage provides several audio files containing examples of the harmonic series. The distribution of frequencies outside equal temperament may appear odd at first; ultimately these tones will be accepted as simply part of the sonority that is the harmonic series.

Process and Sequence

1. Listen carefully to the harmonic series being constructed a single pitch at a time. Notice the spacing between the tones and the sequence of intervals of the series. Work to commit the sequence to memory—both the names of the intervals and the sound quality of the interval sequence should be learned.
2. Through 16 partials, practice recalling the sequence of intervals by writing them—both pitch names and frequencies, and being aware of which harmonics do not conform to equal temperament.
3. Next, practice playing the sequence on a keyboard, remembering the higher intervals do not align with the equal-tempered tuning of the keyboard.
4. Continue to listen to the harmonic series provided, and shift your attention to the perspective of the quality of the 'chord' that is played after the individual pitches of the harmonic series. Listen carefully to the overall quality of this chord; then seek to identify each individual pitch within it.
5. Turn next to an acoustic piano, with lid opened to more easily hear its sounds. Silently depress the keys of the 3rd and 5th harmonics, they play the fundamental with moderate force. Listen for those harmonics to emerge, and create an aural image of their presence within that timbre. Repeat this process with the keys of the 4th, 5th, 6th and 8th harmonics depressed; listen for those harmonics to emerge, and create an aural image of their presence within that timbre. Continue to practice conceptualizing specific pitches/frequencies in order to obtain confidence in spelling the harmonic series and recognizing its pitch succession and overall sound quality.
6. Repeat steps 2 through 5 with a fundamental frequency in the fourth octave (such as A_4), the second octave, and then the fifth octave. Notice partials emerge differently in these octaves. Practice hearing the harmonic series in different octaves until you are comfortable quickly conceptualizing each of the partials of the harmonic series while listening to the timbre; begin to notice the loudness of some harmonics is greater than others.
7. When some comfort in recognizing partials has been accomplished, listen to a single piano note being sustained by depressing only the key of the fundamental, without pedal. Take your focus to the perspective of individual harmonics. You will notice that partials change in loudness level and in pitch during the duration of the sound. Hold one, single harmonic in the center of your focus and follow its changes over the duration of the sound. Do this with several other partials during successive soundings of the fundamental to gain practice at hearing and focusing on spectral components.

This study will prove invaluable to listening inside sounds—that includes a variety of purposes beyond timbre analysis. Remember, the goal of this exercise is to bring the reader to recognize the 'sound quality' (or timbre) of the 'chord' that is created by the harmonic series, in its specific voicing (or spacing) of intervals and pitches. The spacing is important to recognize, as well as the overall quality, and the individual partials. This knowledge will be used as a template against which the reader can learn to identify the partials of a sound's spectrum.

PRAXIS STUDY 7.9 OBSERVING AND NOTATING TIMBRAL CONTENT

Goals and Desired Outcomes

The goal of this study is to deeply explore a sound's timbre to observe and notate the content of its component parts (dynamic envelope, spectrum, spectral envelope). This will be approached in two ways, with one more general in the detail of the data collected than the other.

The reader will gain facility in observing and processing the details within timbres. This opens many opportunities to explore tracks with greater accuracy and detail.

Preparation and Background

Identify an isolated sound of a pitched acoustic instrument from a track; alternatively, you might record a sound yourself. The sound should have with a duration of at least three seconds (up to five seconds), and a sustain of at least two seconds (the longer the duration the easier it will be to execute this study at first). Import the sound into a DAW or other device that will allow you to readily loop or replay the sound with few distractions of the logistics of locating the sound.

Figures 7.8 and 7.9 are examples of timbre analysis graphs. Create two X-Y graphs using the vertical axis from these Figures. The timeline of the first graph should be divided into 0.5 second segments each containing two boxes, and the second graph should be divided into 0.1 second increments of one box per tenth of second; the duration of these timelines will match the duration of the sound you selected.

We will now analyze your selected sound's timbre by using the following sequence of activities.

Process and Sequence

A. Begin with the graph with a timeline divided by half-second increments. This will bring observations that are more gestural than detailed.

1. During the first hearing(s), listen to the example to establish the length of the timeline.
    Internalize the half-second increments as pulses on a metric grid; this will allow you to identify the time locations of the timbre's content.

2. Bring your attention to the activity of each graph tier; estimate the highest and lowest levels within the content of each tier; exceed these by 20% to establish y-axis boundaries of each tier. The amount of space you allocate for each tier should allow the smallest increment of activity you notate to be clear in its levels of activity and speed of activity (dynamic contour, spectral content, spectral envelope).

3. You now have a sense of the qualities within this sound. Methodically work through notating each tier of the graph: pitch definition, loudness contour, spectral content, spectral envelope. By engaging pitch definition first, it can provide guidance to access the content of the other segments. Moving between tiers during the process of data collection may be more productive than collecting all of the data in one before proceeding to the next.

4. Explore the timbre by moving between percepts. Make approximated observations for the levels and contours/content of each tier: pitch definition, dynamic envelope, spectral content, and spectral envelope. Hearing spectral envelope may be challenging during your first analyses.

5. Complete as much of this graph as you can, and return to it the next day to correct your observations and to attempt to find any missing information.

B. Turn now to the graph divided into 0.1 second timeline increments. This will bring more detail to your observations. Use the same sound as part A.; this time seek greater precision and document all detail you perceive.

1. Refer back to Praxis Study 6.1 to remember how to identify and recall tenths of seconds.

Internalize the 0.1 second increments as pulses on a metric grid; they will pass quickly. Patterns or activities such as loudness contour changes or spectrum changes may create points of reference for you to calculate these pulses more clearly. Remembering the purpose of being aware of this time pulsation is to identify the time locations of the timbre's content can allow one to relax into the impossibility of keeping track of each pulsation.

2. Expand the Y-axis of each tier by 30% to allow more room to notate detail of that component.

3. Now methodically work through notating each tier of the graph, adding precision to the levels previously identified and adding more detail to the contours representing changes of levels. Move between tiers less during this more advanced stage of data collection. Focus your attention deeply within the component parts of timbre to observe these details; frequent shifts to other components interfere with the intense focus needed to fully explore these subtle percepts. The top to bottom sequence of pitch definition, loudness contour, spectral content, spectral envelope is typically most effective for data collection.

4. It is common to be tempted to look at the visual display of the DAW or other software or device you are using for playback. Remember, what you see is not always aligned with what you hear. Remember also, and more importantly, the goal of this study is to develop skill in observing timbre through hearing it. Use of visuals to explore the sound and perhaps to validate observations can be helpful to supplement the primary, aural experience.

5. Return to this sound and graph the next day to check your observations, and to add more detail that is revealed with these fresh hearings of the sound.

Repeat the processes of parts A and B on another sound that is dissimilar to this sound.

PRAXIS STUDY 7.10 DESCRIBING THE PHYSICAL CONTENT OF TIMBRE

Goals and Desired Outcomes

The goal of this study is to bring the reader to describe the content of timbre. Developing this approach allows one to clearly separate describing the components of timbre in an objective manner from engaging the character of the timbre. This keeps the objective and measurable aspects of its internal content clear, and allows the character of timbre—in its interpretive qualities—to be articulated as an overall quality. Descriptions of timbral character will appear in Chapters 9 and 10.

Preparation and Background

This study should be performed on a number of sounds. Each description should have a different level of detail ranging from very general to quite detailed. Data collection for descriptions should range from describing both the gestural and the detailed graphs from Praxis Study 7.9, to collecting more superficial observations from a few listenings without creating graphs to illustrate data.

Repeat this exercise at least three times; each time create descriptions spanning the gestural and detailed graphs, and the less detailed observations collected without graphs. Always have clarity in the level of detail you are seeking. Begin with describing the general graph you created in 7.9; next describe one of the detailed graphs you completed in 7.9. Third, identify an isolated sound of at least 4 seconds duration and describe its basic content without the aid of completing a graph.

Each of these approaches has value, and can be appropriate support for a recording analysis depending on its goals.

Process and Sequence

1. Make note of the duration of the sound in beats (or portions) and/or in clock time.
2. Make general observations of the pitch definition of the sound.

    Does it start with a burst of noise (like a piano)? Are there areas where the sound changes in pitch quality? Is the sound mostly pitched or mostly noise-like? How does the pitch sensation change against time?

3. Next describe the dynamic envelope.

    Describe the general shape of the contour first, then add a nuance you might deem desirable. What is the speed of the attack and initial decay? What is the level of the attack? What is the sustain level in relation to the attack? How does the sound end? How does the dynamic envelope change in level during the sound's duration?

4. Describe the spectrum of the sound.

    Is it dominated by harmonics? Between which harmonics are overtones present? Is the spectrum dense with many partials or does it have a simple, open spacing? Is there a different spectrum during the onset than in the body of the sound? Does the spectrum change with dynamic shifts?

5. Describe the spectral envelope.

    What partials are prominent? Is the fundamental louder than the remainder of the spectrum? Which partials change in loudness over the duration of the sound? Can you identify partials that enter and exit the spectrum? How does the spectrum change over time?

6. Practice talking about sound in this way when you seek to identify sounds or discuss their timbre. Ask yourself: what I am hearing related to the actual sound wave? What are its current qualities, or how are those qualities changing?

7. Continue to practice describing sounds this way without first creating a sound quality evaluation graph, and then after creating the graph. Work with a variety of sound sources and sound durations.

This process will bring you to be able to quickly evaluate timbres in a meaningful way, and to provide clear information on timbres.

PRAXIS STUDY 8.1 LOCATING IMAGES IN STEREO LOCATION AND RECOGNIZING THEIR WIDTHS

Goals and Desired Outcomes

The goal of this praxis study is to bring the reader to hear the locations and widths of stereo images. The widths of images can be elusive, especially during beginning studies.

The stereo location graph will help the listener conceptualize and visualize the locations of phantom images, and to bring attention to the lateral space they occupy. With practice, the graphing process step may no longer be needed to accurately perceive stereo imaging of sources.

Preparation and Background

This study and all others related to Chapter 8 assume the listener has a sound playback system that produces a stable, complete sound stage from an established point of audition. Avoid headphone listening as it will distort the very spatial cues you are now attempting to learn.

Identify a record that begins with four to six sound sources through its introduction and first major section. The track should place sources across the entire left-to-right stereo field, including the area around 15°-left and 15°-right of center; at this time avoid tracks that group sources largely in the center.

The stereo location graph of Figure 8.4 will aid the reader in determining the widths and placements of sources. Copy the Y-axis on a sheet of quadrille paper, taking care to distribute the markings of degrees accurately around the center point. Divide the X-axis in half as two song sections, rather than time; the first half represents the introduction and the other the first major section (perhaps a verse or chorus).

PROCESS AND SEQUENCE

1. Make certain you are positioned in the 'sweet spot' at the apex of an equilateral triangle with the two speakers, and that your speakers are balanced and your system functioning well.
2. List the sound sources in the introduction and first sections of your track. Compile a list of sources that perform in the introduction and place it under the timeline area; compile another list under the area of the first section listing its sources. Identify the sources that perform in both sections, and have them at the top of the list.
3. Select a prominent instrument that appears in both sections. Note the center of its image during the introduction and lightly mark it on the graph on the first line from the left; next note the center of its image during the next section and lightly mark that spot on the first line from the left in the next section.

   The center of images naturally draws our attention, and our previous experiences bring our thoughts of location to stop here. It is imperative to records, however, to determine the widths of images. Gradually shift your attention off the center outwards to the left, notice the sound maintaining its image until you reach the edge where the image stops; repeat this to the right to determine that edge. At times it is helpful to deliberately place attention outside the image, so you can verify where the image is not; closing the gap between where you know the image is and isn't can bring more certainty to the process of exploring the edges of phantom images.

   The distance between these two edges is the width of the source. It can be calculated precisely by degrees left and right. Some images have soft or blurry edges that make this difficult, but a great many are clear. Most times when edges are difficult to hear it is because other sounds are occupying the same area or the sound has an environment that spreads its image so as to disguise its boundaries.

   One at a time, find the edges of these two images of the same sound within the two sections. Mark those edges on the graph to expand on the light markings you previously made. Note, it is possible that the two are the same, but it is also possible the size or location have shifted in the next section.

   Source images can shift size and locations within sections as well. If a sound changes within a section acknowledge that it changes in your notes, so you can remain aware that your graph is not a complete representation of the track. During this exercise it is not necessary to account for all detail. This graph is not intended as a representation of the track, but rather is a study in identifying width and location attributes. That you notice changes that cannot be incorporated is an accomplishment. It is important to acknowledge when detail is willfully omitted in the event it is needed at a later time.
4. Turn now to another sound that is present in both sections and repeat the process of 3, locating the sound on the next line of the sections.
5. Continue similarly plotting sounds from your list onto the graph until all sounds are present in each section.

6. Check your observations of individual sounds, and then shift perspectives to the composite texture where you can hold all sounds equally in your attention. Listen to the aggregate texture of all sources in the stereo field while examining your graph.

   Notice areas where images overlap and areas where images gather and are superimposed. Are any sources isolated in their placement? Are there any voids with no image present? Notice the sizes of sources. In comparing them, are some sounds noticeably larger or smaller than others?

7. Listen carefully across the panorama of the stereo field, bringing the focus of your attention to each degree position within the 90° span. Consciously get a sense of the presence or absence of images in every location.

PRAXIS STUDY 8.2 IDENTIFYING SOUND SOURCE POSITIONS AND PLACING THEM ON THE STEREO LOCATION GRAPH

Goals and Desired Outcomes

The goal of this study is to become acquainted with the process of placing sounds on the stereo location graph, identifying the positions of sounds in time and in the lateral dimension of space.

The graphing process will reveal sound source widths and locations, placing them in time will reveal any changes that occur within and between sections and any time attributed created by changes or movements of locations and widths. Unlike the previous study, here you should incorporate all nuance of image size and location you are able to perceive.

Preparation and Background

For this exercise, identify a record that displays significant changes in lateral locations of sound sources and that contains at least eight sources.

Identify (1) the sound that grabs your attention as exhibiting a noticeable amount of change, (2) the sound with the widest spread image and the sound with the narrowest image (perhaps a point source), (3) the lead vocal, and (4) the bass part (bass guitar, keyboard, or other). Graph these sounds throughout the first major sections of the track so as to include both a verse and a chorus using the graph format of Figure 8.5 (note the graph contains two tiers so sounds can be distributed for clarity). Distribute these listed sources on one tier of the graph and all of the other sources on the other tier.

Process and Sequence

The following sequence is usually the most direct and effective process for transcribing stereo sound location.

1. During the initial hearing(s), listen to the example to establish the length of the timeline. Create the stereo location X-Y graph with two tiers in the X-axis to distribute sources into two groups to allow their images to be more clearly visible. To start, allocate one square for every 10° on the X-axis and

one square per measure for the timeline; after drafting the graph these can be adjusted if more or less area is required to clearly represent the material.

2. Next, observe the presence of the selected instrumentation (instruments and voice) that were identified above and sketch the presence of the sound sources against the completed timeline by marking their entrances and exits.

    Now create a list all of other instruments and voices, and sketch their presence against the timeline by marking their entrances and exits.

3. With a clear sense of the sources present in the track, consider which you want to observe first. Plan the order of graphing the identified sources (perhaps the lead vocal first, perhaps the bass); follow this by organizing the order of graphing the remaining sources.

4. Now begin plotting the stereo location of the four identified sound sources onto their tier. Use Praxis Study 8.1 as a guide. The locations of spread images are placed within boundaries. Maintain focus on the source until you have clearly defined its position and size. Work from what you know for certain (where the image *is* and where it is *not*) to gradually remove your doubt and confusion.

    The locations of point-source images are plotted as single lines; while these sources are often the simplest to precisely locate, they are uncommon.

5. Once you have established the locations of several sources, it is possible to use these as a reference to aid in identifying the locations of other sources; compare the locations and sizes of the additional sound sources to the ones already established. This can also help the listener maintain a focus on image size and location, by adding some comparison of their relationships to other sources.

6. Incorporate all changes in width and location; Figure 8.5 illustrates several of these. Sudden changes appear as instantaneous changes on the graph, the piano part contains a number of sudden changes of widths and locations. Gradual movements occur over time and illustrate speed and amount of change, as in Lennon's vocal line.

7. This graph is complete when the smallest audible detail has been incorporated into the graph. Remain aware of the level of stereo location detail needed to reach the goals you have established for your analysis.

Work with several more tracks of different styles. As you gain experience in making these observations, records with more instruments should be examined and longer sections of those works can be examined.

PRAXIS STUDY 8.3 DEVELOPING DISTANCE PERCEPTION

Goals and Desired Outcomes

The goal of this study is to become acquainted with localizing sounds by listening to timbral detail. The listener will acquire a sense of their point of audition and sense of personal space, and learn to make generalized observations of distance location.

Preparation and Background

In these initial observations of distance placement, the listener will bring attention to their sense of occupying a position in space and to localize sounds from a sense of their physical proximity to their point of audition. While timbral detail forms the basis for these assessments, this percept is not commonly consciously processed in everyday listening.

You may return to one of the tracks you used in Praxis Study 8.1 or 8.2, or select a new track that begins with only a few sources for the following sequence. If you are accomplished on an instrument or otherwise know an instrument's timbre intimately well, a track with this instrument's timbre presented clearly will be a helpful example for you.

Process and Sequence

1. While listening to the introduction and the first major section of the track, focus on your point of audition as it relates to the sound sources. Scan the texture for a sense of their relative location to your position; notice the sense of distance between your position and the sources—the space between you and the collective group of sources, as well as the space that separates you and individual instruments and voices.

    You will want to remind yourself that distance perception is unrelated to loudness, and is not related to reverberation and environments except in those atypical instances where they mask or diminish timbral detail.

2. Now, bring your attention to the amount of detail within one of the source's timbre and notice the amount of low-amplitude partials that are present and how the timbre might subtly shift during its duration. Do this twice more, each time holding another source in the center of your attention.

    Next, ask yourself each of these questions while observing each sound source:

    > Is the level of timbre detail relatively typical or common, or is it unusual in some way— heightened detail, or diminished detail from what one might expect?
    > Does this timbral detail bring the sound to a distance common for staged performance?
    > Does the timbral detail instead bring the sound within arm's reach?
    > Is it uncomfortably close, within your own sense of space or area of proximity?
    > Is the sound at a conversation distance, or slightly beyond but still in the same room?
    > Does the timbral detail take the sound out of the area and room you occupy?

3. Bring your attention to a single instrument or the lead vocal—whichever one you might have the most prior experience in listening to its qualities.

    Listening to its timbral detail, ask yourself: can I reach out and touch the instrument/performer? Is the instrument well within arms-length, or is it at a distance twice your arm's length?

    If these seem not to apply to the sound, perhaps the sound is at a further distance. It is also possible you are having trouble bringing attention to the inner workings of timbre and to hearing its detail. You may wish to return to the praxis studies on observing timbral content.

4. Listen to several additional sources and ask these same questions. Keep track of these general observations. Return to this the next day and compare results.

5. Remember to bring attention to timbral detail in assessing distance positions and location relationships. Remember distance perception is unrelated to loudness, and is related to reverberation and environments only when they mask or diminish timbral detail.

PRAXIS STUDY 8.4 OBSERVING DISTANCE, AND CREATING A DISTANCE LOCATION GRAPH

Goals and Desired Outcomes

This study builds on 8.3, with a goal of observing sound source distance positions, using placement in the distance zones of Figures 8.6 and 8.7. The listener will acquire substantial accuracy in placing sources within these zones through consideration of proxemics and timbral detail of sources.

Preparation and Background

Before starting your observations of distance placements of sources, review the descriptions of territorial distance zones in Chapter 8. Precise placement of sources in distance position from the listener's point of audition is accomplished using three observations: amount of timbral detail, sense of territorial zones surrounding the listener, and positioning sounds in distance positions relative to one another. This triangulation uses timbral detail to identify position of sources within a distance zone and then at a specific region of that zone; any further alteration of positioning with a zone will be the result of positioning sources relative to one another.

For this exercise, identify a record that displays significantly different in distance positions of sound sources, and that contains at least 8 sources.

Identify (1) the sound that appears closest to the point of audition, (2) the sound that appears furthest from the point of audition, (3) the lead vocal, (4) the bass part, and (5) each instrument of the drum set. Prepare to graph all these sounds using the graph format of Figure 8.7; the graph's timeline should run throughout the first major sections of the track so as to include both a verse and a chorus. Distribute the distance zones vertically with sufficient space in each area to accommodate significant activity. This will be the default draft format of the graph that may often be altered to omit unused zones in the final copy.

Process and Sequence

1. During the initial hearing(s), listen to the excerpt to establish the length of the timeline. Draft the X-Y graph with the Y-axis divided into zones of six boxes each; after drafting the graph these can be adjusted if more or less area is required to clearly represent the material.
2. Next, observe the presence of the identified instruments and voice and sketch their presence against the completed timeline by lightly marking their entrances and exits.
3. With a clear sense of the sources present in the track, consider which you want to observe first. Plan the order of graphing the identified sources (perhaps the lead vocal first, perhaps the bass). This ordering is to establish a work plan that sets an intention for each listening step; it can be altered whenever desired or as necessary to accomplish any tasks and to conform to unexpected findings.
4. Now begin plotting the distance positions of one sound source. You may use Praxis Study 8.4 as a preliminary step, or you may move directly into placing sounds within distance zones by their

timbral detail. Rely on timbral detail to determine the distance zone of the source. Consider the amount of timbral detail again, to locate the sound within the front or the rear half of the zone; consider the qualities of the zone to help this observation, then place the sound closer to the front edge or the rear edge of that half-zone. Continue cycling these steps until you have focused the precise position of the source.

Repeat this for each sound source listed above.

5. Once sounds are located within their zone, this positioning can be refined further by comparing source distance positions relative to one another. This is accomplished by holding sources equally in attention while comparing and observing their timbral detail and distance zone positions.

This process can aid in keeping the listener focused on distance positioning, and can also make changes of distance more apparent.

6. Sources can change distance positions in the mix. This often happens between structural divisions of a track, though it can occur at any time and change may be anywhere from subtle to pronounced.

Examine each source carefully at the transitional point or area between sections, and recalculate distance placements.

7. Work with several more tracks of different styles.

As you gain experience in making these observations, records with more instruments should be examined and longer sections of those works should be examined.

This graph is capable of illustrating the subtlest detail of distance placement and gradual or sudden changes of distance over time; it may also be used in a more general way by not incorporating a division in each zone. Distance location is, however, important to the character of the track, and a substantial amount of detail may contribute significantly to analysis—often an area not anticipated in the goals established for an analysis.

PRAXIS STUDY 8.5 MAKING DETAILED AND GENERAL OBSERVATIONS OF SOUND SOURCE IMAGES, AND PLACING IMAGES OF SOURCES ON PROXIMATE AND SCALED SOUND STAGE DIAGRAMS

Goals and Desired Outcomes

The goal of this study is to accurately observe the positioning of sound sources on the sound stage, and to notate them on a sound stage diagram (representing the stereo image and the distance position of sound sources). This may be either an approximation of source locations on the sound stage (the proximate diagram) or a detailed representation of sound source imaging (the scaled sound stage diagram).

The listener will acquire the ability to place sounds on both a proximate and a scaled sound stage diagram, with as much detail as needed for an analysis.

Preparation and Background

Review Figures 8.8 and 8.9 and related explanations. The proximate sound stage (Figure 8.8) is used for sketching approximate sound source image placements in lateral and distance dimensions; it may stand alone in some analyses, or it may function as preparation to determine the more precise localizations and detailed observations within the scaled sound stage diagram (Figure 8.9). The proximate sound stage may also be used for preliminary observations before creating stereo location or distance placement graphs.

It is important to remember both diagrams represent a span of time; changes of image size or location within that time span are difficult to account for, and will be more clearly shown on a stereo location or distance X-Y graph. Rarely can a single diagram (either proximate or scaled) accurately represent all sections of a track; even dating back to the early stereo tracks, image positioning will typically change (to some variable degree) between verses, choruses and bridges. Diagrams representing structural sections may successfully account for their different 'scenes of sound relationships.'

To execute this study, select one of the tracks you examined for stereo location and one of the tracks you examined for distance. Create a proximate diagram matching Figure 8.8 and a scaled diagram matching Figure 8.9; be certain to maintain proportions (NOTE: PDF versions of these diagrams are on the book's webpage). Make copies of each of these diagrams to use as this study progresses (and other uses in the future).

Process and Sequence

A.

1. Revisit one of the stereo location graphs you completed earlier. Transfer the sources images from the stereo location graph onto a proximate sound stage diagram without concern for distance; take notice of the challenges in notating any shifts of images.

    Listen to the distance cues for these sources, and make general observations; now shift the stereo images to that approximate distance placement.

2. Listen more intently to the distance locations of sources in the introduction, and sketch them onto a proximate sound stage dedicated to the introduction; add approximate lateral location and image widths from the stereo location graph to this distance information to localize the source by both dimensions.

    Repeat this process for the next major section (verse or chorus?). Focus your attention on listening for general distance position, then take your attention to the stereo location graph to begin to 'hear' the image in position, with both lateral and distance dimensions. Sketch all sources onto the diagram.

    Repeat this process for the next major section (verse or chorus?). This time, when listening for distance position, also include in your awareness the stereo location graph to begin to 'hear' the image in position, with both lateral and distance dimensions. Begin to experience the sound stage as a two dimensional space.

    Continue this process until all sounds in each section are included in their respective diagrams.

3. The next day or a few hours later, return to these proximate stage diagrams.

    Listen again to the track and observe greater detail in distance positions.

    Transfer all of the sources in the introduction onto a scaled sound stage diagram. Note the degree of detail expected for distance placement and for stereo image placement. Work deliberately and methodically to hear the details of distance, and incorporate those cues into the diagram.

    Repeat this process for each of the next two sections. Notice how the process of adding detail to previous more general observations is at times fluid, and at other times challenging; try to identify matters that give you pause, and work with them.

B.

1. Recall one of the distance graphs you completed earlier. Transfer the source images onto a single proximate sound stage diagram without concern for stereo location; make notice of the challenges in notating any images that shift distance locations.

    Listen to the stereo location cues for these sources, and make general observations to shift the distance images to an approximate stereo image placement. Sketch out this single diagram to represent all three sections, and note its general qualities.

2. The processes from A (above) will be repeated here, except distance positions have been predetermined by your previous work. Here in the proximate diagrams you will add the stereo image width and location placements to the distance observations from the X-Y graphs. Work through each of the three structural sections carefully and separately.

3. The next day or a few hours later, return to these proximate stage diagrams.

    Listen again to the track and observe greater detail in stereo image positions.

    Transfer all of the sources in the introduction onto a scaled sound stage diagram. Note the degree of detail expected for distance placement and for stereo image placement. Work methodically to hear the details of stereo location, and to incorporate them into the diagram. Listen to distance placements again and refine your observations.

    Repeat this process for each of the next two sections.

Beginning with a more thorough observation of one element (stereo location or distance position) before including a second will facilitate crafting these sound stages—especially during your initial analyses. Later you may be able to craft a proximate sound stage in both dimensions in quick alternation, almost simultaneously. Focus on elements individually, alternating between them.

At this point you will have four proximate diagrams and three scaled diagrams of each track. The observations contained in these diagrams are rich in data (some more than others), comparing the diagrams will generate additional pertinent observations. All of these will be of use during the evaluation processes.

PRAXIS STUDY 8.6 HEARING SURFACE ATTRIBUTES OF ENVIRONMENTS: REVERBERATION DURATION, DENSITY AND CONTOUR

Goals and Desired Outcomes

The goal of this study is to be able to perceive and observe the surface attributes of reverb.

After working through this exercise, the reader/listener will be able to identify the duration of the reverb time (RT60), to recognize the density of reverberation reflections as a relative value, and will be able to determine its loudness contour.

Preparation and Background

Separating the reverberation/environment from its bond with the source is the first step toward identifying any of its attributes. This is accomplished by directing attention inside the source timbre to recognize the qualities of reverberation, as have been described throughout Chapter 8. Here we will begin with those qualities that are (typically) most apparent. Remember, we are seeking to observe and identify the attributes of the environment (or reverb), and not the sound source+space bonded quality. This holds for the following studies as well.

Identify a sound with pronounced reverberation for your first experiences with this study; it is best if this sound has little else occurring at the same time, or that it be isolated on the sound stage. A sound that ends abruptly and is clearly audible in the track's texture will reveal its surface characteristics most readily. If possible, edit this sound into a loop, so you can listen to it repeatedly without spending much time cuing the sound.

Process and Sequence

1. In listening to the sound pass through its duration, focus your attention to prepare for the moment the sound will cease and the reverb tail will continue. Listen to the sound several times until you recognize this point, and until you come to expect when it will occur.

    Having established this point, listen to the reverberation from when the sound ceases until the reverb has faded to silence. This is the reverb tail.

    Time the duration of this reverb tail—from the end of the source to the point where the reverb sound has ceased. To do this, you can use the pulse of the track to determine the place in the meter where the sound stopped, use a timing device to determine the duration, or you might use the skill in calculating tenths of seconds learned earlier. Other possibilities exist including using the time-line on a DAW; as before, discovering answers by using one's eyes should be verified and validated by listening.

2. To observe the overall shape (loudness contour) of the reverberation, bring your attention to the tail again. Get a sense of the loudness at its beginning and its shape; note if there is a steady decay, a decay that gradually increases or decreases in speed, a decay with some temporary small increases in level or sustained levels, or some other shape. Take a pencil and sketch out this shape, generalizing time and loudness levels—this general observation of shape may be a starting-point for refining the contour, or it may be an end in itself.

    You may wish to create a timeline of the duration identified above, and then divide it into tenths of seconds. This allows the contour to be placed against time. The loudness of the contour might be graphed against an axis of starting and ending levels with 25, 50 and 75 percent of full volume being marked as a vertical axis. This would provide some exacting detail to the contour.

3. Direct your attention again to the reverb tail. Within that tail is a collection of reflections that are it-erations of the original sound. The spacing of these iterations establishes the number of reflections in any given time period, and the density of the reverb. Conceptually similar to the vertical density of frequencies within pitch areas, this horizontal density takes place over time. Reverberation den-sity can be observed as a relative value, spanning a continuum from sparse to very dense.

    Observe the quality of the reflections taking place; seek to consider if the number of reflections is few and provide a sense of transparency, or many that provide a sense of much activity. Added experiences listening to reverberation will provide a richer sense of this observation; expect your understanding of this to grow with more exposure to the sounds of additional reverbs. Identify a relative value of the density of reflections; you might wish to return to this in a few days—or in a few months, after hearing and learning about many additional reverb tails.

4. Repeat these processes with a similar sound that is equally exposed in a track. Become aware of what aspects come more easily to you and which are more challenging. Give added attention to those that need work.

5. Next, execute these processes with a sound that does not contain an abrupt end to its sound; note the results and challenges.

6. Finally, work through these processes with a sound that is one of several appearing within the same texture. Note the results and challenges as less of the reverb is exposed to clear examination.

The detail contained in the values for these attributes can vary widely—a continuum from relative values to accounting for all nuance of variables. The degree of detail should be adequate to support your analysis; often general observations are sufficient in providing distinguishing qualities.

It is important to note here: this is a set of observations on a single presentation of the source. A source's many pitches or variety of performance intensities have the potential to reveal different information present within the reverb. Surface information does not often vary considerably between notes, but the following attributes (especially frequency response) hold the potential for significant variation.

PRAXIS STUDY 8.7 LISTENING INSIDE ENVIRONMENTS: FREQUENCY RESPONSE AND RATIO OF DIRECT TO REVERBERANT SOUND

Goals and Desired Outcomes

The goal of this study is to be able to perceive and observe how reverb frequency content may differ from the frequency content of the original sound, and to perceive the loudness relationship between the direct sound and the reverberation.

After working through this exercise, the reader/listener will be better able to identify the spectral content of the reverb and note differences within the reverberation as compared to the direct sound; the listener should also be better able to identify the loudness relationship or balance between the direct sound and the reverberation.

Preparation and Background

Separating the reverberation/environment from its bond with the source is a critical first step toward identifying these attributes. The listener must be able to distinguish between the direct, original sound source and the sound of the reverberation (the space of the source).

Return to the first sound you observed in study 8.6; plan to return to each of the other sounds you observed in that study. Your experiences will deepen as you learn more about the same sound(s).

Process and Sequence

1. Listen to the sound, bringing your attention to the reverb tail. (a) Experience again how it lapses over time as an overall quality that is distinct from its quality when bonded with the source. (b) Shift now to experiencing that moment when the source+space ceases and the reverberation takes over. (c) Next listen to the sound's timbre, noticing the first source+space portion of the sound as one timbre and the reverb as another; listen to these qualities several times, bringing your attention to the frequency content of the first portion and the frequency content of the reverberation. The skills you obtained in hearing the spectrum of timbres are useful here.

2. The process of identifying the frequency response of the reverb is a comparison of the reverb's timbral spectrum and the spectrum of the original sound (that contains both the reverb and the direct sound of the source). (a) To begin, identify the dominant traits of the original sound's spectrum—the frequency areas of pronounced activity and of characteristic frequency traits; take note of these by

sketching their presence against pitch registers or otherwise identifying the relative frequency areas you hear. (b) Next identify the spectral content of the reverb; note the frequency levels or bands present at any time in its duration (here we are summarizing all activity over time into one observation of its time span), and sketch their presence as you did for the original sound. (c) Take note of the differences; the reverberation will often accentuate some frequency bands from the original source, and often will attenuate some frequency bands; further, some frequency ranges may not be present at all within the reverberation.

3. This is a set of observations on a single presentation of the source, at one pitch level and one performance intensity of source timbre. The source's many different pitches and gradations of timbre have the potential to reveal different information of the reverb.

    Repeat these steps on various notes, as this sound source plays different pitches and/or changes its timbre. Each hold the potential for significant variation. Compare observations of these different sounds to determine the aggregate collection of frequency levels/bands that are accentuated or attenuated within the reverberation

4. The frequency response of the reverberation is calculated by examining the results of all of these observations. The response can be noted by those frequencies that are emphasized and those that are eliminated or attenuated. It may also be observed as the band between the lowest and highest frequency levels it contains.

    It is important to remember the reverb is the object of concern, not the source+reverb portion. Comparing frequency response of reverberation to that of the direct sound identifies their differences; observing the spectral qualities of the reverb alone allows one to identify its unique qualities.

    For more detailed observations, the changes of spectrum over time might be examined.

5. The comparing of the direct sound to the reflected sound that was just the focus of frequency response now continues. Here we observe the loudness level of the reverberation as compared to the loudness level of the source+space.

    Make a general assessment of this loudness relationship. Identify the reverb's level compared to the direct sound as, for example: about as loud, slightly softer, moderately softer, distinctly softer, or much softer. The reader might find other terms are more meaningful to their work.

6. Repeat these processes with the similar sound you previously examined for surface details.

7. Execute these processes again with the other sounds from before: a sound that does not contain an abrupt end to its sound and a sound that is one of several appearing within the same texture. Again, note your successes and those aspects that challenged you. Assess how you might refine your skills in further study.

PRAXIS STUDY 8.8 HEARING MICROTIMING WITHIN THE EARLY TIME FIELD AND PRE-DELAY; EXPERIENCING THE TIMBRE OF TIME

Goals and Desired Outcomes

The goal of this study is to bring the listener to perceive and observe the microtimings of the pre-delay and of the early time field. To realize this, the listener will need to experience how microtime increments are heard as timbre rather than rhythmic iterations.

By learning these skills, the reader/listener will be able to observe the general qualities of the early time field reflections.

Preparation and Background

The early time field reflections are those reflections in an environment that exist before the reverberant sound is formed. It is unreasonable to expect the listener to identify specific microtiming increments in nearly all environments within tracks—even after much skill has been reached. They are typically masked by their being blended with the source timbre or by other sound sources. Still, their impact is present in the environment, and even a general impression of the arrival time gap can illuminate the observation of environment attributes.

We are able to process the sound quality produced by closely spaced iterations of sound as timbral qualities. We may use these timbral qualities to identify microtime units down to several milliseconds; by getting to know the timbres of microtime units we learn their spacings in time. This information can be used in observing the attributes of the early time field.

This exercise will require the reader/listener have access to some audio equipment and recorded sounds. Regrettably, it is nearly impossible to hear these qualities well enough to learn them by working directly with records. The reader is asked to secure (1) a recording of a percussion sound with a fast attack and little decay (such as claves, a woodblock or a high drum with little ring), (2) a way to replay the sound with time delay (this could be a DAW, a delay unit, or related software or plug-in), and (3) a way to route sound separately to two speakers.

Process and Sequence

1. Route the original sound to one loudspeaker and the processed signal to the other loudspeaker.
2. Set your device to delay the signal at an audible echo. Begin repeating the sound while slowly changing time increments to successively shorter values to get a sense of condensing time; the amount of delay is not as important as the experience of hearing time getting compressed.

    Begin to repeat the sound again, this time set the delay at 100 ms; get a sense of this delay, then double it to 200 ms; listen again at 100 ms, then listen at 50 ms. Now listen at 150 ms. Alternate listening to these delays until you are getting comfortable you may be able to recognize each delay increment.

    Now, route both signals—the sound file and its delay—to both speakers to establish a center image. Repeat the alternating listening to these delay times. Conceptualize the time delays as timbres—as an overall quality of loudness envelope, spectrum and spectral envelope. Listen deeply, alternating your attention toward time and toward timbre; with practice you will notice notice how the two percepts can blend.
3. As confidence is obtained in being able to accurately judge these time units, move to multiples of 25 ms: 50 ms, 25 ms, 75 ms and 100 ms. Route the signals to separate channels again. Repeat the sequence in Step 2, ending with repetitions with both signals routed to both channels equally.

    Give this a rest for a day, and listen to steps 2 and 3 again.
4. The following day repeat step 3 with repetitions in multiples of 10 ms (10, 20, 30, 40 and 50 ms). Work to get a sense of time and a sense of timbre; note your success, point of confusion and areas of difficulty, and seek to address them directly.
5. Continue to work with smaller time units a systematic manner, comparing the qualities of the time relationships of each listening to previous and successive material.

    Incorporate the delay times of 5 ms, 15 ms, 25 ms and 35 ms within the last sequence. Continue to focus attention on hearing the subtle differences of time/timbre with the sounds only routed to both speakers. This blending is important to the sensation of time as timbre. Attempt to learn the unique sound of these time-related timbres.

6. This will all take effort over some time (perhaps weeks of 10–20 minute sessions per day), but is worth the effort. Continue moving to smaller time units, until consistency has been achieved at being able to accurately judge time increments of 2 or 3 ms.

7. This knowledge of the sound of time/timbre opens one to recognizing the arrival gap—the time before the environment begins sounding—and to the density, timing and overall duration of the early time field.

Here, the objective is to be aware of the early moments of the environment—the number of reflections present and a general sense of the spacing of reflections. Much of this is masked in most sounds; still, by following a source through the entire track or section one will often identify a sound that has little going on simultaneously.

Attention then is brought to the attack of the sound, and when the environment begins. This is not easily learned or recognized, but the skill is latent within one who has worked through the above steps. The early time field provides information on the placement of the source within its host environment. It also provides a sense of spaciousness around the source. This is sometimes heard as a width of the reverberation *around* the direct sound of the sound source—which is explored next.

PRAXIS STUDY 8.9 OBSERVING THE ATTRIBUTES OF DEPTH AS CREATED BY ENVIRONMENTS AND REVERBERATION, AND OF THE POTENTIAL OF ADDED WIDTH TO IMAGES (CREATED BY EXTENDING THE ENVIRONMENT AROUND THE IMAGE OF THE DIRECT SOUND).

Goals and Desired Outcomes

The goal of this study is to bring the listener to perceive and observe the spatial qualities of environments and reverberation in their potential to establish depth (sense of the back wall) and width (sense of side walls) to stereo images.

By learning these skills, the reader/listener will be able to observe these subtle attributes that extend and add spaciousness to sound source images.

Preparation and Background

A. Reverberation and environment cues will at times establish a sense of depth to the environment of the source+space image, and the source and space remain bonded; this attribute of depth is not always prominent, though at times it is integral to the source (see the backing vocal and hi-hat sounds Figure 9.8).

B. Reverberation and environment cues have the potential to extend the width of an image. This often occurs with a sense that the source is bonded with the space, and the image has a singular quality. Alternatively, the two may have a sense of their own width, and the reverberation is heard as being *around* the direct sound of the sound source—the space has more presence and its sonic identity is clearly separated from the direct sound (see the lead vocal and organ sounds in Figure 9.8). This is another way environments can be separated from source bonding.

Return to the tracks you examined for stereo location and for distance. Prepare to listen carefully to several; seek to identify (1) a source that appears to have considerable depth, and (2) a source with a broad overall image with a center that is a concentrated impression of the instrument/voice.

Process and Sequence

A.

1. Return to one of your sound stage diagrams, and listen to the sources for depth of images, a sense of the back wall. With this sense of depth will be an indication of the amount of distance present behind the source—this sense of depth is not always strong and may be entirely concealed. Choose a sound with a prominent sense of depth.

2. Seek to conceptualize the placement of the rear wall. This can be an estimation based on extending the listener to source distance by a perceived proportion, such as the rear wall is double the distance between listener and source. It can also be based on the type of reverberation; larger spaces tend to have reverberation with more low frequencies than smaller spaces, and the further away the rear wall, the longer the path of reflection, bringing reverberation to have the potential to be less dense than smaller spaces.

3. Generalize the rear wall of the source's host environment in this way, and place it on the sound stage by extending the back of the image to the location of the rear wall you are conceptualizing.

   This placement of the rear wall will not be precise; it is a general observation even if included in the scaled sound stage.

4. Next seek another source that contains an apparent sense of depth, and repeat this process. As you go through several additional sources, it will become apparent that some sources extend further than others; you may also notice that the density of reverberation may at times play a role in depth.

B.

1. Return to one of your stereo location graphs or sound stage diagrams, and listen to the sources for width of images; search for an image that is wider than the width of the source itself, that has been broadened by its environment. It is not common to have an environment clearly wider than the source itself, but when this occurs it is often a significant feature of the track.

   Search the tracks you have already examined for stereo location for such a sound source; if you do not find one, listen to tracks that emphasize acoustic sound sources to continue your search. This image containing separated room and source sounds can be a sense that the source is within a room and the room extends around the source; the added width can create a sense of side walls or of a larger space within which the source is contained.

2. Once a source is identified, create a stereo location graph or a sound stage diagram. You may sketch the source on either.

3. First, locate the outer boundaries of the entire image—which will be the edges of the environment—using the process from stereo location, Praxis Study 8.1 and 8.2. Mark those edges on the graph or diagram; these can be precise locations or more general placements.

4. Next bring your attention to the direct sound of the source, or the core image. Recognize the different attributes of the source and of the reverberation, as developed in the previous several studies. Locate the position where these changes occur—these positions will be the image edges of the instrument or voice *within* the environment. If these edges are not clearly apparent, the source and space are bonded well, and the source is not displaying this characteristic.

PRAXIS STUDY 9.1 RECOGNIZING REFERENCE DYNAMIC LEVEL

Goals and Desired Outcomes

The goal of this study is to be able to recognize the reference dynamic level of a track.

After working through this exercise, the reader/listener will have some clarity of the experience of the RDL and also how to recognize it. Ultimately the reader will arrive at a well define dynamic level that embodies the RDL.

Preparation and Background

Select a recording you know well for your initial attempts at determining the reference dynamic level of a piece of music. It would be best for the track to have a duration of approximately three minutes or less.

This performance intensity of the track is its reference dynamic level (RDL). The goal, here, is to experience the RDL as a singular, overall sense of energy, tension, motion, expression, exertion and intensity of the track. Perceived holistically, considered without interference from verbalization, as a single level of intensity that embodies the track.

Prepare to direct attention toward nonverbal reflection and a sense of inherent recognition, one that is based on *musical thought* and aural imagery and aesthetic expression. This contrasts markedly from the natural tendency of many analysts to engage the rational *thinking mind* of processing, calculating, problem solving, and to otherwise search out a universal formula for this percept—a process that quickly and prematurely brings verbalization, and does not access the inherent character of and the intensity of the RDL.

Process and Sequence

1. Before listening to the track, spend some time thinking about the overall character of the piece; consider its overall energy level as a performance intensity, consider concept or message, and other important aspects of the track as a presence of exertion, energy and emotion.

    Do not attempt to verbalize these; rather apply a general dynamic marking (it will be refined or redefined later) to these thoughts. Recall the threshold between *mp* and *mf* marks the transition between withholding and expending energy.

    It is important this be a reduction of the track to a single impression or object, void of the passage of time.
2. Next, listen to the track several times. Listen as if it is a new experience, and a singular expression; seek to listen to the overall expression, not the subtle details of passing moments.

    At the end of several hearings, let the last one fade. In the silence after the track, allow the listening experience of the track to dissipate into a single awareness and reflect on the impression that remains. Reflect on this presence of the track that lingers in your memory and sensibilities, in your nonverbal consciousness and awareness. The goal is to not 'make sense of it' but rather to open to recognize it for its level of energy and expression, intensity and exertion.

3. Seek to reconsider the observations you made from memory. Listen again to the track, and focus on its overall expression and impression. Reconsider your idea of the RDL with each new hearing of the recording.

   Within this impression and conception is a sense of the amount of energy and the level of exertion of the performance, and the speed of tensional motion and the magnitude of intensity of its expression, supplemented with the drama and meanings of the lyrics and the spatial attributes and other characteristics of the recording. These all coalesce into an impression of the performance intensity of the track; how this happens and the proportions of these factors and attributes is unique to each track—it is part of the mystery of how tracks (and music and lyrics) work.

4. Gradually focus in to attempt to define a precise dynamic level for the RDL. Begin this process by working from the extreme levels—*ppp* and *fff*—asking if the level exists in those areas. Eliminate dynamic areas where the RDL is obviously not present. This will focus your efforts.

5. Once the dynamic area has been determined, work to define a precise level by asking if the RDL is below 50 percent in the level, or above. Continue to work toward a specific level by narrowing the area further.

6. Leave the example and your answer for a period of time (several hours or several days). Listen to the song again. Reconsider the RDL previously defined.

For tracks one does not know well, many hearings may be required before initial observations can be made. A deep sense of knowing brings access to the impressions that allow the RDL to be recognized.

PRAXIS STUDY 9.2 HEARING OVERALL LOUDNESS AND CREATING THE TRACK LOUDNESS CONTOUR GRAPH

Goals and Desired Outcomes

The goal of this study is to bring the listener to experience and recognize the sensation of the overall loudness of the track, and to trace that loudness throughout the duration of the track.

After working through this exercise, the reader/listener will be better able to create the track loudness contour graph, and to accurately perceive the overall loudness of the track as it changes over time, and as it relates to the track's RDL.

Preparation and Background

The entire track will be notated for its overall loudness contour on a "track loudness contour graph," as in Figure 9.2. Attention must be deliberately directed to the overall dynamic/loudness level, and remain at that level of perspective throughout the entire record; the single sensation of loudness (that is established by the combination of all sounds) is the focus of this exercise.

Select a track with a duration of no more than three minutes; it may be the same track you used for the previous RDL study or any other track you already know well.

To aid in developing this skill, initial attempts at this exercise should use a track with distinct and sudden changes of dynamic level. Afterward, repeat the exercise using tracks containing mostly changes of loudness levels that are smaller (or subtler) and more gradual. Seek to determine the general shape of the loudness contour first, then work to grasp the subtlest details.

Process and Sequence

1. During the first hearing(s), listen to the example to establish the length of the timeline. At the same time, become acquainted with the character of the track to begin formulating an idea of its RDL.
2. Check the timeline for accuracy and identify structural divisions.

   Begin to establish the RDL by working through the previous exercise.
3. Notice the activity of the program dynamic contour for boundaries of loudness level and of the speed of changing levels.

   The boundary of speed will establish the smallest time unit required to accurately plot the smallest significant change of loudness.

   The boundary of levels of loudness will establish the upper and lower thresholds of the vertical axis. Next, determine the smallest increment of the Y-axis required to plot the smallest change of the dynamic contour.
4. Draw the axes of the graph using the information from 3.

   Begin plotting the dynamic contour on the graph. Determine several loudness levels (at specific locations) to use as references and to provide some guidance as your observations progress. Some that can often be effective are: the starting and ending levels of the track, starting and ending levels of major sections, the loudest and softest levels, dramatic peaks and lulls, and levels before and after silences; each track may have its own unique moments when overall loudness level is important and noticeable.
5. Relate the perceived loudness levels to the RDL; shift the contour levels as needed.

   Begin to fill in the contour between references, and add more references as significant observations unfold. Focus on the contour, speed, and amount of loudness level changes in order to complete the plotting of the contour.

PRAXIS STUDY 9.3 PERCEIVING LOUDNESS BALANCE: THE CONTRASTING PERCEPTS OF LOUDNESS RELATIONSHIPS AND THE LEVELS AND CONTOURS OF INDIVIDUAL SOURCES; RELATING SOURCE LEVELS TO THE RDL AND FORMATTING LOUDNESS BALANCE AGAINST A TIMELINE.

Goals and Desired Outcomes

The goals of this study are to bring the reader/listener (1) to perceive and observe the loudness contours of individual sources and the loudness level relationships between sources, (2) to relate those levels and relationships to the RDL and (3) to craft a loudness balance graph. The goal is to more accurately hear loudness, and to not confuse loudness with prominence or any other percept.

After working through this exercise, the listener/analyst will be better able to make accurate observations of loudness balance and loudness levels of sources as they relate to the track's RDL, and ultimately be able to construct a loudness balance graph at any desired level of detail.

Preparation and Background

To allow this study to take less preparation time and to perhaps be more meaningful, you may wish to use a track from Praxis Study 9.1or 9.2, where you have already determined the RDL. A track with fewer sources will be easiest to observe during your first attempts at this study; tracks with more sources can be engaged after some facility has been acquired.

Create a timeline of the introduction through the first verse and the first chorus; add a Y-axis of dynamic areas, using 5 squares per area. The graph should appear like that of Figure 9.3.

Process and Sequence

1. During the first hearing(s), establish the length and structural divisions of the timeline.
    Next, get a sense of the loudest and softest general dynamic levels of the loudest and softest sources.
2. Establish a complete list of sound sources (instruments and voices) or a list of the sources of interest for the analysis; sketch the presence of the sound sources against the completed timeline. A key may now be created that assigns each sound source a distinguishing number.
3. Determine the reference dynamic level of the track (using the above process study), and add it to the graph.
4. Observe the activity of the loudness levels of the sound sources (instruments and voices) for highest and lowest boundary levels and for fastest speed of change. The boundary of speed will establish the smallest time unit required to accurately chart the smallest significant change of loudness level. The boundary of levels will establish the smallest Y-axis increment required to clearly illustrate the smallest change of dynamics.
    Recopy the graph to adjust to these boundaries.
5. Begin plotting the loudness contours of each instrument and voice on the graph. Start with a prominent sound and work toward any sounds that blend into the texture.
    Keeping the RDL clearly in mind, establish the beginning dynamic levels of each sound source. Next, determine other prominent points of reference. Use the points of reference to judge the activity of the preceding and following material. Focus attention toward contours, speed of changes, and amounts of level changes to complete the plotting of the dynamic contours.
6. You should periodically shift your focus and level of perspective to compare the dynamic levels of each sound source to others. This shift must be conscious and intentional. This will aid in observing the loudness contours and will keep you focused on the relationships of loudness levels of the various sources.
    The evaluation is complete when the smallest significant detail necessary to meet the goals of your analysis have been incorporated into the graph. Note that Figure 9.4 contains significantly less detail than Figure 9.3, yet it is adequate to illustrate the points under discussion.

It is important to remain focused on the actual loudness of instruments, making certain your attention is not drawn to other aspects of sound.

As you gain experience and confidence in making these evaluations, tracks with more instruments should be examined and longer sections of the works should be evaluated. You will begin to notice subtle changes as striking differences as your skill level improves.

PRAXIS STUDY 9.4 OBSERVING AND NOTATING PERFORMANCE INTENSITY; GRAPHING PERFORMANCE INTENSITY AND LOUDNESS BALANCE AS TWO TIERS OF AN X-Y GRAPH

Goals and Desired Outcomes

The goals of this study are to bring the listener (1) to observe the performance intensity levels and contours of individual sources and (2) to relate those levels to the loudness levels of sources.

With the support of this study, one will be better able to observe and understand the timbral states and changes brought about by the expression, energy and exertion of performance and recognize their relationship to the actual loudness levels of the sources as they appear within the track.

Preparation and Background

To assist your first execution of this study, return to a loudness balance graph you completed earlier. A tier also containing the performance intensity of sources will be added above what was previously completed; the graph will be formatted as the one in Figure 9.5.

Should it best serve an analysis, the performance intensity of sources can be plotted without reference to loudness balance. Observations and data collection can flex and conform to the objectives of an analysis; all sources may be examined, though alternatively only those sources that are pertinent to the analysis might be included.

The perceived level of exertion or expending of energy of the performance—in its potential to shift at any moment in time—is graphed for sound sources. This level is reflected in the timbre that was present at the time the sound source was recorded.

Process and Sequence

1. The following sequence assumes a loudness balance graph has been completed, as in the previous study. This will generate the graph's timeline; should loudness balance not be pursued, a timeline should be determined at this stage.
2. Focusing on one source at a time, performance intensity is next observed for each selected sound source; the resulting contours are added to the performance intensity graph or tier of the combined graph.
3. Begin listening to the general performance intensity of each sound source in order to acquaint yourself with them. Then, establish the beginning performance intensity levels of each sound source.
4. Bring your attention next to a select source. Observe its intensity (timbre) within its first line; notice if the exertion or energy is uniform or if there are point of increased or decreased exertion or energy. Now, transfer these impressions of exertion/intensity into dynamic levels (as described in Chapter 9).

To continue, first notice the overall intensity level of its first phrase and sketch that contour, then move to the next phrase, and so forth. After sketching the general contour and levels, you may choose to move to another source or return to the beginning and listen more deeply, adding detail and nuance of expression and intensity to what you previously observed. The level of detail is (again) a variable you must determine based on need for the analysis. In initial studies, though, observing these subtle cues will be an experience worthy of your effort, as timbral qualities of performance and expression will emerge with richness—perhaps more than you have previously experienced.

Your attention should shift intentionally to contours, then to speed of changes, then to amounts of intensity changes, and finally to how these relate to time and rhythm of the line. This sequence might take any order—the point is to remain on task, and to seek specific information in these steps. If one wishes, the purpose of a hearing might be to randomly explore the sound to discover qualities; that can be valuable, especially if it is followed by a focused listening to what has been revealed.

5. Repeat the previous step for each sound source you wish to observe.
6. Once you have graphed all sources you wish to observe, you have the data needed to compare the performance intensity and loudness balance tiers. In doing so you might learn significant information about how the voices and instruments were shaped by the record, and how the sources contribute to the track—this was illustrated in Chapter 10.

As you gain experience in making these evaluations, all of the sound sources of tracks with many instruments could be examined for longer sections; that data often reveals significant dimensions of a track.

PRAXIS STUDY 9.5 DESCRIBING TIMBRE BY OBSERVING AND EVALUATING TIMBRAL CONTENT, CHARACTER AND CONTEXT: GENERATING A DESCRIPTION OF A TIMBRE

Goals and Desired Outcomes

The goal of this study is to bring the listener to observe and then describe a timbre—attributes of a timbre that provide the source with its unique, inherent qualities.

With the support of this study, the reader/listener will be better able to articulate the attributes and associated values within the content, character and context of a timbre.

Preparation and Background

These timbre attributes are significant, prominent, substantive or in some other way embody the core attributes that provide the timbre with distinctive traits, and also elements that interact with those attributes and in some way are supportive of central qualities, provide important elaboration, intrinsic ornamentation to the timbre's core content or character, or that represent its context.

This study should first focus on a single sound, one isolated timbre performed by a specific source (instrument or voice). As fluidity is gained, one might observe an instrument or voice performing throughout its range, or throughout a track to identify traits that bring it to be unique.

For now, identify an isolated sound of a pitched acoustic instrument from a track, or you may record a sound yourself; select a sound with a duration of at least three seconds (up to five seconds), with a sustain of at least two seconds (the longer the duration the easier it will be to execute this study at first).

Import the sound into a DAW or other device that will allow you to readily loop or replay the sound with few distractions of the logistics of locating the sound. This sound could be the same sound you used for the timbral content study 7.9.

Create a timbre typology table using Table 9.5 as a point of departure. The table required to observe the traits of your timbre may differ from this; replace some of the attributes in Table 9.5 with others as needed. Many alternative attributes are listed in Table 9.4 (though this is not a definitive list).

Process and Sequence

1. Observe the physical content of the timbre; note or notate the onset traits, dynamic contour (sustain and decay), spectral content, and spectral envelope attributes. Start with the most prominent qualities and add detail as you are able.

   If comparing a number of timbres from the same source, makes observations to identify formant regions or traits of dynamic contour (including the onset) that are consistent and distinguishing.

   Note the traits of the content attributes on the typology table. Later you may choose to graph them for more detail. Beginning with a more general approach will speed this process and may provide all the detail needed for a cursory analysis.

2. Identify the context of the timbre: its source (generic type or a specific instrument/voice), associations, meanings, its conformity to the texture, etc.

   Note the traits of the contextual attributes on the typology table.

3. Identify the character attributes that are prominent; incorporate them into the typology table; observe the traits within those attributes and add them to the table. Listen more deeply to the timbre for its expression, energy, intensity, and its mood or emotion. Observe whether or not a defining trait arises—what it is and the magnitude of its influence.

   Compile the attributes and their traits on the typology table.

   A substantial level of detail is not required for this table to be of value.

4. Evaluate the prominent attributes of all three categories (content, context, character) in the table; note those that dominate and those that support others, how attributes interact, align and provide conflicting tendencies. Identify substantive and ornamental attributes, and their level of influence in the timbre.

   The more detail that is present in 1, 2 and 3, the more the description might represent the unique traits of the timbre.

This may be a lengthy process—especially during the first few times one undertakes the study. With practice, one is able to bring attention to what is prominent in the timbre, with greater ease and accuracy—and ultimately, significant detail and nuance is revealed as well.

PRAXIS STUDY 9.6 LISTENING WITH OPEN AWARENESS, AND REMAINING WITHIN THE WINDOW OF THE PRESENT

Goals and Desired Outcomes

The goal of this exercise is to develop the skill of non-directed deep listening. Non-directed deep listening holds a position of open awareness within which one's attention remains within the present moment and is not distracted.

With this skill the listener/analyst will possess the discipline and concentration to practice the open listening technique that allows sounds to arrive and leave without consciously imposed associations, meanings or functions.

Preparation and Background

When we hear sounds we are prone to naming them (a bird, door slam, airplane), perhaps we become aware of the feeling that accompanies the sound (laughter, pleasant; pneumatic drill, unpleasant). We anticipate what might follow the door slam, and project it may be unpleasant. In context of the track we name instruments or materials, have a sense of our like or dislike of it, have a sense of what we expect to follow, or perhaps what we want to hear next; we get a sense of motion and tension, and other conventions. In all these instances, the listening process is reaching out into the sounds arriving and engaging them.

Non-directed deep listening seeks to suspend all of this. In open listening we wait for the sounds to arrive to us, in their own time, and with their unique abstract qualities. We seek to listen as if the past and the future are irrelevant to what is happening at this moment; neither memories of prior events nor anticipation of what may follow exist within this open window of the present. We will not imagine, name, or analyze sounds. We will suspend our personal biases and prejudices, and our sense of the track as musically (or otherwise) unfolding over time—all sounds will arise and pass away in their own way.

Staying in the present requires concentration, attention and discipline so as not to be distracted. To stay in the present moment, thoughts cannot intrude on the listening experience—whatever these thoughts might be.

This study may be performed (1) listening to a record or (2) by listening to the sounds of real-life occurrences around you. It will prove helpful to perform this exercise several times, in each of these approaches.

(1) Identify a track with slowly paced sounds and some silences interspersed (this will allow your attention to reset from time to time, which is helpful when learning this technique) and (2) find a quiet space for listening to the sounds of the world around you, or do this at a quieter time of day. As you become more experienced you can repeat the study in situations that are busier—whether listening to a track, or listening to the world around you.

Process and Sequence

1. Minimize distraction.

   As with all other listening experiences, closed eyes heighten the sense of sound. The distractions of sight are eliminated. Sitting comfortably and still, in a posture that is relaxed and alert, you will be prepared to listen without physical distractions.

2. Begin to listen—to either a track, or to the world around you. Just listen, avoid language or imagining visuals.

   Do not scan for sounds; wait for them.

   Cultivate a position of awareness in allowing sound to reach your position. Release a sense of reaching out to listen. Envision sound arriving at your position, as opposed to actively seeking it out.

   Maintain awareness, while openly being prepared for the sound to find you.

3. When a sound arrives resist the impulse to identify it or to judge it; observe its unfolding.

   Allow it to arise and to pass without naming it, without analyzing its content or functions, without engaging any sense of feelings, and without sensing what is to follow. Let the sound be the sound; do not influence (distort) it with your thoughts. Resist the impulse to avoid listening fully to sounds you (personally) find unpleasant; accept them into your full attention without judgement.

4. Hold the present clearly within your listening process.

    Clear thoughts of sounds from your awareness as soon as they have passed. Be fully aware and present with what you are hearing at each moment—do not hold on to sounds (the memory of a sound) after it has stopped or you will not be aware of the present.

5. Experience the succession of sounds that each have their own place in time, within your awareness that is progressing from one moment to the next moment seamlessly.

    Any sense of a seam within this ongoing string of passing moments is likely to be a lapse of holding sounds in memory, wondering (or anticipating) what will follow, or from distractions.

    Beware of engaging rhythm; rhythm requires memory of durations and patterns of durations. Memory holds what has passed, not what is present.

6. Cultivate concentration, attention and remaining aware of listening. Remain aware of listening to what is happening now, at this moment.

    Work to stay attentive. This is not at all easy. Distractions have countless sources, and a mere moment of distraction brings another moment you were not present while listening.

7. Do not be discouraged, though. When you recognize you have slipped out of awareness of listening, recognize the accomplishment of noticing.

    Maintaining awareness of the present opens the listening experience to receive all that arrives with equal attention.

8. Repeat this entire process with the alternate experience—listen to the world around you, if your first exercise listened to a track, or listen to a track if your first experience did not.

9. One more step can be of assistance. Listen to a well-known track several times in succession; each time listen as if you have never heard the track before. Following this approach, it is common for something to emerge that has not been previously experienced.

PRAXIS STUDY 9.7 CULTIVATING AWARENESS OF BEING WITHIN YOUR SUBJECTIVE VANTAGE AND OF LISTENING FROM A MORE OBJECTIVE POSITION; SHIFTING BETWEEN THESE POSITIONS WITH INTENTION AND WITH ATTENTION TO WHAT EACH OFFERS

Goals and Desired Outcomes

The goal of this study is to (1) cultivate the ability to be present within your subjective vantage, (2) to learn your personal biases and preferences (and aversions) and (3) to step out of these into a more neutral, objective listening vantage.

With this skill, the listener will be able to observe tracks—elements and domains and/or their confluence—and to evaluate them to understand what about them speaks to one personally. With intention, one can then switch to the more objective, neutral vantage that is more suitable for academic discourse, and that has commonality with the audience you wish to engage.

Preparation and Background

For this study it is helpful to revisit Praxis Study 5.6.A and B. There we worked with approaching observations from a neutral unbiased position related to the music and lyrics domains and elements. Here we may expand our awareness to all that is within the track, and also to the overall qualities of the track.

Select two tracks for this study: one you know well and one track that is new to you. Listen to each of these tracks to perform study 5.6.A on a few elements of the music and recording domains. This will remind you of the position of neutral observation for unbiased data collection, and will prepare for the process and sequence that follows.

Process and Sequence

1. Returning to study 5.6.B, review the outline you created of your biases, preferences, knowledge and skill levels. As time has now passed, consider revising the outline.

2. Listen to the track you know well, and allow yourself to be immersed in what grabs your attention, what speaks to you and interests you.

   Notice where your attention wishes to go, and things in the track that are unappealing to you — sounds, materials, affects, whatever.

3. Listen again, several times. Reflect on each experience. Consider: what do you remember? What did you find fascinating? What was meaningful or of interest to you? What traits and what materials emerged?

   Allow yourself to experience and settle into what it is to be listening from your personal, subjective vantage — honoring your tastes, tendencies and preferences. Inside this subjective vantage you can learn a great deal about yourself and your music listening tendencies.

4. Repeat 2 and 3 for the track that is new to you. Notice similarities in what emerges, and ask if they are track-specific or related to personal preferences. Note differences between the two tracks, and explore which are track-specific and which represent your subjective vantage.

5. Listen to the track you know well again, this time seeking a position that is more unbiased, objective, relevant for academic discourse. Settle into this frame of mind, into this objective and detached listening perspective.

6. Listen to the track again, this time listening from your subjective vantage. Notice the difference in this listening approach compared to the unbiased approach of 5. Consider the differences between the experiences, and consider the different overall quality of each experience.

7. Next, listen to the track again. Bring your attention to focus on one specific aspect — the lead vocal performance, for example.

   This time practice shifting between the two approaches while listening to the track. Listen objectively from the beginning through the first chorus; listen from your subjective vantage to the next verse and chorus; then return to listening objectively for the remainder of the track.

8. Repeat steps 5, 6, and 7 with any other track — perhaps one that you do not know well. The goal here is to experience the shift between your subjective vantage and your analytic position, and then acquire the ability to make this shift with intention at the time you wish it to happen.

9. Repeat steps 5, 6, and 7 with a track that you are particularly fond of or that holds meaning to you. This is a test of how well you can shift into the objective position.

Bibliography

Altman, Rick. 1992. "Introduction: Four and a Half Fallacies." In *Sound Theory, Sound Practice*, edited by Rick Altman, 35–45. New York: Routledge.

American National Standards Institute (ANSI). 1994. "American National Standard Acoustical Terminology." American National Standards Institute, ANSI S1.1–1994 (R1999).

Ashby, Patricia. 2011. *Understanding Phonetics*. New York: Routledge.

Baars, Bernard. 1993. *A Cognitive Theory of Consciousness*. New York: Cambridge University Press.

Baars, Bernard, and Stan Franklin. 2003. "How conscious experience and working memory interact." *Trends in Cognitive Sciences*, vol. 7, no. 4, 166–172.

Baars, Bernard, and Nicole M. Gage. 2013. *Fundamentals of Cognitive Neuroscience: A Beginner's Guide*. Boston: Elsevier.

Bartlett, Bruce and Jenny Bartlett. 2017. *Practical Recording Techniques*, 7th edition. New York: Routledge.

Bayley, Amanda. 2010. "Introduction." In *Recorded Music: Performance, Culture and Technology*, edited by Amanda Bayley, 1–11. Cambridge: Cambridge University Press.

The Beatles. 2000. *The Beatles Anthology*. San Francisco: Chronicle.

Bent, Ian. 1987. *Analysis*. New York: W. W. Norton.

Berendt, Joachim-Ernst. 1992. *The Third Ear: On Listening to the World*, translated by Tim Nevill. New York: Henry Holt.

Bjerke, Kristoffer Yddal. 2010. "Timbral Relationships and Microrhythmic Tension: Shaping the Groove Experience Through Sound." In *Musical Rhythm in the Age of Digital Reproduction*, edited by Anne Danielsen, 85–101. New York: Routledge.

Björnberg, Alf. 2007. "On Aeolian Counterpoint in Contemporary Popular Music." In *Critical Essays in Popular Musicology*, edited by Allan F. Moore, 275–282. Aldershot: Ashgate.

Blauert, Jens. 1983. *Spatial Hearing: The Psychophysics of Human Sound Localization*, translated by John Allen. Cambridge, MA: The MIT Press.

Blaukopf, Kurt. 1971. "Space in Electronic Music." *Music and Technology, Stockholm Meeting, June 8–12, 1970*, 157–172. New York: Unipub.

Bourbon, Andrew and Simon Zagorski-Thomas. 2017. "The Ecological Approach to Mixing Audio: Agency, Activity and Environment in the Process of Audio Staging." *Journal on the Art of Record Production*, Issue 11, March 2017. http://arpjournal.com/the-ecological-approach-to-mixing-audio-agency-activity-and-environment-in-the-process-of-audio-staging/. Accessed February 28, 2018.

Bowman, Rob. 2003. "Determining the Role of Performance in the Articulation of Meaning: The Case of 'Try a Little Tenderness.'" In *Analyzing Popular Music*, edited by Allan Moore, 103–130. Cambridge: Cambridge University Press.

Brackett, David. 2000. *Interpreting Popular Music*. Berkeley: University of California Press.

Bradby, Barbara and Torode, Brian. 2000. "Pity Peggy Sue." In *Reading Pop: Approaches to Textual Analysis in Popular Music*, edited by Richard Middleton, 203–227. Oxford: Oxford University Press.

Bregman, Albert. 1990. *Auditory Scene Analysis: The Perceptual Organization of Sound*. Cambridge, MA: MIT Press.

Brøvig-Hanssen, Ragnhild. 2010. "Opaque Mediation: The Cut-and-Paste Groove in DJ Food's 'Break.'" In *Musical Rhythm in the Age of Digital Reproduction*, edited by Anne Danielsen, 159–175. New York: Routledge.

———. 2019. "Listening To or Through Technology: Opaque and Transparent Mediation." In *Critical Approaches to the Production of Music and Sound*, edited by Samantha Bennett and Eliot Bates, 195–210. New York: Bloomsbury.

Brøvig-Hanssen, Ragnhild and Anne Danielsen. 2013. "The Naturalised and the Surreal: Changes in the Perception of Popular Music Sound." *Organized Sound*, vol. 18, no. 1, pp. 71–80.

Brøvig-Hanssen, Ragnhild and Anne Danielsen. 2016. *Digital Signatures: The Impact of Digitalization on Popular Music Sound*. Cambridge: The MIT Press.

Brøvig-Hanssen, Ragnhild and Anne Danielsen. 2017. "Music Production: Recording Technologies and Acousmatic Listening." In *The Routledge Companion to Music Cognition*, edited by Richard Ashley and Renee Timmers, 191–201. New York: Routledge.

Bull, Michael, editor. 2013. *Sound Studies*. New York: Routledge.

Burnham, Scott. 2001. "How Music Matters: Poetic Content Revisited." In *Rethinking Music*, edited by Nicholas Cook and Mark Everist, 193–216. Oxford: Oxford University Press.

Burns, Lori. 1997. "'Joanie' Get Angry: k.d. lang's Feminist Revision." *Understanding Rock: Essays in Musical Analysis*, edited by John Covach and Graeme Boone, 93–112. Oxford: Oxford University Press.

——. 2005. "Meaning in a Popular Song: The Representation of Masochistic Desire in Sarah McLachlan's 'Ice.'" In *Engaging Music: Essays in Music Analysis*, edited by Deborah Stein, 136–148. Oxford: Oxford University Press.

——. 2008. "Analytic Methodologies for Rock Music." In *Expressions in Pop-Rock Music: Critical and Analytical Essays*, 2nd edition, edited by Walter Everett, 63–92. New York: Routledge.

——. 2010. "Vocal Authority and Listener Engagement: Musical and Narrative Expressive Strategies in the Songs of Female Pop-Rock Artists, 1993–95." In *Sounding Out Pop: Analytical Essays in Popular Music*, edited by Mark Spicer and John Covach, 154–192. Ann Arbor: The University of Michigan Press.

——. 2019. "Interpreting Transmedia and Multimodal Narratives: Seven Wilson's "The Raven That Refused to Sing."" In *The Routledge Companion to Popular Music Analysis: Expanding Approaches*, edited by Ciro Scotto, Kenneth Smith, and John Brackett, 95–113. New York: Routledge.

Burns, Lori, Mark Lafrance, and Laura Hawley. 2008. "Embodied Subjectivities in the Lyrical and Musical Expression of PJ Harvey and Björk." *Music Theory Online*, Vol 14, no. 4. http://mtosmt.org/classic/mto.08.14.4/mto.08.14.4.burns_lafrance_hawley.html. Accessed June 6, 2012.

Buser, Pierre, and Michel Imbert. 1992. *Audition*. Cambridge: MIT Press.

Butler, David. 1992. *The Musician's Guide to Perception and Cognition*. New York: Schirmer.

Byrne, David. 2012. *How Music Works*. San Francisco: McSweeney's.

Cage, John. 1961. *Silence*. Middletown, CT: Wesleyan University Press.

Camilleri, Lelio. 2010. "Shaping Sounds, Shaping Spaces." *Popular Music*, vol. 29, no. 2, 199–211.

Chanan, Michael. 1995. *Repeated Takes: A Short History of Recording and its Effects on Music*. London: Verso.

Chion, Michel. 1994. *Audiovision: Sound on Screen*. New York: Columbia University Press.

——. 2012. "The Three Modes of Listening." In *The Sound Studies Reader*, edited by Jonathan Stern, 48–53. New York: Routledge.

Cirque Apple Creation Partnership, The. 2016. *Creating "While My Guitar Gently Weeps" (LOVE Version)*. https://thebeatles.com/news/watch-evocative-new-music-video-while-my-guitar-gently-weeps-love-version. Accessed April 12, 2019.

Chomsky, Noam and Morris Halle. 1968. *The Sound Pattern of English*. New York: Harper & Row.

Clarke, Eric. 2005. *Ways of Listening: An Ecological Approach to the Perception of Musical Meaning*. Oxford: Oxford University Press.

——. 2011. "Music Perception and Musical Consciousness." In *Music and Consciousness: Philosophical, Psychological, and Cultural Perspectives*, edited by David Clarke and Eric Clarke, 193–214. Oxford: Oxford University Press.

Clifton, Thomas. 1983. *Music as Heard: A Study in Applied Phenomenology*. New Haven: Yale University Press.

Cogan, Robert. 1984. *New Images of Musical Sound*. Cambridge, MA: Harvard University Press.

Cogan, Robert and Pozzi Escot. 1984. *Sonic Design: The Nature of Sound and Music*. Cambridge, MA: Publication Contact International.

Cone, Edward. 1989. "Analysis Today." In *Music: A View from Delft*, edited by Robert Morgan, 39–54. Chicago: University of Chicago Press.

Cook, Nicholas. 1987. *A Guide to Musical Analysis*. New York: W. W. Norton.

Cook, Nicholas. 2009. "Methods of Analysing Recordings." In *The Cambridge Companion to Recorded Music*, edited by Nicholas Cook, Eric Clarke, Daniel Leech-Wilkinson and John Rink, 221–245. Cambridge: Cambridge University Press.

Cook, Nicholas and Daniel Leech-Wilkinson. 2009. "A musicologist's guide to Sonic Visualiser." http://charm.rhul.ac.uk/analysing/p9_1.html. Accessed December 9, 2017.

Cooper, Grosvenor and Leonard Meyer. 1960. *The Rhythmic Structure of Music*. Chicago: University of Chicago Press.

Covach, John. 1997. "We Won't Get Fooled Again: Rock Music and Musical Analysis." In *Keeping Score: Music, Disciplinarity, Culture*, edited by David Schwartz, Anahid Kassabian, and Lawrence Siegel, 75–89. Charlottesville: University of Virginia Press.

——. 2001. "Popular Music, Unpopular Musicology." In *Rethinking Music*, edited by Nicholas Cook and Mark Everist, 452–470. Oxford: Oxford University Press.

——. 2018. "Analyzing Texture in Rock Music: Stratification, Coordination, Position, and Perspective." *Pop weiter denken: Neue Anstöße aus Jazz Studies, Philosophie, Musiktheorie und Geschichte, Beiträge zur Populärmusikforschung 44*, edited by Ralf von Appen and André Doehring, 53–72. Bielefeld: Transcript Verlag.

——. 2019. "Yes, the Psychedelic Cover, and 'Every Little Thing.'" In *The Routledge Companion to Popular Music Analysis: Expanding Approaches*, edited by Ciro Scotto, Kenneth Smith, and John Brackett, 277–290. New York: Routledge.

Covach, John and Graeme Boone, editors. 1997. *Understanding Rock: Essays in Musical Analysis*. Oxford: Oxford University Press.

Covach, John and Andrew Flory. 2015. *What's That Sound? An Introduction to Rock and Its History*, 4th edition. New York: W. W. Norton.

Cox, Christoph and Daniel Warner. 2004. "Glossary." In *Audio Culture: Readings in Modern Music*, edited by Christoph Cox and Daniel Warner, 413. London: Continuum.

Dack, John. 2010. "From Sound to Music, From Recording to Theory." In *Recorded Music: Performance, Culture and Technology*, edited by Amanda Bayley, 271–290. Cambridge: Cambridge University Press.

Damasio, Antonio. 2003. *Looking for Spinoza: Joy, Sorrow and the Feeling Brain*. Orlando: Harcourt.

Danielsen, Anne. 1998. "His Name Was Prince: A Study of *Diamonds and Pearls*." *Popular Music*, vol. 16, no. 3, 275–291.

——. 2006. *Presence and Pleasure: The Funk Grooves of James Brown and Parliament*. Middletown, CT: Wesleyan University Press.

——. 2010a. "Introduction: Rhythm in the Age of Digital Reproduction." In *Musical Rhythm in the Age of Digital Reproduction*, edited by Anne Danielsen, 1–16. New York: Routledge.

——. 2010b. "Here, There and Everywhere: Three Accounts of Pulse in D'Angelo's 'Left and Right.'" In *Musical Rhythm in the Age of Digital Reproduction*, edited by Anne Danielsen, 19–36. New York: Routledge.

——. 2012. "The Sound of Crossover: Microrhythm and Sonic Pleasure in Michael Jackson's 'Don't Stop Till You Get Enough'." *Popular Music and Society*, vol. 35, no. 2, 151–168.

——. 2015. "Metrical Ambiguity or Microrhythmic Flexibility? Analysing groove in 'Nasty Girl' by Destiny's Child." In *Song Interpretation in 21ˢᵗ Century Pop Music*, edited by Ralf von Appen, André Doehring, Dietrich Helms, and Allan F. Moore, 53–71. Surrey: Ashgate.

——. 2019. "Pulse as Dynamic Attending: Analysing Beat Bin Metre in Neo Soul Grooves." In *The Routledge Companion to Popular Music Analysis: Expanding Approaches*, edited by Ciro Scotto, Kenneth Smith, and John Brackett, 179–189. New York: Routledge.

Day, Aidan. 1988. *Jokerman*. Oxford: Blackwell.

Dennett, Daniel. 1991. *Consciousness Explained*. London: Penguin.

Deutsch, Diana. 2012a. "Grouping Mechanisms in Music." In *The Psychology of Music*, 3rd edition, edited by D. Deutsch, 183–247. Boston: Academic Press.

Deutsch, Diana. 2012b. *The Psychology of Music*, 3rd edition. Boston: Academic Press.

Deutsch, Diana, Trevor Henthorn and Mark Dolson. 2004. "Absolute Pitch, Speech and Tone Language: Some Experiments and a Proposed Framework." *Music Perception*, 21, 339–356.

Deutsch, Diana, Trevor Henthorn, Elizabeth Marvin, and HongShuai Xu. 2006. "Absolute Pitch Among American and Chinese Conservatory Students: Prevalence Differences, and Evidence of Speech-Related Critical Period." *Journal of the Acoustical Society of America*, vol. 119, no. 2, 719–722.

Dockwray, Ruth. 2017. "Proxemic Interaction in Popular Music Recordings." In *Perspectives On Music Production: Mixing Music*, edited by Russ Hepworth-Sawyer and Jay Hodgson, 53–61. New York: Routledge.

Dockwray, Ruth and Allan Moore. 2010. "Configuring the Sound-Box 1965–1972." *Popular Music*. vol. 29, no. 2, 181–197.

Doll, Christopher. 2019. "Some Practical Issues in the Aesthetic Analysis of Popular Music." In *The Routledge Companion to Popular Music Analysis: Expanding Approaches*, edited by Ciro Scotto, Kenneth Smith, and John Brackett, 3–14. New York: Routledge.

Dowling, W. Jay, and Dane Harwood. 1986. *Music Cognition*. New York: Academic Press.

Doyle, Peter. 2005. *Echo & Reverb: Fabricating Space in Popular Music Recording 1900–1960*. Middletown, CT: Wesleyan University Press.

Durant, Alan. 1984. *Conditions of Music*. London: Macmillan.

Dylan, Bob. 2016. *Bob Dylan: The Lyrics 1961–2012*. New York: Simon & Schuster.

Edelman, Gerald M. 1992. *Bright Air, Brilliant Fire: On the Matter of Mind*. New York: Basic.

——. 2006. *Second Nature: Brain Science and Human Knowledge*. New Haven: Yale University Press.

Eisenberg, Evan. 1987. *The Recording Angel: The Experience of Music from Aristotle to Zappa*. New York: Penguin.

——. 2005. *Recording Angel: Music, Records and Culture from Aristotle to Zappa*, 2ⁿᵈ edition. New Haven: Yale University Press.

Eliot, T.S. 1943. *Four Quartets*. New York: Houghton Mifflin Harcourt.

Emerick, Geoff, and Howard Massey. 2006. *Here, There and Everywhere: My Life Recording the Music of The Beatles*. New York: Gotham.

Erickson, Robert. 1975. *Sound Structure in Music*. Berkeley, CA: University of California Press.

Everest, F. Alton and Ken Pohlmann. 2015. *Master Handbook of Acoustics*, 6th edition. New York: McGraw Hill.

Everett, Walter. 1986. "Fantastic Remembrance in John Lennon's 'Strawberry Fields Forever' and 'Julia'." *The Musical Quarterly*, vol. 72, no. 3, 360–393.

——. 1999. *The Beatles as Musicians: Revolver through the Anthology*. Oxford: Oxford University Press.

——. 2000. "Confessions from Blueberry Hell, or, Pitch Can Be a Sticky Substance." In *Expression in Pop-Rock Music: Critical and Analytical Essays*, edited by Walter Everett, 269–345. New York: Garland.

——. 2001. *The Beatles as Musicians: The Quarry Men through Rubber Soul*. Oxford: Oxford University Press.

——. 2007a. "Fantastic Remembrance in John Lennon's 'Strawberry Fields Forever' and 'Julia'." *Critical Essays in Popular Musicology*, edited by Allan Moore, 391–416. Aldershot: Ashgate.

——. 2007b. "Making Sense of Rock's Tonal Systems." In *Critical Essays in Popular Musicology*, edited by Allan Moore, 301–335. Aldershot: Ashgate.

——. 2008. "Pitch Down the Middle." In *Expression in Pop-Rock Music: Critical and Analytical Essays*, 2nd edition, edited by Walter Everett, 111–174. New York: Routledge.

——. 2009a. "Any Time at All: the Beatles' Free Phrase Rhythms." *The Cambridge Companion to the Beatles*, edited by Kenneth Womack. Cambridge: Cambridge University Press.

——. 2009b. *The Foundations of Rock: From Blue Suede Shoes to Suite Judy Blue Eyes*. Oxford: Oxford University Press.

Fales, Cornelia. 2002. "The Paradox of Timbre." *Ethnomusicology*, vol. 46, no. 2, 56–95.

Fink, Robert, Melinda Latour and Zachary Wallmark. 2018. "Introduction: Chasing the Dragon: In Search of Tone in Popular Music." In *The Relentless Pursuit of Tone: Timbre in Popular Music*, edited by Robert Fink, Melinda Latour, and Zachary Wallmark, 1–17. New York: Oxford University Press.

Ford, Phil. 2019. "Style as Analysis." In *The Routledge Companion to Popular Music Analysis: Expanding Approaches*, edited by Ciro Scotto, Kenneth Smith, and John Brackett, 15–28. New York: Routledge.

Frith, Simon. 1983. *Sound Effects: Youth, Leisure and the Politics of Rock'n'Roll*. London: Constable.

——. 1987. "Towards an Aesthetic of Popular Music." In *Music and Society: The Politics of Composition, Performance and Reception*, edited by Richard Leppert and Susan McClary, 133–149. Cambridge: Cambridge University Press.

——. 1988. *Music for Pleasure: Essays in the Sociology of Pop*. Cambridge: Polity Press.

——. 1996. *Performing Rites: On the Value of Popular Music*. Cambridge, MA: Harvard University Press.

Gardner, Mark. 1969. Distance Estimation of 0° or apparent 0°-oriented Speech Signals in Anechoic Space." *Journal of the Acoustical Society of America*, vol. 45, no. 1, 47–53.

Gibson, David. 2005. *The Art of Mixing*, 2nd edition. Boston: Course Technology.

Gibson, James. 2015. *The Ecological Approach to Visual Perception*, Classic Edition. New York: Psychology Press, first published 1979, 1986.

Goldstein, Richard. 1969. *The Poetry of Rock*. New York: Bantam.

Gracyk, Theodore. 1996. *Rhythm and Noise: An Aesthetics of Rock*. Durham: Duke University Press.

Griffiths, Dai. 2000. "The Tributaries of 'The River.'" In *Reading Pop: Approaches to Textual Analysis in Popular Music*, edited by Richard Middleton, 192–202.

——. 2003. "From Lyric to Anti-Lyric: Analyzing the Words in Pop Song." In *Analyzing Popular Music*, edited by Allan F. Moore, 39–59. Cambridge: Cambridge University Press.

——. 2004. *OK Computer*. New York: Continuum International.

——. 2013. "Words to Songs and the Internet: A Comparative Study of Transcriptions of Words to the Song 'Midnight Train to Georgia,' Recorded by Gladys Knight and the Pips in 1973." *Popular Music and Society*, vol. 36, no.2, 234–273.

Hall, Edward T. 1969. *The Hidden Dimension*. New York, Anchor.

Handel, Stephen. 1993. *Listening: An Introduction to the Perception of Auditory Events*. Cambridge, MA: The MIT Press.

Harley, Robert. 2015. *The Complete Guide to High-End Audio*, 5th edition. Albuquerque, NM: Acapella.

Harvey, Steve. 2016. "P&E Wing: It's High Time for Hi-Res Audio," *Pro Sound News*, August 2016; 20, 26.

Hawkins, Stan. 2000. "Prince: Harmonic Analysis of 'Anna Stesia.'" In *Reading Pop: Approaches to Textual Analysis in Popular Music*, edited by Richard Middleton, 27–57. Oxford: Oxford University Press.

——. 2003. "Feel the Beat Come Down: House Music as Rhetoric." In *Analyzing Popular Music*, edited by Allan Moore, 80–102. Cambridge: Cambridge University Press.

Heidemann, Kate. 2016. "A System for Describing Vocal Timbre in Popular Song." *Music Theory Online*, vol. 22, no. 1. http://mtosmt.org/issues/mto.16.22.1/mto.16.22.1.heidemann.html. Accessed December 11, 2017.

Hennion, Antoine. 1990. "The Production of Success: An Antimusicology of the Pop Song." In *On Record: Rock, Pop and the Written Word*, edited by Simon Frith and Andrew Goodwin, 185–206. London: Routledge.

Hertsgaard, Mark. 1995. *A Day in the Life: The Music and Artistry of the Beatles*. New York: Delacorte.

Higgins, Kathleen Marie. 2012. *The Music Between Us: Is Music a Universal Language*. Chicago: University of Chicago Press.

Himes, Geoffrey. 2005. *Born in the U.S.A.* New York: Bloomsbury.

Hodgson, Jay. 2010. *Understanding Records: A Field Guide to Recording Practice*. New York: Continuum.

——. 2017. "Mix as Auditory Response." In *Mixing Music*, edited by Russ Hepworth-Sawyer and Jay Hodgson, 216–225. New York: Focal Press.

Høier, Svein. 2012. "The Relevance of Point of Audition in Television Sound: Rethinking a Problematic Term." *Journal of Sonic Studies*, vol. 3, no. 1. http://journal.sonicstudies.org/vol03/nr01/a04. Accessed January 27, 2018.

Horton, Julian. 2001. "Postmodernism and the Critique of Musical Analysis," *The Musical Quarterly*, 85, pp. 342–366.

Howard, David and Jamie Angus. 2017. *Acoustics and Psychoacoustics*, 5th edition. New York: Routledge.

Imberty, Michel. 1979. *Entendre la musique: Sémantique psychologique de la musique*. Paris: Dunod.

International Phonetic Association. 2015. https://internationalphoneticassociation.org/content/ipa-chart. Accessed July 16, 2016.

Jairazbhoy, Nazir. 1977. "The 'Objective' and Subjective View of Music Transcriptions." *Ethnomusicology*, vol. 21, no. 2, 263–273.

James, William. 1890. *The Principles of Psychology, Vol. 1*. New York: Cosimo, 2007.

Juslin, Patrik. 2019. *Musical Emotions Explained: Unlocking the Secrets of Musical Affect*. Oxford: Oxford University Press.

Kane, Brian. 2014. *Sound Unseen: Acousmatic Sound in Theory and Practice*. New York: Oxford University Press.

Kassabian, Anahid. 2008. "Rethinking Point of Audition in *The Cell*." *Lowering the Boom: Critical Studies in Film Sound*, edited by Jay Beck and Tony Grajeda. Urbana: University of Illinois Press.

Kennett, Chris. 2003. "Is Anybody Listening?" In *Analyzing Popular Music*, edited by Allan F. Moore, 196–217. Cambridge: Cambridge University Press.

King, Richard. 2017. *Recording Orchestra and Other Classical Ensembles*. New York: Routledge.

Koffka, Kurt. 1935 (2013). *Principles of Gestalt Psychology*. New York: Routledge (The International Library of Psychology – Cognitive Psychology Book 7).

Koozin, Timothy. 2008. "Fumbling Towards Ecstasy: Voice Leading, Tonal Structure, and the Theme of Self-Realization in the Music of Sarah McLachlan." In *Expression in Pop-Rock Music: Critical and Analytical Essays*, 2edition, edited by Walter Everett, 267–284. New York: Routledge.

Kramer, Jonathan. 1988. *The Time of Music*. New York: Schirmer Books.

Krause, Bernie. 2015. *Voices of the Wild: Animal Songs, Human Din, and the Call to Save Natural Soundscapes*. New Haven: Yale University Press.

Krims, Adam. 2007. "What Does It Mean To Analyse Popular Music?" In *Critical Essays in Popular Musicology*, edited by Allan F. Moore, 185–213. Aldershot: Ashgate.

Krumhansl, Carol. 1989. "Why is Musical Timbre so hard to understand?" In *Structure and Perception of Electroacoustic Sound and Music*, edited by Sören Nielzén and Olle Olsson, 43–53. Amsterdam: Elsevier.

Lacasse, Serge. 2000a. *'Listen to My Voice': The Evocative Power of Vocal Staging in Recorded Rock Music and Other Forms of Vocal Expression*. PhD thesis, University of Liverpool.

———. 2000b. "Voice and Sound Processing: Examples of *Mise en Scène* of Voice in Recorded Rock Music." *Popular Musicology Online*. http://popular-musicology-online.com/issues/05/lacasse.html. Accessed October 28, 2017.

———. 2002. "Ontological Misunderstandings in Music Analysis: Composition, Performance, Phonography." Canadian University Music Society Conference, University of Toronto. 26 May 2002.

Lacasse, Serge. 2005. "Persona, Emotions and Technology: the Phonographic Staging of the Popular Music Voice," *The Proceedings of the 2005 Art of Record Production Conference* (www.artofrecordproduction.com).

———. 2006. "Composition, Performance, Phonographie: Un Malentendu Ontologique En Analyse Musicale?" In *Groove: enquête sur les phénomènes musicaux contemporains: mélanges à la mémoire de Roger Chamberland*, edited by Patrick Roy and Serge Lacasse, 65–78. Québec: Les Presses de l'Université Laval.

———. 2007. "Intertextuality and Hypertextuality in Recorded Popular Music." In *Critical Essays in Popular Musicology*, edited by Allan F. Moore, 148–170. Aldershot: Ashgate.

———. 2010a. "The Phonographic Voice: Paralinguistic Features and Phonographic Staging in Popular Music Singing." In *Recorded Music: Performance, Culture and Technology*, edited by Amanda Bayley, 225–251. Cambridge: Cambridge University Press.

———. 2010b. "Slave to the Supradiegetic Rhythm: A Microrhythmic Analysis of Creaky Voice in Sia's 'Breath Me.'" In *Musical Rhythm in the Age of Digital Reproduction*, edited by Anne Danielsen, 141–155. New York: Routledge.

LaRue, Jan. 2011. *Guidelines for Style Analysis*, expanded 2nd edition. Detroit: Harmonie Park.

Lavengood, Megan. 2017. *A New Approach to the Analysis of Timbre*. PhD Dissertation, The City University of New York.

Lefford, M. Nyssim. 2014. "Producing and Its Effect on Vocal Recordings." In *Digital Da Vinci*, edited by Newton Lee, 29–77. New York: Springer.

Lerdahl, Fred. 1987. "Timbral Hierarchies." *Contemporary Music Review*, vol. 2, 135–160.

Lerdahl, Fred and Ray Jackendoff. 1983. *A Generative Theory of Tonal Music*. Cambridge, MA: MIT Press.

Levitin, Daniel J. 1994. "Absolute Memory for Musical Pitch: Evidence from the Production of Learned Melodies." *Perception & Psychoacoustics*, 56, 414–423.

———. 2006. *This is Your Brain on Music: The Science of a Human Obsession*. New York: Dutton.

Levitin, Daniel J. & Perry R. Cook. 1996. "Memory for Musical Tempo: Additional Evidence that Auditory Memory is Absolute." *Perception & Psychoacoustics*, 58, 927–935.

Lewis, Michael. 2016. *The Undoing Project: A Friendship That Changed Our Minds*. New York: W. W. Norton.

Lewisohn, Mark. 1988. *The Beatles Recording Sessions: The Official Abbey Road Studio Session Notes 1962–1970*. New York: Harmony.

——. 2003. *The Complete Beatles Chronicle*. London: Hamlyn.

Lora, Doris. 1979. "Musical Pattern Perception." *College Music Symposium*, vol. 19, no. 1, 166–182.

Ludwig, Bob. 2013. "The Loudness Wars: Musical Dynamics Versus Volume." In *Less Noise, More Soul: The Search for Balance in the Art, Technology, and Commerce of Music*, edited by David Flitner, 57–65. Milwaukee, WI: Hal Leonard.

MacDonald, Ian. 2003. *The People's Music*. London: Pimlico.

——. 2005. *Revolution in the Head: The Beatles' Records and the Sixties*, 3rd edition. Chicago: Chicago Review Press.

Maconie, Robin. 2007. *The Way of Music: Aural Training for the Internet Generation*. Lanham, MD: The Scarecrow Press.

Mather, George. 2016. *Foundations of Sensation and Perception*, 3rd edition. New York: Routledge.

McAdams, Stephen. 1999. "Perspectives on the Contribution of Timbre to Musical Structure." *Computer Music Journal*, vol. 23, no. 3, 85–102.

McAdams, Stephen and Bruno Giordano. 2011. "The Perception of Timbre." In *The Oxford Handbook of Music Psychology*, edited by Susan Hallam, Ian Cross, and Michael Thant, 72–80. Oxford: Oxford University Press.

McClary, Susan. 1994. "Same As It Ever Was: Youth Culture and Music." In *Microphone Friends: Youth Music and Youth Culture*, edited by Andrew Ross and Tricia Rose, 29–40. London: Routledge.

——. 2000. *Conventional Wisdom: The Content of Musical Form*. Berkeley: University of California Press.

McClary, Susan and Robert Walser. 1990. "Start Making Sense! Musicology Wrestles with Rock." In *On Record: Rock, Pop and the Written Word*, edited by Simon Frith and Andrew Goodwin, 277–292. London: Routledge.

McGuiness, Andy and Katie Overy. 2011. "Music, Consciousness, and the Brain: Music as Shared Experience of an Embodied Present." In *Music and Consciousness: Philosophical, Psychological and Cultural Perspectives*, edited by David Clarke and Eric Clarke, 246–262. Oxford: Oxford University Press.

Meintjes, Louise. 2003. *Sound of Africa! Making Music Zulu in a South African Studio* Durham & London: Duke University Press.

Mellers, Wilfred. 1973. *Twilight of the Gods: The Music of the Beatles*. New York: Viking.

——. 1985. *A Darker Shade of Pale: A Backdrop to Bob Dylan*. New York: Oxford University Press.

Meyer, Leonard B. 1967. *Music, the Arts, and Ideas: Patterns and Predictions in Twentieth-Century Culture*. Chicago: The University of Chicago Press.

——. 1973. *Explaining Music: Essays and Explorations*. Berkeley: University of California Press.

Middleton, Richard. 1990. *Studying Popular Music*. New York: Open University Press.

——. 2000a. "Popular Music Analysis and Musicology: Bridging the Gap." In *Reading Pop: Approaches to Textual Analysis in Popular Music*, edited by Richard Middleton, 104–121. Oxford: Oxford University Press.

——. 2000b. "Words and Music." In *Reading Pop: Approaches to Textual Analysis in Popular Music*, edited by Richard Middleton, 163–164. Oxford: Oxford University Press.

Møller, Henrik and Christian Sejer Pedersen. 2004. "Hearing at Low and Infrasonic Frequencies." *Noise & Health*, vol. 6, 37–57.

Moore, Allan F. 1992. "Textures of Rock." In *Secondo Convegno Europeo di Analisi Musicale* edited by Rosanna Dalmonte and Mario Baroni, 241–244. Trento: Università degli Studi di Trento.

——. 1997. *The Beatles: Sgt. Pepper's Lonely Hearts Club Band*. Cambridge: Cambridge University Press.

——. 2001. *Rock: The Primary Text*. Aldershot: Ashgate.

——. 2003. "Introduction." In *Analyzing Popular Music*, edited by Allan F. Moore, 1–15. Cambridge: Cambridge University Press.

——. 2004. *Aqualung*. New York: Bloomsbury Academic.

——, editor. 2007a. *Critical Essays in Popular Musicology*. Aldershot: Ashgate.

——. 2007b. "The So-Called 'Flattened Seventh' in Rock." In *Critical Essays in Popular Musicology*, edited by Allan Moore, 281–299. Aldershot: Ashgate.

——. 2010. "The Track." In *Recorded Music: Performance, Culture and Technology*, edited by Amanda Bayley, 252–267. Cambridge: Cambridge University Press.

——. 2012a. *Song Means: Analysing and Interpreting Recorded Popular Song*. Surrey: Ashgate.

——. 2012b. "Beyond a Musicology of Production." In *The Art of Record Production: An Introductory Reader for a New Academic Field*, edited by Simon Frith and Simon Zagorski-Thomas, 99–111. Surrey: Ashgate.

Moore, Allan F. and Ruth Dockwray. 2008. "The Establishment of the Virtual Performance Space in Rock." *Twentieth-Century Music*. vol. 5, no. 2, 219–241.

Moore, Allan F. and Remy Martin. 2019. *Rock: The Primary Text: Developing a Musicology of Rock*, 3rd edition. London: Routledge.

Moore, Allan F., Patricia Schmidt, and Ruth Dockwray. 2009. "The Hermeneutics of Spatialization in Recorded Song." *Twentieth-Century Music*, vol. 6, no. 1, 83–114.

Moore, Brian C. J. 2013. *An Introduction to the Psychology of Hearing*, 6th edition. Leiden: Brill.

Moorefield, Virgil. 2005. *The Producer as Composer: Shaping the Sounds of Popular Music*. Cambridge, MA: The MIT Press.

Moy, Ron. 2007. *Kate Bush and Hounds of Love*. Aldershot: Ashgate.

Moylan, William. 1983. *An Analytical System for Electronic Music*, Doctor of Arts dissertation, Ball State University. Ann Arbor, MI: University Microfilms.

——. 1985. "Aural Analysis of the Characteristics of Timbre." 79th Convention of the Audio Engineering Society, 13 October 1985.

——. 1986. "Aural Analysis of the Spatial Relationships of Sound Sources as Found in Two-Channel Common Practice." 81st Convention of the Audio Engineering Society, Los Angeles, 14 November 1986.

——. 1992. *The Art of Recording: the Creative Resources of Music Production and Audio*. New York: Van Nostrand Reinhold.

——. 2012. "Considering Space in Recorded Music." In *The Art of Record Production: An Introductory Reader for a New Academic Field*, edited by Simon Frith and Simon Zagorski-Thomas, 163–188. Surrey: Ashgate.

——. 2015. *Understanding and Crafting the Mix: The Art of Recording*, 3rd edition. Burlington, MA: Focal Press.

——. 2017. "How to Listen, What to Hear." In *Perspectives On Music Production: Mixing Music*, edited by Russ Hepworth-Sawyer and Jay Hodgson, 24–52. New York: Routledge.

Negus, Keith. 1996. *Popular Music in Theory: An Introduction*. Cambridge: Polity.

Negus, Keith and Michael Pickering. 2002. "Creativity and Musical Experience" in *Popular Music Studies* edited by David Hesmondhalgh and Keith Negus, 178–190. London: Arnold.

Neisser, Ulric. 1967. *Cognitive Psychology*. New York: Appleton-Century-Crofts.

——. 1976. *Cognition and Reality: Principles and Implications of Cognitive Psychology*. New York: Cambridge University Press.

Nelson, Dean. 2017. "Between the Speakers: Discussions on Mixing." In *Mixing Music*, edited by Russ Hepworth-Sawyer and Jay Hodgson, 122–139. New York: Routledge.

Neuhoff, John, editor. 2004a. *Ecological Psychoacoustics*. San Diego: Elsevier.

——. 2004b. "Ecological Psychoacoustics: Introduction and History." In *Ecological Psychoacoustics*, edited by John Neuhoff, 1–4. San Diego: Elsevier.

——. 2004c. "Auditory Motion and Localization." In *Ecological Psychoacoustics*, edited by John Neuhoff, 87–111. San Diego: Elsevier.

——. 2004d. "Interacting Perceptual Dimensions." In *Ecological Psychoacoustics*, edited by John Neuhoff, 249–269. San Diego: Elsevier.

Noë, Alva. 2004. *Action in Perception*. Cambridge: MIT Press.

Oliver, Mary. 1998. *Rules for the Dance: A Handbook for Writing and Reading Metrical Verse*. New York: Houghton Mifflin Company.

Oliveros, Pauline. 2004. "Some Sound Observations." In *Audio Culture: Readings in Modern Music*, edited by Christoph Cox and Daniel Warner, 102–106. London: Continuum.

Parrott, W. Gerrod. 2001. *Emotions in Social Psychology*. Philadelphia: Psychology Press.

Parsons, Alan and Julian Colbeck. 2014. *Alan Parsons' Art & Science of Sound Recording—The Book*. Milwaukee, WI: Hal Leonard.

Pattison, Pat. 2009. *Writing Better Lyrics: The Essential Guide to Powerful Songwriting*, 2nd edition. Cincinnati: Writer's Digest Books.

Pedler, Dominic. 2003. *The Songwriting Secrets of The Beatles*. London: Omnibus.

Perlman, Helen Harris. 1986. *Persona: Social Role and Personality*. Chicago: University of Chicago Press.

Plutchik, Robert, and Henry Kellerman. 1980. "Theories of Emotion." In *Emotion: Theory, Research and Experience Vol. I*. New York: Academic.

Poldy, Carl A. 2001. "Headphones." In *Loudspeaker and Headphone Handbook*, edited by John Borwick, 585–686. Oxford: Focal Press.

Polizzotti, Mark. 2006. *Highway 61 Revisited*. New York: Bloomsbury.

Porcello, Thomas. 2004. "Speaking Sound: Language and the Professionalization of Sound-Recording Engineers." *Social Studies of Science*, vol. 34, no. 5, 733–758.

Poyatos, Fernando. 1993. *Paralanguage: a Linguistic and Interdisciplinary Approach to Interactive Speech and Sound*. Amsterdam: John Benjamins.

Reetz, Henning and Allard Jongman. 2008. *Phonetics: Transcription, Production, Acoustics, and Perception*. West Sussex: Wiley-Blackwell.

Ricks, Christopher. 2003. *Dylan's Visions of Sin*. London: Penguin.

Roederer, Juan. 2008. *Introduction to the Physics and Psychophysics of Music*, 4th edition. New York: Springer.

Roholt, Tiger. 2014. *Groove: A Phenomenology of Rhythmic Nuance*. New York: Bloomsbury.

Rosenblum, Lawrence, A. Paige Wuestefeld, and Krista Anderson. 1996. "Auditory Reachability: An Affordance Approach to the Perception of Sound Source Distance." *Ecological Psychology*, vol. 8, 1–24.

Rossing, Thomas. 1990. *The Science of Sound*, 2nd edition. Reading, MA: Addison-Wesley.

Rumsey, Francis. 2001. *Spatial Audio*. Boston: Focal Press.

Rumsey, Francis and Tim McCormick. 2014. *Sound and Recording: Applications and Theory*. Boston: Focal Press.

Ryan, Kevin and Brian Kehew. 2006. *Recording the Beatles: The Studio Equipment and Techniques Used to Create Their Classic Albums*. Houston, Texas: Curvebender.

Sacks, Oliver. 2008. *Musicophilia: Tales of Music and the Brain*. New York: Vintage.

Savage, Roger. 2015. *Hermeneutics and Music Criticism*. New York: Routledge.

Schaeffer, Pierre. 1952. *À la recherche d'une musique concrete*. Paris: Editions du Seuil.

——. 1966. *Traité des objets musicaux*. Paris: Editions du Seuil.

——. 2004. "Acousmatics." In *Audio Culture: Readings in Modern Music*, edited by Christoph Cox and Daniel Warner. London: Continuum.

——. 2012. *In Search of a Concrete Music*, translated by Christine North and John Dack. Berkeley: University of California Press.

Schafer, R. Murray. 1971. *The New Soundscape*. London and Vienna: Universal Edition.

——. 1977. *The Tuning of the World*. Toronto: McClelland and Stewart.

——. 1994. *The Soundscape: Our Sonic Environment and the Tuning of the World*. Rochester, VT: Destiny.

——. 2004. "The Music of the Environment." In *Audio Culture: Readings in Modern Music*, edited by Christoph Cox and Daniel Warner, 29–39. London: Continuum.

Schenker, Heinrich. 1979. *Free Composition (Der freie Satz)*, 2nd edition, translated and edited by E. Oster. New York: Longman.

Schlauch, Robert. 2004. "Loudness." In *Ecological Psychoacoustics*, edited by John Neuhoff, 317–345. San Diego: Elsevier.

Schoenberg, Arnold. 1966. *Harmonielehre*, 7th edition. Vienna: Universal Edition.

Schnupp, Jan, Israel Nelken, and Andrew King. 2012. *Auditory Neuroscience: Making Sense of Sound*. Cambridge: The MIT Press.

Schoenberg, Arnold. 1966. *Harmonielehre*, 7th edition. Vienna: Universal Edition.

Schwenke Wyile, Andrea. 2003. "The Value of Singularity in First- and Restricted Third-Person Engaging Narration," *Children's Literature*, vol. 31, pp. 116–141.

Shinn-Cunningham, Barbara, Scott Santarelli & Norbert Kopco. 2000. "Tori of Confusion: Binaural Localization Cues for Sources Within Reach of a Listener." *Journal of the Acoustical Society of America*, vol. 107, no. 3, 1627–1639.

Shuker, Roy. 2002. *Popular Music: The Key Concepts*. London and New York: Routledge.

Slawson, Wayne. 1985. *Sound Color*. Berkeley: University of California Press.

Sloboda, John. 2005. *Exploring the Musical Mind: Cognition, Emotion, Ability, Function*. Oxford: Oxford University Press.

Smalley, Denis. 1997. "Spectromorphology: Explaining Sound-Shapes." *Organised Sound*, vol. 2, no. 2, 107–126.

——. 2007. "Space-Form and the Acousmatic Image." *Organised Sound*, vol. 12, no. 1, 35–58.

Snyder, Bob. 2000. *Music and Memory: An Introduction*. Cambridge, MA: The MIT Press.

Sommer, Robert. 1969. *Personal Space: The Behavioral Basis of Design*. Englewood Cliffs, NJ: Prentice Hall.

Song, Yading, Simon Dixon, Marcus Pearce, and Andrea Halpern. 2016. "Perceived and Induced Emotion Responses to Popular Music." *Music Perception: An Interdisciplinary Journal*, vol. 33, no. 4, 472–492.

Spencer, Peter and Peter Temko. 1994. *Practical Approach to the Study of Form in Music*. Prospect Heights, IL: Waveland.

Spicer, Mark and John Covach, editors. 2010. *Sounding Out Pop: Analytical Essays in Popular Music*. Ann Arbor: The University of Michigan Press.

Stanyek, Jason. 2014. "Forum on Transcription." *Twentieth-Century Music*, vol. 11, no. 1, 101–161.

Stein, Deborah. 2005. "Introduction to Musical Ambiguity." In *Engaging Music: Essays in Music Analysis*, edited by Deborah Stein, 77–88. Oxford: Oxford University Press.

Stephenson, Ken. 2002. *What to Listen for in Rock: A Stylistic Analysis*. New Haven: Yale University Press.

Sterne, Jonathan, editor. 2012. *The Sound Studies Reader*, New York: Routledge.

Tagg, Philip. 1987. "Musicology and the Semiotics of Popular Music." *Semiotica*, vol. 66, nos 1–3, pp. 279–298.

——. 2000. "Analyzing Popular Music: Theory, Method and Practice." In *Reading Pop: Approaches to Textual Analysis in Popular Music*, edited by Richard Middleton, 71–103. Oxford: Oxford University Press.

——. 2013. *Music's Meanings: A Modern Musicology for Non-Musos*. New York: The Mass Media Music Scholars' Press.

——. 2016. *Everyday Tonality II: Towards a Tonal Theory of What Most People Hear*. New York: The Mass Media Music Scholars' Press.

Tan, Siu-Lan, Peter Pfordresher, and Rom Harré. 2018. *Psychology of Music: From Sound to Significance*. New York: Routledge.

Tenney, James. 1986. *Meta ≠ Hodos and META Meta ≠ Hodos: A Phenomenology of 20th Century Musical Materials and an Approach to the Study of Form*. Oakland, CA: Frog Peak Music.

Théberge, Paul. 1989. "The 'Sound' of Music: Technological Rationalization and the Production of Popular Music." *New Formulations*, 8, 99–111.

——. 1997. *Any Sound You Can Imagine: Making Music/Consuming Technology*. Middletown, CT: Wesleyan University Press.

——. 2018. "The Sound of Nowhere: Reverb and the Construction of Sonic Space." In *The Relentless Pursuit of Tone: Timbre in Popular Music*, edited by Robert Fink, Melinda Latour, and Zachary Wallmark, 323–344. New York: Oxford University Press.

Tomasello, Michael. 1999. *The Cultural Origins of Human Cognition.* Cambridge: Harvard University Press.

Vernon, Philip. 1934. "Auditory Perception I. The Gestalt Approach." *British Journal of Psychology*, vol. 25, 123–139.

Wallmark, Zachary. 2014. *Appraising Timbre: Embodiment and Affect at the Threshold of Music and Noise.* PhD Dissertation, University of California, Los Angeles.

Walser, Robert. 1993. *Running with the Devil: Power, Gender, and Madness in Heavy Metal Music.* Hanover, NH: Wesleyan University Press.

——. 2003. "Popular Music Analysis: Ten Apothegms and Four Instances." In *Analyzing Popular Music*, edited by Allan F. Moore, 16–38. Cambridge: Cambridge University Press.

Watson, Jada and Lori Burns. 2010. "Resisting Exile and Asserting Musical Voice: the Dixie Chicks are 'Not Ready to Make Nice.'" *Popular Music*, vol. 23, no. 3, 325–350.

Wessel, David. 1979. "Timbre Space as a Musical Control Structure." *Computer Music Journal*, vol. 3, no. 2, 45–52.

White, John. 1994. *Comprehensive Musical Analysis.* Landham, MD: Scarecrow Press.

Whitesell, Lloyd. 2008. *The Music of Joni Mitchell.* Oxford: Oxford University Press.

Williams, Alan. 1980. "Is Sound Recording Like a Language?" *Cinema/Sound*, edited by Rick Altman, 51–66. Yale French Studies, No. 60.

Williams, Paul. 1990. *Performing Artist: The Music of Bob Dylan, Vol. 1 1960–1973.* Novato, CA: Underwood-Miller.

Winer, Ethan. 2012. *The Audio Expert: Everything You Need to Know About Audio.* Boston: Focal Press.

Winkler, Peter. 1997. "Writing Ghost Notes: The Poetics and Politics of Transcription." In *Keeping Score: Music, Disciplinarity, Culture*, edited by David Schwartz, Anahid Kassabian, and Lawrence Siegel, 169–203. Charlottesville: University of Virginia Press.

——. 2000. "Randy Newman's Americana." In *Reading Pop: Approaches to Textual Analysis in Popular Music*, edited by Richard Middleton, 27–57. Oxford: Oxford University Press.

Yavas, Mehmet. 2016. *Applied English Phonology*, 3rd edition. West Sussex: Wiley-Blackwell.

Zagorski-Thomas, Simon. 2010. "The Stadium in Your Bedroom: Functional Staging, Authenticity and the Audience-led Aesthetic in Record Production." *Popular Music*, vol. 29, no. 2, 251–266.

——. 2014. *The Musicology of Record Production.* Cambridge: Cambridge University Press.

——. 2015. "An Analysis of Space, Gesture and Interaction in King of Leon's 'Sex on Fire.'" In *Song Interpretation in 21st Century Pop Music*, edited by Ralf von Appen, André Doehring, Dietrich Helms, and Allan F. Moore, 115–132. Surrey: Ashgate.

Zak III, Albin. 2001. *The Poetics of Rock: Cutting Tracks, Making Records.* Berkeley, CA: University of California Press.

——. 2005. "Bob Dylan and Jimi Hendrix: Juxtaposition and Transformation 'All along the Watchtower.'" *Journal of the American Musicological Society*, vol. 57, no. 3, pp. 599–644.

——. 2010. "Painting the Sonic Canvas: Electronic Mediation as Musical Style." In *Recorded Music: Performance, Culture and Technology*, edited by Amanda Bayley, 307–324. Cambridge: Cambridge University Press.

Zbikowski, Lawrence. 2002. *Conceptualizing Music: Cognitive Structure, Theory, and Analysis.* Oxford: Oxford University Press.

——. 2011. "Music, Language, and Kinds of Consciousness." In *Music and Consciousness: Philosophical, Psychological, and Cultural Perspectives*, edited by David Clarke and Eric Clarke, 179–192. Oxford: Oxford University Press.

Zieliński, Sławomir, Francis Rumsey and Søren Bech. 2008. "On Some Biases Encountered in Modern Audio Quality Listening Tests—A Review," *Journal of the Audio Engineering Society*, vol. 58, no. 6, pp. 427–451.

Zsiga, Elizabeth. 2013. *The Sounds of Language: An Introduction to Phonetics and Phonology.* West Sussex: Wiley-Blackwell.

Discography

Adams, Bryan. 1991. "(Everything I Do) I Do It for You." A&M.

Adele. 2015. "Hello." *25*. XL Recordings, Columbia. 88875175952.

Amos, Tori. 1992. "Crucify." *Little Earthquakes*. Atlantic.

——. 2001. "97 Bonnie and Clyde." *Strange Little Girls*. Atco/Atlantic.

Beach Boys. 1966. *Pet Sounds*. Capitol. CDP 548421.

Beatles, The. 1963. "I Want to Hold Your Hand." *1*. Parlophone, Apple, Capitol. CDP 7243 5 29325 2 8.

——. 1964. "Every Little Thing." *Beatles for Sale*. Parlophone, Capitol.

——. 1964. "A Hard Day's Night." *A Hard Day's Night*. Parlophone, Capitol.

——. 1965. "Nowhere Man." *Rubber Soul*. Parlophone, Capitol.

——. 1965. "We Can Work It Out." Parlophone, Capitol.

——. 1966. "Eleanor Rigby." *Revolver*. Parlophone, Capitol. CDP 7 46441 2.

——. 1966. "I'm Only Sleeping." *Revolver*. Parlophone, Capitol.

——. 1966. "Paperback Writer." Parlophone, Capitol.

——. 1966. "She Said She Said." *Revolver*. Parlophone, Capitol.

——. 1966. "Tomorrow Never Knows." *Revolver*. Parlophone, Capitol.

——. 1967. "All You Need is Love." Parlophone, Capitol.

——. 1967. "Blue Jay Way." *Magical Mystery Tour*. Parlophone, Capitol. CDP 7 48062 2.

——. 1967. "A Day in the Life." *Sgt. Pepper's Lonely Hearts Club Band*. Parlophone, Capitol. CDP 7 46442 2.

——. 1967, 1987. "Lucy in the Sky With Diamonds." *Sgt. Pepper's Lonely Hearts Club Band*. Parlophone, Capitol. CDP 7 46442 2.

——. 1967. "Penny Lane." Parlophone, Capitol.

——. 1967. "Sgt. Pepper's Lonely Hearts Club Band." *Sgt. Pepper's Lonely Hearts Club Band*. Parlophone, Capitol. CDP 7 46442 2.

——. 1967. *Sgt. Pepper's Lonely Hearts Club Band*. Parlophone, Capitol. CDP 7 46442 2.

——. 1967, 1987. "Strawberry Fields Forever." *Magical Mystery Tour*. Parlophone, Capitol. CDP 7 48062 2.

——. 1967. "With a Little Help from My Friends." *Sgt. Pepper's Lonely Hearts Club Band*. Parlophone, Capitol.

——. 1968, 1987. "While My Guitar Gently Weeps." *The Beatles* (White Album). Parlophone, Apple. CDP 7 46443 2.

——. 1968, 2000. "Hey Jude." *1*. Parlophone, Apple, Capitol. CDP 7243 5 29325 2 8.

——. 1969, 1987. "Come Together." *Abbey Road*. Parlophone, Apple. CDP 7 46446 2.

——. 1969, 1987. "Here Comes the Sun." *Abbey Road*. Parlophone, Apple. CDP 7 46446 2.

——. 1969, 1987. "Maxwell's Silver Hammer." *Abbey Road*. Parlophone, Apple. CDP 7 46446 2.

——. 1969, 1987. "She Came in Through the Bathroom Window." *Abbey Road*. Parlophone, Apple. CDP 7 46446 2.

——. 1969, 1987. "Something." *Abbey Road*. Parlophone, Apple. CDP 7 46446 2.

——. 1969, 1987. "The End." *Abbey Road*. Parlophone, Apple. CDP 7 46446 2.

——. 1969, 1987. "You Never Give Me Your Money." *Abbey Road*. Parlophone, Apple. CDP 7 46446 2.

——. 1970, 1987. "Let It Be." *Let It Be*. Parlophone, Apple. CDP 7 46447 2.

——. 1988. "Let It Be." *Past Masters, Volume Two*. EMI, Apple. CDP 7 90044 2.

——. 1999. "Lucy in the Sky with Diamonds," *Yellow Submarine*. EMI, Apple. CDP 7243 5 21481 2 7.

——. 2000. "Let It Be." *1*. Parlophone, Apple, Capitol. CDP 7243 5 29325 2 8.

——. 2003. "Let It Be." *Let It Be . . . Naked*. Apple, EMI. CDP 7243 5 95713 2 4.

——. 2006. "Strawberry Fields Forever." *LOVE*. Apple, Capitol, Parlophone. 0946 3 79810 2 3/9 2.

——. 2006. "While My Guitar Gently Weeps." *LOVE*. Apple, Capitol, Parlophone. 0946 380 789 2.

——. 2009. "Let It Be." *Let It Be*. EMI Records. 0946 3 82472 2 7.

——. 2009. "Strawberry Fields Forever." *Magical Mystery Tour*. EMI Records. 5099969946028 (mono).

——. 2009. "Strawberry Fields Forever." *Magical Mystery Tour*. EMI Records. 0946 3 82465 2 7 (stereo).

———. 2009. "While My Guitar Gently Weeps." *The Beatles*. EMI Records 5099968495725 (mono).

———. 2009. "While My Guitar Gently Weeps." *The Beatles*. EMI Records. 0946 3 824662 (stereo).

———. 2017. "Strawberry Fields Forever." *Sgt. Pepper's Lonely Hearts Club Band*. Calderstone, Universal, Apple. 0602557455434 (mono).

———. 2018. "While My Guitar Gently Weeps." *The Beatles*. Apple. LC01846.

Björk. 1995. "Army of Me." *Post*. One Little Indian.

———. 2001. "Cocoon." *Vespertine*. One Little Indian, Polydor. 5 016958 046026.

Bowie, David. 1969, 2015. "Space Oddity." *David Bowie*. Parlophone. RPR2-218988.

———. 1971. "Changes." *Hunky Dory*. RCA. PCD1-4623.

———. 1972. *The Rise and Fall of Ziggy Stardust and the Spiders from Mars*. RCA/Rykodisc. RCD 10134.

———. 1974. "Diamond Dogs." *Diamond Dogs*. Parlophone. RP2-219044.

———. 1983. "Let's Dance." *Let's Dance*. Parlophone, EMI.

Brown, James. 1967. "Cold Sweat." King.

———. 1974. "Funky President (People Its Bad)." *Reality*. Polydor.

Bush, Kate. 1982. "Get Out of My House." *The Dreaming*. EMI.

———. 1989. "This Woman's Work." *The Sensual World*. Columbia. CK 44164.

Byrds, The. 1965. "Mr. Tambourine Man." *Mr. Tambourine Man*. Columbia.

Chapman, Tracy. 1988. "Fast Car." *Tracy Chapman*. Elektra. E2 60774.

Cocker, Joe. 1968. "With a Little Help from My Friends." *With a Little Help from My Friends*. A&M.

Collins, Phil. 1981. "In the Air Tonight." *Face Value*. Atlantic. SD 16029.

Cooke, Sam. 1964. "Try a Little Tenderness." *At the Copa*. RCA Victor.

Costello, Elvis. 1983. "Pills and Soap." *Punch the Clock*. Columbia.

Crosby, Bing. 1933. "Try a Little Tenderness." *Sweet Georgia Brown*. Brunswick.

———. 1944. "I'll Be Seeing You." Decca.

Crosby, Stills & Nash. 1969. "Wooden Ships." *Crosby, Stills & Nash*. Atlantic.

Crosby, Stills, Nash & Young. 1970. "Woodstock." *Déjà Vu*. Atlantic.

Dixie Chicks. 2006. "Not Ready to Make Nice." *Taking the Long Way*. Columbia, Open Wide. 82876 80739 2.

Dylan, Bob. 1963. "A Hard Rain's A-Gonna Fall." *The Freewheelin' Bob Dylan*. Columbia. CK 8786.

———. 1963. "Blowing in the Wind." *The Freewheelin' Bob Dylan*. Columbia. CK 8786.

———. 1964. "Boots of Spanish Leather." *The Times They Are A-Changin'*. Columbia. CK 8905.

———. 1965. "Like a Rolling Stone." *Highway 61 Revisited*. CK 9189.

———. 1965. "Mr. Tambourine Man." *Bringing It All Back Home*. Columbia. CK 92401.

———. 1966. "I Want You." *Blonde on Blonde*. Columbia. CGK 841.

———. 1967. "All Along the Watchtower." *John Wesley Harding*. Columbia. CK 9604.

———. 1970. "If Not For You." *New Morning*. Columbia. 8869734002.

———. 1975. "Simple Twist of Fate." *Blood on the Tracks*. Columbia. CK 33235.

———. 1975. "Tangled Up in Blue." *Blood on the Tracks*. Columbia. CK 33235.

———. 1976. "Hurricane." *Desire*. Columbia. CK 92393.

———. 1986. "Brownsville Girl." *Knocked Out Loaded*. Columbia. 467040 2.

———. 1997. *Time Out of Mind*. Columbia. CK 68556.

———. 1997. "Can't Wait." *Time Out of Mind*. Columbia. CK 68556.

———. 1997. "Cold Irons Bound." *Time Out of Mind*. Columbia. CK 68556.

———. 1997. "Dirt Road Blues." *Time Out of Mind*. Columbia. CK 68556.

———. 1997. "Highlands." *Time Out of Mind*. Columbia. CK 68556.

———. 1997. "Love Sick." *Time Out of Mind*. Columbia. CK 68556.

———. 1997. "Make You Feel My Love." *Time Out of Mind*. Columbia. CK 68556.

———. 1997. "Million Miles." *Time Out of Mind*. Columbia. CK 68556.

———. 1997. "Not Dark Yet." *Time Out of Mind*. Columbia. CK 68556.

———. 1997. "Standing in the Doorway." *Time Out of Mind*. Columbia. CK 68556.

———. 1997. "'Til I Fell In Love With You." *Time Out of Mind*. Columbia. CK 68556.

———. 1997. "Tryin' To Get To Heaven." *Time Out of Mind*. Columbia. CK 68556.

Eminem. 1999. "97 Bonnie and Clyde." *The Slim Shady LP*. Interscope.

Franklin, Aretha. 1962. "Try a Little Tenderness." *The Tender, the Moving, the Swinging Aretha Franklin*. Columbia.

———. 1967. "I Never Loved a Man (The Way I Love You)." *I Never Loved a Man The Way I Love You*. Atlantic. R2 71934.

Gabriel, Peter. 1980. "Intruder." *Peter Gabriel*. Mercury. SRM 1-3848.

———. 1986. "In Your Eyes." *So*. Geffen. 9 24088-2.

———. 1986. *So*. Geffen. 9 24088-2.

Harrison, George. 1970. "If Not For You." *All Things Must Pass*. EMI. CDP 7243 5 30474 2 9.

Holiday, Billie. 1944. "I'll Be Seeing You." London, Commodore.

Jackson, Michael. 1982. "Thriller." *Thriller*. Epic. EK 38112.

Jimi Hendrix Experience, The. 1967. "Purple Haze." *Are You Experienced*. Reprise.

———. 1968. "All Along the Watchtower." *Electric Ladyland*. Reprise. 6307-2.

Kingsmen, The. 1963. "Louie Louie." Wand Records.

Led Zeppelin. 1969. "Whole Lotta Love." *Led Zeppelin II*. Atlantic. 48236.

Lennon, John. 1971. "Imagine." *Imagine*. Apple. CDP 7 46641 2.

Linkin Park. 2007. "Valentine's Day." *Minutes to Midnight*. Warner Brothers. 44477-2.

Madonna. 1986. "Where's the Party." *True Blue*. Sire, Warner Brothers.

McLachlan, Sarah. 1993. "Ice." *Fumbling Towards Ecstasy*. Nettwerk.

McKenzie, Scott. 1967. "San Francisco (Be Sure to Wear Flowers in Your Hair)." Columbia. 2757.

Mitchell, Joni. 1970. "Woodstock." *Ladies of the Canyon*. Reprise. 7599-27450-2.

———. 1971. "The Last Time I Saw Richard." *Blue*. Warner Bros. 2038-2.

Morissette, Alanis. 1995. "You Oughta Know." Jagged Little Pill. Maverick. 45901-2.

Paul, Les and Mary Ford. 1955. "Falling in Love With Love." *Les and Mary*. Capitol.

Perry, Katy. 2010. "Firework." Teenage Dream. Capitol. 509996 84601 2 9.

Petty, Tom and the Heartbreakers. 1985. "Don't Come Around Here No More." *Southern Accents*. MCA.

Pink Floyd. 1973. *The Dark Side of the Moon*. Capitol. CDP 7 46001 2.

Presley, Elvis. 1954, 1992. "Blue Moon." *The King of Rock 'n' Roll: The Complete 50's Masters*, disc 1. RCA, 66050-2.

Prince. 1982. "1999." *1999*. Warner Bros. 29820-0.

———. 1988. "Anna Stesia." *Lovesexy*. Paisley Park. 9 25720-2.

Prince and the New Power Generation. 1991. *Diamonds and Pearls*. Paisley Park, Warner Bros.

Prince and the Revolution. 1984. "When Doves Cry." *Purple Rain*. Warner Bros. 25110-1.

———. 1986. "Kiss." *Parade*. Warner Bros.

Redding, Otis. 1966. "Try a Little Tenderness." *The Otis Redding Dictionary of Soul: Complete and Unbelievable*. Volt Records.

Reznor, Trent and Atticus Ross. 2011. "Immigrant Song." *The Girl with the Dragon Tattoo*. Null. 002.

Rolling Stones, The. 1965. "(I Can't Get No) Satisfaction." London.

———. 1968. "Sympathy for the Devil." *Beggars Banquet*. London. 75392, CD 539.

Simon, Paul. 1973. "Something So Right." *There Goes Rhymin' Simon*. Columbia.

———. 1986. "The Boy in the Bubble." *Graceland*. Warner Bros. W2-25447.

———. 1986. "Crazy Love, Vol. II." *Graceland*. Warner Bros. W2-25447.

———. 1986. "Graceland." *Graceland*. Warner Bros. W2-25447.

———. 1986. "Gumboots." *Graceland*. Warner Bros. W2-25447.

———. 1986. "Under African Skies." *Graceland*. Warner Bros. W2-25447.

Springsteen, Bruce. 1980. "The River." *The River*. Columbia. C2K 36854.

———. 1984. "Born in the U.S.A." *Born in the U.S.A*. Columbia. CK 38653.

Stewart, Rod. 1971. "Every Picture Tells A Story." *Every Picture Tells A Story*. Mercury. 314 558 060-2.

Sting. 1993. "Fields of Gold." *Ten Summoner's Tales*. A&M. 31454 0070 2.

U2. 1984. *The Unforgettable Fire*. Island. B0013201-02.

———. 1987. "Bullet the Blue Sky." *The Joshua Tree*. Island. A2-90581.

Williams, Hank. 1951. "Hey Good Lookin'." MGM.

Yes. 1969. "Every Little Thing." *Yes*. Atlantic.

Acknowledgements

William Moylan and Taylor & Francis Group would like to thank the following parties for providing permission for the use of the materials listed below:

Strawberry Fields Forever
Words and Music by John Lennon and Paul McCartney
Copyright © 1967 Sony/ATV Music Publishing LLC
Copyright Renewed
All Rights Administered by Sony/ATV Music Publishing LLC, 424 Church Street, Suite 1200, Nashville, TN 37219
International Copyright Secured   All Rights Reserved
Reprinted by Permission of Hal Leonard LLC

Figure 5.6
Figure from *Presence and Pleasure: The Funk Grooves of James Brown and Parliament* © 2006 by Anne Danielsen. Published by Wesleyan University Press. Used by permission.

Figure 9.1
Image courtesy of API (Automated Processes, Inc.).

Index

abstraction 31, 61–2, 70, 82, 114, 119–20, 236, 242

accelerando 72

accompaniment 3, 14, 43–4, 71, 84, 95, 145; and distance 309; and lead vocal 97–100; and lyrics 185; and music elements 162, 166, 173; and observations 161; and recording elements 197, 420, 422; and spatial elements 468, 472–3, 477; and spatial staging 326, 330; and stereo location 299; and timbre 436, 439, 484

acousmatic listening 241–3, 257, 259, 267, 288, 310, 325, 400, 404, 434, 484

acoustic ecology 24, 361, 404

acoustic waves 50, 375

activities 209–12, 240, 254, 272, 422

Adams, Bryan 187n6; "(Everything I Do) I Do It for You" 187n6

Adele 137–8, 331"Hello" 137–8, 331

Aeolian mode 75, 77–8, 80, 83

aesthetics 18–19, 25, 49, 57, 62, 156, 270; and confluence 376, 384; and deep listening 406; and illusion of space 287–8, 291, 293, 323–4, 326, 331, 337, 339; and loudness 358–60, 371, 397; and recording domain 223; and sound of place 346, 352; and sound source timbres 391; and spatial elements 459

affects 22–4, 58, 90, 92, 256, 262, 282–3; and confluence 357; identification with 423–4; and loudness 372; and lyrics domain 125; and observations 171, 180; and recording domain 216; and recording elements 431; and sound source timbres 391

Africa 82

agency 134–6

albums 458–62

algorithms 155, 274, 392

alternants 130–1, 180–1, 313, 316

Altman, Rick 223

ambience 39, 44, 61, 195, 292, 337, 341, 345, 350–2, 376, 401, 468–9

American National Standards Institute (ANSI) 239

Amos, Tori 80, 162, 180; "97 Bonnie and Clyde" 180; "Crucify" 80, 162

amplifiers 223–6, 228, 232, 265

amplitude 154, 238–41, 260, 265, 270, 273–4, 283; and illusion of space 296–9, 306, 315–17, 335, 343; and loudness 358, 360, 365, 367, 375, 389; and praxis studies 500, 507, 518; and recording domain 194, 196,

201, 219, 226, 228; and recording elements 431, 438, 445; and spatial elements 452

anthropology 24

antiphony 113, 117, 462

architecture 24

area studies 16

arrangement 69, 87, 93–7, 150, 152, 166–9, 176, 262, 272

arrival time gap 343, 347–9, 352, 475, 526–7

art music 12, 14, 22, 31, 40, 70, 72, 75, 78, 99, 268, 414

artificial double tracking (ADT) 265, 336, 472

Ashby, Patricia 141n15

assemblages 134, 270, 386

attention 1–3, 10–11, 13, 19–20, 28–9, 35, 54, 59–65, *see also* listening with attention

attenuation 228, 297, 306, 310, 335, 346–9, 351–2, 475, 525

attributes 209–11, 240, 254, 259, 261, 272, 279; and confluence 389, 392–4; and illusion of space 299, 305–6, 321–2, 329, 332–3, 341–2; and loudness 341–2, 397; and recording domain 213–14; and recording elements 418, 420; and sound of place 342–3, 345–8, 351, 353; and sound source timbres 391; and spatial properties 291–5

auditory images *see* aural images

auditory scene analysis 323

aural images 49, 57, 252, 258–9, 261, 323, 426–7; and confluence 375, 386, 389; and crystallized form 398–9, 401; and deep listening 402–3, 407–8; and evaluation stage 419; and loudness 362–3, 391, 398, 406; and memory 400–1, 407; and praxis studies 511, 529; and recording elements 483, 486; and sound of place 335, 337, 341; and sound source timbre 263; and spatial properties 292–3; and spatial staging 322, 327–8, 331; and stereo location 298–300; and subjective vantage 408; and timbral signatures 283; and timbre analysis 267–8, 270; and timbre of track 395–6

auto-tuning 77, 132

backing vocals 71, 166, 300, 326, 453–4, 456–8

bandwidths 4, 96, 167, 196, 199, 229, 245; and loudness 358; and pitch 247–9, 251–2; and praxis studies 495, 508–9; and recording elements 260, 437, 444; and timbre 256, 265

basic-level 40, 54, 90, 149–50, 160, 177, 180; basic-level categorization 54, 66n10, 67n14, 292–3; and confluence 377, 379, 383; and illusion of space

Printed in the United States
by Baker & Taylor Publisher Services